Food Mycology

A Multifaceted Approach to Fungi and Food

MYCOLOGY SERIES

Editor
J. W. Bennett
Professor
Department of Plant Biology and Pathology
Rutgers University
New Brunswick, New Jersey

Founding Editor
Paul A. Lemke

1. *Viruses and Plasmids in Fungi,* edited by Paul A. Lemke
2. *The Fungal Community: Its Organization and Role in the Ecosystem,* edited by Donald T. Wicklow and George C. Carroll
3. *Fungi Pathogenic for Humans and Animals (in three parts),* edited by Dexter H. Howard
4. *Fungal Differentiation: A Contemporary Synthesis,* edited by John E. Smith
5. *Secondary Metabolism and Differentiation in Fungi,* edited by Joan W. Bennett and Alex Ciegler
6. *Fungal Protoplasts,* edited by John F. Peberdy and Lajos Ferenczy
7. *Viruses of Fungi and Simple Eukaryotes,* edited by Yigal Koltin and Michael J. Leibowitz
8. *Molecular Industrial Mycology: Systems and Applications for Filamentous Fungi,* edited by Sally A. Leong and Randy M. Berka
9. *The Fungal Community: Its Organization and Role in the Ecosystem, Second Edition,* edited by George C. Carroll and Donald T. Wicklow
10. *Stress Tolerance of Fungi,* edited by D. H. Jennings
11. *Metal Ions in Fungi,* edited by Gü'fcnther Winkelmann and Dennis R. Winge
12. *Anaerobic Fungi: Biology, Ecology, and Function,* edited by Douglas O. Mountfort and Colin G. Orpin
13. *Fungal Genetics: Principles and Practice,* edited by Cees J. Bos
14. *Fungal Pathogenesis: Principles and Clinical Applications,* edited by Richard A. Calderone and Ronald L. Cihlar
15. *Molecular Biology of Fungal Development,* edited by Heinz D. Osiewacz
16. *Pathogenic Fungi in Humans and Animals: Second Edition,* edited by Dexter H. Howard
17. *Fungi in Ecosystem Processes,* John Dighton
18. *Genomics of Plants and Fungi,* edited by Rolf A. Prade and Hans J. Bohnert
19. *Clavicipitalean Fungi: Evolutionary Biology, Chemistry, Biocontrol, and Cultural Impacts,* edited by James F. White Jr., Charles W. Bacon, Nigel L. Hywel-Jones, and Joseph W. Spatafora
20. *Handbook of Fungal Biotechnology, Second Edition,* edited by Dilip K. Arora
21. *Fungal Biotechnology in Agricultural, Food, and Environmental Applications,* edited by Dilip K. Arora
22. *Handbook of Industrial Mycology,* edited by Zhiqiang An

23. *The Fungal Community: Its Organization and Role in the Ecosystem, Third Edition,* edited by John Dighton, James F. White, and Peter Oudemans
24. *Fungi: Experimental Methods in Biology,* Ramesh Maheshwari
25. *Food Mycology: A Multifaceted Approach to Fungi and Food,* edited by Jan Dijksterhuis and Robert A. Samson

Food Mycology

A Multifaceted Approach to Fungi and Food

Edited by

Jan Dijksterhuis

CBS Fungal Biodiversity Centre
The Netherlands

Robert A. Samson

CBS Fungal Biodiversity Centre
The Netherlands

CRC Press
Taylor & Francis Group
Boca Raton London New York

CRC Press is an imprint of the
Taylor & Francis Group, an **informa** business

CRC Press
Taylor & Francis Group
6000 Broken Sound Parkway NW, Suite 300
Boca Raton, FL 33487-2742

ISBN 13: 978-0-8493-9818-6 (hbk)

Library of Congress Cataloging-in-Publication Data

Food mycology : a multifaceted approach to fungi and food / edited by Jan Dijksterhuis and Robert A. Samson.
p. cm. -- (Mycology series)
Includes bibliographical references and index.
ISBN-13: 978-0-8493-9818-6 (alk. paper)
ISBN-10: 0-8493-9818-5 (alk. paper)
1. Fungi. 2. Mycology. 3. Food--Microbiology. I. Dijksterhuis, Jan. II. Samson, Robert A. III. Title. IV. Series.

QK603.F66 2007
664.001'5795--dc22 2007013028

Visit the Taylor & Francis Web site at
http://www.taylorandfrancis.com

and the CRC Press Web site at
http://www.crcpress.com

Contents

List of contributors

Aldred, David, Applied Mycology Group, Biotechnology Centre, Cranfield University, Silsoe, Bedford MK45 4DT, U.K. E-mail: d.aldred@cranfield.ac.uk.

Amsing, Jos G. M., Applied Plant Research, Mushroom Unit, Wageningen University and Research Center, P.O. Box 6042, 5966 AA Horst, The Netherlands. E-mail: Jos.Amsing@wur.nl.

Andersen, Birgitte, Center For Microbial Biotechnology, Biocentrum-DTU, Technical University of Denmark, DK-2800 Kgs. Lyngby, Denmark. E-mail: ba@biocentrum.dtu.dk.

Baar, Jacqueline, Applied Plant Research, Mushroom Unit, Wageningen University and Research Center, P.O. Box 6042, 5966 AA Horst, The Netherlands. E-mail: Jacqueline.Baar@wur.nl.

Chitarra, Gilma Silva, Laboratory for Food Microbiology, University of Wageningen, The Netherlands. Present Address: Rua Rio de Janeiro 832, Bairro Nova Várzea Grande,Várzea Grande- MT CEP 78 135 710, Brazil. E-mail: gilmachitarra@yahoo.com.br.

Corrochano, Luis M., Departamento de Genética, Facultad de Biología, Universidad de Sevilla, Apartado 1095, 41080 Sevilla, Spain.

Cristescu, Simona M., Department of Molecular and Laser Physics, Institute for Molecules and Materials, Radboud University Nijmegen, Toernooiveld 1, 6525 ED Nijmegen, The Netherlands. E-mail: simona@science.ru.nl.

Dijksterhuis, Jan, Applied and Industrial Mycology, CBS Fungal Diversity Centre, Uppsalalaan 8, 3584 CT, Utrecht, The Netherlands. E-mail: dijksterhuis@cbs.knaw.nl.

Frisvad, Jens C., Centre for Microbial Biotechnology, BioCentrum-DTU, Building 221, Technical University of Denmark, 2800 Kgs. Lyngby, Denmark. E-mail: jcf@biocentrum.dtu.dk.

Geisen, Rolf, Federal Research Centre for Nutrition and Food, Location Karlsruhe, Haid-und-Neu-Str. 9, 76131 Karlsruhe, Germany. E-mail: Rolf.Geisen@bfel.de.

Harren, Frans J. M., Department of Molecular and Laser Physics, Institute for Molecules and Materials, Radboud University Nijmegen, Toernooiveld 1, 6525 ED Nijmegen, The Netherlands. E-mail: F.Harren@science.ru.nl.

Jongh, Wian A. de, Center for Microbial Biotechnology BioCentrum-DTU Søltofts Plads 221, DK-2800 Kgs. Lyngby, Denmark. Present Address: Pharmexa A/S, Kogle Allé 6, DK-2970 Hørsholm, Denmark. E-mail: WDJ@pharmexa.com.

Karlshøj, Kristian, Center for Microbial Biotechnology, BioCentrum-DTU, Technical University of Denmark, Kgs. Lyngby, Denmark. E-mail: kk@biocentrum.dtu.dk.

Kolattukudy, Pappachan E., Burnett College of Biomedical Sciences, Biomolecular Science Center, University of Central Florida, 12722 Research Parkway, Orlando, FL 32826, USA. E-mail: pkolattu@mail.ucf.edu.

Larsen, Thomas Ostenfeld, Center for Microbial Biotechnology, BioCentrum-DTU, Technical University of Denmark, Kgs. Lyngby, Denmark. E-mail: tol@biocentrum.dtu.dk.

Leong, Su-lin L., Food Science Australia, PO Box 52, North Ryde NSW 1670; School of Agriculture, Food and Wine, University of Adelaide; Cooperative Research Centre for Viticulture, Australia. E-mail: su-lin.leong@gmail.com.

Magan, Naresh, Applied Mycology Group, Biotechnology Centre, Cranfield University, Silsoe, Bedford MK45 4DT, United Kingdom. E-mail: n.magan@cranfield.ac.uk.

McCartney, Alastair, Plant-Pathogen Interactions Division, Rothamsted Research, Harpenden, Herts. AL5 2JQ, United Kingdom. E-mail:alastair.mccartney@bbsrc.ac.uk.

Nielsen, Jens, Center for Microbial Biotechnology BioCentrum-DTU Søltofts Plads 221, DK-2800 Kgs. Lyngby, Denmark. E-mail: jn@biocentrum.dtu.dk.

Nielsen, Per Væggemose, Center for Microbial Biotechnology, BioCentrum-DTU, Technical University of Denmark, Kgs. Lyngby, Denmark. E-mail: Pvn@biocentrum.dtu.dk.

Nout, Rob M. J., Laboratory of Food Microbiology, Wageningen University, P.O. Box 8129, 6700 EV Wageningen, The Netherlands. E-mail: Rob.Nout@wur.nl.

Paradi, Istvan, Applied Plant Research, Mushroom Unit, Wageningen University and Research Center, P.O. Box 6042, 5966 AA Horst, The Netherlands.

Prusky, Dov, Department of Postharvest Science, Institute of Food Technology, Agricultural Research Organization, Bet Dagan, Israel. E-mail: dovprusk@volcani.agri.gov.il.

Rahardjo, Yovita S. P., DSM Food Specialties, P.O. Box 1, 2600 MA Delft, The Netherlands. E-mail: Yovita.Rahardjo@DSM.com.

Rinzema, Arjen, Wageningen University, Food and Bioprocess Engineering Group, P. O. Box 8129, 6700 EV Wageningen, The Netherlands. E-mail: Arjen.Rinzema@wur.nl.

Samson, Robert A., Applied and Industrial Mycology, CBS Fungal Diversity Centre, P.O. Box 85167, 3508 AD Utrecht, The Netherlands. E-mail: samson@cbs.knaw.nl.

Scholtmeijer, Karin, Microbiology, Institute of Biomembranes, University of Utrecht, Padualaan 8, 3584 CH Utrecht, The Netherlands.

Stark, Jacques, DSM Food Specialties, P.O. Box 1, 2600 MA Delft, The Netherlands. E-mail: Jacques.Stark@DSM.com.

Straatsma, Gerben, Applied Plant Research, Mushroom Unit, Wageningen University and Research Center, P.O. Box 6042, 5966 AA Horst, The Netherlands. E-mail: Gerben.Straatsma@wur.nl.

Thrane, Ulf, Centre for Microbial Biotechnology, BioCentrum-DTU, Building 221, Technical University of Denmark, 2800 Kgs. Lyngby, Denmark. E-mail: ut@biocentrum.dtu.dk.

Ugalde, Unai, Unidad de Bioquímica II, Facultad de Química, Universidad del País Vasco, Apartado 1072, 20080 San Sebastian, Spain. E-mail: qppugmau@ehu.es.

Vries, Ronald P. de, Microbiology, Institute of Biomembranes, University of Utrecht, Padualaan 8, 3584 CH Utrecht, The Netherlands. E-mail: r.p.devries@bio.uu.nl.

West, Jon, Plant Pathogen Interactions, Rothamsted Research, Harpenden, Herts., AL5 2JQ, United Kingdom. jon.west@bbsrc.ac.uk.

Woltering, Ernst, J., Wageningen University and Research Center, Agrotechnology & Food Innovations (A&F B.V.), P.O. Box 17, 6700 AA Wageningen, The Netherlands. E-mail: Ernst.Woltering@wur.nl.

Wösten, Han A. B., Microbiology, Institute of Biomembranes, University of Utrecht, Padualaan 8, 3584 CH Utrecht, The Netherlands. E-mail: H.A.B.Wosten@bio.uu.nl.

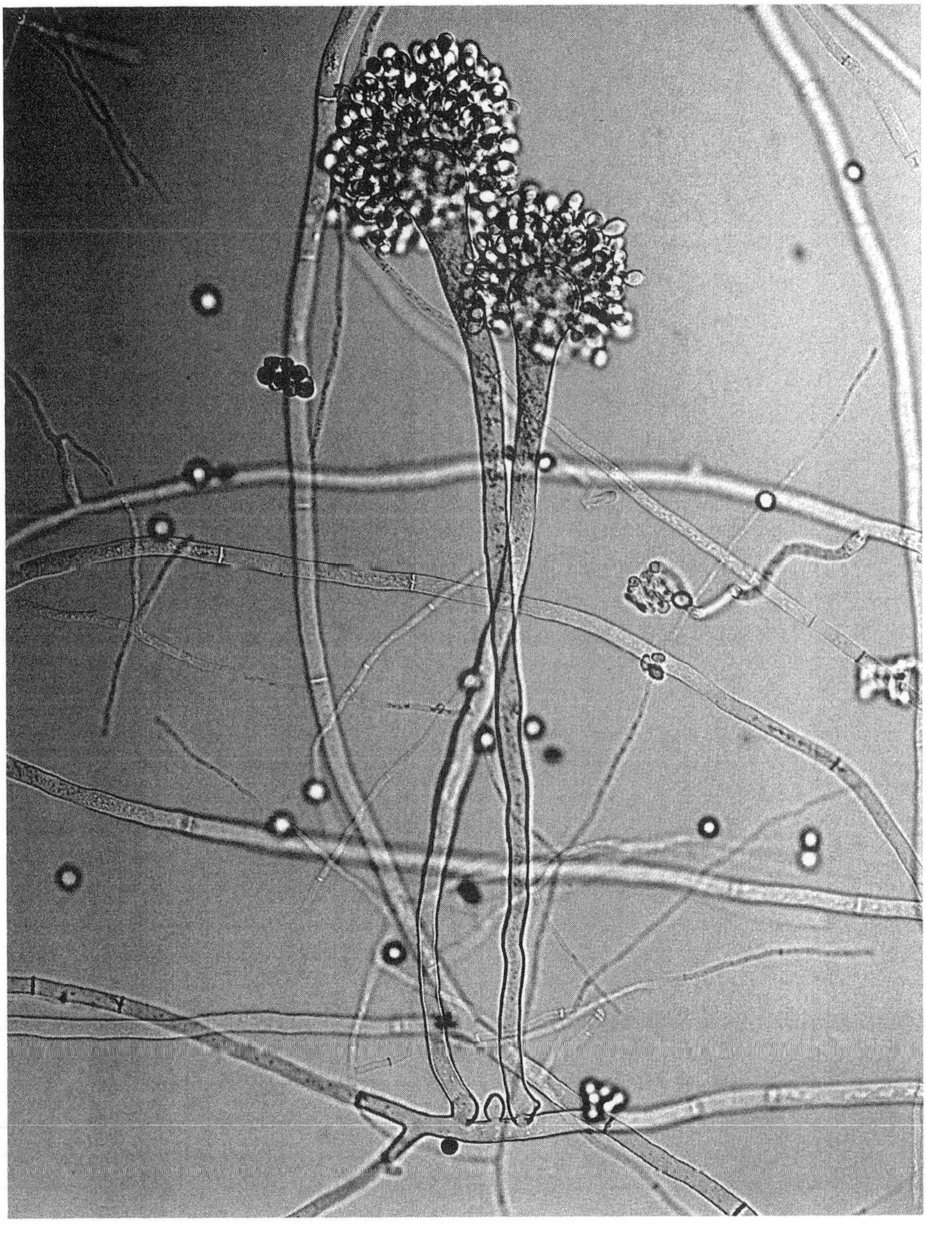

Preface

Fungi and food are strongly related in many ways. Firstly, they spoil a considerable part of all the food we produce and store. In food they can produce toxic compounds that threaten our health. On the other hand, these microorganisms are used for centuries to ferment bread, soft cheeses, soybeans, alcoholic drinks and many other products. In this book many of these aspects are highlighted including subjects ranging from post-harvest infection, molecular detection of fungi and the association between specific fungi and food products. We have divided the chapters in six parts, in which we deal with the fungi in living crops, as propagules, and the important fact that fungi produce mycotoxins, other metabolites and enzymes. Much emphasis has been given to fungal spoilage including various aspects such as ecology, growth and detection. Finally, we complete this book with the fungi as food.

The preparation of this book was unthinkable without the authors who spent so much time writing these excellent chapters and we would like to thank them for their contribution.

Jan Dijksterhuis and Robert A. Samson
Utrecht, January 2007

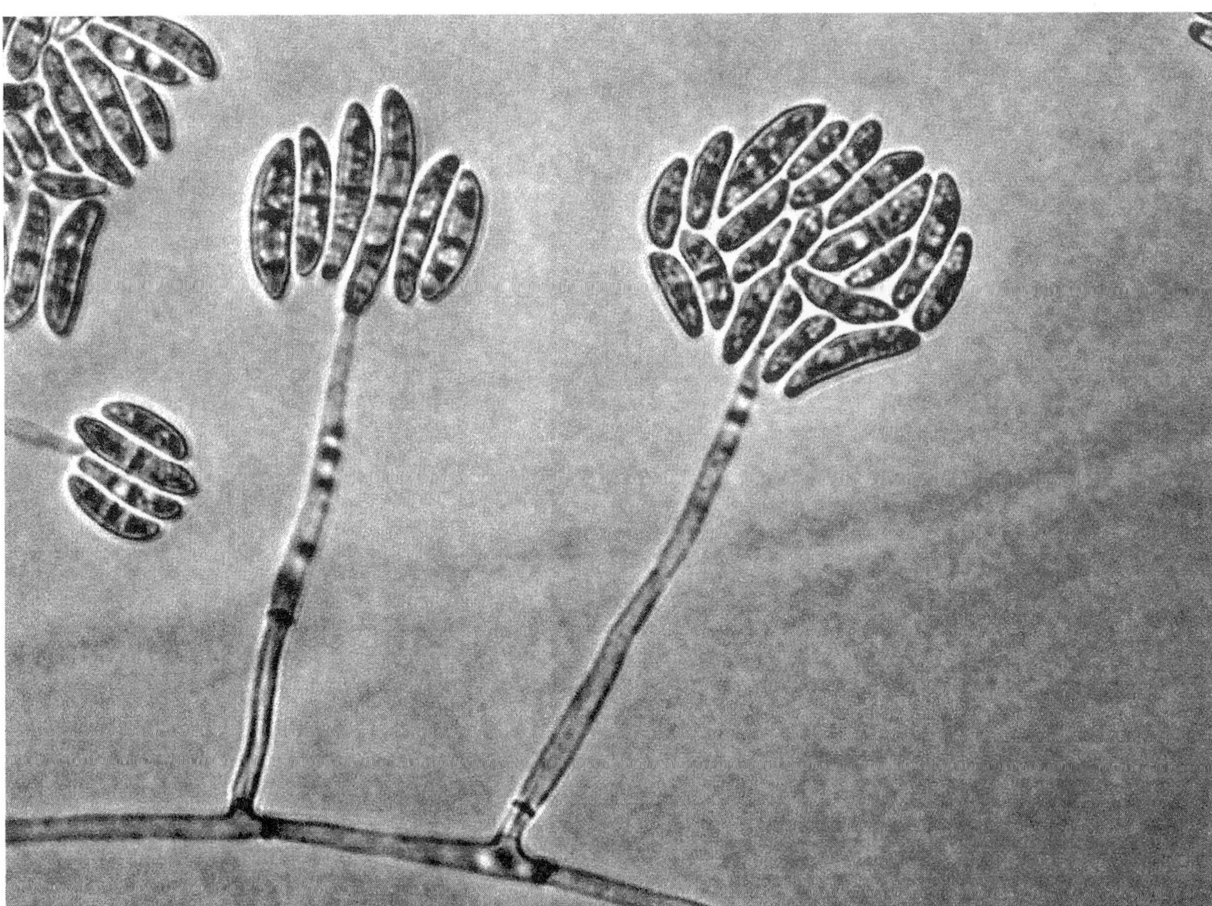

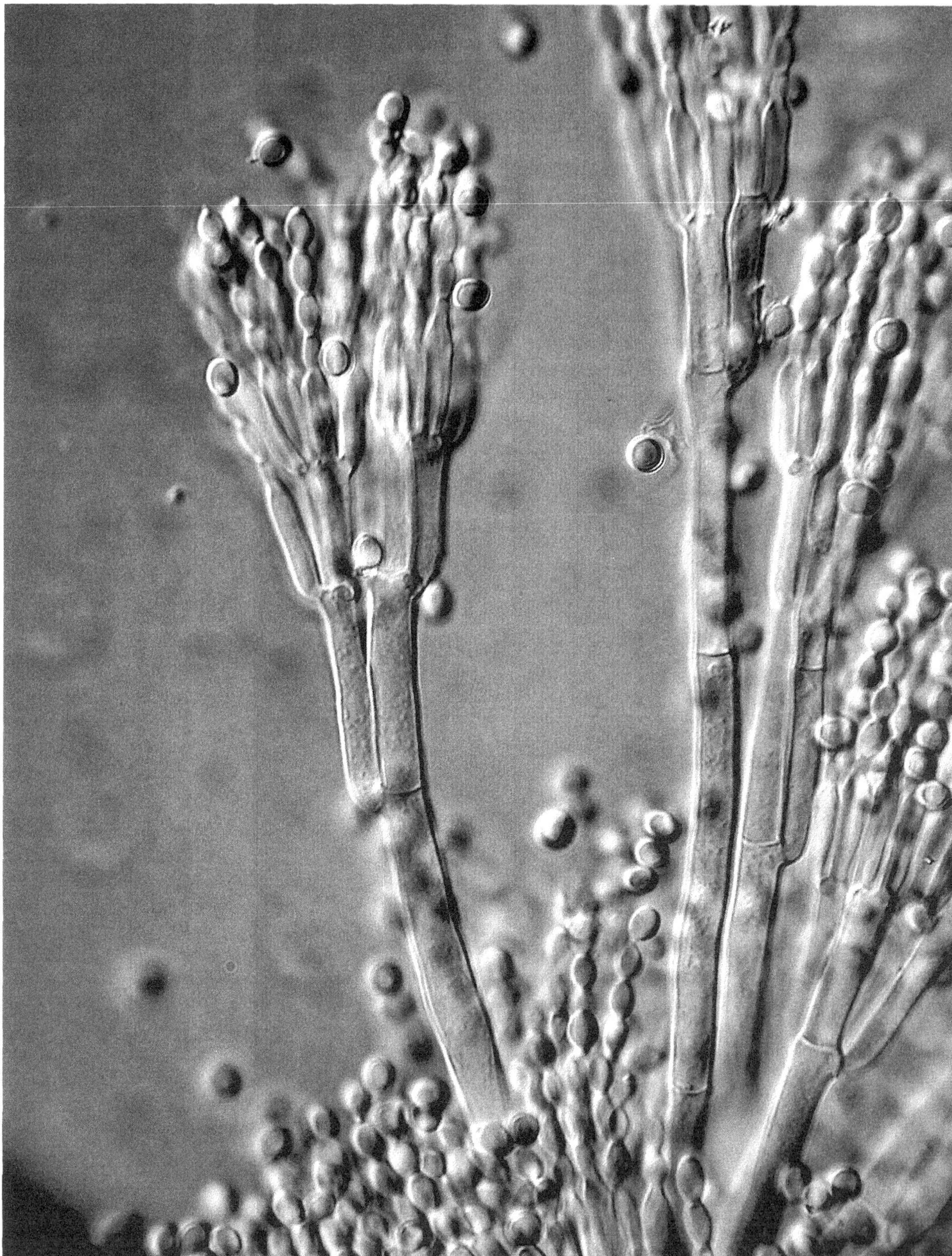

Part 1

FUNGI AND LIVING CROPS

Food products can be either a living crop or a processed matrix that contains high amounts of nutrients. Development of fungi on a living crop is essentially a plant-pathogenic relationship between a fungus and a plant host. With good reason losses of living crops are designated as the result of a postharvest disease. The establishment of these infections includes complex patterns of communication between host and fungus. For example, with respect to the timing of infection where fungal structures await fruit ripening before they penetrate the barriers of the cell wall. Penetration itself is done by specialised structures called appressoria, that can build up pressures large enough to drill into the sturdy plant cell wall solely by mechanical force. The pressures measured inside these cells are the highest ever reported for a living cell.

Other fungi can only enter the crop when little wounds are present, but the damage these fungi can do to crops is devastating, while wounds are nearly always afflicted on the crop after handling or by the action of insects. Losses due to infection of oranges and apples by *Penicillium* species are enormous. Inside the host, enzymes are released that macerate the host cell walls and destroy the food product.

The first two chapters in this book address the postharvest diseases of food products. Prusky and Kolattukudy address in Chapter 1 many aspects of host and fungus including recognition, penetration and the role of many enzymes released by fungi. The chapter ends with possible new approaches to control of postharvest diseases. Cristescu, Woltering and Harren describe in Chapter 2 a brand-new technique called laser-based photo acoustic spectroscopy following fungus-host interactions in real time by monitoring ethylene, a plant hormone. Fast detection of postharvest problems may offer new possibilities to minimalise losses.

Chapter 1

Cross-talk between host and fungus in postharvest situations and its effect on symptom development

Dov Prusky[1] and Pappachan E. Kolattukudy[2]
[1]*Department of Postharvest Science, Institute of Food Technology, Agricultural Research Organization, Bet Dagan, Israel;* [2]*Burnett College of Biomedical Sciences, Biomolecular Science Center, University of Central Florida, BMS Building Room 136, Orlando, FL 32826, U.S.A.*

POSTHARVEST MYCOLOGY AND THE MAJOR LOSSES

The fruit and vegetable production industry has undergone major structural changes during recent years owing to new consumer expectations. The pressure to diminish fungicide residues on fruits and vegetables at all points along the supply chain presents even more problems than that of other agricultural products. At present, growers have to conform with regulations that limit undesirable biocide residues while, at the same time, choosing treatments that will maintain the quality of their produce. Losses from postharvest disease can be as high as 25%; they may result from poor handling during harvesting, processing, storage, and/or transportation to the point of sale. In tropical countries losses may be as high as 50% because at elevated temperatures postharvest senescence is accelerated. After harvest, rapid physiological processes are initiated that result in the breakdown of the host resistance mechanism and lead to enhanced development of rots. In fruits, the physiological changes that occur during ripening serve as a signal for initiation of fungal attack and colonization.

Despite the magnitude of the problem, the development of new approaches to disease control has not always received priority. This is partly because the abundance of the food supply in developed countries has masked the severity of postharvest losses, but is mostly due to the difficulty of devising treatments that prolong produce quality while at the same time satisfying consumer demand for reduced biocide residues.

The requirement to improve produce quality and to reduce postharvest disease within the limitations imposed by the new marketing controls has stimulated revision of the old techniques and the development of new protocols. These aim at a more holistic approach whereby the chemical control of disease is not the only means considered. Since senescence results in the activation of infections, and improper handling and storage encourages disease development, the new approaches encompass improvements in produce handling and storage, in combination with techniques to enhance host resistance. In the present chapter we will refer firstly to the mechanism of fungal pathogenicity and host-pathogen communication, and secondly to specific cases in which new approaches have resulted in improved quality of the stored produce.

MECHANISMS OF HOST SURFACE PENETRATION AND SUBSEQUENT COLONIZATION

A number of postharvest pathogens start their disease cycle with a conidium landing on the host surface. The fungus must have evolved strategies to recognize a suitable host, to penetrate and invade plant tissues and to overcome host defences. To perform these tasks, the fungus is capable of perceiving chemical and physical signals from various different host plants and of responding with the appropriate metabolic activities required for pathogenic development. Communication between the fungal conidium and the plant surface begins as soon as the conidium lands on the plant. Some aspects of this interaction are specific to the host whereas others are relatively nonspecific, and depend only on the lipophilic nature of the plant cuticle. Fungal conidia are often covered with a lipophilic self-inhibitor when they arrive on the plant surface. Diffusion of the self-inhibitor into the hydrophobic plant cuticle relieves the self-inhibition and allows germination of the conidia. This concept was demonstrated with the conidia of *Magnaporthe grisea* (Hegde and Kolattukudy, 1997): the conidial surface material was recovered by washing with organic solvents and was found to inhibit conidial germination in a dose-dependent manner; this inhibition was reversed by plant-surface wax. *Colletotrichum gloeosporioides* self-inhibitors, although they have not been identified, are known to be lipophilic (Tsurushima *et al.*, 1995) and the self-inhibition is probably relieved by diffusion of the inhibitor into the host cuticle.

How self-inhibitors prevent conidial germination is not known, but they would be expected to cause suppression of early gene expression (Chitarra *et al.*, 2005). Since the *cam* (calmodulin) gene of *M. grisea* was found to be expressed very early during the conidial interaction with the host (Liu and Kolattukudy, 1999), it was chosen as a test gene to examine the effects of self-inhibitors. *Cam* gene promoter-driven expression of green fluorescent protein (GFP) reporter gene in *M. grisea* was inhibited by self-inhibitors whose effect was reversed when the self-inhibition was relieved by the addition of plant-surface wax. Surface attachment was required for *cam* promoter-driven GFP expression and appressorium formation, and both of these were inhibited by concanavalin that inhibits conidial surface attachment.

Beside the self-inhibitory conidial factors, it has been known for some time that fungal conidia require contact with a hard surface before they can be induced to germinate and to differentiate into appressoria. The molecular basis of this requirement has not been elucidated. A differential display approach was used to identify some of the fungal genes of *C. gloeosporioides* that are induced by contact with a hard surface (*chip* genes). One such gene was identified as that which encodes a 16.2-kDa ubiquitin-conjugating enzyme; this gene complemented the *ubc5* yeast mutant (Liu and Kolattukudy, 1998). Thus, one role of contact with a hard surface is to induce ubiquitin-dependent protein degradation, which is involved in conidial germination and appressorial differentiation. Two other *Colletotrichum* hard-surface-induced protein genes (*chip* genes) that were discovered by differential display encode CHIP2 and CHIP3, two novel proteins of 65 and 64 kDa, respectively. CHIP2 contains a putative nuclear localization signal, a leucine zipper motif and a heptad repeat region which might dimerize into a coiled-coil structure. The targets of this putative transcription factor and its biological function are unknown. CHIP3 contains nine transmembrane domains. Although induction of *Chip2* and *Chip3* by hard surface contact was confirmed, their biological functions remain unknown (Kim *et al.*, 2000a).

Also, Ca^{+2} calmodulin signaling is probably activated by contact of the conidia with a hard surface (Kim *et al.*, 1998). The *C. gloeosporioides* calmodulin gene (*cam*) showed almost 90% identity with other fungal *cam* genes. The 1.3-kb *cam* transcript level was elevated more than tenfold by contact of the conidia with a hard surface for one hour and, furthermore, a calmodulin antagonist severely inhibited germination and appressorium formation (Kim *et al.*, 1998). Thus, the *cam* gene product seems to be involved in the induction of conidial germination and appressorial differentiation. Involvement of calmodulin signaling in germination and appressorium formation would involve calmodulin kinase (CaMK). *cam*K transcript was also obtained from a cDNA library prepared from hard-surface-induced transcripts isolated from *C. gloeosporioides* conidia. The identity of the *cam*K gene was confirmed by demonstrating CaMK activity of the cloned gene product expressed in *E. coli* (Kolattukudy *et al.*, 2000). Involvement of CaMK in germination and appressorium formation was strongly suggested by the finding that the CaMK selective inhibitor, KN93, inhibited phosphorylation of proteins that were found to be associated with hard-surface treatment of the fungal conidia.

Kim *et al.*, (2002) also reported another novel gene (*Chip* 6) that was induced by contact of *C. gloeosporioides* with a hard surface; it encodes a sterol glycosyl transferase, as confirmed by the measurement of glycosyl transferase activity of the gene product expressed in *E. coli*. This glycosyl transferase was identified as a novel pathogenesis gene, since its disruption caused a drastic decrease in the virulence, although the mutants grew normally and formed normal-looking appressoria (Kim *et al.*, 2002). This suggests that conidia of postharvest pathogens sense and react to various stimuli on the fruit, even before penetration. In light of the importance of volatiles produced by fruits, it will be of interest to search for their effect on the initial stages of pathogenicity.

The biotrophic stage

After invading the host, fungi use various strategies to gain access to host nutrients. Whereas necrotrophs quickly kill plant cells in order to feed subsequently as saprotrophs, other fungi maintain biotrophic relationships with their hosts either transiently or until sporulation. Most of the postharvest pathogens are considered to be necrotrophs, e.g., *Botrytis cinerea, Alternaria alternata, Penicillium* spp. The biotrophic lifestyle is realized in a remarkable range of ways: intercellular (*Cladosporium fulvum*); subcuticular (*Venturia inaequalis*); inter- and intracellular (*Claviceps purpurea, Ustilago maydis*, monokaryotic rust fungi); extracellular with haustoria within epidermal cells (powdery mildews); intercellular with haustoria within parenchyma cells (dikaryotic rust fungi and downy mildews). A transient type of biotrophy followed by necrotrophy is observed in the so-called hemibiotrophic fungi (*M. grisea, Phytophthora infestans* and *Colletotrichum* spp.) (Mendgen and Hahn, 2001). These are regarded as the hemibiotrophic fungi members of the genus *Colletotrichum* initially grow within the cell walls of host epidermal cells leading to the formation of long-term biotrophic or quiescent infections. After penetration, the intracellular infection vesicle and the primary hyphae colonize only a few host cells, and both are surrounded by a matrix that separates the fungal cell wall from the invaginated host plasma membrane (Mendgen and Hahn, 2001). This matrix is extracytoplasmic and is connected to the plant apoplast. It seems that the existence of a matrix layer is crucial for the biotrophic life style. Within the interfacial matrix, a fungal glycoprotein, encoded by *CIH1*, was identified. The protein was shown to be present uniquely at this interface in the biotrophic stage of hemibiotrophic *Colletotrichum* spp.; its expression was switched off at the onset of necrotrophic development. The completion of the biotrophic stage, which is a quiescent stage, might be the result of a host signal accompanied by a signal transduction process that leads to the initiation of processes leading to the destruction of the plant cell.

Induction of penetrating structures

The first host barrier to be breached is the cuticle, which covers all aerial parts of the plant. The cuticle consists of cutin, an insoluble polyester composed mainly of two families of hydroxy and hydroxy-epoxy fatty acids, derived from the most common cell fatty acids: one derived from C_{16} fatty acids and the other from C_{18} unsaturated fatty acids. The monomers of some plants are mainly of the C_{16} family, whereas others are mixtures of the C_{16} and C_{18} families. There is a layer (associated with the cutin layer) consisting of complex mixture of soluble lipids, collectively called waxes, that are distinctly different from cell lipids. The most common major components of cuticular waxes are hydrocarbons and their oxygenated derivatives such as secondary alcohols and ketones, very long-chain fatty acids, aldehydes and alcohols, and wax esters composed of very-long-chain fatty alcohols and fatty acids. Very-long-chain β-diketones and pentacyclic triterpenes are sometimes major components of waxes, particularly on stems and fruits of some plants. As if to provide chemical and metabolic stability, the cuticle-associated waxes, which function at the surface exposed to oxygen and other atmosphere components as well as to microbes, are mostly saturated and are usually hard for microbes to absorb and metabolize (Kolattukudy, 1996).

If the signals at the plant surface are perceived as favorable by the fungi, conidia germinate and the germ tube differentiates into an infection structure called an appressorium, which produces the infection peg that penetrates into the host. In other fungi where no signals are perceived or appressorium are not produced, their germ tubes penetrate the cuticle directly or through wounds.

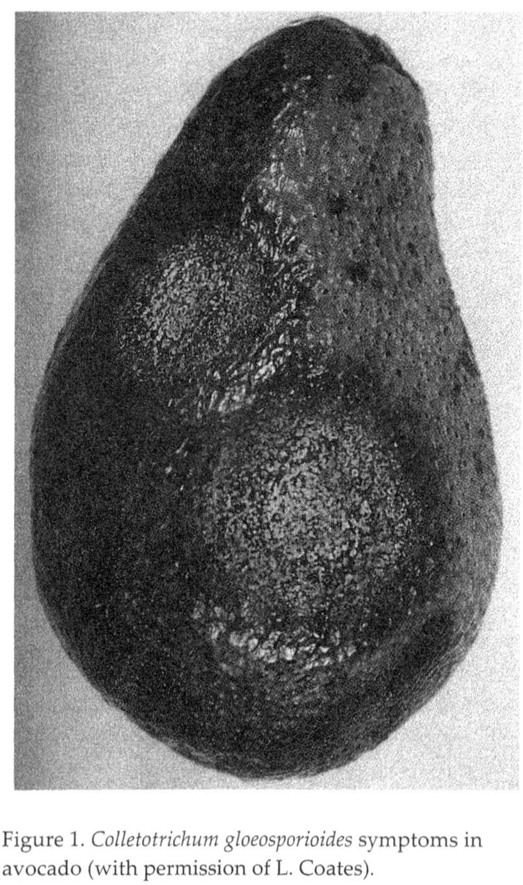

Figure 1. *Colletotrichum gloeosporioides* symptoms in avocado (with permission of L. Coates).

The nature of the plant signals that trigger the programmed differentiation process is poorly understood; topographical features or chemical signals at the plant surface could be involved. Experimental evidence obtained in recent years shows that physical signals are involved in some cases and chemical signals in others. The most clearly established examples of each are reviewed below.

After the diffusive removal of self-inhibitors into the cuticle, the conidia are primed by the surface contact to respond to the chemical signals from the host; host surface wax induces conidial germination and appressorium formation. Avocado fruit surface wax induces germination and appressorium formation of *C. gloeosporioides* conidia (Podila *et al.*, 1993; Prusky and Saka, 1989) (Figures 1-2). This induction is specific for the host wax; waxes from other plants do not elicit this response and avocado wax does not show any biological activity with conidia from *Colletotrichum* species that infect other plants (Podila *et al.*, 1993). The long-chain primary alcohols from avocado wax were shown to be the components that induce germination and appressorium formation of *C. gloeosporioides* conidia, but the terpenoids found in plant wax also induce germination and appressorium formation (Kolattukudy *et al.*, 2000).

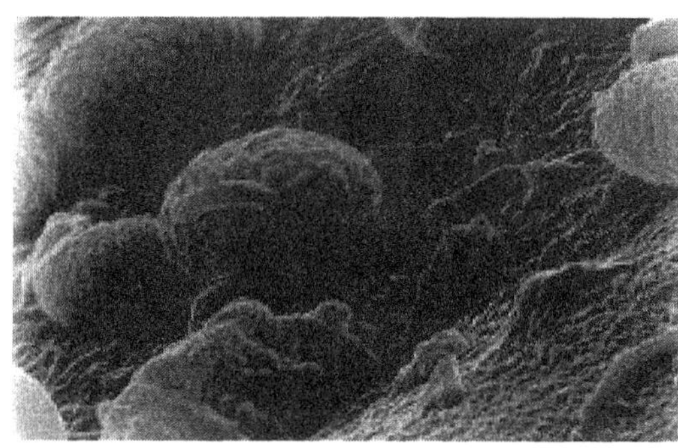

Figure 2. Appressoria formation on avocado fruits by *C. gloeosporioides*.

In many cases germinated appressoria remain dormant on fruit until the fruit ripens, when the fungus causes major damage. This type of pathogen includes *Colletotrichum* spp., *Botrytis, Alternaria, Fusarium* and others. The nature of the host signal that prompts latent anthracnose fungi to attack the host when the host ripens remained unknown for a long time. Ethylene, the fruit ripening hormone, was found to be a potent stimulator of germination and appressoria formation in anthracnose fungal spores (Flaishman and Kolattukudy, 1994): remarkably, this volatile agent that emanates from the ripening fruit caused branching of the germ tubes and formation of up to six appressoria by each conidium. This biological activity is unique to *Colletotrichum* species that infect climacteric fruits which emit ethylene when they ripen. Transgenic tomato fruits, from plants engineered to prevent ethylene production, did not permit germination and appressorium formation by *C. gloeosporioides* and thus were not attacked by the pathogen. The increase in anthracnose associated with the use of ethylene treatment to improve fruit color might be caused by the biological activity of ethylene on dormant conidia on the fruits. However, the effect of ethylene requires further study, since other reports suggest an effect of the hormone on *C. gloeosporioides* appressorium formation but not on the activation of quiescent infections in avocado fruits (Prusky *et al.*, 1996).

To study the genes expressed during appressoria formation, a subtractive library approach was used to isolate cDNA representing transcripts produced during this period. This approach yielded four clones that represented genes expressed uniquely during appressorium formation. Two of them would generate cys-rich peptides with 26 and 27 amino acid residues that showed homology to metallothionins. These genes may be developmentally regulated genes that might also serve in a response to heavy metal stress. Another clone, which represented a transcript uniquely expressed during appressorium formation, encoded a 22-kDa protein that immunocytochemical examination showed to be present in the appressorial wall, probably in a glycosylated form (Hwang and Kolattukudy, 1995). Another gene uniquely expressed during appressorium formation would encode a 20 kDa protein (Hwang *et al.*, 1995). A mutant in which this gene was disrupted failed to infect avocado and tomato fruits, even though it formed normal-looking appressoria. This protein, which was found to be associated with appressorial walls, was also found in deeper layers of infected tissue at the infection front, where appressorium-like structures were found. Probably this gene product is necessary for penetration through the host tissue.

One interesting observation on the response to appressoria formation is that the CaMK inhibitor KN93 inhibited germination and appressorium formation. This inhibitor also inhibited melanin synthesis and thus even the appressorium-like structures formed in the presence of KN93 could not attain the normal shape and structure that require melanization for their formation (Kim *et al.*, 1998). Inhibition of melanization probably occurred at the polyketide synthase step and would have been distinctly different from the inhibition of germination and differentiation. These processes require Ca^{+2}/CaMK signalling, therefore a Ca^{+2} chelator applied during the early hard surface contact stage would inhibit germination and appressorium formation. Ca^{+2} release from internal stores, mediated by IP_3 generated by phospholipase C, participates in the early signalling involved in germination and appressorium formation, therefore these processes are severely inhibited by the phospholipase C selective inhibitor U73122. Thus, Ca^{+2}, CaM and CaMK signaling play critical roles in the early phase of the interaction between *C. gloeosporioides* and its host, and probably also in other host-fungus systems.

A second means of penetration by postharvest pathogens is involved in the opportunistic type of infection, in which the pathogens penetrate through a natural wound or one that occurs mainly after harvest or following storage stresses (wounding, chilling injury, high temperature stress, etc.). This type of penetration may be exploited by the same pathogens that penetrate directly as well as others, such as *Penicillium, Rhizopus, Mycosphaerella,* etc., that require a breached or weakened cuticle. However, although pathogens may differ in their initial mechanisms of penetration, the colonization mechanism of the pathogens that penetrate by means of the two different processes are essentially the same. Both direct penetration and wound penetration are discussed below for various postharvest pathogens.

The role of esterases and lipases

Penetration by postharvest pathogens through wounds in fruits resulted in earlier appearance of symptoms than direct penetration. This may indicate that even though the wax layers do not seem to pose a serious barrier to penetration, removal of the wax and direct wounding may increase the incidence of infection. In the case of *Monilinia* in peaches, it was reported that increased thickness of the cuticle/wax layer modulated the susceptibility to fungal attack (Bostock *et al.*, 1998). Since most of fruits and vegetables have thick cuticles it has been suggested that pathogens might secrete surfactants in the form of proteins or other metabolites that reduce surface hydrophobicity and dissolve the wax layer, thereby providing access to the underlying cutin polymer (Kars and van Kan, 2004). In *Botrytis*, the polysaccharide that covers the *B. cinerea* germ tubes might act as a surfactant or, alternatively, the host surface tension might be reduced enzymatically. Production of acids from cutin seems to be the result of the activity of a fungal cutinase (van den Ende and Linskins, 1974; Kolattukudy, 1984, 1985). Based on crystal structure, cutinase falls into a special class of lyases/esterases (Martinez, *et al.*, 1992, 1994; Edgmond and de Vlieg, 2000; Kolattukudy, 2001). A variety of esterases is produced by many fungi and they have received diverse names, often based on the substrate used, but precise identification of the class of enzyme lacks. In case of the fungi *Alternaria* (Köller *et al.*, 1995; Berto *et al.*, 1997; Fan and Köller, 1998), and *Colletotrichum* (Pascholati *et al.*, 1993) cutinases seem to be involved in fungal adhesion of the plant surface. *B. cinerea* produces an extracellular triacylglycerol lipase able to hydrolyse unsaturated long chain acid esters (Commenil *et al.*, 1995), known to be components of cutin and waxes. Lipase production was induced *in vitro* by wax esters and free fatty acids (Commenil *et al.*, 1998). These lipases have cutinolytic activity and play a role in the modification of the waxes and the cuticle and in the adhesion of conidia to the plant surface. Antibodies to the lipase inhibited infection and adhesion of *Botrytis* and also the infection of *Alternaria brassicicola* (Berto *et al.*, 1999).

The role of cutinases

A cutin degrading enzyme was purified in the early 1970s (Purdy and Kolattukudy, 1975) and characterized as a polyesterase that uses a catalytic triad involving active serine for catalysis (Köller and Kolattukudy, 1982). Recent X-ray crystallographic studies on recombinant cutinase suggest that cutinases form a unique class of enzymes that constitutes a bridge between esterases and lipases (Martinez *et al.*, 1992). The small amounts of constitutively expressed cutinase present on the conidia of pathogenic fungi (Köller *et al.*, 1982) could release small amounts of cutin monomers that might help in the differentiation of infection structures (Francis *et al.*, 1996; Gilbert *et al.*, 1996) and might also transcriptionally activate an inducible cutinase gene that causes production of cutinase that helps the infection peg to penetrate the cuticle (Kolattukudy, 1985).

Induction of cutinase by cutin hydrolysate was discovered (Lin and Kolattukudy, 1978) soon after cutinase was first purified (Purdy and Kolattukudy, 1975). More recently, after the cutinase gene had been cloned, regulation of expression of cutinase gene was investigated (Kämper *et al.*, 1994). When the presence of multiple cutinase genes was recognized, the regulation of individual genes and the transcription factors that regulate each gene could be examined (Li and Kolattukudy, 1997). Thus it was found that a constitutively expressed cutinase gene in *Fusarium solani* forma speciales *pisi* was regulated by a transcription factor distinctly different from the one involved in the regulation of the inducible gene (Li *et al.*, 2002). The repressor that binds the palindrome-sequence that overlaps the binding site of the activator of the inducible gene was incapable of binding the constitutively expressed gene as a result of two different nucleotides in the promoter region (Li *et al.*, 2002). The cutinase transcription factor α (CTFα) which activates the inducible cutinase gene has a promoter element that is involved in the response to cutin, which is also able to transactivate CTFα promoter in a yeast system (Yang, Z., Kang, T.J., Liu, S. and Kolattukudy, P.E., manuscript in preparation).

Although it has been debated for almost a century whether cutinase plays a crucial role in fungal penetration during pathogenesis, there is now overwhelming evidence that this is indeed the case (Kolattukudy, 1985, 1996; Gevens and Nicholson, 2000) as is summarized below:

Plant pathogens produce and secrete cutinase targeted at the penetration point, and cutinase is produced during the actual infection of the host. (Podila *et al.*, 1995; Shaykh *et al.*, 1977).

Inhibition of cutinase by chemicals or antibodies, including monoclonal antibodies, prevents infection (Kolattukudy, 1985; Salinas, 1992). In a field study, spraying of an active serine-directed cutinase inhibitor was found to protect papaya fruits against anthracnose (Kolattukudy, 1987). Cutinase-deficient mutants have significantly reduced virulence, but their infectivity can be restored by the application of exogenous cutinase (Danzig *et al.*, 1986; Dickman and Patil, 1986).

Pathogens that cannot infect a host without a breached cuticle (wound) can be enabled to infect an intact host by genetically engineering to provide them with cutinase-producing capability (Dickman *et al.*, 1989). Disruption of cutinase gene in *F. solani* f. sp. *pisi* caused a drastic loss in virulence (Rogers *et al.*, 1994).

Deletion-mutant of individual cutinase genes do not always lead to a significant reduction in virulence (Sweigard *et al.*, 1992; Köller *et al.*, 1995; Crowhurst *et al.*, 1997; van Kan *et al.*, 1997) suggesting that not all cutinase activity is removed as found in case of *Magnaportha grisea* (Sweigard *et al.*, 1992). The existence of multiple cutinase genes, including some that may be expressed only in planta, would make the single gene knock-out approach inappropriate for assessing the role of cutinase in pathogenesis.

Analysis of the recently finished sequence of the genome of the rice blast organism, *M. grisea*, has identified eight genes that encode putative cutinases, several of which "are significantly upregulated during infection-related development" (Dean *et al.*, 2005). The cutinase gene *CUT1*, previously disrupted to investigate the role of this enzyme in plant infection (Sweigard *et al.*, 1992), "is not among the genes differentially regulated during appressorium formation" (Dean *et al.*, 2005). Thus, approaches based on single gene disruption cannot be used to draw firm conclusions about the role of enzymatic cutin degradation in pathogenesis.

The role of pectinases in the initial stages of penetration

Over the years several studies have dealt with the secretion of cell-wall-degrading enzymes (CWDEs) by postharvest pathogens in the early stages of infection. Swelling of the anticlinal epidermal cell wall (Mansfield and Richardson, 1981) suggested active involvement of CWDEs in penetration. In *B. cinerea* endopolygalacturonase (PG) was detected in ungerminated conidia (Verhoeff and Warren, 1972) and two PG isozymes were detected during the infection process (Van den Heuvel and Waterreus, 1985). However, mutants of *B. cinerea* and *C. gloeosporioides* in pectinolytic genes were capable to penetrate intact host tissue (Yakoby *et al.*, 2000a).

Pathogens have evolved multiple pectin-degrading enzymes including pectate lyases. For example, *N. haematococca* has at least four pectate lyase genes (Guo *et al.*, 1996). These include constitutively expressed lyase (Guo, *et al.*, 1995), pectin-inducible pectate lyase gene (Gonzalez-Candelás and Kolattukudy, 1992), and a lyase gene that is expressed only when the pathogen is within the host (Guo *et al.*, 1996). The latter in combination with the presence of multiple genes make disruption of individual genes unsuitable to elucidate the role of lyases in pathogenesis. This was illustrated in *Nectria haematococca* where disruption of either the pectin-inducible lyase gene or the host-inducible lyase gene did not reduce virulence but when both were disrupted there was a dramatic decrease in virulence (Rogers *et al.*, 2000).

Pathogenesis-related genes have long been known to be expressed only when the pathogen is inside the host, but the host signals that trigger the expression of such genes were only identified very recently (Yang *et al.*, 2005). The *pel*D_gene of *N. haematococca* was known to be induced only in its host (pea plants), and fractionation of pea seedling extract revealed that two soluble amino acids, homoserine and asparagines, were the activating principle. The presence of these amino acids in seedlings correlated strongly with the sensitivity to *Nectria* attack. Thus, the pathogen probably coevolved with the host to use the two soluble host components to induce pathogenesis genes.

THE NECROTROPHIC STAGE

Once the host barriers have been overcome and the initial penetration has taken place, the pathogen switches from the biotrophic to the necrotrophic stage. The changes include the transformation from a quiescent to an active infection in which cell death occurs and initial symptoms are observed. Recent data suggest that it is the initial invasion of plant tissue by the various pathogens that triggers processes that activate pathogenicity factors involved in host colonization. These include a modulation of the host environment and the activation of a mechanism of cell death induction. This implies that diffusible factors that have a direct or indirect phytotoxic activity are released by the pathogen. The inducing factors may be low-molecular-weight secreted by the fungus or proteins that are secreted to the environment by the infected plant.

Modulation of environmental pH

Tissue pH is an important parameter in aqueous environments, since it affects the activities of enzymes and determines the expression of virulence genes inside the host. Many pathogens can thrive over a wide range of pH, but generate enzymes and other products only at those pH levels at which such products will function efficiently. These products include molecules that leave the organism's internal homeostasis system, as permeases, secreted enzymes and exported metabolites, (Denison, 2000; Prusky and Yakoby, 2003). A change in the ambient pH during fungal attack may be a critical factor in the expression of pathogenicity factors (Eshel *et al.*, 2002b; Prusky *et al.*, 2001, 2003; Yakoby *et al.*, 2000b).

Analysis of the endoglucanases *AaK*1 from *A. alternata* indicated that gene expression was maximal at pHs higher than 6.0, i.e., at values similar to those found in the decayed tissues in which maximal virulence is observed (Eshel *et al.*, 2002b). *C. gloeosporioides pel*B and the encoded PL are expressed and secreted at pHs higher than 5.7, a value similar to those present in decaying tissue (Prusky *et al.*, 1989; Yakoby *et al.*, 2000b, 2001). In *C. gloeosporioides*, the transcription factor *pac1* that is expressed during alkalinization conditions, follows a pattern similar to that of *pel*B, suggesting the presence of a regulatory mechanism for the control of secreted enzymes (Drori *et al.*, 2003). Similarly, for acidic environments, both *pg*1 and *pg*5 expression of *F. oxysporum* was enhanced (Caracuel *et al.*, 2003), and a sequence corresponding to the PacC consensus recognition site was found in *pg*1. A mutant of *F. oxysporum*, pacC^c carrying a dominant (constitutively expressed) PacC expression exhibited significantly diminished virulence in tomato, which has an acidic pH (5.5-6.0), This suggests that PacC is a negative regulator of genes important for fungal attack under acidic conditions (Caracuel *et al.*, 2003). The endoPG family of *B. cinerea* has been found to be differentially expressed under different pHs. Although no evidence for a PacC homologue has yet been found in *B. cinerea*, the sequence corresponding to the PacC consensus recognition site has been found in all endoPG genes (Wubben *et al.*, 1999, 2000; Manteau *et al.*, 2003). Analysis of the transcript levels of *Penicillium expansum* PG (*pepg*1) found the highest accumulation at pH 4.0 and only minor expression at pH values higher than 5 (Prusky *et al.*, 2003). Also in *Sclerotinia sclerotiorum*, the expression of endoPg, *pg*1, was specific to acidic pH conditions. Transcription of the *pac*C homologue *pac*1 declined during acidification, concomitantly with an increase in *pg1* expression; this gene was found to contain the PacC consensus recognition site in its promoter (Rollins and Dickman, 2001). Other putative virulence factors in case of *B. cinerea,* including oxalic acid, laccase and protease also were released in a pH-regulated manner, (pH range 3.1-6.0), which are close to the average pH values of the potential host tissue (Manteau *et al.*, 2003;

Movahedi *et al.*, 1990; Movahedi and Heale, 1990a, b). Protease inhibitors inhibited the colonization by pathogenic fungi such as *Mycosphaerella, Fusarium, Botrytis, Alternaria* (Vernekar *et al.*, 1999; Ye at al., 2001; Ye and Ng, 2002), which implies that proteases are important as virulence factors (Caracuel *et al.*, 2003; Manteau *et al.*, 2003; Poussereau *et al.*, 2001a,b; Rollins and Dickman, 2001). Large gene families of CWDEs as endoPGs in *B. cinerea* (van der Cruyssen *et al.*, 1994; Wubben *et al.*, 1999) and *S. sclerotiorum* (Lumsden, 1976; Rollins and Dickman, 2001), and of glucanases in *A. alternata* (Eshel *et al.*, 2002a) may be differentially expressed in correlation to the different hosts of the pathogen. The differential expression of endoPG by *B. cinerea* (ten Have *et al.*, 2001) correlated with the pH situation in two crops, apple and courgette, characterized by low and neutral pH, respectively. The expression of *Bcpg2* was negatively modulated by low ambient pH (Wubben *et al.*, 2000), which might explain its lack of expression in apple fruits, whereas *Bcpg3* expression was induced at low pH in liquid cultures (Wubben *et al.*, 2000), and it occurs in apple fruits. In addition, Manteau *et al.* (2003), examined the difference between the *B. cinerea* isolates – 630 from grapevine (approximate pH 3.5) and T8 from tomato (approximate pH 6.0) – for expression of putative virulence factors. T8 displayed a higher PG expression at the tomato pH, whereas isolated 630 secreted more laccase. This fine-tuning of enzyme induction and secretion in response to the ambient pH, not only in the host but also in fungal isolates, further demonstrates the importance of the specific regulatory system controlled by environmental pH.

Effect of the pathogen on ambient pH

The pathogen itself can dynamically alter the local pH to fit its enzymatic arsenal, with the level of pathogenicity being related to the efficiency of the pH change (Prusky *et al.*, 2001). This ability lies behind the terms "alkaline fungi" and "acidic fungi ."

Alkaline fungi – Ambient alkalization by fungi is achieved through the active secretion of ammonia produced as a result of proteases activity and deamination of amino acids. The pathogenicity of *C. gloeosporioides* and expression of the virulence factor PL-B both depend on pH increase. It was noticed by Yakoby *et al.* (2000b, 2001) that the accumulation of pectate lyase (PL) *in vitro* was accompanied by an increase in the pH of the medium from 3.8 to 7.0. In addition, avocado fruits naturally contribute to its alkalinization by increasing its pH during ripening from 5.2 to 6.0 (Prusky *et al.*, 2001, 2003; Yakoby *et al.*, 2000b, 2001). In the case of polyphage pathogens such as *A. alternata,* ammonia accumulation (threefold to tenfold increase) and pH increase (by 0.2 to 2.4 pH units) were detected in several of their hosts: tomato, pepper, melon, cherry and persimmon (Eshel *et al.*, 2002b).

Acidic fungi – Other postharvest pathogens, such as *P. expansum* (Hadas *et al.*, 2004), *P. digitatum, P. italicum, B. cinerea* (Prusky *et al.*, 2003) and *S. sclerotiorum* (Bateman and Beer, 1965) use tissue acidification in their attack realized by the accumulation of organic acids and/or H^+ excretion. *S. sclerotiorum* and *B. cinerea* decrease the host pH by secreting significant amounts of oxalic acid (Manteau *et al.*, 2003; Rollins and Dickman, 2001); gluconic and citric acids are mainly secreted by *Penicillium* (Prusky *et al.*, 2003) and *Aspergillus* (Ruijter *et al.*, 1999). *P. expansum* acidifies the tissue to pH levels of 3.5 to 4.0, at which *pepg*1 transcription was found to be significantly enhanced (Prusky *et al.*, 2003). Acidic-pH-specific expression of other members of the PG family was found in *S. sclerotiorum* (Rollins and Dickman, 2001), and of *Bcpg*3 in *B. cinerea* (Wubben *et al.*, 2000). However, acids also may act directly as a virulence factor in case of *S. sclerotiorum* as mutants lacking oxalate secretion were non-pathogenic (Godoy *et al.*, 1990). Oxalic, citric and gluconic acids exhibited strong Ca^{2+} chelating activities that weaken the plant cell wall by altering its mineral balance, and thereby affect the stability of cell membranes and cell wall pectate polymers (Cunningham and Kuiack, 1992). Also, oxalate may be directly toxic through suppression of the plant's oxidative burst (Cessna *et al.*, 2000) which would inhibit the activity of plant-produced polyphenol oxidase (Magro *et al.*, 1984; Marciano *et al.*, 1983). Taken together, these results suggest that environmental acidification is important for fungal attack.

Reactive oxygen species production as a factor for enhancing the necrotrophic stage

A frequent consequence of the occurrence of biotic stress is a perturbation in the production of reactive oxygen species (ROS), which results in changes in the redox potential of the organism. In plant-pathogen interactions, both the plant and the pathogen are involved in ROS production (Mayer *et al.*, 2001), and both possess an extensive antioxidative machinery that can moderate the damaging effects.

Most postharvest pathogens become necrotrophic during the period of colonization, and under these conditions a virtually instantaneous burst of oxidative activity occurs in plant tissues during maceration (Goodman *et al.*, 2002). This is not a pathogen-derived reaction, but a manifestation of physical damage to the host. Infection of plant tissue with *B. cinerea* provides evidence (Edlich *et al.*, 1989, Urbanek *et al.*, 1996; von Tiedemann, 1997, Govrin and Levine, 2000) that the generation of ROSs assists the colonization of the plant tissues during the necrotrophic stage. Thus, the resistive response that is expressed for host defense is utilized by the pathogen to enhance colonization.

B. cinerea does produce H_2O_2 (Bratt *et al.*, 1988) as a result of oxidase activity including glucose, xylose, galactose or ascorbate oxidase, which are commonly produced by many fungi. However, the glucose oxidase isolated by Liu *et al.* (1998) appears to be an intracellular enzyme that differs from the typical secreted sugar oxidases of other pathogens. This localization of the enzyme, and the development of a knock-out mutant of *B. cinerea* for a putative secreted glucose oxidase *bcgod*1, make it very unlikely that this enzyme would be important in creating conditions for pathogenicity.

In a different system, which acts when *P. expansum* attacks fruits, H_2O_2 and gluconic acid are produced as a result of the activity of glucose oxidase (GOX) during the pathogenic activity. Natural *P. expansum* isolates with increased pathogenicity accumulated greater amounts of gluconic acid and H_2O_2 than isolates with reduced pathogenicity (Hadas *et al.*, 2004). Reactive oxygen species resulting from GOX activity were easily detected in the decayed apple tissue, and specifically in the hyphae, but it is still not clear what was the contribution of the H_2O_2 to the necrotrophic stage of *P. expansum*.

Pathogens possess an array of enzymes for protection against ROSs during the infection process. These include guaiacol peroxidase, ascorbic peroxides, glutathione peroxidase, laccase, catalase and Cu/Zn SOD (Choi *et al.*, 1997; Gil-ad *et al.*, 2000). An extracellular catalase of *B. cinerea* (*Bccat*2) was rapidly upregulated in the presence of H_2O_2, while disruption had increased sensitivity with higher levels of mRNA of another enzyme namely glutathione S-transferase. However, this mutant was still as virulent on tomato leaves suggesting that there is no simple correlation between catalase and virulence.

This is illustrated by Van der Vlugt-Bergmans *et al.* (1997) which could not detect expression of the fungal catalase gene in the plant. The fungus rapidly metabolized H_2O_2, but it is not clear which enzyme(s) were involved (Gil-ad and Mayer, 1999). Gil-ad *et al.* (2001) suggested that the glucan sheath surrounding the mycelium of *B. cinerea* might play a role in protecting the fungus from the host response. It is important to know whether these microbial enzymes are of the extracellular or intracellular form: the latter might serve to protect the fungus from its own ROSs, whereas the former might be involved in protecting the fungus from the plant ROSs.

Comparison of the aggressiveness of six isolates of *B. cinerea* during the infection of bean leaves led von Tiedemann (1997) to conclude that the primary role of ROSs was in the induction of plant cell death. However, Dat *et al.* (2001) suggested that enhanced levels of H_2O_2 are not, in themselves, the direct cause of cell death, but that they trigger a signal transduction cascade that ultimately leads to an active cell death process. This response would be important for defense against biotrophs but would presumably increase the vulnerability of the host to necrotrophs, such as postharvest pathogens.

MECHANISM-ENCODING FACTORS THAT REGULATE THE SIGNAL EXPRESSION OF GENES DURING COLONIZATION

The fungal responses required for the various processes of fungal penetration and development require a network of signal transduction pathways, such as the activation of G proteins, cyclic adenosine monophosphate (cAMP) signaling, and mitogen-activated protein kinase (MAPK) cascades, to communicate the perceived external signal to the fungal genome so that specific genes are activated and expressed, in order to enhance fungal morphogenesis and colonization. Although significant progress has been achieved in studies of several different systems, our knowledge of the types of signals/communication between postharvest pathogens and their hosts is relatively scanty.

Gα subunits of heterotrimeric G proteins

Heterotrimeric guanine nucleotide-binding proteins (G proteins) are involved in regulating a variety of cellular functions in eukaryotic cells; they act as transducers between activated cell-surface receptors and intracellular effectors. In *B. cinerea* two Gα subunits, genes *bcg*1 and *bcg*2, have been identified (Schulze Gronover *et al.*, 2004). Deletion mutants revealed that the Gα subunits affect growth and fungal pathogenicity in different ways. The *bcg*1-product controlled multiple functions, including vegetative growth, pigmentation, proteolytic activity and pathogenicity. Further, it played a major role in the process of colonization of host tissue, since it inhibits lesion spreading after penetration. The *bcg*2 disruption mutants exhibited an infection process similar to that of the wild type (WT), except that the lesions caused by conidia of the mutant spread more slowly.

The host genes affected by these pathogen genes were identified by suppression subtractive hybridization. Among the 22 differentially expressed genes were found several that encoded for unknown proteases, some for enzymes involved in secondary metabolism, and others that encoded cell-wall-degrading enzymes (Schulze Gronover *et al.*, 2004).

cAMP signaling pathway

The cAMP-dependent signaling pathway regulates several important processes in plant pathogenic fungi, such as morphogenesis, differentiation and virulence.

In *B. cinerea*, the *bac* gene encoding adenylate cyclase was cloned and characterized (Klimpel *et al.*, 2002): the BAC protein consists of functional domains typical of adenylate cyclases, such as the "Ras association" motif, the middle leucine-rich repeat regions, the catalytic domain and the C terminus with a putative binding site for the cyclase-associated protein (CAP). Expression studies of BAC indicate that expression occurs as early as the beginning of necrosis development (12 h post-inoculation) and continues until at least 36 hours later, when spreading soft rot lesions start to grow out of the primary necrotic spots (Klimpel *et al.*, 2002).

In light of the suggestion that both BAC and BCG1 positively influence the production of cAMP, the intracellular cAMP levels were measured in the WT, *Δbcg1*, *Δbcg*2 and *Δbac* mutant. Deletion of *bac* resulted in an 85% reduction of the intracellular cAMP level, which remained constant for up to 6 days; the *Δbcg1* showed about 50% reduction of cAMP, which, however, increased after 6 days to a level similar to that in the WT. The aggressiveness of the *Δbac* mutants was significantly reduced and was comparable with that of the *Δbcg* mutants. The spreading of lesions after inoculation was much slower than that caused by the WT strain and, in addition, no conidia developed on the surface of the *Δbac*-infected leaves. All these data suggest that in *B. cinerea* adenylate cyclase plays an important role in vegetative development and aggressiveness.

MAP kinase pathways

Several MAP kinase genes have an important role in vegetative and sexual development of different fungi including osmoregulation and pathogenicity.

In *B. cinerea*, the *pmk*1 homologue, *bmp1*, was cloned and found essential for pathogenesis (Zheng *et al.*, 2000): the *Δbmp1* mutant grew and sporulated as the WT, but was non-pathogenic on carnation and rose flowers and failed to elicit a plant defense response. Conidia from *Δbmp1* mutants still germinated on plant surfaces, but lost their ability to penetrate and macerate the plant tissue. Wounding of a plant did not overcome the penetration defect, which indicates that both the cAMP and the MAP kinases pathways are involved in the infection process.

An MAPK gene was cloned from *N. haematococca*, and immunoblot analysis showed that this kinase was expressed by the pathogen (Li *et al.*, 1997). Disruption of MAPK (CgMBK) in *C. gloeosporioides* resulted in the loss of the pathogen's ability to form appressoria in response to the host signals, and in loss of virulence (Kim *et al.*, 2000b). This kinase involved with two stages of appressorium differentiation namely: (a) the polarized cell division with preferential increase in F-actin in one of the daughter nuclei after the nuclear division and septum formation; and (b) differentiation of the germ tube into an appressorium.

Genes of Ras superfamily

Ras proteins are a superfamily of small GTP-binding proteins that are highly conserved in all eukaryotic organisms and that are involved in several processes of morphogenesis, differentiation, nutrient sensing and pathogenicity. However, few reports indicate the importance of these genes in the pathogenicity of postharvest pathogens.

In *C. lindemuthianum* Dumas *et al.* (2002) and Siriphutthaiwan *et al.* (2003) demonstrated that small G proteins belonging to the Rab subfamily of the Ras superfamily could affect pathogenicity.

In the case of *B. cinerea*, seven small G proteins of the Ras superfamily have been cloned so far. Among them is one Ras-encoding gene homologous to *ras*2 of *S. cerevisiae* and one gene homologous to CLPT1. Deletion of both genes showed their involvement in morphogenesis, conidiation and pathogenesis (Tudzynski and Schulze Gronover, 2004).

MECHANISMS OF HOST COLONIZATION BY POSTHARVEST PATHOGENS

Postharvest pathogens elicit two main types of symptoms: soft and dry rots; decay caused by *Penicillium, Botrytis, Colletotrichum* and *Monilinia* leads to soft rots, whereas in *Alternaria* attacks mainly a dry rot is observed. In some cases, however, a given pathogen, such as *A. alternata* attacking citrus plants, may cause either soft or dry rot. The plant cells are made up of several different types of polysaccharides: the primary cell wall consists of cellulose and hemi-cellulose, whereas the middle lamella has a high concentration of pectin. Pectin, a complex of various polygalacturonans, also extends into the primary wall. Pathogens that affect the pectin and the primary cell wall lead to cell wall maceration and result in soft rots, whereas those that mostly attack the cellulose layer tend to kill the host cells but preserve the structure of the tissue. The occurrence of multiple fungal genes encoding enzymes that can degrade physical barriers on host plants makes it necessary to evaluate results from single gene disruption carefully, as indicated elsewhere in this chapter.

Pectinases

Pectin is a major component of the plant cell wall and consists of three main types of polygalacturonans: homogalacturonan, rhamnogalacturonan I, and rhamnogalacturonan II. Homogalacturonans are built of α-1, 4-linked chains of D-galacturonic acids that can be methylated. Highly methylated homogalacturonan is designated pectin, and that with a low degree of methylation is called pectate. All the enzymes that are able to degrade pectic components are considered pectinases, and several of them are produced during penetration.

Pectin methyl esterase – Most pectinolytic enzymes cannot degrade the highly methylated pectin, and therefore, to enable penetration, the pectin must be demethylated to pectate. Two *B. cinerea* pectin methyl esterases (BcPMEs) with differing molecular masses were described by Marcus and Schejter (1983). In contrast, Reignault *et al.* (1994) identified another two isozymes in a different strain of *B. cinerea* (Bd90), with identical molecular masses, but with differing pI values (pI 7.0 and 7.4). Gene disruption of the pectin methyl esterase gene *Bcpme*1 in the Bd90 strain revealed that *B. cinerea* has more than one pectin methyl esterase-encoding gene. The *Bcpme*1 disrupted mutant showed 75% reduction in PME activity and was less virulent on apples and grapevine (Valette-Collet *et al.*, 2003). A second BcPME isozyme (pI 7.1) was detected in strain Bd90 (Valette-Collet *et al.*, 2003), but it was hypothesized to play a less prominent role than the *Bcpme*1. Different strains of *Penicillium* sp. and *C. gloeosporioides* (Ortega, 1996) were also reported to produce PME, but no reference was made to their possible importance during the initial stages of interaction with the fruit host.

Endopolygalacturonase – Endopolygalacturonases are endo-acting enzymes that catalyze the hydrolysis of homogalacturonan, resulting in substrate fragmentation. These enzymes are not able to hydrolyse highly methylated pectin, but first require the action of pectin methyl esterase to demethylate pectin to pectate. The first BcPG-encoding genes were cloned and characterized by Wubben *et al.* (1999), who described a gene family that encoded for six isozymes of diverse biochemical characteristics. Characterization of five expressed genes revealed that these isozymes differed from each other in specific activity, protein stability, substrate preference and end-products (Kars and Van Kan, 2004). Disruption of *Bcpg*1 resulted in a reduction in virulence on both tomato and apple fruits (Ten Have *et al.*, 1998). Disruption of *Bcpg*2 also played an importance in virulence in other hosts.

For *Alternaria citri*, the production of an extracellular endoPG with a molecular mass of 60 kDa was also demonstrated by Isshiki *et al.* (1997, 2001), and following the disruption of the gene, the fungus showed a dramatic 85% reduction in development of soft rot symptoms, because its penetration and maceration of citrus tissue were inhibited (Isshiki *et al.*, 2001). Hyphae of the wild type, but not of the mutant, could develop by penetration into the fruit peel from the pedicel, through the pectin-rich central axis, which suggests the importance of endoPG as a factor for pathogenesis in this fungus. However, both isolates spread equally in the juice sac region of citrus fruits (Isshiki *et al.*, 2003, 2001). For *Penicillium* the production of endo PG was also described as a factor for pathogenesis. *P. olsonii* (Wagner *et al.*, 2000) secretes at least three different several polygalacturonases (PGs) with molecular masses of about 47 kDa, which include several basic and acidic isoforms. The gene *pg*1 enodes the acid PG activity and *pg*2 the alkalic *P. expansum*, attacking deciduous fruits, also produced a PG with a basic pI and molecular mass of 34 kDa (Yao *et al.*, 1996).

C. gloeosporioides also produced an endopolygalacturonase with a molecular mass of 62 kDa. This enzyme was able to macerate ripe avocado fruit tissue. Unripe tissue contained inhibitory concentrations of a flavonoid inhibitor of the enzyme and was not macerated by the enzyme. This observation indicates a role of this enzyme in fruit pathogenesis.

Exopolylgalacturonase – The exopolygalacturonases cleave monomeric or dimeric glucosyl groups from pectic polysaccharides, thereby providing the fungus with potential nutrients. *B. cinerea* produced exoPG on tomato fruits (Verhoeff and Warren, 1972). Johnston and Williamson (1992) were the first to purify and characterize *B. cinerea* exoPGs, which has molecular masses of 65 and 70 kDa, as confirmed by Rha *et al.* (2001). Secretion of exoPG was detected in cucumber leaves 9 h after inoculation with *B. cinerea*, which suggests that these enzymes play a role in the early stages of infection and subsequent tissue maceration. Apart from these examples in *B. cinerea*, there have been no reports of the production of exoPG by postharvest pathogens.

Pectin lyase and pectate lyase – Pectin lyase is a pectin-degrading enzyme that cleaves homogalacturonan with a high degree of methyl esterification; it is inactive at acidic pH. Pectin isozymes were detected in extracts of ungerminated conidia and in the extracellular matrix of *B. cinerea* germlings (Movahedi and Haele, 1990b; Chilosi and Magro, 1997; Doss, 1999):

pectin lyase was produced soon after inoculation of zucchini fruits, but not in infected apple tissue – which is very acidic (pH 3-4) whereas zucchini has a more neutral pH. Since *B. cinerea* acidifies its environment prior to pectin degradation, pectin lyases are in any case unlikely to contribute significantly to pectin degradation in the early stages of infection by pathogens such as *Botrytis* and *Penicillium*, which acidify the environment. No results have been reported for pathogens that alkalinize the environment, such as *Colletotrichum* and *Alternaria*. Pectin lyase from *Aspergillus aculeatus* was reported to be the main enzyme responsible for the maceration of potato tubers (van den Broek *et al.*, 1997).

Pectate lyase catalyses the cleavage of pectate, which is the unmethylated homogalacturonan. These enzymes are inactive at acidic pH and the presence of Ca^{2+} ions is essential for catalysis. *C. gloeosporioides* produced pectin lyase A (*pnlA*) (Bowen *et al.*, 1995; Templeton *et al.*, 1994), and pectate lyases *pel*B (Wattad *et al.*, 1997), *pel-1* and *pel-2* (Shih *et al.*, 2000), during the colonization of infected tissue. Disruption of *pel*B by homologous recombination yielded independent isolates that did not produce and secrete PL, and that exhibited 25% lower pectate lyase (PL) and pectin lyase activities, but 15% higher polygalacturonase (PG) activity than the wild type.

When *pel*B mutants were inoculated onto avocado fruits, a 36 to 45% reduction in estimated decay diameter was observed: the reduction in virulence by *ΔpelB*, on the one hand, and the lack of effect of Δ*pnl*A, on the other hand (Bowen *et al.*, 1995), highlight the importance of PL as a pathogenicity factor of *C. gloeosporioides*. The presence of multiple pectate lyase genes that are differentially regulated and their role in penetration into the host tissue were briefly discussed above.

Cellulases – The cellulolytic complex comprises, among others, endoglucanase, cellobiohydrolase and β-glucosidase, and it degrades cellulose into cellobiose and glucose. Verhoeff and Warren (1972) did not detect cellulase activity in either ungerminated or germinated conidia of *B. cinerea*. In *A. alternata* five glucanase genes, corresponding to the C, F and K families, were cloned by means of "family-specific" oligonucleotide primers. The genes, present in a single copy, encoded for exoglucanases *AaC1* and *AaC2*, endoxylanase *AaF1*, endoglucanase *AaK1*, and the mixed-linked glucanase *AaMLG1* (Eshel *et al.*, 2002a, b). RT-PCR analysis of RNA extracted from persimmon fruits 2 and 4 days after infection with *A. alternata* (see Figure 3) showed expression of all five glucanase genes.

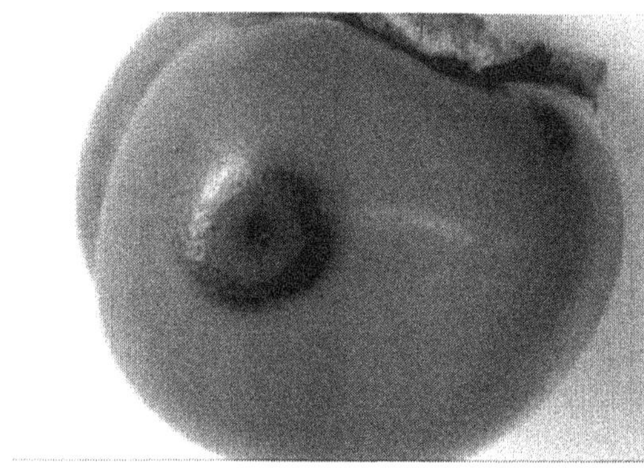

Figure 3. *Alternaria alternata* symptoms in persimmon fruits.

However, transcription levels and enzyme production of the endoglucanase (*AaK1*) and of one exoglucanase (*AaC1*) were enhanced during *A. alternata* growth on cell walls taken from more susceptible fruits, whereas expression of these genes and their enzyme production were significantly reduced in fruits showing resistance to fungal attack. Those results suggest the involvement of endo- and exoglucanase in the development of symptoms elicited by *A. alternata* in resistant and susceptible persimmon fruits (Eshel *et al.*, 2002a, b).

Transporters facilitating fungal attack

In natural environments microorganisms are exposed to a wide variety of antibiotic compounds produced by competing organisms. Some of the postharvest pathogenic organisms have evolved various mechanisms to overcome the natural resistance as in the case of resistance of *B. cinerea* to the phytoalexin resveratrol in grapes. *B. cinerea* has a broad host range and, consequently, is liable to be exposed to many plant defense compounds (Hayashi *et al.*, 2002). Two major classes of membrane transport proteins that have been reported to be involved in transport of fungitoxic compounds are ATP-binding cassette (ABC) transporters and major facilitator (MFS) transporters. ABC transporters are primary transporters that use the energy of ATP hydrolysis to drive transport against a concentration gradient. MFS transporters are secondary transporters that use the proton motive force to transport compounds. Genes encoding ABC and MFS transporters belong to large gene families. Schoonbeek *et al.* (2001) cloned an ABC transported from *B. cinerea* (*Bcatr*B) and its expression was enhanced after treatment of germlings with the grapevine phytoalexin resveratrol, but deletion mutants exhibited increased sensitivity and decreased virulence on tomato or grapevine leaves.

Secondary metabolite degraders: Laccases

Extensive studies have been performed on laccase activity in *B. cinerea*. Laccases are part of a larger group of multicopper enzymes, detected in several fungal species, that act on a variety of polyphenol substrates which have been implicated in a wide range of biological processes that affect fungal virulence (Guetsky *et al.*, 2005). The production of gallic acid-inducible laccase in culture was suppressed by secondary metabolites called cucurbitacins, produced by *Cucurbitacea* (Viterbo *et al.*, 1993). Cucurbitacins protected cucumber fruits and cabbage leaves from infection by *B. cinerea* (Bar-Nun and Mayer, 1990), which suggests that laccase plays an important role in pathogenesis (Viterbo *et al.*, 1993; Staples and Mayer, 1995). Cucurbitacin reduce laccase activity (Viterbo *et al.*, 1993) via repression at the mRNA level (Gonen *et al.*, 1996). Deletion of *Bclcc*1 or *Bclcc*2 did not result in any detectable reduction of virulence (Schouten *et al.*, 2002), suggesting that cucurbitacins do not protect leaves by laccase activity.

Laccase activity in *C. gloeosporioides* degraded the flavonoid epicatechin when used as a substrate (Guetsky *et al.*, 2005). Epicatechin modulates the metabolism of preformed antifungal compounds in avocado and activates quiescent *C. gloeosporioides* in ripening avocado fruits (Prusky, 1996). Extracts of laccase enzyme obtained from decayed tissue and from culture media fully metabolized the epicatechin substrate within 4 and 20 h, respectively. Isolates of *C. gloeosporioides* with reduced laccase activity and no capability to metabolize epicatechin showed no pathogenicity on ripening fruits. In contrast, isolates with enhanced capability to metabolize epicatechin elicited early symptoms of disease in unripe fruits. These results suggest that degradation of epicatechin by laccase, followed by the decline of the preformed antifungal diene compound, may activate the quiescent infections in ripening avocado fruits.

NEW APPROACHES TO CONTROL OF POSTHARVEST DISEASES

Approaches based on fungal biology

Fungal viruses for biocontrol – Fungi, including fruit pathogens are attacked by a range of other organisms, which include other fungi, bacteria and viruses, and some of these agents have been exploited for biocontrol of fruit pathogens. A study of the mycoviruses of *B. cinerea* and *Monilinia fructicola* found that over 70% of strains harbor viruses which, however, have little if any effect on their hosts (Howitt *et al.*, 1995; Tsai *et al.*, 2004). Recently a virus has been discovered in Europe that reduces pathogenicity of *B. cinerea* (Castro *et al.*, 2003) and could be developed as a future biocontrol agent.

Reduced-pathogenicity strains for biocontrol – Classically the use of fungi for biocontrol involves the use of other fungal species to control the target pathogen. A more subtle approach involves the use of strains of the pathogens themselves or of similar species with reduced pathogenicity as a biocontrol. This approach is based on the facts that the biocontrol agent may efficiently displace the target fungus ecologically and physiologically and, at the same time, stimulate host defences in advance of infection (Droby *et al.*, 2004). Several examples have demonstrated these possibilities: preinoculation with a non-aggressive strain of *B. cinerea* reduced the ability of a subsequently inoculated aggressive strain to cause lesions (Weeds *et al.*, 2000); and similarly, reduced-pathogenicity isolates of *C. gloeosporioides* induced by inclusion of a hygromycin resistance cassette induced resistance to pathogenic strains in avocado (Yakoby *et al.*, 2002). This approach could easily be explored for the control of postharvest diseases by means of naturally occurring variants of reduced-pathogenicity strains.

Modulation of pathogenicity factors – Postharvest pathogenic fungi modulate the host pH as a basis for expression of virulence factors during the colonization of the target host tissue (See 3.1). Pathogens may modulate their virulence by local acidification or alkalinization the host tissue, e.g., colonization of acidified citrus and apple tissues by *Penicillium* spp. was enhanced by low pH. A simple approach based in these findings comprises the application of neutralizing solutions, depending on the type of pathogen it is desired to control. This approach is important for the control of postharvest disease because (i) it directly affects the germination of conidia and fungal colonization; and (ii) it affects the efficiency of fungicides that are used to control these diseases. This approach is being used for the control of *A. alternata* in stored mango fruits (Prusky *et* al., 2006).

Approaches based on enhancing host defences

Host resistance in fruit and vegetables is very complex and changes during fruit ripening and senescence. A defence mechanism may be expressed continuously throughout the life of a host, regardless of the presence or absence of a pathogen (constitutive and preformed resistance) or the defence machinery may be turned on at a particular point in time, either by a pathogen or by a sensitising treatment (inducible resistance). The manipulation of these natural resistance processes offers opportunities to develop durable disease management.

Treatments inducing preformed resistance mechanisms – Prusky *et al.* (1991) described the capability of high-CO_2 treatment to increase the level of the preformed antifungal diene that is present in avocado fruits and hence to increase the resistance of ripening avocado fruits to *C. gloeosporioides*. More recently it was also found that short cold treatment (4°C, 24 h) also activated the mechanism of resistance by inducing higher levels of the antifungal diene synthesis (Madi *et al.*, 2003). These treatments that have the advantage of being environmentally benign and residue free should further be analyzed for commercial considerations (Prusky *et al.*, 1991). More recently, Anderson *et al.* (2004) had used potassium silicate applied as trunk injection or a paint to the tree to stimulate the expression of natural defence reaction by accumulation of antifungal phenols and activation of defence related enzymes (chitinase and β-1,3 glucanase).

In summary, novel approaches based on use of biological control and exploitation of host resistance to enhance disease control offer the prospects of sustainable and environmentally benign disease management. However, these approaches still demand investment in long-term research before clear results can be obtained. Until then, the traditional approaches based on the use of synthetic fungicides in combination with physical treatments such as temperature (high and low) and controlled atmosphere should be efficiently applied.

REFERENCES

Anderson, J., Pegg, K., Coates, L., Dann, L., Cooke, T., Smith,T. and Dean, J. (2004). Silicon and disease management in avocados. Talking Avo. 15: 23-25.

Bar-Nun, N., and Mayer, A. M. (1990). Cucurbitacins protect cucumber tissue against infection of *Botrytis cinerea*. Phytochemistry 29:787-792.

Bateman, D. F., and Beer, S. V. (1965). Simultaneous production and synergistic action of oxalic acid and polygalacturonase during pathogenesis by *Sclerotium rolfsii*. Phytopathology 58:204-211.

Berto, P., Belingheri, L., and Dehorter, B. (1997). Production and purification of a novel extracellular lipase from *Alternaria brassicicola*. Biotechnology Letters 19:533-536.

Berto, P., Commenil, P., Belingheri, L., and Dehorter B. (1999). Occurrence of a lipase in spores of *Al-*

ternaria brassicicola with a crucial role in the infection of cauliflower leaves. FEMS Microbiology Letters 180:183-189.

Bostock, R. M., Wilcox, S. M., Wang, G., and Adaskaveg, J. E. (1998). Suppression of *Monilinia fructicola* cutinase production by peach fruit surface phenolic acids. Physiological and Molecular Plant Pathology 54:37-50.

Bowen, K. J., Templeton, D. M., Sharrock, R. K., Crowhurst, N. R., and Rikkerink, A. H. E. (1995). Gene inactivation in the plant pathogen *Glomerella cingulata*: three strategies for the disruption of the pectin lyase gene *pnlA*. Molecular and General Genetics 246:196-205.

Bratt, R. P., Brown, A. E., and Mercer, P. C. (1988). A role for hydrogen peroxide in degradation of flax fibre by *Botrytis cinerea*. Transactions of the British Mycological Society 91:481-488.

Broek L. A. M. van den, Aantrekker, E. D. den, Voragen, A. G. J., Beldman. G., and Vincken, J. P. (1997). Pectin lyase is a key enzyme in the maceration of potato tuber. Journal of the Science of Food and Agriculture 75: 167-172.

Caracuel, Z., Roncero, M. I. G., Espeso, E. A., Gonzales-Verdejo, C. I., Garcia-Maceira, F. I., and Di Pietro, A. (2003). The pH signaling transcription factor PacC controls virulence in the plant pathogen *Fusarium oxysporum*. Molecular Microbiology 48:765-779.

Castro, M., Karmer, K., Valdivia, L., Ortiz, S., and Castillo, A. (2003). A double-stranded RNA mycovirus confers hypovirulence-associated traits to *Botrytis cinerea*. FEMS Microbiology Letters 229:87-91.

Cessna, S., Sears, V., Dickman, M., and Low, P. (2000). Oxalic acid, a pathogenicity factor of *Sclerotinia sclerotiorum*, suppresses the host oxidative burst. Plant Cell 12:2191–2199.

Chilosi, G., and Magro, P. (1997). Pectin lyase and polygalacturonase isoenzyme production by *Botrytis cinerea* during the early stages of infection on different host plants. Journal of Plant Pathology 78:61-69.

Chitarra, G. S., Abee T., Rombouts, F. M., and Dijksterhuis, J. (2005). 1-Octen-3-ol inhibits conidia germination of *Penicillium paneum* despite of mild effects on membrane permeability, respiration, intracellular pH, and changes the protein composition. FEMS Microbiology Ecology 54: 67-75

Choi, G. J., Lee, H. J., and Cho, K. .Y. (1997). Involvement of catalase and superoxide dismutase in resistance of *Botrytis cinerea* to dicarboximide fungicide vinclozolin. Pesticide Biochemistry and Physiology 59:1-10.

Commenil, P., Brunet, L., and Audran, J. C. (1995). The development of the grape berry cuticle in relation to susceptibility to bunch rot disease. Journal of Experimental Botany 48:1599-1607.

Commenil, P. L., Belingheri, L., and Dehorter, B. (1998). Antilipase antibodies prevent infection of tomato leaves by *Botrytis cinerea*. Physiological and Molecular Plant Pathology 52:1-14.

Crowhurst, R. N., Binnie, S. J., Bowen, J. K., Hawthorn, B. T., Plummer, K. M., Rees-George, J., Rikkernik, E. H., and Templeton, M. D. (1997). Effect of disruption of a cutinase gene (cutA) on virulence and tissue specificity of *Fusarium solani* f. sp. *cucurbitae* race 2 toward *Cucurbita maxima* and *C. moschata*. Molecular Plant-Microbe. Interactions 10:335-368.

Cruyssen, G. van der, Meester, E. de, and Kamoen, O. (1994). Expression of polygalacturonase of *Botrytis cinerea* in vitro and in vivo. Mededelingen van de Faculteit Landbouwwetenschappen van de Rijksuniversiteit Gent. 59:895-905.

Cunningham, J. E., and Kuiack, C. (1992). Production of citric and oxalic acids and solubilization of calcium phosphate by *Penicillium bilaii*. Applied and Environmental Microbiology 58:1451-1458.

Dat, J. F., Inze, D., and Van Breusegem, F. (2001). Catalase-deficient tobacco plants: tools for in planta studies on the role of hydrogen peroxide. Redox Report 6:37-42.

Danzig, A. H., Zuckermann, S. H., and Andonov-Roland, M. M. (1986). Isolation of a *Fusarium solani* mutant reduced in cutinase activity and virulence. Journal of Bacteriology 168:911-916.

Dean, R. A., Talbot, N. J., Ebbole, D. J., *et al.* (2005). The genome sequence of the rice blast fungus *Magnaporthe grisea*. Nature 434:980-986.

Denison, S. H. (2000). pH regulation of gene expression in fungi. Fungal Genetics and Biology 29:61-71.

Dickman, M. B., and Patil, S. S. (1986). Cutinase deficient mutants of *Colletotrichum gloeosporioides* are non-pathogenic to papaya fruit. Physiological and Molecular Plant Pathology 28:235-242.

Dickman, M. B., Podila, G. K., and Kolattukudy, P. E. (1989). Insertion of cutinase gene into a wound pathogen enables it to infect intact host. Nature 342:446-448.

Doss, R. P. (1999). Composition and enzymatic activity of the extracellular matrix secreted by germlings of *Botrytis cinerea*. Applied and Environmental Microbiology 65:255-263.

Droby, S., Wisniewski, M. E., Wilson, C. L., and El-Ghaouth, A. (2004). Biologically-based technology for the control of postharvest diseases. *In* Food Microbial Contamination (Wilson, C., and

Droby, S., eds.), CRC Press, Boca Raton, FL., U.S.A., pp. 187-205.

Drori, N., Kramer-Haimovich, H., Rollins, J., Dinoor, A., Okon, Y., Pines, O., and Prusky, D. (2003). External pH and nitrogen source affect secretion of pectate lyase by *Colletotrichum gloeosporioides*. Applied and Environmental Microbiology 69:3258-3262.

Dumas, B., Borel, C., Herbert, C., Maury, J., Jacquet, C., Balsse, R., and Esquerre'-Tugaye', M. T. (2002). Molecular characterization of CLPT1, a SEC4-like Rab/GTPase of the phytopathogenic fungus *Colletotrichum lindemuthianum* which is regulated by carbon source. Gene 272:219-225.

Edgmond, M. R., and Vlieg, J. de (2000). *Fusarium solani pisi* cutinase. Biochimie 82:1015-1021.

Edlich, W., Lorenz, G., Lyr, H., Nega, E., and Pommer, E-H. (1989). New aspects on the infection mechanism of *Botrytis cinerea* Pers. Netherlands Journal of Plant Pathology 95 (Sup 1):53-62.

Ende, G. van den, and Linskens, H. F. (1974). Cutinolytic enzymes in relation to pathogenesis. Annual Reviews of Phytopathology 12:247-258.

Eshel, D., Lichter, A., Dinoor, A., and Prusky, D. (2002a). Characterization of *Alternaria alternata* glucanase genes expressed during infection of resistant and susceptible persimmon fruits. Molecular Plant Pathology 3: 347-358.

Eshel, D., Miyara, I., Ailinng, T., Dinoor, A., and Prusky, D. (2002b). pH regulates endoglucanase expression and virulence of *Alternaria alternata* in persimmon fruits. Molecular Plant-Microbe Interactions 15: 774-779.

Fan, C. Y., and Köller, W. (1998). Diversity of cutinases form plant pathogenic fungi: differential and sequential expression of cutinolytic esterases by *Alternaria brassicicola*. FEMS Microbiology Letters 158:33-38.

Flaishman, M. A., and Kolattukudy, P. E. (1994). Timing of fungal invasion using host's ripening hormone as a signal. Proc. Natl. Acad. Sci. U.S.A. 91:6579-6583.

Francis, S. A., Dewey, F. M., and Gurr, S. J. (1996).The role of cutinase in germling development and infection by *Erysiphe graminis* f. sp. *hordei*. Physiological and Molecular Plant Pathology 49:201-211.

Gevens, A., and Nicholson, R. L. (2000). Cutin composition: A subtle role for fungal cutinase. Physiological and Molecular Plant Pathology 57:43-45.

Gil-ad, N. L., and Mayer, A. M. (1999). Evidence for rapid breakdown of hydrogen peroxide by *Botrytis cinerea*. FEMS Microbiology Letters 176:455-461.

Gil-ad, N. L., Bar-Nun, N., Noy, T., and Mayer, A. M. (2000). Enzymes of *Botrytis cinerea* capable of breaking down hydrogen peroxide. FEMS Microbiology Letters 190:121-126.

Gil-ad, N. L., Bar-Nun, N., and Mayer, A. M. (2001). The possible function of the glucona sheath of *Botrytis cinerea*: effects on the distribution of enzyme activities. FEMS Microbiology Letters 199:109-113.

Gilbert, R. D., Johnson, A. M., and Dean, R. A. (1996). Chemical signals responsible for appressorium formation in the rice blast fungus *Magnaporthe grisea*. Physiological and Molecular Plant Pathology 48:335-346.

Godoy, G., Steadman, J. R., Dickman, M. B., and Dam, R. (1990). Use of mutants to demonstrate the role of oxalic acid in pathogenicity of *Sclerotinia sclerotiorum* on *Phaseolus vulgaris*. Physiological and Molecular Plant Pathology 37: 179-191.

Gonen, L., Viterbo, A., Cantone, F., Staples, R. C., and Mayer, A. M. (1996). Effect of cucurbitacins on mRNA coding for laccase in *Botrytis cinerea*. Phytochemistry 42:321-324.

Gonzalez-Candelás, L., and Kolattukudy, P. E. (1992). Isolation and analysis of a novel inducible pectate lyase gene from the phytopathogenic fungus *Fusarium solani* f. sp. *pisi (Nectria haematococca,* Mating population VI*)*. Journal of Bacteriology 174:6343-6349.

Goodman, B. A., Glidewell, S. M., Arbuckle, C. M. Bernardin, S., Cook, T. R., and Hillman, J. R. (2002). An EPR study of free radical generation during maceration of uncooked vegetables. Journal of the Science of Food and Agriculture 82:1208-1215.

Govrin, E. M., and Levine, A. (2000). The hypersensitive response facilitates plant infection by the necrotrophic pathogen *Botrytis cinerea*. Current Biology 10:751-757.

Guetsky, R., Kobiler, I., Wang, X., Perlman, N., Gollop, N., Avila-Quezada, G., Hadar, I., and Prusky, D. (2005). Metabolism of the flavonoid epicatechin by laccase of *Colletotrichum gloeosporioides* and its effect on pathogenicity on avocado fruits. Phytopathology 95:1341-1348.

Guo, W., González-Candelas, L., and Kolattukudy, P. E. (1995). Cloning of a novel constitutively expressed pectate lyase gene *pel*B from *Fusarium solani* f. sp. *pisi (Nectria haematococca,* mating type VI) and characterization of the gene product expressed in *Pichia pastoris*. Journal of Bacteriology 177:7070-7077.

Guo, W., González-Candelas, L., and Kolattukudy, P. E. (1996). Identification of a novel *pel*D gene expressed uniquely in planta by *Fusarium solani* f. sp. *pisi* (*Nectria haematococca,* mating type VI) and characterization of its protein product as an endo-pectate lyase. Archives of Biochemistry and Biophysics 332:305-312.

Hadas, Y., Goldberg, I., Pines, O., and Prusky, D. (2004). Tissue acidification by *Penicillium* as a mechanism to enhance pathogenicity. 7th European Conference on Fungal Genetics, Copenhagen, April 2004.

Have, A. ten, Mulder, W., Visser, J., and Kan, J. A. L. van (1998). The endopolygalacturonase gene *Bcpg*1 is required for full virulence of *Botrytis cinerea*. Molecular Plant-Microbe Interactions 11:1009-1016.

Have, A. ten, Beuil, W. O., Wubben, J. P., Visser, J., and Kan, J. A. L. van (2001). *Botrytis cinerea* endopolygalacturonase genes are differentially expressed in various plant tissues. Fungal Genetics and Biology 33:97-105.

Hayashi, K., Schoonbeek, H., and De Waard, M. A. (2002). *Bcmfs1*, a novel major facilitator superfamily transporter from *Botrytis cinerea*, provides tolerance towards the natural toxic compounds camptothecin and cercosporin, and towards fungicides. Applied and Environmental Microbiology 68:4996-5004.

Hegde, Y., and Kolattukudy, P. E. (1997). Cuticular waxes relieve self-inhibition of germination and appressorium formation by the conidia of *Magnaporthe grisea*. Physiological and Molecular Pathology 51:75-84.

Heuvel, J. van den, and Waterreus, L. P. (1985). Pectic enzymes associated with phosphate-stimulated infection of French bean leaves by *Botrytis cinerea*. Netherlands Journal Plant Pathology 91:253-264.

Howitt, R. L. J., Beever, R. E., Pearson, M. N., and Foster, R. L. S. (1995). Presence of double-stranded RNA and virus like particles in *Botrytis cinerea*. Mycological Research 99:1472-1478.

Hwang, C.-S., and Kolattukudy, P. E. (1995). Isolation and characterization of genes expressed uniquely during appressorium formation by *Colletotrichum gloeosporioides* conidia induced by the host surface wax. Molecular and General Genetics 247:282-294.

Hwang, C.-S., Flaishman, M. A., and Kolattukudy, P. E. (1995). Cloning of a gene expressed during appressorium formation by *Colletotrichum gloeosporioides* and a marked decrease in virulence by disruption of this gene. Plant Cell 7:183-193.

Isshiki, A., Akimitsu, K., Yamamoto, M., and Yamamoto, H. (2001). Endopolygalacturonase is essential for citrus black rot caused by *Alternaria citri* but not brown spot caused by *Alternaria alternata*. Molecular Plant-Microbe Interactions 14:749-757.

Isshiki, A., Akimitsu, K., Nishio, K., Tsukamoto, M., and Yamamoto, H. (1997). Purification and characterization of an endopolygalacturonase from the rough lemon pathotype of *Alternaria alternata*, the cause of citrus brown spot disease. Physiological and Molecular Plant Pathology 51:155-167.

Isshiki, A., Ohtani, K., Kyo, M., Yamamoto, H., and Akimitsu, K. (2003). Green fluorescent detection of fungal colonization and endopolygalacturonase gene expression in the interaction of *Alternaria citri* with citrus. Phytopathology 93:768-773.

Jennings D. H. (1989). Some perspectives on nitrogen and phosphorous metabolism in fungi. *In* Nitrogen, Phosphorous and Sulphur Utilization by Fungi (Boddy, L., *et al*., eds.), Cambridge: Cambridge University Press, U. K., pp. 1-31.

Johnston, D. J., and Williamson, B. (1992). An immunological study in the induction of polygalacturonases of *Botrytis cinerea*. Mycological Research 96:343-349.

Kämper, J., Kämper, U., Rogers, L. M., and Kolattukudy, P. E. (1994). Identification of regulatory elements in the cutinase promoter from *Fusarium solani* f. sp. *pisi* (*Nectria haematococca*). Journal of Biological Chemistry 269:9195-9204.

Kan, J. A. van, Klooster, J. van't W., Wagemakers, C. A., Dees, D. C., and Vlugt-Bergmans, C. J. van der (1997). Cutinase A of *Botrytis cinerea* is expressed, but not essential, during penetration of gerbera and tomato. Molecular and Plant Microbe Interactions 10:30-38.

Kars, I., and Kan, J. A. L. van (2004). Extracellullar enzymes and metabolites involved in pathogenesis of *Botrytis*. *In* Botrytis: Biology, Pathology and Control (Elad, Y., Williamson, B., Tudzynski, P., and Delen, N., eds.), Kluwer Academic Publishers, Dordrecht, The Netherlands, pp. 99-118.

Kim, Y.-K., Li, D., and Kolattukudy, P. E. (1998). Induction of Ca^{2+}-calmodulin signaling by hard-surface contact primes *Colletotrichum gloeosporioides* conidia to germinate and form appressoria. Journal of Bacteriology 180:5144-5150.

Kim, Y.-K.., Liu, Z.-M., Li, D., and Kolattukudy, P. E. (2000a). Two novel genes induced by hard-surface contact of *Colletotrichum gloeosporioides* conidia. Journal of Bacteriology 182:4688-4695.

Kim, Y.-K., Kawano, T., Li, D., and Kolattukudy, P. E. (2000b). A mitogen-activated protein kinase required for the induction of cytokinesis and appressoria formation by host signals in the conidia of *Colletotrichum gloeosporioides*. Plant Cell 12:1331-1334.

Kim Y.-K., Wang, Y., Liu, Z.-M., and Kolattukudy, P. E. (2002). Identification of a hard surface contact-induced gene in *Colletotrichum gloeosporioides* conidia as a sterol glycosyl transferase, a novel fungal virulence factor. Plant Journal; 30:177-187.

Klimpel, A. Schulze Gronover, C., Williamson, B., Stewart, J. A., and Tudzynski, B. (2002). The adenylate cyclase (BAC) in *Botrytis cinerea* is required for full pathogenicity. Molecular Plant Pathology 3:439-450.

Kolattukudy, P. E. (1984). Cutinases from fungi and pollen. *In* Lipases (Borgstrom, B. and Brockman, H. eds.), Elsevier Science Publishers, Amsterdam, The Netherlands, pp. 471-504.

Kolattukudy, P. E. (1985). Enzymatic penetration of the plant cuticle by fungal pathogens. Annual Reviews of Phytopathology 23:223-250.

Kolattukudy, P. E. (1987). Lipid derived defensive polymers and waxes and their role in plant-microbe interaction. *In* The Biochemistry of Plants Vol. 9 (Stumpf, P.K., ed.), Academic Press, New York, U.S.A, pp. 291-314.

Kolattukudy, P. E. (1996). Biosynthetic pathways of cutin and waxes, and their sensitivity to environmental stresses. *In* Plant Cuticles - an Integrated Functional Approach (Kerstiens, G., ed.), BIOS Scientific Publishers Ltd., Oxford, U.K., pp. 83-108.

Kolattukudy, P. E. (2001). Cutin from plants. *In* Biopolymers, Volume 3a, Wiley-VCH, Münster, Germany, pp. 1-35.

Kolattukudy, P. E., Kim, Y., Li, D., Liu, Z. M., and Rogers, L. (2000). Early molecular communication between *Colletotrichum gloeosporioides* and its host. *In* Host Specificity, Pathology and Host Pathogen Interaction of *Colletotrichum* (Dickman, M., Freeman, S., and Prusky, D., eds.), The American Phytopathological Society, St. Paul, MN, U.S.A., pp. 78-79.

Köller, W., and Kolattukudy, P. E. (1982). Mechanism of action of cutinase: chemical modification of the catalytic triad characteristic for serine hydrolases. Biochemistry 21:3083-3090.

Köller, W., Allan, C. R., and Kolattukudy, P. E. (1982). Role of cutinase and cell wall degrading enzymes in infection of *Pisum sativum* by *Fusarium solani* f. sp. *pisi*. Physiological and Molecular Plant Pathology 20:47-60.

Köller, W., Yao, C., Trial, F., and Perker, D. M. (1995). Role of cutinases in the infection of plants. Canadian Journal of Botany 73 (Supplement 1):1109-1118.

Li, D., and Kolattukudy, P. E. (1997). Cloning of cutinase transcription factor 1, a transactivating protein containing Cys_6Zn_2 binuclear cluster DNA-binding motif. Journal of Biological Chemistry 272:12462-12467.

Li, D., Rogers, L., and Kolattukudy, P. E. (1997). Cloning and expression of cDNA encoding a mitogen-activated protein kinase from a phytopathogenic filamentous fungus. Gene 195: 161-166.

Li, D., Sirakova, T., Rogers, L., Ettinger, W. F., and Kolattukudy, P. E. (2002). Regulation of constitutively expressed and induced cutinase genes by different zinc finger transcription factors in *Fusarium solani* f *pisi* (*Nectria hematococca*). Journal Biological Chemistry 277:7905-7912.

Lin, T.-S., and Kolattukudy, P. E. (1978). Induction of a biopolyester hydrolase (cutinase) by low levels of cutin monomers in *Fusarium solani* f. sp. *pisi*. Journal of Bacteriology 133:942-951.

Liu, Z.-M., and Kolattukudy, P. E. (1998). Identification of a gene product induced by hard-surface contact of *Colletotrichum gloeosporioides* conidia as a ubiquitin-conjugating enzyme by yeast complementation. Journal of Bacteriology 180:3592-3597.

Liu, Z.-M., and Kolattukudy, P. E. (1999). Early expression of the calmodulin gene, which precedes appressorium formation in *Magnaporthe grisea*, is inhibited by self-inhibitors and requires surface attachment. Journal of Bacteriology 181:3571-3577.

Liu, S., Oeljeklaus, S., Gerhardt, B., and Tudzynski, B. (1998). Purification and characterization of glucose oxidase of *Botrytis cinerea*. Physiological and Molecular Plant Pathology 53:123-132.

Lumsden, R. D. (1976). Pectolytic enzymes of *Sclerotinia sclerotiorum* and their localization in infected bean. Canadian Journal of Botany 54:2630-2641.

Madi, L., Xuejun, W., Kobiler, I. Lichter, A., and D. Prusky (2003). Stresses regulate Δ^9-stearoyl ACP desaturase expression, fatty acid composition, antifungal diene level and resistance to *Colletotrichum gloeosporioides* attack. Physiological and Molecular Plant Pathology 62:277-283.

Magro, P., Marciano, P., and Di Lenna, P. (1984). Oxalic acid production and its role in pathogenesis of *Sclerotinia sclerotiorum*. FEMS Microbiological Letters 24:9–12.

Mansfield, J. W., and Richardson, A. (1981). The ultrastructure of interaction between *Botrytis* species and broad leaves. Physiological Plant Pathology 19:41-48.

Manteau, S., Abouna, S., Lambert, B., and Legendre, L. (2003). Differential regulation by ambient pH of putative virulence factors secretion by the phytopathogenic fungus *Botrytis cinerea*. FEMS Microbiology Ecology 43:359-366.

Marciano, P., Di Lenna, P., and Magro, P. (1983). Oxalic acid, cell wall-degrading enzymes and pH in pathogenesis and their significance in the virulence of two *Sclerotinia sclerotiorum* isolates on sunflower. Physiological Plant Pathology 22:339–345.

Marcus, L., and Schejter, A. (1983). Single step chromatographic purification and characterization of the endopolyglacturonases and pectinesterases of the fungus *Botrytis cinerea* Pers. Physiological and Molecular Plant Pathology 36:289-302.

Martinez, C., Geus, P. de, Lauwereys, M., Matthyssens, G., and Cambillau, C. (1992). *Fusarium solani* cutinase is a lipolytic enzyme with a catalytic serine accessible to solvents. Nature 356:615-618.

Martinez, C., Nicolas, A., Tilbeurgh, H. van, Egloff, M. P., Cudrey, C., Verger, R., and Cambillau, C. (1994). Cutinase, a lipolytic enzyme with a preformed oxyanion hole. Biochemistry 33:83-89.

Mayer, A. M., Staples, R. C., and Gil-ad, N. L. (2001). Mechanism of survival of necrotrophic fungal plant pathogens in host expression the hypersensitive response. Phytochemistry 58:33-41.

Mendgen, K., and Hahn, M. (2001). Plant infection and the establishment of fungal biotrophy. Trends in Plant Science 6: 496-498.

Movahedi, S., and Heale, J.B. (1990a). Purification and characterization of an aspartic proteinase secreted by *Botrytis cinerea* Pers ex. Pers in culture and in infected carrots. Physiological and Molecular Plant Pathology 36:289-302.

Movahedi, S., and Heale, J. B. (1990b). The roles of aspartic proteinase and endo-pectin lyase enzymes in the primary stages of infection and pathogenesis of various host tissues by different isolates of *Botrytis cinerea* Pers. Ex. Pers. Physiological and Molecular Plant Pathology 36:303-324.

Movahedi, S., Norey, C. G. Kay, J., and Heale, J. B. (1990). Infection and pathogenesis of cash crops by *Botrytis cinerea*: primary role of an aspartic proteinase. Advances in Experimental Medicine and Biology 306:213-216.

Ortega, J. (1996). Pectolytic enzymes produced by the phytopathogenic fungus *Colletotrichum gloeosporioides*. Texas Journal of Science 48:123-128.

Pascholati, S. F., Deising, H., Leite, B., Anderson, D., and Nicholson, R. L. (1993). Cutinase and nonspecific esterase activities in the conidial mucilage of *Colletotrichum graminicola*. Physiological and Molecular Plant Pathology 42:37-51.

Podila, G. K., Rogers, L. M., and Kolattukudy, P. E. (1993). Chemical signals from avocado surface wax trigger germination and appressorium formation in *Colletotrichum gloeosporioides*. Plant Physiology 103:167-272.

Podila, G. K., Rosen, E., San Francisco, M. J. D., and Kolattukudy, P. E. (1995). Targeted secretion of cutinase in *Fusarium solani* f. sp. *pisi* and *Colletotrichum gloeosporioides*. Phytopathology 85:238-242.

Poussereau, N., Creton, S., Billon-Grand, G., Rascle, C., and Fevre, M. (2001a). Regulation of acp1, encoding a non-aspartyl acid protease expressed during pathogenesis of *Sclerotinia sclerotiorum*. Microbiology 147:717-726.

Poussereau, N., Gente, S., Rascle, C., Billon-Grand, G., and Fevre, M. (2001b). aspS encoding an unusual aspartyl protease from *Sclerotinia sclerotiorum* is expressed during phytopathogenesis. FEMS Microbiology Letters 1:27-32.

Prusky, D., (1996). Pathogen quiescence in postharvest diseases. Annual Review of Phytopathology 34:413-434.

Prusky, D., and Saka, H. (1989). The role of epicuticular wax of avocado fruit in appressoria formation of *Colletotrichum gloeosporioides*. Phytoparasitica 17: 140.

Prusky, D., and Yakoby, N. (2003). Pathogenic fungi: leading or led by ambient pH? Molecular Plant Pathology 4:509-516.

Prusky, D., Kobiler, I., Akerman, M. and Miyara, I. (2006). Effect of acidic solutions and acidic prochloraz on the control of postharvest decay caused by *Alternaria alternata* in mango and persimmon fruit. Postharvest Biology and Technology 42: 134-141

Prusky, D., Gold, S., and Keen, N. T. (1989). Purification and characterization of an endopolygalacturonase produced by *Colletotrichum gloeosporioides*. Physiological and Molecular Plant Pathology 35:121-133.

Prusky, D., Plumbley, R., and Kobiler, I. (1991). Modulation of natural resistance of avocado fruits to *Colletotrichum gloeosporioides* by CO_2. Physiological and Molecular Plant Pathology 39: 325-334.

Prusky, D., Wattad, C., and Koliber, I. (1996). Effect of ethylene on the activation of quiescent infections of *Colletotrichum gloeosporioides* in avocado fruits. Molecular Plant-Microbe Interactions 9:864-868.

Prusky, D., McEvoy, J. L., Leverentz, B., and Conway, W. S. (2001). Local modulation of host pH by *Colletotrichum* species as a mechanism to increase virulence. Molecular Plant-Microbe Interactions 14: 1105-1113.

Prusky, D., McEvoy, J. L., Saftner, R., Conway, W. S., and Jones, R. (2003). The relationship between host acidification and virulence of *Penicillium* spp. on apple and citrus fruit. Phytopathology 94:44-51.

Purdy, R. E., and Kolattukudy, P. E. (1975). Hydrolysis of plant cuticle by plant pathogens. Purification, amino acid composition, and molecular weight of two isozymes of cutinase and a nonspecific esterase from *Fusarium solani* f. *pisi*. Biochemistry 14:2824-2831.

Reignault, P., Mercier, M., Bompeix, G., and Boccara, M. (1994). Pectin methylesterase from *Botrytis cinerea*: physiological, biochemical and immunochemical studies. Microbiology 140:3249-3255.

Rha, E., Park, H. J., Kim, M. O., Chung, Y. R., Lee, C. W., and Kim, J. W. (2001). Expression of exopolygalacturonases in *Botrytis cinerea*. FEMS Microbiology Letters 2001:105-109.

Rogers, L. M., Flaishman, M. A., and Kolattukudy, P. E. (1994). Cutinase gene disruption in *Fusarium solani* f. sp. *pisi* decreases its virulence on pea. Plant Cell 6:935-945.

Rogers, L., Kim, Y., Guo, W., González-Candelas, L., Li, D., and Kolattukudy, P. E. (2000). Requirement for either a host-induced or a pectin-induced pectate lyase for infection of *Pisum sativum* by *Nectria haematococca*. Proceedings of the National Academy of Sciences U.S.A. 97:9813-9818.

Rollins, J. A., and Dickman, M. B. (2001). pH signalling in *Sclerotinia sclerotiorum*: identification of *pac*C/RIM1 homolog. Applied and Environmental Microbiology 67:75-81.

Ruijter, G. J. G., Vondervoort, P. J. I. van de, and Visser, J. (1999). Oxalic acid production by *Aspergillus nige*r: an oxalate-non-producing mutant produces citric acid at pH 5 and in the presence of manganese. Microbiology 145:2569–2576.

Salinas, J. (1992). Function of cutinolytic enzymes in the infection of gerbera flowers by *Botrytis cinerea*. Ph.D. Thesis University of Utrecht, The Netherlands.

Schoonbeek, H., Del Sorbo, G., and Waard, M. A. de (2001). The ABC transporter *BcatrB* affects the sensitivity of *Botrytis cinerea* to the phytoalexin resveratrol and the fungicide fenpiclonil. Molecular Plant-Microbe Interactions 14: 562-571.

Schouten, A., Wagemakers, L. Stefanato, F. L., Kaaij, R. M. van der, and Kan, J. A. L. van (2002). Resveratrol acts as a natural profungicide and induces self-intoxication by a specific laccase. Molecular Microbiology 43:883-894.

Schulze Gronover, C., Schorn, C., and Tudzynski, B. (2004). Identification of *Botrytis cinerea* genes up regulated during infection and controlled by the Gα subunit BCG1 using suppression substractive hybridization (SSH). Molecular Plant-Microbe Interactions 17:537-546.

Shaykh, M., Soliday, C. L., and Kolattukudy P. E. (1977). Proof for the production of cutinase by *Fusarium solani* f. *pisi* during penetration into its host, *Pisum sativum*. Plant Physiology 60:170-172.

Shih, J., Wei, Y., and Goodwin, P. H. (2000). A comparison of the pectate lyase genes, pel-1 and pel-2, of *Colletotrichum gloeosporioides* f. sp. *malvae* and the relationship between their expression in culture and during necrotrophic infection. Gene 243:139-150.

Siriphutthaiwan, P., Herbert, C., Jauneau, A. Esquerre-Tugaye, M. T., and Dumas, B. (2003). Function analysis of CLPT1, a RAB/GTPase from the bean pathogen *Colletotrichum lindemuthianum*. Proceedings of the 22nd Fungal Genetics conference, Asilomar, p. 120.

Staples, R. C., and Mayer, A. M. (1995). Putative virulence factors of *Botrytis cinerea* acting as wound pathogen. FEMS Microbiology Letters 134:1-7.

Sweigard, J. A., Chumley, F. G., and Valent, B. (1992). Disruption of a *Magnaporthe grisea* cutinase gene. Molecular and General Genetics 232:183-190.

Templeton, M. D., Keith, K. R., Bowen, J. K., Crowhurst, R. N., and Rikkerink, E. H. (1994). The pectin lyase-encoding gene (*pnl*) family from *Glomerella cingulata*: Characterization of *pnlA* and its expression in yeast. Gene 142:141-146.

Tiedemann, A. von (1997). Evidence for a primary role of active oxygen species in induction of host cell death during infection of bean leaves with *Botrytis cinerea*. Physiological and Molecular Plant Pathology 50:151-166.

Tsai, P. F., Person, M. N., and Beever, R. E. (2004). Mycoviruses in *Monilinia fructiocola*. Mycological Research 108:907-912.

Tsurushima, T., Ueno, T., Fukami, H., Irie, H., and Inoue, M. (1995). Germination self-inhibitors from *Colletotrichum gloeosporioides* f. sp. *jussiaea*. Molecular Plant-Microbe Interactions 8:652-657.

Tudzynski, B., and Schulze Gronover, C. (2004). Signaling in *Botrytis cinerea*. *In* Botrytis: Biology, Pathology and Control (Elad, Y., Williamson, B., Tudzynski, P., and Delen, N., eds.), Kluwer Academic Publishers, Dordrecht, The Netherlands , pp. 85-97.

Urbaneck, H., Gajewaska, E., Karwowska, R., and Wielanek, M. (1996). Generation of superoxide anion and induction of superoxide dismutase and peroxidase in bean leaves infected with pathogenic fungi. Acta Biochimica Polonica 43:679-685.

Valette-Collet, O., Cimerman, A., Reignault, P., Levis, C., and Boccara, M. (2003). Disruption of *Botrytis cinerea* pectin methylesterase gene *Bcpme*1 reduces virulence on several host plants. Molecular Plant-Microbe Interactions 16:360-367.

Verhoeff, K., and Warren, J. M. (1972). In vitro and in vivo production of cell wall degrading enzymes by *Botrytis cinerea* from tomato. Netherlands Journal of Plant Pathology 78:179-185.

Vernekar, J. V., Ghatge, M. S., and Deshpande, V. V. (1999). Alkaline protease inhibitor: A novel class of antifungal proteins against phytopathogenic fungi. Biochemical and Biophysical Research Communications 262:702-707.

Viterbo, A., Yagen, B., and Mayer, A. M. (1993). Cucurbitacins, 'attack' enzymes and laccase in *Botrytis cinerea*. Phytochemistry 32:61-65.

Vlugt-Bergmans, C. J. B. van der, Wagemakers, C. A. M.,. Dees, D. C. T., and Kan J. A. L. van (1997). Catalase A from *Botrytis cinerea* is not expressed during infection on tomato leaves. Physiological and Molecular Plant Pathology 50:1-15.

Wagner, F., Kusserow, H., and Schafer, W., (2000). Cloning and targeted disruption of two polygalacturonase genes in *Penicillium olsonii*. FEMS Microbiology Letters 186:293-299.

Wattad, C., Kobiler, D., Dinoor, A., and Prusky, D. (1997). Pectate lyase of *Colletotrichum gloeosporioides* attacking avocado fruits: cDNA cloning and involvement in pathogenicity. Physiological and Molecular Plant Pathology 50:197-212.

Weeds, P. L., Beever, R. E., and Long, P. G. (2000). Competition between aggressive and non-aggressive strains of *Botrytis cinerea* (*Botryotinia fuckeliana*) on French bean leaves. Australasian Plant Pathology 29:200-204.

Wubben, J. P., Mulder, W., Have, A. ten, Kan, J. A. L. van, and Visser, J. (1999). Cloning and partial characterization of endopolygalacturonase genes from *Botrytis cinerea*. Applied and Environmental Microbiology 65:1596-1602.

Wubben, J. P., Have, A. ten, Kan, J. A. L. van, and Visser, J. (2000). Regulation of endopolygalacturonase gene expression in *Botrytis cinerea* by galacturonic acid, ambient pH and carbon catabolite repression. Current Genetics 37:152-157.

Yakoby, N., Kobiler, I., Dinoor, A., and Prusky, D. (2000a). pH regulation of pectate lyase secretion modulates the attack of *Colletotrichum gloeosporioides* on avocado fruits. Applied and Environmental Microbiology 66:1026-1030.

Yakoby, N., Freeman, S., Dinoor, A., Keen N. T., and Prusky, D. (2000b). Expression of pectate lyase from *Colletotrichum gloeosporioides* in *C. magna* promotes pathogenicity. Molecular Plant-Microbe Interactions 13:887-891.

Yakoby, N., Beno-Moualem, D., Keen, N. T., Dinoor, A., Pines, O., and Prusky, D. (2001). *Colletotrichum gloeosporioides pel*B, is an important factor in avocado fruit infection. Molecular Plant-Microbe Interactions 14:988-995.

Yakoby, N., Beno-Moualem, D., Kobiler, I., and Prusky, D. (2002). The analysis of fruit protection mechanisms provided by reduced pathogenicity REMI mutants of *Colletotrichum gloeosporioides*. Phytopathology 92: 1196-1201.

Yang, Z., Rogers, L. M., Song, Y., Guo, W., and Kolattukudy, P. E. (2005). Homoserine and asparagine are host signals that trigger *in planta* expression of a pathogenesis gene in *Nectria haematococca*. Proceedings of the National Academy of Sciences U.S.A. 102: 4197-4202.

Yao, C. L., Conway, W. S., and Sams, C. E. (1996). Purification and characterization of a polygalacturonase produced by *Penicillium expansum* in apple fruit. Phytopathology 86:1160-1166.

Ye, X. Y., and Ng, T. B. (2002). A new peptidic protease inhibitor from *Vicia faba* seeds exhibits antifungal, HIV-1 reverse transcriptase inhibiting and mitogenic activities. Journal of Peptide Science 8:565-662.

Ye, X. Y., Ng, T. B., and Rao, P. F. (2001). A Bowman-Birk-type trypsin-chymotrypsin inhibitor from broad beans. Biochemical and Biophysical Research Communications 289:91-96.

Zheng, L., Campbell, M., Murphy, J., and Xu, J. R. (2000). The BMP1 gene is essential for pathogenicity in the gray mold fungus *Botrytis cinerea*. Molecular Plant-Microbe Interactions 13:724-732.

Chapter 2

Real time monitoring of ethylene during fungal-plant interaction by laser-based photoacoustic spectroscopy

Simona M. Cristescu[1], Ernst J. Woltering[2] and Frans J.M. Harren[1]
[1]*Department of Molecular and Laser Physics, Institute for Molecules and Materials, Radboud University Nijmegen, Toernooiveld 1, 6525 ED Nijmegen, The Netherlands;* [2] *Wageningen University and Research Center, Agrotechnology & Food Innovations (A&F B.V.), P.O. Box 17, 6700 AA Wageningen, The Netherlands.*

INTRODUCTION

The quality of agricultural products at the time they get to the consumers strongly depends on the developmental stage at harvest, shipping and storage conditions. It is commercially advantageous that fruit and vegetables have a long shelf life and do not deteriorate immediately after harvest. In many deteriorative processes the plant hormone ethylene plays an important role; by controlling the ethylene production or sensitivity, important benefits can be obtained (Saltveit, 1999).

The plant hormone ethylene

Ethylene is involved in virtually all aspects of the plant life cycle, as well as in the plant's response to many environmental stimuli. In the broadest of terms, ethylene is responsible for signaling changes during germination, growth, flower and fruit development, senescence of plant organs, programmed cell death, the onset of plant defense mechanisms and the action of other plant hormones. Biotic stress (e.g., pathogen attack) and abiotic stress conditions (e.g., wounding, hypoxia, ozone, chilling, and freezing) elicit ethylene synthesis in plants (Abeles *et al.*, 1992; Mattoo and Suttle, 1991).

The elucidation of the ethylene biosynthetic pathway and the molecular cloning of genes encoding the enzymes involved have provided insight into the regulation of ethylene biosynthesis in plants. Plants biosynthesize ethylene via the Yang cycle, wherein methionine is converted to S-adenosylmethionine (SAM) by the enzyme SAM synthase. The conversion of SAM to 1-aminocyclopropane-1-carboxylic acid (ACC) is then catalyzed by the enzyme ACC synthase (ACS). ACC is oxidized to ethylene by ACC oxidase (ACO) (Yang and Hoffman, 1984) (Figure 1). The conversion of SAM to ACC is generally considered to be the rate-limiting step in the synthesis of ethylene (Kende, 1993). Both latter enzymes play a role in the regulation of ethylene biosynthesis and are encoded by small gene families.

Ethylene biosynthesis in microorganisms

In addition to plants, some microorganisms, including phytopathogenic fungi and bacteria, can synthesize ethylene themselves. Except for few fungal species, such as the slime mold *Dictyostelium mucoroides* (Amagai and Maeda, 1992) and *Penicillium citrinum* (Jia, 1999), the ACC pathway for ethylene biosynthesis has not been found operative in microorganisms. Presently, two different ethylene biosynthetic pathways have been established in microorganisms (Fukuda *et al.*, 1993). Ethylene can be produced either from glutamic acid via 2-oxoglutarate as in *Penicillium digitatum* (Fukuda *et al.*, 1989a) and in *Pseudomonas syringae* (Nagahama *et al.*, 1991) or from methionine via 2-keto-4-methylbutyric acid (KMBA) as in *Escherichia coli* (Ince and Knowles, 1986), *Cryptococcus albidus* (Fukuda *et al.*, 1989b),

Colletotrichum musae (Daundasekera *et al.*, 2003) and in *Botrytis cinerea* (Cristescu *et al.*, 2002; Chague *et al.*, 2002) (Figure 1). Additionally, KMBA has been identified as an intermediate in methionine-derived ethylene biosynthesis by microbial cultures in soil (Nazli *et al.*, 2003).

Effect of ethylene on fungal development

It was reported that ethylene has different effects on various phases of fungal development *in vitro*. Exogenous application of ethylene stimulates conidial germination of *B. cinerea, Penicillium expansum, Rhizopus stolonifer* and *Gloeosporium perennans* (Kepczynski and Kepczynska, 1977), *P. digitatum, P. italicum, Thielaviopsis paradoxa* (El-Kazzaz *et al.*, 1983), *Diplodia natalonis* and *Phomopsis citri* (Abeles, 1973). Elad (2002) showed that ethylene did not affect conidial germination and hyphal growth of *B. cinerea* on PDA media (potato dextrose agar), whereas on glass, tomato or bean leaf surfaces both germination rate and germ tube elongation were enhanced. A specific inhibitor of ethylene action in plants, 2,5-norbornadiene (NBD) inhibited growth of hyphae and mycelium and retarded the *B. cinerea* development (Kepczynska, 1989; 1993). A similar inhibitory effect was reported following application of the plant ethylene production inhibitor, aminoethoxyvinylglycine (AVG) a specific inhibitor of ACC synthase (Figure 1), which reduced mycelium growth and sporulation of *B. cinerea*. As ethylene biosynthesis in *B. cinerea* does not involve ACC synthase, the target of AVG is probably some other aminotransferase and the effect may not be related to ethylene biosynthesis.

Many fungi are known to remain dormant at the fruit surface until the fruit ripens, at which time the fungus infects the fruit. In some fungi ethylene was found to play a role as a signaling molecule. For instance, in *Colletotrichum gloeosporioides* and *C. musae* that attack ripe fruit, exposure to ethylene induces germination and appressorium formation. The reception of ethylene by the fungus was supposed to act through a mechanism with similarity to the receptor-mediated effects of ethylene in plants. Sensing of ethylene was blocked by the ethylene perception inhibitors, silver thiosulphate (STS) and NBD, while the ethylene analog propylene (but not methane) could substitute for ethylene. On transgenic tomato fruits, that did not produce ethylene, the fungus was unable to germinate. Upon treatment with ethylene, the spores germinated and produced multiple appressoria and infection lesions (Kolattukudy *et al.*, 1995).

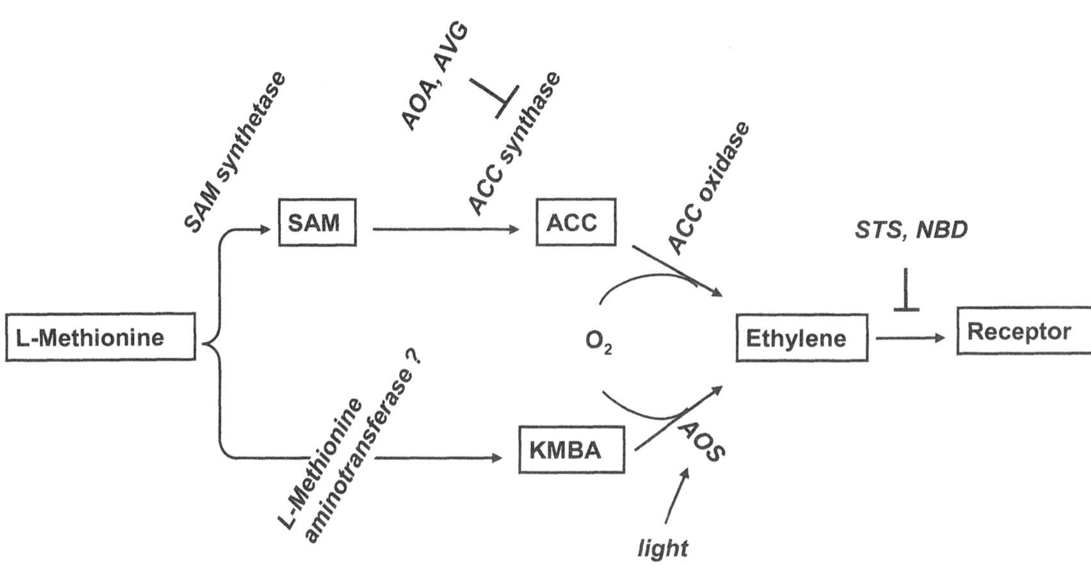

Figure 1. Ethylene biosynthesis in plants and *B. cinerea*.

In *Aspergillus parasiticus* ethylene was found to affect fungal development and aflatoxin synthesis in a dose dependent manner. The effect of ethylene could be influenced by treatment with the plant ethylene binding inhibitor 1-MCP (Roze *et al.*, 2004). Such observations suggest that fungi can sense ethylene by a similar mechanism as plants. In plants ethylene is sensed by two component histidine kinases (Chang and Stadler, 2001) and, recently, a similar type of protein was identified in *B. cinerea* (Catlett *et al.*, 2003). However, the functionality of these proteins has not yet been elucidated.

Host-pathogen interactions and ethylene

Enhanced ethylene production is one of the earliest responses of plants to the perception of a pathogen (Boller, 1991) and ethylene has been associated with both resistance and susceptibility to disease (Knoester *et al.*, 1998). A survey of the literature yields conflicting reports on the role of ethylene in pathogenesis (Johnson and Ecker, 1998). Ethylene may be a stimulus for defense responses by activating the plant defense genes which thereby lead to increasing plant resistance or, conversely, it may play a role in disease symptom development and in the breakdown of internal plant resistance (Boller, 1991; Abeles *et al.*, 1992; Lund *et al.*, 1998). Depending on the type of pathogen and plant species, the role of ethylene can be dramatically different. This diversity can be explained by taking into account the involvement of ethylene in the multiple physiological processes in plants and its interaction with other hormones and pathways. External application of ethylene reduces, increases or does not have any effect on disease incidence depending on the plant-pathogen system (El-Kazzaz *et al.*, 1983; Elad, 1990; Marte *et al.*, 1993).

Ethylene promotes fruit ripening, induces necrosis and chlorosis and accelerates the senescence in plants (Matoo and Suttle, 1991; Abeles *et al.*, 1992). Ripe fruits and senescent or wounded plant organs are more susceptible to *B. cinerea* infection. Increased ethylene production is also characteristic of the hypersensitive response (HR) during the incompatible combination of an avirulent pathogen and its resistant host in which cells at the infection site die to form a necrotic lesion. As a result, a biotrophic pathogen is deprived of nutrients and its growth restricted (Cohn *et al.*, 2001). On the contrary, some so-called necrotrophic pathogens, such as *B. cinerea,* may benefit from HR by using the dead tissues as a food support to further spread and invade healthy living tissues (Govrin and Levine, 2000).

Different defense mechanisms are involved in resistance and each of them has the capacity to withstand infection of certain pathogens (Thomma *et al.*, 2001). The plant defense-related processes activated by ethylene include production of pathogenesis-related (PR) proteins (Rodrigo *et al.*, 1993, van Kan *et al.*, 1995), production of phytoalexins and lignin biosynthesis (Fan *et al.*, 2000), activation of hydrolases (Boller, 1988, Brown and Lee, 1993), the induction of the phenylpropanoid pathway (Chappell *et al.*, 1984) and cell wall alterations (Bell, 1981). Although the synthesis of the pathogenesis-related proteins can also be induced by ethylene-independent pathways (Dixon and Lamb, 1990), biological elicitation of some of them may require a functional ethylene response (Penninckx, *et al.*, 1996). However, enhanced endogenous ethylene production is not always a requirement for the induction of defense responses (Boller, 1991; Bent *et al.*, 1992; Ciardi *et al.*, 2000; Lawton *et al.*, 1995).

Van Loon (1984) suggested that a large part of the plant damage during pathogen infection is caused by autocatalytic ethylene synthesis and not from the direct action of the pathogen. Based on this, it was proposed that exogenous ethylene often increases the disease severity and, moreover, that inhibitors of ethylene synthesis may decrease the fungal infection severity. There are many examples supporting this hypothesis: (i) Increase of ethylene production due to the infection has been correlated with increased plant disease susceptibility in the case of wheat plants infected with *Septoria nodorum* (Hyodo, 1991). (ii) Exogenous ethylene applied to cucumber plants prior to infection increased disease severity in the case of *Colletotrichum lagenarium* (Biles *et al.*, 1990) and for *Verticillium* wilt of tomato (Cronshaw and Pegg, 1976). (iii) Specific inhibitors of ethylene synthesis reduced disease severity to *B. cinerea* infection in rose and carnation flowers, detached leaves of tomato, pepper, French-bean and cucumber (Elad, 1990 and 1993; Boller, 1991). (iv) Cotton plants pretreated with AVG (an inhibitor of ethylene biosynthesis) showed decreased disease severity when infected with *Alternaria* (Bashan, 1994). However, virtually conflicting results are observed with tomato plants pretreated with ethylene or the inhibitor of ethylene perception, 1-methylcyclopropene (MCP). Ethylene pretreatment caused a decreased susceptibility against *B. cinerea* and MCP pretreatment resulted in increased susceptibility (Diaz *et al.*, 2002). These examples show that ethylene may play a different role in disease development depending on the time it is applied or produced with respect to the timing of infection. Ethylene can affect the disease development by its possible direct action on the pathogen and/or indirectly by inducing various modifications in host plant metabolism (Kader, 1985).

Ethylene was not only found to stimulate fungal growth directly (Brown and Lee, 1993), but also to increase the activity of certain abscission-associated enzymes in the plant and, therefore, to predispose the plant to pathogen invasion (Brown and Burns, 1998). Although it was reported by El-Kazzaz *et al.*, (1983) that ethylene stimulates the growth of *B. cinerea* and *P. italicum*, Palou *et al.*, (2003) found that ethylene did not affect aerial mycelial growth on table grapes infected with *B. cinerea*. Neither incidence nor disease severity to *Monilinia fructicola* of stone fruit was affected by exogenous application of ethylene, suggesting that ethylene plays no role in the pathogenicity of this fungus.

An additional approach to study the effect of ethylene on disease development is with ethylene-insensitive mutants (Knoester *et al.*, 1998; Geraats *et al.*, 2002; Hoffman *et al.*, 1999). For example, the ethylene-insensitive *Arabidopsis thaliana* mutant *ein2* displayed enhanced susceptibility to *B. cinerea* (Thomma *et al.*, 1999), but decreased susceptibility to infection with the beet cyst nematode (Wubben *et al.*, 2001). Soybean mutants with reduced ethylene sensitivity developed more severe symptoms than the wild type when infected by necrotrophic pathogens, *Septoria glycines* and *Rhizoctonia solani*, and less severe chlorotic symptoms when infected by biotrophic pathogens, *P. syringae* pv. *glycines* and *P. sojae* (Hoffman *et al.*, 1999). Ethylene-insensitive *A. thaliana* and tomato lines did not display high susceptibility to the bacteria *Pseudomonas* and *Xanthomonas* or the fungal species *Fusarium* and the Oomycete *Peronospora* (Bent, *et al.*, 1992; Lawton *et al.*, 1995; Lund *et al.*, 1998). It was found that ethylene-insensitive tobacco plants Tetr 1 (Tetr 1, expressing the mutant *A. thaliana etr1-1*gene) were also more susceptible than the wild type plants to several other fungi, including *Colletotrichum destructivum* (Chen *et al.*, 2003) and *Chalara elegans* (Knoester *et al.*, 1998; Geraats *et al.*, 2002). Plants that did not produce ethylene developed much larger necrotic areas, while addition of ethylene restricted disease spreading. However, high levels of ethylene are usually destructive to plant growth and health and facilitate their damage by pathogen. During the fungal-plant interaction both the partners can produce ethylene, usually via a different pathway. Since methionine is the common precursor for ethylene biosynthesis in plants and in different microorganisms, it is difficult to determine if ethylene originates from the plant or the fungus. Usually, ethylene production by fungi was studied under *in vitro* conditions and then correlated with the ethylene emission from the infected host. Currently, there is no evidence of ethylene production by a specific fungus *in planta* and it is not known if fungal ethylene may play a role in triggering host ethylene production. Molecular tools to study the biology of fungi (ten Have *et al.*, 2001; Wubben *et al.*, 2000) will enable the role of fungus-produced ethylene in pathogenesis and the isolation of the genes involved in fungal ethylene biosynthesis and their deletion will unequivocally show whether ethylene production by the fungus plays a role in the fungus-plant interaction. In our research group we have developed a new approach to study the pathogen-host interaction by using a laser spectroscopic technique. As application, we chose the tomato fruit – *B. cinerea* system. *B. cinerea* is an important worldwide pathogen that attacks more than 200 plant species and causes extensive crop losses to many field-grown and greenhouse crops. It is visible as the well-known grey mold on fruits, vegetables, ornamentals, trees, shrubs and various types of foods. *B. cinerea* infects many plant species, but also various organs at different developmental stages of a particular plant host. Fruit and stem rot, blossom blight, stem cankers, leaf spots, bulb-, corm-, tuber- and root-rots and twig blight are all manifestations of the fungus in plants. Under humid conditions it can cause massive losses in yield and quality, particularly in wine and fruit production. The losses imposed by this pathogen require the intense use of fungicides worth about € 50-100 million per year in Europe. *B. cinerea* is one of the most ubiquitous and serious fungal diseases of greenhouse tomatoes. Even more research was done on the *B. cinerea* infection of the vegetative plant parts (stem and leaves) than on fruits. It has been shown that *B. cinerea* produces ethylene *in vitro* most probably via pathway for ethylene formation using KMBA (Cristescu *et al.*, 2002; Chague *et al.*, 2002). Therefore, analysis of ethylene emission from the plant-pathogen system, such as tomato-*B. cinerea* becomes a complex phenomenon and more information is required to elucidate the contribution of each organism to the total ethylene produced. We present a laser-based ethylene detector suitable to monitor on-line the ethylene released during the infection process. The instrument allows ethylene emission in a flow-through system with a detection limit down to 10 pptv (pptv = parts-per-trillion volume, $1{:}10^{12}$) (Bijnen *et al.*, 1996) and has relatively high time resolution for measuring the dynamics of ethylene production by *B. cinerea in vitro* and by infected tomatoes. We indicate it as a powerful tool to study the relationship between ethylene released by the fungus *in vitro* and the enhanced ethylene production in *B. cinerea* infected tomato with respect to disease development. Especially for this particular type of interaction when both, the host and the pathogen, are able to produce ethylene, it is difficult to separate their contribution to the total ethylene emission by the host-pathogen system. In combination with an effective broad range of well-characterised chemical inhibitors and pathway intermediates related to ethylene biosynthesis, the use of this technique enables us to produce a comprehensive description of the mode of ethylene formation and action in *B. cinerea*, both *in vitro* and *in vivo*.

Real time monitoring of ethylene

To follow dynamic processes in plants, it is necessary to measure ethylene directly and with high resolution in time. This can be achieved if a flow-through system in line with sampling cuvettes is combined with the extremely sensitive laser-based photoacoustic detector. The schematic diagram of the set-up is presented in Figure 2. The various parts of the system are computer controlled, enabling a fully automated sampling of ethylene production rates of biological tissue for periods up to several weeks. The detector consists of a line-tunable CO_2 laser which emits radiation in the 9-11 μm infrared wavelength region and a photoacoustic cell, through which the laser light is directed for detecting the gas of interest (Harren and Reuss, 1997; Lintel Hekkert *et al.*, 1998). The laser-based ethylene detector is able to distinguish between different gases by making use of their wavelength dependent "fingerprint" absorption. Thanks to its distinct fingerprint-like spectrum in the CO_2 laser wavelength range (Brewer *et al.*, 1982), ethylene can be measured with very high sensitivity, exhibiting a detection limit of three orders of magnitude better than gas chromatography (i.e., 10 $pL{\cdot}L^{-1}$).

Since the past decade we have used the laser-based photoacoustic systems to determine on-line ethylene release in various processes in plants and microorganisms, such as seed germination (Petruzzelli *et al.*, 1995; Thuring *et al.*, 1994), flower senescence (Woltering *et al.*, 1993; Wagstaff *et al.*, 2005), diffusion through aerenchymatous roots (Visser *et al.*, 1997), submergence (Voesenek *et al.*, 1993), fruit ripening (de Vries *et al.*, 1995, 1996), nitrogen fixation by cyanobacteria (Zuckermann *et al.*, 1997a; Staal *et al.*, 2001, 2003), interaction with auxin (van den Bussche *et al.*, 2003), dehydration (Leprince *et al.*, 2000) and circadian rhythm (Thain *et al.*, 2004).

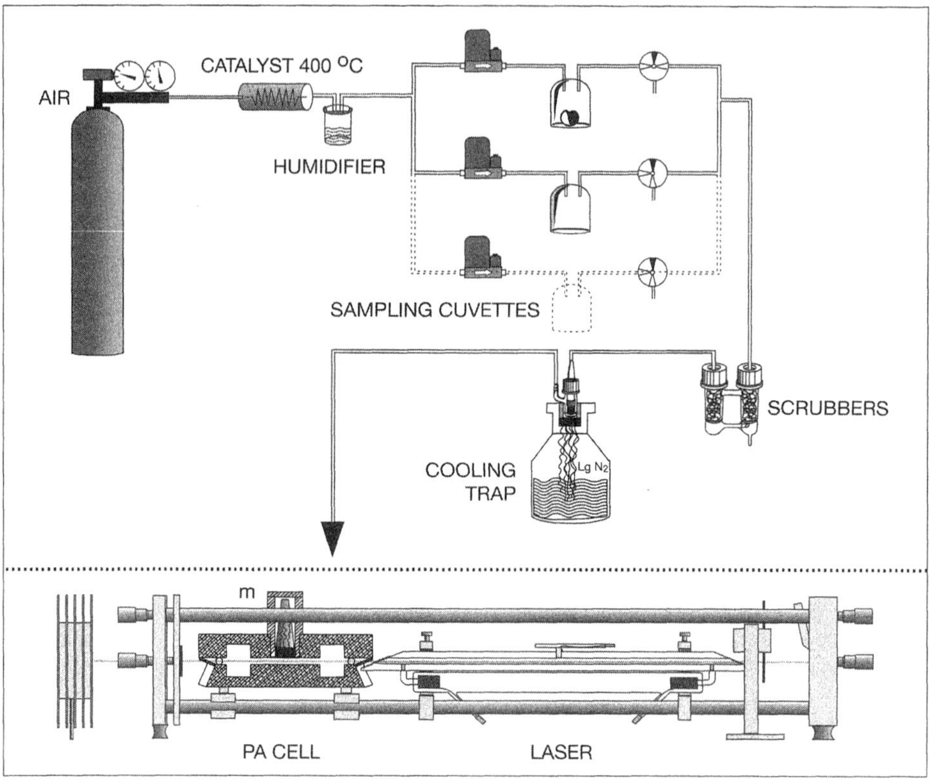

Figure 2. The ethylene detection set-up. Gas flow system (top). Laser-based ethylene detector consisting of a photoacoustic (PA) cell and a CO_2 laser (bottom). Ethylene released by the biological samples (plated fungi or infected fruits) is transported to the PA cell where it absorbs the laser radiation and gives rise to an acoustic wave. The amplitude of this wave is measured with a sensitive microphone (m) and is directly proportional with the ethylene concentration.

Trace gases released by the biological samples (plated fungi or infected fruits) are transported to the photoacoustic cell through a flow system using air as carrier gas. Once inside the photoacoustic cell, traces of ethylene absorb the laser radiation and convert it into heat which will further generate an increase of pressure inside a closed volume. By modulating the laser beam with a chopper, pressure waves (i.e., sound) are generated and detected with a sensitive miniature microphone. The amplitude of the acoustic waves is directly proportional to the concentration of ethylene in the photoacoustic cell.

The overall measuring sequence can be chosen in function of the dynamics of the process under study. It can vary from less than 10 seconds for fast processes up to about one minute.

The gas flow through the measuring system can be controlled using electrical three-way valves that switch a particular gas stream to the photoacoustic cell (on-position) or into the laboratory (off-position). In this way the gas emission from a number of cuvettes (up to 8 per experiment) containing the biological samples can be transported to the photoacoustic cell alternately and at controlled flow rates. The flow is continuously monitored and adjusted by mass flow controllers.

Other interfering gases released by the samples might influence the quantification of ethylene emission, due to the overlap between their spectral absorption and the CO_2 laser wavelengths; therefore, a number of scrubbers and traps are introduced in the measuring system to remove them from the gas flow. A platinum based catalyzer (platinum on Al_2O_3), which operates at minimum 400 °C placed before the entrance of the cuvettes, provides air free of any traces of ethylene (or other hydrocarbons). When monitoring low ethylene concentrations, the CO_2 and water concentrations have to be reduced before entering the photoacoustic cell. Consequently, a scrubber with KOH (moist pellets) is usually used to reduce the CO_2 concentration below 1 ppmv and a tube with $CaCl_2$ (granules) is placed directly after it in order to decrease the water content in the gas flow. Ethanol and some other heavier hydrocarbons are removed by inserting a cooling trap (-150 °C) into the gas flow system just before the photoacoustic cell. In addition, the gas flow was filtered by passing through 0.2 μm millipore filters placed at the inlet and outlet of the sampling cuvettes.

From the obtained emission rates, readings of an empty cuvette are subtracted in order to adjust for externally induced variations (e.g., over hours or days). In most of the cases, the ethylene production from the fungi was related to the emission rate by multiplying the measured value with the flow rate, and expressed in nl h^{-1}. While monitoring the emission by fruits, the rate of ethylene production was expressed in nl h^{-1} g^{-1}fresh weight.

ETHYLENE PRODUCTION AND PERCEPTION BY *BOTRYTIS CINEREA IN VITRO*

Ethylene production by *B. cinerea in vitro*

When plants mature and die, the senescing tissues serve as a food base for many microorganisms. *B. cinerea* is such an organism. This fungus very quickly colonizes wounded, dead or dying stems, leaves, flowers, and fruits. As a necrotrophic pathogen, *B. cinerea* has the ability to kill and macerate the host cells before invading them to obtain the nutrients for its growth. By this mechanism, using the previously colonized dead tissues as a food support, the fungus can spread and invade healthy living tissues (Jarvis, 1977).

The ability of *B. cinerea* to adapt to various environmental conditions has been investigated and different mechanisms of its action in function of the attacked host tissue were proposed (Barkai-Golan *et al.*, 1988; Elad and Eversen, 1995; ten Have *et al.*, 2001; von Tiedemann, 1997; Yang and Hoffman, 1984). However, there is no broad understanding of the "attack strategies" of this fungus. One intriguing part in this scenario is the role of ethylene. Ethylene is generated during the *Botrytis*-host interaction possibly by both organisms as both plants and *B. cinerea* have the ability to produce it. In spite of the large extent of published data dedicated to ethylene production by the *Botrytis*-host system and, in particular by *B. cinerea*, it is not very clear so far how and why the fungus produces ethylene.

Research performed in this field showed that *B. cinerea* is able to produce ethylene *in vitro*. Qadir *et al.* (1997) reported ethylene production from liquid cultures of *B. cinerea* grown in methionine-enriched media. However, using gas chromatography (GC) it is difficult to quantify the ethylene released by *B. cinerea* in the absence of methionine in the basal media and to offer a good description of ethylene emission over time. These data were revealed by using the more sensitive laser-based ethylene detector. We proved that *B. cinerea* produced low, but detectable levels of ethylene, when grown *in vitro* on PDA (potato dextrose agar) media without added methionine. A constant emission of 0.17 ± 0.04 nl h^{-1} was detected during the first 24 h for 160 μl of suspension at $2x10^7$ conidia ml^{-1} concentration. The ethylene production increased to a peak of about 1 ± 0.05 nl h^{-1} after 43 h from plating the conidia on PDA, after which it decreased to 0.2 ± 0.04 nl h^{-1}. As control we used 160 μl of autoclaved conidial suspension and autoclaved hyphae, respectively, plated on PDA (data not shown). In this case, no increase of ethylene emission was found over a period of 3 days. The equivalent ethylene production of the control, representing the background of non-enzymatically produced ethylene, showed a constant level of 0.18 ± 0.05 nl h^{-1} for the autoclaved conidial suspension, 0.18 ± 0.04 nl h^{-1} for autoclaved hyphae, and 0.17 ± 0.03 nl h^{-1} for PDA media alone, respectively. Addition of methionine greatly enhances the ethylene production by *B. cinerea* (Qadir *et al.*, 1997; Cristescu *et al.*, 2002; Chague *et al.*, 2002). We investigated 15 concentrations of methionine and found that a small amount of L-methionine (0.05 m*M*) present in the PDA media already increased ethylene release by three-fold. The pattern of ethylene production, showing a peak after approximately 43 hr, was similar as the pattern observed for the fungus grown on PDA without methionine. The highest ethylene production occurred in the presence of 3 to 15 m*M* L-methionine (around 30 nl h^{-1}). At higher levels it decreases with increasing methionine concentration, probably due to its effect on fungal vitality. These results differ from those reported by Qadir *et al.* (1997) who examined the effect of adding 1, 5, 10, 35 and 50 m*M* L-methionine in PDA and found a maximum ethylene production by fungus grown on PDA supplemented with 35 m*M* L-methionine. This may be caused by the fact that large errors are introduced by the integration method over 7 days ethylene production (Qadir *et al.*, 1997).

Ethylene emission by *B. cinerea* was found to be dependent on the concentration of conidia plated on the growing media. The ethylene release was higher and the peak in ethylene production occurred earlier with increasing conidia concentrations (Figure 3). Radial growth of the fungus was slightly lower for media containing methionine than for the media without methionine, although no significant variations could be observed between media with different methionine concentrations.

Addition of other ethylene precursors than methionine, such as 2-oxoglutarate or glutamate including their co-factors (e.g., ferric ions, L-arginine) in the growing media (either in liquid culture or in PDA) did not stimulate the ethylene released by *B. cinerea* indicating methionine-dependent synthesis (Qadir *et al.*, 1997; Cristescu *et al.*, 2002; Chague *et al.*, 2002).

Using a pharmacological approach, we showed that *B. cinerea* most likely produces ethylene from methionine via the KMBA pathway (Cristescu *et al.*, 2002). Inhibitors of the plant ethylene pathway, such as amino oxyacetic acid (AOA) and aminoethoxyvinylglycine (AVG), had no effect on the ethylene emission from the fungus. Furthermore, using 2,4-dinitrophenylhydrazine as reagent, Chague *et al.*, (2002) tested the presence of KMBA in different growth media. They found that KMBA was present only in the media supplemented with methionine, being produced by the fungus and secreted into the media. Additionally, no ACC synthase or ACC oxidase homologs have been found in *B. cinerea* genome while *Penicillium citrinum* has a functional ACC synthase gene (Jia *et al.*, 1999). These observations show that *B. cinerea* does not use the plant ethylene pathway for ethylene synthesis.

Light microscopic analysis showed that conidial germination occurs within the first 3 hours after harvesting and plating, well before ethylene release becomes substantial. In fact, no ethylene was monitored during conidia dormancy and germination. Ethylene emission increased slowly during hyphal elongation, then very rapidly and at high rates till it reached the maximum production when the fungal hyphae extensively grew and began to branch (Cristescu *et al.*, 2002). Thus, ethylene released by *B. cinerea* is associated with hyphal growth rather than conidial germination. Once it reaches a maximum, ethylene emission by the fungus shows a decline associated with further growth of the fungus (Cristescu *et al.*, 2002). These results were supported by Chague *et al.* (2002) who found that KMBA is produced by young hyphae and not by dormant and germinating conidia. Moreover, older mycelium produces less KMBA and subsequently, less ethylene.

KMBA conversion

Ethylene production by *B. cinerea* is partly dependent on light. We found that ethylene produced by the fungus grown on methionine-enriched media was fivefold higher under low light intensity conditions (5 μmol photons $m^{-2}s^{-1}$) than in dark (Figure 3 inset).

Using the photoacoustic ethylene detector we were able to monitor the dynamics of ethylene released by *B. cinerea* while interchanging the light and dark regimes (Figure 3). The ethylene emission from *B. cinerea* changes in less than 2 minutes after switching from light to dark and/or from dark to light (Figure 3 inset). When the light is switched off, ethylene emission firstly shows a fast exponential decline during the first 30 minutes followed by a slower linear decrease. After switching back to light, ethylene release increases exponentially to levels that occurred before the dark treatment.

These data indicate that KMBA conversion to ethylene is considerably lower in the dark than in light. Similar results were reported by Chague *et al.* (2002). These authors hypothesized that in the dark KMBA accumulates in the medium and, upon switching on the light, the accumulated KMBA is rapidly converted into ethylene. However, their data show that the light-induced ethylene release following 24 h dark incubation is over 20 times less than expected considering the huge amounts that should have accumulated assuming continuous KMBA synthesis. This implicates that KMBA synthesis is apparently also affected by light/dark.

To explain the observed dynamics in fungal ethylene production, we need to consider the mechanism for KMBA conversion to ethylene (Figure 1). A major role in this process is played by the free radicals which are generated in both dark and light conditions. In the dark, the radicals are formed by respiratory processes (chemical). In the light, additional radicals will be generated due to photochemical processes.

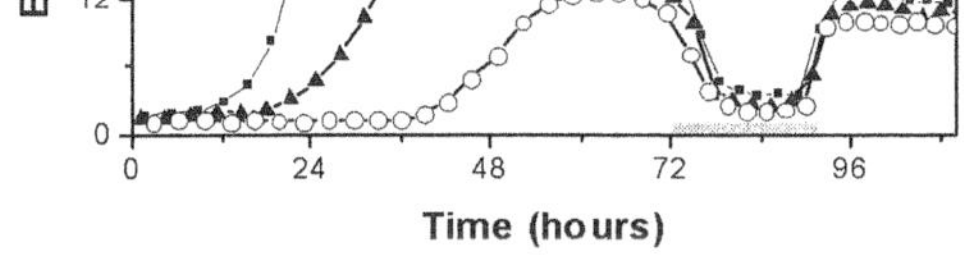

Figure 3. Ethylene released by *B. cinerea in vitro* at different concentrations $1.5x10^8$ (■), $2x10^7$ (▲) and $2x10^5$ (o) conidia ml^{-1} (160 µl) plated on PDA containing 25 m*M* L-methionine in light and dark (grey bar), respectively. Inset: the KMBA conversion to ethylene when switching from light to dark and back to light for $2x10^5$ (o) conidia ml^{-1} starts within 2 minutes.

In *B. cinerea* other sources may be considered for the formation of free radicals such as pigments and/or the reaction of light directly with chemical compounds present in the fungus cells.

Light seems to exert a dual action on ethylene release by *B. cinerea*. Firstly, light produces extra radicals which increase the conversion of KMBA to ethylene and secondly, the KMBA synthesis is mediated by action of a flavin, which may be stimulated by light.

The acceleration of ethylene release *in vitro* at higher conidia concentrations (see Figure 3) also suggests that development of the fungus responds to chemical signals in its neighborhood. This could be related to sensing, e.g., (self-produced) ethylene.

Since the ethylene precursor (methionine) is present in plant tissues, it may be used by the fungus as substrate for ethylene production via the KMBA pathway. The mechanism for ethylene production by *B. cinerea in planta* is rather complex due to the multiple ways to convert the fungus released KMBA. One such possibility may be furnished by the oxidative environment generated during the plant-fungus interaction. For example, the hydroxyl radicals that are produced together with other active oxygen species (AOS) can cause chemical oxidation of KMBA. It has been shown that *B. cinerea* produces hydrogen peroxide (H_2O_2) when grown on autoclaved flax stems (Bratt *et al.*, 1988), possibly due to an oxidase activity, and H_2O_2 can be further converted to superoxide (O_2^-) and hydroxyl (OH•) radicals. In addition, Georgieva *et al.* (2000) reported an enhanced peroxidase activity in the tomato fruit pericarp upon infection. From here it raises naturally the question whether the *B. cinerea* ethylene significantly contributes to the total ethylene production in infected plant and if it plays a role in triggering the plant ethylene production (or manipulates ethylene production to trigger other defense mechanisms).

Ethylene perception by *B. cinerea*

It is likely that *B. cinerea*, like some other fungi, can sense ethylene and change its behavior accordingly. To gain more insight into the role of ethylene on fungal development, we monitored ethylene released by *B. cinerea* grown on medium containing ethylene perception inhibitor silver thiosulphate (STS). We observed that in the presence of STS, the hyphal growth was reduced. Moreover, the increase in ethylene production was delayed and the maximum ethylene level was lower compared to *B. cinerea* growing on PDA without STS. This may indicate that ethylene perception mediates fungal growth (Figure 4).

Role of ethylene in plant-fungal interaction *in vivo*

Infection of tomato fruits with *B. cinerea* resulted in enhanced ethylene release which started to rise before visible decay development. This demonstrates that ethylene can be considered a sensitive marker for early infection in harvested fresh products (Cristescu *et al.*, 2002; Polevaya *et al.*, 2001).

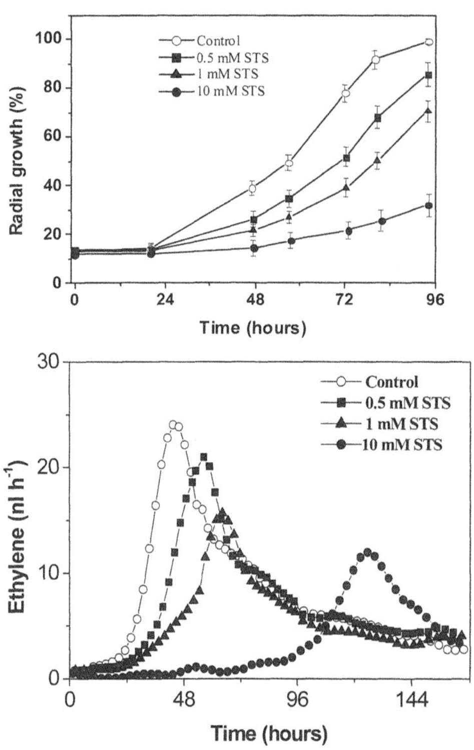

Figure 4. Ethylene production from *B. cinerea* (160 μl at $2x10^7$ conidia/ml) grown on PDA with 25 m*M* methionine without STS (Control) and with different STS concentrations, respectively (bottom). The radial growth of the fungus in these conditions (top).

In a previous work it was shown that higher inoculum concentrations of *B. cinerea* increased infection when applied to flowers and to wounds caused by leaf removal (Eden *et al.*, 1996). We found that also in the case of infected tomato fruit the decay development is dependent on the concentration of inoculum and it is faster for higher conidia concentrations (Cristescu *et al.*, 2002). It is known that ethylene production induces fruit ripening and that ripe fruits and senescent or wounded plant organs are more susceptible to *B. cinerea*. Accordingly, we observed a faster disease development for the fast ripening tomato cultivar Money Maker accompanied by higher levels of ethylene release compared to the slow ripening cultivar Daniela.

We found that exogenous application of ethylene (10, 20 ppbv and 1 ppmv of ethylene in air) did not affect the conidial germination or the hyphal growth of *B. cinerea in vitro*. In this case, the ethylene production by the fungus and its development were similar as in non-treated hyphae. Therefore, the presence of ethylene may represent an (indirect) advantage for the fungus, because it stimulates ripening and softening of the plant tissue and, therefore, facilitates tissue penetration and fungal spread (Diaz *et al.*, 2002).

B. cinerea is able to engage various infection strategies depending on the infected host. For example, infection of tomato leaves with *B. cinerea* occurs in three phases (Benito *et al.*, 1998): (i). primary lesion formation characterized by a necrotic lesion appearance; (ii). quiescent phase when no disease development or fungal growth can be seen and (iii). lesion expansion phase when the fungus colonizes the whole host leaf. Our results indicate that infection in fruits can be described according to a similar scheme. As an example, we present the evolution of the ethylene emission from a tomato fruit, artificially inoculated with *B. cinerea* by four small infections (2 mm deep in the epidermis, 160 μl) at $2x10^7$ and $2x10^5$ conidia ml^{-1} (Figure 5).

The following pattern of ethylene production was repeatedly observed:

(i). Initial inoculation (period 0-12 hours). A small peak in ethylene emission is observed both in *B. cinerea* and in mock infected tomato (not shown). However, ethylene production from *B. cinerea* infected tomato is more pronounced. Spore germination starts within the first 3 hours from inoculation. As no ethylene was produced by the fungus *in vitro* during its germination or dormancy, it is suggested that this first peak is part of the defense response by the attacked host (Ciardi *et al.*, 2000).

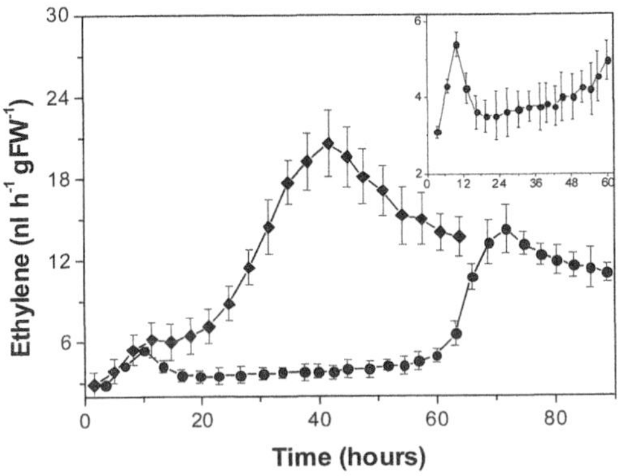

Figure 5. Ethylene production from tomatoes (FW = 80÷100g) infected with *B. cinerea* at $2x10^5$ (•) and $2x10^7$ conidia ml^{-1} (♦), respectively. At 0 h tomatoes were inoculated and immediately placed into cuvettes under continuous air flow of 4 liter h^{-1}. The inset shows the increase of ethylene emission from infected tomato with $2x10^5$ conidia ml^{-1} (■) for the first 2 days. Measurements were stopped when the fruits were completely deteriorated. Data are displayed as the averages of the sampling rate every 3 h (the errors due to averaging were smaller than the symbol size).

According to Robinson *et al.* (2001), this first peak is due to the conversion of the ACC existing in the tomato tissue, followed by increased transcription of the ACC synthase genes which will generate more ACC inside the fruit tissues, presumably used during the last stage of infection.

(ii). Non symptoms stage (period 12-48 hours) when there are no visible disease symptoms. At the beginning of this period the fungus stays dormant; its growth is temporarily stopped due to the complex defense machinery activated by the tomato host. At a certain point the fungus switches to an invasive action. This corresponds to the moment of hyphal elongation which *in vitro* occurs after 18 hours. Soon the hyphae are branching (after 24 h) and ethylene starts to increase slowly.

(iii). Visible disease development (after 48 h). A second peak in ethylene production, much larger than the first one, is recorded. This burst in ethylene release from infected tomato is a clear expression of disease development as a consequence of the tissue damage caused by *B. cinerea*. Senescence is initiated in this stage and the host lost the battle for its survival. The decrease in ethylene emission of the infected tomato after reaching the peak corresponds with an advanced stage of the fungal infection. It was suggested that during the infection process the infected tissue gradually loses the capacity to convert ACC to ethylene (Achilea *et al.*, 1985).

Infection-related ethylene production by both tomato cultivars showed appropriate patterns as for the ethylene released by the fungus *in vitro*, although at much higher values (hundreds fold). Therefore, ethylene generation by infected tomatoes can be considered a likely response of the host to the stress caused by *B. cinerea* infection. Ethylene production by the fungus *in vitro* is very low even in comparison with ethylene released by mock-infected fruits to be considered substantial in inducing fruit spoilage. Moreover, enhanced formation of ethylene by the infected tomato was monitored well before ethylene released by the fungus *in vitro* started to increase. Our cytological analysis indicated that conidia germination and fungal growth inside the fruit were comparable to those *in vitro*. These results indicate that the ethylene emission by the tomato-fungus system is not triggered directly by ethylene production of *B. cinerea*, although it is strongly "synchronised" with the growth rate of the fungus inside the tomato.

The first two stages of disease development, as described above, seem to be critical for the host because it has to generate adequate actions to successfully restrict or stop the fungus, such as production of phytoalexins and other phenolic compounds, pathogenesis-related proteins including those against the cell wall degrading enzymes (CWDEs) (Benito *et al.*, 1998), polygalacturonase inhibiting proteins, AOS (Baker and Orlandi, 1995; Levine,

1994). As for the fungus, it has to cope with the oxidative stress induced by the host and counteract the host produced compounds which inhibit its growth (Pezet at al., 1991). A complex and dynamic pattern of H_2O_2 formation was observed within tomato and bean leaves early after inoculation with *B. cinerea*. In the early stage of the infection the H_2O_2 originates from the host cells as an induced defense reaction, while after penetration of the epidermal cell wall, additional H_2O_2 may be generated directly by the fungal enzymes. A considerable increase in cytosolic H_2O_2 was found between 5-24 h after inoculation of tomato leaves with *B. cinerea* and apoplastic generation, as indicated by NADH peroxidase activity, was enhanced between 24 and 72 h (Patykowski and Urbanek, 2003). It was suggested that *B. cinerea* experiences H_2O_2 stress only in the early stage of infection. At this time, the fungus had penetrated the host, but the infection remained symptomless (Schouten *et al.*, 2002a). As a consequence of the oxidative stress imposed, *B. cinerea* temporarily exhibits a decrease in growth. Recently, Malolepsza and Urbanek (2000) showed that mycelial growth of *B. cinerea* was completely inhibited at 100 m*M* H_2O_2 (not checked if the fungus was killed or just temporarily impaired in growth), while others reported that germination of conidia occurred in the presence of 180 m*M* H_2O_2 and both germination and fungal development was slowed down, but not inhibited by up to 1 *M* H_2O_2. Moreover, *B. cinerea* is able to produce both intra- and extracellular enzymes (i.e., superoxide dismutase, laccase, catalase, different peroxidase to inactivate the H_2O_2 (Gil-ad *et al.*, 2000). The intracellular enzymes may only serve to protect the fungus from its own AOS, while the extracellular ones may be involved in protecting the fungus against the AOS from the host plant. It was recently postulated that *B. cinerea* actively triggers the production of AOS in planta in order to kill host cells, thereby facilitating entry into host. In leaves of *A. thaliana* infected by *B. cinerea*, massive depletion of ascorbic acid levels occurred before visible infection as a result of damage to the antioxidant mechanism (i.e., redox status) represents an early event in the infection process (Muckenschnabel *et al.*, 2002). In the mitochondrial fraction a continuous decrease in activity of ascorbate peroxidase (APX), one of the major H_2O_2-decomposing enzymes in plant cells, was observed in the inoculated leaves (Kuzniak and Sklodowska, 2004). They found that GSH (glutathione) and AA (ascorbate) pools together with the ascorbate-related enzymatic reactions were heavily suppressed once the spreading lesions started to develop.

The three stages described above were observed when low concentrations of the inoculation were used (in the order of 10^5 conidia/ml or lower). For higher concentrations the second stage of non-symptoms is very short and from the ethylene perspective, its production presents a continuous and rapid rise from beginning toward the third stage.

The low ethylene production of the fungus compared to the production of the tomato-fungus system already indicated that the contribution of fungal ethylene to the total ethylene is negligible. However, while invading the host, the fungus may have access to increased amounts of methionine or alternative substrates that increase its ethylene production. Therefore, tomato slices were treated with inhibitors of plant (not fungal) ethylene production and thereafter infected with *B. cinerea*. Inhibition of ethylene biosynthesis in *Botrytis*-infected tomato slices with AOA applied prior to inoculation significantly decreased the ethylene emission. However, it did not block it completely (Cristescu *et al.*, 2002). This remaining activity may be due to either plant ethylene in the case the inhibitor is not 100% effective or, alternatively, it may be due to fungal ethylene production. To determine the efficiency of the inhibitor treatment, the experiments were repeated with AVG, while also a wound control was inserted to determine the efficiency of the applied AVG (Figure 6). AVG-treated slices were severely wounded and their ethylene production compared to wounded non-pretreated slices. Wounded slices produced a significant amount of ethylene that was almost completely blocked by prior AVG treatment. Ethylene production in AVG treated and then *B. cinerea* infected slices was blocked to the same extent, which clearly shows that virtually all the ethylene produced in the plant-pathogen system is derived through ACC and, therefore of plant origin. The fungal growth was slightly reduced in the AVG treated tissues, although it recovered rapidly.

A control experiment where the fungus was allowed to grow on a medium containing autoclaved grounded tomato tissue showed no increased ethylene production by the fungus (unpubl.). To the contrary, ethylene production was consistently lower on tomato medium than on PDA medium. This indicates that tomato tissue is not a good substrate for ethylene production by *B. cinerea*. The presumed chemical conversion of KMBA to ethylene may be suppressed by the presence of, e.g., antioxidants in the tomato medium.

The question which arises is whether ethylene production in *B. cinerea* infected tomato is an autocatalytic process and whether traces of ethylene produced through fungal KMBA may trigger plant ethylene production. Experiments were performed with MCP pre-treated tomatoes that were later inoculated with *B. cinerea*.

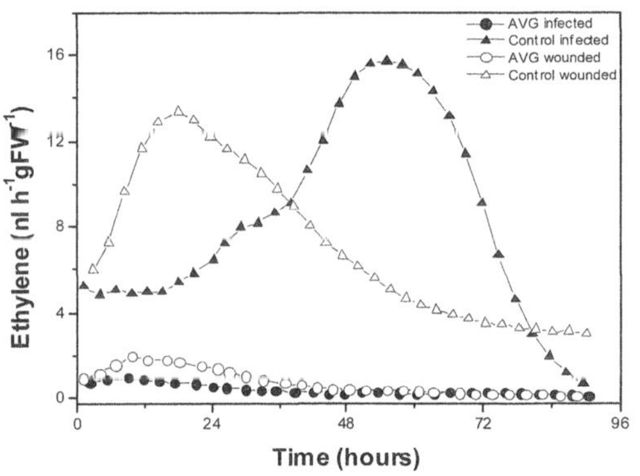

Figure 6. Ethylene production from slices of tomatoes treated with AVG (0.5 m*M*) compared to the non-treated fruits. Following 3 h AVG treatment, slices were either wounded or inoculated with 10^5 conidia/ml.

The experiments revealed that ethylene levels monitored in this case are comparable to or slightly higher than those from non-treated infected tomatoes. In MCP treated tomato, fungal ethylene (if produced at all) will not be able to trigger the plant ethylene production. If the (initially) produced ethylene would be important for further ethylene production, the MCP tomato should show much less ethylene especially at later stages of infection. Because this was not the case, we indicate that ethylene produced by tomato in response to *B. cinerea* infection is rather directly elicited by other mechanisms initiated by the fungus than by autocatalysis. One such mechanism may be the production by the fungus of necrosis- and ethylene-inducing protein (NEP) that directly induces ethylene production in the attacked host. The NEPs and their homologues have been reported in many fungal species, like *Verticillium* (Wang *et al.*, 2004), *Fusarium* (Bailey *et al.*, 1994), oomycete species, including *Phytophthora* and some eubacteria (Qutob *et al.*, 2002; Fellbrich *et al.*, 2002). Whether this is the case in *B. cinerea* infection remains to be elucidated. In this context, ethylene apparently plays a role in plant resistance. Interestingly, we observed a significant increase of disease development in the MCP-treated tomatoes compared to the non-treated fruits as it was also reported by Diaz *et al.* (2002) for disease development in leaves.

ETHYLENE AND RESVERATROL

One of the challenges to the modern agriculture is to deal with the enormous postharvest losses of fresh products which may add up to 30% (Kader *et al.*, 1992). These losses are mainly due to product deterioration as a result of over-ripening, senescence or pathogen attack. The use of low temperature storage, controlled atmosphere (CA) conditions and various pesticides is a common solution to overcome these problems, however, not without human health risks and environmental consequences. New strategies based on exploitation of the natural plant capabilities to improve its defense mechanisms and, hence, resistance may decrease the postharvest losses as a result of pathogen attack. In this context, resveratrol (3, 5, 4′-trihydroxystilbene) takes an important place because of its implications in both phytopathology and human health. This compound is naturally produced as phytoalexin in grapevine, peanut and other plants and it was found to protect the host against fungal infections

(Breuil *et al.*, 1999). These include fungi as *Plasmopara viticola* (Dai *et al.*, 1995), *Phomopsis viticola* (Hoos and Blaich, 1990) or *Rhizopus stolonifer* (Sarig *et al.*, 1997). Because the compound is effective against a broad range of fungal species and the selective accumulation of resveratrol in grape skin, this compound is a good candidate as a natural pesticide against pathogen attack as a result of improvement of the natural resistance of grapes. Due to its antioxidant properties, resveratrol can also have positive effects on fruit conservation during storage as it may slow down the deteriorative processes. Consequently, both endogenous enhancement and exogenous application could be exploited to reduce grape spoilage.

Since it was reported that *B. cinerea* can elicit the production of resveratrol in grapevines (Langcake and Pryce, 1976), many investigations have been carried out on this particular host-pathogen interaction (Stein and Blaich, 1985; Jeandet *et al.*, 1995, Adrian *et al.*, 2000, Breuil *et al.*, 1998; Montero *et al.*, 2003). In the wine industry, the growth of *B. cinerea* on wine grapes has been known as "noble rot" which gives an added effect to the bouquet of certain wines. Nevertheless, the grey mold is more often a severe problem for all grape varieties; the fungus can settle in on immature grapes and during the humid periods early in the season continues to penetrate the grapes causing them to rot.

Resistance of grapevines to *B. cinerea* infection is the result of multiple defense mechanisms consisting mainly of accumulation of phytoalexins, such as resveratrol, and the synthesis of pathogenesis related (PR)-proteins (Derckel *et al.*, 1999). Adrian *et al.* (1998) showed that in the presence of resveratrol conidia germination as well as mycelium growth of *B. cinerea* were significantly reduced. In response to the enhanced levels of resveratrol, the fungus produces blue-copper oxidases known as stilbene oxidases or laccases which were believed to detoxify resveratrol. By means of functional molecular genetic analysis of *B. cinerea* laccases, Schouten *et al.* (2002b) found a resveratrol-induced laccase gene *Bclcc2* which paradoxically is responsible for transforming resveratrol into fungitoxic compounds, thus, producing self-intoxication. In spite of this obvious advantage for the host, in the long term, it seams that the fungus can profit from the expression of this gene, because there are no *B. cinerea* strains reported so far in which this gene is deleted or its expression repressed. Thus, both the fungus and the plant-host can mutually benefit from this mechanism.

Ethylene is involved in the ripening process of many fruits and it may also play a role in pathogenesis. During the ripening phase of climacteric fruits (e.g., apples, tomatoes, etc.) both CO_2 and ethylene are emitted at elevated levels as opposed to non-climacteric fruits (e.g., citrus). In grapes it was reported that the resveratrol content decreases during ripening and, therefore, the fruit become more susceptible to *B. cinerea* infection (Sarig *et al.*, 1997). As non-climacteric fruit, grapes release ethylene at very low production rate (Archbold *et al.*, 1997); almost undetectable with standard procedures. The use of the laser-based ethylene detector, however, enabled us to investigate the dynamics of ethylene evolution in grapes with much improved accuracy. This device was simultaneously complemented by another laser-based instrumentation that uses Laser Desorption (LD) coupled with Laser Resonant Multiphoton Ionisation with Time-of-Flight Mass Spectrometric detection (REMPI-TOFMS) that provided fast and direct analysis of resveratrol in grapes (Montero *et al.*, 2000a, 2000b, Orea *et al.*, 2001). As resulting from this combination, we provided the first report on the real-time monitoring of ethylene by *B. cinerea* infected grapes in association with resveratrol levels (Montero *et al.*, 2003) (Figure 7).

In the case of non-infected grapes, the resveratrol content and the ethylene released showed an opposite behaviour with respect to each other; high resveratrol content corresponds to a low ethylene emission. Interestingly, the resveratrol content from the non-infected fruits was higher than that corresponding to the mock-infected fruits, which drastically decreased to zero during the first day after the buffer inoculation. In correlation, ethylene released by mock-infected grapes increased in the first day up to a certain level and showed higher values compared to the non-infected fruits. For the *Botrytis*-infected fruits, the resveratrol content increased to a maximum on the second day after infection, followed by a fast decrease, most probably due to its oxidation by the *B. cinerea* laccase. At the time the resveratrol content showed a decline, the ethylene emission started to rise. Overall, it seems that in grapes, the resveratrol content is determining the rate of ethylene production.

The relation is not reciprocal because, as it was described previously, continuous exposure to exogenous ethylene did not affect gray mold nesting ability on table grapes artificially inoculated with *B. cinerea* (Palou *et al.*, 2002, 2003). Previous investigations on the production of resveratrol by grapes in response to *Botrytis* infection (Jeandet *et al.*, 1995, Adrian *et al.*, 2000) showed that its elicitation occurred predominantly in the non-infected grapes surrounding the infected ones, while in the latter the resveratrol content was always lower than in the non-infected grapes.

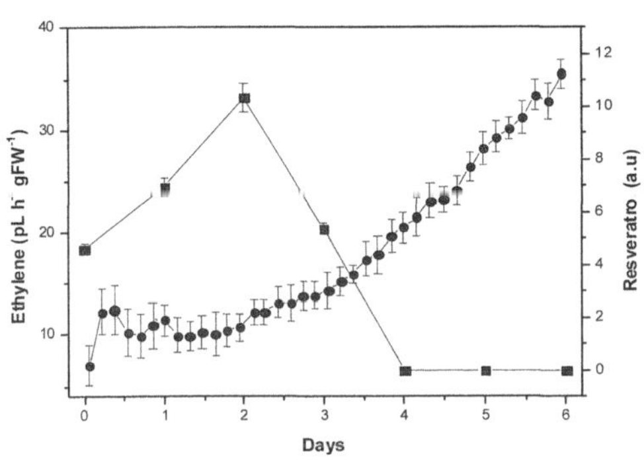

Figure 7. Ethylene production by grapes infected with *B. cinerea* (5 μl of the suspension at 10^3 conidia mL^{-1} per grape inoculated at 0 h) (•) and evolution of *trans*-resveratrol content (■) in grape skin of *Botrytis* infected grapes.

This apparently contradictory result is due to the time of resveratrol analysis which was done several days after the *Botrytis* infection when, as the authors claim, resveratrol could have been already metabolized by the fungus (as it happened in the present case after the second day). In accordance to our study, Paul *et al.* (1998) reported the induction of resveratrol by *B. cinerea* in leaves which reached a maximum yield in the third day after infection, followed by a rapid reduction of the resveratrol content by the fifth day.

The activity of resveratrol as natural pesticide has been investigated by exogenous application on grapes. A short submerge (5 s) of the grapes in 1.6 x 10^{-4} M solution of resveratrol affected the ethylene production in two directions: (1) delaying the enhanced ethylene emission with about 2 days and (2) decreasing the ethylene production of at least threefold. This significant decrease of the ethylene production in the treated grapes can be attributed to the action of trans-resveratrol on different microorganisms (bacteria and fungi) present on the grapes. This hypothesis is supported by a recent work on the effects of *B. cinerea* on grapes (Dorado *et al.*, 2001), performed in similar experimental conditions and with the same variety of grapes as in the present study. According to this report, other microorganisms like bacteria and fungi, distinct from the inoculated *Botrytis,* were developed during the incubation period of grapes and caused the deterioration of the fruits. The identified non-inoculated microorganisms present on grapes were mainly yeasts and imperfect fungi such as *Penicillium, Aspergillus* and *Alternaria* spp. which are known to include ethylene-producing species (Fukuda *et al.*, 1993). This treatment has positive effects on fruit conservation during storage; it doubled the normal shelf-life of grapes at room temperature, maintaining their postharvest quality for 10 days. This result offers a new, simple and inexpensive modality which can be used to improve the shelf-life of fruits and to preserve their natural postharvest quality. Recently, it was reported that exogenous application of resveratrol reduced postharvest decay also in other types of fruits than grapes, such as tomatoes, apples, avocado, pears and peppers (Jimenez *et al.*, 2005).

Preventing water losses is a very important issue during the postharvest period of fresh plant products. In this respect, it is suggested that resveratrol acts like a thin coat on the fruit, which not only protects against the microbial growth, but it also reduces water losses, thus conserving the water content and fruit firmness. Additionally, it was demonstrated that application of resveratrol to several fruits does not alter their organoleptic and biochemical properties (González Ureña *et al.*, 2003).

Microbial contamination of food is one of major problems with risks for the human health. Although some reports intended to show that risks related to using natural chemicals in foods are even greater than the risks from pesticide residues (Pimentel *et al.*, 1996, Swirsky *et al.*, 1997), the lack of toxicity of the resveratrol has been demonstrated. A considerable number of investigations are currently focussed on the health benefits of resveratrol consumption (see Frémont, 2000, German and Walzem, 2000 or Parr and Bolwell, 2000 for recent reviews on this subject) giving it an additional value as candidate for bio-control experiments against *B. cinerea*, as better alternative than the use of harmful chemical pesticides.

CONCLUSIONS

We have shown that a laser-based ethylene detector represents a suitable instrumentation for on-line measurements of ethylene released by fungi *in vitro* and *in vivo*. Moreover, the instrument is a powerful tool for the early detection of traces of ethylene released in the case of infection caused by microorganisms (fungi or bacteria) with a long period of incubation (weeks).

Its high sensitivity and fast time response allow to investigate the temporal and functional relationship between fungal and plant ethylene biosynthesis. In addition, this method eliminates the large data variability which might be generated during the use of the standard instrumentation (e.g., GC-gas chromatography) due to the integration procedure over many hours/days.

This development may lead to future applications in the postharvest technologies based on alternative strategies for fresh produce protection, mainly focused on action on the fungus rather that on inhibition plant-produced ethylene, that usually associate enhanced shelf life with decreased flavor and quality. Among ethylene, other components of biological interest can be monitored in real-time with the laser-based photoacoustic detectors. A trend in fruit storage is to reduce the oxygen level in order to slow down ripening and senescence. In this way, the aerobic respiration is gradually replaced by alcoholic fermentation that leads to production of acetaldehyde and ethanol. Alcoholic fermentation is also connected with stress-signal transduction and the disease-resistance response in plants. Several plant species when exposed to environmental stress such as water deficit, low temperature, ozone exposure or pathogen infection can generate significant amounts of acetaldehyde and ethanol at ambient of even or at elevated oxygen concentrations (Tadege and Kuhlemeier, 1997). Laser-based photoacoustic detectors have been proven to sensitive detection of acetaldehyde and ethanol at and below the part per billion level (Zuckermann *et al.*, 1997b). These compounds were monitored as markers for alcoholic fermentation in (pos)anoxic fruits (Zuckermann *et al.*, 1997b; Oomens *et al.*, 1998), (post)submerged rice seedlings (Boamfa *et al.*, 2003), wheat dough (Tomas *et al.*, 2001), dehydrated radicles of cucumber and pea (Leprince *et al.*, 2000) and poplar trees (Kreuzwieser *et al.*, 2001).

In many cases, due to a disturbed balance between the formation of AOS in plant tissue as a result of stress and the normal scavenging capacity, the plant tissues suffer from, e.g., lipid peroxidation that can cause damage of cell membranes (Halliwell and Gutteridge, 1989). The gaseous endproducts of this process are ethylene and ethane which can be sensitively monitored by laser-based photoacoustics. Ethane was monitored as a result of photo-oxidative damage of chilled cucumber leaves (Santosa *et al.*, 2003), membrane peroxidation in pears (Veltman *et al.*, 1999), lipid peroxidation induced during artificial aging of onion seeds (Klein *et al.*, 2004).

As mentioned previously, the plant recognition of pathogen infection leads to so-called hypersensitive response (HR) indicated by a fast, localized cell death at the site of infection. A synergistic mechanism has been proposed between the AOS and nitric oxide (NO) during the HR response in plants and using a CO laser-based photoacoustic detector, the first in planta and direct measurements of NO emission from plants undergoing various responses to *P. syringae* challenge has been obtained (Mur *et al.*, 2005).

These examples clearly show the high potential of photoacoustic detection systems for studying plant metabolism and emphasize the many possibilities for detailed studies of plant-pathogen interactions.

Acknowledgements

We thank Dr. Marc Staal (NIOO-KNAW, Yerseke, the Netherlands) for helpful comments during the preparation of this manuscript. Our research was supported by EU–FAIR grant ("Fruta Fresca" CT98-4211).

REFERENCES

Abeles, F. B. (1973). Ethylene in plant Biology. Academic Press. New York, p. 302.

Abeles, F. B., Morgan, P. W., and Saltveit, Jr, M. R. (1992). Ethylene in plant biology, 2nd ed. Academic Press, San Diego, CA.

Achilea, O., Fuchs, Y., Chalutz, E., and Rot, I. (1985). The contribution of host and pathogen to ethylene biosynthesis in *Penicillium digitatum*. Physiological Plant Pathology 27:55-63.

Adrian, M., Rajaei, H., Jeandet, P., Veneau, J., and Bessis, R. (1998). Resveratrol oxidation in *Botrytis cinerea* conidia. Phytopathology 88:472-476.

Adrian, M., Jeandet, P., Douillet-Breuil, A. C., Tesson, L., and Bessis, R. (2000). Stilbene content of mature *Vitis vinifera* berries in response to UV-C elicitation. Journal of Agricultural and Food Chemistry 48:6103-6105.

Amagai, A., and Maeda, Y. (1992). The ethylene action in the development of cellular slime molds: an analogy to higher plants. Protoplasma 167:159-168.

Archbold, D. D., Hamilton-Kemp, T. R., Barth, M. M., and Langlois, B. E. (1997). Identificating natural volatile compounds that control gray mold (*Bot. Cin.*) during post-harvest storage of strawberry, blackberry and grape. Journal of Agricultural and Food Chemistry 45:4032-4037.

Bailey, J., Jennings, C., and Anderson, J. D. (1997). The 24-KDa protein from *Fusarium oxysporum f. sp. erythroxyli*: occurrence in related fungi and the effect of growth medium on its production. Canadian Journal of Microbiology 43:45–55.

Baker, C. J., and Orlandi, E. W. (1995). Active oxygen in plant pathogenesis. Annual Review of Phytopathology 33:299-321.

Barkai-Golan, R., Lavy-Meir, G., and Kopeliovitch, E. (1988). Pectolytic and cellulolytic activity of *Botrytis cinerea. Pers.* related to infection of non-ripening tomato mutants. Journal of Phytopathology 123:174-183.

Bashan, Y. (1994). Symptom expression and ethylene production in leaf blight of cotton caused by *alternaria macrospora* and *Alternaria alternate* alone and combined. Canadian Journal of Botany 72:1574-1579.

Bell, A. A. (1981). Biochemical mechanisms of disease resistance. Annual Reviews of Plant Physiology 32:21-81.

Benito, E. P., Have, A. ten, Klooster, J. W. van't, and Kan, J. A. L. van (1998). Fungal and plant gene expression during synchronized infection of tomato leaves by *Botrytis cinerea*. European Journal of Plant Pathology 104:207-220.

Bent, A., Innes, R., Ecker, J., and Staskawicz, B (1992). Disease development in ethylene-insensitive Arabidopsis thaliana infected with virulent and avirulent *Pseudomonas* and *Xanthomonas* pathogens. Molecular Plant-Microbe Interactions 5:372-378.

Bijnen, F. G. C., Reuss, J., and Harren, F: J. M. (1996). Geometrical optimization of a longitudinal resonant photoacoustic cell for sensitive and fast trace gas detection. Rev. Sci. Instr. 67:2914-2923.

Biles, C. L., Abeles, F. B., and Wilson, C. L. (1990). The role of ethylene in anthracnose of cucumber, *Sucumis sativus,* caused by *Colletotrichum lagenarium*. Phytopathology 80:732-736.

Boamfa, E. I., Ram, P. C., Jackson, M. B., Reuss, J., and Harren, F. J. M. (2003). Dynamic aspects of alcoholic fermentation of rice seedlings in response to anaerobiosis and to complete submergence: relationship to submergence tolerance. Annals of Botany 91:279-290.

Boller, T. (1988). Ethylene and the regulation of antifungal hydrolases in plants. Oxford Surveys on Plant Molecular and Cell Biology 5:145-174.

Boller, T. (1991). Ethylene in pathogenesis and disease resistance. *In* A. K. Mattoo and J. C. Suttle (ed.), The plant hormone ethylene (Mattoo, A. K., and Suttle, J. C., eds.), CRC Press, Boca Raton, FL, U.S.A., pp. 293-324.

Bratt, R. P., Brown, A. E., and Mercer, P. C. (1988). A role for hydrogen peroxide in degradation of flax fibre by *Botrytis cinerea.* Transactions of the British Mycological Society 91:481-488

Breuil, A. C., Adrian, M., Pirio, N., Meunier, P., Bessis, R., and Jeandet, P. (1998). Metabolism of stilbene phytoalexins by *Botrytis cinerea*: 1. Characterization of a resveratrol dehydromer. Tetrahedron Letters 39:537-540.

Breuil, A. C., Jeandet, P., Adrian, M., Chopin, F., Pirio, N., Meunier, P., and Bessis, R. (1999). Characterization of a pterostilbene dehydrodimer produced by laccase of *Botrytis cinerea*. Phytopathology 89:298-302.

Brewer, R. J., Bruce, C. W., and Mater, J. L. (1982). Optoacoustic spectroscopy of C_2H_4. Applied Optics 21:4092-4100.

Brown, G. E., and Burns, J. K. (1998). Enhanced activity of abscission enzymes predisposes oranges to invasion by *Diplodia natalensis* during ethyelene degreening. Postharvest Biology and Technology 14:217-227.

Brown, G. E., and Lee, H. S. (1993). Interaction of ethylene with citrus stem-end rot caused by *Diplodia natalensis*. Phytopathology 83:1204-1208.

Bussche, F. van den, Smalle, J., Jie, L., Madeira Saibo, N. J., de Paepe, A., Chaerle, L., Tietz, O., Smets, R., Laarhoven, L. J. J., Harren, F. J. M., Onckelen, H. van, Palme, K., Verbelen, J. P., Straeten, D. van der (2003). The *Arabidopsis* mutant *alh1* illustrates a cross talk between ethene and auxin. Plant Physiology 131:1228-1238.

Catlett, N. L., Yoder, O. C., and Turgeon, B. G. (2003). Whole-genome analysis of two-component signal transduction genes in fungal pathogens. Eukaryotic Cell 2:1151-1161.

Chague, V., Elad, Y., Barakat, R., Tudzynski, P., and Sharon, A. (2002). Ethylene biosynthesis in *Botrytis cinerea*. FEMS Microbiology Ecology 40:143-149.

Chang, C., and Stadler, R. (2001). Ethylene hormone receptor action in *Arabidopsis*. BioEssays 23:619-627.

Chappell, J., Hahlbrock, K., and Boller, T. (1984). Rapid induction of ethylene biosynthesis in cultured parsley cells by fungal elicitor and its relationship to the induction of phenylalanine ammonia-lyase (*Phytophthora megasperma*). Planta 161:475-480.

Chen, N., Goodwin, P. H., and Hsiang, T. (2003). The role of ethylene during the infection of *Nicotiana tabacum* by *Colletotrichum destructivum*. Journal of Experimental Botany 54:2449-2456.

Ciardi, J. A., Tieman, D. M., Lund, S. T., Jones, J. B., Stall, R. E., and Klee, H. L. (2000). Response to *Xanthomonas campestris* pv. *Vesicatoria* in tomato involves regulation of ethylene receptor gene expression. Plant Physiology 123:81-92.

Cohn, J., Sessa, G., and Martin, G. B. (2001). Innate immunity in plants. Current Opinion in Immunology 13:55-62.

Cristescu, S. M., Martinis, D. de, Lintel Hekkert, S. te, Parker, D. H., and Harren, F. J. M. (2002). Ethylene production by *Botrytis cinerea in vitro* and in tomatoes. Applied and Environmental Microbiology 68:5342-5350.

Cronshaw, D. K., and Pegg, G. F. (1976). Ethylene as a toxin synergist in *Verticillium* wilt of tomato. Physiological Plant Pathology 9:33-38.

Dai, G.H., Andary, C., Mondolot-Cosson, L. and Boubals, D. (1995). Histochemical Studies on the Interaction between three species of grapevine, *Vitis Vinifera, V. rupestris* and *V. rotundifolia* and the downy mildew fungus, *Plasmopara viticola*. Physiological and Molecular Plant Pathology 46:177-188.

Daundasekera, M. Joyce, D. C., Aked, J., Abikaram, N. K. B. (2003). Ethylene production by *Colletotrichum musae in vitro*. Physiological and Molecular Plant Pathology 62:21-28.

Derckel, J. P., Baillieul, F., Manteau, S., Audran, J. C., Haye, B., Lambert, B., and Legendre, L. (1999). Differential Induction of Grapevine Defenses by two strains of *Botrytis cinerea*. Biochemistry and Cell biology 89:197-203.

Diaz, J., Have, A. ten, and Kan, J. A. L. van (2002). The role of ethylene and wounding signalling in resistance of tomato to *Botrytis cinerea*. Plant Physiology 129:1341-1351.

Dixon, R. A., and Lamb, C. J. (1990). Molecular communication in interactions between plants and microbial pathogens. Annual Review of Plant Physiology and Plant Molecular Biology 41:339-367.

Dorado, M., Bermejo, E. González, J. L., Sánchez, A., and Luna, N. (2001). Development influence of *Botrytis cinerea* on grapes. Advances in Food Chemistry 23:153-159.

Eden, M. A., Hill, R. A., Beresford, R., and Stewart, A. (1996). The influence of inoculum concentration, relative humidity, and temperature of infection of greenhouse tomatoes by *Botrytis cinerea*. Plant Pathology 45:795-806.

El-Kazzaz, M. K., Sommer, N. F., and Kader, A. A. (1983). Ethylene effects on *in vitro* and *in vivo* growth of certain postharvest fruit-infecting fungi. Phytopathology 73:998-1001.

Elad, Y. (1990). Production of ethylene by tissue of tomato, pepper, French-bean and cucumber in response to infection by *Botrytis cinerea*. Physiollogical Molecular Plant Pathology 36:277-287.

Elad, Y. (1993). Regulators of ethylene biosynthesis or activity as a tool for reducing susceptibility of host plant tissues to infection by *Botrytis cinerea*. Netherlands Journal of Plant Pathology 99:105-113.

Elad, Y. (2002). Ethylene and reactive oxygen species in a plant-pathogen system. Phytoparasitica 30:307.

Elad, Y., and Eversen, K. (1995). Physiological aspects of resistance to *Botrytis cinerea*. Phytopathology 85:637-643.

Fan, X. T., Mattheis, J. P., and Roberts, R. G. (2000). Biosynthesis of phytoalexin in carrot root requires ethylene action. Physiologica Plantarum 110:450-454.

Fellbrich, G., Romanski, A., Varet, A., Blume, B., Brunner, F., Englehardt, S., Felix, G., Kemmerling, B., Krzymowska, M., and Nurnberger, T. (2002). NPP1, a *Phytophthora*-associated trigger of plant defense in *Arabidopsis*. Plant Journal 32:375–390.

Frémont, L. (2000). Biological effects of resveratrol. Life Sciences 66:663-673.

Fukuda, H., Kitajima, H., Fujii, T., Tazaki, M., and Ogawa, T. (1989a). Purification and some properties of novel ethylene-forming enzyme produced by *Penicillium digitatum*. FEMS Microbiology Letters 59:1–6.

Fukuda, H., Takahashi, M., Fujii, T., Tazaki, M., and Ogawa, T. (1989b). An NADH: Fe(III) EDTA oxidoreductase from *Cryptococcus albidus*: an enzyme involved in ethylene production *in vivo*? FEMS Microbiology Letters 60:107–112.

Fukuda, H., Ogawa, T., and Tanase, S. (1993). Ethylene production by microorganisms. *In* Advances in microbial physiology, Vol. 35, (Rose, A. H., ed.), Academic Press Inc., London, U.K., pp. 275-306.

Georgieva, I., Edreva, A., Rodeva, R., Sotirova, V., and Stoimenova, E. (2000). Metabolic changes in tomato fruits and seeds after viral, bacterial and fungal infection. Acta Physiologica Plantarum 22:281-284.

Geraats, B. P. J., Bakker, P. A. H. M., and van Loon, L. C. (2002). Ethylene insensitive impairs resistance to soilborne pathogens in tobacco and *Arabidopsis thaliana*. Molecular Plant-Microbe Interactions 15:1078-1085.

German, J. B., and Walzem, R.L. (2000). The health benefits of wine. Annual Review of Nutrition 20:561-593.

Gil-ad, N. L., Bar-Nun, N., Noy, T., Mayer, A. M. (2000). Enzymes of *Botrytis cinerea* capable of breaking down hydrogen peroxide. FEMS Microbiology Letters 190:121-126.

Gonzáles Ureña, A., Orea, J. M., Montero, C., Jimenez, J. B., Gonzalez, J. L., Sanchez, A., and Dorado, M. (2003). Improving the post-harvest resistance in fruits by external application of *trans*-resveratrol. Journal of Agricultural Food Chemistry 51:82-89.

Govrin, E. M., and Levine, A. (2000). The hypersensitive reaction facilitates plant infection by the necrotophic fungus *Botrytis cinerea*. Current Biology 10:751-757.

Halliwell, B., and Gutteridge, J. M. C. (1989). Free radicals in Biology and medicine. Oxford: Clarendon press: 188-276.

Harren, F. J. M., and Reuss, J. (1997). Photoacoustic spectroscopy. *In* Encyclopedia of applied physics, Vol. 19 (Trigg, G. L., ed.), VCH Publishers, Inc., Weinheim, Germany, pp. 413-435.

Have, A. ten, Breuil, W. O., Wubben, J. P., Visser, J., and Kan, J. A. L. van (2001). *Botrytis cinerea* endopoligalacturonase genes are differentially expressed in various plant tissues. Fungal Genetics and Biology 33:97-105.

Hoffman, T., Schmidt, J. S., Zheng, X., and Bent, A. F. (1999). Isolation of ethylene-insensitive soybean mutants that are altered in pathogen susceptibility and gene-for-gene disease resistance. Plant Physiology 119:935-950.

Hoos, G., and Blaich, R. (1990). Influence of resveratrol on germination of conidia and mycelial growth of *Botrytis cinerea* and *Phomopsis viticola*. Journal of Phytopathology 129:102-110.

Hyodo, H. (1991). Stress/wound ethylene. *In* The plant hormone ethylene (Mattoo, A. K., and Suttle, J. C., eds.), CRC Press, Boca Raton, FL, U.S.A., pp. 65-80.

Ince, J. E., and Knowles, C. J. (1986). Ethylene formation by cell-free extracts of *Escherichia coli*. Archives of Microbiology 146:151-158.

Jarvis, W. R. (1977). *Botryotinia* and *Botrytis* species: taxonomy and pathogenicity. Monogr. Res. Branch Can. Dept. Agric. 15, Harrow, Ontario, Canada.

Jeandet, P., Bessis, R., Sbaghi, M., and Meunier, P. (1995). Production of the phytoalexin Resveratrol by grapes as a response to *Botrytis* attack under natural conditions. Journal of Phytopathology 143:135-139.

Jia, Y. (1999). Synthesis and degradation of 1-aminocyclopropane-1-carboxylic acid by *Penicillium citrinum*. Bioscience Biotechnology and Biochemistry 63:384-387.

Jimenez, J. B., Orea, J. M., Montero, C., Gonzáles Ureña, A., Navas, E., Slowing, K., Goez-Serranillos, M. P., Carretero, E., and de Martinis, D. (2005). Resveratrol treatment controls microbial flora, prolongs shelf life and preserves nutritional quality of fruit. Journal of Agricultural and Food Chemistry 53:1526-1530.

Johnson, P., and Ecker, J. (1998). The ethylene gas signal transduction pathway: A molecular perspective. Annual Review of Genetics 32:227-254.

Kader, A. A. (1985). Ethylene-induced senescence and physiological disorders in harvested horticultural crops. HortScience 20:54-57.

Kader, A. A. (1992). Postharvest biology and technology: an overview. *In* Postharvest technology of horticultural crops 42, 2nd ed., University of California, Oakland, CA, U.S.A., pp. 15–20.

Kan, J. A. L. van, Cozijnsen, T., Danhash, N., and Wit, P. J. G. M. de (1995). Induction of tomato stress protein mRNAs by ethephon, 2, 6-dichloroisonicotic acid and salicylate. Plant Molecular Biology 27:1205-1213.

Kende, H. (1993). Ethylene biosynthesis. Annual Review of Plant Physiology 44:283-307.

Kepczynska, E. (1989). Ethylene requirement during germination of *Botrytis cinerea* spores. Physiology Plantarum 77:369-372.

Kepczynska, E. (1993). Involvement of ethylene in the regulation of growth and development of the fungus *Botrytis cinerea Pers. ex. Fr.* Plant Growth Regulators 13:65-69.

Kepczynski, J., and Kepczynska, E. (1977). Effect of ethylene on germination of fungal spores causing fruit rot. Fruit Science Reports 4:31-35.

Klein, J. D., Santosa, E., Laarhoven, L. J., Boamfa, E. I., Hebbe, Y., and Harren, F. J. M. (2004). Acta Horticulturae 631:39-42.

Knoester, M., Loon, L. C. van, Heuvel, J. van den, Henning, J., Bol, J. F., and Linthorst, H. J. M. (1998). Ethylene-insensitive tobacco lacks nonhost resistance against soil-borne fungi. Proceeding of the National Academy of Sciences USA 95:1933-1937.

Kolattukudy, P. E., Rogers, L. M., Li, D., Hwang, C. S., and Flaishman, M. A. (1995). Surface signaling in pathogenesis. Proceedings of the National Academy of Sciences of the USA 92:4080-4087.

Kreuzwieser, J., Harren, F. J. M., Laarhoven, L. J. J., Boamfa, I., te Lintel-Hekkert, S., Scheerer, U., Hüglin, C., and Rennenberg, H. (2001). Acetaldehyde emission by the leaves of trees - correlation with physiological and environmental parameters. Physiologia Plantarum 113:41-49.

Kuzniak, E., and Sklodowska, M. (2004). The effect of *Botrytis cinerea* infection on the antioxidant profile of mitochondria from tomato leaves. Journal of Experimental Botany 55:605-612.

Langcake, P., and Pryce, R. J. (1976). The production of resveratrol by *Vitis vinifera* and other members of the *Vitaceae* as a response to infection or injury. Physiological Plant Pathology 9:77-86.

Lawton, K., Weymann, K., Friedrich, L., Vernooij, B., Uknes, S., and Ryals, J. (1995). Systemic acquired resistance in *Arabidopsis* requires salicylic acid but not ethylene. Molecular Plant-Microbe Interactions 8:863-870.

Leprince, O., Harren, F. J. M., Buitink, J., Alberda, M., and Hoekstra, F. A. (2000). Metabolic dysfunction and unabated respiration precede the loss of membrane integrity during dehydration germinating radicals. Plant Physiology 122: 597-608.

Levine, A., Tenhaken, R., and Dixon, R. (1994). H_2O_2 from the oxidative burst orchestrates the plant hypersensitive disease resistance response. Cell 79:583-593.

Lintel Hekkert, S., Staal, M. J., Nabben, R. H. M., Zuckermann, H., Persijn, S., Stal, L. J., Voesenek, L. A. C. J., Harren, F. J. M., Reuss, J., and Parker, D.H. (1998). Laser photoacoustic trace gas detection, an extremely sensitive technique applied in biological research. Instruments Science and Technology 26:157-175.

Loon, L. C. van (1984). Regulation of pathogenesis and symptom expression in diseased plants by ethylene. *In* Ethylene: biochemical, physiological and applied aspects (Fuchs, Y., and Chalutz, E., eds.), The Hague: martinus Nijhoff/Dr. W.Junk, pp. 171-180.

Lund, S. T., Stall, R. E., and Klee, H. J. (1998). Ethylene regulates the susceptibility response to pathogen infection in tomato. The Plant Cell 10:371-382.

Malolepsza, U., and Urbanek, H. (2000). The oxidants and antioxidant enzymes in tomato leaves treated with o-hydroxyethylorutin and infected with *Botrytis cinerea*. European Journal of Plant Pathology 106:657–665.

Marte, M., Buonaurio, R., and Dellatorre, G. (1993). Induction of systemic resistance to tobacco powdery mildew by tobacco mosaic virus, tobacco necrosis virus and ethephon. Journal of Phytopathology 138:137-144.

Mattoo, A. K., and Suttle, J. C. (1991). The plant hormone ethylene. CRC Press, Boca Raton, FL.

Montero, C., Bescós, B., Orea, J. M., and González Ureña, A. (2000a). Food chemical analysis by laser desorption and resonant ionization mass spectrometry. Revista Analytical Chemistry 19:1-29.

Montero, C., Orea, J. M., Muñoz, M. S., Lobo, R. F., and González Ureña, A. (2000b). Non volatile analysis in fruits by laser resonant ionization spectrometry: Application to resveratrol in grapes. Applied Physics B 71:601-605.

Montero, C., Cristescu, S., Jimenez, J. B., Orea, J. M., Lintel Hekkert, S. te, Harren, F. J. M., and Gonzalez Urenia, A. (2003). trans-Resveratrol and grape disease resistance. A dynamical study by high-resolution laser-based techniques. Plant Physiology 131:129-138.

Muckenschnabel, I., Goodman, B. A., Williamson, B., Lyon, G. D., and Deighton, N. (2002). Infection of leaves of Arabidopsis thaliana by *Botrytis cinerea*; changes in ascorbic acid, free radicals, and lipid peroxidation products. Journal of Experimental Botany 53:207-214.

Mur, L. A. J., Santosa, I. E., Laarhoven, L. J. J., Holton, N. J., Harren, F. J. M., and Smith, A. R. (2005). Laser photoacoustic detection allows in planta detection of nitric oxide in tobacco following challenge with avirulent and virulent *Pseudomonas syringae* pathovars. Plant Physiology 138:1247-1258.

Nagahama, K., Ogawa, T., Fujii, T., Tazaki, M., Tanase, S., Morino, Y., and Fukuda, H. (1991). Purification and properties of an ethylene-forming enzyme from *Pseudomonas syringae* pv. *phaseolicola* PK2. Journal of General Microbiology 137:2281–2286.

Nazli, Z.-i.-H., Arshad, M., and Khalid, A. (2003). 2-Keto-4-methylthiobutyric acid-dependent biosynthesis of ethylene in soil. Biol. Fertil. Soils 37:130-135.

Oomens, J., Zuckermann, H., Persijn, S., Parker, D. H., and Harren, F. J. M. (1998). CO-laser based photoacoustic trace-gas detection: applications in postharvest physiology. Applied Physics B67:459-466.

Orea, J. M., Montero, C., Jiménez, J. B., and González Ureña, A. (2001). Analysis of *trans*-Resveratrol by laser desorption coupled with resonant ionization spectrometry. Application to *trans*-resveratrol content in vine leaves and grape skin. Analytical Chemistry 73:5921-5929.

Palou, L., Crisosto, C. H., Garner, D., Basinal, L. M., Smilanick, J. L., and Zoffoli, J. P. (2002). Minimum constant sulfur, dioxide emission rates to control gray mold of cold-stored table grapes. American Journal of Enol. Vitic. 53:110-115.

Palou, L., Crisosto, C. H., Garner, D., and Basinal, L. M. (2003). Effect of continuous exposure to exogenous ethylene during cold storage on postharvest decay development and quality attributes of stone fruits and table grapes. Postharvest Biology and Technology 27:243-254.

Parr, A. J. and Bolwell, G. P. (2000). Phenols in the plant and in the man. The potential for possible nutritional enhancement of the diet by modifying the phenols content or profile. Journal of Agricultural and Food Science 80:985-1012.

Patykowski, J., and Urbanek H. (2003). Activity of enzymes related to H2O2 generation and metabolism in leaf apoplastic fraction of tomato leaves infected by *Botrytis cinerea*. Journal of Phytopathology 151:153-161.

Paul, B., Chereyathmanjiyil, A., Masih, I., Chapuis, L., and Benoit, A. (1998). Biological control of *Botrytis cinerea* causing grey mould disease of grapevine and elicitation of stilbene phytoalexin (resveratrol) by a soil bacterium. FEMS Microbiology Letters 165:65-70.

Penninckx, I. A. M., Eggermont, K., Terras, F. R. G., Thomma, B. P. H. J., Samlanx, G. W. de, Buchala, A., Metraux, J. P., Manners, J. M., and Broekaert, W. F. (1996). Pathogen-induced systemic activation of a plant defensin gene in *Arabidopsis* follows a salicylic acid-independent pathway. Plant Cell 8:2309-2323.

Petruzzelli, L, Harren F., Perrone, C., and Reuss, J. (1995). On the role of ethylene in seed germination and early root growth of *Pisum Sativum*. Journal of Plant Physiology 145:83-86.

Pezet, R., Pont, V., and Hoang-Van, K. (1991). Evidence for detoxification of pterostilbene and resveratrol by laccase-like stilbene oxidase produced by *Botrytis cinerea*. Physiol. Mol. Plant Pathol. 39:441-450.

Pimentel, D., Culliney, T. W., and Bashore, T. (1996). Public health risks associated with pesticides and natural toxins in foods. *In* The Radcliffes's IPM World Textbook (Radcliffe, E. B., and Hutchison, W. D., eds.), University of Minnesota, St. Paul, MN, U.S.A.

Polevaya, Y., S. Alkalai-Tuvia, A. Copel, and E. Fallik. (2002). Early detection of grey mould development in tomato after harvest. Postharvest Biology and Technology 25:221-225.

Qadir, A., Hewett, E. W., and Long, P. G. (1997). Ethylene production by *Botrytis cinerea*. Postharvest Biology and Technology 11:85-91.

Qutob, D., Kamoun, S., and Gijzen, M. (2002). Expression of *Phytophthora sojae* necrosis-inducing protein occurs during transition from biotrophy to necrotrophy. Plant Journal 32:361–373.

Rodrigo, I., Vera, P., Tornero, P., Hernandez Yago, J., and Conejero, V. (1993). cDNA cloning of voroid-induced tomato pathogenesis-related protein P23: characterization as a vacuolar antifungal factor. Plant Physiology 102:939-945.

Roze, L. V., Calvo, A. M., Gunterus, A., Beaudry, R., Kall, M., and Linz, J. E. (2004). Ethylene Modulates Development and Toxin Biosynthesis in *Aspergillus* possibly via an ethylene sensor-mediated signaling pathway. Journal of Food Protection 67:438-447.

Saltveit, M. E. (1999). Effect of ethylene on quality of fresh fruits and vegetables. Postharvest Biology and Technology 15:279-292.

Santosa, I. E., Laarhoven L. J. J., Harbinson J., Driscoll S., and Harren F. J. M. (2003). Laser-based trace gas detection of ethane as a result of photo-oxidative damage in chilled cucumber leaves. Review of Scientific Instruments 74:680-683.

Sarig, P., Zutkhi, Y., Monjauze, A., Lisker, N., and BenArie, R. (1997). Phytoalexin elicitation in grape berries and their susceptibility to *Rhizopus stolonifer*. Physiological and Molecular Plant Pathology 50:337-347.

Schouten, A., Tenberge, K. B., Vermeer, J., Stewart, J., Wagemakers, C. A. M., Williamson, B., and Kan, J. A. L. van (2002a). Functional analysis of an extracellular catalase of *Botrytis cinerea*. Molecular Plant Pathology 3:227-238.

Schouten, A., Wagemakers, L., Stefanato, F. L., Kaaij, R. M. van der, and Kan J. A. L. van (2002b). Resveratrol acts as a natural profungicide and induces self-intoxication by a specific laccase. Molecular Microbiology 43:883-894.

Staal, M., Lintel Hekkert, S. te, Harren, F., and Stal, L. J. (2001). Nitrogenase activity in cyanobacteria measured by the acetylene reduction assay: a comparison between batch incubation and online monitoring. Environmental Microbiology 3:343-351.

Staal, M., Lintel-Hekkert, S. te, Harren, F., and Stal, L. J. (2003). Light action spectra of nitrogenase activity in Baltic Sea cyanobacteria. Journal of Phycology 39:668-677.

Stein, U., and Blaich, R. (1985). Investigations on the production of stilbenes and susceptibility to *Botrytis* of Vitis spp. Vitis 24:75-87.

Swirsky Gold, L., Slone, T. H., and Ames, B. N. (1997). Priorization of possible carcinogenic hazards in food. *In* Food Chemical Risk Analysis (Tennant, D. R., ed.), Chapman and Hall., New York, U.S.A., pp. 267-295.

Tadege, M., and Kuhlemeier, C. (1997). Aerobic fermentation during tabacco pollen development. Plant Molecular Biology 35:343-354.

Thain, S. C., Laarhoven, L. J. J., Dowson-Day, M. J., Wang, Z. Y., Tobin, E. M., Harren, F. J. M., Straeten, D. van der, and Millar, A. J. (2004). Circadian rhythms of ethylene emission in Arabidopsis detected by laser photoacoustics. Plant Physiology 136:3751-3761.

Thomma, B., Eggermont, K., Tierens, K. F., and Broekaert, W. F. (1999). Requirement of functional ethylene-insensitive 2 gene for efficient resistance of *Arabidopsis* to infection by *Botrytis cinerea*. Plant Physiology 121:1093-1102.

Thomma, B., Penninckx, I., Broekaert, W. F., and Cammue, B. P. A. (2001). The complexity of disease signaling in Arabidopsis. Current Opinion. in Immunology 13:63-68.

Thuring, J. W. J. F., Harren F. J. M., Nefkens G. H. L., Reuss, J., Titulaer, G. T. M., de Vries, H. S. M., and Zwanenburg, B. (1994). Ethene production by seeds of *Striga hermonthica* induced by germination stimulants. *In* Biology and management of Orobanche (Pieterse, A. H., Verkleij, J. A. C., and ter Borg, S. J., eds.), Royal Tropical Institute, Amsterdam, The Netherlands, pp. 225-236.

Tiedemann, A. von (1997). Evidence for a primary role of active oxygen species in induction of host cell death during infection of bean leaves with *Botrytis cinerea*. Physiological and Molecular Plant Pathology 50:151-166.

Tomas, S. A., and Harren, F. J. M. (2001). Kinetics of ethanol and acetaldehyde production in fermenting wheat dough by laser-based trace gas detection. Food Science and Technology International 7:307-315.

Veltman, R. H., Sanders, M. G., Persijn, S. T., Peppelenbos H. V., and Oosterhaven, J. (1999). Decreased ascorbic acid levels and brown core development in pears (*Pyrus communis* L. cv. Conference). Physiologia Plantarum 107:39-45.

Visser, E. J. W., Nabben, R. H. M., Blom, C. W. P. M. and Voesenek, L. A. C. J. (1997). Growth of primary lateral roots and adventitious roots during conditions of hypoxia and high ethylene concentrations. Plant, Cell and Environment 20:647-653.

Voesenek, L. A. C. J., Banga, M., Thier, R. H., Mudde, C. M., Harren, F. J. M., Barendse, G. W. M., and Blom, C. W. P. M. (1993). Submergence induced ethylene synthesis, entrapment and growth in two plant species with a contrasting flooding resistance. Plant Physiology 103:783-791.

Vries, H. S. M. de, Harren, F. J. M., and Reuss, J. (1995). In situ, real-time monitoring of wound-induced ethylene in cherry tomatoes by two infrared laser-driven systems. Post-Harvest Biology and Technology 6:275-285.

Vries, H. S. M. de, Wasono, M. A. J., Harren, F. J. M., Woltering, E. J., Valk, H. C. P. M. van der, and Reuss, J. (1996). Ethylene and CO2 emission rates and pathways in harvested fruits investigated, in situ, by laser photodeflection and photoacoustic techniques. Post-Harvest Biology and Technology 8:1-10.

Wagstaff, C., Chanasut, U., Harren, F. J. M., Laarhoven, L. J., Thomas, B., Rogers, H. J., and Stead, A. D. (2005). Ethylene and flower longevity in *Alstromeria*: relationship between tepal senescence, abscission and ethylene biosynthesis. Journal of Experimental Botany 56:1007-1016.

Wang, J-Y., Cai, Y., Gou, J-Y., Mao, Y-B., XU, Y-H., Jiang, W-H., and Chen, X-Y. (2004). Applied and Environmental Microbiology 70:4989-4995.

Woltering, E. J., Hout, M. van, Somhorst, D., and Harren, F. (1993). Roles of pollination and short-chain saturated fatty acids in flower senescence. Plant Growth Regulation 12:1-10.

Wubben, J. P., Have, A. ten, Kan, J. A. L. van, and Visser, J. (2000). Regulation of endopolygalacturonase genes expression in *Botrytis cinerea* by galacturonic acid, ambient pH and carbon catabolite repression. Current Genetics 37:152-157.

Wubben, M. J. E, Su, H., Rodermel, S. R., and Baum, T. J. (2001). Susceptibility to the sugar beet cyst nematode is modulated by ethylene signal transduction in *Arabidopsis thaliana.* Molecular Plant-Microbe Interactions 14:1206-1212.

Yang, S. F, and N. E. Hoffman. (1984). Ethylene biosynthesis and its regulation in higher plants. Annual Review of Plant Physiology 35:155-189.

Zuckermann, H., Staal, M., Stal, L. J., Reuss, J., Lintel-Hekkert, S. te, Harren, F. J. M., and Parker, D. H. (1997a). On line monitoring of nitrogenase activity in cyanobacteria by sensitive laser photoacoustic detection of ethylene. Applied and Environmental Microbiology 63:4243-4251.

Zuckermann, H, Harren, F. J. M., Reuss, J., and Parker, D. H. (1997b). Plant Physiology 113:925-932.

Part 2

THE FUNGAL SPORE IN FOOD MYCOLOGY

Fungal spores are the main vehicles of distribution for fungi. The variation in shape and behaviour of these cells in the fungal kingdom is bewildering. They can be formed in large numbers and travel through the air, water, or attached to small animals. Many food products are colonised by the action of single or multicelled spores. Alternatively, fungi are present inside the ingredients of the food at earlier stages of the production chain. Formation of spores takes place on structures that are as variable as spores and the taxonomy of fungi is partly based on the morphology of this apparatus. In Chapter 3, Ugalde and Corrochano describe examples of the process of spore formation in different groups of fungi. Spore germination requires a highly specialized apparatus and its biology is intricating.

In Chapter 4, McCartney and West describe the fate of spores after release from the spore-bearing apparatus as small particles through the air. As everyone can experience many food products are spoiled by fungi from the air, but how do these spores travel through the air and how does the shape of the spore influence the travelling time of the spore is the topic of this chapter.

In Chapter 5, Chitarra and Dijksterhuis describe the behaviour of fungal spores before and after entering of the substrate. How do spores behave as cells when they travel through the air and directly upon landing on the food substrate. They summarize the knowledge about spore germination as an intriguing transition from a static, relatively dormant cell to actively metabolising and growing fungal hyphae.

Some fungal spores are resistant to stress to such an extent that they can be compared to bacterial spores. This unique group of fungi survives pasteurisation and spoil food products after such a treatment. Dijksterhuis summarises the knowledge about these fungi in Chapter 6.

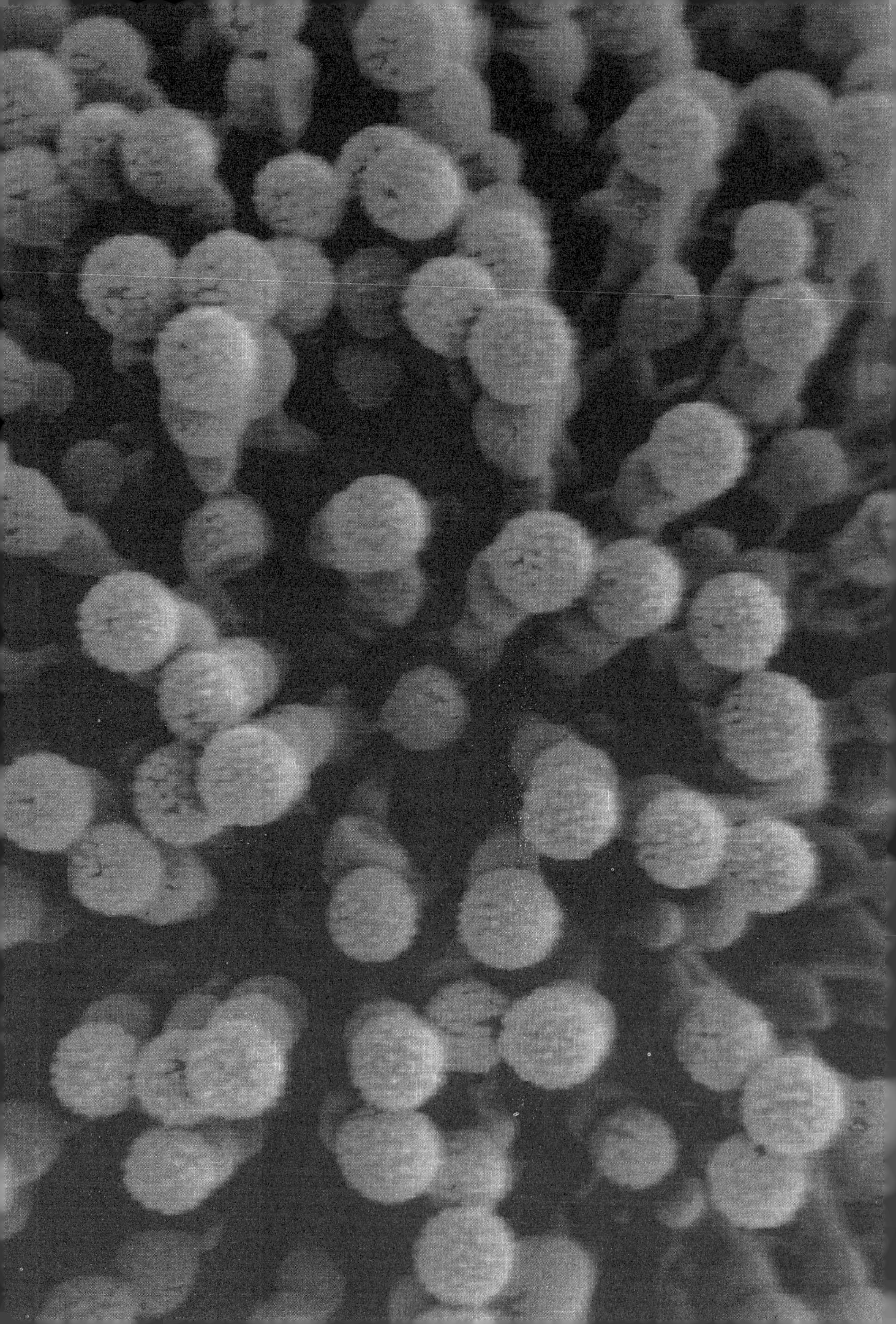

Chapter 3

Spore formation in food-relevant fungi

Unai Ugalde[1] and Luis M. Corrochano[2]
[1]*Unidad de Bioquímica II, Facultad de Química, Universidad del País Vasco, Apartado 1072, 20080 San Sebastian, Spain;* [2]*Departamento de Genética, Facultad de Biología, Universidad de Sevilla, Apartado 1095, 41080 Sevilla, Spain.*

INTRODUCTION

A large number of filamentous fungi are notoriously familiar to most people for their dashing colonisation of foods, often resulting in spoilage, even under cold storage (Fisher, 2002). They also share the ability of producing large numbers of asexual spores. This apparently harmless feature renders them ubiquitous in natural and human environments, including thoroughly sanitised food storage and processing facilities. Indeed, prolific spore production and dispersal is at the very heart of their unwelcome success.

This chapter aims to provide an overview of spore production as well as the stimuli which are involved in triggering this important biological phenomenon. Given the fundamental differences at the phylogenetic, cellular and developmental level between the Zygomycetes, which produce sporangiospores, and other fungal groups, which normally form conidia (Deuteromycetes and Ascomycetes), sporulation in these two groups of organisms will be presented separately.

CONIDIAL FUNGI

The Process of Conidiation

Conidia are cellular propagules which commonly emerge from aerial hyphae at zones which lie behind the growing colony edge, and therefore, no longer participate in vegetative growth. Their purpose is to provide the fungal colony with a means of dispersal in a rapidly changing environment. Hence, conidial production (conidiation) typically relies on relatively simple cellular transformations which can be completed relatively swiftly in every aerial hypha, resulting in a concerted and massive production of spores. In line with the above requirements, conidia rely on a readily available dispersion method: aerial transport. However, they may also be carried by insects and other living organisms with remarkable success (Ngugi and Scherm, 2004).

Conidiation has attracted considerable interest in the food industry (Gray, 1981), since conidia can be used as biotransformation catalysts (Larroche and Gros, 1997) and as inoculum for industrial fermentations (Smith and Calam, 1980). In addition, fungal spores are at the start of food spoilage and decay processes, and are well known for harbouring mycotoxins (Pestka, 1995).

Further, the health risks associated with the presence of fungi and their spores in human living environments is an increasing concern (Nielsen, 2003). In this section, we shall focus on three representatives of the most frequently encountered conidial fungi: *Penicillium*, *Aspergillus* and *Neurospora*.

Conidiation in *Penicillium* species involves the differentiation of the hyphal apex into a specialised reproductive cell called phialide, which undergoes mitotic divisions, each resulting in a new specialised daughter cell: the conidium (Cole and Kendrick, 1969).

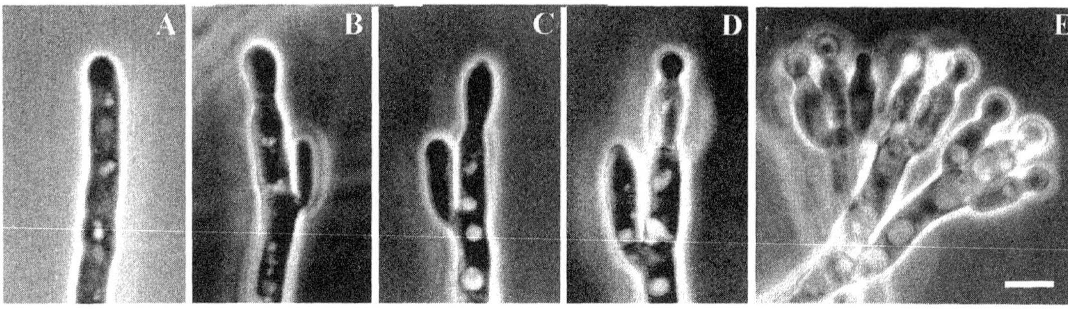

Figure 1. Morphological stages of *Penicillium cyclopium* during conidiation in submerged liquid culture. (A) Stage 1: vegetative hypha. (B) Stage 2: apical cell swelling and subapical branching. (C) Stage 3: phialide formation. (D) Stage 4: conidium formation. (E) Penicilli. Scale bar: 10 μm. Reprinted with Permission, Ref (Roncal and Ugalde, 2003).

The successive cellular and morphogenetic changes occurring throughout conidiation have been studied in liquid cultures, where induction could be effected synchronously (Hadley and Harrold, 1958). Four morphogenetic stages have been identified (Ugalde and Pitt, 1983; Figure 1): Upon induction, apically growing vegetative hyphae, immediately arrest extension. After four hours with no apparent morphological change, the apical cell is delimited by a septum and begins to swell, with the concomitant formation of subapical branches (stage 2). Depending on the species, these branches may themselves septate and branch again, in which case, they are termed *metulae*. After six hours, the apical cell differentiates into a phialide (stage 3), which finally buds at its tip, giving rise to the first conidium (stage 4). The overall time period required to fulfill this process is approximately 7 h. Once the first conidium has formed, a new conidium appears at the tip of the phialide approximately each hour, resulting in a chain of conidia that can surpass one hundred units. Subapical branches, or metulae (stage 2) also give rise to phialides and conidia, resulting in the formation of characteristic branched brush-like structures called penicilli, which are at the origin of the name of the entire genus (Penicillium, the brush). Each penicillus can bear between 5000 and 7000 conidia, and most members of this genus are able to produce 2-3 x 10^6 conidia per square centimetre.

In the Aspergilli, the conidiation process begins with the erection of a conidiophore stalk which emerges by apical extension with a specialised *foot cell* at the level of the substratum. The conidiophore rises as a column of 4-5 μm in diameter, to a height that may surpass 100 μm, depending on the species. The conidiophore tip then swells, giving rise to a spherical structure or *vesicle*. The foot cell, conidiophore and vesicle are not separated by septa. Buds emerge at the vesicle surface, in equal number to the nuclei arising from the multiple nuclear divisions taking place beneath it. The number of buds varies between species. The nuclei migrate into the buds forming *metulae*, from which further divisions give rise to phialides. In some instances, phialides emerge directly from the vesicle. In *Aspergillus nidulans*, each of the 60 or so metulae produced gives rise to two phialides. Each phialide can then produce 100 or more conidia, and each conidiophore can produce up to 10,000 spores. Under favourable sporulation conditions, Aspergilli are capable of producing up to 3 x 10^6 conidia per square centimetre. The stages of conidiation in *A. nidulans* are shown in Figure 2.

The biology of *Neurospora crassa* has been reviewed in detail (Davis, 2000; Perkins and Davis, 2000; Davis and Perkins, 2002), and the stages of its asexual cycle resemble those of the previously described examples. Conidiation in *Neurospora* is induced by environmental cues, including lack of carbon or nitrogen sources or dessication. After transfer to a dry surface the mycelia will grow away from the substrate resulting in the massive development of aerial hyphae. The growth of the aerial hyphae by apical elongation will last for several hours.

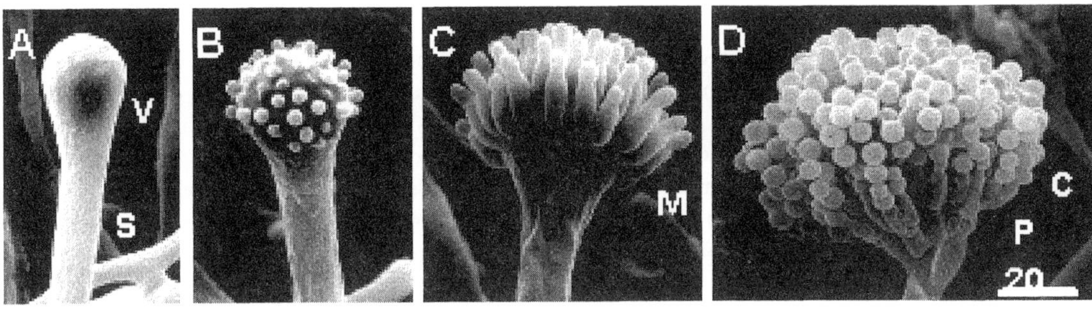

Figure 2. Conidiophore development as observed in the scanning electron microscope. An aerial hypha (S = stalk) swells terminally to a vesicle (V) (A), which nearly synchronously form metulae (M) in a budding-like process (B,C). Metulae produce two to three phialides (P), which continuously generate conidia (C) (D). Reprinted with permission (Fisher, 2002).

Then repeated apical budding will produce conidiophores consisting of long and branched chains of proconidia, each 5-10 µm in length. Proconidia are separated by minor and then major interconidial constrictions along the proconidial chain. When budding growth ceases, new cell wall is laid down between each proconidium. These cross-walls will thicken and redistribute giving rise to a fragile connective thread that will hold the newly separated multinucleated macroconidia for easy dispersal (Figure 3). Measurements of 8 x 10^8 macroconidia per slant (about 6-8 x 10^7, per cm^2) have been reported under laboratory conditions (Lauter *et al.*, 1997). Extra cross-walls may appear in the region of aerial hyphae near the conidiophore resulting in separate hyphal segments, the arthroconidia. Microconidiation is an alternative pathway that results in the formation of uninucleated microconidia. Microconidia are not very common, are formed within the vegetative hyphae and are liberated after breaking the hyphal cell wall. The morphological events and the regulation of the sporulation pathways in *Neurospora* have been reviewed by Springer (1993).

For more examples of conidiation structures in food relevant fungi which are not covered in this chapter, we recommend an extensive review (Samson *et al.*, 2002).

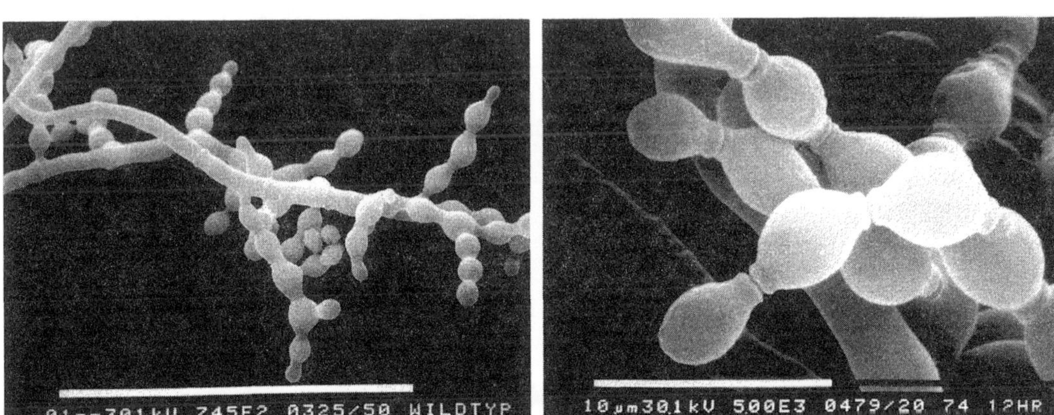

Figure 3. Conidiation in *Neurospora crassa*. Left. A representative region of conidiophores consisting of proconidial chains with cross-walls before separation. Scale bar length 100 µm. Right. Mature conidia with cross-walls are held by connective tissue before separation. Scale bar length 10 µm. Photographs by M. Springer obtained from the Fungal Genetics Stock Center, www.fgsc.net.

Induction of Sporulation

Since conidia are mostly dispersed through the air, the emergence of hyphae from the growth substrate into the atmosphere represents a *natural stimulus* for spore production. Indeed, the production of conidia is most potently stimulated by the emergence to the air, and only under carefully controlled submerged conditions can any form of conidiation be stimulated (Morton, 1961).

The precise mechanism involved in aerial induction has been the subject of much investigation since early studies by Klebs, who postulated that exposure to the aerial medium inflicted physiological stress on hyphae (Klebs, 1896). This hypothesis found considerable backing for many decades, since some stress conditions also induce spore production in submerged culture (see below). The compounded evidence from various model conidial fungi indicates that conidiation induction involves the integration of several environmental cues. Significantly, not all are necessarily stress-related.

A low molecular weight endogenous extracellular factor, the synthesis of which was shown to depend on an active *fluG* gene, was first assigned the function of reporting emergence to the air in *A. nidulans,* (Lee and Adams, 1994). Although the molecule was not identified, and remains unknown to date, the evidence supported the view of a specific signal reporting on the emergence to the air (Adams *et al.*, 1998). Separate studies with *Penicillium cyclopium* (Roncal *et al.*, 2002a) resulted in the identification of conidiogenone, a tetracyclic diterpene which was shown to play that precise role (Figure 4). The molecule is constitutively produced and contains only two functional groups (a hydroxyl and a ketone group) which are essential for biological activity (Roncal *et al.*, 2002a and b). This compound purportedly reports on the emergence to the atmosphere by accumulating at the thin water film remaining at the hyphal surface. Under submerged conditions, conidiogenone is diluted below the threshold concentration required for signalling, and is also gradually converted to an inactive derivative (conidiogenol), thus avoiding equivocal induction by accumulation of the inducer in the bulk liquid surrounding the hypha over prolonged periods. Although the evidence of such a mechanism is currently limited to the abovementioned case, it could be envisaged that other fungi also dispose of similar systems to sense aerial emergence, albeit with variations in the molecules and receptors involved.

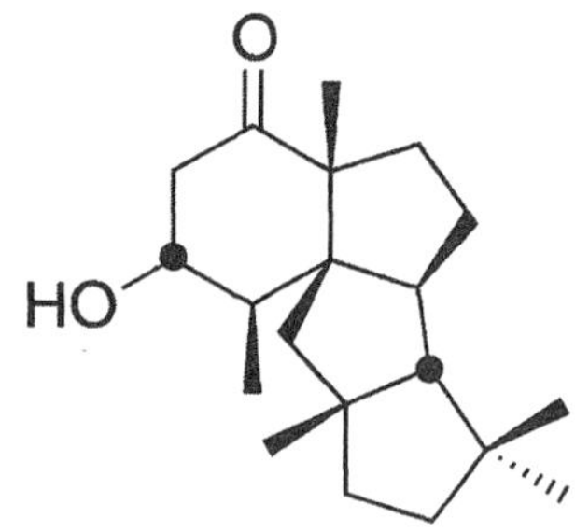

Figure 4. Chemical structure of the conidiation autoregulator conidiogenone.

A good many food relevant fungi, including those which participate in natural food processes, such as *P. roqueforti,* solely require the emergence to the air as natural sporulation stimulus. However, other notable examples, such as *N. crassa* (Lauter *et al.*, 1997), *Trichoderma viride* (Betina, 1995), *Aspergillus* spp. (Mooney and Yager, 1990; Calvo *et al.*, 1999), and a few examples in the *Penicillia* (Pazout *et al.*, 1982), require light to undergo spore production in addition to emergence to the air. In the case of *A. nidulans,* an active *veA* gene (*velvet*) is required for light sensitivity in the red and blue range. Moreover, suppressor mutations that regain sensitivity to light were alleles of the earlier mentioned *fluG* gene (Kim *et al.*, 2002; Yager *et al.*, 1998), and it is currently understood that the *velvet* gene acts as a negative regulator of conidiation.

In *N. crassa* the White Collar-1 (WC-1) protein, which contains FAD as a cofactor, has been identified as the blue light photoreceptor (Froehlich *et al.*, 2002; He *et al.*, 2002). This photoreceptor also is involved in a circadian clock that regulates conidiation and that can be entrained by a light exposure. The genetics and molecular biology of the *Neurospora* circadian clock and its regulation by light has been investigated in detail (Dunlap and Loros, 2004).

The combined evidence supports the view that light would most probably act as a second switch which needs to be turned on to confirm the first stimulus of emergence to the air. In terms of signal transduction, it could be said that light acts downstream from the initial triggering step of aerial emergence, by depressing the onset of conidiation. The biological value of such a confirmatory system lies in the role of light as indicator of an open environment, favouring conidial dispersal.

There are other environmental cues which promote spore production even under submerged conditions. They are all associated with impaired growth, and in these instances, sporulation clearly appears as an alternative survival strategy.

High osmolarity (at levels below those which inhibit growth, and which are used in food preservation) was first identified as an induction stimulus separate from aerial emergence in Penicillia by Morton (1961). Relatively high concentrations (10% w/v) of glucose, and other non-metabolizable sugars and sugar-alcohols induced conidiation in liquid culture, where the aerial stimulus is precluded. Studies with *A. nidulans* also reported similar results by including 0.8 *M* NaCl in the medium (Lee and Adams, 1995). In *A. oryzae*, KCl concentrations higher than 0.1 *M* have been reported to promote the formation of conidia (Song *et al.*, 2001). The mechanism mediating this effect remains unknown, but may involve the participation of two-component system osmosensors similar to those first described in *Saccharomyces cerevisiae* (Maeda *et al.*, 1994). Two component systems are protein complexes of which one component is a signal receiver or sensor, which in turn then affects a second component termed the response regulator, often with DNA binding capacity, that regulates transcription of a specific set of response genes. Such two-component systems have also been found in *A. nidulans* (Appelyard *et al.*, 2000; Furukawa *et al.*, 2002) and *N. crassa* (Alex *et al.*, 1996).

Nutrient limitation is another widespread and much studied sporulation trigger. Considerable evidence has emerged over decades, on the induction of sporulation in liquid cultures when the carbon or nitrogen source to sustain growth is limiting (Bu'Lock, 1975). This is exemplified in *A. nidulans*, where transfer of a mycelium grown in a nutritionally sufficient medium to a medium lacking the carbon or the nitrogen source results in the formation of asexual reproductive structures (Skromne *et al.*, 1995). The same phenomenon also occurs in *Penicillium griseofulvum* following transfer to a nitrogen-free medium (Morton, 1961). In most cases, reduced availability of the nitrogen source is responsible for sporulation induction, although the involvement of carbon limitation on conidiation induction has been also reported (Righelato *et al.*, 1968). The signalling mechanisms involved in this form of induction remain unexplained. An interesting possibility would involve the existence of different nutrient sensors that would report on the cell's nutritional status. The presence of this kind of nutrient sensing receptors was first proposed in *S. cerevisiae* (Özcan *et al.*, 1996). In *N. crassa*, submerged mycelia, which remain vegetative in nutritionally sufficient media, can be induced to conidiate when carbon limitation is imposed (Springer, 1993). However, a mutation in the *rco-3* gene, which encodes a protein with sequence similarity to the members of the sugar transporter gene superfamily, results in a strain that could conidiate in submerged culture without nutrient limitation, suggesting a role for that protein as glucose sensor (Madi *et al.*, 1997).

The sensing of neighbouring colonies and overcrowded conditions through specific signals not only results in a concerted limitation of colony expansion, but also in an enhancement of sporulation. Early studies with *Glomerella cingulata* showed that inoculum loads surpassing 10^6 spores/mL resulted not only in reduced germination, but also in microcycle conidiation (Lingappa and Lingappa, 1969). Similar results have been encountered in studies with *Penicillium paneum* where 1-octen-3-ol has been assigned an autoinhibitor role, which also results in microcycle conidiation (Chitarra *et al.*, 2003).

SPORULATION IN ZYGOMYCETE FUNGI

Zygomycete fungi are characterized by a coenocytic mycelium in which the nuclei are not separated by septa, and a sexual cycle in which gamentangia fuse to form zygospores. The phylum Zygomycota includes the class Zygomycetes and the class Trichomycetes, which are obligate symbionts of arthropods. The Zygomycetes include the order Mucorales, which is divided into several families. A distinguishing feature of the Mucorales is that they make the asexual spores inside receptacles called sporangia. Spore production in the sporangium occurs when the cytoplasm and the nuclei are divided by newly deposited cell walls to form spores, a developmental process very different from conidial production in Ascomycete fungi. Many members of the family Mucoraceae are easily spotted as food contaminants and cause animal diseases, including diseases in humans. Prominent members of the Mucorales are species of the genus *Mucor*, *Rhizopus*, *Rhizomucor*, *Absidia*, *Blakeslea*, and *Phycomyces*. The biology of the Mucorales has been reviewed by Ingold (Ingold, 1978). In addition to the Mucorales, the order Entomophthorales contains insect parasites and pathogens of other animals, including humans. The biology of the zygomycete fungi, particularly those causing human diseases, with detailed descriptions of their appearance and classification methods has been described by Ribes *et al.* (2000). An interesting feature of *Mucor* is the capacity of the sporangiospores to grow as yeast or as multinucleated hyphae depending on the presence of oxygen in the culture medium. The biology of *Mucor* with an emphasis on the mechanisms of dimorphism has been reviewed by Orlowski (1991).

Spore germination and hyphal growth allows zygomycete fungi to expand and colonize the substrate provided that nutrients are available. It is generally believed that mycelial growth will continue until nutrient deprivation and other environmental cues trigger sporangiophore development and the formation of spores for further dispersal. Sporangiophore development and its regulation by environmental signals, most notably light, has been investigated in detail in the Mucoral fungus *Phycomyces blakesleeanus* and will be described in the following sections. It is likely that similar patterns of growth and regulatory mechanisms will be present in other mucoral fungi.

Sporangiophore development in *Phycomyces*
The biology of the zygomycete *Phycomyces blakesleeanus* has been described recently by Cerdá-Olmedo (2001).

Sporangiophores are aerial hyphae, normally unbranched, that form a sporangium filled with spores at their tip. The upward growth of the sporangiophore is supported by fast cytoplasmic streams that carry cellular materials, including the nuclei that will be packed into the spores. Macrophores are giant sporangiophores that grow several centimeters long guided by many stimuli, most notably by light (reviewed by Galland, 2001) and gravity (Galland *et al.*, 2004); their sporangia contain about 10^5 spores. The microphores are dwarf sporangiophores, about 1 mm long, and their sporangia contain about 10^3 spores (Figure 5).

On induction, the formation of macrophores depends on temperature (Thornton, 1973) and on the availability of nutrients, particularly asparagine (Corrochano and Cerdá-Olmedo, 1988) zinc (Hilgenberg and Hofmann, 1977), and oxygen (Galland and Russo, 1979). The composition of the culture medium could have a prominent role in the development of the spores and their requirements for germination. In a related fungus, *Mucor racemosus*, the germination of the spores with different carbon sources depended on the concentration and type of ingredients present in the sporulation medium (Tripp and Paznokas, 1981). Whether a similar effect occurs in *Phycomyces* remains undocumented.

The usual agar cultures form macrophores both in the dark and in light. Under some culture conditions, however, the formation of macrophores is clearly determined by light. Growth on a phosphate-rich medium under periodic alternations of darkness and blue illumination results in bands of macrophores (Bergman, 1972). When *Phycomyces* is grown in

Figure 5. Vegetative development of *Phycomyces*. Left. A sporangiophore primordium growing upwards from the mycelium (photograph by W. Schröder). Right. Two types of sporangiophores of very different sizes (macrophores and microphores) grow out of the mycelium. The sporangia at the top of the sporangiophores contain the spores (photograph by F. Gutiérrez-Corona). Microphores are about 1-2 mm in length.

a closed jar, the formation of macrophores depends on blue illumination (Russo, 1977).

Early reports on the microphores (Burgeff, 1914, 1915; Orban, 1919) were followed by experiments to determine the conditions for their appearance (Rudolph, 1958; Thornton, 1972; Gutiérrez-Corona and Cerdá-Olmedo, 1985; Ortiz-Castellanos and Gutiérrez-Corona, 1988). Microphorogenesis is brought about by unfavourable conditions: high plating densities, scarcity of nitrogen sources, low temperatures, and limited ventilation. Light counteracts these effects: under a defined set of very harsh conditions (Thornton, 1973) microphorogenesis is suppressed by light. Under high-plating density, blue light stimulates macrophorogenesis and inhibits microphorogenesis (Corrochano and Cerdá-Olmedo, 1988). The effect of light on sporangiophore development has been reviewed by Corrochano and Cerdá-Olmedo (1991, 1992).

As in the case of conidial fungi, light acts as an indicator of an open environment, and photomorphogenesis would presumably improve the dispersion of *Phycomyces* spores in nature. Guided by light and other stimuli, macrophores grow into the open air where spores can adhere to or be eaten by passing animals. Microphores, much less onerous to build than macrophores, are formed in the dark, when *Phycomyces* resigns itself to leaving the spores *in situ*.

The effect of light on sporangiophore development

Sporangiophore development in *Phycomyces* is highly synchronized. In a defined set of growth conditions and spore density only vegetative mycelium is detected at the age of 48 h, when the mycelium is ready to develop sporangiophores and is sensitive to blue light. Sporangiophores appear soon thereafter and can be easily collected at the age of 72 h. The maximum number of sporangiophores is obtained at the age of 96 h, and remains constant for several days. The final numbers of macrophores and microphores in the cultures depend on the blue light intensity applied at the age of 48 h (Corrochano and Cerdá-Olmedo, 1988, 1990). The effect of blue light on sporangiophore development follows a two-step stimulus-response curve with thresholds at 10^{-4} J/m^2 and 1 J/m^2 which suggests the presence of different photosystems optimized to operate at different light intensities. The effect of light depends on the product of the exposure time (between 12 s and 3 h) and the intensity, a suggestion that *Phycomyces* counts and remembers the photons received over a long time. The absolute threshold corresponds to the arrival of one photon per μm^2 every 20 min (Corrochano and Cerdá-Olmedo, 1988, 1990). A complex photosensory system with separate transduction pathways for photomicrophorogenesis and photomacrophorogenesis has been proposed based on differences in action spectra and on the effect of mutations in several genes (Corrochano and Cerdá-Olmedo, 1988, Flores *et al.*, 1998; Cerdá-Olmedo and Corrochano, 2001).

Light probably acts through the activation of gene transcription in concert with other stimuli, such as the emergence to the air or the lack of nutrients, resulting in the initiation of phorogenesis. The molecular basis of photophorogenesis remains unknown, though evidence of the involvement of heterotrimeric G proteins and protein phosphorylation (Tsolakis *et al.*, 1999, 2004), pteridines and NO synthase (Maier and Ninnemann, 1995; Maier *et al.*, 2001), and polyamines (Ruiz-Herrera, 1994) have been proposed. In addition, the gene for the heat-shock protein HSP100 is induced by light in mycelia at the onset of sporangiophore development, suggesting a role for this protein in the regulation by light of sporangiophore development (Corrochano, 2002; Rodríguez-Romero and Corrochano, 2004). In addition, ras-type proteins with an ability to bind GTP and key parts of many regulatory networks have also been implicated in sporangiophore development in *Mucor* (Roze *et al.*, 1999).

Conclusion

An overview of spore production and the stimuli triggering the process across a diverse group of organisms, such as those covered in this chapter, reveals many common themes which are all of great biological relevance, and practical value in food microbiology.

Regardless of the phylogenetic group and developmental pathway involved in spore production, the emergence to the atmosphere and lack of nutrients appear as key factors triggering the transition from vegetative growth to the initiation of sporulation. The former stimulus likely involves a series of endogenous signalling compounds, specifically designed as reporters of environmental change.

Light, usually blue light but sometimes red light, is an important environmental factor, which apparently acts as a second confirmatory switch for spore production. Light sometimes promotes sporulation, or may be a decisive factor determining the choice between alternative spore producing programmes. As in the case of *Neurospora*, light may also act as a signal to entrain a circadian clock that is a superimposed regulatory circuit modulating sporulation.

Further clarification of the molecular events responsible for the initiation and regulation of sporulation by environmental signals will yield novel insights into possible control mechanisms to deal with pathogenic and food contaminant fungi.

REFERENCES

Adams, T. H., Wieser, J. K., and Yu, J.-H. (1998). Asexual sporulation in *Aspergillus nidulans*, Microbiology and Molecular Biology Reviews 62:35-54.

Alex, L. A.. Borkovich, K. A., and Simon, M. I. (1996). Hyphal development in *Neurospora crassa*: involvement of a two-component histidine kinase, Proceedings of the National Academy of Sciences U.S.A. 93:3416-3421.

Appelyard, M. V. C. L., McPheat, W. L., and Stark, M. J. R. A. (2000). Novel 'two-component' protein containing histidine kinase and response regulator domains required for sporulation in *Aspergillus nidulans*, Current Genetics 37:364-372.

Bergman, K. (1972). Blue-light control of sporangiophore initiation in *Phycomyces*. Planta 107:53-67.

Betina, V. (1995). Photoinduced conidiation in *Trichoderma viride*, Folia Microbiologica 40:219-224.

Bu'Lock, J. D. (1975). *In* The Filamentous Fungi, Vol. 1 (Smith, J. E., and Berry, D. R., eds.), Edward Arnold, London, U.K., pp. 33-58.

Burgeff, H. (1914). Untersuchungen über Variabilität, Sexualität und Erblichkeit bei *Phycomyces nitens* Kunze I. Flora 107:259-316.

Burgeff, H. (1915). Untersuchungen über Variabilität, Sexualität und Erblichkeit bei *Phycomyces nitens* Kunze II. Flora 108:353-448.

Calvo, A. M., Hinze, M. I. I., Gardner, H. W., and Keller N. P. (1999). Sporogenic effect of polyunsaturated fatty acids on development of *Aspergillus* spp. Applied and Environmental Microbiology 65:3668-3673.

Cerdá-Olmedo, E. (2001). *Phycomyces* and the biology of light and color. FEMS Microbiological Reviews 25:503-512.

Cerdá-Olmedo, E., and Corrochano L. M. (2001). Genetics of *Phycomyces* and its responses to light. *In* Photomovement. Comprehensive Series in Photosciences, Vol. 1. (Häder, D. P., and Lebert, M., eds.), Elsevier, Amsterdam, The Netherlands, pp. 589-620.

Chitarra, G. S., Abee, T., Rombouts, F. M., Posthumus, M. A., and Dijksterhuis J. (2003). Germination of *Penicillium paneum* conidia is regulated by Octen-3-ol, a volatile self-inhibitor. Applied Environmental Microbiology 70:2823-2829.

Cole, G. T., and Kendrick, W. D. (1969). Conidium ontogeny in Hyphomycetes. The phialides of *Phialophora, Penicillium,* and *Ceratocystis*. Canadian Journal of Botany 47:779-789.

Corrochano, L. M. (2002). Photomorphogenesis in *Phycomyces*: differential display of gene expression by PCR with arbitrary primers. Molecular Genetics and Genomics 267:424-428.

Corrochano, L. M., and Cerdá-Olmedo, E. (1988). Photomorphogenesis in *Phycomyces*: dependence on environmental conditions. Planta 174: 309-314.

Corrochano, L. M., and Cerdá-Olmedo, E. (1990). Photomorphogenesis in *Phycomyces*: competence period and stimulus-response relationships. Journal of Photochemistry and Photobiology B5: 255-266.

Corrochano, L. M., and Cerdá-Olmedo, E. (1991). Photomorphogenesis in *Phycomyces* and in other fungi. Photochemistry and Photobiology 54: 319-327.

Corrochano, L. M., and Cerdá-Olmedo, E. (1992). Sex, light and carotenes: the development of *Phycomyces*. Trends in Genetics 8:268-274.

Davis, R. H. (2000). *Neurospora*: contributions of a model organism. Oxford University Press, New York.

Davis, R. H., and Perkins D. D. (2002). *Neurospora*: a model of model microbes, Nature Reviews Genetics 3:397-403.

Dunlap, J. C., and Loros J. J. (2004). The *Neurospora* circadian system, Journal of Biological Rhythms 19:414-424.

Fischer, R. (2002). Conidiation in *Aspergillus nidulans*. *In* Molecular Biology of Fungal Development (Osiewacz, H. D., ed.), Marcel Dekker, New York, U.S.A.

Flores, R., Cerdá-Olmedo, E., and Corrochano L. M. (1998). Separate sensory pathways for photomorphogenesis in *Phycomyces*. Photochemistry and Photobioly 67:467-472.

Froehlich, A. C., Liu, Y., Loros, J. J., and Dunlap, J.C. (2002). White Collar-1, a circadian blue light photoreceptor, binding to the *frequency* promoter, Science 297:815-819.

Furukawa, K., Katsuno, Y., Urao, T., Yave, T., Yamada-Okabe, T., Yamada-Okabe, H., Yamagata, Y., Abe, K., and Nakajima, T. (2002). Isolation and functional analysis of a gene, *tcsB*, encoding a transmembrane hybrid-type histidine kinase from *Aspergillus nidulans*, Applied Environmental Microbiology 68:5304-5310.

Galland, P. (2001). Phototropism in *Phycomyces*. *In* Photomovement. Comprehensive Series in Photosciences Vol. 1. (Häder, D. P., and Lebert, M., eds.), Elsevier, Amsterdam, The Netherlands, pp. 621-657.

Galland, P., Finger, H., and Wallacher Y. (2004). Gravitropism in *Phycomyces*: threshold determination on a clinostat centrifuge. Journal of Plant Physiology 161:733-739.

Galland, P., and Russo, V. E. A. (1979). The role of retinol in the initiation of sporangiophores of *Phycomyces blakesleeanus*. Planta 146:257-262.

Gray, W. D. (1981). Food technology and industrial mycology. Biology Conidial Fungi 2:237-268.

Gutiérrez-Corona, F., and Cerdá-Olmedo, E. (1985). Environmental influences in the development of *Phycomyces* sporangiophores. Experimental Mycology 9:56-63.

Hadley, G., and Harrold, C. E. (1958). The sporulation of *Penicilium notatum* Westling in submerged liquid culture. I. The effect of calcium and nutrients on sporulation. Journal of Experimental Botany 9:408-417.

He, Q., Cheng, P., Yang, Y., Wang, L., Gardner, K. H., and Liu, Y. (2002). White collar-1, a DNA binding transcription factor and a light sensor. Science 297:840-843.

Hilgenberg, W., and Hofmann F. (1977). Tryptophansynthase in *Phycomyces blakesleeanus*. Teil II: Tryptophansynthaseaktivität des Pilzes in Abhängigkeit von Lichtbedingungen und vom Zinkgehalt des Kulturmediums. Physiologica Plantarum 40:235-238.

Ingold, C. T. (1978). The biology of *Mucor* and its allies. Edward Arnold-Camelot Press, Southampton, U.K.

Kim, H. S., Han, K. Y., Kim, K. J., Han, D. M., Jahng, K. Y., and Chae, K. S. (2002) The *veA* gene activates sexual development in *Aspergillus nidulans*, Fungal Genetics and Biology 37:72-80.

Klebs, G. (1896). Die Bedingungen der Fortplanzung bei enigen Algen und Pilzen. Verlag von Gustav Fischer. Jena. Germany.

Larroche, C., and Gros, J. B. (1997). Special transformation processes using fungal spores and immobilized cells. Advances in Biochemical Engineering and Biotechnology 55:179-220.

Lauter, F. R., Yamashiro, C. T., and Yanofsky, C. (1997). Light stimulation of conidiation in *Neurospora crassa*: studies with the wild-type strain and mutants *wc-1*, *wc-2* and *acon-2*, Journal of Photochemistry Photobiology B: 37:203-211.

Lee, B. N., and Adams, T. H. (1995). *fluG* and *flbA* function interdependently to initiate conidiophore development in *Aspergillus nidulans* through *brlA* activation. EMBO Journal 15:299-309.

Lee, B. N., and Adams, T. H. (1994). The *Aspergillus nidulans fluG* gene is required for production of an extracellular developmental signal and is related to prokaryotic glutamine synthetase I. Genes and Development 8:641-651.

Lingappa, B. T., and Lingappa, Y. (1969). Role of auto-inhibitors on mycelial growth and dimorphism of *Glomerella cingulata*. Journal of General Microbiology 56:35-45.

Madi, L., McBride, S. A., Bailey, L. A., and Ebbole, D. J. (1997). *rco-3*, a gene involved in glucose transport and conidiation in *Neurospora crassa*, Genetics 146:499-508.

Maeda, T., Wurgler-Murphy, S. M., Saito, H. A. (1994). Two-component system that regulates an osmosensing MAP kinase cascade in yeast. Nature 369:242-245.

Maier, J., Hecker, R., Rockel, P., and Ninnemann H. (2001). Role of nitric oxide synthase in the light-induced development of sporangiophores in *Phycomyces blakesleeanus*. Plant Physiology 126:1323-1330.

Maier, J., and Ninnemann H. (1995). Inhibition of light-dependent photomorphogenesis of sporangiophores from *Phycomyces blakesleeanus* by application of pteridine biosynthesis inhibitors. Photochemistry and Photobiology 61:206-209.

Mooney, J. L., and Yager, L. N. (1990). Light is required for conidiation in *Aspergillus nidulans*, Genes and Development 4:1473-1482.

Morton, A. G. (1961). The induction of sporulation in mould fungi. Proceedings Royal Microscopical Society B 153:548-569.

Ngugi, H. K., and Scherm, H. (2004). Pollen mimicry during infection of blueberry flowers by conidia of *Monilinia vaccinii-corymbosi*. Physiology and Molecular Plant Pathology 64:113-123.

Nielsen, K. F. (2003). Mycotoxin production by indoor molds. Fungal Genetics and Biology 39:103–117.

Orban, G. (1919). Untersuchungen über die Sexualität von *Phycomyces nitens*. Beih Bot Centralbl 36:1-59.

Orlowski, M. (1991). *Mucor* dimorphism. Microbiological Reviews 55:234-258.

Ortiz-Castellanos, M. L., and Gutiérrez-Corona J. F. (1988). The sensitive period for light and temperature regulation of sporangiophore development in *Phycomyces*. Planta 174:305-308.

Özcan, S., Dover, J., Rosenwald, A. G., Wölfl, S., and Johnston, M. (1996). Two glucose transporters in *S. cerevisiae* are glucose sensors that generate a signal for induction of gene expression. Proceeding of the National Academy of Sciences U.S.A. 93:12428-12432.

Pazout, J., Pazoutova, S., and Vancura, V. (1982). Effects of light, phosphate and oxygen on ethylene formation and conidiation in surface cultures of *Penicillium cyclopium* Westling. Current Microbiology 7:133-136.

Perkins, D. D., and Davis R. H. (2000). *Neurospora* at the millennium, Fungal Genetics and Biology 3:153-167.

Pestka, J. (1995). Fungal toxins in raw and fermented meats. *In* Fermented Meats (Campbell-Platt, G., and Cook, P. E., eds.), Blackie Academic and Professional, London, U.K., pp. 194-216.

Ribes, J. A., Vanover-Sams, C. L., and Baker D. J. (2000). Zygomycetes in human disease. Clinical Microbiology Reviews 13:236-301.

Righelato, R. C., Trinci, A. P. J., Pirt, S. J., and Peat, A. (1968). The influence of maintenance energy and growth rate on the metabolic activity, morphology and conidiation of *Penicillium chrysogenum*. Journal of General Microbiology 50:399-412.

Rodríguez-Romero, J., and Corrochano, L. M. (2004). The gene for the heat-shock protein HSP100 is induced by blue light and heat shock in the fungus *Phycomyces blakesleeanus*. Current Genetics 46:295-303.

Roncal, T., Cordobés, S., Sterner, O., and Ugalde, U. (2002a). Conidiation in *Penicillium cyclopium* is induced by conidiogenone, an endogenous diterpene. Eukaryotic Cell 1:823-829.

Roncal, T., Cordobés, S. Ugalde, U. He, Y. and Sterner, O. (2002b). Novel diterpenes with potent conidiation activity. Tetrahedron Letters 43: 6799-6802.

Roncal, T., and Ugalde, U. (2003). Conidiation induction in *Penicillium*. Research in Microbiology 154:539-546.

Roze, L. V., Mahanti, N., Mehigh, R., McConnell, D. G., and Linz J. E. (1999). Evidence that MRas1 and MRas3 proteins are associated with distinct cellular functions during growth and morphogenesis in the fungus *Mucor racemosus*. Fungal Genetics and Biology 28:171-189.

Rudolph, H. (1958). Entwicklungsphysiologische Untersuchungen an den Sporangiophoren von *Phycomyces blakesleeanus*. Biologisches Zentralblatt 77:385-437.

Ruiz-Herrera, J. (1994). Polyamines, DNA methylation, and fungal differentiation. Critical Reviews in Microbiology 20:143-150.

Russo, V. E. A. (1977). The role of blue light in synchronization of growth and inhibition of differentiation of stage I sporangiophore of *Phycomyces blakesleeanus*. Plant Science Letters 10:373-380.

Samson, R. A, Hoekstra, E. S., and Frisvad, J. C. (2002). Introduction to food and airborne fungi. Centraalburea voor Schimmelcultures. Utrecht. The Netherlands.

Skromne, I., Sanchez, O., and Aguirre, J. (1995). Starvation stress modulates the expression of the *Aspergillus nidulans brlA* regulatory gene. Microbiology 141:21-28.

Smith, G. M., and Calam, C. T. (1980). Variations in inocula and their influence on the productivity of antibiotic fermentations. Biotechnology Letters 2:261-266.

Song, M. H., Nah, J. Y., Han, Y. S., Han, D. M., and Chae, K., S. (2001). Promotion of conidial head formation in *Aspergillus oryzae* by a salt. Biotechnology Letters 23:689-691.

Springer, M. L. (1993). Genetic control of fungal differentiation: the three sporulation pathways of *Neurospora crassa*. Bioessays 15 (1993) 365-374.

Thornton, R. M. (1972). Alternative fruiting pathways in *Phycomyces*. Plant Physiology 49:194-197.

Thornton, R. M. (1973). New Photoresponses of *Phycomyces*. Plant Physiology 51:570-576.

Tsolakis, G., Parashi, E., Galland, P., and Kotzabasis K. (1999). Blue light signaling chains in *Phycomyces*: phototransduction of carotenogenesis and morphogenesis involves distinct protein kinase/phosphatase elements. Fungal Genetics and Biology 28:201-213.

Tripp, M. L., and Paznokas J. L. (1981). Relationship between sporulation medium and germination ability of *Mucor racemosus* sporangiospores. Journal of General Microbiology 127:35-43.

Tsolakis, G., Moschonas, N. K., Galland, P., and Kotzabasis, K. (2004). Involvement of G proteins in the mycelial photoresponses of *Phycomyces*. Photochemistry and Photobiology 79:360-370.

Ugalde, U., and Pitt, D. (1983). Morphology and calcium-induced conidiation of *Penicillium cyclopium* in submerged culture. Transactions of the British Mycological Society 80:319-325.

Yager, L. N., Lee, H.-O., Nagle, D. L., and Zimmerman, J. E. (1998). Analysis of *fluG* mutations that affect light-dependent conidiation in *Aspergillus nidulans*. Genetics 149:1777-1786.

Chapter 4

Dispersal of fungal spores through the air

Alastair McCartney and Jon West
Plant Pathogen Interactions, Rothamsted Research, Harpenden, Herts., AL5 2JQ, United Kingdom.

INTRODUCTION

Fungal spore dispersal can rarely be considered alone and is usually a combination of at least some of the key stages in the aerobiology pathway (Edmonds and Benninghoff, 1973), i.e., source, take-off, dispersal, deposition and effect. In this chapter we will consider the three steps needed to transport a spore, through the air, from one place another, namely: how the spore gets into the air, how it is transported through the air and how it is deposited at its final destination. Many examples of spore dispersal studies, used here, have been made in outdoor environments but the same principles apply to indoor applications to identify sources of microbial contamination in food processing situations.

Air flow considerations

Like all fluids, the flow of air can be in one or two modes: "laminar" where the air molecules follow parallel paths; and "turbulent" where the flow is more chaotic and the molecules follow different paths, although in the same general direction. Laminar flow is usually associated with low velocities and smooth surfaces, and rarely occurs outside wind tunnels or other specialised facilities (Grace, 1977). Therefore in most environments, especially outdoors, air flow is turbulent, and it is the effects of turbulence that are largely responsible for the dispersal of spores carried in the air. However, when air flows over a surface, friction slows it down so that the airspeed decreases as the surface is approached (Figure 1). The area of transition from free air flow to the surface is known as the boundary layer. Very close to the surface the air flow becomes laminar (laminar sub-layer) and air speed is almost zero (Grace, 1977). The thickness of this "boundary layer" depends on the nature of the flow over the surface and the structure of the surface itself. Air flow over surfaces has been extensively studied, and flow over natural surfaces such as leaves is discussed by Monteith and Unsworth (1990). The existence of the surface boundary layer has consequences for spore release and dispersal and is discussed below.

On a different scale, in the atmosphere, wind speed increases with distance above the surface as frictional forces have a decreasing effect on atmospheric flow. This layer is called the "planetary boundary layer" and extends from the surface to where friction-induced turbulence is effectively zero (Figure 1).

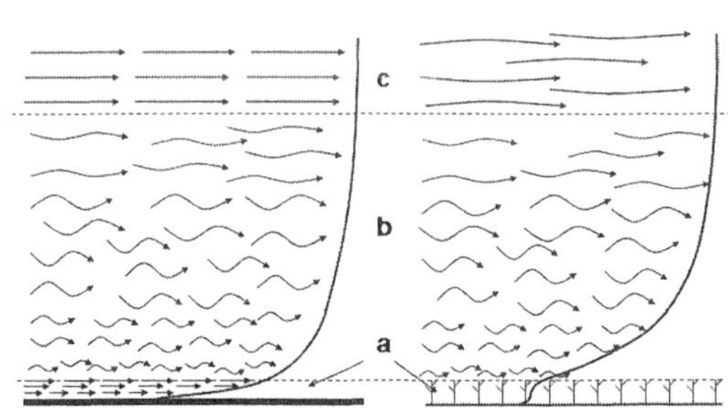

Figure 1. Left: Laminar air flow over a flat plate: (a) laminar sub-layer close to the surface; (b) turbulent boundary layer; (c) laminar free air flow. Right: Atmospheric flow over a crop: (a) surface layer within crop; (b) planetary boundary layer (turbulent); (c) pressure gradient air flow. The lines represent the wind speed profiles above the plate and the crop. Adapted from Grace (1977) and McCartney and Fitt (1985).

Sometimes the boundary layer is defined by a well-marked temperature inversion, at other times no clearly marked delineation exists and turbulence decreases gradually with increasing height. However, in the presence of large scale convection there may be significant vertical transport and the boundary layer can "break down" (Pasquill and Smith, 1983). Synoptic weather fronts and flow over mountain ranges can also cause boundary layer breakdown. The depth of the boundary layer changes in response to changes at the surface and is typically between 400 and 2000 m during the day and from a few tens of metres to about 400 m at night. Air flow within the planetary boundary layer can be very complex and is influenced not only by the physical nature of the underlying surface but also by thermal effects such as large scale convection. The nature of air flow in the planetary boundary layer had been extensively studied and the reader is referred to standard texts on atmospheric dispersal such as the publication by Pasquill and Smith (1983).

The dispersive ability of wind depends on its turbulence structure: the larger the vertical component of turbulent eddies, the greater the potential for spores to be transported into the atmosphere. Eddy structure is influenced by the thermal stability of the layer. Under neutral stability a rising air parcel remains in thermal equilibrium with the surrounding air and turbulence is dominated by friction, here the magnitude of the vertical and horizontal fluctuations in wind speed are similar (Figure 2) (Monteith and Unsworth, 1990). In contrast, in unstable conditions a rising air parcel tends to continue to rise and vertical motion is enhanced (Figure 2). Unstable conditions occur when the surface is heated, usually during the day. Conversely, under stable stratification, for example during a clear night with light winds, a rising air parcel becomes cooler than the surrounding air due to expansion as pressure decreases and so it tries to descend, thus repressing vertical motion. In this case, the vertical fluctuations in wind speed are smaller than the horizontal fluctuations (Figure 2). During the day, the air near the ground is often unstable, thus spores released during the day are more likely to be more efficiently dispersed than spores released at night. Once spores are transported above the planetary boundary layer, they have the potential to be dispersed over very large distances in large scale atmospheric motion.

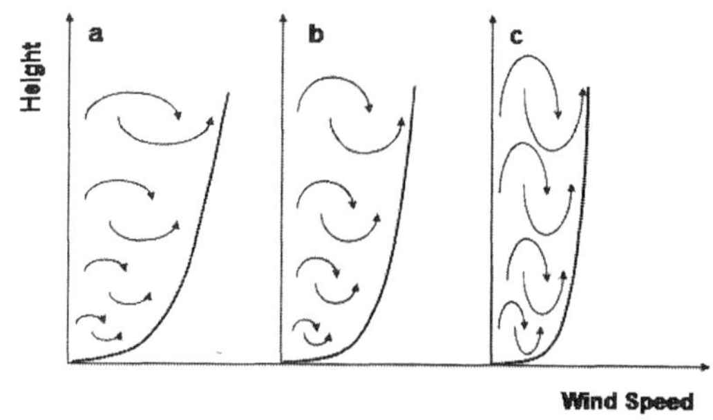

Figure 2. The influence of atmospheric thermal stability on wind speed and turbulence. (a) Stable conditions: vertical air motion is suppressed; (b) neutral stability: vertical and horizontal fluctuations are of a similar magnitude; (c) unstable conditions: vertical mixing is enhanced. Adapted from Thom (1975).

AERODYNAMIC CHARACTERISTICS OF FUNGAL SPORES

As fungal spores are much denser than air, they will naturally fall through the air under the force of gravity. The rate at which they fall plays an important role in the dispersal and deposition of airborne spores. Spores that fall quickly will tend to be less efficiently dispersed and more readily deposited than those that fall slowly. Any object falling through the air will eventually reach a steady speed, v_s, called the "settling speed", "fall speed", or "terminal velocity", when the forces of gravity are balanced by drag and lift forces that tend to slow the object down. The settling speed of a spore depends on its physical properties: mass, size and shape. However, environmental factors such as temperature or humidity can have small effects by altering the density of air or the spore itself. The gravitational forces acting on a spore are determined by its mass, while the drag forces depend on the size and shape of the spore. For objects such as winged seeds, lift forces can become important as the seed rotates or glides, but fungal spores are usually small enough for viscous drag to be dominant. Fungal spores occur in a wide range of shapes and sizes (see plates 6 and 7, Gregory, 1973). Many fungal spores have compact shapes and can be approximated to spheres (e.g., *Aspergillus* spp., diameter 2-3 μm; *Penicillium* sp., diameter ≈ 5 μm; *Puccinia striiformis,* diameter 20-25 μm) or ellipsoids (e.g., *Sclerotinia sclerotiorum* ascospores, 8 μm long, x 3 μm diameter; *Blumeria graminis* conidia, 30 μm long x 10 μm diameter). Other spore types are more elongated and behave more like cylinders (e.g., *Helminthosporium* sp., 80 μm long x 15 μm diameter) or fibres (*Claviceps purpurea,* ≈1 μm diameter, 80-120 μm long). Some have more complex shapes, for example the conidia of some *Alternaria* sp. are club shaped. The settling speeds of fungal spores range from less than 0.1 cm s^{-1} (*Aspergillus fumigatus* spores ~ 0.03 cm s^{-1}) to over 2 cm s^{-1} (*Helminthosporium sativum* conidia 2.0-2.78 cm s^{-1}) (Gregory, 1973). For a given species the settling speed of the spores can generally be estimated only within ± 20% due to natural variation in spore sizes and moisture content, which can be affected by the ambient relative humidity.

Although v_s for many fungal spores has been measured experimentally (Gregory, 1973), it can often be estimated from physical principles if the shape and density of the spore are known (Chamberlain, 1975; McCartney, 1990; McCartney, 1997). In this approach the gravitational forces acting on the particle are equated to the drag (and if appropriate the lift) forces acting on the spore. For spherical spores of diameter d, v_s, can be calculated from Stokes' law (Chamberlain, 1975):

$$v_s = \frac{d^2 g \rho}{18 \nu \rho_a} \qquad (1)$$

where g is the acceleration due to gravity (9.81 m s^{-2}), ρ and ρ_a are the densities of the spore and air, respectively, and ν is the kinematic viscosity of air. The density of spores depends on the species and can vary with relative humidity, but many spores have densities close to that of water (Gregory, 1973). The settling speed (cm s^{-1}) of a spherical spore that has the same density as water and a diameter of d μm falling through air at 20 °C is:

$$v_s = 0.00308\, d^2 \qquad (2)$$

For non-spherical spores v_s can be estimated from that of a spherical spore of the same volume, v_{ss}, by dividing by a shape factor, α ($v_s = v_{ss}/\alpha$). Shape factors have been evaluated for a number of simple shapes such as ellipsoids, cylinders and discs (Mercer, 1973; Chamberlain, 1975). McCartney *et al.* (1993) showed that v_s for *Alternaria* sp. conidia could be estimated from that of cylinders with the density of water and the same length, L, and mean diameter, d, of the spores. At 20 °C, v_s (cm s^{-1}) of such spores is approximated by:

$$v_s = \frac{0.00404\, d^2 (L/d)^{2/3}}{\alpha} \qquad (3)$$

where d is in μm and the shape factor, α, is a function of L/d. For cylinders with L/d values up to about 5, $\alpha = 0.087(L/d) + 0.97$ (Chamberlain, 1975). For cylinders with large values of L/d, v_s may depend only on d (Mercer, 1973). Using the shape factor values for glass fibres given by Mercer (1973) v_s (cm s^{-1}), at 20 °C, of long thin spores, such as *Claviceps purpurea* ascospores, with L/d between 50 and 150 can be estimated to within about 2% (assuming that they have the same density as water):

$$v_s = 0.0117 d^2 \qquad (4)$$

where d is in μm. Table 1 illustrates the relationship between fall speed and particle diameter for spheres, spheroids, cylinders and fibres.

Fungal spores may also be dispersed in clusters or chains (McCartney, 1997). The value of v_s for a cluster of spores is usually larger than that for a single spore, but less than that for a sphere of the same volume. Ferrandino and Aylor (1984) found that the settling speed of clusters of *Uromyces phaseoli,* and *Lycopodium clavatum* spores and *Ambrosia elatior* pollen could be estimated from:

$$v_{sn} = 0.98 v_s n^{0.53} \qquad (5)$$

where n was the number of spores in the cluster and v_s was the settling speed of a single spore. Equation 5 also described the relationship between v_s for of a single *Blumeria graminis* conidium and clusters of conidia (McCartney and Bainbridge, 1987). But, v_{sn} for chains of *Alternaria* sp. conidia were better described by Equation 3 with L the length of the chain and d the mean diameter (McCartney *et al.*, 1993).

The settling speed of a spore clearly influences its potential for dispersal. However, a spore's aerodynamic properties also affect deposition processes such as rate of sedimentation and efficiency of impaction (see below). The aerodynamic characteristics of a spore can be summarised using the concept of an "aerodynamic diameter," d_a. This is the diameter of a sphere with the density of water that has the same aerodynamic behaviour as the spore. The aerodynamic diameter of a spore can be calculated from Equation 1 by setting $\rho = 1$ g cm^{-3} and solving for d. In air at 20 °C, d_a (in μm) for a spore of settling speed v_s is:

$$d_a = 18.02\sqrt{v_s} \qquad (6)$$

when v_s is measured in cm s^{-1}.

SPORE RELEASE

As in any forms of air transport, fungal spore dispersal has three distinct phases: take-off, flight and landing. Thus, before spores can be dispersed they need to become airborne, and because spores are very small, this entails escaping the boundary layer of nearly still air on the surface on which they are growing (Figure 1). Spores can be passively released into the air by for example gusts of wind, mechanical disturbance (e.g., animal movement or rain tapping a leaf) or by rain splash, but many fungi have developed mechanisms that actively release their spores into the air (Ingold, 1971). These mechanisms are complex and varied and have been discussed at length by several authors (Ingold, 1971, Lacey, 1986, Lacey, 1996, Ingold, 1999). Most of the fungi that employ an active spore release mechanism are basidiomycetes and ascomycetes (Ingold, 1999) although fungi in other genera also actively liberate spores. Ballistospore discharge in basidiomycetes rarely projects spores further than a few mm (Ingold, 1999), whereas ascospore discharge in ascomycetes usually propel spores 0.5–2 cm, but distances up to 40 cm have been reported for some species (Lacey, 1996). Many active release mechanisms require water, for example the "squirt-gun" mechanism common in many ascomycetes, but in some fungi or Oomycetes, spore liberation takes place under dry conditions (Lacey, 1996). For example, sporangiophores of *Phytophthora infestans* and *Peronospora tabacina* twist in response to changes in relative humidity with sufficient violence to release sporangia.

Active spore release is often driven by environmental factors such as temperature, humidity and light, but is often related to water requirements. Ascospores are usually released after wetting by rain or dew, as water is needed for the release mechanism (Lacey, 1986, 1996). For example, periods of *Pyrenopeziza brassicae* ascospore release in oilseed rape crops are associated with rain, but spore release can continue for up to five days without further rainfall as the crop debris, on which the fruiting bodies develop, continues to respond to wetting and drying cycles caused by dew (McCartney and Lacey, 1990). Similarly, ascospores of *Leptosphaeria maculans* (phoma stem canker of oilseed rape) are released after rain, and can exhibit a diurnal periodicity with most spores being released between 10:00 and 12:00, possibly due to changes in relative humidity (West *et al.*, 2002). Other fungal groups that use active spore release, such as some *Entomophthorales,* also exhibit diurnal periodicities in spore concentrations. Conidia of *Erynia neoaphidis,* a pathogen of aphids, tend to be released during the night or in the early morning (01:00–07:00) when environmental conditions favour spore production (Hemmati *et al.*, 2001).

In contrast to fungi that have developed active spore release mechanisms, many fungi rely on external physical forces to release their spores into the air. Wind can release spores directly by blowing them off surfaces or by dislodging them by shaking the surface on which the fungus is growing (Bainbridge and Legg, 1976). Many fungi have evolved spore bearing structures that hold the spores away from the surface to enhance their chances of being blown off. Powdery mildews, such as *Blumeria graminis,* produce conidia in chains, the oldest spores being raised away from the leaf by progressively produced spores. Spores are removed by wind when the aerodynamic forces acting on the spore exceed the attachment forces (Aylor and Parlange, 1975), but these forces are not known for most fungi. The wind speeds needed to remove spores are probably relatively large (Grace, 1977), for example, conidia of *Blumeria graminis* are only released in wind speeds exceeding about 0.5 m s^{-1} (Hammett and Manners, 1974), while wind speeds exceeding 5 m s^{-1} are needed to remove conidia of *Drechslera maydi* (southern leaf spot of maize) (Aylor, 1975). It is therefore likely that in many environments spore release by wind takes place only in gusts when speeds are sufficient to remove spores. Thus wind intermittency (turbulence) probably plays an important role in spore removal (Aylor, 1978; Aylor *et al.*, 1981). The importance of gusts in the removing spores has been demonstrated in wind tunnel experiments using conidia of *Passalora personata* (late leaf spot of groundnut) (Wadia *et al.*, 1998).

Spores can also be released into the air by other mechanical actions such as crop disturbance by machinery: combined harvesters release large numbers of spores into the air. Mechanical disturbance can also be responsible for spore release in industrial and indoor environments, for example waste composting and processing cork oak (Avila and Lacey, 1974; Lacey *et al.*, 1992; Lacey, 1997). Spores released in such environments can represent potential health risks to workers. Cleaning operations in food factories have been shown to generate aerosols containing microorganisms (Burfoot *et al.*, 2003) and such activities could easily remove and disperse fungal contaminants growing on surfaces. Rain drops falling on leaves or other surfaces dislodge "dry" spores to allow them to be dispersed by wind. Some plant pathogen spores can be released into the air in this manner, for example late leaf spot of groundnut (*Passalora personata*) (Wadia *et al.*, 1998) and brown (*Puccinia recondita*) and yellow (*P. striiformis*) rust of wheat (Geagea *et al.*, 2000). Spores of some puff-balls (*Lycoperdaceae*) and earth stars (*Geastraceae*) can be ejected into the air when raindrops strike their ripe fruiting bodies.

Rain or spray can remove spores directly from surfaces in run-off water or in splash droplets (Madden, 1992). Raindrops striking surfaces can remove spores by incorporating them in the droplets produced by splash. The spores of many plant pathogens can only be dispersed by water (usually splash) because they are contained in mucilage which prevents dispersal by wind (Fitt *et al.*, 1989). The droplets produced by rain splash range in size from a few μm to up to 1-2 mm, but spores tend to be carried in droplets greater than about 50 μm and most spore-carrying droplets tend to be between about 300 and 700 μm (Fitt *et al.*, 1989). As the fall speed of most spore carrying droplets is relatively large (between 1 and 3 m s^{-1}) they tend to be quickly deposited and are therefore not efficiently dispersed by wind. However, the smaller droplets can evaporate leaving the spores effectively airborne.

DISPERSAL

Once spores have become airborne they can be transported by wind. If air flow were steady and non-turbulent, then the distance, x, a spore would travel could be calculated simply from its fall speed, v_s, the wind speed, u and the height, h, from which it was released:

$$x = \frac{h.u}{v_s} \qquad (7)$$

Unfortunately, natural winds (and many indoor air flows) are turbulent, which causes the concentration of spores in a spore plume to be diluted as the plume expands downwind, in a manner analogous to a smoke plume. Because of this it difficult to define the "dispersal distance" for windborne spores. Spore dispersal is therefore often described in terms of a spore concentration or dispersal gradient that describes how concentration changes away from the spore source (Gregory, 1973). Patterns of spore concentration gradients round spore sources are complex, but, spore concentrations measured in one direction away from the source decrease monotonically with distance

(Figure 3). Several different functions have been used to describe dispersal gradients (Minogue, 1986; Fitt *et al.*, 1987; McCartney and Fitt, 1998).

Two simple functions have been extensively used to describe dispersal gradients: the negative exponential function:

$$C = C_0 \exp(-\alpha x) \quad (8)$$

and the inverse power function:

$$C = Ax^{-\beta} \quad (9)$$

Where C is spore concentration, x is the distance from the source and C_0, α, A and β are constants. The coefficients α and β determine the rate of decrease in spore concentration with distance, and are sometimes referred to as "dispersal gradients". Although both functions appear to behave in a similar manner, they are fundamentally different (Figure 3). In the negative exponential, concentration decreases by half over fixed distances (half distance, $d_{½}$ = 0.693α, analogous to "half life" in radioactive decay). The idea of a "half distance" is frequently used to describe spore dispersal gradients (Table 2). In the inverse power law, the rate of decrease in concentration decreases with distance from the source giving a "long tailed" distribution.

The negative exponential equation is more appropriate when concentration depletion is predominantly by deposition (as in crop canopies), but the inverse power equation is more suited when turbulent mixing is dominant (McCartney and Fitt, 1998). Thus, a power law is usually more appropriate when describing dispersal over long distances (Brown and Hovmøller, 2002).

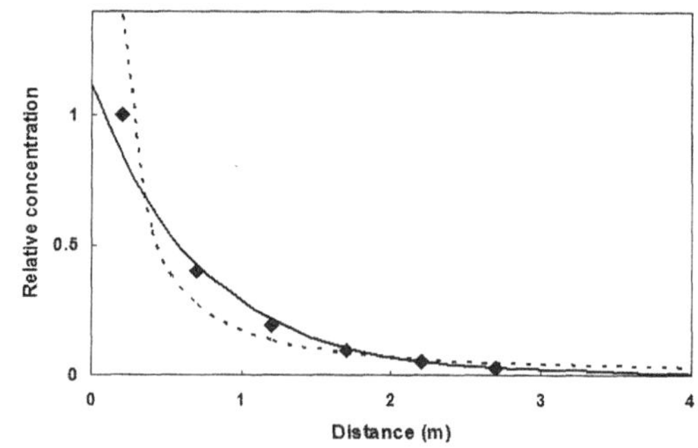

Figure 3. Spore dispersal gradients. Concentrations of airborne *Puccinia striiformis* (yellow rust) conidia measured downwind of an infected patch in a wheat field (◆). Solid line: negative exponential function fitted to the measured concentrations; broken line: inverse power function fitted to the measured concentrations (West and McCartney, unpublished data).

Table 1. Relationship between fall speed (v_s) and particle diameter for spheres, spheroids, cylinders and fibres. The density of the particle was assumed to be that of water. Fall speeds are given in cm s^{-1} when diameters are in μm. Shape factors were taken from Chamberlain (1975) and Mercer (1973). The shape factors for glass fibres estimated from Mercer (1973) were used to calculate the relationships for fibres. The ration of fall speed/ fall speed of a sphere of the same volume (v_s / v_{ss}) is given in the third column, and the aerodynamic diameter, d_a, in the fourth column. These relationships only hold for particles with d_a between about 1 and 60 μm.

Shape	v_s (cm s^{-1})	v_s / v_{ss}	d_a
Sphere	0.00308 d^2	1	D
Spheroid (height/diameter = 0.75)	0.00251 d^2	0.99	0.90 d
Spheroid (height/diameter = 0.5)	0.00186 d^2	0.96	0.78 d
Spheroid (height/diameter = 0.2)	0.00085 d^2	0.80	0.53 d
Spheroid (height/diameter = 0.1)	0.00045 d^2	0.67	0.38 d
Cylinder (length/diameter = 1)	0.00382 d^2	0.95	1.11 d
Cylinder (length/diameter = 2)	0.00561 d^2	0.87	1.35 d
Cylinder (length/diameter = 5)	0.00841 d^2	0.71	1.65 d
Fibre (length/diameter = 50)	0.0115 d^2	0.21	1.93 d
Fibre (length/diameter = 100)	0.0118 d^2	0.14	1.95 d
Fibre (length/diameter = 150)	0.0119 d^2	0.10	1.96 d

Table 2: Spore concentration or deposition gradients: values are given as half distances[1]. The entries are ordered in approximately increasing aerodynamic diameter.

Species[2]	Half distance[1] (m)	Spore shape[3] and size (μm)	Comments	Source
Pyrenopeziza brassicae	7-10	rounded cylinder ~12 x 2.4	Light leaf spot of oilseed rape, concentration downwind from field edge.	McCartney *et al.* (1986)
Scerotinia sclerotiorum	0.2–0.9	ellipsoid ~8 x 3	Ascospore deposition downwind of pasture plots inoculated with *S. sclerotiorum* sclerotia.	Bourdôt *et al.* (2001)
Bovista plumbea	5-8	ovoid 4-5.5 x 5-6.5	Grey puffball.	Fitt *et al.* (1987)
Cladosporium sp.	32-110	ellipsoid 10-20x3-4	Spores, concentration downwind from a wheat crop.	From: Eversmeyer and Kramer (1992)
Cryphonectria parasitica	43	ellipsoid 7-12 x 3-5.5	Chestnut blight.	Fitt *et al.* (1987)
Gibberella zeae	11-33	curved fusoid 19-24 x 3-4	Fusarium head blight of wheat, concentration downwind of infected wheat plots.	de Luna *et al.* (2002)
Ustilago scitaminea	1.1–4.6	spheroid 5.5-7.5	Smut of sugar cane, deposition to the ground close to infected plants.	Hoy *et al.* (1991)
Ustilago violacea	1.4	spheroid 4-10	Anther smut of white campion, deposition to flowers and ground traps.	Roche *et al.* (1995)
Venturia inaequalis	8–11	ellipsoid ~12 x 5	Apple scab, ascospores, estimated from average spore concentrations measured over a season.	From Holb *et al.* (2004)
Botrytis cinerea	1.5 2.1	ellipsoid ~13 x 7	Gray mould of snap beans, deposition to leaves during crop bloom.	From: Johnson and Powelson (1983)
Podaxis pistillaris	6-7	ovoid 9-12 x 10-14	Desert shaggy main mushroom.	Fitt *et al.* (1987)
Phaeoisariopsis personata	0.4-1	rounded cylinder 18-60 x 6-10	Late leaf spot of groundnut, deposition estimated from "trap" plants.	Savary and Van Santen (1992)
Blumeria graminis	~1.8	ellipsoid 25-40 x 8-10	Barley powdery mildew, within a barley canopy.	From: Bainbridge and Stedman (1979)
Blumeria graminis	20-89	ellipsoid 30 x10	Wheat powdery mildew, concentration downwind from a wheat crop.	From: Eversmeyer and Kramer (1992)
20 μm drops	0.25–2.5	spheroid 20	Deposition to horizontal collectors in a barley crop.	McCartney and Bainbridge (1984)
Tilletia tritici	5.5–7.5	spheroid 14 25	Bunt of wheat.	Fitt *et al.* (1987)
Puccinia recondita	0.5-1.2	ellipsoid 16-34 x 13-25	Wheat rust, deposition to "trap" wheat plants in a winter barley crop.	Aylor (1987)
Puccinia recondita	12.5-24	ellipsoid 16-34 x 13-25	Brown rust of wheat, concentration downwind from a wheat crop.	From: Eversmeyer and Kramer (1992)

Puccinia striiformis	0.5	spheroid 20-25	Yellow rust of wheat, concentration downwind of an infected patch in a wheat crop.	Figure 3
Alternaria linicola	1.6–2.3	club shape 60-220 x 15-21	Alternaria blight of linseed, concentration downwind of line source in a linseed crop.	Vloutoglou *et al.* (1995)

[1] Half distances were estimated from original data when not quoted in the source.
[2] Latin binomials are currently accepted usage, and may differ from those quoted in the original source.
[3] Sizes of spheroids are given as the diameter; for ellipsoids and other shapes, the size is given as length x diameter.

When both functions have been compared using measured spore or pollen gradients, both models often fit equally well (Gregory, 1968, Fitt *et al.*, 1987, Ferrandino, 1996). Local environment can play an important role. Within crops, gradients are generally relatively steep (Table 1), for example $d_{½}$ values for 20 μm droplets measured within a barley crop ranged between about 0.5 and 2.5 m depending on the canopy structure (McCartney and Bainbridge, 1984).

Gradients at the edges of crops tend to be shallower (Table 2): $d_{½}$ values measured for oilseed rape pollen (v_s value about 1.6 cm s^{-1} compared to about 1.2 cm s^{-1} for 20 μm drops) dispersing from the edge of an oilseed rape field were between about 2 and 8 m (McCartney and Lacey, 1991). Spore size (aerodynamic diameter) can also affect dispersal gradients: *Pyrenopeziza brassicae* ascospores, which are much smaller than oilseed rape pollen grains (v_s about 0.03 cm s^{-1}), had $d_{½}$ values of about 9 m when measured downwind from the same oilseed rape field as the oilseed rape pollen (McCartney *et al.*, 1986). Other functions have been used to describe spore dispersal gradients (McCartney and Fitt, 1998) and recently Bullock and Clarke (2000) have suggested a combined exponential and inverse power equation to describe wind-borne seed dispersal:

$$C = A(ae^{-\alpha x} + bx^{-\beta}) \qquad (10)$$

This equation allows for two different components of dispersal: a steep short distance gradient and a flatter long distance "tail". This could also be used for spores, but this equation requires two parameters to describe the shape of the gradient. Unfortunately, there has been little systematic work on the influence of environmental factors, surface structures or spore characteristics on dispersal gradient parameters. Thus it is difficult to estimate *a priori* dispersal gradient shapes.

Because of their turbulent nature, wind can also rapidly transport spores vertically into the atmosphere, where they have the potential to disperse over a large distance. Vertical transport is most likely, when turbulence is high, for example during unstable atmospheric conditions that occur during sunny afternoons (Figure 2). Such conditions also favour the passive removal of spores and escape from plant canopies. Spore concentrations decrease with height if the source of spores is local and at ground level, but concentrations can increase with height if the spores have a distant source. Spore aerodynamic diameter can also affect the potential for vertical transport of spores: concentrations of ascospores of *P. brassicae* (d_a about 3.1 μm) decreased less quickly with height than much larger oilseed rape pollen grains (d_a about 23 μm) (Figure 4). Although most fungal spores probably travel relatively short distances, once they have been mixed into the planetary boundary layer (Figure 1) they can travel long distances. Hirst *et al.* (1967) found measurable concentrations of fungal spores between 500 and 1000 m above the North Sea many kilometres from the nearest source, and spores and pollen grains, which must have been produced in distant continents, have been found in air samples taken in Antarctica (Marshall, 1996). The introduction of new plant diseases into countries has been attributed to long distance transport of spores, although such events are probably rare (Brown and Hovmøller, 2002). Natural or man-made events that enhance vertical air movement, such as bush fires, may be responsible for individual long distance transport events, for example the spread of viable bacteria and fungal spores from the Yucatan in Mexico to Texas and from Southeast Asia to Hawaii (Mims and Mims, 2004). Long distance transport of spores need not take place in a single step: plant pathogen inoculum can spread over continental distances in multiple "jumps". In most years, tobacco blue mould (caused by *Peronospora tabacina*) spreads from Cuba up the eastern states of the USA in a series of dispersal events (Davis and Main, 1986). The North American Plant Disease Forecast Centre, North Carolina State University, Raleigh, NC, provides an Internet-based blue mould disease risk forecasting system for tobacco and cucurbit growers (http://www.ces.ncsu.edu/depts/pp/ blue mold/index.html, Main *et al.*, 2001). The forecasting system uses atmospheric dispersal models to predict *P. tabacina* spore transport events.

Empirical descriptions of spore dispersal gradients have limited applications as it is difficult to estimate gradient parameters for conditions different from those in which they were measured.

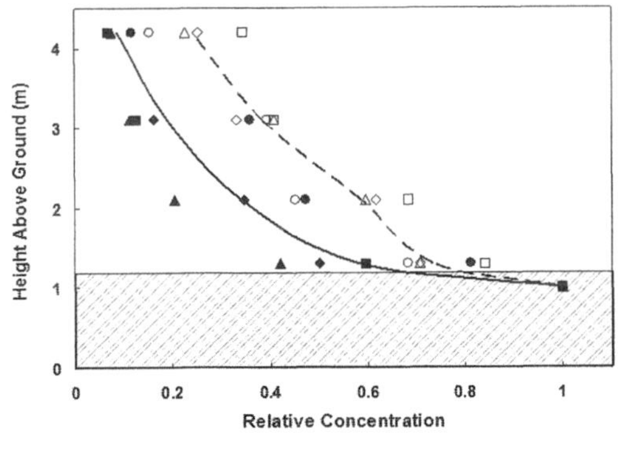

Figure 4. Vertical transport of spores. Relative concentration profiles of *Pyrenopeziza brassicae* ascospores and oilseed rape pollen measured simultaneously above an oilseed rape crop: filled symbols pollen; open symbols ascospores. The solid line is the mean profile for the pollen and the broken line is that for the ascospores (McCartney and Lacey (1991) and McCartney, unpublished data).

As a result, atmospheric dispersal models that are used to calculate air pollution dispersal have been adapted to estimate spore dispersal patterns (McCartney and Fitt, 1998). Average pollution concentrations downwind of point and line sources can be estimated using Gaussian Plume dispersal models (Pasquill and Smith, 1983). Some atmospheric dispersal models that have been developed for use in air pollution regulation and emergency planning are of this type (Caputo *et al.*, 2003). For example, the US Environmental Protection Agency developed the AEROMOD model for regulatory purposes (USEPA, 1999). Gaussian Plume models assume that air pollutant concentration profile distributions are Gaussian in both crosswind and vertical directions (Figure 5). The values of the standard deviations of the crosswind (σ_y) and vertical (σ_z) distributions are dependant on downwind distance and determine the downwind gradients. As Gaussian plume models have been used for many years, much effort has been spent on parameterising σ_y and σ_z for different atmospheric conditions (Pasquill and Smith, 1983). Gaussian Plume models were developed in the 1980s to assess the risk of the aerial transmission of Foot and Mouth disease in farm animals (Blackall and Gloster, 1981; Gloster, 1983a) and Newcastle Disease in poultry (Gloster, 1983b). This type of model was also used in the management of the 2001 Foot and Mouth outbreak in the UK (Mikkelsen, 2003).

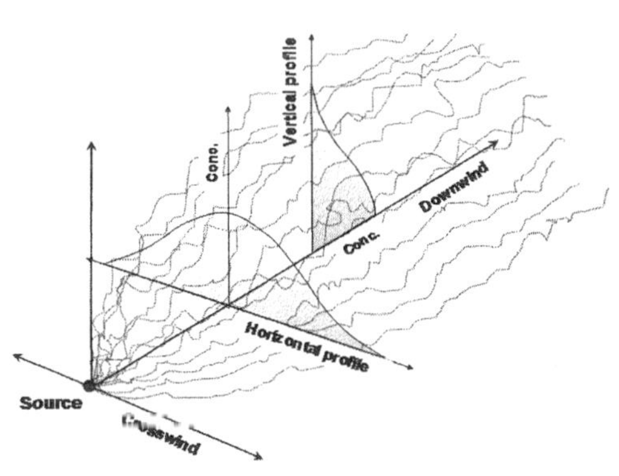

Figure 5. Spore dispersal downwind of a ground level point source. Gaussian plume models assume that the spore concentrations in the crosswind and vertical directions are distributed according to a Gaussian distribution.

Air parcel trajectory analysis has been used in tracing potential long distance spread of plant pathogens (Aylor, 1986, Davis, 1987, Brown and Hovmøller, 2002). Trajectory analysis uses information on wind fields and atmospheric temperature profiles to account for large scale movement of air parcels due to changes in wind direction, and is frequently used in air pollution analysis (Stohl, 1998). Web-based trajectory models are available from the USA National Ocean and Atmosphere Administration (HYSPLIT model http://www.arl.noaa. gov) and the British Atmospheric Data Centre (NERC Centres for Atmospheric Science, http://badc.nerc.ac.uk/ community/). Trajectory analysis can be combined with the Gaussian Plume approach to take account of spore dispersal within the air parcel (Aylor, 1986; Davis, 1987; Aylor, 1999). The spore plume is treated as an expanding "puff" travelling along the path of the trajectory. The spore concentrations in the "puff" are assumed to follow a Gaussian distribution in the vertical, horizontal and downwind directions, unless constrained by atmospheric structures such as inversions. The rate of "puff" expansion is determined by how σ_z, σ_y and σ_x (the standard deviations of the Gaussian distributions) change with distance. Gaussian "puff" models for spore dispersal also allow for spore loss through deposition by sedimentation and washout (Aylor, 1999). The model used by the North American Plant Disease Forecasting Centre for their Internet-based tobacco and cucurbit disease risk forecasting system is a Gaussian "puff" trajectory model (Main *et al.*, 2001). The NOAA HYSPLIT model is used to estimate the risk of inoculum movement from infected to uninfected areas. A Gaussian "puff" model was also one of the dispersal models used to analyse the Foot and Mouth disease outbreak in the UK in 2001 (Mikkelsen *et al.*, 2003).

Other models, based on physical principles, have been used to describe airborne spore dispersal (see for example: Aylor, 1990; McCartney, 1997; McCartney and Fitt, 1998; Aylor *et al.*, 2003).

EAD models spore dispersal is assumed to be analogous to molecular diffusion; while LS models estimate the trajectories of "individual spores" allowing for the effects of turbulence.

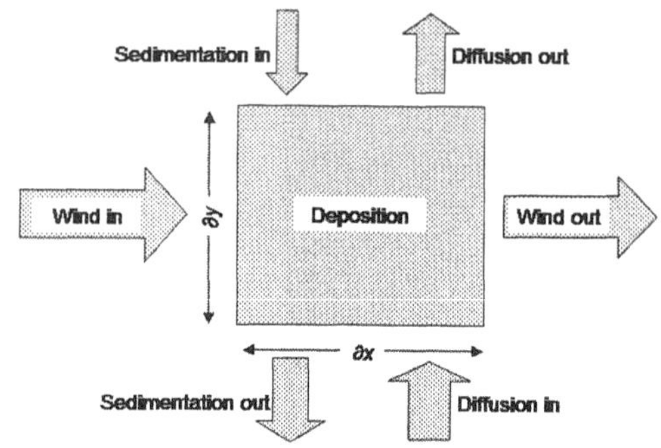

$$u\frac{\partial C}{\partial x} = \frac{\partial}{\partial z}\left(K\frac{\partial C}{\partial z}\right) + v_s\frac{\partial C}{\partial z} + S(x,y)$$

Figure 6. Eulerian advection-diffusion models are based on the number balance of spores entering and leaving a small volume of air. Schematic diagram for an infinite line source of spores (assumes that cross-wind diffusion ignored and downwind diffusion negligible compared to downwind transport). The difference between the rate the spores carried into and out of the volume by wind is balanced by the rate at which spores enter and leave the volume vertically (sedimentation and diffusion) and the rate at which spores are deposited within the volume (deposition). The rate of diffusion is proportional to the vertical concentration gradient (∂C/∂z).

EAD models are based on the number balance of spores entering and leaving a small volume of air (Figure 6). This can be illustrated by considering an infinite line source of spores at right angles to the wind. In this case, cross-wind and downwind diffusion can be ignored (advection is assumed to be the predominant mechanism for transporting spores downwind). For a small volume of air, the difference between the rate at which spores enter or leave the volume horizontally by wind is balanced by the rates at which spores leave the volume vertically by diffusion or by sedimentation and the rate at which spores are removed from the volume by deposition (Figure 6). The rate of vertical diffusion is assumed to be proportional to the vertical concentration gradient ($K_z\partial C/\partial z$) and determined by a diffusion coefficient, K_z, which is a function of height. Dispersal from point and area sources can also be described by EAD models, but these require terms to describe horizontal diffusion and advection (Yao *et al.*, 1997, Aylor, 1999). EAD models usually need to be solved by numerical methods, and diffusion coefficients, wind speeds and spore deposition rates need to be defined for all points in the model space. Diffusion coefficients and wind speeds can be measured directly or estimated theoretically for different atmospheric conditions (Yao *et al.*, 1997; D'Amours, 1998). Deposition rates depend on spore aerodynamic properties and the nature of the surface (see next section). EAD models implicitly assume that the size of the eddies that cause diffusion are small compared with the size of the dispersing plume. This may not be the case in some situations, such as in plant canopies, which may explain why EAD models have been found to overestimate concentrations of plant pathogen spores close to the spore source (Legg and Powell, 1979; Aylor and Ferrandino, 1989). However, EAD models are useful when the dominant eddies are small compared with the vertical width of the plume (Aylor, 1999), such as dispersal downwind from the edge of a field (Yao *et al.*, 1997). Xu and Burfoot (2000) have applied EDA-based models to the application, by fogging, of agrochemicals to potatoes in storage.

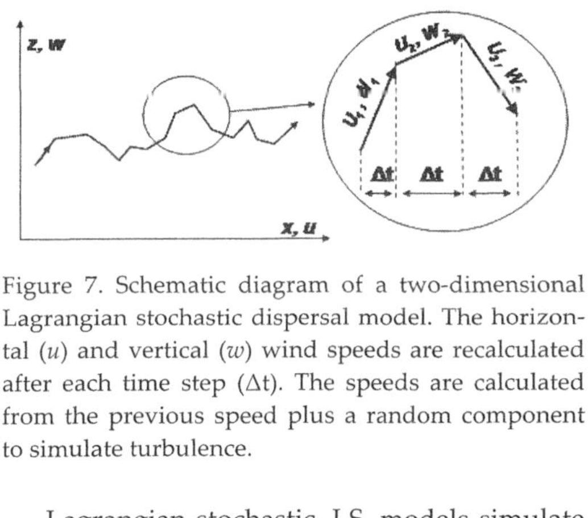

Figure 7. Schematic diagram of a two-dimensional Lagrangian stochastic dispersal model. The horizontal (*u*) and vertical (*w*) wind speeds are recalculated after each time step (Δt). The speeds are calculated from the previous speed plus a random component to simulate turbulence.

Lagrangian stochastic, LS, models simulate the paths of individual spores as a pseudo-random walk (Figure 7). Spore paths are simulated as series of discrete steps determined partly by a correlation between successive velocities (the velocity "memory") and partly by a random element that represents the effects of turbulence (Figure 7). The formulation of LS models has been reviewed by Wilson and Sawford (1996). Because LS models simulate the flights of "individual spores", they are useful for estimating dispersal close to sources (Aylor, 1989, 1999) and have been used to calculate the escape of *Venturia inaequalis* (apple scab) ascospores from the ground (Aylor and Flesch, 2001) and *Phytopthora infestans* (late blight) sporangia from potato canopies. The dispersal of pollen from maize crops has also been investigated using LS models (Aylor *et al.*, 2003; Jaroz *et al.*, 2004). The effects of gust release on spore dispersal can be simulated using LS models, by only starting spore trajectory simulations when the wind speed exceeds the spore release threshold. LS simulations suggest that deposition near the source may be enhanced by gust release due to more efficient impaction at the higher wind speeds in which the spores are travelling (see below) (Legg, 1983). LS models have the potential to describe spore dispersal in a wide range of environments as long as the flow fields can adequately be described (mean flow and turbulence statistics). Mean flow fields in indoor environments can be calculated using sophisticated computational fluid dynamics (CFD) programs that solve the continuity equations for mass and momentum (Burfoot *et al.*, 1999). For example such programs have been used to study the air flow in potato storage facilities (Xu *et al.*, 2002). The mean flows predicted by CFD programs can be combined with appropriate descriptions of turbulent fluctuations to form the basis for LS particle dispersal models (Burfoot *et al.*, 1999, Riddle *et al.*, 2004). This approach has recently been used to model the dispersal of airborne microbial particles from cleaning operations in an enclosed room (Harral and Burfoot, 2005). In this study dispersal of aerosol particles generated by a boot scrubber, in a room ventilated by ceiling air ducts, was simulated using two different CDF/ Lagrangian modelling approaches (Gosman and Ioannides, 1981; Reynolds, 1998). Both modelling approaches predicted the general pattern of particle dispersal within the room, but the Reynolds model more accurately predicted particle clearance times. The Reynolds model simulates the effects of velocity fluctuations more accurately than the Gosman and Ioannides model, and therefore may be more applicable to modelling the dispersal of microorganisms indoors. As our understanding of indoor and outdoor air flow increases, the accuracy and applicability of LS models should increase.

Atmospheric dispersal models are becoming increasingly more sophisticated, for example some can describe dispersal over complex terrain (e.g., Aloyan, 2004; Wang and Ostoja-Starzewski, 2004). These new approaches could help in understanding the influence of landscape on spore dispersal. As noted above, the combination of computational fluid dynamic models with dispersal models should allow spore dispersal to be modelled in urban (Riddle *et al.*, 2004) and indoor environments (Harral and Burfoot, 2005).

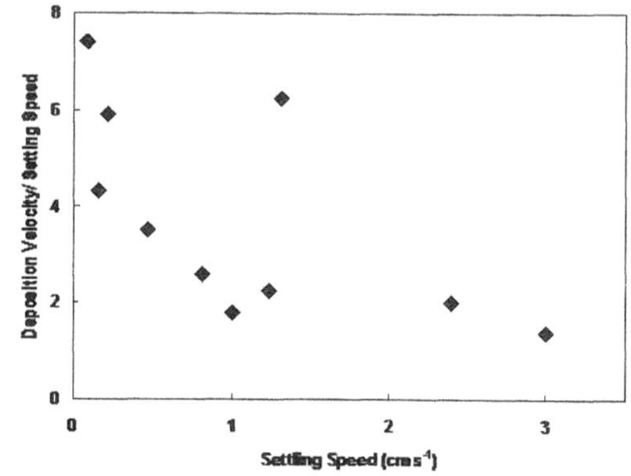

Figure 8. Enhancement of deposition by turbulent diffusion. The ratio of deposition velocity/ settling speed (v_d/v_s) plotted against settling speed for spore and pollen deposition to microscope slides 20 cm above a barley crop. The ratio decreases with increasing spore size showing that turbulence has a smaller effect on large spores. Adapted from McCartney *et al.* (1985).

DEPOSITION

The rate of deposition of spores, (D, number per unit area per unit time) from the atmosphere to a horizontal surface is proportional to the concentration of spores above the surface, C, and is given by:

$$D = v_d C \qquad (11)$$

the constant of proportionality, v_d, is called the deposition velocity (Chamberlain, 1975). If atmospheric air flow was non-turbulent, v_d would have the same value as the spore settling speed, v_s. However, in turbulent flow diffusion can enhance deposition rates and so deposition velocities tend to be about 2-5 times larger than v_s (McCartney and Fitt, 1985). Deposition velocities also tend to increase with wind speed and turbulence (Davidson *et al.*, 1982; Callander and Unsworth, 1983). However, the effects of turbulence on spore deposition tend to decrease with increasing aerodynamic diameter (Figure 8). There have been several models developed to estimate particle deposition velocities from surface characteristics, most using an EAD approach (see McCartney and Fitt, 1985; Ferrandino and Aylor, 1985) and more recently using LS models (Reynolds, 2000).

Spores not only settle on surfaces, they can also be impacted onto an object, such as a leaf, or to the effects of inertia (Figure 9).

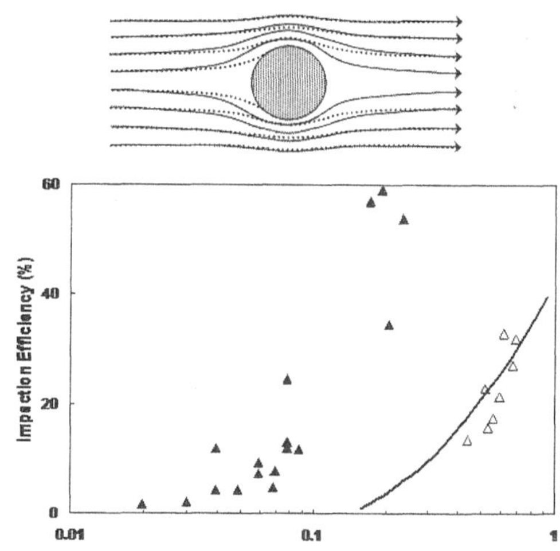

Figure 9. Top: Inertial impaction of spores. Air flow around a vertical cylinder: air streamlines are shown as solid lines (—); spore trajectories (··········) cannot follow the streamlines exactly, so some spores may strike the cylinder. Bottom: Impaction efficiencies measured in a barley crop for *Blumeria graminis* conidia as a function of Stokes' number (see text). Solid symbols (▲): measurements made inside the crop; open symbols (△): measurements made at the top of the crop. The solid line is the relationship between efficiency and Stokes' number for low turbulence flow (Chamberlain, 1975). Adapted from McCartney (1991).

The rate of dep sition by impaction, I, is proportional to the wind speed, u and the spore concentration, C:

$$I = CuE \tag{12}$$

the constant of proportionality, E, is called the efficiency of impaction, and increases with spore aerodynamic diameter and wind speed, but decreases as the size of the object impacted upon increases (Chamberlain, 1975). In laminar flow the efficiency of impaction is a non-linear function of the particle Stokes' number, St, defined as:

$$St = \frac{v_s u}{gL} \tag{13}$$

where g is gravitational acceleration (9.81 m^2s^{-1}) L is a characteristic length of the object (e.g., width of a leaf or diameter of a stem). Aylor (1982) gives the following functional representation for the relationship between E and St:

$$E = \frac{0.86}{1 + 0.442 St^{-1.967}} \tag{14}$$

McCartney and Bainbridge (1987) found that impaction efficiencies for *Blumeria graminis* (barley powdery mildew) conidia measured in a barley crop were significantly larger than those calculated from Equation 14 using measured mean wind speeds (Figure 9). They attributed this to the effects of releasing spores only in gusts. Spores released in gusts will be carried in air travelling faster than the mean wind, consequently their impaction efficiencies will be larger than that for spores travelling at the mean wind speed (Aylor *et al.*, 1981). Model calculations suggest that enhanced impaction due to gusts could decrease dispersal distance close to the source (McCartney, 1987).

Deposition to individual surfaces, such as leaf elements within vegetation canopies, has often been treated as a combination of two processes: gravitational settling and inertial impaction (McCartney and Fitt, 1985). Non-horizontal and non-vertical objects are resolved into areas projected horizontally (along the mean wind direction) and vertically to determine the proportions of the total deposition that are by sedimentation (horizontal) and impaction (vertical). Thus, for an object at an angle, θ, to the horizontal the deposition rate is given by:

$$D_\theta = D_0 \cos(\theta) + D_{90} \sin(\theta) \tag{15}$$

where D_0 and D_{90} are the deposition rates on equivalent horizontal and vertical surfaces. This assumption may not strictly be valid as the variable boundary layer on the sloping surface has a vertical component that may affect sedimentation. However, Equation 15 was found to describe deposition on sloping surfaces in a wheat canopy (McCartney and Aylor, 1987) and thus may be adequate for practical purposes.

CONCLUSIONS

Airborne transfer of microorganisms is now seen as significant route for contamination in many sectors of the food industry (Burfoot *et al.*, 2000). Thus, an understanding of the physical and biological processes involved in the aerial transport of such organisms is needed to assess the risks of contamination and to develop appropriate strategies to avoid contamination. In this chapter we have attempted to highlight the physical and biological nature of the dispersal of fungal spores through the air. Although many of the examples cited have been related to the spread of plant pathogenic fungi, the mechanisms involved are equally applicable to dispersal in other environments.

Recent advances in fluid dynamics (Reynolds, 1998) combined with new methods for detecting and identifying fungal contaminants (Ward *et al.*, 2004) offer new opportunities for in-depth studies of the spread of contaminants in indoor and outdoor environments. Such studies, in turn, will lead to a better understanding of the role dispersal plays in fungal contamination and allow improvements to be made in both the risk assessment and the management of contamination.

REFERENCES

Aloyan, A. E. (2004). Numerical modelling of minor gas constituents and aerosols in the atmosphere. Ecological Modelling 179:163-175.

Avila, R., and Lacey, J. (1974). The role of *Penicillium frequentans* in suberosis (respiratory disease in the cork industry). Clinical Allergy 4:109-117.

Aylor, D. E. (1975). Force required to detach conidia of *Helminthosporium maydis*. Plant Physiology 55:99-101.

Aylor, D. E. (1978). Dispersal in time and space: aerial pathogens. *In* Plant Disease; an Advanced Treatise (Horsfall, J. G., and Cowling, E. B., eds.), Academic Press, New York, U.S.A., pp. 159-158.

Aylor, D. E. (1982). Modelling spore dispersal in a barley crop. Agricultural Meteorology 26:215-219.

Aylor, D. E. (1986). A framework for examining inter-regional aerial transport of fungal spores. Agricultural and Forest Meteorology 38:263-288.

Aylor, D. E. (1987). Deposition gradients of urediniospores of *Puccinia recondita* near a source. Phytopathology 77:1442-1448.

Aylor, D. E. (1989). Aerial spore dispersal close to a focus of disease. Agriculture and Forest Meteorology 47:109-122.

Aylor, D. E. (1990). The role of intermittent wind in the dispersal of plant pathogens. Annual Review of Phytopathology 28:73-92.

Aylor, D. E. (1999). Biophysical scaling and passive dispersal of fungus spores: relationship to integrated pest management strategies. Agricultural and Forest Meteorology 97:275-292.

Aylor, D. E., and Parlange, J-Y. (1975). Ventilation required to entrain small particles from leaves. Plant Physiology 56:97-99.

Aylor, D. E., McCartney, H. A., and Bainbridge, A. (1981). Deposition of particles liberated in gusts of wind. Journal of Applied Meteorology 10:1212-1221.

Aylor, D. E., and Ferrandino, F. J. (1989). Dispersion of spores released from an elevated line source within a wheat canopy. Boundary-Layer Meteorology 46:251-273.

Aylor, D. E., and Flesch, T. K. (2001). Estimating spore release rates using a Lagrangian stochastic simulation model. Journal Applied Meteorology 40: 1196-1208.

Aylor, D. E. Schultes, N. P., and Shields, E. J. (2003). An aerobiological framework for assessing cross-pollination in maize. Agricultural and Forest Meteorology, 119:111-129.

Bainbridge, A., and Legg, B. J. (1976). Release of barley-mildew conidia from shaken leaves. Transactions of the British Mycological Society 66:97-99.

Bainbridge, A., and Stedman, O. J. (1979). Dispersal of *Erysiphe graminis* and *Lycopodium clavatum* spores near to the source in a barley crop. Annals of Applied Biology 91:187-198.

Blackall, R. M., and Gloster, J. (1981). Forecasting the airborne spread of foot and mouth disease. Weather 36:162-167.

Bourdôt, G. W., Hurrell, G. A., Saville, D. J., and Jong, M. D. de (2001). Risk analysis of *Sclerotinia sclerotiorum* for biological control of *Cirsium arvense* in pasture: ascospore dispersal. Biocontrol Science and Technology 11:119-139.

Brown, J. K., and Hovmøller, M. S. (2002). Aerial dispersal of pathogens on the global and continental scales and its impact on plant disease. Science 297:537-541.

Bullock, J. M., and Clarke, R. T. (2000). Long distance seed dispersal by wind: measuring and modelling the tail of the curve. Oecologia 124:506-521.

Burfoot, D., Hall, K., Brown, K., and Xu, Y. (1999). Fogging for the disinfection of food processing in factories and equipment. Trends in Food Science and Technology 20:205-210.

Burfoot, D., Reavell, S., Tuck, C., and Wilkinson, D. (2003). Generation and dispersion of droplets from cleaning equipment used in the chilled food industry. Journal of Food Engineering 58:343-353.

Callander, B. A., and Unsworth, M. H. (1983). Deposition of 30-60μm diameter droplets in winter wheat. Agricultural Meteorology 28:109-127.

Caputo, M., Giménez, M., and Schlamp, M. (2003). Intercomparison of atmospheric dispersion models. Atmospheric Environment 37:2435-2449.

Chamberlain, A. C. (1975). The movement of particles in plant communities. *In* Vegetation and the Atmosphere Vol. 1, (Monteith, J. L., ed.) Academic Press, London, U.K., pp. 155-203.

D'Amours, R. (1998). Modeling the ETEX Plume dispersion with the Canadian emergency response model. Atmospheric Environment 32:4335-4341.

Davidson, C. I., Miller, J. M., and Pleskow, M. A. (1982). The influence of surface structure on predicted particle dry deposition to natural grass canopies. Water, Air and Soil Pollution 18:25-43.

Davis, J. M. (1987). Modeling the long-range transport of plant pathogens in the atmosphere. Annual Review of Phytopathology 25:169-188.

Davis, J. M., and Main, C. E. (1986). Applying atmospheric trajectory analysis to problems in epidemiology. Plant Disease 70:490-497.

De Luna, L., Bujold, I., Carisse, O., and Paulitz, T. C. (2002). Ascospore gradients of *Gibberella zeae* from overwintered inoculum in wheat fields. Canadian Journal of Plant Pathology 24:457-464.

Edmonds, R. L., and Benninghoff, W. S. (1973). Aerobiology and its Modern Applications, Report No. 3, Aerobiology Component, US Component of the International Biological Program.

Eversmeyer, M. G., and Kramer, C. L. (1992). Local dispersal and deposition of fungal spores from a wheat canopy. Grana 31: 53-59.

Ferrandino, F. J. (1996). Length scale of disease spread: fact or artefact of experimental geometry. Phytopathology 86:807-811.

Ferrandino, F. J., and Aylor, D. E. (1984). Settling speed of clusters of spores. Phytopathology 74:969-972.

Ferrandino, F. J., and Aylor, D. E. (1985). An explicit equation for deposition velocity. Boundary Layer Meteorology 31:197-201.

Fitt, B. D. L., Gregory, P. H., Todd, A. D., McCartney, H. A., and Macdonald, J. M. (1987). Spore dispersal and plant disease gradients: a comparison between two empirical models. Journal of Phytopathology 118:227-242.

Fitt, B. D. F, McCartney, H. A., and Walklate, P. J. (1989). The role of rain in the dispersal of pathogen inoculum. Annual Review of Phytopathology 27:241-470.

Geagea, L., Huber, L., Sache, I., Flura, D., McCartney, H. A., and Fitt, B. D. L. (2000). Influence of simulated rain on dispersal of rust spores from infected wheat seedlings. Agricultural and Forest Meteorology 101:53-66.

Gloster, J. (1983a). Forecasting the airborne spread of foot and mouth disease and Newcastle disease. Philosophical Transactions of the Royal Society, London B302:535-541.

Gloster, J. (1983b). Analysis of two outbreaks of Newcastle disease. Agricultural Meteorology 28:177-189.

Gosman, A. D., and Ionnides, E. (1981). Aspects of computer simulation of liquid fuelled combustors. Journal of Energy 7:482-490.

Grace, J. (1977). Plant Response to Wind. Academic Press, London.

Gregory, P. H. (1968). Interpreting plant disease dispersal gradients. Annual Review of Phytopathology 6:189-212.

Gregory, P. H. (1973). Microbiology of the Atmosphere, Second Edition. Leonard Hill, Aylesbury.

Hammett, K. R. W., and Manners, J. G. (1974). Conidium liberation in *Erysiphe graminis* III: wind tunnel studies. Transactions of the British Mycological Society 62:267-282.

Harral, B. and Burfoot, D. (2005). A comparison of two models for predicting the movements of air borne particles from cleaning operations. Journal of Food Engineering 69:443-451.

Hemmati, F., Pell, J. K., McCartney, H. A., and Deadman, M. L. (2001). Airborne concentrations of conidia of *Erynia neoaphidis* above cereal fields. Mycological Research 105:485-489.

Hirst, J. M., Stedman, O. J., and Hurst, G. W. (1967). Long distance spore transport: vertical sections of spore clouds over the sea. Journal of General Microbiology 48:357-377.

Holb, I. J., Heijne, B., Withagen, J. C. M., and Jeger, M. J. (2004). Dispersal of *Venturia inaequalis* ascospores and disease gradients form a defined inoculum source. Journal of Phytopathology 152: 639-646.

Hoy, J. W., Grisham, M. P. and Chao, C. P. (1991). Production of sori and dispersal of teliospores of *Ustilago scitaminea* in Louisiana. Phytopathology 81:574-579.

Ingold, C. T. (1971). Fungal Spores: their Liberation and Dispersal. Clarendon Press, Oxford.

Ingold, C. T. (1999). Active liberation of reproductive units in terrestrial fungi. The Mycologist 13 (part 3):113-116.

Jaroz, N., Loubet, B., and Huber, L. (2004). Modelling airborne concentration and deposition rate of maize pollen. Atmospheric Environment 38:5555-5566.

Johnson, K. B., and Powelson, M. L. (1983). Analysis of spore dispersal gradients of *Botrytis cinerea* and grey mold disease in snap beans. Phytopathology 73:741-746.

Lacey, J. (1986). Water availability and fungal reproduction: patterns of spore production, liberation and dispersal. *In* Water Fungi and Plants (Ayres, P. G., and Boddy, L., eds.), Cambridge University Press, Cambridge, U.K., pp. 65-86.

Lacey, J. (1996). Spore dispersal – its role in ecology and disease: the British contribution to fungal aerobiology. Mycological Research 100:641-660.

Lacey, J. (1997). Actinomycetes in composts. Annals of Agricultural and Environmental Medicines 4:113-121.

Lacey, J., Williamson, P. A. M., and Crook, B. (1992). Microbial emissions from composts made for mushroom production and from domestic waste. *In* Composting and Compost Quality Assurance Criteria (Jackson, D. V., Merillot, J. M., and L'Hermite, P., eds.), Luxembourg, office for Official Publications of the European Community. EUR14254, pp. 117-130.

Legg, B. J. (1983). Movement of plant pathogens in the crop canopy. Philosophical Transactions of the Royal Society London B302:559-574.

Legg, B. J., and Powell, F. A. (1979) Spore dispersal in a barley crop: a mathematical model. Agricultural Meteorology 20:47-67.

Madden, L. V. (1992). Rainfall and dispersal of fungal spores. Advances in Plant Pathology 8:39-79.

Main, C. E., Keever, T., Holmes, G. J., and Davis, J. M. (2001). Forecasting long-range transport of downy mildew spores and plant disease epidemics. APS*net*, May 2001, http://www.apsnet.org/online/feature/.

Marshall, W. A. (1996). Biological particles over Antarctica. Nature 383:680.

McCartney, H. A. (1987). Deposition of *Erysiphe graminis* conidia on a barley crop, II: consequences for spore dispersal. Journal of Phytopathology 118:258-264.

McCartney, H. A. (1990). Dispersal mechanisms through the air. *In* Dispersal in Agricultural Habitats (Bunce, R. G. H., and Howard, D. C., eds.), Belhaven Press, London, U.K., pp. 133-158.

McCartney, H. A. (1991). Airborne dissemination of plant fungal pathogens. Journal of Applied Bacteriology (suppl) 70:S39-S48.

McCartney, H. A. (1997). Physical factors in the dispersal of aerobiological particles. *In* Aerobiology (Agashe, S. N., ed.), Oxford & IBH Publishing Co., New Delhi, India, pp. 439-450.

McCartney, H. A., and Aylor, D. E. (1987). Relative contribution of sedimentation and impaction to deposition of particles in crop a canopy. Agriculture and Forest Meteorology 40:343-358.

McCartney, H. A., and Bainbridge, A. (1984). Deposition gradients close to a point source. Phytopathologische Zeitschrift 109:219-236.

McCartney, H. A., and Bainbridge, A. (1987). Deposition of *Erysiphe graminis* conidia on a barley crop, I: sedimentation and impaction. Journal of Phytopathology 118:243-257.

McCartney, H. A., Bainbridge, A., and Stedman, O. J. (1985). Spore deposition velocities measured over a barley crop. Phytopathologische Zeitschrift 114: 224-233.

McCartney, H. A., and Fitt, B. D. L. (1985). Construction of dispersal models. *In* Advances in Plant Pathology, Vol.3: Mathematical Modelling of Crop Disease (Gilligan, C. A., ed.), Academic Press, London, U.K., pp. 107-143.

McCartney, H. A., and Fitt, B. D. L. (1998). Dispersal of fungal plant pathogens: mechanisms, gradients and spatial patterns. *In* Plant Disease Epidemiology (Jones, G., ed.), Kluwer Academic Publishers, London, U.K., pp. 138-160.

McCartney, H. A., and Lacey, M. E. (1990). The production and release of ascospores of *Pyrenopeziza brassicae* Sutton et Rawlinson on oilseed rape. Plant Pathology 39:17-32.

McCartney, H. A., and Lacey, M. E. (1991). Wind dispersal of pollen from crops of oilseed rape (*Brassica napus* L). Journal of Aerosol Science 22:467-477.

McCartney, H. A., Lacey, M. E., and Rawlinson, C. J. (1986). Dispersal of *Pyrenopeziza brassicae* spores from an oil-seed rape crop. Journal of Agricultural Science, Cambridge 107:299-305.

McCartney, H. A., Schmechel, D., and Lacey, M. E. (1993). Aerodynamic diameter of conidia of *Alternaria* species. Plant Pathology 42:280-286.

Mercer, T. T. (1973). Aerosol Technology in hazard Evaluation. Academic Press, New York.

Mikkelsen, T., Alexandersen, S., Astrup, P., Champion, H. J., Donaldson, A. I., Dunkerley, F. N., Gloster, J., Sørensen, J. H., and Thykier-Nielsen, S. (2003). Investigation of airborne foot and mouth disease virus transmission during low-wind conditions in the early phase of the UK 2001 epidemic. Atmospheric Chemistry and Physics 3:2101-2110.

Mims, S. A., and Mims, F. M. (2004). Fungal spores are transported long distances in smoke from biomass fires. Atmospheric Environment 38:651-655.

Minogue, K. P. (1986). Disease gradients and the spread of disease. *In* Plant Disease Epidemiology, Vol. 1, Population Dynamics and Management (Leonard, K. J., and Fry, W. E., eds.), Macmillan, New York, U.S.A., pp. 285-310.

Monteith, J. L., and Unsworth, M. H. (1990). Principles of Environmental Physics, Second Edition. Edward Arnold, London.

Pasquill, F., and Smith, F. B. (1983). Atmospheric Diffusion, Third Edition. Ellis Horwood, Chichester.

Reynolds, A. M. (1998). Modelling particle dispersion within a ventilated airspace. Fluid Dynamics Research 22:139-152.

Reynolds, A. M. (2000). Prediction of particle deposition on to rough surfaces. Agricultural and Forest Meteorology 104:107-118.

Riddle, A, Carrithers, D. Sharpe, A., McHugh, C., and Stocker, J. (2004). Comparison between FLUENT and ADMS for atmospheric dispersion modelling. Atmospheric Environment 38:1029-1038.

Roche, B. M., Alexander, H. M., and Maltby, A. D. (1995). Dispersal and disease gradients of anther-smut infection of *Silene alba* at different life stages. Ecology 76:1863-1871.

Savary, S., and van Santen, G. (1992). Effect of crop age on primary gradients of late leaf spot (*Cercosporidium personatum*) on groundnut. Plant Pathology 41:265-273.

Stohl, A. (1998). Computation, accuracy and applications of trajectories – a review and bibliography. Atmospheric Environment 32:947-966.

Thom, A. S. (1975). Momentum, mass and heat exchange of plant communities. *In* Vegetation and the Atmosphere, Vol. 1 (Monteith, J. L., ed.), Academic Press, London, UK, pp. 57-109.

USEPA (1999). Revised draft user's guide for the AEOMOD meteorological processor (aermet). EPA, 273 pp.

Vloutoglou, I, Fitt, B. D. L., and Lucas, J. L. (1995). Periodicity and gradients in dispersal of *Alternaria linicola* in linseed crops. European Journal of Plant Pathology 1010:639-653.

Wadia, K. D. R., McCartney, H. A., and Butler, D. R. (1998). Dispersal *of Phaeoisariopsis personata* conidia from groundnut by wind and rain. Mycological Research 102:355-360.

Wang, G., and Ostoja-Starzewski (2004). A numerical study of plume dispersion motivated by a mesoscale atmospheric flow over a complex terrain. Applied Mathematical Modelling 28:957-981.

West, J. S., Jedryczka, M., Leech, P. K., Dakowska, S., Huang, Y. J. Steed, J. M., and Fitt, B. D. L. (2002). Biology of *Leptosphaeria maculans* ascospore release in England and Poland. *IOBC/WPRS* Bulletin 25:21-29.

Wilson, J. D., and Sawford, B. L. (1996). Review of Lagrangian stochastic models for trajectories in the turbulent atmosphere. Boundary-Layer Meteorology 78:191-210.

Xu, Y and Burfoot, D. (2000). Modelling the application of chemicals in box potato stores. Pest management Science 56:111-119

Xu, Y., Burfoot, D., and Huxtable, P. (2002). Improving the quality of stored potatoes using computer modelling. Computers in Electronics and Agriculture 34:159-171.

Yao, C, Arya, S. P., Davis, J, and Main, C. E. (1997). A numerical model of the transport and diffusion of *Peronospora tabacina* spores in the evolving atmospheric boundary layer. Atmospheric Environment 31:1709-1714.

Chapter 5

The germinating spore as a contaminating vehicle

Gilma Silva Chitarra[1$] and Jan Dijksterhuis[2]

[1]*Laboratory for Food Microbiology, University of Wageningen, The Netherlands;* [2]*Applied and Industrial Mycology, CBS Fungal Biodiversity Centre, Uppsalalaan 8, 3584 CT, Utrecht, The Netherlands.*

[$]*Present Address: Rua Rio de Janeiro 832, Bairro Nova Várzea Grande,Várzea Grande- MT CEP 78 135 710, Brazil.*

INTRODUCTION

Fungi can be found in a wide variety of environments, such as in seeds, plants, soil, water, insects, food and food products, and animal products. Phytopathogenic fungi cause damage to the living crops upon storage as *Colletotrichum* that causes anthracnose disease in several fruits and vegetables such as banana, avocado, papaya, and tomato, decreasing their commercial values. Fungal infection of grain, nuts and fruits is often preceded by physical damage caused by insect invasion or mechanical injury during harvest. Fungal growth reduces the nutritional value of storage grains and animal feed and can result in the production of mycotoxins (D'Mello and MacDonald, 1997). Mycotoxins are poisonous, often carcinogenic secondary metabolites of fungi, which are associated with certain disorders in animals and humans (for *Fusarium* on grain see for instance, D'Mello *et al.*, 1998; Reid *et al.*, 1999).

Food products also become contaminated during processing and handling operations. Processed food can be considered as a complex often plant-based medium that fungi colonise and spoil. Fungal species associated with particular foods correlate with the characteristics and properties of the product (Dijksterhuis and Samson, 2002; Filtenborg *et al.*, 2004). The primary cause for the deterioration of rye bread for example are the fungi *Penicillium roqueforti, P. paneum, P. carneum* and *Paecilomyces variotii.* Contaminated commodities, such as cereals, can deteriorate during storage, resulting in enhanced contamination levels of whole wheat flour (Weidenborner *et al.*, 2000). In food products, the issue of mycotoxins requires continuous attention, but more recently fungal spores are also increasingly recognized as aeroallergen sources (Green *et al.*, 2005). Fungal contamination and the toxic metabolites it forms cause massive economic losses of food. There is a great interest among agricultural, food industrial and medical disciplines to prevent or control fungal contamination. These include different techniques that manipulate the physical environment of the fungus including acidification, increase of the osmotic potential, drying, cooled storage, pasteurisation and the use of modified atmospheres. Some fungal species are able to grow at such adverse conditions and are able to thrive at situations that are meant to be free of spoilage.

Contamination and colonisation of the food products is often by means of survival vehicles including airborne spores. Fungi are known for their capability to produce sexual and/or asexual spores as agents of reproduction, dispersal and survival. Some fungal species predominantly form sexual spores as *Talaromyces* species even without the need of different mating types (homothallic) and ascospores are produced in high numbers, while there is only restricted production of asexual spores. Alternatively, many fungal species do not have a well recognised sexual stage and are designated as the Deuteromycetes (mitosporic fungi). This group includes many members of genera as *Aspergillus, Penicillium* and *Fusarium,* which are very relevant fungi for food situations (Dijksterhuis and Samson, 2002). Spores play an important role in the life cycle of fungi acting as dispersal or survival spores. Dispersal spores are separated completely from the parent mycelium by different factors to facilitate migration to a new site. They have a moderate capacity for survival in a resting state (dormancy). They are also capable to germinate readily in the presence of nutrients or favourable environmental conditions (Griffin, 1994). In case of *Aspergillus* and *Penicillium*, conidia are formed in chains on specialised spore-forming cells (phialides). Mature conidia have to survive in dry conditions during dispersion through the air current (Dijksterhuis and Samson, 2002). In contrast, survival spores are often produced in lower numbers and may not be separated from the parent mycelium (Carlisle *et al.*, 1994). As an example, thick-walled chlamydospores are produced by, e.g., *Mucor racemosus, F. culmorum* and *Paecilomyces variotii* and typically produced between hyphal cells. Besides, many ascospores are formed inside closed or open fruit bodies (ascomata) that reside within the mycelium and not on specialised structures (conidiophores) that enable the spores to be distributed by air- or water currents. Many fungal species are able to produce different types of spores within one colony as is the case with for example *Fusarium* species (microconidia, macroconidia and chlamydospores) and *Eurotium* species (conidia and ascospores) (Samson *et al.*, 2004).

As is stated above, fungal contamination of foods and food products and colonisation and infection of plants and animals is usually initiated by contact of the host with spores (conidia). Contamination by the external environment, e.g., air, water, walls and floors for instance is considered to be the main source of contamination of beef carcasses with *Penicillium, Aspergillus, Mucor* and *Cladosporium* species (Ismail *et al.*, 1995). Additionally spores can be brought on the crop or food product via an encounter with organisms (insects, mites). The germination process is the beginning of fungal colonisation into food and on plants or animals. It involves the initiation of biochemical activities, with an increase of the metabolic rates and induction of morphological changes (Griffin, 1994; D'Enfert and Fontaine, 1997). A better understanding of spore survival and the different processes of spore germination could lead to novel techniques to prevent food spoilage. This chapter describes the germination process of fungal spores and the relation between germination and fungal contamination, mycotoxin production, control methods and the mode of action of antifungal agents. The problem of fungal contamination can be partially confronted with the use of fungal inhibitors of germination and hyphal growth, but spores are less sensitive to different compounds. It is here, that the terms fungistatic and fungicidal have a different meaning. Germination of a spore includes a continuous change from a "stasis"-like situation towards a vegetatively growing hyphal cell expressing processes as active metabolism, expanding cell mass and nuclear division.

LANDING, ADHESION AND WETTING OF THE CONIDIA

The first events of fungal colonisation are the landing of the spores on the substratum and subsequent hydration. Airborne spores are cells that have to deal with drying and rewetting and certainly will possess mechanisms that address the redistribution of cell components that accompany these changes. In *Magnaporthe grisea*, the conidia that are transported through the air have a collapsed appearance as a result of dehydration and this stage is regarded as a normal part of the life cycle of the cell and not as an artefact due to preparation of the cells. After rewetting these conidia retained their turgid shape (Howard, 1993). A similar feature is visible with dry rust spores immediately after contact with the leaf surface (Deising *et al.*, 1992). Upon landing, attachment of the spore is important especially in case of the colonisation of plant surfaces, which often have a hydrophobic nature.

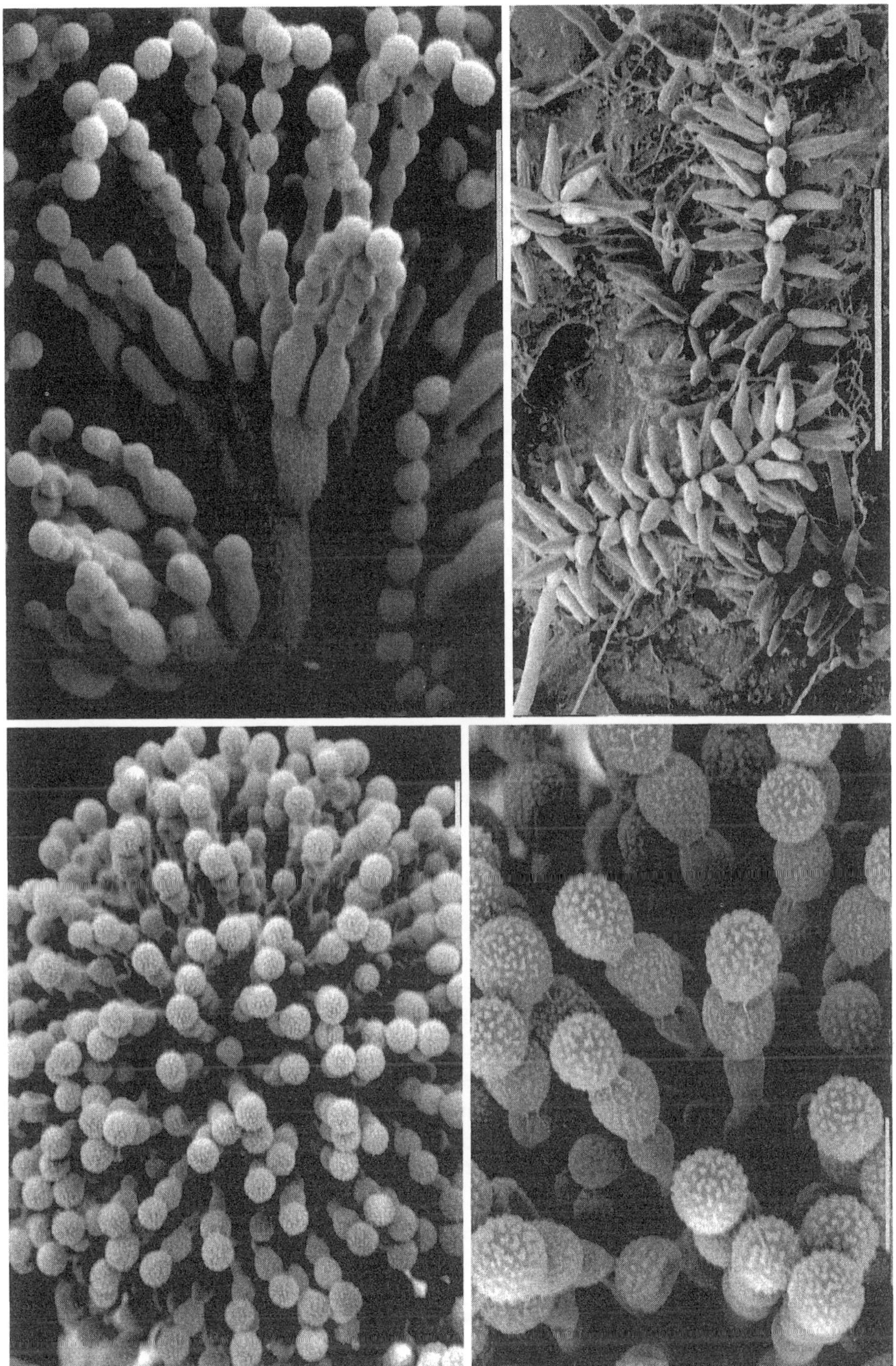

Figure 1. Formation of spores on specialized structures. Top left, a conidiophore of *Penicillium bialovaiense* where spore-forming cells (phialides) are clearly visible. Top right, multicelled spores of the fungus *Helminthosporium solani* formed on a spore-bearing structure that is located on potato skin. Bottom left, *Aspergilllus oryzae* forms numerous conidia on a conidiophore where many strings of conidia form a spherical structure. Bottom right, detail of the previous micrograph with visible ornamentation on the conidia.

Some spores possess a droplet of adhesive material as conidia of *M. grisea* that bear a spherical droplet on the tip that literally glues the spore to the leaf surface when it is transported to it (Hamer *et al.*, 1988).

Other spores do not have these appendages and attachment must be reached by other means. In case of urediniospores of *Uromyces viciae-fabae,* 10-25% of the spores were able to adhere immediately to different surfaces and were not removed by sterile distilled water (Clement *et al.*, 1993). Attachment was higher on hydrophobic surfaces and increased after more than 20 min, which indicates that during development of the spore the attachment to different surfaces increased indicating a "vulnerable" stage where spores can be removed easier. Immediate adhesion of *Botrytis cinerea* conidia was studied by Doss *et al.* (1993) and it was found that dry conidia adhered for 15% to tomato cuticle, but after vapour hydration for 2 min, 93% of the conidia were adhered (were resistant to a jet of nitrogen gas). Washing with water resulted in 37% adhered spores on the cuticle (and 9% on glass). The process was not influenced by lectins or proteases, but detergents had a strong inhibiting effect on adhesion to both hydrophobic substrata polystyrene and tomato. The authors conclude that adhesion is a two-step process with "passive" adhesion mediated by hydrophobic interactions (and also occurring with killed conidia) and subsequent stronger delayed adhesion during swelling and germ tube formation. Upon contact and wetting of the urediniospores *Uromyces viciae-fabae* material accumulation between the spore-cuticle (in this case bean leaves) interface became apparent (Deising *et al.*, 1992). This material dubbed "the adhesion pad" plays a role in firmer attachment to and degradation of the (wax) cuticle of the host. The tenacity of adhesion to artificial surfaces of uredospores of *U. appendiculatus* was correlated with the extent of hydrophobicity (Terhune and Hoch, 1993), which was measured after more than 30 min. The black rot fungus of grape, *Phyllosticta ampelicida,* exhibits a complex adhesion pattern towards different substrates including electrochemical attraction (spores have a polyanionic surface, i.e., it is negatively charged) and hydrophobic interactions. Remarkably, attachment was a prerequisite in this fungus for subsequent germination, indicating that these phenomena were linked (Kuo and Hoch, 1996). In a number of cases, immediate attachment to the substrate was seen (within seconds) and factors of the imbibing solution (as acidity) were vital here.

Filonov (2001) confirmed that *B. cinerea* conidia become more firmly attached to apple skin during swelling and germination. A short ultrasonication treatment removed >95% of the conidia immediately after addition. During the first 4 h 80% of the spores were recovered, but during the formation of germ tubes attachment to the substrate had increased strongly with 70% attachment after the treatment after 24 h. Filonov (2003) further studied the adhesion and germination of conidia of different fungal species on polycarbonate membranes. Adhesion was assayed after 24 h which means that spores could have germinated very well and that adhesion of germlings is assayed in such a case. From these studies it became clear that adhesion/germination of *Penicillium expansum, claviforme* and *roqueforti* as well as *Botrytis cinerea* was markedly higher and influenced through by the presence of acetate esters. It is tempting to assume that the fruit rotters, *P. expansum* and *B. cinerea,* react on volatile constituents of fruits, while adhesion of only these fungi had increased on apple skin with additional esters present in the air. It was also clear that fresh wounds on apples captured 80-100% of the spores compared to approx. 20% on the skin and that the age of the wound was correlated with the firmness of attachment after 4h. Wounds of 24 h old age exhibited 40% recovery of *B. cinerea* conidia after sonication treatment at 150 W for 10 s while fresh wounds did not show recovery. In case of *P. expansum* this was 60% vs. 20% of the conidia. This observation is very relevant for postharvest problems while disinfection of wounds directly after formation is in fact very important (see also Filonov, 2004).

The act of wetting alone leads to changes in the conidium as is observed with *Neurospora crassa* (Bonnen and Brambl, 1983). They observed an increase in the fraction of polyribosomes correlated with water harvested conidia while cells obtained in an isoparaffinic hydrocarbon fluid had the same levels of these structures as dry-harvested conidia. This shows that cellular constituents change immediately after contact of the cytoplasm with water. In this case protein synthesis might occur quickly after wetting of the cells. Incubation of sporangiospores of *R. oligosporus* after a 2 hour storage period in buffer (pH 4) showed metabolisation of cFDA (carboxyfluorescein diacetate) and germ tube formation in a subpopulation of the cells after 4 hours (Thanh and Nout, 2004), which indicate that spores develop under very poor nutrient conditions upon wetting.

FURTHER STAGES OF GERMINATION

In general water and nutrients are important requirements for proper germination. Many fungal species need external addition of these nutrients for optimal germination, other species often related to plant (leaf) surfaces do germinate in distilled water and have internal deposits of nutrients (as rust-fungi do, e.g., *Uromyces vignae,* see Dijksterhuis, 2003). Leaf pathogenic fungi as *Colletotrichum* species and *Magnaporthe grisea* need a hard surface as one of the requirements for germination and appressorium formation. Similarly, *Botrytis cinerea* germinates not only on glass surfaces that are hydrophic, but also on rich media. In the latter case conidia rapidly germinate with long germ tubes that soon branch (Doehlemann *et al.*, 2006).

Addition of phosphate, amino acids, glucose and combinations of the compounds resulted in increased germination in case of sporangiospores of *Rhizopus oligosporus* (Thanh and Nout, 2004; Thanh *et al.*, 2005). Besides, also physical factors can invoke germination; sporangiospores of *Phycomyces blakesleeanus* are activated to germinate by a heat treatment at 50 °C (van Assche *et al.*, 1972). When proper nutrients are available, the spores continue to develop, which results in isotropic growth also designated as swelling, which is observed in numerous fungal species. *Fusarium culmorum* and *Rhizopus* spores require a carbon and nitrogen source for development. *Penicillium griseofulvum* and *Aspergillus nidulans* conidia need glucose for germination (D'Enfert and Fontaine, 1997; Osherov and May, 2001). In addition, other low molecular weight nutrients as for example inorganic salts can activate germination (Griffin, 1994). Uptake and metabolisation of the probe carboxyfluorescein diacetate (cFDA) was strongly increased after the introduction of dried sporangiospores of *R. oligosporus* in malt extract at 37 °C (Thanh and Nout, 2004; Thanh *et al.*, 2005). This was interpreted as a monitor of the beginning of the germination process. After long drying periods (11 months) the spores did show no colony formation on 2% glucose alone (<1%). More complex media as malt extract, peptone, yeast extract and glucose/peptone medium resulted in much higher numbers of germinated spores (33-36% of the spores) and colonies. The use of the fluorescent probes propidium iodide (PI) indicated that a large subpopulation of the dried spores show PI-related membrane permeabilisation and DNA staining (Thanh *et al.*, 2006). However, the dye TOTO-1 was not observed inside the cells. This is remarkable while both PI and TOTO-staining inside the cell was regarded as an indicator of cell death. When dried spores were pre-inoculated in malt extract broth, the majority of the spores stained with cFDA and therefore were metabolically active. This is evidence for a regeneration of a damaged cell population. The fraction of PI positive-TOTO negative (thus damaged) spores increased with storage time. The requirements of germinating spores may clearly differ from other stages of the fungal lifecycle. With *R. oryzae*, sporangiospores germinate readily in malt pepton medium, but germination decreases below a pH of 4.8 (J. Dijksterhuis, unpublished results). Optimal germination was observed at 30 °C, while the highest radius of colonies is observed in case at 35 °C.

The first obvious change in spore morphology in many fungal species is isotropic growth, also designated as swelling which is observed in the case of *Penicillium* and *Aspergillus* species (M.R. van Leeuwen, CBS, unpublished results) and *Fusarium culmorum* macroconidia (Chitarra *et al.*, 2005a), where the spore starts to swell and consequently increases its volume. Swelling is not merely water uptake, it is also characterised by changes in the composition of the cell, cell wall growth, and increase in dry weight (Bartnicki-García and Lippman, 1977). Isotropic growth is accompanied by numerous metabolic activities including respiration, RNA and protein synthesis (van Etten *et al.*, 1983; Ojha and Barja, 2003), and degradation of trehalose into glucose (Osherov and May, 2001).

Following swelling, cell wall deposition becomes polarized, and the extension occurs at a restricted area at the tip of the developing germ tube (Parton *et al.*, 1997). Momany (2002) distinguished different stages of spore germination. First initiation included breaking of dormancy and the start of isotropic growth. A phase of isotropic growth, which is roughly between 3 and 7 hours is followed by the establishment of an area of polarised growth, which includes the proper positioning of cell wall deposition and directioning of the vesicle transport machinery in which the cytoskeleton, different proteins and the plasma membrane show a precise interplay (Cheng *et al.*, 2001). This cooperation results in the outgrowth of a germ tube and later, the formation of a branching mycelium. Extensive studies have been carried out on the germination of unicellular spores, e.g., *Colletotrichum, Aspergillus, Penicillium* and *Rhizopus,* but hardly anything is known about germination of multicellular conidia. (Bourret, 1986; Breeuwer *et al.*, 1997; Marin *et al.*, 1998; Chaky *et al.*, 2001; Leandro *et al.*, 2001).

The different stages of conidial development seem to be linked to the different stages of the cell cycle (Harris, 1999). Isotrophic growth takes place in *A. nidulans* until the first mitosis. After mitosis an axis of polarity is established and maintained in the emergence and elongation of the germ tube (Momany, 2002). Mitosis is also associated with septum formation at the base of the emerged germ tube. Between *A. fumigatus* and *A. nidulans* interesting differences were observed in timing of polarity establishment related to the mitotic state. Pear shape (germ tube emergence) was observed in 22% of the conidia before the first mitosis in case of *A. fumigatus* and not with *A. nidulans* (Momany and Taylor, 2000). Similar differences are observed with septation and the emergence of a second germ tube and these morphogenetic changes are also related to the nutrient status of the medium. The authors mention that a critical size of cell volume could be an important factor in septum formation. Remarkably, Dijksterhuis *et al.* (unpublished results) have found that a gradual decrease occurs in conidia of different *Penicillium* species with respect to fluid phase viscosity of the cytoplasm to a level that is typical for vegetative cells. In such a case a global and physical parameter might induce cell changes.

SIGNALLING DURING EARLY GERMINATION

Different signalling factors are involved with germination of conidia. In sporangiospores of *Pilobolus longipus,* glucose resulted in a rise of cAMP before germination (Bourret, 1986) and the role of this signalling pathway was also recognised in yeast ascospores (Thevelein, 1984). Fillinger *et al.* (2002) studied the role of adenylate cyclase in the cAMP signalling pathway as well as the downstream kinases schA and pkaA during germination of conidia of *A. nidulans*. A double mutant of pkaA and cyaA (the adenylate cyclase) and the single cyaAΔ mutant exhibited delayed conidial germination (30% in 15 h), but certainly not a complete arrest. Trehalose degradation was blocked in the cyaAΔ and the schaAΔpkaAΔ mutants. This indicates that individual signalling elements of the adenylate cyclase sequence play a role in different aspects of germination, but that they have several targets, which results in the operation of a signalling network. Changing the activity of the signal mediator Ras to dominant activity led to blocked germ tube formation and resulted in prolonged swelling and multiple nuclei. This Ras-pathway operates independent of the adenylate cyclase pathway/network and dominant activity of Ras results in a defect of polarity establishment. Ras is a member of the small GTPase family and plays an important role in the communication inside different signalling networks in the cell. Two different GTPase types, a ras and a rho type were studied in the dimorphic fungus *Penicillium marneffei* (Boyce *et al.*, 2005). For this study dominant negative and dominant positive transformants were used, while a deletion could not be generated (lethal?). The dominant negative $rasA^{D125A}$ and the dominant activated $rasA^{G19V}$ both showed less germination after 12 h and the authors state that an increase proportion of the cells are misshapen. This indicates that activity of these molecules above and below a certain level has a bearing on germination. Dominant activation of a rho GTPase and CDC42 homologue $cflA^{G14V}$ did undo the effect of $rasA^{D125A}$ and resulted in high germination again. Another rho GTPase named *cflB* was deleted and showed some disturbance in conidial germination including not complete germination after 12 h, but a somewhat higher incidence of secondary germ tubes at that stage (Boyce *et al.*, 2003).

Zuber *et al.* (2003) studied the effect of alterations of the G-protein α-subunit on germination of conidia at 25 °C. This is also a molecule that cycles between a GTP-bound active (signalling) state and a non-signalling Gα-subunit. Germination rate was lowest (25% at 27 h) in the Δ*gasC* mutant, also delayed in the $gasC^{G207R}$, which is non-signalling (25% at 15 h), but was accelerated in the dominant active $gasC^{G45R}$ compared to the wildtype (35% vs. 25% in 10 h. These features remained similar under carbon poor situations.

Similarly, in *A. nidulans,* conidia showed enhanced germination with a constitutively active $ganB^{Q208L}$ mutant and was lowest in the $ganB^{G207R}$, which is kept in the inactive state (Chang *et al.*, 2004).

Calcium is a factor that plays an important signalling role in cells that settle on a surface. These can be Oomycetes as *Phytophthora parasitica* (Warburton and Deacon, 1998) or fungal species as *Phyllosticta ampelicida* (Shaw and Hoch, 2000) or *Colletotrichum gloeosporioides* (Kim *et al.*, 1999). The latter two species need a hard surface to germinate and also this feature has to be communicated into the cell. Doehlemann *et al.* (2006) have studied germination in *Botrytis cinerea* and found that a disrupted Gα3 subunit reduced fructose dependent germination to approximately 20% of the cells over a long period, but this mutant germinated like the wildtype on a hydrophobic polypropylene surface. Deletion of the MAP kinase BMP1 resulted in no germination at all on hydrophobic surfaces. Buhr and Dickman (1997) observe maximum expression of serine-threonine kinase, calmodulin and protein kinase C prior to germ tube morphogenesis of *C. trifolii,* which illustrates that during germination many different factors play a role and a complex interplay of signalling routes may depict this picture to the cell.

COMPATIBLE SOLUTES IN CONIDIA

Accumulation of compatible solutes inside living cells is thought to protect cells against osmotic stresses and it also is observed after an oxidative or heat shock in germinating conidia of *Aspergillus nidulans* (Fillinger *et al.*, 2001). These compounds do not disturb the functioning of proteins and other biomolecules and the complexes formed by them when they are present in high amounts inside the cell, henceforth the name *compatible* solutes (see also Dijksterhuis and Samson, 2002). Trehalose, an α-1,4 non-reducing disaccharide including two linked glucose moieties (α-D-glucopyranosyl-α-D-glucopyrano-side) is an important compatible solute and protects both membranes and proteins (Crowe *et al.*, 1984; Hottiger *et al.*, 1994; Prestrelski *et al.*, 1993; Wolkers *et al.*, 1998) against drying and heat. Trehalose is synthesised in yeast cells from glucose by the action of trehalose-6-phosphatase (encoded by *TPS1*) which links two phosphorylated glucose molecules to each other via a UTP-bound energizing step. The resulting trehalose-6-P is dephosphorylated by means of a trehalose-6-phosphate phosphatase (*TPS2*).

Germination of conidia is associated with a degradation of the trehalose pool from 1,2 pg per spore, to zero within 120 min and this phenomenon is observed with different types of spores (Thevelein, 1984; D'Enfert *et al.*, 1999; Dijksterhuis *et al.*, 2002). Assuming a cell diameter of approximately 3,0 μm (based on conidia of *A. niger* studied by Tiedt (1993) and a density of conidia above 1,0 gr/ml (otherwise the cells could not be centrifuged so quickly) this results in a weight of one spore of minimally 20 pg meaning that trehalose accumulates to maximally 6,7 % of the cell wet weight. In *A. nidulans* trehalose degradation is performed by an neutral cAMP-activated and calcium-dependent neutral trehalase. Germinating conidia of 2 h show 80% decrease in colony forming units after 20 min at 50 °C. A mutant defective of neutral trehalase activity retained levels of trehalose at the maximum level and showed nearly unchanged germination and full heat tolerance for at least 40 min at this temperature (D'Enfert *et al.*, 1999). Conidia of *A. nidulans* that had germinated for three hours at 30 °C and shifted to 50 °C showed accumulation of trehalose within 30 min to a level of maximally 0,8 pg/spore (Fillinger *et al.*, 2001). Lower accumulation was observed after addition of 100 m*M* H_2O_2. A tpsAΔ strain of *A. nidulans* was, unable to produce trehalose and, surprisingly, the wildtype and mutant showed a similar sensitivity for the stressors and the accumulation of trehalose did not increase the survival of the germlings during short-term exposure. However, germinating conidia of the mutant showed very low colony formation when stored for approx. 15 hours at 44 °C, while the wildtype showed no significant decrease. Sustained storage of conidia at 20°C showed gradual decrease in viability during a period of 20 days in case of the mutant, but with maximal germination by the wild type after 50 days. Of course, trehalose only can perform its function when it is present in the cell, it cannot restore the damage done to the germling after a heat shock. Yeast cells that were treated with a short heat shock showed a increased *acquired* tolerance to subsequent heat treatment that was associated with the presence of trehalose (De Virgilio *et al.*, 1994). Combined, these data strongly suggest that the presence of trehalose provides protection against different types of stress and also plays an important role in the longevity of the life of the conidium.

Surprisingly, the *A. nidulans* conidia also contain 0,8 pg mannitol per spore (4% wet weight) which is degraded to zero in 3 hours during early germination. Also in conidia of the related fungus *A. niger*, mannitol is an important compatible solute (Ruijter *et al.*, 2003). Mannitol is produced by the action of two enzymes mediating a reduction and a phosphatase activity from fructose 6-phosphate via mannitol 1-phosphate. Conidia of *A. niger* contain 10,9% dry weight mannitol and assuming a percentage of water inside the spores of 50% or more this would be approx. 5,5% mannitol (wet weight). For a comparison, stress-resistant ascospores of the fungus *Talaromyces macrosporus* contain approximately 38% water and are regarded as very dense (Dijksterhuis *et al.*, 2002). The *A. niger* conidia contain a somewhat smaller quantity of trehalose (3,6% dry weight). The ΔmpdA strain of *A. niger* that is deficient for the mannitol 1-phosphate dehydrogenase has more trehalose (11.5% dry weight) and reduced mannitol levels (4.0% dry weight). Mutant conidia show 90% viability loss after nearly 1 hour of heating at 50 °C while the wildtype survives easily 2 h at this temperature. Further, conidia are more sensitive for a freeze-thaw step, lyophilization and a hypo-chlorite treatment. Interestingly, there is no difference in long-term storage between the wildtype and the mutant, which indicated that trehalose and not mannitol plays an important function in this respect. These data combined may suggest that only a combination of trehalose and mannitol, that are both present in conidia of the two species in approximately equal amounts, give protection against different stressors. The data may also suggest that another cell mechanism is connected with the processes of compatible solute accumulation, for instance, the expression of heat shock related proteins.

Hallsworth and Magan (1994, 1996) provided clear evidence that the growth conditions of the spore-delivering culture in the fungal species *Metarrhizium anisopliae, Beauveria bassiana* and *Paecilomyces farinosus* strongly influences the accumulation and composition of compatible solutes inside the conidia. These species are insect pathogens and the spores of these fungi were used for biocontrol of insect pests and proper storage and survival of the conidia was of great importance. In control situations (in this case Saboraud Dextrose Agar), mannitol was the most dominant solute in the three species (Hallsworth and Magan, 1996). The presence of trehalose, glycerol or starch in the growth medium of the three species highly influenced the internal composition of the spores with mannitol, glycerol, erithr(e)itol and trehalose as main players with total solute levels between 10 and 20% dry weight (Hallsworth and Magan, 1994). These authors observed that glycerol and erythritol dominance inside the spores was correlated with (faster) germination in case of lower water activities (Hallsworth and Magan, 1995). *A. nidulans* conidia were also tested (Hallsworth *et al.*, 2003) from PDA (potato dextrose agar) media with excess glycerol or KCl. All growth conditions resulted in mannitol levels of 4.4-4.6% dry weight, but glycerol containing medium also showed 6.3 and 2.7% glycerol and erythreitol, two other important compatible solutes in fungi. The KCl conidia showed intermediate levels of the latter (0.35 and 0.64%), and 0.084% and 0.21% in PDA grown cells. The authors observed protection of germination of conidia that contained high levels of ethanol and erithreitol in the presence of ethanol and NaCl (up to 7.5 and 16% wt/vol, respectively). *Penicillium chrysogenum* was grown on pearl barley by Ballio *et al.* (1964) and the harvested conidia showed 10% and 8.3% mannitol and trehalose, respectively (dry weight) and 3.0 and 2.7% glycerol and erythreitol which confirm the typical levels of these solutes inside this type of spores. Table 1 summarizes very shortly the functions correlated with the different compatible solutes discussed till now.

Table 1. Compatible solutes and their function in conidia

Compatible solute	Function inside spore
Mannitol/Trehalose	Protection against heat
Trehalose	Longevity
Glycerol/Erithreitol	Protection during germination at low α_w

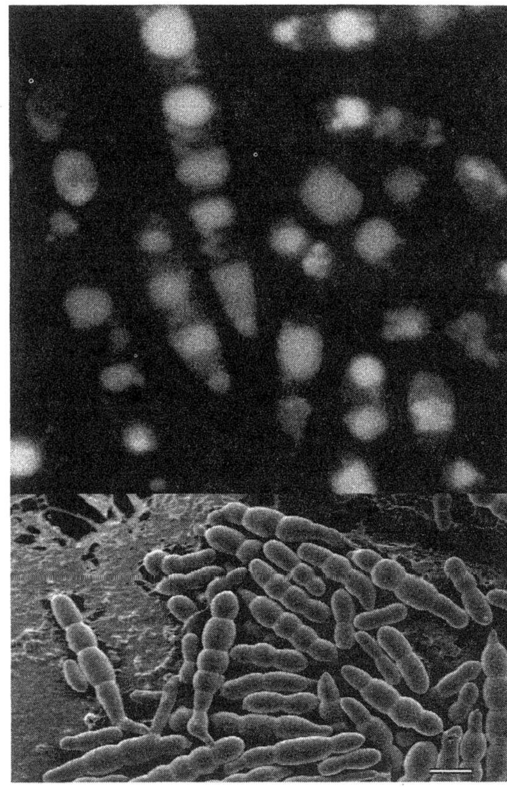

Figure 2. Multi-compartment macroconidia of the fungus *Fusarium culmorum*. Top, cells stained with a fluorescent dye (cFDA-SE), which is used for the measurement of the internal pH of the cells. The cells are studied with the confocal microscope. Bottom, cryo-electron microscopy of germinating macroconidia that clearly show isotropic growth of the individual compartments.

MULTICELLULAR CONIDIA AND INTERNAL pH.

Multicellular conidia are formed by a large number of fungi among them are a number of species very relevant in food situations. *Fusarium* species are important food related (cereals) and plant pathogenic mycotoxin forming fungi. *Alternaria* species are often observed on cereals and form mycotoxins and are related to allergenic reactions. *Helminthosporium solani* causes silver scab on stored potatoes and *Magnaporthe grisea* is the most important pathogen of rice. Germination of these multicellular conidia is only scarcely studied and one could argue if the different cells of such a spore behave as individual conidia "glued" together or do exhibit a certain differentiation. Is there any ecological reason to produce these structures? It was known that in conidia *M. grisea* and *F. culmorum* germ tubes developed preferably from apical cells and seldom from middle cells (Jelitto, 1999; Atkinson *et al.*, 2002; Chitarra *et al.*, 2005a). The last authors studied *F. culmorum* as a model system of multicellular conidia by means of FRIM (fluorescence ratio imaging) where the internal pH inside spores was followed during germination.

The pH_{in} for *F. culmorum* (pH_{in} 6.4; Chitarra *et al.*, 2005a) was higher than *P. paneum* (pH_{in} 5.4; Chitarra *et al.*, 2005b), but both within the pH range previously reported for *Rhizopus oligosporus,* which is between pH 5 to 6.5 (Breeuwer *et al.*, 1997). The germination of *P. paneum* conidia and *F. culmorum* macroconidia under optimal conditions included both isotropic growth and an increase of intracellular pH of both types of spores and no statistical significant differences among compartments of *F. culmorum* was observed during early stages of swelling. During further stages of swelling and germ tube formation, differentiation between the compartments of the macroconidium was statistically sound. After the first stages of swelling, the ungerminated middle compartments of *F. culmorum* showed a significant decrease in pH_{in}, which was not associated with an increase in the fraction of vacuoles in the cells. The latter was observed with *M. grisea* conidia (Atkinson *et al.*, 2002). The monitoring of the internal pH in multicelled conidia showed higher pH values in apical cells and certainly inside the germ tubes. The pH_{in} in germ tubes of *F. culmorum* (>7.2) was in agreement with the pH_{in} of *M. grisea* germ tubes (7.4) incubated in complete nutrient medium stained with the dye SNARF-1 (Jelitto, 1999) or with the cytoplasmic pH found with dextran conjugated dye in hyphae of *Neurospora crassa* (Parton *et al.*, 1997). Changes in pH may be associated with differentiation processes as has been reported elsewhere (Inouye, 1985; Stewart *et al.*, 1988). These studies do not establish whether an increase in pH_{in} occurs prior to or if it is a product of the metabolic changes occurring inside the cell. Controversial results of internal pH gradient in tip growth hypha have been previously reported (Roncal *et al.*, 1993; Jelitto *et al.*, 1994; Robson *et al.*, 1996; Parton *et al.*, 1997).

Taken together these observations suggest that differentiation occurs between the different compartments. The inclination of the apical cells to germinate over middle cells may be started by the difference between the surface-to-volume ratio of apical cells. This may facilitate differences in, for instance, transport processes and henceforth introduce an asymmetry in the development of the conidium. The consistent germination pattern of the macroconidium was changed after treatments of the conidia with sublethal doses of nystatin. Apical cells were preferably targeted by the compound and the distorted germination was counteracted by an increased germination of the middle compartments. This shift indicates a way of communication between the cells and a type of "apical dominance" may be alleviated from the middle cells. The ecological function of such differentiation and communication may be sought in the more versatile response of these cells towards adverse conditions for germination. When the first germination fails, another attempt can be made later and if conditions remain unattractive for colonization, the middle cells may differentiate further to long-survival spores as chlamydospores and this was already observed decades ago (French and Nielsen, 1966; Schneider and Seaman, 1974).

FUNGAL GERMINATION AND SELF INHIBITORS

Fungi produce substances during growth that influence their own development. These can be inhibitory substances and then are named self-inhibitors. These compounds inhibit germination of spores or growth of hyphae. For example, germination of spores of *Rhizopus oryzae* shows lower germination when they are present in higher densities. Germination lowers from 67 to 22% after 4 hours of incubation when the density of the spores increases from 10^6 to 7×10^7 spores/ml (J. Dijksterhuis, unpublished results). Self-inhibitors have been characterised in many fungal (and non-fungal) genera *Puccinia, Uromyces, Colletotrichum, Dictyostelium, Fusarium* and *Aspergillus* and can be volatile or non-volatile (see for instance, Allen, 1955, Bacon and Sussman, 1973, Barrios-Gonzalos *et al.*, 1989). Various self-inhibitors have been isolated and identified after extraction from culture filtrates of fungi (Table 2).

Self-inhibitors also can influence other fungal processes, for example, mycosporine-alanine produced by *C. graminicola* prevents appressorium formation (Leite and Nicholson, 1992, 1993). The self-inhibitors produced by *Glomerella cingulata* and *Dictyostelium discoideum* (not strictly a fungus, but this illustrates that concept of self-inhibition might be widely spread) inhibit protein synthesis (Bacon and Sussman, 1973; Lingappa *et al.*, 1973). Self-inhibitors must inhibit spore germination in a reversible manner, after removal of the compound from the spore or its environment, germination is initiated (see also Chapter 1 of this book). The major function of self-inhibitors is stated as prevention of premature germination of spores directly after formation when they are located at conidiophores, inside fruiting bodies or on pustules (in case of rust-fungi) and before spore dispersion. This mechanism guarantees that spores only germinate after dispersal into the environment that favour outgrowth to establish a mycelium.

Breeuwer *et al.* (1997) studied the mode of action of the self-inhibitory compound nonanoic acid in sporangiospores of *Rhizopus oligosporus*. Nonanoic acid results in both a decrease in internal pH and a lower number of metabolic active cells, but this effect is transient and restoration of the internal pH to normal levels occurs at a concentration of 1 m*M*. The mode of action of this compound is compared to that of weak organic acids that are used as food preservatives, like sorbate, propionate and acetate. Also spores of other fungal species show similar phenomena in the presence of nonanoic acid.

The intracellular pH of macroconidia of *Fusarium culmorum* fluctuated between 5.4 and 6.5 in the presence of nonanoic acid during a period of 90 minutes at an extracellular pH of 4.0 (Chitarra *et al.*, 2005a). The disturbed or fluctuated intracellular pH was recovered twice, indicating that macroconidia had energy

Table 2: Self-inhibitors from fungi

Fungal species	Chemical compound	References
Aspergillus niger		Krishnan, 1954; Barrios-Gonzales, 1989
Anisogramma anomala		Stone *et al.*, 1994
Blastocladiella emersonii		Adelman and Lovett, 1974
Colletotrichum capsici		Louis *et al.*, 1988
Colletotrichum gloeosporioides	Gloeosporone	Lax *et al.*, 1985
Colletotrichum graminicola	Microsporine-alanine	Leite and Nicholson, 1992
Dictyostelium discoideum	N,N-dimethylguanosine	Bacon *et al.*, 1973
Fusarium oxysporum	Nonanoic acid	Garrett and Robinson, 1969
Geotrichum candidum		Steele, 1973
Glomerella cingulata		Lingappa *et al.*, 1973
Hemileia vastatrix	Free organic acid	Musumeci *et al.*, 1974
Microsporum gypseum		Page and Stock, 1971
Penicillium griseofulvum		Fletcher and Morton, 1970
Peronospora tabacina	5-Isobutyroxy-β-ionone	Leppik *et al.*, 1972 Page and Stock, 1971
Puccinia graminis var *tritici*	Coumarins and phenolic acids Methyl-*cis*-ferulate	Sumere *et al.*, 1957; Macko *et al.*, 1971a
Puccinia helianthi; P. antirrhini	Methyl-3,4 dimethoxycinnamate	Macko *et al.*, 1971b
Syncephalastrum racemosum	Nonanoic acid	Hobot and Gull, 1980
Tiletia caries	Trimethylalanine	Trione, 1973
Uromyces phaseoli var *typica*	Aspartic and glutamic acid	Wilson, 1958; Stone *et al.*, 1994; Steele, 1973

enough to pump excess protons out of the cell.

In addition, swelling and germ tube formation of the conidia of *Penicillium paneum* was inhibited and transient collapse of the internal pH of the spores was also observed (Chitarra *et al.*, 2005b).

Recently, a volatile self-inhibitor, 1-octen-3-ol, was identified in case of the fungus *P. paneum* (Chitarra *et al.*, 2004; Chitarra *et al.*, 2005b) that blocked swelling and germination of conidia at a millimolar (4 m*M*) concentration range and approximately 70% of the conidia had the same size as freshly harvested conidia after 4 h, while 80% of the control cells was clearly swollen. 1-Octen-3-ol was initially identified in and above dense suspensions of conidia. Small droplets of very dense (10^9 spores/ml) conidial suspensions placed on thin agar layers showed less than 10% germination after 24 h, which indicates a clear crowding effect. There was some entering of the fluorescent indicators PI and TOTO into the conidia in the presence of 1-octen-3-ol (in case of 20 and 10%) which indicates a mild permeabilization of the plasma membrane. In addition oxygen consumption was slightly lowered and a transient drop in internal pH was observed. Taken together these observations suggest that 1-octen-3-ol has a mild systemic effect on the developing conidial cells. Surprisingly, there were notable differences in the composition of the protein population of treated cells after 5 h compared to the controls. So, despite its mild physiological effects, a profound influence on protein expression was observed.

1-Octen-3-ol also inhibits other fungal life stages including radial growth of the mycelium of different fungal species. Further, microcycle conidiation was observed in the presence of the compound. One could suggest that 1-octen-3-ol acts as a fungal hormone during development of the fungal thallus. Physiologically, 1-octen-3-ol is a product of the enzymatic breakdown of linoleic acid by the enzyme lipoxygenase and a hydroperoxide. In *Pleurotus pulmonarius,* linoleic acid splits in two compounds, 10-HPOD (10-hydroperoxyocta-decadienoic acid) a precursor of 1-octen-3-ol and 13-HPOD (13-hydroxyperoxy-*cis*-9,*trans*-11-octadecadienoic acid)(Assaf *et al.*, 1997; Kuribayashi *et al.*, 2002). Together with 1-octen-3-ol, a non-volatile metabolite, 10-oxo-*trans*-8-decenoic acid (ODA) is formed in this process, which is stated to have an influence on the development of the mushroom. It stimulates growth of the mycelium, stipe elongation, and fruiting initiation during mushroom development and it has been regarded as a growth regulating substance (GRS) produced by gills (Mau *et al.*, 1992; Champavier *et al.*, 2000). Other linoleic acid derivatives play a role in sporulation phenomena in *Emericella (Aspergillus) nidulans*, which suggest that polyunsaturated lipid compounds and their degradation products are remarkably important in development of fungi.

Further investigation of the role of self-inhibitors may reveal novel methods for the inhibition of fungal development in food.

ANTIFUNGAL COMPOUNDS

Germination and growth of fungi in food and feed is discouraged by the introduction of different adverse conditions as the use of altered gas composition, low water activities and the presence of organic acids or a combination of these factors. Low oxygen pressure and organic acids for instance are used to preserve grass forage in silos during ensilage. These conditions lower the metabolism of fungi and prevent their growth as is also the result of lowering of the water activity of a medium to α = 0.65-0.86. However, some osmotolerant and xerophilic fungi that are able to grow in the presence of high concentrations of sugar and salt cause spoilage in these conditions (Dijksterhuis and Samson, 2002), but with many commodities including seeds, grains, beans and peas prevention of fungal spoilage is successful due to their low water activity if properly dried and well stored.

The main antimicrobial food preservatives are weak organic acids and esters (propionate, sorbate, benzoate and benzoate esters (parabens), organic acid acidulants (lactic, citric, malic, and acetic acids), inorganic acid preservatives (sulfite), mineral acids (phosphoric and hydrochloric acids) and other compounds as natamycin (Britt *et al.*, 1974; Kabara and

Eklund, 1991; Gould, 1996; Stark, 2003). Propionate is a highly effective fungal inhibitor used in cheese and bakery product industries. Sec-butylamine is commonly used in its free form to preserve fruits against damage by storage fungi such as *Penicillium* and *Aspergillus*. In addition, sorbate prevents fungal growth and decreases mycotoxin biosynthesis by inhibiting the biological pathways responsible for their production. Nowadays, a wide range of antifungal agents is used in combating biodeterioration; prevention or treatment of fungal disease of plants and treatment of diseases in animals and humans (Table 3). The mode of action of these compounds is variable, but an important part of the compounds have plasma membrane and cell wall related targets.

Novel antibiotics against bacteria compounds are actively search at and one potential family of antifungal compounds are the iturins (A-E) that are produced by *Bacillus subtilis*. Iturins are cyclic lipopeptides characterised by the presence of seven α-amino acids (Isogai *et al.*, 1982; Latoud *et al.*, 1990). Iturins interact with sterols in the cytoplasmic membrane and are similar to the antifungal polyene amphotericin B (Maget-Dana *et al.*, 1985; Latoud *et al.*; 1990). Other lipopeptides that belong to the iturin group are the bacillomycins D, F, and L, and mycosubtilin (Bland, 1996; Moyne *et al.*, 2001). Iturin A reduced the fungal population on seed with variation among the fungal species with respect to their sensitivity, but it is not able to inhibit aflatoxin production of *Aspergillus flavus* (Klich *et al.*, 1993, 1994). For instance, *Rhizopus* sp. was known previously not to be sensitive to it (Gould, 1996). This may be explained by the low ergosterol content of the *Rhizopus* sp. membrane (Groll *et al.*, 1998). Fungal inhibition was observed in case of postharvest fungal spoilage of peaches and the role of iturins during biological control with *Bacilli* studied (Gueldner *et al.*, 1988). An iturin-like compound inhibited the germination of *Penicillium paneum* conidia (Chitarra *et al.*, 2003). Fluorescence microscopy and FCM revealed that the PI was able to label damaged cells, indicating the permeabilisation of *P. paneum* conidiospores membrane after exposure to the HCl precipitate.

EPILOGUE

The fungal spore is a resting phase and as such is not very reactive on antifungal compounds. Killing of spores with other methods than heat is a very difficult task. Germination of spores (conidia), however, is a gradual development from resistant and not-responsive cells to germ-tube bearing cells via a number of stages. Knowledge about the sensitivity of these different phases to antifungal compounds is vital to evaluate the potential of fungal spores to form "spoilage time bombs" when the antifungal compound is inactivated due to its stability or diffusion.

REFERENCES

Adelman, T. G., and Lovett, J. S. (1974). Evidence for a ribosome associated translation inhibitor during differentiation of *Blastocladiella emersonii*. Biochemistry and Biophysics Acta 335:236-245.

Table 3: Antifungal compounds and their mode of action

Synthetic antifungals	Mode of action
Benzimidazoles; Griseofulvin	Mitosis
5-Fluorocytosine	Nucleic acid synthesis
Acylalamines	RNA polymerase I
Kasugamycin; Sordarins	Protein synthesis
Carboxamides; Strobilurins	Respiration
Fosetyl-Al	Phosphate metabolism
Imidazoles; Triazoles; Thiocarbamates	Ergosterol synthesis
Nystatin; Amphotericin B; Natamycin	Plasma membrane
Polyoxin; Nikkomycins; Echinocandins	Cell wall synthesis

Allen, P. J. (1955). The role of a self-inhibitor in the germination of rust uredospores. Phytopathology 215:259-266.

Assaf, S., Hadar, Y., and Dosoretz, C. G. (1997). 1-octen-3-ol and 13-hydroperoxylinoleate are products of distinct pathways in the oxidative breakdown of linoleic acid by *Pleurotus pulmonarius*. Enzyme and Microbial Technology 21:484-490.

Assche, J. A. van, Carlier, A. R., and Dekeersmaeker, H. I. (1972). Trehalase activity in dormant and activated spores of *Phycomyces blakesleeanus*. Planta 103:327-333.

Atkinson, H., Daniels, A., and Read, N. D. (2002). Live-cell imaging of endocytosis during conidial germination in the rice blast fungus, *Magnaporthe grisea*. Fungal Genetics and Biology 37: 233-244.

Bacon, C. W., and Sussman, A. S. (1973). Effects of the self-inhibitor of *Dictyostelium discoideum* on spore metabolism. Journal of General Microbiology 76:331-344.

Bacon, C. W., Sussman, A.S., and Paul, A.G. (1973). Identification of a self-inhibitor from spores of *Dictyostelium discoideum*. Journal of Bacteriology 113:1061-1063.

Ballio, A., Vittorio V. di, and Russi S. (1964). The isolation of trehalose and polyols from the conidia of *Penicillium chrysogenum*. Archives of Biochemistry and Biophysics 107:177-183.

Barrios-Gonzales, J., Martinez, C., Aguilera, A., and Raimbault, M. (1989). Germination of concentrated suspensions of spores from *Aspergillus niger*. Biotechnology Letters 11:551-554.

Bartnicki-García, S., and Lippman, E. (1977). Polarization of cell wall synthesis during spore germination of *Mucor rouxi*. Experimental Mycology 1:230-240.

Bland, J. M. (1996). The first synthesis of a member of the iturin family, the antifungal cyclic lipopeptide, iturin-A2. Journal of Organic Chemistry 61:5663-5664.

Bonnen, A., and Brambl, R. (1983). Germination physiology of *Neurospora crassa* conidia. Experimental Mycology 7:197-207.

Bourret, J. A. (1986). Evidence that a glucose-mediated rise in cyclic AMP triggers germination of *Pilobolus longipes* spores. Experimental Mycology 10:60-66.

Boyce, J. B., Hynes, M. J., and Andrianopolis, A. (2003). Control of morphogenesis and actin localization by the *Penicillium marneffei* RAC homolog. Journal of Cell Science 116:1249-1260.

Boyce, J. B., Hynes, M. J., and Andrianopolis, A. (2005). The Ras and Rho GTPases genetically intereact to co-ordinately regulate cell polarity during development in *Penicillium marneffei*. Molecular Microbiology 55:1487-1501.

Breeuwer, P., Reu, J. C. d., Drocourt, J. L., Rombouts, F. M., and Abee, T. (1997). Nonanoic acid, a fungal self-inhibitior, prevents germination of *Rhizopus oligosporus* sporangiospores by dissipation of the pH gradient. Applied Environmental Microbiology 63:178-185.

Britt, D. G., Huber, J. T., and Rogers, A.L. (1974). Fungal growth and acid production during fermentation and refermentation of organic acid treated corn silages. Journal of Dairy Science 58:532-539.

Buhr, T. L., and Dickman, M. B. (1997). Gene expression analysis during conidial germ tube and appresorium development in *Colletotrichum trifolii*. Applied and Environmental Microbiology 63: 2378-2383.

Carlisle, M. J., Watkinson, S. C., and Gooday, G. W. (1994). Spores, dormancy and dispersal. *In* The Fungi, 2nd ed (Carlisle, M. J., Watkinson, S. C., and Gooday, G. W., eds.), Academic Press, California, U.S.A, pp. 185-240.

Chaky, J., Anderson, K., Moss, M., and Vaillancourt, L. (2001). Surface hydrophobicity and surface rigidity induce spore germination in *Colletotrichum graminicola*. Phytopathology 91:558-564.

Champavier, Y., Pommier, M., Arpin, N., Voiland, A., and Pellon, G. (2000). 10-oxo-trans-8-decenoic acid (ODA): production, biological activities, and comparison with other hormone-like substances in *Agaricus bisporus*. Enzyme and Microbial Technology 26:243-251.

Chang, M. H., Chae, K. S., Han, D. M., and Jahng, K. Y. (2004). The GanB Galpha-protein negatively regulates asexual sporulation and plays a positive role in conidial germination in *Aspergillus nidulans*. Genetics 167:1305-1315.

Cheng, J., Park, T. S., Fischl, A. S., and Ye, X. S. (2001). Cell cycle progression and cell polarity require sphingolipid biosynthesis in *Aspergillus nidulans*. Molecular Cellular Biology 21:6198-6209.

Chitarra, G. S., Breeuwer, P., Nout, M. J. R., Aelst, A. van, Rombouts, F. M., and Abee, T. (2003). An antifungal compound produced by *Bacillus subtilis* YM 10-20 inhibits germination of *Penicillium roqueforti* conidiospores. Journal of Applied Microbiology 94:159-166.

Chitarra, G. S., Abee, T., Rombouts, F. M., Posthumus, M. A., and Dijksterhuis, J. (2004). Germination of *Penicillium paneum* conidia is regulated by a volatile self-inhibitor. Applied and Environmental Microbiology 70:2823-2829.

Chitarra, G. S., Breeuwer, P., Rombouts, F. M., Abee, T., and Dijksterhuis, J. (2005a). Differentiation inside multicelled macroconidia of *Fusarium culmorum* during early germination. Fungal Genetics and Biology 42:694-703.

Chitarra, G. S., Abee, T., Rombouts, F. M., and Dijksterhuis, J. (2005b). 1-Octen-3-ol has mild effects on membrane permeability, respiration and intracellular pH, but blocks germination and changes the protein composition of *Penicillium paneum* conidia. FEMS Microbiology Ecology 54:67-75.

Clement, J. A., Martin, S. G., Porter, R., Butt, T. M., and Beckett, A. (1993). Germination and the role of extracellular matrix in adhesion of urediniospores of *Uromyces viciae-fabae* to synthetic surfaces. Mycological Research 97:585-593.

Crowe, J. H., Crowe, L. M., and Chapman, D. (1984). Preservation of membranes in anhydrobiotic organisms: the role of trehalose. Science 223:701-703.

D'Enfert, C., and Fontaine, T. (1997). Molecular characterization of the *Aspergillus nidulans* treA gene encoding an acid trehalase required for growth on trehalase. Molecular Microbiology 24:203-216.

D'Enfert, C., Bonini, B. M., Zapella, P. D. A., Fontaine, T., Da Silva, A. M., and Terenzi, H. F. (1999). Neutral trehalase catalyse intracellular trehalose breakdown in the filamentous fungus *Aspergillus nidulans* and *Neurospora crassa*. Molecular Microbiology 32:471-483.

Deising, H., Nicholson, R. L., Haug, M., Howard, R. J., and Mendgen, K. (1992). Adhesion pad formation and the involvement of cutinase and esterases in the attachment of uredospores to the host cuticle. The Plant Cell 4:1101-1111.

De Virgilio, C., Hottiger, T., Dominguez, J., Boller, T., and Wiemken, A. (1994). The role of trehalose for the acquisition of thermotolerance in yeast. I. Genetic evidence that trehalose is a thermoprotectant. European Journal of Biochemistry 219:179-186.

Dijksterhuis, J., and Samson, R. A. (2002). Food and crop spoilage on storage. *In* The Mycota XI, Agricultural Applications (Kempken, F., ed.), Springer Verlag, Berlin, Germany, pp. 39-52.

Dijksterhuis, J., Driel, K. G. A. van, Sanders, M. G., Molenaar, D., Houbraken, J. A. M. P., Samson, R. A., and Kets, E. P. W. (2002). Trehalose degradation and glucose efflux precede cell ejection during germination of heat-resistant ascospores of *Talaromyces macrosporus*. Archives of Microbiology 178:1-7.

Dijksterhuis, J. (2003). Confocal microscopy of Spitzenkörper dynamics during growth and differentiation of rust fungi. Protoplasma 222:53-59.

D'Mello, J. P. F., and Macdonald, A. M. C. (1997). Mycotoxins Animal Feed Science Technology. 69:155-166.

D'Mello, J. P. F., Macdonald, A. M. C., Postel, D., Dijksma, W. T. P., Dujardin, A., and Placinta, C. M. (1998). Pesticide use and mycotoxin production in *Fusarium* and *Aspergillus* phytopathogens. European Journal of Plant Pathology. 104:741-751.

Doehlemann, G., Berndt, P., and Hahn, M. (2006). Different signalling pathways involving a Gα protein, cAMP and a MAPkinase control germination of *Botrytis cinerea* conidia. Molecular Microbiology 59:821-835.

Doss, R. P., Potter, S. W., Chastagner, G. A., and Christian, J. K. (1993). Adhesion of nongerminated *Botrytis cinerea* conidia to several substrata. Applied and Environmental Microbiology 59: 1786-1791.

Etten, J. L. van, Dahberg, K. R., and Russo, G. M. (1983). Fungal spore germination. *In* Fungal Differentiation (Smith, J. E., ed.), Dekker, New York, U.S.A., pp. 235-266.

Fillinger, S., Chaveroce, M. K., Dijck, P. van, Vries, R. de, Ruijter, G., Thevelein, J., and D'Enfert, C. (2001). Trehalose is required for the acquisition of tolerance to a variety of stresses in the filamentous fungus *Aspergillus nidulans*. Microbiology 147:1851-1862.

Fillinger, S., Chaveroce, M. K., Shimize K., Keller, N. and D'Enfert, C. (2002). cAMP and ras signalling independently control spore germination in the filamentous fungus *Aspergillus nidulans*. Molecular Microbiology 44:1001-1016.

Filonov, A. B. (2001). A procedure for quantifying adhesion of conidia of *Botrytis cinerea* to the skin of apple fruit. Canadian Journal of Microbiology 47:787-791.

Filonov, A. B. (2003). Germination and adhesion of fungal conidia on polycarbonate membranes and on apple fruit exposed to mycoactive acetate esters. Canadian Journal of Microbiology 49:130-138.

Filonov, A. B. (2004). Adhesion of decay-causing fungal conidia in wounds of *Malus domestica* 'Golden Delicious' apple fruit is influenced by wound age. Canadian Journal of Botany 82:265-272.

Filtenborg, O., Frisvad, J. C., and Samson, R. A. (2004). Specific association of fungi to foods and influence of physical environmental factors. *In* Introduction to food and airborne fungi, 7th edition (Samson, R. A., Hoekstra, E. H., and Frisvad,

J. C., eds.), Centraalbureau voor Schimmelcultures, Utrecht, The Netherlands, pp. 306-320.

Fletcher, J., and Morton, A. G. (1970). Physiology of germination of *Penicillium griseofulvum* conidia Transactions of the Britisch Mycological Society 54:65-81.

French, E. R., and Nielsen, W. (1966). Production of macroconidia of *Fusarium oxysporum* f. *batatas* and their conversion to chlamydospores. Phytopathology 88:879-884.

Garret, M. K., and Robinson, P. M. (1969). A stable inhibitor of spore germination produced by fungi. Archieves Mikrobiology 67:370-377.

Gould, G. W. (1996). Methods for preservation and extension of shelf life. International Journal of Food Microbiology 33:51-64.

Green, B. J., Sercombe, J. K., and Tovey, E. R. (2005). Fungal fragments and undocumented conidia function as new aerollergen sources. Journal of Allergy and Clinical Immunology 115:1043-1048.

Griffin, D. H. (1994). Spore dormancy and germination, *In* Fungal Physiology, 2nd ed., I. N. Y. John Wiley & Sons, pp. 375-398.

Groll, A. H., Lucca, A. J. De, and Walsh, T. J. (1998). Emerging targets for the development of novel antifungal therapeutics. Trends in Microbiology 6:117-124.

Gueldner, R. C., Reilly, C. C., Pusey, P. L., Costello,C. C., Arrendale, R. F., Cox, R. H., Himmelsbach, D. S., Crumley, F. G., and Cutler, H. G. (1988). Isolation and identification of iturins as antifungal peptides in biological control of peach brown rot with *Bacillus subtilis*. Journal of Agricultural and Food Chemistry 36:366-370.

Hallsworth, J. E., and Magan N. (1994). Effect of carbohydrate type and concentration on polyhydroxy alcohol and trehalose content of conidia of three entomopathogenic fungi. Microbiology 140:2705-2713.

Hallsworth, J. E., and Magan N. (1995). Manipulation of intracellular glycerol and erythritol enhances germination of conidia at low water availability. Microbiology 141:1109-1115.

Hallsworth, J. E., and Magan N. (1996). Culture, age, temperature, and pH affect the polyol and trehalose contents of fungal propagules. Applied and Environmental Microbiology 62:2345-2442.

Hallsworth J. E., Prior, B. A., Nomura Y., Iwahara, M., and Timmis K. N. (2003). Compatible solutes protect against chaotrope (ethanol)-induced nonosmotic water stress. Applied and Environmental Microbiology 69:7032-7034.

Hamer, J. E., Howard, R. J., Chumley, F. G., and Valent, B. (1988). A mechanism of surface attachment in spores of a plant pathogenic fungus. Science 239: 288-290.

Harris, S. D. (1999). Morphogenesis is coordinated with nuclear division in germinating *Aspergillus nidulans* conidiospores. Microbiology 145:2747-2756.

Hobot, J. E., and Gull, K. (1980). The identification of a self inhibition from *Syncephalatrum racemosus* and its effects upon sporangiospore germination. Antonie van Leeuwenhoek. 46:435-441.

Hottiger, T., Vergilio, C. De, Hall, M. N., Boller, T., and Wiemken, A. (1994). The role of trehalose for the acquisition of thermotolerance in yeast. II. Physiological concentrations of trehalose increase the stability of proteins in vivo. European Journal of Biochemistry 219:187-193.

Howard, R. J. (1993). During a contribution at the BMS-meeting held at Portsmouth, UK.

Inouye, K. (1985). Measurements of intracellular pH and its relevance to cell differentiation in *Dictyostelium discoideum*. Journal of Cell Science 76:235-245.

Ismail, M. A., Elala, A. H. A., Nassar, A., and Michail, D. G. (1995). Fungal contamination of beef carcasses and the environment in a slaughterhouse. Food Microbiology 12:441-445.

Isogai, A., Takayama, S., Murakoshi, S., and Suzuki, A. (1982). Structure of beta-amino acids in antibiotics iturin A Screened for use against phytopathogenic fungi. Tetrahedron Letters 23:3065-3068.

Jelitto, T. C., Page, H. A., and Read, N. D. (1994). Role of external signals in regulating the prepenetration phase of infection by the rice blast fungus, *Magnaporthe grisea*. Planta 194:471-477.

Jelitto, T. C. (1999). Confocal ratio imaging of cytolasmic pH during germ tube growth and appressorium induction by *Magnaporthe grisea*. New Phytopathologist 144:499-506.

Kabara, J. J., and Eklund, T. (1991). Organic acids and esters. *In* Food Preservatives (Russell, G. W., ed.), Blackie, Avi in USA and imprint of Van Nostrand Reinhold, New York., Glasgow and London, U.K., p. 290.

Kim, Y. K., Li, D., and Kolattukudy, P. E. (1999). Induction of Ca^{2+}-calmodulin signaling by hard-surface contact primes *Colletotrichum gloeosporioides* conidia to germinate and form appressoria. Journal of Bacteriology 180:5144-5150.

Klich, M. A., Lax A. R., Bland, J. M., and Scharfenstein Jr, L. L. (1993). Influence of iturin A on mycelial weight and aflatoxin production by *Aspergillus flavus* and *Aspergillus parasiticus* in shake culture. Mycopathologia 123:35-38.

Klich, M. A., Arthur, K. S., Lax, A. R., and Bland, J. M. (1994). Iturin A: a potential new fungicide for stored grains. Mycopathologia 127:123-127.

Krishnan, P. S. (1954). Some observations on the growth of *Aspergillus niger* from spore inoculum. Applied Microbiology 2:303-308.

Kuo, K., and Hoch, H. C. (1996). Germination of *Phyllosticta ampelicida* pycnidiospores: prerequisite of adhesion to the substratum and the relationship of substratum wettability. Fungal Genetics and Biology 20:18-29.

Kuribayashi, T., Kaise, H., Uno, C., Hara, T., Hayakawa, T., and Joh, T. (2002). Purification and characterization of lipoxygenase from *Pleurotus ostreatus*. Journal of Agricultural and Food Chemistry 50:1247-1253.

Latoud, C., Peypoux, P., and Michel, G. (1990). Interaction of iturin A, a lipopeptide antibiotic, with *Saccharomyces cerevisiae* cells: influence of the sterol membrane composition. Canadian Journal of Microbiology 36:384-389.

Lax, A. R., Templeton, G. E., and Meyer, W. L. (1985). Isolation, purification, and biological activity of a self-inhibitor from conidia of *Colletotrichum gloeosporioides*. Phytopathology 75:386-390.

Leandro, L. F. S., Gleason, M. L., Wegulo, S. M., Dixon, P. M., and Nutter, F. W., Jr. (2001). Survival and sporulation of *Colletotrichum acutatum* on symptomless strawberry leaves. Phytopathology 91: 659-664.

Leite, B., and Nicholson, R. L. (1992). Mycosporine-alanine: A self-inhibitor of germination from the conidial mucilage of *Colletotrichum graminicola*. Experimental Mycology 16:76-86.

Leite, B., and Nicholson, P. (1993). A volatile self-inhibitor from *Colletotrichum graminicola*. Mycologia 85:945-951.

Leppik, R. A., Hollomon, D. W., and Bottomley, W. (1972). Quiesone: An inhibitor of the germination of *Peronospora tabacina* conidia. Phytochemistry 11:2055-2063.

Lingappa, B. T., Lingappa, Y., and Bell, E. (1973). A self-inhibitor of protein synthesis in the conidia of *Glomerella cingulata*. Archives of Mikrobiology 94:97-107.

Louis, I. A., Chew A., and Lim, G. (1988). Influence of spore density and extracellular conidial matrix on spore germination in *Colletrotrichum capsici*. Transactions of the Britisch Mycological Society 91:694-697.

Macko, V., Staples, R. C., Allen, P. J. , and Renwick, J. A. A. (1971a). Identification of the germination self-inhibitor from wheat stem rust uredospores. Science 173:835-836.

Macko, V., Staples, R. C., and Renwick, J. A. A. (1971b). Germination self-inhibitor of sunflower and snapdragon rust uredospores. Phytophatology 61:902.

Maget-Dana, R., Ptak, M., Peypoux, F., and Michel, G. (1985). Pore-forming properties of iturin A, a lipopeptide antibiotic. Biochimica et Biophysica Acta 815:405-409.

Marin, S., Sanchis, V., Saenz, R., Ramos, A.J.,Vinas, I., and Magan, N. (1998). Ecological determinants for germination and growth of some *Aspergillus* and *Penicillium* spp. from maize grain. Journal of Applied Microbiology 84:25-36.

Mau, J. L., Beelman, R. B., and Ziegler, G. R. (1992). Effect of 10-oxo-trans-8-decenoic acid on growth of *Agaricus bisporus*. Phytochemistry 31:4059-4064.

Momany, M. (2002). Polarity in filamentous fungi: establishment, maintenance and new axes. Current Opinion in Microbiology 5:580-585.

Momany, M., and Taylor, I. (2000). Landmarks in the early duplication cycles of *Aspergillus fumigatus* and *Aspergillus nidulans*: polarity, germ tube emergence and septation. Microbiology 146: 3279-3284.

Moyne, A. L., Shelby, R., Cleveland, T. E., and Tuzun, S. (2001). Bacillomycin D: an iturin with antifungal activity against *Aspergillus flavus*. Journal of Applied Microbiology 90:622-629.

Musumeci, M. R., Moraes, W. B. C., and Staples, R. C. (1974). A self-inhibitor in uredospores of coffee rust fungus. Phytophatology 64:71-73.

Ojha, M., and Barja, F. (2003). Spacial and cellular localization of calcium-dependent protease(CDPII) in *Allomyces arbuscula*. Journal of Cell Science 116:1095-1105.

Osherov, N., and May, G. S. (2001). The molecular mechanisms of conidial germination. FEMS Microbiology Letters 34:1-8.

Page, W. J., and Stock, J. J. (1971). Regulation and self-inhibitor of *Microsporum gryseum* macroconidia germination. Journal of Bacteriology 108:276-281.

Parton, R. M., Fisher, S., Malho, R., Papasouliotis, O., Jellito, T. C., Leonard, L., and Read, N. D. (1997). Pronounced cytoplasmic pH gradients are not required for tip growth in plant and fungal cells. Journal of Cell Science 110:1187-1198.

Prestrelski, S. J., Tedeschi, N., Arakawa, T., Carpenter, J. F. (1993). Dehydration-induced conformational transitions in protein and their inhibition by stabilizers. Biophysical Journal 65:661-671.

Reid, L. M., Nicol, R. W., Ouellet, T., Savard, M., Miller, J. D., Young, J. C., Stewart, D. W., and Schaafsma, A. W. (1999). Interaction of *Fusarium graminearum* and *F. moniliforme* in maize ears: disease progress, fungal biomass, and mycotoxin accumulation. Phytopathology 89:1028-1037.

Robson, G. D., Prebble, E., Rickers, A., Hosking, S., Denning, D. W., Trinci, A. P. J., and Robertson, W. (1996). Polarized growth of fungal hyphae is defined by an alkaline pH gradient. Fungal Genetics and Biology 20:289-298.

Roncal, T., Ugalde, U. O. and Irastorza, A. (1993). Calcium-induced conidiation in *Penicillium cyclospium*: calcium triggers cytosolic alkalinization at the hyphal tip. Journal of Bacteriology 175:879-886.

Ruijter, G .J. G., Bax, M., Patel, H., Flitter, S.J., Vondervoort, P. J. I. van de, Vries, R. P. de, Kuyk, P. A. van, and Visser, J. (2003). Mannitol is required for stess tolerance in *Aspergillus niger* conidiospores. Eucaryotic Cell 4:690-698.

Samson, R. A., Hoekstra, E. S., Frisvad, J. C., and Filtenborg, O. (eds) (2004). Introduction to Food- and Airborne Fungi. Seventh Edition, Centraalbureau voor Schimmelcultures, Utrecht, The Netherlands.

Schneider, E. F., and Seaman, W. L. (1974). Development of conidial chlamydospores of *Fusarium sulphureum* in distilled water. Canadian Journal of Microbiology 20:247-254.

Shaw, B. D., and Hoch, H.C. (2000). Ca^{2+} regulation of *Phyllosticta ampelicida* pycnidiospore germination and appressorium formation. Fungal Genetics and Biology 31:43-53.

Stark, J. (2003). Natamycin, an effective fungicide for food and beverages. *In* Natural antimicrobials for the minimal processing of foods (Roller. S., ed.), Woodhead Publishing Limited, Londen, U.K., pp. 82-97.

Steele, S. D. (1973). Self-inhibition of arthrospore germination in *Geotrichum candidum Canadian* Journal of Microbiology 19:943-947.

Stewart, L., Gow, N. A. R., and Bowen, D. V. (1988). Cytoplasmic alkalinization during germ tube formation in *Candida albicans*. Journal of General Microbiology 134:1079-1087.

Stone, J., Pinketon, J. N., and Johnson, K. B. (1994). Axenic culture of *Anisogramma anomala*: evidence for self-inhibitor of ascospore germination and colony growth. Mycologia 86:674-683.

Sumere, C. F. v., Preter, C. v S., Vining, L. C., and Ledingham, G. A. (1957). Coumarins and phenolic acids in the uredospores of wheat stem rust. Canadian Journal of Microbiology 3:847-862.

Terhune, B. T., and Hoch H. C. (1993). Substrate hydrophobicity and adhesion of *Uromyces* urediospores and germlings. Experimental Mycology 17:241-252.

Thanh, N. V., and Nout, M. J. (2004). Dormancy, activation and viability of *Rhizopus oligosporus* sporangiospores. International Journal of Food Microbiology 92:171-179.

Thanh, N. V., Rombouts, F. M., and Nout, M. J. (2005). Effect of individual amino acids and glucose on activation and germination of *Rhizopus oligosporus* sporangiospores in tempe starter. Journal of Applied Microbiology 99:1204-1214.

Thanh, N. V., Rombouts, F. M., and Nout, M. J. (2007). Viability and physiological state transitions of *Rhizopus oligosporus* sporangiospores in tempe starter culture. Antonie Van Leeuwenhoek 91:35-44.

Thevelein, J. M. (1984). Cyclic-AMP content and trehalase activation in vegetative cells and ascospores of yeast. Archives of Microbiology 138:64-67.

Tiedt, L. R., (1993). An electron microscopical study of conidiogenesis and wall formation of conidia of *Aspergillus niger*. Mycological Research 97:1459-1462.

Trione, E. J. (1973) The physiology of germination of *Tilletia* teliospores. Phytopathology 63:643-648.

Warburton, A. J., and Deacon, J. W. (1998). Transmembrane Ca2+ fluxes associated with zoospore encystment and cyst germination by the phytopathogen *Phytophthora parasitica*. Fungal Genetics and Biology 25:54-62.

Weidenborner, M., Wieczorek, C., Appel, S., and Kunz, B. (2000). Whole wheat and white wheat flour — the mycobiota and potential mycotoxins Food Microbiology 17:103-107.

Wilson, E. M. (1958). Aspartic and glutamic acid as self-inhibitors of uredospore germination. Phytophatology 48:595-600.

Wolkers, W. F., Kilsdonk, M. G. van, and Hoekstra, F. A. (1998). Dehydration-induced conformational changes of poly-L-lysine as influenced by drying rate and carbohydrates. Biochimica Biophysica Acta 1425:127-136.

Zuber, S. Hynes, M. J., and Andrianopoulos, A. (2003). The G-protein α-subunit GasC plays a major role in germination in the dimorphic fungus *Penicillium marneffei*. Genetics 164:487-499.

Chapter 6

Heat-resistant ascospores

Jan Dijksterhuis
Applied and Industrial Mycology, CBS Fungal Biodiversity Centre, Uppsalalaan 8, 3584 CT, Utrecht, The Netherlands.

INTRODUCTION

Fungal spoilage is a cause of immense food losses during storage of crops and food products. In order to overcome food spoilage by fungi, mankind has developed an array of methods to make food, which can be regarded as a rich and complex medium, difficult to colonize. This is done, among others, by means of lowering of the water activity by addition of sugars, salt or after drying. In other cases, food products are kept under lowered temperatures. A third possibility is the killing of fungi present in the food matrix by a transient heat treatment (pasteurisation). Currently, novel techniques are introduced as high-pressure treatment or storage of a food product under a modified atmosphere. Although heat treatment is a reliable and historical method, a class of fungal organisms is able to survive prolonged periods of heat and these organisms cause a restricted, but consistent, loss of food products. This is especially interesting with modern trends to use minimally processed food where the heat treatment is minimalised in favour of the organoleptic properties of the product and vitamin content.

HEAT RESISTANCE IN FUNGI

Vegetative (yeast) cells usually have little heat resistance. When heated in beer (pH 4.0, 5% ethanol) *Saccharomyces carlsbergensis, Hansenula anomala* and *Pichia membranaefaciens* had D_{49}-values of less than 2 min. *S. willianus*, was more heat-resistant with a D_{49} around 15 min (Tsang and Ingledew, 1982). For *Zygosaccharomyces bailii, Saccharomyces bisporus, H. anomala* and *Z. rouxii* decimal reduction was observed after heating for 1 min in a buffer (pH 5,5) supplied with 400 g/l sucrose, at 56, 55, 54 and 50°C, respectively (Baggerman and Samson, 1988). Yeast species that were heated in grapefruit serum showed similar characteristics (Parish, 1991). The presence of sucrose, sodium chloride or glycerol in the heating menstruum was somewhat protective during heating, while sorbate and benzoate usually reduced the heat resistance of the yeast cells (Beuchat, 1981; Agab and Collins, 1992). Living (and growing) fungal cells exhibit after a sudden increase in temperature of approximately 10 °C above the optimal growth temperature, a phenomenon called heat shock. Increases in temperature lead to a number of changes including protein denaturation, cell cycle arrest and changes in the fluidity of the membrane. One of the most deleterious factors in heat shocked cells is the disturbance (unfolding) of the protein structure that leads to the formation of aggregates of proteins that affect the functioning of the cell lethally (Riezman, 2004). Accumulation of compatible solutes (glycerol, trehalose, mannitol) that protect the cellular membrane and the proteins (Crowe *et al.*, 1984; Prestrelski *et al.*, 1993) is one of the answers of the cell to heat shock. In addition, heat shock proteins play an important role in cooperation with compatible solutes in yeasts (Elliott *et al.*, 1996), but these are scarcely studied in fungal survival structures. Recently, the genome of the filamentous fungus *Aspergillus fumigatus* was sequenced (Nierman *et al.*, 2005) and 323 genes showed higher expression at 48 °C compared to 37 °C. The most strongly upregulated genes included three proteins related to compatible solute synthesis and degradation and nine heat shock proteins. This also illustrates the role of different factors during the heat-shock response.

Fungal survival structures as conidia, sclerotia, chlamydospores and ascospores can be regarded as more or less heat resistant when they are compared to actively growing fungal cells. They do not exhibit the metabolism of the vegetative cells, but prevail in a dormant state and often are encompassed by a thicker cell wall.

Conidia are non-motile asexual spores of many fungal species that are often dispersed by air and water and which exhibit some dormancy, namely that they germinate only when proper nutrients are present in the medium. Conidia of the fungal species *Aspergillus niger, Penicillium chrysogenum, Wallemia sebi, Eurotium rubrum* and *P. glabrum*, exhibit a decimal reduction in buffer in the range of 56 and 62 °C (Baggerman and Samson, 1988) and *P. roqueforti, P. expansum, P. citrinum* and *A. flavus* have D-values between 3.5 and 230 min at 54-56 °C (Bröker *et al.*, 1987a). The age of the spores at the time of harvest as well as the composition of the growing and heating media influence the heat resistance. Furthermore, conidia could restore from a heat treatment while *P. expansum* showed recovery during a 3-day storage in aerated water at 23 °C after a heat treatment at 54 °C compared to conidia that were plated out directly (Baldy *et al.*, 1970). Numbers of colonies were 20-fold higher after this treatment, indicating that the situation after heating has an effect on the number of survivors.

Ascospores are widely formed sexual spores within the Ascomycetes with often a high survival capability. In an investigation of 20 yeast strains from soft drinks and fruit products, mainly *Saccharomyces cerevisiae, S. bailii* (now *Z. bailii*) and *S. chevalieri* strains, the D_{60}-values of ascospores were 25 to 350 times higher than those of the corresponding vegetative cells (Put and de Jong, 1980). In a pH 4,5 buffer *S. cerevisiae, S. chevalieri* and *S. bailii* ascospores exhibited D_{60}-values of 22.5, 13 and 10 min (Put and de Jong, 1982).

In general, ascospores of filamentous fungi are more heat resistant than mycelia and conidia and more resistant than yeast ascospores. A number of observations suggest that thick-walled hyphal fragments and chlamydospores (*Paecilomyces variotii, Fusarium* spp) and sclerotia (e.g., in *Eupenicillium*) can perform high thermo-resistance (Splittstoesser and King, 1984). But by far the most heat-resistant fungal structures known to date are ascospores produced by some members of the genera *Byssochlamys, Neosartorya* and *Talaromyces*. Ascospores of these fungi are extraordinary heat resistant and show D_{90} values of several minutes (Beuchat, 1986). In fact, these spores belong to the most resilient eukaryotic structures observed hitherto. A decimal reduction time of 1.5-11 min is observed at 90 °C among different species (Scholte *et al.*, 2001, Dijksterhuis and Samson, 2006). Ascospores of the fungus *Talaromyces macrosporus* are able to survive at 85 °C for 100 min, which makes these fungal spores as resilient as some bacterial spores (e.g., *Bacillus subtilis*).

One has to realise that heating of cells in a low moisture environment is much less effective. Even yeast cells show very high inactivation times at low humidities. When heated on aluminium foil at a relative humidity of 33-38%, *P. membranaefaciens* and *Rhodotorula rubra* cells showed little survival after 5 min at 110 °C (Scott and Bernard, 1985). With D_{110}-values of 1.3, 1.8, 2.9 and 3.6 min, respectively, *Debaryomyces hansenii, Kloeckera apiculata, Lodderomyces elongisporus* and *H. anomala* cells were more resistant.

HEAT-RESISTANT FUNGI

Heat-resistant fungi can survive pasteurising heat treatments of especially high-acid food products (e.g., fruits, see Silva and Gibbs, 2004). Subsequent germination causes spoilage of canned and pasteurised fruit products. Fungi that are associated with product recalls that cause a damage of millions of dollars in the fruit-juice branch are *Byssochlamys nivea (fulva), Talaromyces flavus (macrosporus), Neosartorya fischeri, Eupenicillium brefeldianum* (as reviewed by Tournas, 1994). Despite several decades of research these fungi still represent problems in the food branch. Heat-resistant fungi are basically soil-borne fungi and fruits that develop in contact with soil (like strawberries and pineapple) are more prone to contamination. The fungus *Talaromyces flavus* is found to have a worldwide distribution and was isolated from soil samples from 16 different countries including Bermuda, Tasmania, Pakistan and Finland (Fravel and Adams, 1986)

The first description of the heat-resistant nature of these fungi in literature was the isolation of *Byssochlamys nivea* from processed fruit by Olliver and Rendle in the 1930s (Olliver and Rendle, 1934). In 1963, a heat-resistant *Aspergillus* (teleomorph, *Neosartorya fischeri*) was isolated from canned strawberries (Kavanagh *et al.*, 1963) and later two heat-resistant *Penicillium*-like species were isolated from flash-pasteurised apple juice (van der Spuy *et al.*, 1975) which became later known as *Talaromyces (flavus* and *macrosporus*). In addition, other fungal species are now identified as heat-resistant (e.g., *Talaromyces trachyspermus*, Enigl *et al.*, 1993; *Talaromyces helicus* and *stipitatus*, Dijksterhuis and Samson, 2006; *Byssochlamys spectabilis*, J. Houbraken, unpublished results).

INACTIVATION OF HEAT-RESISTANT FUNGI

Methods

Heat resistance of cells is expressed by means of the D- and Z-values, as is well documented, but the exact measurement and calculation of these values is not so self-evident. Throughout different studies various attempts are made to heat ascospore solutions instantaneously, which is important for an accurate D-value acquisition. These attempts include small sealed vessels that are plunged into water or oil baths (King and Whitehand, 1990), or a two-phase slug flow heat exchanger (King, 1997) or spiral steel capillary tubes (Engel and Teuber, 1991). A simpler method is the addition of a small aliquots of ascospore suspension to a larger preheated volume of fluid. Dijksterhuis *et al.* (2002) used small diameter Teflon tubing for sampling of the heated solution without opening the water bath. The *D-value*, the decimal reduction time, is the time (usually in min) that is needed to inactivate 90% of the microorganisms at a given temperature. Ideally, a linear curve will be obtained when the number of surviving microorganisms, on a logarithmic scale, is plotted against time. However, many cases of non-linear death-rate kinetics are observed (Put and de Jong, 1982; Splittstoesser and King, 1984, King and Halbrook, 1987). King *et al.* (1979) have provided a mathematical method for these situations. They use the formula: $(\log N_0 - \log N_t)^{\alpha} = kt + C$, where α is a term that addresses the non-linearity of the death kinetics, and k equals a rate constant and 1/k is a measure for the D-value at the used temperature. Apart from non-linear survivor plots, tailing is a phenomenon where a small subpopulation of spores seems to be extra resistant to heat (Bayne and Michener, 1979; Casella *et al.*, 1990). One cause of tailing can be that a small subpopulation escapes proper heating. Fujikawa and Itoh (1996) modelled the non-linear thermal inactivation of *A. niger* conidia at 60 °C that included a shoulder, an accelerated decline and a tail. A later report of Fujikawa *et al.* (2000) indicated that the conidia that adhered to the inner wall of test tubes were the cause of tailing in a test tube system. To compare heat-inactivation rates at different temperatures, a second parameter, the z-value, is used. It denotes the increase or decrease in temperature needed, respectively, to decrease or to increase the *D-value* by a factor of 10. For *A. niger* conidia for example, a D_{59} of 3.3 min with a z-value of 4.9 °C implies a D_{54} of approximately 33 min (Baggerman and Samson, 1988).

Homogeneity of the ascospore suspension is important for heat-resistance measurements;

Table 1. Heat resistance of ascospores at different temperatures and medium composition

Fungal species	T	D-value	Medium	Reference
Byssochlamys fulva	86°	13–14	Grape juice	1
		8	Tomato juice	2
B. nivea	85°	1.3–4.5	Buffer pH 3.5	3
		34.6	15° Brix Strawberry pulp	4
	88°	8–9 sec	Ringer solution	5
	90°	1.5	Tomato juice	2
B. spectabilis	85°	ca. 70	Buffer, pH 6,8	6
Eurotium herbariorum	70°	1.1–4.6	Grape juice, 65° Brix	7
Eupenicillium javanicum	85°	3.7	15° Brix strawberry pulp	4
Monascus ruber	80°	1.7–2.0	Buffers (pH 3.0 ; pH 7.0)	8
		0.9–1.0	In brine	
Neosartorya fischeri	85°	13.2	Apple juice	9
		10.1	Grape juice	9
		10–60	In ACES-buffer, 10 m*M*, pH 6.8	10
		10.4	Buffer pH 7.0	9
		14.5	15° Brix strawberry pulp	4
		15.1	15° Brix apple juice	11
		19.6–29.5	Dionized water, pineapple juice and concentrate	12
		35.3	Buffer pH 7.0	13
	88°	1.4	Apple juice	14
		4.2–16.2	Heated fruit fillings	15
		12.4–17.0	Dionized water, pineapple juice and concentrate	12
	90°	4.4–6.6	Tomato juice	2
	91°	<2	Heated fruit fillings	15
N. pseudofischeri	95°	20 sec		6
Talaromyces flavus	85°	3.3	15° Brix strawberry pulp	4
(macrosporus)	85°	39	Buffer pH 5.0, glucose, 16°	16
		20–26	Buffer pH 5.0, glucose	17
	88°	7.8	Apple juice	14
		7.1–22.3	Heated fruit fillings	15
	90°	2-8	Buffer pH 5.0, glucose	17
		6.2	Buffer pH 5.0, glucose	9
		6.0	Buffer pH 5.0, glucose. Slug flow heat exchanger	9
		2.7–4.1	Organic acids	18
		2.5–11.1	Sugar content (0-60° Brix)	18
		5.2–7.1	PH 3.6-6.6	18
	91°	2.1–11.7	Heated fruit fillings	15
T. helicus	70°	Ca. 20		19
T. macrosporus	85°	30–100	In ACES-buffer, 10 m*M*, pH 6.8	20
T. stipitatus	72°	Ca. 85		19
T. trachyspermus	85°	45 sec		16
Xeromyces bisporus	82.2°	2.3		21

References: 1. Michener and King (1974). 2. Kotzekidou (1997). 3. Aragão (1989). 4. Casella *et al.* (1990). 5. Engel and Tueber (1991). 6. J. Houbraken, unpublished data. 7. Splittstoesser *et al.* (1989). 8. Panagou *et al.* (2002). 9. Conner and Beuchat (1987b). 10. J. Dijksterhuis, unpublished data, strain CBS 133.64. 11. Gumerato (1995). 12. Tournas and Traxler (1994). 13. Rajashekhara *et al.* (1996). 14. Scott and Bernard (1987). 15. Beuchat (1986). 16. King (1997). 17. King and Halbrook (1987). 18. King and Whitehand (1990). 19. J. Eleveld, unpublished data. 20. Dijksterhuis and Teunissen (2004). 21. Pitt and Hocking (1982). This is a revised table from: Dijksterhuis, J., and Samson, R. A. (2006). Activation of ascospores by novel food preservation techniques. *In* Advances in Food Mycology (Hocking, A. D., Pitt, J. I., Samson, R. A., and Thrane, U., eds.). Advances in Experimental Medicine and Biology 571:247-260. Table 1, with kind permission by Springer Science and Business Media.

no other cell types may mask or deform the results obtained.

In some studies pre-treatments are done to kill other cell types in case of very heat-resistant cells (Reyns *et al.*, 2003). For instance *Byssochlamys* (anamorph, *Peacilomyces*) forms ascospores, conidia, chlamydospores and hyphal fragments in one culture. Further, viable counts of suspensions will show deviations of thermal death kinetics when the cells of the population stick together. Asci (ascus; the structures that contain 8 ascospores and give the Ascomycetes their name) must be broken effectively to measure proper thermal death kinetics of single ascospores. Michener and King (1974) use a French press treatment for the particularly resistant asci of *Byssochlamys*.

To enumerate the surviving colonies, in many studies, the stain Bengal Rose (typically 10 mg/L) is added to the agar medium. The stain causes some more distinct boundaries of the fungal colony (King and Halbrook, 1987) and in time a nice purple dot is formed beneath the centre of the colonies; these factors facilitate the counting of the colonies.

Conditions that affect heat resistance

The majority of the studies on heat-resistant spores in the past is done with respect to factors that correlate with the extent of heat resistance. A large part of these studies is summarised in Table 1 (Modified after Dijksterhuis and Samson, 2006). Heat resistance varies with different external and endogenous factors.

External factors include the presence of sugar (Beuchat, 1988a; King and Whitehand, 1990; Splittstoesser and Splittstoesser, 1977) in the heating menstruum, the influence of pH and the presence of organic acids. The presence of sugars invariably has a protective action on the survival of spores within experiments, while, between studies sometimes confusing values are seen. Several organic acids counteract heat resistance of ascospores, but only at low pHs (lower than 4). This is most prominent for fumaric acid (Beuchat, 1988b; Conner and Beuchat, 1987a; Splittstoesser and Splittstoesser, 1977). Conner and Beuchat (1987a) state that pH, type, molarity of the undissociated form of the organic acid act synergistically with heat to inactivate ascospores. Benzoic and sorbic acid had also clear effects on *T. flavus* and *N. fischeri* (Beuchat, 1988b; Rajashekhara *et al.*, 1998). Tartaric and malic acid however correlated with a higher heat resistance of spores of *B. fulva*. The combination of different factors may lead to some puzzling variations in heat resistance observed in literature. For instance, *N. fischeri* is more heat resistant in apple juice compared to grape juice (Conner and Beuchat, 1987a) and both *B. fulva* and *N. fischeri* exhibit a markedly lower heat resistance in cranberry juice compared to grape, apple or tomato juice (Splittstoesser and Splittstoesser, 1977). *N. fischeri* in 0.1 *M* phosphate buffer (pH 7.0) exhibited a far higher heat resistance than in grape jellies with large amounts of sugar cane at pH 3.1-3.3 (Beuchat and Kuhn, 1997). *B. nivea* and *fulva* and *N. fischeri* were approx. twice as heat resistant in tomato juice (pH of 4.2) as in phosphate buffer (pH 7.0, 0.1 *M*, Kotzekidou, 1997).

Endogenous factors that influence heat resistance are the age of the culture for *N. fischeri*, *T. flavus* and *B. nivea* (Beuchat, 1988a; Casella *et al.*, 1990; Conner and Beuchat, 1987b; Dijksterhuis and Teunissen, 2004) that positively correlates with the resilience of the spores, the composition of the medium and the growth temperature of the fungus (Conner and Beuchat, 1987b; King and Whitehand, 1990). The latter reported higher heat resistance of *T. macrosporus* when the fungus was grown on solid medium. *T. flavus* showed higher heat resistance when grown on oatmeal agar compared to malt extract agar. Finally, the individual isolates used showed variation in *B. nivea*, *T. macrosporus* and *flavus* and *N. fischeri* (Bayne and Michener, 1979; Beuchat, 1986; King and Whitehand, 1990). To our knowledge, the only study on the nature of heat resistance was done by Conner *et al.* (1987). They studied younger (11 days) and older (25 days) ascospores of *N. fischeri* with different heat resistance (D_{82} of approx. 23 and >60 min, respectively). Ascospores showed changes in the inner cell wall region at the lateral ridge during aging. They observed qualitative differences in extractable proteins, but did not see changes in fatty acid or lipid content. Older spores contained 2.8% (dry weight) of mannitol and 0.6% of trehalose which could not be measured inside 11-day-old spores. Polyols and disaccharides may play an important role in heat protection. Heat resistance of ascospores not only increased with the age of the fungal culture, but also harvested and washed ascospores showed maturation in case of *T. macrosporus* (i.e., increase of heat resistance in time, Dijksterhuis and Teunissen, 2004) when stored at 30° C. This phenomenon did not occur at lower temperatures (10° C) suggesting a temperature-dependent acquisition of resistance.

More recent reports on heat-resistant ascospores include the fungus *Monascus ruber* from brine containing thermally processed canned green olives (Panagou *et al.*, 2002). The spores are moderately heat-resistant in a phosphate buffer (pH 7.0; D_{75} = 7.1 min) and not much higher in citrate buffer (pH 3.0; D_{75} = 7.8 min). In brine (pH 3.8; 5.6% NaCl) the D_{70} was somewhat increased, but the D_{75} markedly lowered (4.9 min). In brine with higher salt content (10.5% potassium chloride) the D_{70} is markedly higher and the D_{75} similar as in the buffer situations. The D_{80} is lower (0.9-1.0 min) in the salty environments than in buffer (1.7-2.0 min). Taken together these results show a complex behaviour of the spores with respect to the presence of salt in the bathing medium. At lower temperatures salt is correlated with an increase of heat resistance, but this effect inverses at higher temperatures. The author suggests that potassium chloride might have a detrimental effect if damage occurs to the spores. While other heat-resistant fungi show initial activation of germination by heat, this behaviour is not reported in this study.

GERMINATION OF ASCOSPORES AFTER HEAT ACTIVATION

The phenomenon of dormancy

Dormancy can be regarded as a form of hypometabolism, which includes any rest period or reversible interruption of the phenotypic development (Sussman and Halvorson, 1966). Exogenous dormancy (quiescence) includes delayed development due to physical or chemical conditions of the environment. Constitutive dormancy is a condition in which development is delayed as an innate property such as a barrier to the penetration of nutrients or water, a metabolic block, or as a result of the action of a self-inhibitory compound. There are clear indications that ascospores of the heat-resistant fungi described here exhibit constitutive dormancy and need a robust physical signal (as heat, but also high pressure) for breaking of dormancy. In most of the studies done on heat-resistant fungi, activation of spores to germinate is observed where the number of viable counts after a short heat treatment is increased several log cycles (e.g., *Byssochlamys nivea* (5 min, 75 °C) , Yates *et al.*, 1968; *Eurotium herbariorum* (15 min, 60 °C), Splitstoesser *et al.*, 1989; *Neurospora tetrasperma,* (5 min, 65 °C), Lingappa and Sussman, 1959; *Talaromyces flavus*, Katan, 1985; *Neosartorya fischeri,* Beuchat, 1986; Gomez *et al.*, 1989; *T. macrosporus*, (7-10 min, 85 °C, Dijksterhuis and Teunissen, 2004). The fungus *T. flavus* shows heat activation at 80 and 85 °C and activation is followed by killing during prolonged treatment at the higher temperature (Beuchat, 1986). At lower temperatures activation fails and only low numbers of germinated spores are observed. Recent observations at our laboratory indicate that with *T. macrosporus* no activation was observed on Malt Extract Broth below 60 °C during a one-hour activation period (J. Dijksterhuis, unpublished results). When suspensions of *N. fischeri* were heated at different temperatures for 10 min, a very steep activation (nearly 1000 times) occurred between 60 and 70 °C (Gomez *et al.*, 1989). When the cells were dry-heated at relatively low humidities (50%) at 95 °C for 60 min this traject was starting at lower temperatures (50 °C) and completed at 65 °C. Both *T. flavus* and *N. fischeri* isolates show marked variations in the temperature and the extent of heat activation (Katan, 1985; Splittstoesser *et al.*, 1993).

The speed of activation

Remarkably, the speed of activation increases with higher temperatures in the case of *T. flavus*, although *N. fischeri* exhibits constant rates of heat activation between 70-85 °C (Beuchat, 1986; King and Halbrook, 1987). In our laboratory we observed that ascospores of *T. macrosporus* added to preheated buffer of pH 6,8 were fully activated within 2 min. Kikoku (2003) activated ascospores of *T. macrosporus* in a citrate-phosphate buffer (pH 6.5) within 100 s at 81 and 82.5 °C. Shorter activation times were observed at higher temperatures, namely 60 s at 86.5 and 87 °C, and 35 s at 91 °C. From these data the author calculated rate constants of heat activation (expressed as k), which range from 1.2 to 4.1/min between 81 and 91 °C. At 84 °C no difference in k was observed between pH 3.5 or 6.5 (2.9 and 2.8/ min, respectively) and also in phosphate buffer (pH 6.6) the k value was 2.8/min. However in grape juice (5 °Brix) a very high k value was observed (7.7/min). Thus, the presence of the sugars or organic acids or some other compound in the fruit juice resulted in a very rapid activation at this temperature (100% in 20 s). The activation energy (Ea) of a certain reaction was calculated using the Arrhenius plot where ln (0.23303.k) is plotted against 1/T. If activation is confined to a conformational or chemical change of one compound in the ascospore, for instance a receptor protein or structures inside the (plasma) membrane, the Ea reflects the energy needed to convert 1 mole of such a compound. Changes in proteinaceous compounds do need a different energy absorption than lipid compounds and then the E_a could give clues about the nature of activation. In case of systemic changes inside the ascospore, many different alterations absorb the energy delivered by the heat and the Ea calculated does not give much information. Recent findings suggest that heat activation includes the release of a protein, alterations of different fractions inside the cell wall and an increase in the permeability of the ascospore cell wall (J. Dijksterhuis, unpublished results). All these different processes will make a precise indication of the activation factor with the Arrhenius plot more difficult.

Dry heat treatments of ascospores

Apart from heat, also a drying treatment can result in activation, but differences seem to occur between species. For *N. fischeri* the dormant state can be broken by a drying treatment of 18h at 40 °C (Beuchat, 1992), but *T. flavus* ascospores did not show a release of dormancy. After a subsequent storage period under dry conditions (a_w = 0.23) up to 30 months or more, the *T. macrosporus* ascospores did not need heat activation anymore and germinated after wetting. Remarkably, in case of storage in blueberry, grape or strawberry fruit powders, full activation was obtained within 8 months of storage, after 20 months in pineapple powder, but only partial activation after 30 months in apple powders. As stated earlier, heating at 50% r.h. (dry heat treatment) at 95 °C (for 30 or 60 min) activated *N. fischeri* ascospores, but the temperature of the wetting or recovery buffer was crucial for the viable count obtained (Gomez *et al.*, 1989, 1993). The latter could hint into a de-activation phenomenon where ascospores are reversibly brought back into the dormant state. It could also hint into the direction of imbibitional damage to the cells where rewetting in a cold fluid may cause the occurrence of different phases in the membrane at the same time resulting in leakage and cell damage (van Bilsen *et al.*, 1994). Glycerol had a protective action on the ascospores after drying during recovery in ice water, which indeed suggests that these resilient ascospores undergo imbibitional damage. Surprisingly, water vapour alleviated the detrimental effect of temperature on recovery.

Later research of the same group (Gomez *et al.*, 1994) indicated that the ascospores were more heat resistant as the relative humidity had decreased up to four orders in magnitude compared to the resistance during wet heat.

Recovery in hot buffer (80 °C) was higher than in cold buffer (0 °C) only in ascospores that were heated at 30 and 40% RH and subsequently vapour treated, but this "recovery" was interpreted as a result of heat activation.

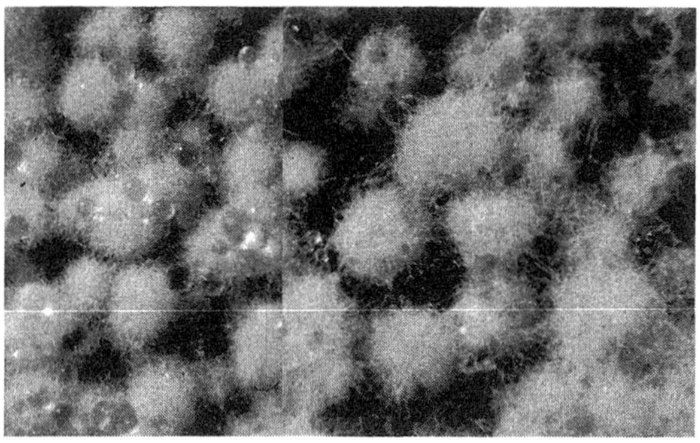

Figure 1. Formation of fruit bodies of the fungus *Talaromyces macrosporus*; numerous yellow coloured ascomata (fruiting bodies) can be observed after 7 (left) and 14 days (right) after inoculation of the fungus.

Can spores be de-activated?

T. macrosporus is a suitable model system for the study of heat-resistant ascospore biology. The fungus has a worldwide distribution and is a homothallic fungus (forms sexual fruit bodies without the need of different mating types) on oatmeal agar at 30 °C (Figure 1). Ascomata and ascospores can be harvested by a simple procedure to a dense homogenous suspension. The fungus also produces Penicillia with elongated phialides and conidia (Figure 2). Ascospores do not germinate when left in malt extract broth for prolonged times. As stated above, a 7-10 min treatment at 85 °C, however, results in activation and synchronisation of germination of the majority of the cells. HPLC studies showed that these cells contain very high concentrations of trehalose, up to 15-20% of the wet cell weight (that is 24-32% of the dry weight, Dijksterhuis *et al.*, 2002). The low water content of the spores (38%) introduces a very high viscosity inside the spores which is recently measured by means of EPR (Electron Paramagnetic Resonance) studies and cryo-electron microscopy (Dijksterhuis *et al.*, 2007). Furthermore these cells are encompassed by a very thick ornamented cell wall (Figure 3) that is covered with numerous spikes (Figure 4).

A sudden lowering of the temperature or a reduction of the water content might introduce a glass transition situation inside the cell. The glassy state is an amorphous phase characterized by very low movement speeds of the molecules inside the cytoplasm. This state inside the cell can be considered as a "biological glass," a term introduced by Buitink (2000). Thoroughly dried glasses, especially those containing trehalose are characterised by a high melting temperature (Wolkers *et al.*, 1998).

Could activated ascospores resume dormancy again when the cytoplasm is brought into a glassy state, for instance, by plunging of the cells in liquid nitrogen (a sudden lowering of the temperature) or controlled drying below 3% water? We observed that heat-activated spores when plunged in nitrogen or kept at –20 °C directly germinated upon introduction into conducive conditions (Dijksterhuis and Samson, 2006). Further, ascospores were dried according to the procedures used at the CBS, and ascospores remained dormant after drying and could be effectively activated by a heat treatment after resuspension in buffer.

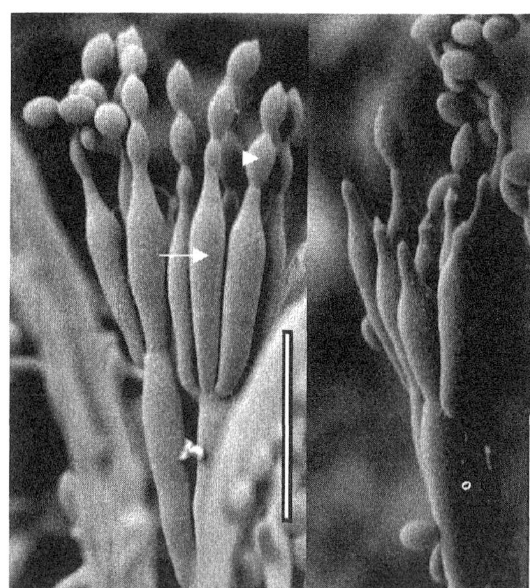

Figure 2. Two examples of conidiophores formed by *T. macrosporus* as viewed by means of cryo-SEM. Not the elongated phialides (arrow) and conidia (arrowhead).

Activated spores, however, germinated direct-ly upon rewetting, but these cells showed similar tolerance to the drying treatment as dormant cells. In addition, these cells survived another heat treatment very well and no additional increase or decrease of cell numbers occurred compared with undried activated cells.

This indicated that these cells had retained their activated state, but still showed heat resistance. Both drying tolerance and heat resistance had decreased markedly after keeping the cells at 2h at 30 °C. These combined observations suggest that important phase transitions of the inner cell may not change the status of activation.

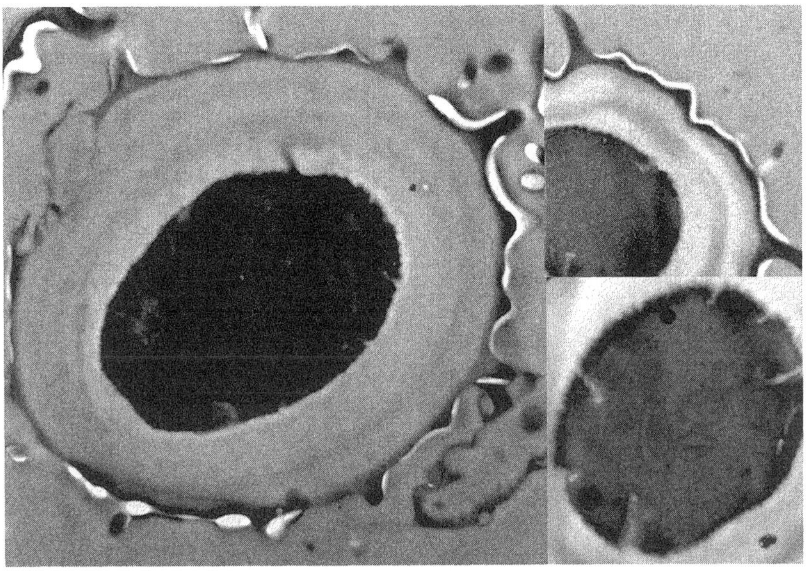

Figure 3. The ascospores of *T. macrosporus* as observed by means of transmission electron microscopy. It clearly illustrates the very thick outer cell wall of the spore which contains different layers and ornamentations (top right) and deep indentations of the cell wall into the cell (lower right). Photographs taken by Kenneth van Driel, CBS, Utrecht.

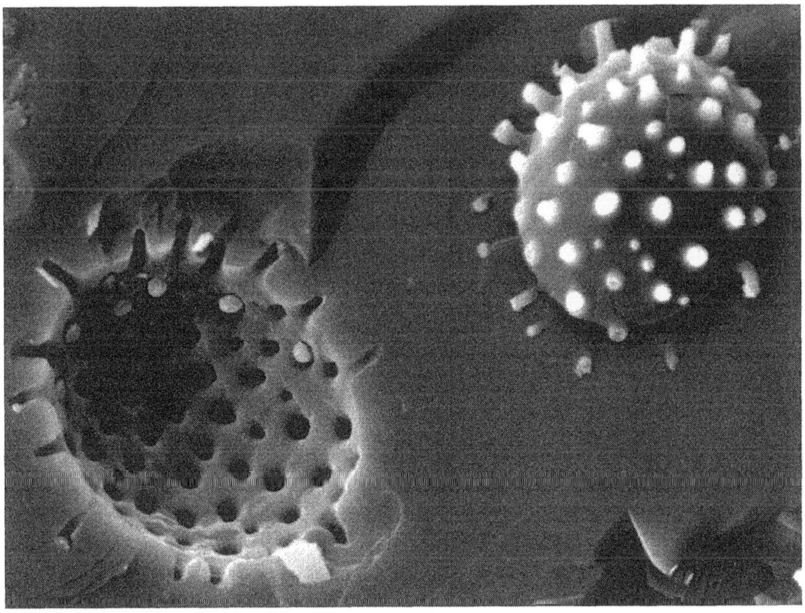

Figure 4. The ornamentations are clearly visible on this freeze break cryo-scanning electron micrograph. Photograph taken by Jaap Nijsse, Plant Physiology, Wageningen.

An ecological function for heat resistance?

The ecological function of heat activation is unknown, but could be related to survival of fungal cells after fires followed by a rapid colonisation of soils. *Talaromyces* is a fungal genus isolated from soil worldwide (Fravel and Adams, 1986) and postfire ascomycetes are described before as *Daldinia loculata,* a fungus that produces fruit bodies on trees killed by fire (Johannesson *et al.*, 2000; Guidot *et al.*, 2003). Remarkably, this fungus produces ascospores that germinate through a stage including a sudden opening of the thick outer cell wall, which enables the germ tube to grow out (Beckett, 1976a,b), which is similar to the phenomenon of prosilition in case of *T. macrosporus* (see below).

Indeed, activated germinating spores could quickly fill empty ecological niches that are formed as a result of the fire. Fire-related germination is well known with plant seeds (e.g., *Eucalyptus*), but attempts at our laboratory to activate *T. macrosporus* with smoke extracts, as is observed in case of plant seeds (Flematti *et al.*, 2004) failed. It is also possible that heat activates spores because it results in changes in the cell that are also addressed by the still hidden activation signal in nature.

A last possibility might be that the dormancy of these ascospores works as a clock to enable the spores to survive long periods and henceforth these spores can be regarded as "dispersed" in time.

BREAKING THE DORMANCY BY ULTRA-HIGH PRESSURE

Ascospores can be activated by a novel non-thermal activation method

Heat treatment of food products is a classical and very important method for preservation, but it also influences firmness, taste and vitamin content. Modern consumers tend to prefer "fresh" food and therefore shorter treatments at lower temperatures gain popularity. In addition, a new generation of non-thermal preservation techniques is under development (Barbosa-Cànovas *et al.*, 1998; Smelt, 1998). One of these alternative methods is the application of high pressure that results in death of microorganisms, but structure and vitamin content of the food stay relatively unchanged.

Vegetative (growing) microbial cells are sensitive to high-pressure treatments at 200-400 MPa (0,1 MPa = 1 bar, Barbosa-Cànovas *et al.*, 1998). Fungal and bacterial spores are inactivated at much higher pressures (Maggi *et al.*, 1994; Butz *et al.*, 1996; Wuytack *et al.*, 1998) and ascospores of *B. nivea* are inactivated above 600 MPa combined with high temperatures (60 °C, Butz *et al.*, 1996), but these treatments were more effective when the treatments were repeated shortly after each other (designated as oscillatory treatments, Palou *et al.*, 1998). Since pressurization is already commercially used to pasteurize fruit juices and other products, pressure resistance of these fungi is of relevance for the food industry.

Two independent recent studies showed that ascospores of *T. macrosporus* were activated by high-pressure treatments (Reyns *et al.*, 2003; Dijksterhuis and Teunissen, 2004). Both studies showed that even a very short treatment at high pressure caused maximal activation. Dijksterhuis and Teunissen (2004) observed full dormancy after 200 MPa treatments and activation of part of the spores (up to 7% of the cells) between 400 and 800 MPa, while Reyns *et al.* (2003) observed partial activation at 200 MPa and full activation at 600 MPa following 15 seconds of treatment. The following aspects seem important for this difference. Firstly, the cultivation of the fungus was different including temperature, medium and harvesting time of the ascospores. Dijksterhuis and Teunissen (2004) report that the age of the fungal culture during harvesting of the ascospores correlated with the heat resistance and that considerable increase of heat resistance occurred between 20 and 40 days of culturing.

So, the combined results of the papers hint in the direction that the extent of maturity of the ascospores may also influence the ability of the cells to remain dormant (see Dijksterhuis and Samson, 2006). Further, Reyns *et al.* (2003)

had a pre-treatment of the ascospore suspension for 20 min at 65 °C to kill the vegetative cells. In a buffer at a pH 6,8; this treatment will not activate ascospores, but 70 °C gives a strong increase of germination (J. Dijksterhuis, unpublished results). At lower pH (3.0), Reyns *et al.* (2003) observed that near complete activation occurs after the heat treatment at 65 °C. At our laboratory we observed activation at even lower temperatures in low pH solutions (J. Dijksterhuis, unpublished results). Thirdly, the actual high-pressure treatments are done in a buffer (Dijksterhuis and Teunissen, 2004, 10 m*M* ACES, pH 6,8) or in distilled water (Reyns *et al.*, 2003). It is known that these treatments lower the acidity of the medium temporarily, which will be less extensive in a buffer and also this could have an effect on activation. Dijksterhuis and Teunissen (2004) performed cryo-electron microscopy on the spores and observed changes of the cell wall after short treatments at 600 MPa, where Reyns *et al.* (2003) illustrated that treated spores collapsed after air drying, while untreated spores maintained their shape. These observations indicate that structural changes occur in the cell wall and that these have a direct influence of the process of activation.

All these factors learn that activation of the ascospores of *T. macrosporus* is influenced by many different cues and that profound knowledge will be needed of activation and germination in order to tackle the growing problem of heat-resistant fungi in food products.

The relation between different treatments

Ascospores of *Eurotium herbariorum* (also sometimes designated as *E. repens*) are mildly heat resistant (Splittstoesser *et al.*, 1989). Eicher and Ludwig (2002) showed that 8% of these spores were activated from dormancy after 60 min at 200 MPa (and not above or below these pressure values, H. Ludwig, personal communication) and that 50% germinated after 8 min at 60 °C. After 18 hours, approx. 15% of the cells showed signs of germination at room temperature. Ascospores that were heat activated (15 min, 60 °C) were more sensitive to a subsequent high-pressure treatment at 500 MPa and the effect was much stronger when the heat treatment was applied immediately after pressurization. The spores recovered ten-fold during storage at 20 °C in an isotonic salt solution; a phenomenon designated by the authors as "re-stabilisation." When pressurisation was first (500 MPa, 30 min), heat activation did not enhance germination, but slightly reduced viability. A pause between these treatments reduced the counts further.

These relations between treatments indicate that germination and inactivation are complex phenomena ruled by many different parameters. In case of *T. macrosporus,* full heat activation occurred after short treatments at 800 MPa (Dijksterhuis and Teunissen, 2004). However, the experiments of Reyns *et al.* (2003) show clear heat sensitivity of ascospores for the activation heat treatment (30 min, 80 °C) after all pressure treatments at 600 MPa and 700 MPa.

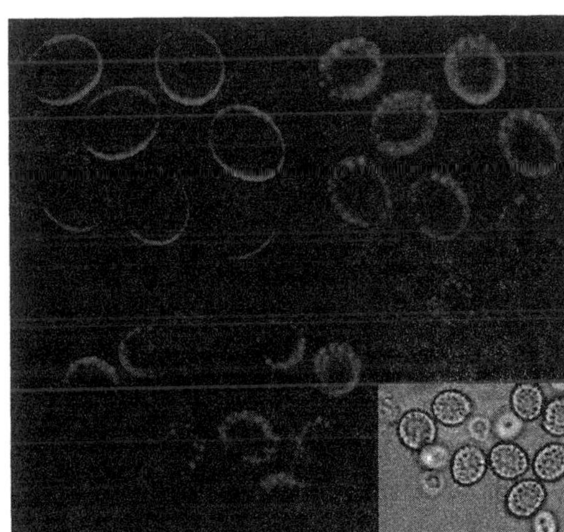

Figure 5. Ascospores of *T. macrosporus* exhibit a strong autofluorescence in a broad range of excitation wavelengths. Top panel shows two focal planes through the spores, illustrating the spikes on the surface of the cell wall. The lower panel shows prosilited ascospores with ruptured cell walls where ejection has taken place (for an image with transmitted light, see the inset).

HEAT-RESISTANT ASCOSPORES OF *TALAROMYCES MACROSPORUS*

Heat activation is as earlier stated a process that includes different mechanisms including a change in permeability in the ascospore cell wall (J. Dijksterhuis *et al.*, in preparation) and a possible influx of water into the inner cell. Upon heat treatment, the disaccharide trehalose is degraded by a high trehalase activity and the product of hydrolysis, glucose, is accumulated inside the spore, but resides inside the cell for only a very short time (Dijksterhuis *et al.*, 2002). Then it is released into the bathing medium till no glucose could be detected inside the spore after breaking of the cells. These observations, indicate a massive release of glucose from the germinating cell. It is not clear if glucose transporters are involved in this process, the spores show only very low respiration during these stages of germination (Dijksterhuis *et al.*, 2002).

After 150 min or more the inner cell jumps through the outer cell wall, which becomes ruptured. The emptied outer cell wall remains attached to the protoplast which is encompassed by the inner layer of the ascospore cell wall by means of a third, fibrillar layer (Dijksterhuis *et al.*, 2007). This process is very sudden, it only takes a second or less, and is termed prosilition (Latin: *prosilire*, to jump out). After this remarkable phenomenon, the respiration of the cells increases strongly and isotrophic growth (swelling) is followed by polarised growth (germ tube formation). The time needed from heat activation to germ tube formation is approx. 7 hours following the heat treatment. Figure 5 shows prosilited and unprosilited cells by means of the fluorescence microscope.

Recently, prosilition was also confirmed in other species of *Talaromyces* namely *T. stipitatus, T. helicus* and *T. bacillosporus*. However, ascospores of *Neosartorya* species seem to germinate by a slow separation of the two shell-like ornamented halves and subsequent formation of a germ tube. Apparently, the two genera show different modes of ascospore germination.

HEAT-RESISTANT ASCOSPORES OF *NEUROSPORA*

Earlier studies on *Neurospora tetrasperma*

One of the most important model systems in fungal biology is *Neurospora crassa* and the very related *N. tetrasperma* forms ascospores that survive temperatures at and above 60 °C for long times (> 1 h) and these ascospores do need a "heat flash" (65 °C, 5 min) for activation of germination (Lingappa and Sussman, 1959). These ascospores are not relevant in the food context, but research on these spores is done in the past and summarised here. After heat activation the spores can be best grown at 27 °C. During this incubation the heat resistance of the spores decreases. The latter is also the case when the exospore (the thin outer layer of the spore cell wall, which is actively shed during germination) is removed by using Clorox, but it was not clear if the compound itself did have side effects (Lingappa and Sussman, 1959). Remarkably, chemical compounds including furfural and phenetyl-alcohol were able to break the dormancy of these spores (Eilers and Sussman, 1970). It was observed that furfural was taken up by the ascospores and that the compound bound to the cell wall of the spores. In 1976 Sussman hypothesized that these compounds may act by causing an alteration in lipid moieties of the spore and thereby break dormancy.

Upon activation trehalose inside the ascopores is rapidly degraded, which was at that time interpreted as carbon storage compounds usage for energy metabolism and biosynthesis. Hecker and Sussman (1973) suggested that the trehalase needed for this degradation was associated with the inner cell wall of the ascospore which was based on immunofluorescence studies (where the cytoplasm had a strong autofluorescence) and the observation that trehalase was stabilised, that means protected from drying followed by a heat treatment, by cell wall preparations of the spores (and also by a fraction of it). Belmans *et al.* (1983) observed that the activation temperature of *N. tetrasperma* ascospores shifted upwards with 4K/100 MPa (1000 bar) during high-pressure treatment.

They argued that this is typical for proteins and not for lipids, which have melting characteristics from 20 K/100 MPa and concluded that the activation trigger in these spores is transduced by a protein conformational change.

Later studies

During sexual development in *N. crassa* (perithecium development and also ascospore germination) the amount and also composition of the fatty acid population in cultures changes (Goodrich-Tanrikulu *et al.*, 1998). In developing asci oleate (18:1) becomes the predominant fatty acid, while vegetative tissue contains mostly linoleate (18:2). The amount of polar lipids inside ascospores before heat activation was 8 pg/spore on a total fatty acid content of 44 pg/spore. In the first 6 hours after activation, the amount of linoleate increases markedly in the polar lipids. The content of triacylglycerol inside spores increases from 60% of the total to a maximum of 80-90% between 3 and 12 h after activation (after 4-6 the first hyphal tips appear from the spores in *N. tetrasperma*). Upon activation, ascospores show a biphasic increase in respiration. The first phase lasts 90 min and is independent of (mitochondrial and cytoplasmic) protein synthesis. Radioactive methionine is incorporated into the spores only after 90 min when the first signs of germ tube formation occur (Hill *et al.*, 1992) and the second stage of increase in respiration starts. However, it is possible that the spores do not take up the labelled compound while the cell wall is not permeable. Isolated mRNA was hybridised with cDNA for mitochondrial ATPase. In dormant cells some transcripts were present, but had disappeared at 30-90 min after activation. Then transcription obviously started as judged by reappearance of mRNA. In another study, germinating ascospores were exposed to a heat shock. Only cells reacted with accumulation of hsp30 after 480 min, cells at 90-300 min did not react (Plesofsky-Vig *et al.*, 1992).

All these data suggest that ascospores germinate through and initial stage characterised by lower respiration and biosynthesis of proteins and nucleic acids. Recent studies on *T. macrosporus* ascospores show that also these spores exhibit such a stage, which is also characterised by high internal viscosity of the cytoplasm, which drops suddenly after prosilition (Dijksterhuis *et al.*, 2007). Obviously, the very stress resistant ascospore needs time to renormalize to a vegetative cell as the cell constituents become "less protected."

FUTURE DEVELOPMENTS

Heat-resistant ascospores remain a concern in food industry due to minimal processing of food products. More recently, these fungi are also reported from raw pectin and dairy products, which may mean that their occurrence is increasing. More research has to be done on the precise mechanism of germination to find alternative methods to prevent spoilage with these fungi. Apart from this, these unique resistant eukaryotic cells may give new insights in unique ways of protection of biological compounds.

Acknowledgements

The authors thank Kenneth van Driel, CBS, for transmission electron microscopy on ascospore, Jaap Nijsse (now at Unilever Research Laboratory, Vlaardingen, The Netherlands) for help during cryo-planing.

REFERENCES

Agab, M. A., and Collins, M. (1992). Effects of treatments environment (temperature, pH, water activity) on the heat resistance of yeasts. Journal of Food Science and Technology 29:5-9.

Aragão, G. M. F. (1989). Identificação e determinação da resistência térmica de fungos filamentosos termoresistentes isolados da polpa de morango. Master Thesis. Universidade de Campinas, Brazil.

Baggerman, W. I., and Samson, R. A. (1988). Heat resistance of fungal spores. *In* Introduction to, food-borne fungi 3rd edition (Samson, R. A., Reenen-Hoekstra, E. S. van, eds.). Centraalbureau voor Schimmelcultures, Baarn, The Netherlands.

Baldy, R. W., Sommer, N. F., and Buckley, P. M. (1970). Recovery of viability and radiation resistance by heat-injured conidia of *Penicillium expansum* Lk. Ex Thom. Journal of Bacteriology 102:514-520.

Barbosa-Cànovas, G. V., Pothakamury, U., Palou, E., and Swanson, B. (1998). High hydrostatic pressure food processing. *In* Nonthermal preservation of foods (Barbosa-Cànovas, G. V., Pothakamury, U., Palou, E., and Swanson, B., eds.), Marcel Dekker, Inc., New York, U.S.A, pp. 9-52.

Bayne, H. G., and Michener, H. D. (1979). Heat resistance of *Byssochlamys* ascospores. Applied and Environmental Microbiology 37:449-453.

Beckett, A. (1976a). Ultrastructural studies on exogenously dormant ascospores of *Daldinia concentrica*. Canadian Journal of Botany 54:689-697.

Beckett, A. (1976b). Ultrastructural studies on germinating ascospores of *Daldinia concentrica*. Canadian Journal of Botany 54:698-705.

Belmans, D. L., Laere, A. J. van, and Assche, J. A. van (1983). Effect of n-alcohols and high pressure on the heat activation of *Neurospora tetrasperma* ascospores. Archives of Microbiology 134:49-51.

Beuchat, L. R. (1981). Effects of potassium sorbate and sodium benzoate on inactivating yeasts in broths containing sodium chloride and sucrose. Journal of Food Protection 44:765-769.

Beuchat, L. R. (1986). Extraordinary heat resistance of *Talaromyces flavus* and *Neosartorya fischeri* ascospores in fruit products. Journal of Food Science 51:1506-1510.

Beuchat, L. R. (1988a). Thermal tolerance of *Talaromyces flavus* ascospores as affected by growth medium and temperature, age and sugar content in the inactivation medium. Transactions of the British Mycological Society 90:359-374.

Beuchat, L. R. (1988b). Influence of organic acids on heat resistance characteristics of *Talaromyces flavus* ascospores. International Journal of Food Microbiology 6:97-105.

Beuchat, L. R. (1992). Survival of *Neosartorya fischeri* and *Talaromyces flavus* ascospores in fruit powders. Letters in Applied Microbiology 14:238-240.

Beuchat, L. R., and Kuhn, G. D. (1997). Thermal sensitivity of *Neosartorya fischeri* ascospores in regular and reduced-sugar grape jelly. Journal of Food Protection 60:1577-1579.

Bilsen, D. van, Hoekstra, F. A., Crowe, L. M. and Crowe, J. H. (1994) Altered phase behaviour in membranes of aging dry pollen may cause imbibitional leakage. Plant Physiology 104:1193-1199.

Bröker, U., Spicher, G., and Ahrens, E. (1987a). Zur Frage der Hitzeresistenz der Erreger der Schimmelbildung bei Backwaren. 2. Mitteilung: Einfluß endogener Faktoren auf die Hitzeresistenz von Schimmelsporen. Getreide, Mehl und Brot 41:278-284.

Bröker, U., Spicher, G., and Ahrens, E. (1987b). Zur Frage der Hitzeresistenz der Erreger der Schimmelbildung bei Backwaren. 3. Mitteilung: Einfluß exogener Faktoren auf die Hitzeresistenz von Schimmelsporen. Getreide, Mehl und Brot 41:344-355.

Buitink, J. (2000). Biological Glasses: Nature's way to preserve life. Ph.D. Thesis, University of Wageningen, The Netherlands, pp. 5-17.

Butz, P., Funtenberger, S., Haberditzl, T., and Tausher, B. (1996). High pressure inactivation of *Byssochlamys nivea* ascospores and other heat resistant moulds. Lebensmittel Wissenshaft und Technologie 29:404-410.

Casella, M. L. A., Matasci, F., and Schmidt-Lorenz, W. (1990). Influence of age, growth medium, and temperature on heat resistance of *Byssochlamys nivea* ascospores. Lebensmittel Wissenshaft und Technologie 23:404-411.

Conner, D. R., and Beuchat, L. R. (1987a). Heat resistance of *Neosartorya fischeri* as affected by sporulation and heating medium. International Journal of Food Microbiology 4:303-312.

Conner, D. R., and Beuchat, L. R. (1987b). Efficacy of media for promoting ascospore formation by *Neosartorya fischeri*, and the influence of age and culture temperature on heat resistance of ascospores. Food Microbiology 4:229-238.

Conner, D. R., Beuchat, L. R., and Chang, C. J. (1987). Age-related changes in ultrastructure and chemical composition associated with changes in heat resistance of *Neosartorya fischeri* ascospores. Transactions of the Britisch Mycological Society 89:539-550.

Crowe, J. H., Crowe, L. M., and Chapman, D. (1984). Preservation of membranes in anhydrobiotic organisms: the role of trehalose. Science 223:701-703.

Dijksterhuis, J., Driel, K. G. A. van, Sanders, M. G., Molenaar, D., Houbraken, J. A. M. P., Samson, R. A., and Kets, E. P. W. (2002). Trehalose degradation and glucose efflux precede cell ejection during germination of heat-resistant ascospores of *Talaromyces macrosporus*. Archives of Microbiology 178:1-7.

Dijksterhuis, J., and Teunissen, P. G. M. (2004). Dormant ascospores of *Talaromyces macrosporus* are activated to germinate after treatment with ultra high pressure. Journal of Applied Microbiology 96:162-169.

Dijksterhuis, J., and Samson, R. A. (2006). Activation of stress-resistant ascospores by novel food preservation techniques. Advances in Experimental Medicine and Biology 517:247-260.

Dijksterhuis, J., Nijsse, J., Hoekstra, F. A., and Golovina, E. A. (2007). High viscosity and anisotropy characterize the cytoplasm of fungal dormant stress-resistant spores. Eukaryotic Cell 6:157-170.

Eicher, R., and Ludwig, H. (2002). Influence of activation and germination on high pressure inactivation of ascospores of the mould *Eurotium repens*. Comparative Biochemistry and Physiology Part A 131:595-604.

Eilers, F. I., and Sussman, A. S. (1970). Furfural uptake by *Neurospora* ascospores. Planta 94:265-272.

Elliot, B., Haltiwanger R. S., and Futcher, B. (1996). Synergy between trehalose and Hsp104 for thermotolerance. Genetics144:923-933.

Engel, G., and Teuber, M. (1991). Heat resistance of ascospores of *Byssochlamys nivea* in milk and cream. International Journal of Food Microbiology 12:225-234.

Enigl, D. C., King, A. D., and Török, T. (1993). *Talaromyces trachyspermis,* a heat-resistant mold isolated from fruit juice. Journal of Food Protection 56:1039-1042.

Flematti, G. R., Ghisalberti, E. L., Dixon, K. W. and Trengove, R. D. (2004). A compound from smoke that promotes seed germination. Science 305: 977.

Fravel, D. R., and Adams, P. B. (1986). Estimation of United States and world distribution of *Talaromyces flavus*. Mycologia 78:684-686.

Fujikawa, H., and Itoh, T. (1996). Tailing if thermal inactivation curve of *Aspergillus niger* spores. Applied and Environmental Microbiology 62:3745-3749.

Fujikawa, H., Morozumi S., Smerage, G. H., and Teixeira, A. A. (2000). Comparison of capillary and test tube procedures for analysis of thermal inactivation kinetics of mold spores. Journal of Food Protection 63:1404-1409.

Goodrich-Tanrikulu, M., Howe, K., Stafford, A., and Nelson, M. A. (1998). Changes in fatty acid composition of *Neurospora crassa* accompany sexual development and ascospore germination. Microbiology 144:1713-20.

Gomez, M. M., Busta, F. F. and Pflug, I. J. (1989). Effect of the post-dry heat treatment temperature on the recovery of ascospores of *Neosartorya fischeri*. Letters in Applied Microbiology 8:59-62.

Gomez, M. M., Pflug, I. J. and Busta, F. F. (1993). Factors affecting recovery of *Neosartorya fischeri* after exposure to dry heat. Journal of Parenteral Science and Technology 47:300-305.

Gomez, M. M., Pflug, I. J. and Busta, F. F. (1994). Resistance of *Neosartorya fischeri* to wet and dry heat. Journal of Pharmaceutical Science and Technology 48:16-23.

Guidot, A., Johannesson, H., Dahlberg, A. and Stenlid, J. (2003). Parental tracking in the postfire wood decay ascomycete *Daldinia loculate* using highly variable nuclear gene loci. Molecular Ecology 12:1717-1730.

Gumerato, H. F. (1995). Desenvolvimento de um program de computador para identificação de alguns fungos comuns em alimentos e determinação de resistência térmica de *Neosartorya fischeri* isolado de maçãs. Master Thesis. Universidade de Campinas, Brazil.

Hecker, L. I. and Sussman, A. S. (1973). Localization of trehalase on the ascospores of Neurospora. Journal of Bacteriology 115:592-599.

Hill, E. P., Plesofsky-Vig, N., Paulson, A., and Brambl, R. (1992). Respiration and gene expression in germinating ascospores of *Neurospora tetrasperma.* FEMS Microbiology Letters 69:111-115.

Johannesson, H. S., Johannesson, K. H. P., and Stenlid, J. (2000). Development of primer sets to amplify fragments of conserved genes for use in population studies of the fungus *Daldinia loculata*. Molecular Ecology 9:375–378.

Katan, T. (1985). Heat activation of dormant ascospores of *Talaromyces flavus.* Transactions of the Britisch Mycological Society 84:748-750.

Kavanagh, J., Larchet, N., and Stuart, M. (1963). Occurrence of a heat-resistant species of *Aspergillus* in canned strawberries. Nature 198:1322.

Kikoku, Y. (2003). Heat activation characteristics of *Talaromyces* ascospores. Journal of Food Science 68:2331-2335.

King, A. D. (1997). Heat resistance of *Talaromyces flavus* ascospores as determined by a two phase slug flow heat exchanger. International Journal of Food Microbiology 35:147-151.

King, A. D., and Halbrook, U. (1987). Ascospore heat resistance and control measures for *Talaromyces flavus* isolated from fruit juice concentrate. Journal of Food Science 52:1252-1254.

King, A. D., and Whitehand, L. C. (1990). Alteration of *Talaromyces flavus* heat resistance by growth conditions and heating medium composition. Journal of Food Science 55:830-832.

King, A. D., Bayne, H. G., and Alderton, G. (1979). Nonlogarithmic death rate calculations for *Byssochlamys fulva* and other microorganisms. Applied and Environmental Microbiology 37:596-600.

Kotzekidou, P. (1997). Heat resistance of *Byssochlamys nivea, Byssochlamys fulva* and *Neosartorya fischeri* isolated from canned tomato paste. Journal of Food Science 62:410-412.

Lingappa, Y., and Sussman, A. S. (1959). Changes in the heat resistance of ascospores of *Neurospora* upon germination. American Journal of Botany 49:671-678.

Maggi, A., Gola, S. , Spotti, E., Rovere, P., and Mutti, P. (1994). High-pressure treatments of ascospores of heat-resistant moulds and patulin in apricot nectar and water. Industria conserve 69:26-29.

Michener, H. D., and King, A. D. (1974). Preparation of free heat-resistant ascospores from *Byssochlamys* asci. Applied Microbiology 27:671-673.

Nierman, W. C. *et al.* (2005). Genomic sequence of the pathogenic and allergenic filamentous fungus *Aspergillus fumigatus*. Nature 438: 1151-1156.

Olliver, M., and Rendle, T. (1934). A new problem in fruit preservation. Studies on *Byssochlamys fulva* and its effects on tissues of processed fruit. Journal of the Society of Chemistry and Industry 53: 166-172.

Palou, E., Lopez-Malo, A., Barbosa-Cánovas, G. V., Welti-Chanes, J., Davidson, P. M., and Swanson, B. G. (1998). Effect of oscillatory high hydrostatic pressure treatments on *Byssochlamys nivea* ascospores suspended in fruit juice concentrates. Letters in Applied Microbiology 27:375-378.

Panagou, E .Z., Katsaboxalis, C. Z., and Nychas, G-J. E. (2002). Heat resistance of *Monascus ruber* ascospores isolated from thermally processed green olives of the Conservolea variety. International Journal of Food Microbiology. 76:11-18.

Parish, M. E. (1991). Microbiological concerns in citrus juice processing. Food Technology 45:128-133.

Pitt, J. I., and Hocking, A. D. (1982). Food spoilage fungi. I. *Xeromyces bisporus* Fraser. CSIRO Food Res. Q.42:1-6.

Plesofsky-Vig, N., Paulson, A., Hill, E. P., Glasser L., and Brambl, R. (1992). Heat shock gene expression in germinating ascospores of *Neurospora tetrasperma*. FEMS Microbiology Letters 69:117-122.

Prestrelski, S. J., Tedeschi, N., Arakawa, T., and Carpenter, J. F. (1993). Dehydration-induced conformational transitions in protein and their inhibition by stabilizers. Biophysical Journal 65:661-671.

Put, H. M. C., and Jong, J. de (1980). The heat resistance of selected yeasts causing spoilage of canned soft drinks and fruit products. *In* Biology and Activities of Yeasts (Skinner, F. A., Passmore, S. M., and Davenport, R. R., eds.). Soc. Appl. Bacteriol. Symp. Series No. 9. Academic Press, London, U.K.

Put, H. M. C., and Jong, J. de (1982). The heat resistance of ascospores of four *Saccharomyces* spp. isolated from spoiled heat-processed soft drinks and fruit products. Journal of Applied Bacteriology 52:235-243.

Rajashekhara, E., Suresh, E. R., and Ethiraj, S. (1996). Influence of different heating media on thermal resistance of *Neosartorya fischeri* isolated from papaya fruit. Journal of Applied Bacteriology 81:337-340.

Rajashekhara, E., Suresh, E. R., and Ethiraj, S. (1998). Thermal death rate of ascospores of *Neosartorya fischeri* ATCC 200957 in the presence of organic acids and preservatives in fruit juices. Journal of Food Protection 61:1358-1362.

Rajashekhara, E., Suresh, E. R., and Ethiraj, S. (2000). Modulation of thermal resistance of ascospores of *Neosartorya fischeri* by acidulants and preservatives in mango and grape juice. Food Microbiology 17:269-275.

Reyns, K. M. F. A., Verbeke, E. A. V., and Michiels, C. E. W. (2003). Activation and inactivation of *Talaromyces macrosporus* ascospores by high hydrostatic pressure. Journal of Food Protection 66:1035-1042.

Riezman, H. (2004). Why do cells require heat shock to survive heat shock proteins to survive heat stress? Cell Cycle 3:61-63.

Scholte, R. P. M., Samson, R. A., and Dijksterhuis, J. (2001). Spoilage fungi in the industrial processing of food. *In* Introduction to Food- and Airborne Fungi (Samson, R. A., Hoekstra, E. S., Frisvad, J. C., and Filtenborg, O., eds.). Sixth Edition, Centraalbureau voor Schimmelcultures, Utrecht, The Netherlands, pp. 339-356.

Scott, V. N., and Bernard, D. T. (1985). Resistance of yeasts to dry heat. Journal of Food Science 50:1754-1755.

Scott, V. N., and Bernard, D. T. (1987). Heat resistance of *Talaromyces flavus* and *Neosartorya fischeri* from commercial fruit juices. Journal of Food Protection 50:18-20.

Silva, F. V. M., and Gibbs, P. (2004). Target selection in designing pasteurisation processes for shelf-stable high-acid fruit products. Critical Reviews in Food Science and Nutrition 44:1-8.

Smelt, J. P. P. M. (1998). Recent advances in the microbiology of high pressure processing. Trends in Food Science and Technology 9:152-158.

Splittstoesser, D. F., and Splittstoesser, C. M. (1977). Ascospores of *Byssochlamys fulva* compared with those of a heat resistant *Aspergillus*. Journal of Food Science 4: 685-688.

Splittstoesser, D. F., and King, A. D. (1984). Enumeration of *Byssochlamys* and other heat resistant moulds. *In* Compendium of Methods for the Microbiological Examination of Foods, 2nd Edition (Speck, M. L., ed.), American Public Health Ass., Washington DC, U.S.A.

Splittstoesser, D. F., Lammers, J. M., Downing, D. L., and Churey, J. J. (1989). Heat resistance of *Eurotium herbariorum*, a xerophilic mold. Journal of Food Science 54:683-685.

Splittstoesser, D. F., Nielsen, P. V., and Churey, J. J. (1993). Detection of viable ascospores of *Neosartorya*. Journal of Food Protection 56:599-603.

Spuy, J. E. van der, Matthee, F. N., and Crafford, D. J. A. (1975). The heat resistance of moulds *Penicillium vermiculatum* Dangeard and *Penicillium brefeldianum* Dodge in apple juice. Phytophylactica 7:105-108.

Sussman, A. S., and Halvorson, H. O. (1966). Spores, their Dormancy and Germination. Harper & Row, New York, USA.

Sussman, A. S. (1976). Activators of fungal spore germination. *In* The fungal spore. Form and function (Weber, D. J., and Hess, W. M., eds.). John Wiley & Sons, New York, U.S.A, pp. 101-139.

Tournas, V. (1994). Heat-resistant fungi of importance to the food and beverage industry. Critical Reviews in Microbiology 20:243-263.

Tournas, V., and Traxler, R. W. (1994). Heat resistance of *Neosartorya fischeri* strain isolated from pineapple juice frozen concentrate. Journal of Food Protection 57:814-816.

Tsang, E. W. T., and Ingledew, W. M. (1982). Studies on the heat resistance of wild yeasts and bacteria in beer. American Society of Brewing Chemists Journal 40:1-8.

Wolkers, W. F., Oldenhof, H., Alberda, M., and Hoekstra, F. (1998). A Fourier transform infrared microspectroscopy study of sugar glasses: application to anhydrobiotic higher plant cells. Biochimica et Biophysica Acta 1379:83-96.

Wuytack, E. Y., Boven, S., and Michiels, C. W. (1998) Comparative study of pressure-induced germination of *Bacillus subtilis* spores at low and high pressures. Applied and Environmental Microbiology 64:3220-3223.

Yates, A. R., Seaman, A., Woodbine, M. (1968). Ascospore germination in *Byssochlamys nivea*. Canadian Journal of Microbiology 14: 319.

Part 3

FUNGI AND MYCOTOXINS

Safety of food products remains in the centre of attention and a considerable part of all food is lost due to spoilage. The release of toxic components in food during spoilage is one of the most important aspects of food mycology. Fungi are very avid producers of many different complex secondary metabolites and among them are some of the most carcinogenic compounds produced in nature. Unfortunately, more and more of these compounds are identified to date. Many debates have taken place on the topic of the lowest allowed concentrations of certain toxic compounds in food. Magan and Aldred discuss in Chapter 7 the influence of the environment on fungi to bring them to mycotoxin production. They highlight possible reasons why fungi do produce these compounds and address the question how the presence of other fungal species does alter the production pattern. In Chapter 8, Frisvad, Thrane and Samson provide a list of mycotoxin producers. The basic thrust in this chapter is the observation that fungal species produce a very species specific cocktail of secondary metabolites that can be either toxic or non-toxic. Therefore, taxonomy of the fungal species must be very sound. In literature there are many examples of mycotoxin producers that are wrongly labelled in the light of the current knowledge. This chapter addresses this problem and delivers an updated list of which fungus produces which toxin.

Chapter 7

Why do fungi produce mycotoxins?

Naresh Magan and David Aldred
Applied Mycology Group, Biotechnology Centre, Cranfield University, Silsoe, Bedford MK45 4DT, U.K.

INTRODUCTION

Fungi are ubiquitous in nature and have evolved over time to colonise a wide range of ecosystems. Part of this evolution process has been the development of the ability to produce a range of extracellular chemicals known as secondary metabolites. Many fungi have been responsible for the production of very useful secondary metabolites with pharmaceutical use (e.g., penicillin, cyclosporin, the statin group) as well as those which are considered to be toxic. This group is known collectively as *mycotoxins*. Two of these are classed as 2B carcinogens (aflatoxin, ochratoxin). Others such as the trichothecenes, zearalenone, patulin and fumonisins are important agents in relation to contamination of human food and animal feed.

Secondary metabolites may be defined as those products produced by microorganisms (and other "lower" organisms) that are not directly essential for growth (Betina, 1994). They may therefore be further defined as those metabolic products that have no *known* role in the "internal economy of the producer" (Williams, 1994). This contrasts with primary metabolism which may be defined as: "a summation of the interrelated enzyme catalysed reactions which are essential to growth by providing energy, synthetic intermediates and key macromolecules" (Betina, 1989).

Secondary metabolites, including mycotoxins, have previously been considered to be somewhat "exotic" and poorly understood chemical substances. An understanding of this enigmatic group of chemicals is especially important in the case of the mycotoxins, because of the special threat they can pose in the human food production chain. Some studies refer to these as extrolites, of which hundreds can be produced by different spoilage fungi and which have been successfully used to aid in identification and differentiation of related species by using micro-extraction techniques (Frisvad and Filtenborg, 1983, 1989; Smedsgaard, 1997).

Secondary metabolites are produced by both major groups of microorganisms – the bacteria and fungi – with the most well-known examples produced by the actinomycetes (bacteria) and the ascomycetes and deuteromycetes (fungi). Many hundreds of secondary metabolites are known, with many more undoubtedly still to be discovered. Indeed, individual microbial species may be capable of producing large numbers of metabolites, and the profile of production may change under different growth conditions, such as nutrient status and water availability. Figure 1 shows patterns of secondary metabolite production for the phyllosphere-dwelling fungus *Epicoccum nigrum* when grown on a range of different solid substrates.

Secondary metabolites from the "higher" fungi (basidiomycetes) also exist, but are less well known because of the general difficulty of culturing this class of fungi in the laboratory. These metabolites may represent a largely untapped resource of useful chemical substances (Aldred *et al.*, 2005).

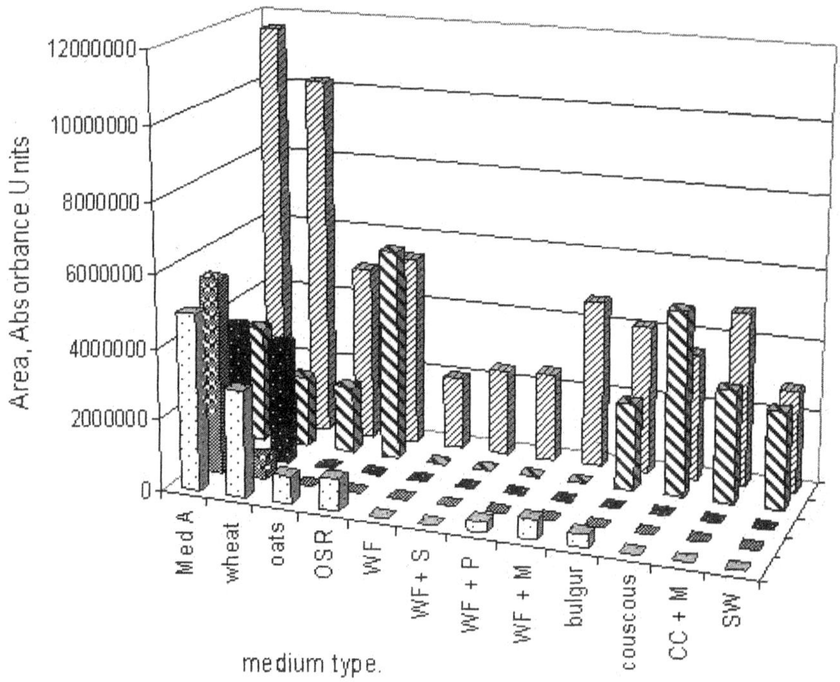

Figure 1. Profile of secondary metabolites obtained from *Epicoccum nigrum* when grown on a range of solid substrates and analyzed by HPLC with photodiode array detection. Secondary metabolites are quantified by mean peak area (absorbance units). Key to substrate treatments: Med A: Medium A (confidential growth medium developed by Xenova Ltd.) OSR: oil seed rape, WF: wheat flakes, S: trace nutrient supplement, P: perlite (moisture retainer), CC: couscous, M: millet, SW: shredded wheat (breakfast cereal) (adapted from Aldred *et al.*, 2005).

ROLE OF SECONDARY METABOLITES

There has for decades been a debate about what, when and why microorganisms produce secondary metabolites. There is still significant discussion about the origin, purpose and significance of products of secondary metabolism. Classically, primary metabolism has been defined as "a summation of the interrelated enzyme catalysed reactions which are essential to growth by providing energy, synthetic intermediates and key macromolecules" (Betina, 1989). They are normally bioactive molecules formed from unique, often elaborate biosynthetic pathways using the intermediates of primary metabolism, e.g., amino acids, sugars and nucleosides and central pathway intermediates (Vining, 1986). The enzymes of secondary metabolites are unique to these processes with substrate specialities markedly different from those of primary metabolism. Production is by ordered sets of genes associated with special regulatory mechanisms that control both timing and level of gene expression. Control mechanisms are usually integrated with the physiology of the producer organism.

The greatest area of debate has been the role that such secondary metabolites serve to the producing organisms (Aharonowitz and Cohen, 1985). Some have suggested them to be "reserve products" or "detoxification products" (Vining, 1986), although others have quite rightly pointed out that they are extremely toxic (Haslam, 1986), and it is certainly unclear which chemicals are being detoxified in these suggested processes. Others workers have put forward the idea that secondary metabolites are produced when conditions no longer allow balanced growth. For example, under conditions of specific nutrient depletion, they have been described as "shunt" or "overflow" metabolites that reduce abnormal concentrations of normal cellular constituents in adverse conditions (Haslam, 1985a,b; 1986). Similarly Ratledge (1993) suggested that the products themselves are largely irrelevant, but that the processes involved in production allow the continuation of metabolic activity and the turnover of enzyme systems under abnormal growth conditions. However, Williams *et al.* (1989) extended the view that primary metabolism is itself far too finely tuned for these ideas to be acceptable.

The more modern view of secondary metabolites including mycotoxins is that the products may confer some selective advantage to producers in the natural state (Williams, 1994). Vining (1986) adapted this view and has described their functionality as falling into two groups: (a) extrinsic functions – those that impinge on growth and reproduction of other microorganisms in the immediate environment; (b) intrinsic factors – those that beneficially affect growth, physiology and reproduction of the producer organism. Secondary metabolites with extrinsic functions would include the antibiotics classes and the mycotoxins. Secondary metabolites with intrinsic functions may, for example, be concerned with the sequestering of limiting nutrients from the immediate environment.

In fact, much has been written on the view that secondary metabolism is a process that *must* in some way benefit the producer, and the most pertinent points are summarised below:

- □ Secondary metabolites are frequently produced by long, complex biosynthetic pathways (often 10-40 steps) and are extremely expensive energetically. They must therefore have been actively selected for, indicating an intrinsic competitive benefit to the producer (Williams, 1994).
- □ Binding of secondary metabolites to receptor molecules is often extremely sophisticated with a complementarity approaching that of enzyme-substrate binding. This also points to evolutionary selection of advantageous characteristics (Maplestone *et al.*, 1992).
- □ All antibiotics so far studied are produced by genes organised into clusters often including regulatory and auto-resistance genes. Gene clustering is usually strong evidence of evolutionary selection of genes (Stone and Williams, 1992).
- □ Secondary metabolism is exhibited by lower organisms lacking an immune system. It can therefore be viewed as replacing such a system in these organisms.
- □ Microorganisms living in harsh environments (where there is little competition), and nutritionally rich environments (where there is always an excess of nutrients) generally do not produce secondary metabolites (Vining, 1990; Demain, 1992).
- □ The general observation that secondary metabolism is regulated to commence when balanced growth is no longer possible may reflect a specific "switching on" of various survival strategies at a time when the environment is becoming hostile and competition is at its greatest (Maplestone, 1992).

Janzen (1977) extends these ideas of competitive advantage even further by suggesting that toxic metabolites (e.g., mycotoxins) and other "objectionable" substances are elaborated onto substrates by microorganisms to prevent their utilisation by other organisms (including man). Lillehoj (1982) has further suggested the importance of ecological imbalance as a stimulus for secondary metabolite production and cites the example of high mycotoxin production in monoculture crops both pre- and post-harvest.

Secondary metabolite production is energetically expensive and thus is physiologically regulated in response to environmental factors (Vining, 1990). It is usually suggested that in liquid submerged culture secondary metabolism often occurs during the idiophase when active growth as ceased, often due to nutrient exhaustion. Others have reported specific regulation exerted by the presence of readily utilizable carbon sources such as glucose and nitrogen sources such as ammonia and phosphate. This mechanism is known as catabolite repression. The mechanism appears to operate by repression of the enzymes of secondary metabolites, therefore working at the level of

DNA transcription. In laboratory systems this type of regulation is lifted when the substance is exhausted and results in the onset of production as a direct response to changes in environmental conditions. This is very important since it indicates that for some secondary metabolites (perhaps including mycotoxins) production is specifically switched on by the initiation of stress (e.g., nutrient depletion) conditions. This is one of the most compelling pieces of evidence which supports the idea that secondary metabolism is initiated for the purpose of conferring a competitive advantage to the producer. The key point here is that energy expensive biosynthetic processes are being switched on at a point where nutrients are becoming scarce. The most plausible explanation for this observation is that the chemicals so produced play a vital role in the survival of the organism. The fact that we often fail to recognise the specific role played by these substances probably reflects our lack of understanding of the ecology and physiology of microorganisms in natural conditions.

However, not all secondary metabolite systems respond to nutritional signals in this way, and many are probably regulated by other signals which may be very specific to the role played by the metabolite once it is formed. Further, the traditional view of secondary metabolites only being produced during later growth is certainly not supported by studies with mycotoxigenic species. For example, natural strains can be found that produce a mycotoxin concomitant with growth, e.g., aflatoxin producing strains of *A. flavus* (Gendloff *et al.*, 1992). In fact, there is a significant body of evidence that suggests that fungi produce mycotoxins rapidly and during the linear phase of active growth. This is further influenced by the prevailing environmental factors and nutritional status of the food matrix.

MYCOTOXIN PRODUCTION BY FUNGI IN RELATION TO ENVIRONMENTAL STRESS

The ability of fungal species to colonise and occupy specific ecological niches is determined by their capacity to effectively compete against other microorganisms which make up the community. Fungi compete very directly with each other for resources by releasing extracellular enzymes into their immediate environment. This will further be impacted upon by other important factors such as temperature, water availability, gas balance and pH. Stress factors may be long-lived or transient resulting in a dynamic fungal community structure in a state of flux. To overcome such changes fungi use different primary strategies to survive and prosper in different ecological niches. They can use combative (C-selected) strategies, e.g., *Trichoderma* species, xylariales and basidiomycetes, which maximise occupation and exploitation of food matrices in relatively unstressed and undisturbed conditions and where recalcitrant compounds such as lignin needs to be degraded; stress (S-selected) strategies which allow survival and endurance of continuous environmental stress, e.g., *Zygosaccharomyces rouxii, Eurotium, Penicillium* and *Aspergillus* species; and ruderal (R-selected) strategies, e.g., *Rhizopus, Mucor* species, characterised by species with a high reproductive potential and shortlife span which facilitates successful exploitation in severely disturbed but nutrient-rich conditions. These three can merge to result in secondary strategies (C-R, S-R, C-S, C-S-R) which form part of a continuum with some transition between them. The main attributes of these three primary groupings are summarised in Figure 2. A key component of fungi which use S- and C-selected strategies to occupy ecological niches is secondary metabolite production. When trying to understand the reasons why fungi produce such metabolites, this must be borne in mind. Furthermore, interactions between fungi, occupation and dominance in a specific niche could all be influenced by whether mycotoxins are produced. Thus, the possible role of such secondary metabolites must be considered in a wider ecological context.

Many studies on production of mycotoxins by spoilage fungi, especially in culture or in naturally contaminated food products have involved quantification at a single point in time.

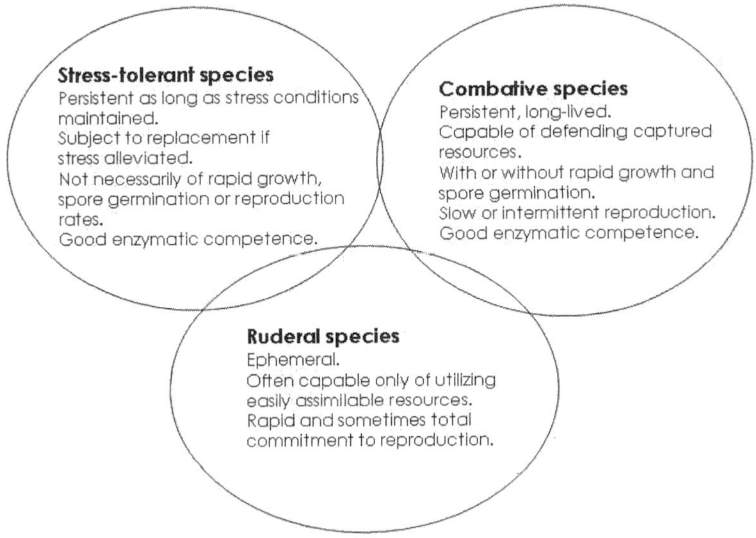

Figure 2. Summary of attributes of fungi in relation to the three major ecological strategies employed by fungi (adapted from Cooke and Rayner, 1984).

This often gives no idea of the temporal production rates and the conditions under which they were produced on often nutritionally heterogeneous food matrices. In nature, for fungi to compete effectively in a mixed microbial community they often have to be very competitive to become established. Thus, rapid early production of mycotoxins could be one way in which they are able to spatially retain a competitive edge in a range of fluctuating conditions. Recent studies by Puel *et al.* (2005) show that mycophenolic acid kinetics by *Byssochlamys nivea* was correlated with biomass with rapid early production in the first 10 days, increasing over 30 days before decreasing, probably due to nutrient exhaustion. These studies were conducted in defined liquid culture only. However, they do not support the view that mycotoxins are produced solely in late exponential and stationary phase. They must be produced earlier for an ecological reason as part of an overall strategy to colonise food matrices.

If fungi colonise a matrix where there is no competition then they still often produce mycotoxins very rapidly, in spite of the traditional view that they are only produced when the nutrients have been exhausted. Thus colonisation and mycotoxin production are related but often there are differential optima for these two parameters. For example, *P. verrucosum* grows optimally at 25 °C and 0.98 water activity (a_w) while it produced OTA (ochratoxin A) optimally at 25 °C and 0.95 a_w indicative of the influence of water stress. Figure 3 shows the temporal production of ochratoxin by a strain of *P. verrucosum* on irradiated wheat grain under marginal and optimum temperature and water availability levels. Recent studies of *Aspergillus carbonarius,* a member of the *Aspergillus* section *Nigri* group, on grape-based media also showed that optimum growth was at 30-35 °C and 0.99-0.98 a_w (Mitchell *et al.*, 2004).

In contrast, many European isolates produced OTA optimally at 15-20 °C and 0.95 a_w. They also produce OTA very rapidly (within 5-10 days) after initial growth, again suggesting an important role in primary and secondary strategies for colonisation of the food matrix (Belli *et al.*, 2004; Esteban *et al.*, 2004). This suggests that when a species is impacted by fluctuations in environmental stress, particularly drought stress, then conditions are marginal for growth, and these fungi are often stimulated to produce higher titres of mycotoxins, even in the absence of competition

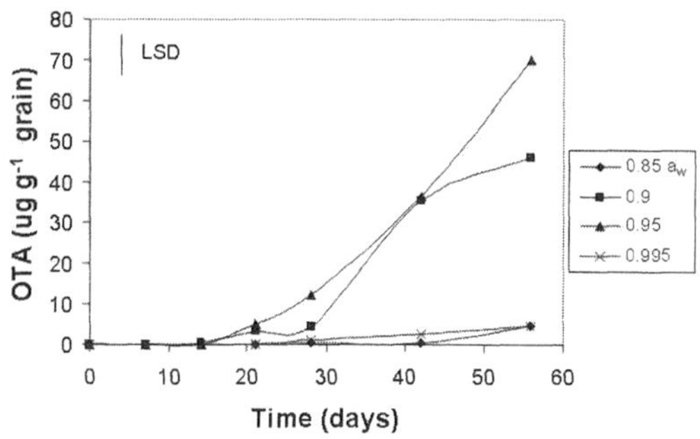

Figure 3. Temporal production of ochratoxin A by a strain of *Penicillium verrucosum* (OTA11) on wheat grain at 25 °C under four different water availabilities.

for space in a particular niche. Recent studies at Cranfield have examined the partitioning of OTA between spores, biomass and medium for *A. ochraceus* and *A. carbonarius* species from Europe. These studies showed that while for *A. ochraceus*, 60% of the OTA was in the biomass, and about 20% was in the spores and the medium. In contrast for many *A. carbonarius* isolates under optimum temperature and water availability conditions (20 °C, 0.98-0.95 a_w) at least 60% of OTA was in the spores, 30% in the mycelial biomass and 20% in the medium (Mitchell, Magan and Aldred, unpublished data). This suggests that different mycotoxigenic species may have different patterns of partitioning of such metabolites into mycelium, spores and substrate. Perhaps abiotic or biotic stress impacts physiologically in a similar way and results in expression of genes or gene products for the biosynthesis of mycotoxins.

Other stresses, especially the application of fungicides to crops, have also been shown to result in an impact on mycotoxigenic species and influence mycotoxin production. Fungicides and fungistats can effectively inhibit germination and growth of fungi which produce mycotoxins, e.g., *Fusarium culmorum, F. graminearum, P. verrucosum.* However, where control is only partial and allows some growth then sometimes a stimulation in mycotoxin production has been observed. For example, Magan *et al.* (2002) found that suboptimal concentrations of triazole fungicides stimulated production of DON (deoxynivalenol) by *F. culmorum* isolates from different parts of Europe, especially when combined with reduced water availability. *F. graminearum* strains have also been shown to be stimulated to produce more DON on wheat grain compared to untreated controls (Ramirez *et al.*, 2004), although some liquid culture studies with addition of fungicides suggest that this does not occur (Nicholson, 2004). However, the liquid fermentation studies did not consider interactions with other abiotic stresses and they are very different from those in natural ecosystems where active colonisation of a solid substrate is required by foraging hyphae colonising the food matrix and releasing enzymes as well as mycotoxins to enable effective establishment to occur. Such fundamental differences in growth conditions can be expected to impact significantly on secondary metabolism production patterns.

Food preservatives are predominantly fungistats as apposed to fungicides and their use is being reduced because of consumer pressure. However, it has been shown that while at the recommended levels they are effective in bakery products, at intermediate levels (<0.3%) they can result in stimulation of growth and in some cases mycotoxin production. Recent work suggests that interaction between environmental factors and calcium proprionate and potassium sorbate can result in a differential effect on growth and ochratoxin production by *P. verrucosum* on bread analogues (Arroyo *et al.*, 2005). Similar results have been obtained by Marin *et al.* (2002) in relation to a range of spoilage moulds. This suggests that where substitution is made with antioxidants or natural preservatives, e.g., essential oils, then potential for such stimulation must be considered as well as impacts on shelf-life of food products.

There is some correlation between hydrolytic enzyme production, growth and mycotoxin production as the latter is dependent on extracellular enzyme production, establishment and colonisation. Figure 4 shows the increase in specific activity of one hydrolytic enzyme (n-acetyl-β-D-Glucosaminidase) in wheat-based media in relation to colonisation and OTA production by *P. verrucosum.* This shows that enzymes are produced rapidly during the first 5-10 days, followed by production of OTA. Thus a sequential series of events occurs to enable the fungus to become established and then perhaps retain possession of territory gained.

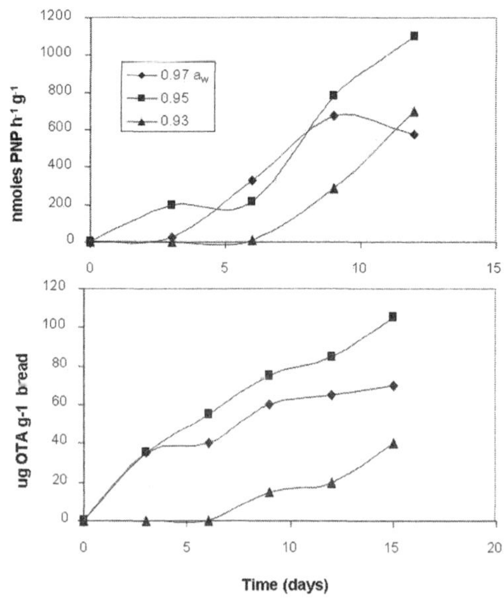

Figure 4. Comparison of hydrolytic enzyme production (n-acetyl-β-D-Glucosaaminidase) and ochratoxin production by a strain of *P. verrucosum* growing on a wheat matrix at different water availabilities.

Recently, some elegant studies have been conducted in relating molecular monitoring of the impact of environmental conditions influencing induction of OTA biosynthesis genes in *Penicillium nordicum* (Geisen, 2004). Real time PCR (polymerase chain reaction) specific for OTA polyketide synthase gene (*otapks*PN) for *P. nordicum* demonstrated the induction of the transcription factor correlates with biosynthesis of OTA. Thus the induction of the molecular signal can be used to support the production kinetics of mycotoxins in the environment. The kinetics using this approach supports the early synthesis of OTA, within 3-5 days. The effect of temperature, pH, and ionic solute concentration all showed a parallel expression of *otapks*PN gene and OTA production. Interestingly, while this occurred for pH and ionic solute stress, there was less relationship between expression and OTA production and temperature (15 to 30 °C). It was speculated that temperature has less effect on *otapks*PN gene expression although it does have an impact on OTA production. The maximum amount of ionic stress imposed was 50 g/l which only equates to about 0.975 water activity. *Penicillium* species, especially *P. verrucosum* grow and produce OTA over a much wider range than this (Sanchis and Magan, 2004; Cairns-Fuller, 2004; Cairns-Fuller *et al.*, 2005). Thus an opportunity does exist to use this approach to confirm the role of mycotoxins in ecological competitiveness during colonisation of food matrices and gain territorial advantage.

It is interesting to note that sometimes fungi are able to break down mycotoxins if necessary and synthesise more when required. For example, *A. carbonarius* produces OTA over 5-15 days then there is a decrease followed by another cycle of production. Abrunhosa *et al.* (2002) found that a range of fungi isolated from grapes could degrade OTA produced by *A. carbonarius*. Varga *et al.* (2002) showed that OTA can be degraded by *Aspergillus* species. Recently, Varga *et al.* (2005) have also demonstrated the capacity of *Rhizopus* species to degrade a range of mycotoxins. For example, they could degrade 95% of OTA in 16 days, even on moist wheat. This suggests that depending on the ecological conditions and the strategy for competitive exclusion of competitors, enzyme systems may be activated to enable such degradation pathways to be instituted.

ECOLOGICAL INTERACTIONS BETWEEN SPOILAGE FUNGI AND MYCOTOXIN PRODUCTION

Mycotoxigenic moulds do not occur in an ecological niche alone. They are often part of a diverse community of microorganisms in which they need to compete effectively, and sometimes exclude other species from colonising territory by using a range of strategies as detailed previously. One such strategy must include the rapid production of the necessary hydrolytic enzymes to gain access to the nutrients and then produce activate synthesis of mycotoxins to maintain a hold on territory.

Recent innovative studies have shown that *Penicillium* species can produce quorum sensing (QS) inhibitors which can affect the functioning of bacteria such as *Pseudomonas aeruginosa*. Microorganisms exist in mixed consortia and this study has shown that penicillic acid and patulin were two metabolites which were biologically active and significantly affected the QS-controlled gene expression of *Ps. aeruginosa* (Rasmussen *et al.*, 2005). This suggests that quite complex interactions might occur between microorganisms and that fungal secondary metabolites may have a much wider influence in mixed communities than thought previously. Many Gram-negative bacteria produce N-acetyl homoserine lactones (AHLs) which coordinate expression of virulence in response to surrounding bacterial populations as QS compounds. The potential interaction of fungal secondary metabolites with such QS AHLs needs to be investigated in more detail under different environmental stress conditions as Rasmussen *et al.* (2005) showed that 33 of 50 extracts from *Penicillium* species produced QS inhibitor compounds.

Studies have been conducted to examine the interactions between fungal species on grain substrates and the outcome of such interactions on dominance/competitiveness. Of course, environmental factors, carbon sources and the ability to capture and assimilate them will further influence these interactions within an ecosystem influencing the outcome of competition (Cooke and Whipps, 1993). Recently, in detailed studies of the maize ecosystem Marin *et al.* (2004) suggested that different strategies are used by spoilage fungi depending on the moisture content of the maize. For example, some species which have adapted to xerophilic conditions, e.g., *Eurotium* species, use specific strategies to colonise partially dried maize (13-14% m.c.). In contrast, wet drying maize (>18% m.c.) is colonised by a range of *Aspergillus*, *Penicillium* species, some mycotoxigenic species. They are involved in primary resource capture, combat and defence of niche resulting in secondary resource capture. Primary resource capture involves utilization of readily available carbon sources, especially water soluble sugars. This promotes access and utilization via secondary resource capture of less readily available resources such as cellulose, hemicelluloses and occupation of territory. Combat involves defence of the colonised territory by an already established occupant. The combative species are able to keep captured resources by storage of lipid bodies, glycogen or compatible solutes with or without fast growth and spore germination, slow or periodic reproduction, and enzymatic competence and the production of mycotoxins.

An Index of Dominance to classify interactions and outcome of competition has been used widely to obtain information on the reasons why certain species are more competitive over others under different environmental conditions (Magan *et al.*, 2003). More recently, the number of carbon sources a fungus can assimilate has been used as a determinant of colonisation potential. The number relative to other fungal species has been used to develop a Niche Overlap Index (NOI) based on the number of common carbon sources utilized, divided by the total number utilised. The ability to use some C-sources which other species cannot use over a range of conditions may provide an advantage in producing secondary metabolites as part of maintaining a competitive edge over other species. This has shown that mycotoxigenic species such as *P. verrucosum* and *Fusarium* section Liseola species are able to compete effectively against other spoilage species such as *Alternaria* and *Cladosporium* species, partially by the ability to utilise carbon sources over a range of conditions but also the ability to produce mycotoxins. For example, *Fusarium* section Liseola species are considered to have endophytic phases of growth in maize tissue. Fumonisin (FB1) production may be an important component for becoming established and also retaining its niche. Growth of a range of species including *A. alternata*, *P. expansum*, *Botrytis cinerea* and *F. graminearum* can be inhibited by FB1 at different concentrations, although generally fumonisin producers are not affected. However, field studies with co-inoculation of fumonisin producers and non-producers of *F. verticillioides* and *F. graminearum* in field infection of maize found that interactions between these two species did not directly affect mycotoxin production or accumulation. Such studies suggest that FB1 production may not be directly associated with a competition strategy.

Other studies however suggest that interactions between mycotoxigenic species and other mycoflora in a niche is further affected by environmental factors resulting in a fluid situation where interactions can vary and dominant species may change. For example, fumonisin production by *F. verticillioides* and *F. proliferatum* was inhibited by the presence of some species (*A. flavus, A. ochraceus, P. implicatum*) regardless of water availability or temperature. However, interactions with *A. niger* resulted in a significant stimulation in FB1 production on maize grain (Marin *et al.*, 1998). Furthermore, co-culture of both *Fusarium* species resulted in a stimulation of FB1 production than when they were grown individually. This could be due to the production of synthesis precursors or to the fact that competition may encourage these spoilage fungi to produce mycotoxins at an earlier stage. It can be speculated that perhaps some mycotoxigenic species may produce QS-like compounds which could influence the activity of other species competing for the same ecological niche. Other studies of different mycotoxin-producing species have shown that when *Fusarium* section Liseola species were competing with *F. graminearum*, FB1 accumulation decreased at 15 °C regardless of moisture content. However, at 25 °C, *F. verticillioides* produced higher amounts of FB1 in the presence of *F. graminearum* than when cultured alone. In contrast, *F. proliferatum* produced less than in the absence of the competitor. There was, however, no effect on production of zearalenone by *F. graminearum*, although there was some stimulation under specific environmental conditions (15 °C, 0.98 a_w). DON production was significantly stimulated at slightly reduced water availability (0.98 a_w; Velluti *et al.*, 2000a,b).

Table 1 summarises some studies where interaction with species has resulted in a stimulation or inhibition of mycotoxin production by other competing species on grain matrices.

Studies have examined interactions between *A. ochraceus* and other spoilage fungi and found that the effect of interactions on growth and OTA production was more pronounced *in vitro* than *in situ* on maize grain (Lee and Magan, 1999, 2000).

Table 1. Interactions between mycotoxigenic fungi and other species and outcomes on mycotoxin production

Species interactions	Effects on mycotoxins	Reference
Fusarium verticillioides + *Trichoderma viride*	Reduce Fumonisin B1 by 85%	Yates *et al.* (1999)
F. verticillioides + *Aspergillus niger, A. flavus, A. ochraceus, Penicillium implicatum*	Inhibited Fumonisin B1, except at 0.98 a_w	Marin *et al.* (1998)
F. verticillioides + *F. proliferatum*	Enhanced Fumonisin B1 (co-culture on maize)	Marin *et al.* (1998)
F. verticillioides + *F. graminearum*	< Fumonisin at 15°C; > at 25°C, deoxynivalenol at 0.98 a_w, no effect on zearalenone	Velluti *et al.* (2002a and b)
A. flavus + different fungi	Inhibition of aflatoxin depending (maize grain) on species	Cuero *et al.* (1988)
A. ochraceus + other species	Stimulation in some cases *in vitro*; inhibition of ochratoxin in maize;	Lee and Magan (1999; 2000)
A. flavus + *Hyphopichia burtonii* + *Bacillus amylofaciens*	Stimulation of aflatoxin (maize grain)	Cuero *et al.* (1987)
F. culmorum + grain fungi	Predominantly inhibition of DON (wheat grain)	Magan *et al.* (2003)
P. verrucosum + *Alternaria alternata, F.culmorum, Eurotium repens, P. aurantiogriseum, A. ochraceus*	Some species inhibited; others stimulated OTA on wheat grain	Cairns *et al.* (2003)

Interaction of *Fusarium* section Lisolea species and *A. parasiticus* showed that there was no effect on FB1 production by the former species. However, the *Fusaria* were competitive and inhibited aflatoxin production over a range of environmental conditions (Marin *et al.*, 2001). Field studies of co-inoculation with these same species (*F. verticillioides, A. flavus*) also resulted in lower aflatoxin accumulation than when the aflatoxigenic species was used alone (Zummo and Scott, 1992).

Recent studies have examined the interaction between *F. culmorum*, and a range of other fungi which colonise ripening ears and grain during harvesting to examine the impact on DON and NIV (nivalenoll) production.

Table 2. Effect of interactions between *F. culmorum* and other species on deoxynivalenol and nivalenol (ng g^{-1} grain) production on irradiated wheat grain at two water activity levels at 25 °C. Key to fungi: F, *Fusarium*; A, *Alternaria*; M, *Microdochium*; *M. majus*, *M. nivale* var *majus*; P, *Penicillium* (from Magan *et al.*, 2003)

	Mycotoxin			
	Deoxynivalenol		**Nivalenol**	
Water activity	*0.995*	*0.955*	*0.995*	*0.955*
F. culmorum	7669	447	289	298
F. culmorum + *C. herbarum*	634	0	316	412
F. culmorum + *M. nivale*	451	444	0	288
F. culmorum + *M. majus*	0	440	292	0
F. culmorum + *P. verrucosum*	3264	450	0	0

LSD (P=0.05: DON=180; NIV=123).

Table 2 shows that on wheat grain the presence of fungi such as *Cladosporium herbarum, Alternaria tenuissima, Microdochium nivale* and *M. nivale* var. *majus* and the mycotoxigenic species *P. verrucosum* resulted in inhibition of DON production but in some cases increased NIV production. This suggests that where a range of mycotoxins are produced there may be a switch from one synthesising pathway to another as part of the strategy to maintain a competitive edge (Magan *et al.*, 2003).

Although DON is produced in higher concentrations than NIV, the latter is in fact more toxic. Similar results were obtained when some essential oils were examined to control growth and DON/NIV production by *F. culmorum* (Hope *et al.*, 2003). They found that some essential oils could inhibit growth and DON production under some environmental factors. However, there was a switch to nivalenol and sometimes other trichothecenes. It was thus suggested that in the stored grain ecosystem complex interactions occurred between germination, establishment and mycotoxin production and environmental factors which were influenced by the presence and activity of other species (Magan *et al.*, 2003).

Metabolomic production profiles for mycotoxins and other secondary metabolites may indeed be a method for such fungi to respond in a fluid way to changes in competition from other species alone or interactions with fluxes of changing abiotic factors. Recently, Aldred *et al.* (2005) showed that ecological niches from which fungi are isolated significantly influence metabolomic profiles and titres of individual metabolites. Production was very different for species from damp rainforest ecosystems and from the phyllosphere. Thus the production of mycotoxins and secondary metabolites in general must be seen in an ecological context.

CONCLUSIONS

Some mycotoxins appear to be produced in response to environmental change, usually due to the onset of stress conditions.

- □ Evidence suggests that mycotoxins produced by fungi (and other micro-organisms) confer a competitive advantage to the producer. This role may not always be clear (especially in laboratory situations), and probably operates only within natural ecosystems.
- □ Under stress conditions the need for competitive advantage is increased and this is when mycotoxins probably become important.
- □ In natural systems the ecological milieu determines the community structure and the predominant species may be partially dependent on stress tolerance and the abil-

ity to produce a battery of secondary metabolites. This array of metabolites could have different functions against other microorganisms, mites and nematodes.

- □ The use of alternative, more natural, preservatives for food preservation requires careful consideration of concentrations to prevent stimulating growth and mycotoxin production in such ecosystems.

REFERENCES

Abrunhosa, L., Serra, R., and Venancio, A. (2002). Biodegradation of ochratoxin A by fungi isolated from grapes. Journal of Agriculture and Food Chemistry 50:7493-7496.

Aharonowitz, Y. and Cohen, B. (1985). Bioactive microbial secondary metabolites. *In* Biotechnology: Potential and Limitations (Silver, S., ed.), Berlin: Springer Verlag.

Aldred, D., Penn, J., and Magan, N. (2005). Water availability and metabolomic profiles of *Epicoccum nigrum* and *Sarophorum palmicola* grown in solid substrate fermentation systems. The Mycologist 19:18-23.

Arroyo, M., Aldred, D., and Magan, N. (2005). Environmental factors and weak organic acid interactions have differential effects on control of growth and ochratoxin A production by *Penicillium verrucosum* isolates in bread. International Journal of Food Microbiology 98:223-231.

Bellí, N., Marín, S., Sanchis, V., and Ramos, A. J. (2004). Influence of water activity and temperature on growth of isolates of *Aspergillus* section *Nigri* obtained from grapes. International Journal of Food Microbiology 96:19-27.

Betina, V. (1989). Mycotoxins: chemical, biological and environmental apects. *In* Bioactive Molecules, Vol. 9., Oxford: Elsevier.

Betina, V. (1994). Microbial primary and secondary metabolism. Progress in Industrial Microbiology 30:1-15.

Cairns, V., Hope, R., and Magan, N. (2003). Environmental factors and competing mycoflora affect growth an ochratoxin production by *Penicillium verrucosum* on wheat grain. Aspects of Applied Biology 68: 81-90.

Cairns-Fuller, V. (2004). Ecology and dynamics of ochratoxin production and control in grain. PhD. Thesis, Institute of BioScience and Technology, Cranfield University.

Cairns-Fuller, V., Aldred, D., and Magan, N. (2005). Water, temperature and gas composition interactions affect growth and ochratoxin A production by isolates of *Penicillium verrucosum* on wheat grain. Journal of Applied Microbiology 99:1215-1221.

Cooke, R. C., and Rayner, A. D. M. (1984). Ecology of saprotrophic fungi. Longman, London and New York, 415 pp.

Cooke, R., and Whipps, J. M. (1993). *Ecophysiology of fungi*. Blackwell, Oxford.

Cuero, R., Lacey, J., and Smith, J. E. (1987). Stimulation by *Hyphopichia burtonii* and *Bacillus amyloliquefaciens* of aflatoxin production by *Aspergillus flavus* in irradiated maize and rice grains. Applied and Environmental Microbiology 53: 1142-1146.

Cuero, R., Lacey, J., and Smith, J. E. (1988). Mycotoxin production by *Aspergillus flavus* and *Fusarium graminearum* in irradiated maize grains in the presence of other fungi. Journal of Food Protection 51:452-456.

Demain, A. L. (1992). Microbial secondary metabolism: a new theoretical frontier for academia, a new opportunity for industry. *In* Secondary Metabolites: their function and evolution. Ciba Foundation Symposium 171, Chichester:Wiley, U.K., pp. 3-23

Esteban, A., Abarca, M. L., Bragulat, M. R., and Cabañes, F. J. (2004). Effects of temperature and incubation time on production of ochratoxin A by black aspergilli. Research in Microbiology 155:861-866.

Frisvad, J. C., and Filtenborg, O. (1983). Classification of terviticillate penicillia based on profiles of mycotoxins and other secondary metabolites. Applied and Environmental Microbiology 46:1301-1310.

Frisvad, J. C., and Filtenborg, O. (1989). Terviticillate penicillia: chemotaxonomy and mycotoxin production. Mycologia 81:836-861.

Geisen, R. (2004). Molecular monitoring of environmental conditions influencing the induction of ochratoxin A biosynthesis genes in *Penicillium nordicum*. Molecular and Nutritional Food Research 48:532-540.

Gendloff, E. H., Chu, F. S., and Leonard, T. J. (1992). Variation in regulation of aflatoxin biosynthesis in isolates of *Aspergillus flavus*. Experientia 48: 84-87.

Haslam, E. (1985a). Secondary metabolism – overflow metabolism an interpretation? *In* Metabolites and Metabolism, Oxford University Press, Oxford, U.K., pp. 117-147.

Haslam, E. (1985b) Some general characteristics of secondary metabolism. *In* Metabolites and Metabolism, Oxford University Press, Oxford, U.K., pp. 148-153.

Haslam, E. (1986). Secondary metabolism – fact and fiction. Natural Products Report 3: 217-249.

Hope, R., Jestoi, M., and Magan, N. (2003). Multi-target environmental approach for control of growth and toxin production by *Fusarium culmorum* using essential oils and antioxidants. *In* Advances in Stored Product Protection (Credland, P., Armitage, D. M., Bell, C. H., and Cogan, P. M., eds.) Cabi International, U.K., pp. 486-492.

Hope, R., Aldred, D., and Magan, N. (2005). Comparison of the effect of environmental factors on deoxynivalenol production by *F. culmorum* and *F. graminearum* on wheat grain. Letters in Applied Microbiology 40:295-300.

Janzen, D. H. (1977). Why fruits rot, seeds mould and meat spoils. American Naturalist 111:691-713.

Lee, H. B., and Magan, N. (1999). Environmental influences on in vitro interspecific interactions between *A. ochraceus* and other maize spoilage fungi on growth and ochratoxin production. Mycopathologia 146:43-47.

Lee, H. B., and Magan, N. (2000). Impact of environment and interspecific interactions between spoilage fungi and *Aspergillus ochraceus* and ochratoxin production on maize. International Journal of Food Microbiology 61:11-16.

Lillehoj, E. B. (1982). Evolutionary basis and ecological role of toxic microbial metabolites. Journal of Theoretical Biology 97:325-327.

Luchese, R. G., and Harrison, W. F. (1993). Biosynthesis of afaltoxin – the role of nutritional factors. Journal of Applied Bacteriology 74:5-14.

Magan, N., Hope, R., Colleate, A., and Baxter, E. S. (2002). Relationship between growth and mycotoxin production by *Fusarium* species, biocides and environment. European Journal of Plant Pathology 108:685-690.

Magan, N., Hope, R., Cairns, V., and Aldred, D. (2003). Post-harvest fungal ecology: impact of fungal growth and mycotoxin accumulation in stored grain. European Journal of Plant Pathology 109: 723-730.

Maplestone, R. A., Stone, M. J. and Dudley, W. H. (1992). The evolutionary role of secondary metabolites- a review. Gene 115:151-157.

Marin, S., Sanchis, V., Ramos, A. G., and Magan, N. (1998). Environmental factors, interspecific interactions, and niche overlap between *Fusarium moniliforme* and *F. proliferatum* and *Fusarium graminearum, Aspergillus* and *Penicillium* spp. isolated from maize Mycological Research 102: 831-837.

Marin, S., Albareda, X., Ramos, A. J., and Sanchis, V. (2001). Impact of environment and interactions of *Fusarium verticillioides* and *F. proliferatum* with *Aspergillus parasiticus* on fumonisin B1 and aflatoxins on maize grain. Journal of Science of Food and Agriculture 81:1060-1068.

Marin, S., Guynot, M. E., Neira, P., Bernado, M., Sanchis, V., and Ramos, A. J. (2002). Risk assessment of the use of sub-optimal levels of weak acid preservatives in the control of mould growth in bakery products. International Journal of Food Microbiology 79:203-211.

Marin, S., Magan, N., Ramos, A. J., and Sanchis, V. (2004). Fumonisin-producing strains of Fusarium: A review of their ecophysiology. Journal of Food Protection 67:1792-1805.

Mitchell, D., Parra, R., Aldred, D., and Magan, N. (2004). Water and temperature relations of growth and ochratoxin A production by *Aspergillus carbonarius* strains from grapes in Europe and Israel. Journal of Applied Microbiology 97:439-445.

Nicholson, P. (2004). Rapid detection of mycotoxigenic fungi in plants. *In* Mycotoxins in Food: detection and control (Magan, N., and Olsen, M., eds.), Woodhead Publishing Ltd, Cambridge, England, pp. 111-136.

Puel, O., Tadrist, S., Galtiere, P., Oswald, I. P., and Delaforge, M. (2005). *Bysocchlamys nivea* as a source of mycophenolic acid. Applied and Environmental Microbiology 71:550-553.

Ramirez, M. L., Chulze, S. N., and Magan, N. (2004). Impact of environmental factors and fungicides on growth and deoxynivalenol production by *Fusarium graminearum* isolates from Argentinian wheat. Crop Protection 23: 117-123.

Rasmussen, T. B., Skindersoe, M. E., Bjarnsholt, T., Phipps, R. K., Christensen, K. B, Jensen, P. O., Andersen, J. B., Koch, B., Larsen, T. O., Hentzer, M., Eberl, L., Hoiby, N., and Givskov, M. (2005). Identity and effects of quorum sensing inhibitors produced by *Penicillium* species. Microbiology 151:1325-1340.

Ratledge, C. (1993). Secondary metabolites and fungal biochemistry. Journal of Chemical Technology 56:203-224.

Sanchis, V., and Magan, N. (2004). Environmental profiles for growth and mycotoxin production. *In* Mycotoxins in food: detection and control (Magan, N., and Olsen, M., eds.), Woodhead Publishing Ltd., Cambridge, U.K.

Smedsgaard, J. (1997). Micro-scale extraction procedure for standardized screening of fungal metabolite production cultures. Journal of Chromatography A 760:264-270.

Stone, M. J. and Williams, D. H. (1992). On the evolution of functional secondary metabolites (natural products). Molecular Biology 6:29-34.

Varga, J., Rigo, K., and Teren, J. (2002). Degradation of ochratoxin A by *Aspergillus* species. International Journal of Food Microbiology 59:1-7.

Varga, J., Peteri, Z., Tabori, K., Teren, J., and Vagvolgyi, C. (2005). Degradation of ochratoxin A and other mycotoxins by *Rhizopus* species. International Journal of Food Microbiology 99:321-328.

Velluti, A. S., Marin, S., Betucci, A., Ramos, A. J. and Sanchis, V. (2000a). The effect of fungal competition on colonisation of maize grain by *Fusarium moniliforme, F. proliferatum* and *F. graminearum* and on fumonisin B1 and zearalenone formation. International Journal of Food Microbiology 59:59-66.

Velluti, A. S., Marin, S., Gonzalez, R., Ramos, A. J. and Sanchis, V. (2000b). Fumonisin B1, zearalenone and deoxynivalenol production by *Fusarium moniliforme, F. proliferatum* and *F. graminearum* in mixed cultures on irradiated maize kernels. Journal of Science of Food and Agriculture 81:88-94.

Vining, L. C. (1986). Secondary metabolism. *In* Biotechnology 4 (Microbial products II), (Pape, H., and Rehm, H. J., eds.), VCH: Weinheim, Germany, pp. 19-38.

Vining, L. C. (1990). Functions of secondary metabolism. Annual Review of Microbiology 44:395-427.

Williams, D. H., Stone, M. J. and Hauck, P. R. (1989). Why are secondary metabolites (natural products) biosynthesised? Journal of Natural Products 52:1189-1208.

Williams, D. H. (1994). Functional roles of secondary metabolites. *In* Discovering Drugs from Nature. Conference Documentation, 8-9 December 1994. London: IBC Technical Services.

Yates, I. E., Meredith, F., Smart, W., Bacon, C. W., and Jaworski, A. J. (1999). *Trichoderma viride* suppresses fumonisin B_1 production by *Fusarium monoliforme*. Journal of Food Protection 62:1326-1332.

Zummo, N., and Scott, G. E. (1992). Interaction of *Fusarium moniliforme* and *Aspergillus flavus* on kernel infection and aflatoxin contamination in maize ears. Plant Disease 76:71-773.

Chapter 8

Mycotoxin producers

Jens C. Frisvad[1], Ulf Thrane[1], and Robert A. Samson[2]
[1] *Centre for Microbial Biotechnology, BioCentrum-DTU, Building 221, Technical University of Denmark, 2800 Kgs. Lyngby, Denmark;* [2] *CBS Fungal Biodiversity Centre, P.O. Box 85167, 3508 AD Utrecht, The Netherlands.*

INTRODUCTION

Mycotoxins are extrolites produced by filamentous microfungi that can cause disease in vertebrate animals when introduced via a natural route: ingested, absorbed through the skin or inhaled. This definition excludes fungal toxins which are active against bacteria, archaea, algae, protozoa, and non-vertebrate animals such as insects, spiders, worms, snails and jellyfish. It also excludes toxins produced by Basidiomycetes, i.e., mushrooms and related fungi, because these toxins are ingested by eating fruit bodies, a problem different from the ingestion of toxins produced by microfungi. The definition of microfungi is not rigorous, but understood here to refer principally to Ascomycetous fungi, including those with no sexual stage. The subkingdom Zygomycotina apparently contains few mycotoxin producers, including genera such as *Rhizopus* and *Mucor*, are not necessarily excluded from potential mycotoxin producers, but compounds of sufficient toxicity to be termed mycotoxins have not been found in these genera, except for rhizonin A and B from *Rhizopus microsporus* (Jennessen *et al.*, 2005).

All fungi produce extrolites, but fortunately only some of these are toxic to vertebrates. Some mycotoxins are acutely toxic, some are chronically toxic and some are both. Furthermore it is possible that mixtures of mycotoxins act synergistically or additively, so a mycological examination of the mycobiota, to species level, is very important, as different species produce different profiles of extrolites (Frisvad and Samson, 2004a). Prevention of mycotoxin formation should rely on accurately identified fungi and determination of the mycotoxins and other extrolites the fungi produce. Knowledge of the physiology of the different species is important, not only regarding growth lag phase and growth rate, but also regarding conidium production and further extrolite production. Mycotoxins and other extrolites may be produced in large amounts on some media and not on other media, so a detailed knowledge of the influence of the substrate on mycotoxin production is particularly important. Many mycotoxins may be produced under artificial laboratory conditions, but not necessarily on food, feedstuff or other substrates. Extrolites other than mycotoxins may also be a serious problem, a problem usually overlooked. For example penicillin is produced by *P. nalgiovense* and *P. chrysogenum*, species often found or deliberately used as starter cultures, in mould-fermented salami products. The contamination of foods with penicillin and other pharmaceutical compounds should also be avoided. Furthermore allergy problems, maybe initiated by living in water damaged buildings, may later be a problem if mouldy foods with the same fungal species are encountered. In the present chapter we will focus on the qualitative relationship between fungal species and their mycotoxin production, in worst case situations.

Table 1. Mycotoxins and other bioactive extrolites which have been named after a misidentified fungus or named after a synonym (Frisvad, 1989; Marasas *et al.*, 1984; Frisvad *et al.*, 2004c).

Extrolite	Producer originally identified as	Correct name
Communesin	*P. commune*	*P. marinum*
Compactin	*P. brevicompactum*	*P. solitum*
Cyclopaldic acid	*P. cyclopium*	*P. commune*
Cyclopenin	*P. cyclopium*	*P. solitum*
Cyclopiamide	*P. cyclopium*	*P. griseofulvum*
Cyclopiamin	*P. cyclopium*	*P. griseofulvum*
Cyclopiazonic acid	*P. cyclopium*	*P. griseofulvum*
Dehydrocyclopeptin	*P. cyclopium*	*P. solitum*
Moniliformin	*Fusarium moniliforme*	*F. verticillioides*
Neosolaniol	*F. solani*	*F. sporotrichioides*
Nivalenol	*F. nivale*	*F. kyushuense*
Ochratoxin	*Aspergillus ochraceus*	*Aspergillus westerdijkiae*
Patulin	*P. patulum*	*P. griseofulvum* (synonym)
Patulodin	*P. patulum*	*P. solitum*
Puberuline	*P. puberulum*	*P. cyclopium*
Rubratoxin	*P. rubrum*	*P. crateriforme*
Sterigmatocystin	*Sterigmatocystis* sp.	*Aspergillus versicolor*
Terrestric acid	*P. terrestre*	*P. crustosum* (synonym)
Vermiculin	*P. vermiculatum*	*Talaromyces flavus*
Verrucosidin	*P. verrucosum* var. *cyclopium*	*P. polonicum*
Verruculogen	*P. verruculosum*	*P. brasilianum*
Viridicatin	*P. viridicatum*	*P. crustosum*
Viridicatol	*P. viridicatum*	*P. crustosum*
Viridicatumtoxin	*P. viridicatum* (*P. expansum*)	*P. aethiopicum*

Important and less important producers of the different mycotoxins will be listed and discussed and examples of misidentified mycotoxin producers or misidentified mycotoxins will also be listed.

Mycotoxins have often been named after the fungus which was first found to produce them (Table 1). A large number of connections between fungal species and mycotoxins and antibiotics have been reported, but unfortunately many of these reports are incorrect (Frisvad, 1989).

Sometimes the fungi reported to produce mycotoxins are erroneous. Some examples of this are the following: Drusch and Ragab (2003) listed "*P. patulinum*" and "*P. clavatus*" as producers of patulin, and Bhatnagar *et al.* (2002) listed "*P. niger*" as an ochratoxin A producer. These taxa never existed and are incorrect combinations of genus and species or names that have never been described. Bhatnagar *et al.* (2002) incorrectly list *P. viridicatum* as producing ochratoxin A in a table, while using *P. verruculosum* as the species name in the text.

The correct name of the ochratoxin producing fungus was neither of these but actually *P. verrucosum*, the most important producer of ochratoxin A in *Penicillium*.

Below we provide an overview of the species producing the important mycotoxins and also list the most important misconceptions concerning species and their associated mycotoxins.

LIST OF MYCOTOXINS

Aflatoxins

Aflatoxins are the most potent natural compound carcinogens known (JECFA, 1997) affecting all vertebrate animal species, including man. Four compounds are commonly produced in foods: aflatoxins B_1, B_2, G_1 and G_2, but other biotransformed aflatoxins may occur for example in milk, such as aflatoxin M_1 and M_2 (Cole and Cox, 1981).

Aspergillus flavus is the most common species producing aflatoxins (Sargeant *et al.*, 1961), occurring in most kinds of foods in tropical countries. This species is very common on maize, peanuts and cottonseed, and produces only B aflatoxins. It has been estimated that only about 30-40% of known isolates produce aflatoxin, but in this estimate other closely related species has not been taken into account. Furthermore several other extrolites are produced by *A. flavus*, including kojic acid, cyclopiazonic acid, aspergillic acid and β-nitropropionic acid, so accurate identification of both the fungus and its profile of extrolites is important, if a prevention program should be successful. Species phenotypically quite similar to *A. flavus*, such as *A. nomius* (Kurtzman *et al.*, 1987; *A. zhaoqingensis*, Sun and Qi, 1991) and *A. bombycis* (Peterson *et al.*, 2001) produce B and G type aflatoxins, but more data are needed to establish their relative importance.

A. parasiticus occurs rather commonly in peanuts, but is apparently quite rare in other foods. It is more restricted geographically as compared to *A. flavus*. *A. parasiticus* produces both B and G aflatoxins (Sargeant *et al.*, 1961), and virtually all known isolates are toxigenic. This species also produce kojic acid and aspergillic acid, but it is not known whether compounds such as parasiticolide and other extrolites act synergistically with the aflatoxins.

Two species closely related to *A. parasiticus* also produce B and G aflatoxins: *A. toxicarius* (Murakami *et al.*, 1966; Murakami, 1971) and *A. parvisclerotigenus* (Saito and Tsuruta, 1993; Frisvad *et al.*, 2005a). *A. flavus* var. *columnaris* was reported to produce only aflatoxin B_2 (van Walbeek *et al.*, 1968). The culture ex-type of this taxon does not produce aflatoxins, and the taxonomic position of this variety is not fully elucidated, but it is probably a naturally occurring mutant of *A. flavus*.

A. pseudotamarii (Ito *et al.*, 2001) is another effective producer of B and G type aflatoxin for which importance for mycotoxins occurring in foods is unknown. The closely related species *A. tamarii* is not able to produce aflatoxins, despite several reports claiming this (Goto *et al.*, 1996; Klich *et al.*, 2000).

A. ochraceoroseus (Frisvad *et al.*, 1999; Klich *et al.*, 2000; Frisvad *et al.*, 2005a) and *A. rambellii* (Frisvad *et al.*, 2005a) are effective producers of aflatoxin B_1, but they have never been found in foods.

Emericella astellata (Frisvad *et al.*, 2004a) and *E. venezuelensis* (Frisvad and Samson, 2004b) are also producers of type B aflatoxins, but they have not only been found in foods yet.

Frisvad *et al.* (2006a) provide a list of the most important characteristics of the aflatoxin producing species.

Incorrect— The list of species that have been (incorrectly) reported to produce aflatoxins is long and includes *Absidia butleri* and *Absidia glauca* (Swelim *et al.*, 1994), *Alternaria cheiranthi* (Swelim *et al.*, 1994) *A. flavo-fuscus, A. glaucus* (Hanssen and Jung, 1966), *A. niger* (Kulik and Holaday, 1966), *A. oryzae* (El-Hag and Morse, 1976; Adebajo, 1992; El-Kady *et al.*, 1994; Atalla *et al.*, 2003; Drusch and Ragab, 2003), *A. ostianus* (Scott *et al.*, 1967), *A. sulphureus* (Scott *et al.*, 1970; Barr and Downey, 1975), *A. tamarii* (Lalithakumari and Govindaswami, 1970; El-Kady *et al.*, 1994; Goto *et al.*, 1996; 1997; Klich *et al.*, 2000), *A. terreus* (Sripathomswat and Thasnakorn, 1981), *A. terricola* (Moubasher *et al.*, 1977), *A. wentii* (Schroeder and Verrett, 1969), *A. zonatus* (El-Kady *et al.*, 1994), *Cephalosporium curticeps* and *C. rosea-griseum* (Swelim *et al.*, 1994), *Cladosporium cladosporioides* and *C. sphaerospermum* (Swelim *et al.*, 1994), *Cunninghamella echinulata* (Swelim *et al.*, 1994), *Emericella nidulans* (as *A. nidulans*) (Hanssen and Jung, 1973), *Emer. rugulosa* (as *A. rugulosus*) (Schroeder and Kelton, 1975), *Eurotium chevalieri, Eur. intermedium, Eur. repens* and *Eur. rubrum* (Kulik and Holaday, 1966; Leitao *et al.*, 1989; El-Kady *et al.*, 1994), *Mucor circinelloides* (Swelim *et al.*, 1994), *M. griseocyanus* (Swelim *et al.*, 1994), *M. mucedo* (Hanssen, 1969; Hanssen and Jung, 1973), *Penicillium citrinum* (Kulik and Holaday, 1966) "*P. citromyces strictum*" (Kulik and Holaday, 1966), *P. digitatum* (Hanssen and Jung, 1966), *P. frequentans* (Kulik and Holaday, 1966), *P. expansum* or *P. glaucum* (Hanssen

Table 2. Producers of mycotoxins

Aflatoxin	*Aspergillus bombycis, Aspergillus flavus, Aspergillus nomius, Aspergillus ochraceoroseus, Aspergillus parasiticus, Aspergillus parvisclerotigenus, Aspergillus pseudotamarii, Aspergillus rambellii, Aspergillus toxicarius, Emericella astellata, Emericella olivicola, Emericella venezuelensis*
Antibiotic Y	*Fusarium avenaceum, Fusarium chlamydosporum, Fusarium lateritium, Fusarium tricinctum*
Beauvericin	*Beauveria bassiana, Fusarium acuminatum, Fusarium avenaceum, Fusarium dlaminii, Fusarium equiseti, Fusarium longipes, Fusarium nygamai, Fusarium oxysporum, Fusarium poae, Fusarium proliferatum, Fusarium sambucinum, Fusarium sporotrichioides, Fusarium subglutinans, Fusarium verticillioides, Isaria fumosorosea*
Butenolide	*Fusarium avenaceum, Fusarium crookwellense, Fusarium culmorum, Fusarium graminearum, Fusarium poae, Fusarium sambucinum, Fusarium sporotrichioides, Fusarium tricinctum, Fusarium venenatum*
Citreoviridin	*Aspergillus terreus, Eupenicillium cinnamopurpureum, Penicillium citreonigrum, Penicillium manginii, Penicillium miczynskii, Penicillium smithii*
Citrinin	*Aspergillus terreus chemotype II, Aspergillus carneus, Aspergillus niveus, Blennoria* sp., *Clavariopsis aquatica, Monascus ruber, Penicillium chrzaczszii, Penicillium citrinum, Penicillium expansum, Penicillium manginii, Penicillium odoratum, Penicillium radicicola, Penicillium verrucosum, Penicillium westlingii*
Culmorin	*Fusarium crookwellense, Fusarium culmorum, Fusarium graminearum, Fusarium langsethiae, Fusarium poae, Fusarium sporotrichioides*
Cyclochlorotine and islanditoxin	*Penicillium islandicum*
Cyclopiazonic acid	*Aspergillus flavus, Aspergillus lentulus, Aspergillus oryzae, Aspergillus parvisclerotigenus, Aspergillus pseudotamarii, Aspergillus tamarii, Penicillium camemberti, Penicillium commune, Penicillium dipodomyicola, Penicillium griseofulvum, Penicillium palitans*
Enniatins	*Fusarium acuminatum, Fusarium avenaceum, Fusarium langsethiae, Fusarium lateritium, Fusarium poae, Fusarium sambucinum, Fusarium sporotrichioides, Halosarpeia* sp., *Verticillium hemipterigenum*
Ergot alkaloids	*Claviceps fusiformis, Claviceps paspali, Claviceps purpurea*
Fumigaclavines	*Aspergillus fumigatus, Aspergillus tamarii (?)*
Fumonisins	*Fusarium anthophilum, Fusarium dlamini, Fusarium napiforme, Fusarium nygamai, Fusarium proliferatum, Fusarium thapsinum, Fusarium verticillioides*
Fumonisin-like compounds	*Alternaria arborescens, Aspergillus cf. fumigatus, Aspergillus niger, Cochliobolus heterostrophus, Paecilomyces variotii*
Fusaproliferin	*Fusarium globosum, Fusarium guttiforme, Fusarium proliferatum, Fusarium pseudocircinatum, Fusarium pseudonygamai, Fusarium subglutinans, Fusarium verticillioides*
Janthitrems	*Eupenicillium shearii, Eupenicillium tularense, P. cf. janthinellum*
Lupinopsis toxin	*Phomopsis leptostromiformis*
Moniliformin	*Fusarium avenaceum, Fusarium napiforme, Fusarium nygamai, Fusarium oxysporum, Fusarium proliferatum, Fusarium subglutinans, Fusarium tricinctum, Fusarium thapsinum, Fusarium verticillioides*
Mycophenolic acid	*Byssochlamys nivea, Penicillium bialowiezense, Penicillium brevicompactum, Penicillium carneum, Penicillium roqueforti, Septoria nodorum*
β-nitropropionic acid	*Arthrinium aureum, Arthrinium phaerospermum, Arthrinium sacchari, Arthrinium saccharicola, Arthrinium sereanis, Arthrinium terminalis, Aspergillus flavus, Aspergillus oryzae, Aspergillus sojae, Penicillium atrovenetum*
Ochratoxin A	*Aspergillus carbonarius, Aspergillus cretensis, Aspergillus flocculosus, Aspergillus lacticoffeatus, Aspergillus niger, Aspergillus ochraceus, Aspergillus pseudoelegans, Aspergillus roseoglobulosum, Aspergillus sclerotioniger, Aspergillus sclerotiorum, Aspergillus steynii, Aspergillus sulphureus, Aspergillus westerdijkiae, Neopetromyces muricatus, Penicillium nordicum, Penicillium verrucosum, Petromyces albertensis, Petromyces alliaceus*

Patulin	*Aspergillus clavatonanica, Aspergillus clavatus, Aspergillus giganteus, Aspergillus longivesica, Aspergillus terreus (?), Byssochlamys nivea, Penicillium carneum, Penicillium clavigerum, Penicillium concentricum, Penicillium coprobium, Penicillium dipodomyicola, Penicillium expansum, Penicillium formosanum, Penicillium gladioli, Penicillium glandicola, Penicillium griseofulvum, Penicillium marinum, Penicillium paneum, Penicillium sclerotigenum, Penicillium vulpinum*
Penicillic acid	*Aspergillus auricomus, Aspergillus bridgeri, Aspergillus cretensis, Aspergillus flocculosus, Aspergillus insulicola, Aspergillus melleus, Aspergillus neobridgeri, Aspergillus ochraceus, Aspergillus ostianus, Aspergillus persii, Aspergillus petrakii, Aspergillus pseudoelegans, Aspergillus roseoglobulosus, Aspergillus sclerotiorum, Aspergillus sulphureus, Aspergillus westerdijkiae, Neopetromyces muricatus, Penicillium aurantiogriseum, Penicillium brasilianum, Penicillium carneum, Penicillium cyclopium, Penicillium fennelliae, Penicillium freii, Penicillium matriti, Penicillium polonicum, Penicillium radicicola, Penicillium tulipae, Penicillium viridicatum*
Penitrem A	*Penicillium clavigerum, Penicillium crustosum, Penicillium glandicola, Penicillium janczewskii, Penicillium melanoconidium, Penicillium tulipae*
Phomopsin	*Phomopsis leptostromiformis, PR toxin, Penicillium chrysogenum, Penicillium roqueforti*
Roquefortine C	*Penicillium albocoremium, Penicillium atramentosum, Penicillium allii, Penicillium carneum, Penicillium chrysogenum, Penicillium concentricum, Penicillium confertum, Penicillium coprobium, Penicillium coprophilum, Penicillium crustosum, Penicillium expansum, Penicillium flavigenum, Penicillium glandicola, Penicillium griseofulvum, Penicillium hirsutum, Penicillium hordei, Penicillium marinum, Penicillium melanoconidium, Penicillium paneum, Penicillium persicinum, Penicillium radicicola, Penicillium roqueforti, Penicillium sclerotigenum, Penicillium tulipae, Penicillium venetum, Penicillium vulpinum*
Rubratoxin	*Penicillium crateriforme*
Satratoxins	*Stachybotrys chartarum, Stachybotrys chlorohalonata*
Secalonic acid D	*Aspergillus aculeatus, Claviceps purpurea, Penicillium oxalicum, Phoma terrestris*
Sporidesmin	*Pithomyces chartarum, Pithomyces maydicus*
Sterigmatocystin (see also aflatoxin producers)	*Aspergillus multicolor, Aspergillus ochraceoroseus, Aspergillus rambellii, Aspergillus versicolor, Bipolaris sorokiniana (?), Chaetomium thielavioideum, Chaetomium* spp., *Emericella nidulans, Emericella* spp., *Humicola fuscoatra, Monocillium nordinii*
Tenuazonic acid	*Alternaria citri, Alternaria japonica, Alternaria kikuchiana, Alternaria longipes, Alternaria mali, Alternaria oryzae, Alternaria solani, Alternaria tenuissima, Phoma sorghina*
Trichothecenes	*Fusarium crookwellense, Fusarium culmorum, Fusarium equiseti, Fusarium graminearum, Fusarium langsethiae, Fusarium poae, Fusarium pseudograminearum, Fusarium sambucinum, Fusarium sporotrichioides, Fusarium venenatum*
Verrucosidin	*Penicillium aurantiogriseum, Penicillium melanoconidium, Penicillium polonicum*
Verruculogen and fumitremorgins	*Aspergillus caespitosus, Aspergillus fumigatus, Neosartorya fischeri, Penicillium brasilianum, Penicillium graminicola, Penicillium mononematosum*
Viriditoxin	*Aspergillus viridinutans, Paecilomyces variotii, Penicillium mononematosum*
Xanthomegnin	*Aspergillus auricomus, Aspergillus bridgeri, Aspergillus elegans, Aspergillus flocculosus, Aspergillus insulicola, Aspergillus melleus, Aspergillus neobridgeri, Aspergillus ochraceus, Aspergillus ostianus, Aspergillus persii, Aspergillus petrakii, Aspergillus roseoglobulosus, Aspergillus sclerotiorum, Aspergillus steynii, Aspergillus sulphureus, Aspergillus westerdijkiae, Microsporon cookei, Neopetromyces muricatus, Penicillium cyclopium, Penicillium freii, Penicillium janthinellum, Penicillium mariaecrucis, Penicillium melanoconidium, Penicillium tricolor, Penicillium viridicatum, Trichophyton megninii, Trichophyton mentagrophytes, Trichophyton rubrum, Trichophyton violaceum*
Zearalenone	*Fusarium crookwellense, Fusarium culmorum, Fusarium equiseti, Fusarium graminearum*

and Jung, 1973), *P. notatum* (Swelim *et al.*, 1994), *P. oxalicum* (Swelim *et al.*, 1994), *P. puberulum* (Hodges *et al.*, 1964), *P. roqueforti* (Swelim *et al.*, 1994), *P. variabile* (Kulik and Holaday, 1966), *Rhizopus* sp. (Kulik and Holaday, 1966; van Walbeek *et al.*, 1968), *Rhizopus nigricans* (Swelim *et al.*, 1994), *Scopulariopsis brevicaulis* (Swelim *et al.*, 1994), *Syncephalastrum racemosum* (Swelim *et al.*, 1994) and the bacterium *Streptomyces* sp. (Mishra and Murthy,

1968). The early reports on aflatoxin production were rejected by Langone and van Vunakis (1976), Rehm (1972), Bösenberg and Becker, (1972), Frank (1972), Mislivec *et al.*, (1968), Parrish *et al.* (1966); Wilson *et al.*, 1968, Hesseltine *et al.*, (1966), Scott (1965) and Rabie and Terblanche (1967). The first report that *A. oryzae* was able to produce aflatoxin was published by El-Hag and Morse (1976). However, the culture of *A. oryzae* they used was shown to be contaminated by an aflatoxin producing *A. parasiticus* (Fennell, 1976). Despite the fact that this problem was solved, others reported that *A. oryzae* produces aflatoxin. Floccose strains of *A. flavus* and *A. nomius* may superficially look like *A. oryzae*, so this macromorphological resemblance may have been the reason for later erroneous reports of aflatoxin production by this species. Since *A. oryzae* is a domesticated form of *A. flavus*, the former species will not be isolated from natural sources, except if they escape the soy sauce production plants and similar factories and contaminate the immediate surroundings. None of these species produce aflatoxins, and many of these names are not accepted as valid species in any case.

Antibiotic Y

Originally called lateropyrone (Bushnell *et al.*, 1984), antibiotic Y has significant antibiotic properties towards phytopathogenic bacteria, but apparently low cell toxicity (Golinski *et al.*, 1986). Producers of antibiotic Y are very common in agricultural, and antibiotic Y and its producers have also been found in apples, cherries and wheat (Andersen and Thrane, 2006).

The main producers are *Fusarium avenaceum* and *F. tricinctum*. The former occurs frequently in cereal grain, fruits and vegetables, while the latter is very frequent on cereal grain especially in temperate climate.

F. lateritium and *F. chlamydosporum* produce antibiotic Y, but their importance in fruits in temperate regions and cereal grains in warmer climates need to be examined.

Beauvericin

Beauvericin was originally found in insect-associated fungi such as *Beauveria bassiana* and *Isaria fumosorosea*, but have also been detected among several *Fusarium* species occurring on food (Logrieco *et al.*, 1998). The species listed as producers of beauvericin (Table 2) may have to be revised as many changes have occurred in *Fusarium* taxonomy in the last decade. *F. subglutinans*, *F. proliferatum* and *F. oxysporum* are often found on corn and fruits and are consistent producers of beauvericin.

Several other species of *Fusarium* have been reported to produce beauvericin in low amounts, including *F. nygamai*, *F. dlaminii* and *F. verticillioides* from cereals and fruits. *F. avenaceum*, *F. poae* and *F. sporotrichioides* from cereal grains, fruits and vegetables are also known to produce beauvericin in low amounts (Morrison *et al.*, 2002; Thrane *et al.*, 2004). Furthermore, *F. sambucinum* and some strains of *F. acuminatum*, *F. equiseti* and *F. longipes* from agricultural products have been reported as weak producers of beauvericin (Logrieco *et al.*, 1998).

Butenolide

Butenolide, or more correctly 4-acetamido-2-buten-4-olide, has been associated with cattle diseases (Yates *et al.*, 1969; Marasas *et al.*, 1984). There are no reports of butenolide occurring in foods, but it is still considered a toxin of potential importance due to the reported synergistic effect with enniatin B (Hershenhorn *et al.*, 1992).

The cereal associated Fusaria, *F. sporotrichioides*, *F. graminearum* and *F. culmorum* are considered the most important producers of 4-acetamido-2-buten-4-olide. Other potential producers of 4-acetamido-2-buten-4-olide are *F. avenaceum*, *F. poae* and *F. tricinctum* which are frequently found in cereal grains together with *F. crookwellense*, *F. sambucinum* and *F. venenatum*. Most of these Fusaria can be found in potatoes and related vegetables.

Citreoviridin

The toxicology of citreoviridin is not well elucidated, but it has been claimed to be involved in acute cardiac beriberi (Ueno, 1974). It has occasionally been associated with yellow rice disease in cases where *P. citreonigrum* (formerly called *P. citreoviride*) was the dominant yellow pigment producing fungus.

Eupenicillium ochrosalmoneum has been found in cereals in the United States and in Slovakia (Labuda and Tancinova, 2003) and is an efficient producer of citreoviridin.

P. smithii, P. miczynskii and *P. manginii* (Frisvad and Filtenborg, 1990) and *Aspergillus terreus* (Bauer and Gedek, 1979; Franck and Gehrken, 1980) has most often been recovered from soil and only rarely from foods.

Incorrect — *P. pulvillorum* (Nagel *et al.*, 1972) reported to produce citreoviridin was actually a *P. manginii* (Pitt, 1979). Both *P. charlesii* (Cole *et al.*, 1981) and *P. fellutanum* (Locci *et al.*, 1965), reported to produce citreoviridin were shown to be *P. citreonigrum* (Frisvad, 1989).

Citrinin

Citrinin is a potent nephrotoxin, occurring in different cereal crops, peanuts and meat products (Reddy and Berndt, 1991). Producers of citrinin are widespread and common in foods. The major source of citrinin in cereals is *P. verrucosum* (Frisvad *et al.*, 2004b; 2005b) while isolates of *P. expansum* isolated from pomaceous fruits often produce citrinin (Harwig *et al.*, 1973; Andersen *et al.*, 2004). Citrinin producing *P. radicicola* has been found in onions, carrots and potatoes (Overy and Frisvad, 2003). *P. citrinum* is a fungus of worldwide distribution, and all isolates produce citrinin (Hetherington and Raistrick, 1931; Frisvad and Filtenborg, 1990).

The fungus used for producing red rice, *Monascus ruber* has been reported to produce citrinin (Blanc *et al.*, 1995).

Several other fungi are able to produce citrinin, but these species are uncommon in foods. These include *A. terreus* (Sankawa *et al.*, 1983), *A. carneus* (Chien *et al.*, 1977), a *Blennoria* species and *Clavariopsis aquatica* (Broadbent, 1966), *P. chrzaczszii* (Pollock, 1947), *P. manginii* (Frisvad and Filtenborg, 1990), *P. odoratum* (Nakajima and Nozawa, 1979) and *P. westlingii* (Frisvad and Filtenborg, 1990).

Incorrect — Citrinin production by *A. oryzae* or *P. camemberti* (Bennett and Klich, 2003) was based on misinformation in another review of Manabe (2001). Citrinin production by *A. candidus* was mentioned by Timonin and Rouatt (1944), but was rejected by Raper and Fennell (1965). Many other species have been claimed to produce citrinin, including *A. ochraceus* (Mantle and McHugh, 1993), *A. wentii* (Abu-Seidah, 2003) and *Eurotium pseudoglaucum* (El-Kady *et al.*, 1994), but either producer or mycotoxins or both may have been misidentified in those cases.

Culmorin

Culmorin has a synergistic effect with deoxynivalenol towards caterpillars, but a low toxicity on its own (Pedersen and Miller, 1999; Dowd *et al.*, 1989). Culmorins in addition to deoxynivalenol and acetyl-deoxynivalenol have been detected in cereals (Ghebremeskel and Langseth, 2000).

The cereal borne *F. culmorum, F. graminearum, F. poae* and *F. langsethiae* are the producers of culmorin, but *F. crookwellense* and *F. sporotrichioides* have also be listed as producers (Thrane *et al.*, 2004).

Cyclochlorotine

Cyclochlorotine and islanditoxin are chlorine containing cyclic peptides from *Penicillium islandicum* that have been associated with yellowed rice toxicosis in connection with the yellow anthraquinones luteoskyrin and rugulosin, also produced by *P. islandicum* (Enomoto and Ueno, 1974). These cyclic peptides have not been found in any other species yet, but may be important in boiled rice that is kept too long without being refrigerated.

Cyclopiazonic acid

Cyclopiazonic acid (Holzapfel, 1968; Antony *et al.*, 2003) is a potent organ damaging calcium chelating mycotoxin that produces focal necrosis in most vertebrate inner organs. It was originally believed that aflatoxins were responsible for all the toxic effects of *Aspergillus flavus* contaminated peanuts to turkeys in turkey X disease, but it has later been shown that cyclopiazonic acid had a severe effect on the muscles and bones of the turkeys (Dorner *et al.*, 1983; Jand *et al.*, 2005).

Aspergillus flavus and the domesticated form *A. oryzae* often produce large amounts of cyclopiazonic acid. *A. flavus* is common on oil seeds, nuts, peanuts and cereals, but may also produce aflatoxin on dried fruits (Huang *et al.*, 1994; Pitt and Hocking, 1997).

P. commune and its domesticated *P. camemberti*, and the closely related species *P. palitans* are common on cheese and meat products and may produce cyclopiazonic acid in these products (Hermansen *et al.*, 1984; Pitt *et al.*, 1986; Polonelli *et al.*, 1987; Frisvad *et al.*, 2004c).

P. griseofulvum is also a major producer of cyclopiazonic acid, and may occur in long stored cereals and cereal products, such as pasta (Pitt and Hocking, 1997).

Other producers of cyclopiazonic acid in *Aspergillus* include *A. parvisclerotigenus, A. pseudotamarii* and *A. tamarii*, but the role of these fungi concerning CPA production is not clear. *P. dipodomyicola*, another good producer of CPA, has until now only been found rarely in foods.

Incorrect – Cyclopiazonic acid was originally isolated from and named after *P. cyclopium* CSIR 1082, but this fungus was re-identified as *P. griseofulvum* (Hermansen *et al.*, 1984; Frisvad, 1989). Despite this, most reviews still cite *P. cyclopium* or *P. aurantiogriseum* as producers of CPA (Scott, 1994; Bhatnagar, 2002; Bennett and Klich, 2003). Another example of an error being often cited is the claimed production of cyclopiazonic acid by *Aspergillus versicolor* (Ohmomo *et al.*, 1973; cited by Bhatnagar *et al.*, 2002) even though Domsch *et al.* (1980) and Frisvad (1989) had stated that the isolate described by Ohmomo *et al.* (1973) was correctly identified as *A. oryzae*, a known producer of cyclopiazonic acid (Orth, 1977). *P. hirsutum, P. viridicatum, P. chrysogenum, P. nalgiovense, A. nidulans* and *A. wentii* have also wrongly been claimed to produce cyclopiazonic acid (El-Banna *et al.*, 1987; Cole *et al.*, 2003; Abu-Seidah, 2003).

Enniatins

Enniatins are a group of more than 15 related cyclic peptides which have antibiotic and ionophoric activities (Kamyar *et al.*, 2004). Both enniatins and the related beauvericins have been detected in agricultural products (Jestoi *et al.*, 2004).

F. avenaceum and *F. sambucinum* are the most important enniatin producers in cereals and other agricultural plants used for food (Morrison *et al.*, 2002).

Other maybe less important producers of enniatins are the cereal-borne *F. langsethiae, F. poae* and *F. sporotrichioides* and the fruit-borne *F. lateritium* and *F. acuminatum* isolated from herbs. Other fungi than *Fusarium* producing enniatins are listed in Table 2. These have not been isolated from foods.

Ergot alkaloids

Ergot alkaloids are common in sclerotia of *Claviceps*, rather often occurring in whole rye, used for rye-bread production. These sclerotia are often removed before milling of the rye.

Claviceps purpurea, C. fusiformis and *C. paspali* are the major sources of ergot alkaloids (Blum, 1995). *C. purpurea* produces ergopeptides, *C. paspali* produces lysergic acid amides and *C. fusiformis* produces clavine type alkaloids, but all species produce chanoclavine I, agroclavine and elymoclavine. Several Penicillia and Aspergilli can produce clavine type alkaloids also, including agroclavine, but the role of clavines in mycotoxicology is not well known.

Fumigaclavines

Fumigaclavines are toxic clavine-like alkaloids produced by *A. fumigatus* (Cole *et al.*, 1977) and *A. tamarii* (Janardhanan *et al.*, 1984). Fumigaclavine production by *A. tamarii* has not been confirmed, but the common occurrence of *A. fumigatus* indicates that this may be an important mycotoxin in some cases.

Fumonisins

The fumonisins were discovered after being ignored or in the late 1980s there has been a lot of attention to these highly carcinogenic compounds. Several reviews on the chemistry, toxicology and mycology have been published (Marasas *et al.*, 2001; Weidenbörner, 2001).

F. verticillioides (formerly known as *F. moniliforme* (Seifert *et al.*, 2003) and *F. proliferatum* are important sources of fumonisins in corn. These species and fumonisins in crops have been reported from all over the world in numerous papers and book chapters.

Other fumonisin producing species are *Fusarium nygamai, F. napiforme, F. thapsinum, F. anthophilum* and *F. dlaminii* from millet, sorghum and rice.

Fusaproliferin

Fusaproliferin is a recent discovered mycotoxin which shows teratogenic and pathological effects in cell assays (Bryden *et al.*, 2001). Fusaproliferin has been detected in natural samples together with beauvericin and fumonisin (Munkvold *et al.*, 1998).

F. proliferatum and *F. subglutinans* are major sources in cereals, especially corn. These fungi and fusaproliferin have been detected in Europe, North America and South Africa (Wu *et al.*, 2003). Few strains of *F. globosum, F. guttiforme, F. pseudocircinatum, F. pseudonygamai* and *F. verticillioides* have been found to produce fusaproliferin.

Janthitrems and shearinins

Janthitrems are very toxic tremorgenic toxins produced by *Eupenicillium tularense, E. shearii* and *P.* cf. *janthinellum*, normally do not grow to a significant extent in foods. On the other hand *P. tularense* has recently been demonstrated to produce janthitrems in tomatoes (Andersen and Frisvad, 2004), so maybe these mycotoxins may occur sporadically.

Lupinopsis toxin is produced by *Phomopsis leptostromiformis* and is a hepatotoxin (Marasas *et al.*, 1984).

Moniliformin

Moniliformin is cytotoxic, inhibits protein synthesis and enzymes, causes chromosome damages and induces heart failure in mammals and poultry (Bryden *et al.*, 2001). Moniliformin has been found worldwide in cereal samples.

In corn, *F. proliferatum* and *F. subglutinans* are the main producers of moniliformin, where as *F. avenaceum* and *F. tricinctum* are the key sources in cereals grown in temperate climate.

In sorghum, millet and rice *F. napiforme, F. nygamai, F. verticillioides* and *F. thapsinum* may be responsible for moniliformin production. Some strains of *F. oxysporum* produce a significant amount of moniliformin under laboratory condition; however, there is no detailed information on a possible production in vegetables and fruits. An overview of other minor sources has been published (Schütt *et al.*, 1998).

Mycophenolic acid

Despite having a low acute toxicity, mycophenolic acid may be a very important indirect mycotoxin as it is highly immunosuppressive, thereby paving the way for bacterial and fungal infections (Bentley, 2000). Mycophenolic acid occurs quite frequently in foods (Lafont *et al.*, 1979, Lopez-Diaz *et al.*, 1996; Overy and Frisvad, 2005). It is not acutely toxic, but mycophenolic acid has been reported to be strongly immunosuppressive and is therefore used in organ transplantation to avoid rejection (Bentley, 2000). It is not unlikely that mycophenolic acid could lower the immune system if ingested often and thereby pave the way for bacterial infections.

P. brevicompactum and *P. bialowiezense* are ubiquitous species and may produce mycophenolic acid in, for example, ginger (Overy and Frisvad, 2005). Two other major species producing mycophenolic acid are *P. roqueforti* and *P. carneum*. Both species can grow in a lactic acid bacterium environment and other acidic substrates with a certain amount of carbondioxide. The same appears to be the case for a fifth important producer *Byssochlamys nivea* (Puel *et al.*, 2005). Mycophenolic acid has been found to occur naturally in blue cheeses (Lafont *et al.*, 1979).

The soil-borne species *P. fagi* also produces mycophenolic acid (Frisvad and Filtenborg, 1990, as *P. raciborskii*) and *Septoria nodorum* (Devys *et al.*, 1980) is another source that is probably not important as a food contaminant.

Incorrect — P. viridicatum was once regarded as a mycophenolic acid producer (Burton, 1949), but this has never been reconfirmed.

β-nitropropionic acid (BNP)

β-nitropropionic acid has been reported to be involved in sugar cane poisoning of children, but may potentially also cause other intoxications, as producers are widespread (Ming *et al.*, 1995; Burdock *et al.*, 2001). Furthermore BNP has been found in miso and shoyu and katsuobushi and it can be produced by *A. oryzae* (Nakamura and Shimoda, 1954) when artificially inoculated on cheese, peanuts, etc. Unfortunately this has not been tested for *A. flavus*, because the production of BNP by *A. oryzae* on peanuts indicates that *A. flavus* may be able to produce this mycotoxin in combination with aflatoxin B_1, cyclopiazonic acid and kojic acid. The possible synergistic effect of these mycotoxins on mammals is unknown.

BNP producing species from sugar cane were found to be *Arthrinium phaerospermum* and *Art. sacchari*. *Art. terminalis*, *Art. saccharicola*, *Art. aureum* and *Art. sereanis* can also produce BNP (Burdock *et al.*, 2001).

A. flavus (Bush *et al.*, 1951) may turn out to be an important producer of this mycotoxin in foods, but there are no surveys that include analytical determination of BNP alongside cyclopiazonic acid and aflatoxin B_1.

A. oryzae and *A. sojae* can produce BNP in miso and shoyu. It is probably more important that their wildtype forms, *A. flavus* and *A. parasiticus*, respectively, may produce BNP in foods. More research is needed in this area.

P. atrovenetum is another authenticated producer of BNP, but this fungus is only found in soil (Raistrick and Stössl, 1958).

Incorrect — *P. cyclopium*, *P. chrysogenum*, *A. wentii*, *Eurotium* spp., and *A. candidus* have been reported as producers of BNP (Brookes *et al.*, 1963; Burrows and Turner, 1966; Burdock *et al.*, 2001), but these identifications are doubtful. The report that *A. candidus* produced β-nitropropionic (Kinoshita *et al.*, 1968) was based on a white spored mutant of *A. flavus* (Frisvad, 1989).

Ochratoxin A (OTA)

Ochratoxin A is a nephrotoxin affecting all tested animal species, though effects in man have been difficult to establish unequivocally. It is listed as a probable human carcinogen (Class 2B) (JECFA, 2001). Links between OTA and Balkan Endemic Nephropathy have long been sought, but not established (JECFA, 2001). The toxicology of ochratoxin A has recently been reviewed (Ringot *et al.*, 2006).

Major producers of ochratoxin A are found among *Aspergillus* sections *Flavi*, *Circumdati* and *Nigri* and in *Penicillium* series *Verrucosa* (Frisvad and Samson, 2000; Frisvad *et al.*, 2004b; Samson *et al.*, 2004b; Frisvad *et al.*, 2004c). *A. ochraceus* (van der Merwe *et al.*, 1965), occurs in stored cereals (Pitt and Hocking, 1997) and coffee (Taniwaki *et al.*, 2003). *A. ochraceus* has been shown to consist of two species (Varga *et al.*, 2000a,b; Frisvad *et al.*, 2004b), and the second and new species, *A. westerdijkiae*, produces large amounts of ochratoxin A consistently. The producer of ochratoxin A first discovered, NRRL 3174, has been designated as the type culture of *A. westerdijkiae* (Frisvad *et al.*, 2004b). This is interesting as *A. westerdijkiae* is both a more efficient and consistent producer of ochratoxin than *A. ochraceus*, and is maybe more prevalent in coffee than *A. ochraceus*. The ex type culture of *A. ochraceus* CBS 108.08 only produce trace amounts of ochratoxin A.

A. carbonarius (Horie, 1995) is a major OTA producer. It is occurring in grapes and grape products, including grape juice, wines and dried vine fruits (International Agency for Research on Cancer (IARC), 2002) and occasionally occurs on coffee beans (Abarca *et al.*, 2004).

Petromyces alliaceus (Lai *et al.*, 1970), now placed in *Aspergillus* section *Flavi*, produces large amounts of ochratoxin A in pure culture, and OTA produced by this fungus has been found in figs in California (Bayman *et al.*, 2002).

A. steynii, from the *Aspergillus* section *Circumdati*, is also a very efficient producer of OTA, and has been found in green coffee beans, mouldy soy beans and rice (Frisvad *et al.*, 2004b). As with *A. westerdijkiae*, *A. steynii* may have been identified as *A. ochraceus* earlier, so the relative abundance of these three species is difficult to evaluate at present.

A. niger (Abarca *et al.*, 1994) is an extremely common species, but only few strains appear to be producers of OTA, so this species may be of much less importance than *A. carbonarius* in grapes, wine and green coffee beans. It can be of major importance, however, as for example an isolate of *A. niger* NRRL 337, referred to as the "food fungus," produce large amounts of OTA in pure culture. This fungus is used for fermentation of potato peel waste, etc. and used for animal feed (Schuster *et al.*, 2002).

P. verrucosum (Frisvad, 1985b; Pitt, 1987) is the major producer of ochratoxin A in stored cereals (Lund and Frisvad, 2003).

P. nordicum (Larsen *et al.*, 2001) is the main OTA producer found in meat products such as salami and ham. Both OTA producing *Penicillium* species have been found on cheese also, but have only been reported to be of high occurrence on Swiss hard cheeses (as *P. casei*, Staub, 1911). The ex type culture of *P. casei* is a *P. verrucosum* (Larsen *et al.*, 2001).

Several other Aspergilli can produce ochratoxin A in large amounts, but they appear to be relatively rare. In *Aspergillus* section *Circumdati* (formerly the *Aspergillus ochraceus* group), the following species can produce ochratoxin A: *Aspergillus cretensis, A. flocculosus, A. pseudoelegans, A. roseoglobulosus, A. sclerotiorum, A. sulphureus* and *Neopetromyces muricatus* (Frisvad *et al.*, 2004b). According to Ciegler (1972) and Hesseltine *et al.* (1972), *A. melleus, A. ostianus, A. persii* and *A. petrakii* may produce trace amounts of OTA, but this has not been confirmed since those papers. Strains of these species reported to produce large amounts of OTA were reidentified by Frisvad *et al.* (2004b). In *Aspergillus* section *Flavi, Petromyces albertensis* produces ochratoxin A. In *Aspergillus* section *Nigri, A. lacticoffeatus* and *A. sclerotioniger* produce ochratoxin A (Samson *et al.*, 2004b).

Incorrect — A very large number of species have been claimed to produce ochratoxin A and are in need of revision. Some of these misidentified isolates will be mentioned here. Of the *Penicillium* species, *P. viridicatum* was the name cited for many years as the major ochratoxin A producer, but it was shown by Frisvad and Filtenborg (1983), Frisvad (1985a) and Pitt (1987) that *P. verrucosum* was the correct name for this fungus, the only species that produces ochratoxin A in cereals in Europe. The closely related *P. nordicum*, which occurs on dried meat in Europe, was mentioned as producing ochratoxin A by Frisvad and Filtenborg (1983) and Land and Hult (1987), but not accepted as a separate species until the publication of Larsen *et al.* (2001). *P. verrucosum* has been correctly cited as the main *Penicillium* species producing ochratoxin A for a number of years now, but in a series of recent reviews and papers *P. viridicatum* and *P. verruculosum* (no doubt mistaken for *P. verrucosum*) have been mentioned again (Mantle and McHugh, 1993; Bhatnagar *et al.*, 2002; Czerwiecki *et al.*, 2002a,b). In the latter two papers *P. chrysogenum, P. cyclopium, P. griseofulvum, P. solitum, Aspergillus flavus, A. versicolor* and *Eurotium glaucum* were listed assigned ochratoxin A producers.

The strain of *P. solitum* reported by Mantle and McHugh (1993) to produce ochratoxin A was assigned more recently to *P. polonicum*, but either species produces ochratoxin A (Lund and Frisvad, 1994; 2003). These isolates were contaminated by *P. verrucosum*.

The reports by Czerwiecki *et al.* (2002 a,b) have to be rejected as the fungi have been discarded, so it will never be possible to check the results.

The following species were listed as ochratoxin A producers by Varga *et al.* (2001): *Aspergillus auricomus, A. fumigatus, A. glaucus, A. melleus, A. ostianus, A. petrakii, A. repens, A. sydowii, A. terreus, A. ustus, A. versicolor, A. wentii, Penicillium aurantiogriseum, P. canescens, P. chrysogenum, P. commune, P. corylophilum, P. cyaneum, P. expansum, P. fuscum, P. hirayamae, P. implicatum, P. janczewskii, P. melinii, P. miczynskii, P. montanense, P. purpurescens, P. purpurogenum, P. raistrickii, P. sclerotiorum, P. spinulosum, P. simplicissimum, P. variabile* and *P. verruculosum*. These species have been reexamined and until now there are no signs of ochratoxin A production in those species.

In the *Handbook of Fungal Secondary Metabolites* (Cole and Schweikert, 2003a,b; Cole *et al.*, 2003), only two of the species cited as producing ochratoxin A are correct: *A. ochraceus* and *A. sulphureus*. The others mentioned are not.

Patulin

Patulin is generally very toxic for both prokaryotes and eukaryotes, but the toxicity for humans has not been conclusively demonstrated. Several countries in European and the United States now have set limits on the level of patulin in apple juice.

P. expansum is by far the most important source of patulin. This species also produces the toxic compounds chaetoglobosin A and C, communesins, roquefortine C and often also citrinin (Andersen *et al.*, 2004).

Byssochlamys nivea may be present in pasteurized fruit juices and may produce patulin and mycophenolic acid (Puel *et al.*, 2005).

P. griseofulvum is a very efficient producer of high levels of patulin in pure culture, and it may potentially produce patulin in cereals, pasta and similar products.

P. carneum may potentially produce patulin in beer, wine, meat products and rye bread as it has been found in those substrates (Frisvad and Samson, 2004b), but there are not any reports yet on patulin production by this particular species in those foods. *P. carneum* also produces mycophenolic acid, roquefortine C and penitrem A (Frisvad *et al.*, 2004c).

P. paneum occurs in rye bread (Frisvad and Samson, 2004b), but again actual production of patulin in this product has not been reported yet.

P. sclerotigenum is common in yams and has the ability to produce patulin in laboratory cultures.

P. dipodomyicola is an efficient producer of patulin and has been found in rice in Australia and in chicked feed mixture in Slovakia, but it may be of sporadic occurrence.

The coprophilous fungi *P. concentricum, P. clavigerum, P. coprobium, P. formosanum, P. glandicola, P. vulpinum, A. clavatus, A. longivesica* and *A. giganteus* are very efficient producers of patulin in the laboratory, but it is only *A. clavatus* that may play any role in human health, as it may be present in beer malt (Lopez-Diaz and Flannigan, 1997). *A. terreus, P. novae-zeelandiae, P. marinum, P. melinii* and other soil-borne fungi may produce patulin in pure culture, but are less likely to occur in any foods.

Incorrect — A number of species in different genera, notably *Penicillium, Aspergillus* and *Byssochlamys*, produce patulin. Among the most efficient producers of patulin are *A. clavatus, A. giganteus, A. terreus, Byssochlamys nivea, P. carneum, P. dipodomyicola, P. expansum, P. griseofulvum, P. marinum, P. paneum* and several dung associated Penicillia (Frisvad, 1989; Frisvad *et al.*, 2004b). It is not, however, produced by species in all of the 42 genera listed by Steiman *et al.* (1989) and Okeke *et al.* (1993). These papers include erroneous statements that *Alternaria alternata, Fusarium culmorum, Mucor hiemalis, Trichothecium roseum* and many others produce patulin. The production of patulin by *Alternaria alternata* was later reported by Laidou *et al.* (2001), and mentioned in a review by Drusch and Ragab (2003). However patulin was not found in hundreds of analyses of *Alternaria* extracts (Montemurro and Visconti, 1992), or in extracts from more than 200 *Alternaria* cultures tested by us at the Technical University of Denmark (B. Andersen, personal communication).

Penicillic acid

Penicillic acid (Alsberg and Black, 1913) and dehydropenicillic acid (Obana *et al.*, 1995) are also small toxic polyketides, but their major role in mycotoxicology may be in their possible synergistic toxic effect with ochratoxin A (Lindenfelser *et al.*, 1973; Stoev *et al.*, 2001) and possible additive or synergistic effect with the naphthiumiuoquinones hepatotoxins xanthomegnin, viomellein and vioxanthin.

Penicillic acid is likely to co-occur with ochratoxin A, xanthomegnin, viomellein and vioxanthin in members of the *Aspergillus ochraceus* group (*Aspergillus* section *Circumdati*) and members of *Penicillium* series *Viridicata* which often co-occur with *P. verrucosum*. The former *Aspergillus* group species often occur in coffee and the latter Penicillia are common in cereals. The major sources of penicillic acid are *P. aurantiogriseum, P. cyclopium* (Birkinshaw *et al.*, 1936), *P. melanoconidium* and *P. polonicum* (Lund and Frisvad, 1994; Frisvad and Samson, 2004b) and all members of *Aspergillus* section *Circumdati* (Frisvad and Samson, 2000). Furthermore it is produced by *P. tulipae* and *P. radicicola*, which are occasionally found on onions, carrots and potatoes (Overy and Frisvad, 2003).

Penicillic acid has been found in *P. carneum* (Frisvad and Samson, 2004b), and it is also produced by several soil-borne Penicillia including *P. brasilianum* (Frisvad and Filtenborg, 1990) and *P. fennelliae* (van Eijk, 1969).

Incorrect — Production reported by *P. roqueforti* (Karow *et al.*, 1944, as *P. suavolens*; Moubasher *et al.*, 1978; Olivigni and Bullerman, 1978) is now considered to be due to the similar species *P. carneum* (Boysen *et al.*, 1996). Also penicillic acid production by *P. commune* (Ciegler and Kurtzman, 1972), *P. chrysogenum* (Leistner and Pitt, 1977) and several other species of *Penicillium* could not be confirmed.

Penitrem A

Penitrem A is a mycotoxic indol-terpene with tremorgenic properties. It has first of all been implicated in mycotoxicoses of animals (Rundberget and Wilkins, 2002), but has also been suspected to be implicated in tremors in humans (Cole *et al.*, 1983).

P. crustosum is the most important producer of penitrem A (Pitt, 1979). This species is of worldwide distribution and often found in foods (El-Banna *et al.*, 1988). This mycotoxin is produced by all isolates of *P. crustosum* examined (Sonjak *et al.*, 2005). *P. melanoconidium* is common in cereals (Frisvad and Samson, 2004b), but it is not known whether this species can produce penitrem A in infected cereals.

P. glandicola, P. clavigerum, and *P. janczewskii* are further producers of penitrem A (Frisvad and Samson, 2004b; Frisvad and Filtenborg, 1990), but they have not been recovered from foods more than sporadically. Other species which do produce penitrem A include *P. carneum* and *P. tulipae* (Frisvad *et al.*, 2004b).

Incorrect — Many species have been claimed to produce penitrem A, but most were misidentifications of *P. crustosum* (Pitt, 1979; Frisvad, 1989). Names given to isolates that were in fact *P. crustosum* include *P. cyclopium, P. verrucosum* var. *cyclopium, P. verrucosum* var. *melanochlorum, P. viridicatum, P. commune, P. lanosum, P. lanosocoeruleum, P. granulatum, P. griseum, P. martensii, P. palitans* and *P. piceum* (Frisvad, 1989).

Phomopsin

Phomopsin and other metabolites have been reported from *Phoma* and *Phomopsis* and these may also be important mycotoxins (Bhatnagar *et al.*, 2002).

PR toxin

PR toxin is a mycotoxin that is acutely toxic and has DNA and protein damaging properties (Moule *et al.*, 1980; Arnold *et al.*, 1987). Even though it is unstable in cheese (Teuber and Engel, 1983), it may be produced in silage and other substrates.

Major sources. P. roqueforti is the major source of PR toxin. It has been reported also from *P. chrysogenum* (Frisvad and Samson, 2004b).

Roquefortine C

The status of roquefortine C as a mycotoxin has been questioned, but it is a very widespread fungal secondary metabolite, and produced by a large number of species. The acute toxicity of roquefortine C is not very high (Cole and Cox, 1981), but it has been reported as a neurotoxin.

P. albocoremium, P. atramentosum, P. allii, P. carneum, P. chrysogenum, P. crustosum, P. expansum, P. griseofulvum, P. hirsutum, P. hordei, P. melanoconidium, P. paneum, P. radicicola, P. roqueforti, P. sclerotigenum, P. tulipae and *P. venetum* are all producers that have been found in foods, but the natural occurrence of roquefortine C has only been reported rarely.

P. concentricum, P. confertum, P. coprobium, P. coprophilum, P. flavigenum, P. glandicola, P. marinum, P. persicinum and *P. vulpinum* are not likely to occur in foods.

Rubratoxin

Rubratoxin is a potent hepatotoxin (Engelhardt and Carlton, 1991) and is of particular interest as it has been implicated in severe liver damage in three Canadian boys, who drank rhubarb wine contaminated with *P. crateriforme*. One of the boys needed to have the liver transplanted (Richer *et al.*, 1997).

Major producers. P. crateriforme is the only known major producer of rubratoxin A and B (Frisvad, 1989).

Incorrect — Rubratoxins are hepatoxic mycotoxins known to be produced only by the rare species *P. crateriforme* (Frisvad, 1989). Rubratoxins are not produced by *P. rubrum, P. purpurogenum* or *Aspergillus ochraceus* as reported by Moss *et al.* (1968), Natori *et al.* (1970) and Abu-Seidah (2003).

Satratoxins

Stachybotrys spp. are first of all of importance for indoor climate, but stachybotrytoxicosis was one of the first horse mycotoxicosis to be reported on (Rodricks and Eppley, 1974).

Stachybotrys chartarum and *S. chlorohalonata* are the two important fungi producing cyclic trichothecenes (satratoxins) and toxic atranones (Andersen *et al.*, 2003; Jarvis, 2003).

Secalonic acid D

The toxicological data on secalonic acid D and F are somewhat unclear (Ehrlich *et al.*, 1982; Reddy and Reddy, 1991), so the significance of this metabolite in human and animal health is somewhat uncertain.

Claviceps purpurea, P. oxalicum, Phoma terrestris and *A. aculeatus* produce large amounts of secalonic acid D and F in pure culture. Secalonic acid D has been found to occur in grain dust in the United States (Reddy and Reddy, 1991).

Sporidesmin is produced by *Pithomyces chartarum* and causes facial eczema in sheep (Atherton *et al.*, 1974). *Pithomyces chartarum* and *P. maydicus* have reported to produce sporidesmin and related compounds.

Sterigmatocystin

Sterigmatocystin is a possible carcinogen, but may be important as is can be produced in rather large amounts on cheese and occasionally in cereals.

The major source of sterigmatocystin in foods is *A. versicolor*. This fungus is common on cheese, but may also occur on other substrates (Pitt and Hocking, 1997).

A large number of species are able to produce sterigmatocystin, including *Chaetomium* spp., *Emericella* spp., *Monocillium nordinii* and *Humicola fuscoatra* (Joshi *et al.*, 2002). These species are probably not likely to contaminate foods.

Although sterigmatocystin is a precursor of aflatoxins (Frisvad, 1989), only *A. ochraceoroseus* (Frisvad *et al.*, 1999; Klich *et al.*, 2000), and some *Emericella* species accumulate both sterigmatocystin and aflatoxin (Frisvad *et al.*, 2004a; Frisvad and Samson, 2004a). Members of *Aspergillus* section *Flavi*, which includes the major aflatoxin producers, efficiently convert sterigmatocystin into 3-methoxysterigmato-cystin and then into aflatoxins (Frisvad *et al.*, 1999).

Incorrect — A large number of *Aspergillus* species have been reported to produce sterigmatocystin incorrectly except for those cited above. Production of sterigmatocystin by *Penicillium* species has not been reported, apart from an obscure reference to *P. luteum* in Dean (1963). However, Wilson *et al.* (2002) claimed that *P. camemberti, P. commune* and *P. griseofulvum* produce sterigmatocystin. Perhaps they have mistaken sterigmatocystin for cyclopiazonic acid. Three *Eurotium* species have been claimed to produce sterigmatocystin (Schroeder and Kelton, 1975), but this was only based on unconfirmed thin layer chromatography assays. Unfortunately the strains used were not placed in a culture collection.

Tenuazonic acid

Tenuazonic acid is regarded as the most toxic of the secondary metabolites from *Alternaria* (Blaney, 1991), but it is also produced by a *Phoma* species.

Phoma sorghina appears to be the most important producer as it has been associated with onyalai, a haematological disease (Steyn and Rabie, 1976). Species in the *Alternaria tenuissima* species group often produce tenuazonic acid, but it has not been found in *A. alternata sensu stricto. A. citri, A. japonica, A. gaisen, A. longipes, A. mali, A. oryzae,* and *A. solani* have also been reported to produce tenuazonic acid (Sivanesan, 1991).

Trichothecenes

More than 200 trichothecenes have been identified and the non-macrocyclic ones are among the most important mycotoxins. Trichothecenes are haematotoxic and immunosuppressive; in animals vomiting, feed refusal and diarrhoea are typical symptoms. Skin oedema in humans has also been observed. A report from a European Union working group on trichothecenes in food has been published (Schothorst and van Egmond, 2004).

Deoxynivalenol (DON) and **acetylated derivatives (3ADON, 15ADON)**

These are by far the most important trichothecenes. Numerous reports on worldwide occurrence have been published and several international symposia and workshops have had DON in focus (Larsen *et al.*, 2004).

F. graminearum and *F. culmorum* are consistent producers of DON, especially in cereals. Within both species strains have been subgrouped into DON/ADON and NIV/FX producers (see below) based on the major metabolites in their trichothecene profile, but intermediates have also been found (Nielsen and Thrane, 2001). Recently, *F. graminearum* has been divided into nine phylogenetic species (O'Donnell *et al.*, 2004); however, in the present context this species concept will not be used as a correlation to existing mycotoxicological literature is impossible at this stage.

Production of DON by *F. pseudograminearum* has been reported, but this species is restricted to warmer climate and less frequent detected.

Nivalenol (NIV) and **Fusarenon X (FX, 4ANIV)**

These occur in the same commodities as DON and are in many cases covered by the same surveys due to the high degree of familiarity. NIV are often detected in much lower concentrations than DON, but are considered to be more toxic.

F. graminearum is a well-known producer of NIV and FX in cereals. In temperate climate *F. poae*, which is a consistent producer of NIV (Thrane *et al.*, 2004), maybe responsible for NIV in cereals.

NIV producing *F. culmorum* strains are less frequent than DON producers, but other NIV producers such as *F. equiseti* and *F. crookwellense* are also present in some cereal samples and in vegetables. In potatoes NIV producing *F. venenatum* strains have been detected (Nielsen and Thrane, 2001).

Incorrect — Trichothecenes were especially troublesome after the introduction of capillary gas chromatography coupled to mass spectrometry (MS). In the last decade the introduction of liquid chromatography combined with atmospheric ionization MS provided more reliable detection methods for these mycotoxins. Because immunochemical methods have also been improved in the latter years they now can be considered valid. However TLC and HPLC (unless combined with immunoaffinity cleanup) based methods must be considered totally invalid, especially as many authors have "forgotten" or neglected crucial but very time consuming cleanup steps.

Marasas *et al.* (1984) showed that *F. nivale*, which gave nivalenol its name, does not produce trichothecenes. However, the original isolate was a *F. kyushuense*, while the correct name for the species is *Microdochium nivale*. This name was still incorrectly mentioned as a trichothecene producer in a recent review (Bhatnagar *et al.*, 2002). It has even been claimed recently that *Aspergillus* species (*A. oryzae, A. terreus, A. parasiticus* and *A. versicolor*) produce nivalenol, deoxynivalenol and T-2 toxin (Atalla *et al.*, 2003). *A. parasiticus* was claimed to produce very high amounts of deoxynivalenol and T-2 toxin after growth on wheat held at 80% relative humidity for 1 to 2 months. These data are totally implausible and have to be rejected as false.

T-2 toxin

T-2 toxin is one of the most toxic trichothecenes, whereas the derivative **HT-2 toxin** is less toxic. Due to structural similarity these toxins are often included in the same analytical method.

F. sporotrichioides and *F. langsethiae*, frequently isolated from cereals, are consistent producers of T-2 and HT-2 (Thrane *et al.*, 2004).

Only few T-2 and HT-2 producing strains of *F. poae* and *F. sambucinum* have been found (Nielsen and Thrane, 2001; Thrane *et al.*, 2004).

Diacetoxyscirpenol (DAS) and monoacetylated derivatives (MAS)

These are a fourth group of important trichothecenes in food.

Major sources. F. venenatum is a high-yield producer of DAS and is frequently isolated from cereals and potatoes (Nielsen and Thrane, 2001). In addition *F. poae* is also a good producer of DAS.

Minor sources. F. equiseti does produce DAS and MAS in high amounts, but this species is infrequently isolated from cereals and vegetables. *F. sporotrichioides* and *F. langsethiae* do also produce DAS and MAS; however, much less than T-2 toxin (Thrane *et al.*, 2004). *F. sambucinum* have also been reported as DAS and MAS producers and may be related to DAS occurrence in potatoes (Ellner, 2002).

Verrucosidin

Verrucosidin is one of the mycotoxins from members of *Penicillium* series *Viridicata* that have claimed to cause mycotoxicosis in animals (Burka *et al.*, 1983). *P. polonicum, P. aurantiogriseum* and *P. melanoconidium* are the major known sources of verrucosidin (Frisvad and Samson *et al.*, 2004a).

Verruculogen and fumitremorgins

Verruculogen is an extremely toxic tremorgenic mycotoxin, but it may not be very common in foods. *Neosartorya fischeri* may be present in heat-treated foods, but *N. glabra* and allied species are much more common in foods, and the latter species do not produce verruculogen.

Major sources. A. fumigatus and *Neosartorya fischeri* are the major sources that may be of some significance in feedstuffs and heat-treated foods. These species produce many other toxic compounds including gliotoxin, fumigaclavins, and tryptoquivalins (Cole *et al.*, 1977; Cole and Cox, 1981; Panaccione and Coyle, 2005).

Minor sources. A. caespitosus, P. mononematosum and *P. brasilianum* are efficient producers of verruculogen and fumitremorgins, but are very rare in foods and feeds.

Viriditoxin

Viriditoxin is a toxic compound of unknown relevance in mycotoxicology. It is produced by *Paecilomyces variotii, A. viridinutans* and *P. mononematosum* (Frisvad *et al.*, 2004c).

Xanthomegnin, viomellein and vioxanthin

These toxins have been reported to cause experimental mycotoxicosis in pigs and they apparently are more toxic to the liver than to kidneys in mammals (Zimmerman *et al.*, 1979). They have been found to be naturally occurring in cereals (Hald *et al.*, 1983; Scudamore *et al.*, 1986).

P. cyclopium, P. freii, P. melanoconidium, P. tricolor and *P. viridicatum* are common in cereals and *A. ochraceus, A. westerdijkiae* and possibly *A. steynii* are common in green coffee beans and are occasionally found in grapes and on rice.

P. janthinellum, P. mariaecrucis are soilborne forms producing these hepatotoxins (Frisvad and Filtenborg, 1990), and *Trichophyton, Microsporum* and *Epidermophyton* spp. are good producers of xanthomegnin, viomellein and vioxanthin, but they are fungi causing superficial skin mycosis of mammals and never occur in foods.

Incorrect — These naphthoquinones are not produced by *P. crustosum* as else reported by Hald *et al.* (1983), by *P. oxalicum* as reported by Lee and Skau (1981) or by *A. nidulans, A. flavus, A. oryzae* or *A. terreus* as reported by Abu-Seidah (2003).

Zearalenone

Zearalenone has for many years been related to hyperoestrogenism in swine and possible effects in humans has also been reported. Derivatives of zearalenone have been used as growth promoters in livestock; however, this is now banned in the Europian Union (Launay *et al.*, 2004). Zearalenone and its derivatives have been reviewed recently (Hagler Jr *et al.*, 2001).

F. graminearum and *F. culmorum* are the most pronounced producers of zearalenone and several derivatives hereof. They occur frequently in cereals all over the world. Recently, *F. graminearum* has been divided into nine phylogenetic species (O'Donnell *et al.*, 2004); however, in the present context this species concept will not be used as a correlation to existing mycotoxicological literature as this is impossible at this stage.

Under laboratory conditions *Fusarium equiseti* does produce a number of zearalenone derivatives in high amounts, but little is known about the production under natural conditions. *F. crookwellense* does also produce zearalenone.

REFERENCES

Abarca, M. L., Bragulat, M. R., Castella, G., and Cabañes, F. J. (1994). Ochratoxin A production by strains of *Aspergillus niger* var. *niger*. Applied and Environmental Microbiology 60:2650-2652.

Abarca, M.L., Accensi, F., Cano, J., and Cabañes, F. J. (2004). Taxonomy and significance of black Aspergilli. Antonie van Leeuwenhoek 86:33-49.

Abu-Seidah, A. A. (2003). Secondary metabolites as co-markers in the taxonomy of Aspergilli. Acta Microbiologica Polonica 52:15-23.

Adebajo, L. O. (1992). Spoilage moulds and aflatoxins from poultry feeds. Nahrung 36:523-529.

Alsberg, C. L., and Black, O. F. (1913). Contributions to the study of maize deterioration; biochemical and toxicological investigations of *Penicillium puberulum* and *Penicillium stoloniferum*. Bulletin Bur. Anim. Ind. U.S. Dept. Agric. 270:1 47.

Andersen B., and Frisvad, J. C. (2004). Natural occurrence of fungi and fungal metabolites in moldy tomatoes. Journal of Agricultural and Food Chemistry 52:7507-7513.

Andersen, B., and Thrane, U. (2006). Food-borne fungi in fruit and cereals and their production of mycotoxins. *In* Advances in Food Mycology (Hocking, A. D., Pitt, J. I., Samson, R. A., and Thrane, U., eds.), Advances in Experimental Medicine and Biology 571, Springer, New York, pp. 137-152.

Andersen, B., Nielsen K. F., Thrane, U., Szaro, T., Taylor, J. W., and Jarvis, B. B. (2003). Molecular and phenotypic descriptions of *Stachybotrys chlorohalonata* sp. nov. and two chemotypes of *Stachybotrys chartarum* found in water-damaged buildings. Mycologia 95:1227-1238.

Andersen, B., Smedsgaard, J., and Frisvad, J. C. (2004). *Penicillium expansum*: Consistent production of patulin, chaetoglobosins and other secondary metabolites in culture and their natural occurrence in fruit products. Journal of Agricultural and Food Chemistry 52:2421-2429.

Antony, M., Shukla, Y., and Janardhanan, K. K. (2003). Potential risk of acute hepatotoxicity of kodo poisoning due to exposure to cyclopiazonic acid. Journal of Ethnopharmacology 87:211-214.

Arnold, D. L., Scott, P. M., McGuire, P. F., Harwig, J. (1987). Acute toxicity studies on roquefortine C and PR-toxin, metabolites of *Penicillium roqueforti* in the mouse. Food Cosmetics Toxicology 16:369-371.

Atalla, M. M., Hassanein, N. M., El-Beih, A. A., and Youssef, Y. (2003). Mycotoxin production in wheat grains by different Aspergilli in relation to different humidities and storage periods. Nahrung 47:6-10.

Atherton, L. G., Brewer, D., and Taylor, A. (1974). *Pithomyces chartarum*: A fungal parameter in the aetiology of some diseases of domestic animals. *In* Mycotoxins (Purchase, I. F. H., ed.), Elsevier, Amsterdam, pp. 29-68.

Barr, J. G., and Downey, G. A. (1975). A multiple inoculation technique for the screening of fungal isolates for the evaluation of growth and mycotoxin production on agar substrates. Journal of the Science of Food and Agriculture 26:1561-1566.

Bauer, J, and Gedek, B. 1979. Zum Vorkommen toxinbildender Schimmelpilze in Cerealien underschiedlicher Herkunft. Landwirtschaftliche Forschung, Sonderheft 35:562-569.

Bayman, P., Baker, J. L., Doster, M. A., Michailides, T. J., and Mahoney, N. E. (2002). Ochratoxin production by the *Aspergillus ochraceus* group and *Aspergillus alliaceus*. Applied Environmental Microbiology 68:2326-2329.

Bennett, J. W., and Klich M. A. (2003). Mycotoxins, Clinical Microbiology Reviews 16:497-516.

Bentley, R. (2000). Mycophenolic acid: A one hundred Odyssey from antibiotic to immunosuppressant. Chemical Reviews 100:3801-3825.

Bhatnagar, D., Yu, J., and Ehrlich, K. C. (2002). Toxins of filamentous fungi. *In* Fungal allergy and pathogenicity (Breitenbach, M., Crameri, R., and Lehrer, S. B., eds.), Chemical Immunology 81, S. Karger A. G., Basel, Switzerland, pp. 167-206.

Birkinshaw, J. H., Oxford, A. E., an Raistrick, H. (1936). Penicillic acid, a metabolic product of *Penicillium puberulum* Bainier and *P. Cyclopium* Westling. Biochemical Journal 30: 394–411.

Blanc, P. J., Loret M. O., and Goma G. (1995). Production of citrinin by various species of *Monascus*. Biotechnology Letters 17: 210-213.

Blaney, B. J. (1991). *Fusarium* and *Alternaria* toxins. *In* Fungi and mycotoxins in stored products (Champ, B. R., Highley, E., Hocking, A. D., and Pitt, J. I., eds.), ACIAR Proceedings No. 36, Australian Centre for International Agricultural Research, Canberra, Australia, pp. 86-98.

Blum, M. S. (1995). The toxic action of marine and terrestrial alkaloids. Alaken, Inc., Fort Collins, Colorado.

Bösenberg, H., and Becker, E. (1972). Bildung Aflatoxin-ähnlicher Susbtanzen durch *Penicillium* Arten. Zeitschrift Lebensmittel Untersuchungen Forschung 150:153-154.

Boysen, M., Skouboe, P., Frisvad, J. C., and Rossen, L. (1996). Reclassification of the *Penicillium roqueforti* group into three species on the basis of molecular genetic and biochemical profiles. Microbiology 142:541-549.

Broadbent, D. (1966). Antibiotics produced by fungi. Botanical Review 32:219-242.

Brookes, D., Kidd, B. K, and Turner, W. B. (1963). Avenaciolide, an antifungal lactone from *Aspergillus avenaceus*. Journal of the Chemical Society 1963:5385-5391.

Bryden,W. L., Logrieco, A., Abbas, H. K., Porter J. K., Vesonder, R. F., Richard, J. L., and Cole, R. J. (2001). Other significant *Fusarium* mycotoxins. *In* Fusarium. Paul E. Nelson Memorial Symposium (Summerell, B. E., Leslie, J. F., Backhouse, D., Bryden, W. L., and Burgess, L. W., eds.), APS Press, St. Paul, MN, USA, pp. 360-392.

Burdock, G. A., Carabin, I. G., and Soni, M. G. (2001). Safety assessment of β-nitropropionic acid: a monograph in support of an acceptable daily intake in humans. Food Chemistry 75:1-27.

Burka, L. T., Ganguli, M., and Wilson, B. J. (1983). Verrucosidin, a tremorgen from *Penicillium verrucosum* var. *cyclopium*. Journal of the Chemical Society, Chemical Communications 1983:544-545.

Burrows, B. F., and Turner, W. B. (1966). 1-amino-2-nitrocyclopentane carboxylic acid. A new naturally occurring nitro-compound. Journal of the Chemical Society C 1966:255-260.

Burton, H. S. (1949). Antibiotics from Penicillia. Britisch Journal of Experimental Pathology 30:151-158.

Bush, M. T., Touster, O., and Brockman, J. E. (1951). The production of β-nitropropionic acid by a strain of *Aspergillus flavus*. Journal of Biological Chemistry 188:685-694.

Bushnell, G. W., Li, Y. L., and Poulton, G. A. (1984). Pyrones. X. Lateropyrone, a new antibiotic from the fungus *Fusarium lateritium* Nees. Canadian Journal of Chemistry 62:2101-2106.

Chien, M. M., Schiff, Jr., P. L., Slatkin, D. J., and Knapp, J. E. (1977). Metabolites of aspergilli. III. The isolation of citrinin, dihydrocitrinone and sclerin from *Aspergillus carneus*. Journal of Natural Products (Lloydia) 40:301-302.

Ciegler, A. (1972). Bioproduction of ochratoxin A and penicillic acid by members of the *Aspergillus ochraceus* group. Canadian Journal of Microbiology 18:631-636.

Ciegler, A., and Kurtzman, C. P. (1972). Penicillic acid production by blue-eye fungi on various agricultural commodities. Applied Microbiology 20:761-764.

Cole, R. J., and Cox, R. H. (1981). Handbook of toxic fungal metabolites. Academic Press, New York.

Cole, R. J., and Schweikert, M. A. (2003a). Handbook of Secondary Fungal Metabolites. Vol. 1. Academic Press, New York.

Cole, R. J., and Schweikert, M. A. (2003b). Handbook of Secondary Fungal Metabolites. Vol. 2. Academic Press, New York.

Cole, R. J., Kirksey, J. W., Dorner, J. W., Wilson, D. M., Johnson, J. C., Jr., Johnson, A. N., Bedell, D. M., Springer, J. P., Chexal, K. K., Clardy, J. J, and Cox, R. H. (1977). Mycotoxins produced by *Aspergillus fumigatus* species isolated from molded silage. Journal of Agricultural and Food Chemistry 25:826-830.

Cole, R. J., Dorner, J. W., Cox, R. H., Hill, R. A., Cluter, H. G., and Wells, J. M. (1981). Isolation of citreoviridin from *Penicillium charlesii* cultures and molded peacan fragments. Applied and Environmental Microbiology 42:677-681.

Cole, R. J., Dorner, J. W., and Cox, R. H. (1983). 2 classes of alkaloid mycotoxins produced by *Penicillium crustosum* Thom isolates from contaminated beer. Journal of Agricultural and Food Chemistry 31:655-657.

Cole, R. J., Jarvis, B. B., and Schweikert, M. A., (2003). Handbook of Secondary Fungal Metabolites. Vol. 3. Academic Press, New York.

Czerwiecki, L., Czajkowska, D., and Witkowska-Gwiazdowska, A., (2002a). On ochratoxin A and fungal flora in Polish cereals from conventional and ecological farms. Part 1: Occurrence of ochratoxin A and fungi in cereals in 1997. Food Additives and Contaminants 19:470-477.

Czerwiecki, L., Czajkowska, D., and Witkowska-Gwiazdowska, A., (2002b). On ochratoxin A and fungal flora in Polish cereals from conventional and ecological farms. Part 2: Occurrence of ochratoxin A and fungi in cereals in 1998. Food Additives and Contaminants 19:1051-1057.

Dean, F. M. (1963). Naturally Occurring Oxygen Compounds. Butterworth, London, p. 526.

Devys, M., Bousquet, J. F., Kollman, A., and Barbier, M. (1980). Dihydroisocoumarines et acide mycophenolique du milieu de culture du champignon phytopathogene *Septoria nodorum*. Phytochemistry 19:2221-2222.

Domsch, K. H., Gams, W., and Anderson, T.-H. (1980). Compendium of Soil Fungi. Academic Press, London.

Dorner, J. W., Cole, R. J., Lomax, L. G., Gosser, H. S., and Diener, U. L. (1983). Cyclopiazonic acid production by *Aspergillus flavus* and its effect on broiler chickens. Applied and Environmental Microbiology 46:698-703.

Dowd, P. F., Miller, J. D., and Greenhalgh, R. (1989). Toxicity and interactions of some *Fusarium graminearum* metabolites to caterpillars. Mycologia 81:646-650.

Drusch, S., and Ragab, W. (2003). Mycotoxins in fruits, fruit juices, and dried fruits. Journal of Food Protection 66:1514-1527.

Ehrlich, K. C., Lee, L. S., Ciegler, A., and Palmgren, M. S. (1982). Secalonic acid D: a natural contaminant of corn dust. Applied and Environmental Microbiology 44:1007-1008.

Eijk, G. W. van (1969). Isolation and identification of orsellinic acid and penicillic acid produced by *Penicillium fennelliae* Stolk. Antonie van Leeuwenhoek 35:497-504.

El-Banna, A. A., Pitt, J. I., Leistner, L. (1987). Production of mycotoxins by *Penicillium* species. Systematic and Applied Microbiology 10:42-46.

El-Banna, A. A., and Leistner, L. (1988) Production of penitrem A by *Penicillium crustosum* isolated from foodstuffs, International Journal of Food Microbiology 7:9-17.

El-Hag, N., and Morse, R. E. (1976). Aflatoxin production by a variant of *Aspergillus oryzae* (NRRL 1988) on cowpeas (*Vigna sinensis*), Science 192:1345-1346.

El-Kady, I., El-Maraghy, S., and Zihri, A. N. (1994). Mycotoxin producing potential of some isolates of *Aspergillus flavus* and *Eurotium* groups from meat products. Microbiological Research 149:297-307.

Ellner, F. M. (2002). Mycotoxins in potato tubers infected by *Fusarium sambucinum*. Mycotoxin Research 18:57-61.

Engelhardt, J. A., and Carlton, W. W. (1991). Rubratoxins. *In* Mycotoxins and Phytoalexins (Sharma, R. P., and Salunkhe, D. K., eds.), CRC Press, Boca Raton, FL, USA, pp. 259-289.

Enomoto, M., and Ueno, I. (1974). *Penicillium islandicum* (toxic yellowed rice) – luteoskyrin – islanditoxin cyclochlorotine. *In* Mycotoxins (Purchase, I. F. H., ed.), Elsevier, Amsterdam, The Netherlands, pp. 303-326.

Fennell, D. I. (1976) *Aspergillus oryzae* (NRRL strain 1988): a clarification, Science 194:1188.

Franck, B., and Gehrken, H.-P. (1980). Citreoviridin aus *Aspergillus terreus*. Angew. Chem. Int. Ed. 19:461-462.

Frank, H. K. (1972). Zweifel über das Vorkommen von Aflatoxin bei der Gattung *Penicillium*. Zeitschrift Lebensmittel Untersuchungen und Forschung 150:151-153.

Frisvad, J. C. (1985a). Profiles of primary and secondary metabolites of value in classification of *Penicillium viridicatum* and related species. *In* Advances in *Penicillium* and *Aspergillus* systematics (Samson, R. A., and Pitt, J. I., eds.), Plenum Press, New York, U.S.A, pp. 311-325.

Frisvad, J. C. (1985b). Classification of asymmetric Penicillia using expressions of differentiation. *In* Advances in *Penicillium* and *Aspergillus* systematics (Samson, R. A., and Pitt, J. I., eds.), Plenum Press, New York, U.S.A, pp. 327-333.

Frisvad, J. C. (1989). The connection between the Penicillia and Aspergilli and mycotoxins with special emphasis on misidentified isolates. Archives of Environmental Contamination and Toxicology 18:452-467.

Frisvad, J. C., and Filtenborg, O. (1983). Classification of terverticillate Penicillia based on profiles of mycotoxins and other secondary metabolites. Applied and Environmental Microbiology 46:1301-1310.

Frisvad, J. C., and Filtenborg, O. (1990). Revision of *Penicillium* subgenus *Furcatum* based on secondary metabolites and conventional characters. *In* Modern concepts in Penicillium and Aspergillus classification (Samson, R. A., and Pitt, J. I., eds.), Plenum Press, New York, U.S.A, pp. 159-170.

Frisvad, J. C., and Samson, R. A. (2000). *Neopetromyces* gen. nov. and an overview of teleomorphs of *Aspergillus* subgenus *Circumdati*. Studies in Mycology 45:201-207.

Frisvad, J. C., and Samson, R. A. (2004a). Polyphasic taxonomy of *Penicillium* subgenus *Penicillium*. A guide to identification of food and air-borne terverticillate Penicilllia and their mycotoxins. Studies in Mycology 49:1-173.

Frisvad, J. C., and Samson, R. A. (2004b). *Emericella venezuelensis*, a new species with stellate ascospores producing sterigmatocystin and aflatoxin B_1. Systematic and Applied Microbiology 27:672-680.

Frisvad, J. C., Houbraken, J., and Samson, R. A. (1999). *Aspergillus* species and aflatoxin production: a reappraisal. *In* Food Microbiology and Food Safety into the Next Millennium (Tuijtelaars, A. C. J., Samson, R. A., Rombouts, F. M., and Notermans, S., eds.), Foundation Food Micro '99, Zeist, The Netherlands, pp. 125-126.

Frisvad, J. C., Samson, R. A., and Smedsgaard, J. (2004a). *Emericella astellata*, a new producer of aflatoxin B_1, B_2 and sterigmatocystin. Letters in Applied Microbiology 38:440-445.

Frisvad, J. C., Frank, J. M., Houbraken, J. A. M. P., Kuijpers, A. F. A., and Samson, R. A. (2004b). New ochratoxin producing species of *Aspergillus* section *Circumdati*. Studies in Mycology 50:23-43.

Frisvad, J. C., Smedsgaard, J., Larsen, T. O, Samson, R. A. (2004c). Mycotoxins, drugs and other extrolites produced by species in *Penicillium* subgenus *Penicillium*. Studies in Mycology 49:201-242.

Frisvad, J. C., Skouboe, P., Samson, R. A. (2005a). Taxonomic comparison of three different groups of aflatoxin producers and a new efficient producer of aflatoxin B_1, sterigmatocystin and 3-O-methylsterigmatocystin, *Aspergillus rambellii* sp. nov. Systematic and Applied Microbiology 28:442-453.

Frisvad, J. C., Lund, F., and Elmholt, S. (2005b). Ochratoxin A producing *Penicillium verrucosum* isolates from cereals reveal large AFLP fingerprinting variability. Journal of Applied Microbiology 98:684-692.

Frisvad, J. C., Thrane, U., Samson, R. A., and Pitt, J. I. (2006a). Important mycotoxins and the fungi which produce them. *In* Advances in Food Mycology (Hocking, A. D., Pitt, J. I., Samson, R. A., and Thrane, U., eds.), Advances in Experimental Medicine and Biology 571, Springer, New York, U.S.A., pp. 3-31.

Frisvad, J. C., Nielsen, K. F., and Samson, R. A. (2006b). Recommendations concerning the chronic problem of misidentification of mycotoxinogenic fungi associated with foods and feeds. *In* Advances in Food Mycology (Hocking, A. D., Pitt, J. I., Samson, R. A., and Thrane, U., eds.), Advances in Experimental Medicine and Biology 571, Springer, New York, York, pp. 33-46.

Ghebremeskel, M., and Langseth, W. (2000). The occurrence of culmorin and hydroxy-culmorins in cereals. Mycopathologia 152:103-108.

Golinski, P., Wnuk, S., Chełkowski, J., Visconti, A., and Schollenberger, M. (1986). Antibiotic Y: Biosynthesis by *Fusarium avenaceum* (Corda ex Fries) Sacc., isolation, and some physiochemical and biological properties. Applied and Environmental Microbiology 51:743-745.

Goto, T., Wicklow, D. T., and Ito, Y. (1996). Aflatoxin and cyclopiazonic acid production by a sclerotium-producing *Aspergillus tamarii* strain. Applied and Environmental Microbiology 62:4036-4038.

Goto, T., Ito, Y., Peterson, S. W., and Wicklow, D. T. (1997). Mycotoxin producing ability of *Aspergillus tamarii*. Mycotoxins 44:17-20.

Hagler, Jr, W. M., Towers, N. R., Mirocha, C. J., Eppley, R. M., and Bryden, W. L. (2001). Zearalenone: mycotoxin or mycoestrogen? *In* Fusarium. Paul E. Nelson Memorial Symposium (Summerell, B. A., Leslie, J. F., Backhouse, D., Bryden, W. L., and Burgess, L. W., eds.), APS Press, St. Paul, MN, U.S.A., pp. 321-331.

Hald, B., Christensen, D. H., and Krogh, P. (1983). Natural occurrence of the mycotoxin viomellein in barley and the associated quinone-producing penicillia. Applied and Environmental Microbiology 46:1311-1317.

Hanssen, E. (1969). Schädigung von Lebensmitteln durch Aflatoxin B. Naturwissenschaften 56: 90.

Hanssen, E., and Jung, M. (1973). Control of aflatoxins in the food industry. Pure Applied Chemistry 35:239-250.

Harwig, J., Chen, Y. K., Kennedy, B. P. C., and Scott, P. M. (1973). Occurrence of patulin and patulin-producing strains of *Penicillium expansum* in natural rots of apple in Canada. Canadian Institute of Food Science and Technology Journal 6:22-25.

Hermansen, K., Frisvad, J. C., Emborg, C., and Hansen, J. (1984). Cyclopiazonic acid production by submerged cultures of *Penicillium* and *Aspergillus* strains. FEMS Microbiology Letters 21:253-261.

Hershenhorn, J., Park, S. H., Stierle, A., and Strobel, G. A. (1992). *Fusarium avenaceum* as a novel pathogen of spotted knapweed and its phytotoxins, acetamido-butenolide and enniatin B. Plant Science 86:155-160.

Hesseltine, C. W., Shotwell, O. L., Ellis, J. J., and Stubblefield, R. D. (1966). Aflatoxin production by *Aspergillus flavus*. Bacteriological Reviews 30:795-805.

Hesseltine, C. W., Vandegraft, E. E., Fennell, D. I., Smith, M., and Shotwell, O. (1972). Aspergilli as ochratoxin producers. Mycologia 64:539-550.

Hetherington, A. C., and Raistrick, H. (1931). Studies in the biochemistry of microorganisms part XIV. On the production and chemical constitution of a new yellow colouring matter, citrinin, produced from glucose by *Penicillium citrinum* Thom. Philosophical Transactions of the Royal Society Series B 220:269-295.

Hodges, F. A., Zust, J. R., Smith, H. R., Nelson, A. A., Armbrecht, B. H., and Campbell, A. D. (1964). Mycotoxins: aflatoxin produced by *Penicillium puberulum*. Science 145:1439.

Holzapfel, C. W. (1968). The isolation and structure of cyclopiazonic acid, a toxic product from *Penicillium cyclopium* Westling. Tetrahedron 24:2101-2119.

Horie, Y. (1995). Productivity of ochratoxin A of *Aspergillus carbonarius* in *Aspergillus* section *Nigri*. Nippon Kingakkai Kaiho 36:73-76.

Huang, X., Dorner, J. W., and Chu, F. S. (1994). Production of aflatoxin and cyclopiazonic acid by various Aspergilli: an ELISA approach. Mycotox in Research 10:101-106.

Ito, Y., Peterson, S. W., Wicklow, D. T., and Goto, T. (2001). *Aspergillus pseudotamarii*, a new aflatoxin producing species in *Aspergillus* section *Flavi*, Mycological Research 105:233-239.

Janardhanan, K. K., Sattar, A., and Husain, A. (1984). Production of fumigaclavine A by *Aspergillus tamarii* Kita. Canadian Journal of Microbiology 30:247-250.

Jand, S. K., Kaur, P., and Sharma, N. S. (2005). Mycoses and mycotoxicosis in poultry. Indian Journal Animal Science 75:465-476.

Jarvis, B. B. (2003). *Stachybotrys chartarum*: a fungus for our time. Phytochemistry 64:53-60.

JECFA (Joint FAO/WHO Expert Committee on Food Additives). 2001. Safety evaluation of certain mycotoxins in foods. Fifty-sixth report of the Joint FAO/WHO Expert Committee on Food Additives. FAO Food and Nutrition Report No. 74.

Jennessen, J., Nielsen, K. F., Houbraken, J., Lyhne, E. K., Schnürer, J., Frisvad, J. C., and Samson, R. A. (2005). Secondary metabolite and mycotoxin production by the *Rhizopus microsporus* group. Journal of Agricultural and Food Chemistry 53:1833-1840.

Jestoi, M., Rokka, M., Yli-Mattila, T., Parikka, P., Rizzo, A., and Peltonen, K. (2004). Presence and concentrations of the *Fusarium*-related mycotoxins beauvericin, enniatins and moniliformin in finnish grain samples. Food Additives and Contaminants 21:794-802.

Joshi, B. K., Gloer, J. B., and Wicklow, D. T. (2002). Bioactive natural products from a sclerotium-colonizing isolate of *Humicola fuscoatra*. Journal of Natural Products 65:1734-1737.

Kamyar, M., Rawnduzi, P., Studenik, C. R., Kouri, K., and Lemmens-Gruber, R. (2004). Investigation of the electrophysiological properties of enniatins. Archives of Biochemistry and Biophysics 429:215-223.

Karow, E. O., Woodruff, H. B., and Forster, J. W. (1944). Penicillic acid from *Aspergillus ochraceus*, *Penicillium thomi* and *Penicillium suavolens*. Archives of Biochemistry 5:279-282.

Kinoshita, R., Ishiko, T., Sugiyama, S., Seto, T., Igarasi, S., and Goetz, I. E. (1968). Mycotoxins in fermented foods. Cancer Research 28:2296-2311.

Klich, M., Mullaney, E. J., Daly, C. B., and Cary, J. W. (2000). Molecular and physiological aspects of aflatoxin and sterigmatocystin biosynthesis by *Aspergillus tamarii* and *A. ochraceoroseus*. Applied Microbiology and Biotechnology 53:605-609.

Kulik, M. M., and Holaday, C. E. (1966). Aflatoxin: a metabolic product of several fungi. Mycopathologia et Mycologia Applicata 30:137-140.

Kurtzman, C. P., Horn, B. W., and Hesseltine, C. W. (1987). *Aspergillus nomius*, a new aflatoxin-producing species related to *Aspergillus flavus* and *Aspergillus parasiticus*. Antonie van Leeuwenhoek 53:147-158.

Labuda, R., and Tancinova, D. (2003). *Eupenicillium ochrosalmoneum*, a rare species isolated from a pig feed mixture in Slovakia. Biologia 58:1123-1126.

Lafont, O., Debeaupuis, J.-P., Gaillardin, M., and Payen, J. (1979). Production of mycophenolic acid by *Penicillium roqueforti* strains. Applied and Environmnetal Microbiology 39:365-368.

Lai, M., Semeniuk, G., and Hesseltine, C. W. (1970). Conditions for production of ochratoxin A by *Aspergillus* species in a synthetic medium. Applied Microbiology 19:542-544.

Laidou, I. A., Thanassoulopoulos, C. C., and Liakopoulou-Kyriakidis, M. (2001). Diffusion of patulin in the flesh of pears inoculated with four post-harvest pathogens. Journal of Phytopathology 149:457-461.

Lalithakumari, D., and Govindaswarni, C. V. (1970). Role of aflatoxins in groundnut seed spoilage. Current Science 39:308-309.

Land, C. J., and Hult, K. (1987). Mycotoxin production by some wood-associated *Penicillium* spp., Letters in Applied Microbiology 4:41-44.

Langone, J. J., van Vunakis, H. (1976). Aflatoxin B_1, specific antibodies and their use in radioimmunoassay. Journal of the National Cancer Institute 56:591-595.

Larsen, J. C., Hunt, J., Perrin, I., and Ruckenbauer, P. (2004). Workshop on trichothecenes with a focus on DON: summary report. Toxicology Letters 153:1-22.

Larsen, T. O., Svendsen, A., and Smedsgaard, J. (2001). Biochemical characterization of ochratoxin A-producing strains of the genus *Penicillium*. Applied and Environmental Microbiology 67:3630-3635.

Launay, F. M., Ribeiro, L., Alves, P., Vozikis,V., Tsitsamis, S., Alfredsson,G., Sterk, S. S., Blokland, M., Iitia, A., Lovgren, T., Tuomola, M., Gordon, A., and Kennedy, D. G. (2004). Prevalence of zeranol, taleranol and *Fusarium* spp. toxins in urine: implications for the control of zeranol abuse in the European Union. Food Additives and Contaminants 21:833-839.

Lee, L. S., and Skau, D. B. (1981). Thin layer chromatographic analysis of mycotoxins: a review of recent literature. Journal of Liquid Chromatography 4, Suppl. 1:43-62.

Leistner, L., and Pitt, J. I. (1977). Miscellaneous Penicillium toxins. *In* Mycotoxins in human and animal health (Rodricks, J. V., Hesseltine, C. W., and

Mehlman, M. A., eds.), Pathotox Publishers, Park Forest South, pp. 639-653.

Leitao, J., Le Bars, J., Bailly, J. R. (1989). Production of aflatoxin B_1 by *Aspergillus ruber* Thom and Church. Mycopathologia 108:135-138.

Lindenfelser, L. A., Lillehoj, E. B., and Milburn, M. S. (1973). Ochratoxin and penicillic acid in tumorigenic and acute toxicity tests with mice. Developments in Industry and Microbiology 14:331-336.

Locci, R., Merlini, L., Nasini, G., and Locci, J. R. (1965). On a strain of *Penicillium fellutanum* Biourge producing a yellow flourescent pigment. Giornale Microbiologia 13:271-277.

Logrieco, A., Moretti, A., Castella, G., Kostecki, M., Golinski, P., Ritieni, A., and Chełkowski, J. (1998). Beauvericin production by *Fusarium* species. Applied and Environmental Microbiology 64:3084-3088.

Lopez-Diaz, T. M., and Flannigan, B. (1997). Mycotoxins of *Aspergillus clavatus*: toxicity of cytochalasin E, patulin, and extracts of contaminated barley malt. Journal of Food Protection 60:1381-1385.

Lopez-Diaz, T. M., Roman-Blanco, C., Garcia-Arias, M. T., Garcia-Fernández, M. C., and Garcia-López, M. L. (1996). Mycotoxins in two Spanish cheese varieties. International Journal of Food Microbiology 30:391-395.

Lund, F., and Frisvad, J. C. (1994). Chemotaxonomy of *Penicillium aurantiogriseum* and related species. Mycological Research 98:481-492.

Lund, F., and Frisvad, J. C. (2003). *Penicillium verrucosum* in cereals indicates production of ochratoxin A. Journal of Applied Microbiology 95:1117-1123.

Manabe, M. (2001). Fermented foods and mycotoxins. Mycotoxins 51:25-28.

Mantle, P. G., and McHugh, K. M. (1993). Nephrotoxic fungi in foods from nephropathy households in Bulgaria. Mycological Research 97:205-212.

Marasas, W. F. O., Nelson, P. E., and Toussoun, T. A. (1984). Toxigenic *Fusarium* species. Identity and mycotoxicology. The Pennsylvania State University Press, University Park and London, pp. 1-328.

Marasas, W. F. O., Miller, J. D., Riley, R. T., and Visconti, A. (2001). Fumonisins - occurrence, toxicology, metabolism and risk assessment. *In* Fusarium. Paul E. Nelson Memorial Symposium (Summerell, B. A., Leslie, J. F., Backhouse, D., Bryden, W. L., and Burgess, L. W., eds.), APS Press, St. Paul, MN, U.S.A., pp. 332-359.

Merwe, K. J. van der, Steyn, P. S., Fourie, Scott., L., and Theron, J. J. (1965). Ochratoxin A, a toxic metabolite produced by *Aspergillus ochraceus* Wilh. Nature 205:1112-1113.

Ming, L. (1995). Moldy sugarcane poisoning – a case report with a brief review. Journal Toxicology. Clinical Toxicology 33:363-367.

Mishra, S. K., and Murthy, H. S. R. (1968). An extra fungal source of aflatoxins. Current Science (India) 37:406.

Mislivec, P. B., Hunter, S. H., and Tuite, J. (1968). Assay for aflatoxin production by the genera *Aspergillus* and *Penicillium*. Applied Microbiology 16:1053-1055.

Montemurro, N., and Visconti, A. (1992). *Alternaria* metabolites - chemical and biological data. *In* Alternaria. *Biology, Plant Diseases and Metabolites* (Chełkowski, J., and Visconti, A., eds.), Elsevier, Amsterdam, The Netherlands, pp. 449-557.

Morrison, E., Kosiak, B., Ritieni, A., Aastveit, A. H., Uhlig, S., and Bernhoft, A. (2002). Mycotoxin production by *Fusarium avenaceum* strains isolated from Norwegian grain and the cytotoxicity of rice culture extracts to porcine epithelial cells. Journal of Agricultural and Food Chemistry 50: 3070-3075.

Moss, M. O., Robinson, F. V., and Wood, A. B. (1968). Rubratoxin B, a toxic metabolite of *Penicillium rubrum*. Chemistry and Industry 1968:587-588.

Moubasher, A. H., El-Khady, I. A., and Shoriet, A. (1977). Toxigenic aspergilli isolated from different sources in Egypt. Annale de Nutrition et Aliments 31:407-415.

Moubasher, A. H., Abdel-Kader, M. I. A., and El-Kady, I. A. (1978) Toxigenic fungi isolated from Roquefort cheese. Mycopathologia 66:187-190.

Moule, Y., Moreau, S., Aujard, C. (1980). Induction of cross-links between DNA and protein by PR-toxin, a mycotoxin from *Penicillium roqueforti*. Mutation Research 77:79-89.

Munkvold, G., Stahr, H. M., Logrieco, A., Moretti, A., and Ritieni, A. (1998). Occurrence of fusaproliferin and beauvericin in *Fusarium*-contaminated livestock feed in Iowa. Applied and Environmental Microbiology 64:3923-3926.

Murakami, H. (1971). Classification of the koji mould. Journal of General and Applied Microbiology 17:281-308.

Murakami, H., Ouwaki, K, and Takase, S. (1966). An aflatoxin strain, ATCC 15517. Journal of General and Applied Microbiology 12:195-206.

Nagel, D. W., Steyn, P. S., and Scott, D. B. (1972). Production of citreoviridin by *Penicillium pulvillorum*. Phytochemistry 11:627-630.

Nakajima, S., and Nozawa, K. (1979). Isolation in high yield of citrinin from *Penicillium odoratum* and of mycophenolic acid from *Penicillium brun-*

neo-stoloniferum. Journal of Natural Products 42:423-426.

Nakamura, S., and Shimoda, C. (1954). Studies on an antibiotic substance oryzacidin, produced by *Aspergillus oryzae*. V. Existence of β-nitropropionic acid. Journal of the Agricultural Chemistry Society Japan 28:909-913.

Natori, S., Sakaki, S., Kurata, M., Udagawa, S., Ichinoe, M., Saito, M., Umeda, M., and Ohtsubo, K., (1970). Production of rubratoxin B by *Penicillium purpurogenum*, Applied Microbiology 19:613-617.

Nielsen, K. F., and Thrane, U. (2001). Fast methods for screening of trichothecenes in fungal cultures using gas chromatography-tandem mass spectrometry. Journal of Chromatography A 929:75-87.

Obana, H., Kumeda, Y., and Nishimune, T. (1995). *Aspergillus ochraceus* production of 5,6-dihydropenicillic acid in culture and foods. Journal of Food Protection 58:519-523.

O'Donnell, K., Ward, T. J., Geiser, D. M., Kistler, H. C., and Aoki, T. (2004). Genealogical concordance between the mating type locus and seven other nuclear genes supports formal recognition of nine phylogenetically distinct species within the *Fusarium graminearum* clade. Fungal Genetics and Biology 41:600-623.

Ohmomo, S., Sugita, M., and Abe, M. (1973). Isolation of cyclopiazonic acid, cyclopiazonic acid imine and bissecodehydrocyclopiazonic acid from the cultures of *Aspergillus versicolor* (Vuill.) Tiraboschi. Journal of the Agricultural and Chemical Society of Japan 47:57 93.

Okeke, B., Seigle-Murandi, F., Steiman, R., Benoit-Guyod, J.-L., and Kaouadjii, M. (1993). Identification of mycotoxin-producing fungal strains: a step in the isolation of compounds active against rice fungal diseases, Journal of Agricultural and Food Chemistry 41:1731-1735.

Olivigni, F. J., and Bullerman, L. B. (1978). Production of penicillic acid and patulin by an atypical *Penicillium roqueforti* isolate. Applied Microbiology 35:435-438.

Orth, R. (1977). Mycotoxins of *Aspergillus oryzae* strains for use in the food industry as starters and enzyme producing molds. Annale de Nutrition et Aliments 31:617-624.

Overy, D. P., and Frisvad, J. C. (2003). New *Penicillium* species associated with bulbs and root vegetables. Systematic and Applied Microbiology 26: 631-639.

Overy, D. P., and Frisvad, J. C. (2005). Mycotoxin production and postharvest storage rot of ginger (*Zingiber officinale*) by *Penicillium brevicompactum*. Journal of Food Protection 68:607-609.

Panaccione, D. G., and Coyle, C. M. (2005). Abundant respirable ergot alkaloids from the common airborne fungus *Aspergillus fumigatus*. Applied and Environmental Microbiology 71:3106-3111.

Parrish, F. W., Wiley, B. J., Simmons, E. G., and Lang, L. (1966). Production of aflatoxins and kojic acid by species of *Aspergillus* and *Penicillium*. Applied Microbiology 14:139.

Pedersen, P. B., and Miller, J. D. (1999). The fungal metabolite culmorin and related compounds. Natural Toxins 7:305-309.

Peterson, S. W., Ito, Y., Horn, B. W., and Goto, T. (2001). *Aspergillus bombycis*, a new aflatoxigenic species and genetic variation in its sibling species, *A. nomius*. Mycologia 93:689-703.

Pitt, J. I. (1979). *Penicillium crustosum* and *P. simplicissimum*, the correct names for two common species producing tremorgenic mycotoxins. Mycologia 71:1166-1177.

Pitt, J. I. (1987). *Penicillium viridicatum*, *P. verrucosum*, and the production of ochratoxin A. Applied and Environmental Microbiology 53:266-269.

Pitt, J. I., and Hocking, A. D. (1997). Fungi and Food Spoilage. 2nd ed. Blackie Academic and Professional, London.

Pitt, J. I., Cruickshank, R. H., and Leistner, L. (1986). *Penicillium commune, P. camembertii*, the origin of white cheese moulds, and the production of cyclopiazonic acid. Food Microbiology 3:363-371.

Pollock, A. V. (1947). Production of citrinin by five species of *Penicillium*. Nature 160:331-332.

Polonelli, L., Morace, G., Rosa, R., Castagnola, M., and Frisvad, J. C. (1987). Antigenic characterization of *Penicillium camemberti* and related common cheese contaminants. Applied and Environmental Microbiology 53:872-878.

Puel, O., Tadrist, S. Galtier, P., Oswald, I. P., Delaforge, M. (2005). *Byssochlamys nivea* as a source of mycophenolic acid. Applied and Environmental Microbiology 71:550-553.

Rabie, C. J., and Terblanche, M. (1967). Influence of temperature on the toxicity of different isolates of *Aspergillus wentii*. South African Journal Agricultural Science 10:263-266.

Raistrick, H., and Stössl, A. (1958). Studies in the biochemistry of microorganisms 104. Metabolites of *Penicillium atrovenetum* G. Smith. β-nitropropionic acid, a major metabolite. Biochemical Journal 68:647-653.

Raper, K. B., and Fennell, D. I. (1965). *The genus* Aspergillus. Williams and Wilkins, Baltimore.

Raper, K. B., and Thom, C. (1949). *A Manual of the Penicillia*. Williams and Wilkins, Baltimore.

Reddy, R. V., and Berndt, W. O. (1991). Citrinin. *In* Mycotoxins and Phytoalexins (Sharma, R. P., and

Salunkhe, D. K., eds.), CRC Press, Boca Raton, FL, U.S.A., pp. 237-250.

Reddy, C. S., and Reddy, R. V. 1991. Secalonic acids. *In* Mycotoxins and Phytoalexins (Sharma, R. P., and Salunkhe, D. K., eds.), CRC Press, Boca Raton, FL, U.S.A, pp. 167-190.

Rehm, H. J. (1972). Mykotoxine in Lebensmitteln. VI. Mitteilung. Aflatoxinbildung verschiedene Pilzarten. Zeitschrift für Lebensmittel Untersuchungen und Forschung 150:146-151.

Richer, L., Sigalet, D., Kneteman, N., Jones, A., Scott, R. B., Ashbourne, R., Sigler, L., Frisvad, J, and Smith, L. (1997). Fulminant hepatic failure following ingestion of moldy homemade rhubarb wine. Gastenterology 112: A1366.

Ringot, D., Chango, A., Schneider, Y. J., and Larondelle, Y. (2006). Toxicokinetics and toxicodynamics of ochratoxin A, an update. Chemico-Biological Interactions 159:18-46.

Rodricks, J. V., and Eppley, R. M. (1974). *Stachybotrys* and stachybotryotoxicosis. *In* Mycotoxins (Purchase, I. F. H., ed.), Elsevier, Amsterdam, pp. 181-197.

Rundberget, T., and Wilkins, A. L. (2002). Thomitrems A and E, two indole-alkaloid isoprenoids from *Penicillium crustosum* Thom. Phytochemistry 61:979-985.

Saito, M., and Tsuruta, O. (1993). A new variety of Aspergillus flavus from tropical soil in Thailand and its aflatoxin productivity. Proceedings of the Japanese Association of Mycotoxicology 37:31–36.

Samson, R. A., Hoekstra, E. S., and Frisvad, J. C., eds., (2004a). Introduction to Food- and Airborne Fungi, 7th Edition, Centraalbureau voor Schimmelcultures, Utrecht, 389 pp.

Samson, R. A., Houbraken, J. A. M. P., Kuijpers, A. F. A., Frank, J. M., and Frisvad, J. C. (2004b). New ochratoxin A or sclerotium producing species in *Aspergillus* section *Nigri*. Studies in Mycology 50:45-61.

Sankawa, U., Ebizuka, Y., Noguchi, H., Isikawa, Y., Kitaghawa, S., Yamamoto, Y., Kobayashi, T., and Iitak, Y. (1983). Biosynthesis of citrinin in *Aspergillus terreus*. Tetrahedron 39:3583-3591.

Sargeant, K., Sheridan, A., O'Kelley, J., and Carnaghan, R. B. A. (1961). Toxicity associated with certain samples of groundnut. Nature 192:1096-1097.

Schothorst, R. C., and Egmond, H. P. van (2004). Report from SCOOP task 3.2.10 "collection of occurrence data of *Fusarium* toxins in food and assessment of dietary intake by the population of EU member states"; Subtask: trichothecenes. Toxicological Letters 153:133-143.

Schroeder, H. W., and Kelton, W. H. (1975). Production of sterigmatocystin by some species of the genus *Aspergillus* and its toxicity to chicken embryos, Applied Microbiology 30:589-591.

Schroeder, H. W, and Verrett, M. J. (1969). Production of aflatoxin by *Aspergillus wentii* Wehmer. Canadian Journal of Microbiology 15:895-899.

Schuster, E., Dunn-Coleman, N., Frisvad, J. C., Dijck, P. W. M. van (2002). On the safety of *Aspergillus niger* - a review. Applied Microbiological Biotechnology 59:426-435.

Schütt, F., Nirenberg, H. I., and Deml, G. (1998). Moniliformin production in the genus *Fusarium*. Mycotoxin Research 14:35-40.

Scott, D. B. (1965). Toxigenic fungi isolated from cereal and legume products. Mycopathol. Mycol. Appl. 25:213-222.

Scott, P. M. (1994). *Penicillium* and *Aspergillus* toxins, *In* Mycotoxins in Grain. Compounds other than Aflatoxin (Miller, J. D., and Trenholm, H. L., eds.), Eagan Press, St. Paul, Minnesota, U.S.A, pp. 261-285.

Scott, P. M., Walbeek, W. van, and Forgacs, J. (1967). Formation of aflatoxins by *Aspergillus ostianus* Wehmer. Applied Microbiology 15:945.

Scott, P. M., Lawrence, J. W., and Walbeek, W. van (1970). Detection of mycotoxins by thin layer chromatography: Application to screening of fungal extracts. Applied Microbiology 20:839-842.

Scudamore, K. A., Atkin, P., and Buckle, A. E. (1986). Natural occurrence of the naphthoquinone mycotoxins, xanthomegnin, viomellein and vioxanthin in cereals and animal foodstuffs. Journal of Stored Product Research 22:81-84.

Seifert, K. A., Aoki, T., Baayen, R. P., Brayford, D., Burgess, L. W., Chulze, S., Gams, W., Geiser, D., de Gruyter, J., Leslie, J. F., Logrieco, A., Marasas, W. F. O., Nirenberg, H. I., O'Donnell, K., Rheeder, J. P., Samuels, G. J., Summerell, B. A., Thrane, U., and Waalwijk, C. (2003). The name *Fusarium moniliforme* should no longer be used. Mycological Research 107:643-644.

Sivanesan, A. (1991). The taxonomy and biology of dematiaceous hyphomycetes and their mycotoxins. *In* Fungi and mycotoxins in stored products (Champ, B. R., Highley, E., Hocking, A. D., and Pitt, J. I., eds.), ACIAR Proceedings No. 36, Australian Centre for International Agricultural Research, Canberra, Australia, pp. 47-64.

Sonjak, S., Frisvad, J. C., and Gunde-Cimerman, N. (2005). Comparison of secondary metabolite production by *Penicillium crustosum* strains, isolated from Arctic and other various ecological niches. FEMS Microbiology Ecology 53:51-60.

Sripathomswat, N., and Thasnakorn, P. (1981). Survey of aflatoxin-producing fungi in certain fermented foods and beverages in Thailand. Mycopathologia 73:83-88.

Staub, W. (1911). *Penicillium casei* n. sp. als Ursache die rotbraunen Rinderfarbung bei Emmenthaler Käsen. Centralblatt für Bakteriologie (II) 31:454.

Steiman, R., Seigle-Murandi, F., Sage, L., and Krivobok S. (1989). Production of patulin by micromycetes. Mycopathologia 105:129-133.

Steyn, P. S, and Rabie, C. J. (1976). Characterisation of magnesium and calcium tenuazonate from *Phoma sorghina*. Phytochemistry 15:1977-1979.

Stoev, S. D., Vitanov, S., Anguelov, G., Petkova-Bocharova, T., and Creppy, E. E. (2001). Experimental mycotoxic nephropathy in pigs provoked by a diet containing ochratoxin A and penicillic acid. Veterinary Research Communications 25:205-223.

Sun, Z. M., and Qi, Z. T. (1991). A new aflatoxin producing species of sect. *Flavi* of *Aspergillus*. Acta Mycologica Sinica 10:22-26.

Swelim, M. A., Baka, Z. A. M., El-Dohlob, S. M., Hazzaa, M. M., and El-Sayed, T. I. (1994). Mycoflora of stored poultry fodder in Egypt and their ability to produce aflatoxins. Microbiological Research 149:435-442.

Taniwaki, M. H., Pitt, J. I., Teixeira, A. A., and Iamanaka, B. T. (2003). The source of ochratoxin A in Brazilian coffee and its formation in relation to processing methods. International Journal of Food Microbiology 82:173-179.

Teuber, M., and Engel, G. (1983). Low risk of mycotoxin production in cheese. Microbiology of Aliments and Nutrition 1:193-197.

Thrane, U., Adler, A., Clasen, P.-E., Galvano, F., Langseth, W., Lew, H., Logrieco, A., Nielsen, K. F., and Ritieni, A. (2004). Diversity in metabolite production by *Fusarium langsethiae, Fusarium poae*, and *Fusarium sporotrichioides*. International Journal of Food Microbiology 95:257-266.

Timonin, M. I., and Rouatt, J. W. (1944). Production of citrinin by *Aspergillus* species of the candidus group. Canadian Journal of Public Health 35:80-88.

Ueno, Y. (1974). Citreoviridin from *Penicillium citreoviride* Biourge. *In* Mycotoxins (Purchase, I. F. H., ed.), Elsevier, Amsterdam, The Netherlands, pp. 283-302.

Varga, J., Tóth, B., Rigó, K., Téren, J., Hoekstra, R. F., and Kozakiewics, Z. (2000a). Phylogenetic analysis of *Aspergillus* section *Circumdati* based on sequences of the internal transcribed spacer regions of the 5.8 S rRNA gene. Fungal Genetics and Biology 30:71-80.

Varga, J., Kevei, É., Tóth, B., Kozakiewicz, Z., and Hoekstra, R. F. (2000b). Molecular analysis of variability within the toxigenic *Aspergillus ochraceus* species. Canadian Journal of Microbiology 46:593-599.

Varga, J., Rigó, K., Téren, J., and Mesterházy, Á., (2001). Recent advances in ochratoxin research. I. Production, detection and occurrence of ochratoxins. Cereal Research Communications 29:85-92.

Walbeek, W. van, Scott, P. M., and Thatcher, F. S. (1968). Mycotoxins from some food-borne fungi. Canadian Journal of Microbiology 14:131-137.

Weidenbörner, M. (2001). Food and fumonisins. European Food Research and Technology 212:262-273.

Wilson, B. J., Campbell, T. C., Hayes, A. W., and Hanlin, R. T. (1968). Investigation of reported aflatoxin production by fungi outside the *Aspergillus flavus* group. Applied Microbiology 16: 819-821.

Wilson, D. M., Mutabanhema, W., and Jurjevic, Z., (2002). Biology and ecology of mycotoxigenic *Aspergillus* species as related to economy and health concerns. *In* Mycotoxins and Food Safety (Vries, J. W. de, Trucksess, M. W., and Jackson, L. S., eds.), Kluwer Academic Publishers, Dordrecht, The Netherlands, pp. 3-17.

Wu, X., Leslie, J. F., Thakur, R. A., and Smith, J. S. (2003). Purifiction of fusaproliferin form cultures of *Fusarium subglutinans* by preparative high-performance liquid chromatography. Journal of Agricultural and Food Chemistry 51:383-388.

Yates, S. G., Tookey, H. L., Ellis, J. J., Tallent, W. H., and Wolff, I. A. (1969). Mycotoxins as a possible cause of fescue toxicity. Journal of Agricultural and Food Chemistry 17:437-442.

Zimmerman, J. L., Carlton, W. W., and Tuite, J. (1979). Mycotoxicosis produced by cultural products of an isolate of *Aspergillus ochraceus*. 1. Clinical observations and pathology. Veteri nary Pathology 16:583-592.

Part 4

FUNGI AS HYPERPRODUCERS

The capability of fungi to form so many different metabolites is reflected in the importance of fungi as industrial producers of a number of compounds including chemicals as citrate and penicillin. The latter alone already is of crucial importance of mankind. One can ask if it is possible to direct the fungal metabolism in such a way that they become super-producers of such compounds. De Jongh and Nielsen describe in Chapter 9 different fungal species that are hyperproducers. In more detail, the producers of citrate and penicillin are addressed here. They describe the history of production, the different ways production was increased and also novel developments as metabolic engineering are addressed.

Being plant degraders in nature, fungi produce a great variety of different enzymes that can attack complex biopolymers. Some of these enzymes are of industrial interest and fungi are used for the commercial production of these. In Chapter 10 Wösten, Scholtmeier and de Vries describe the specific problems that appear when fungi become hyperproducers of enzymes. The chapter gives a valuable overview of all the different industrially used enzymes and their producers. Synthesis of enzymes is different from the biosynthesis of metabolites in that proteins are made in association with ribosomes and subsequently travel via the ER, the Golgi- equivalents and excretion vesicles to be released from the hyphal tip. Hyperproduction can induce interesting problems inside the cell and this can be related to the expression of heterologous proteins, proteins from other species.

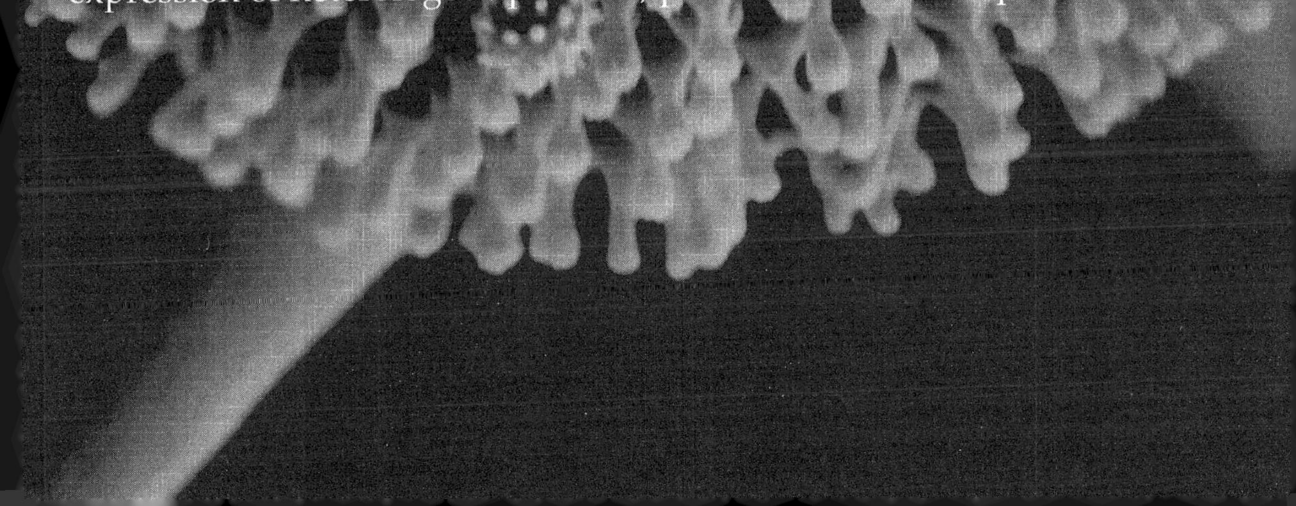

Chapter 9

Filamentous fungi as cell factories for metabolite production

Wian A. de Jongh and Jens Nielsen
Center for Microbial Biotechnology BioCentrum-DTU Søltofts Plads 221, DK-2800 Kgs. Lyngby, Denmark.

INTRODUCTION

Filamentous fungi are extensively used as cell factories for different biotechnological products such as enzymes, pharmaceuticals, as well as primary and secondary metabolites. The two most eminent biotechnological processes are citrate production by *Aspergillus niger* and penicillin production by *P. chrysogenum,* which will be the focus of this chapter. In terms of industrial production, these two metabolite hyperproducers represent billions of U.S.$ in annual revenues.

Hyperproduction can be defined as the ability of an organism to obtain very high yields, in some cases close to stoichiometric conversion of carbon into the product of interest in a relatively short period of time (generally a few days). Such intensive production can only be obtained with a few selected isolates of filamentous fungal species under specific nutrient and morphological conditions.

The original natural isolates would be considered poor producers by modern standards. Intensive efforts have been applied to the improvement of these original isolates, using mostly classical random mutation strain improvement techniques, and more recently genetic and metabolic engineering techniques. Concurrent to the advances in fungal genome manipulation techniques, great strides forward in the development of comprehensive metabolite detection, large scale mRNA transcript quantification and other so-called "omic" techniques have been made. The application of these techniques to strain improvement is already starting to revolutionize the way in which we improve and develop current or novel processes. In this chapter we discuss these developments, but we start with a discussion on the historical development of citrate and penicillin production, two examples where hyperproducing filamentous fungi have been developed.

HISTORY

From a historic perspective, citrate production by *A. niger,* fumarate production by *Rhizopus oryzae,* and penicillin production by *Penicillium chrysogenum* have great importance as the first true examples of industrial scale biotechnological processes. These fungal fermentation processes have led the way towards further industrial applications of biotechnology. Particularly the development of citrate and penicillin production processes resulted in mutual advantages, and led to rapid progress in the field of biotechnology as a whole. In the course of developing these processes many similar problems arose, and hence many of the techniques developed for one process could rapidly be transferred to the other process (see Figure 1 for a timeline overview of key historical developments), e.g., large-scale fermentation technology.

Citrate was first discovered by Scheele in 1784 and was produced mainly from lemons during the nineteenth century. In 1880 it was attempted to produce citrate from glycerol, but this was not economically competitive.

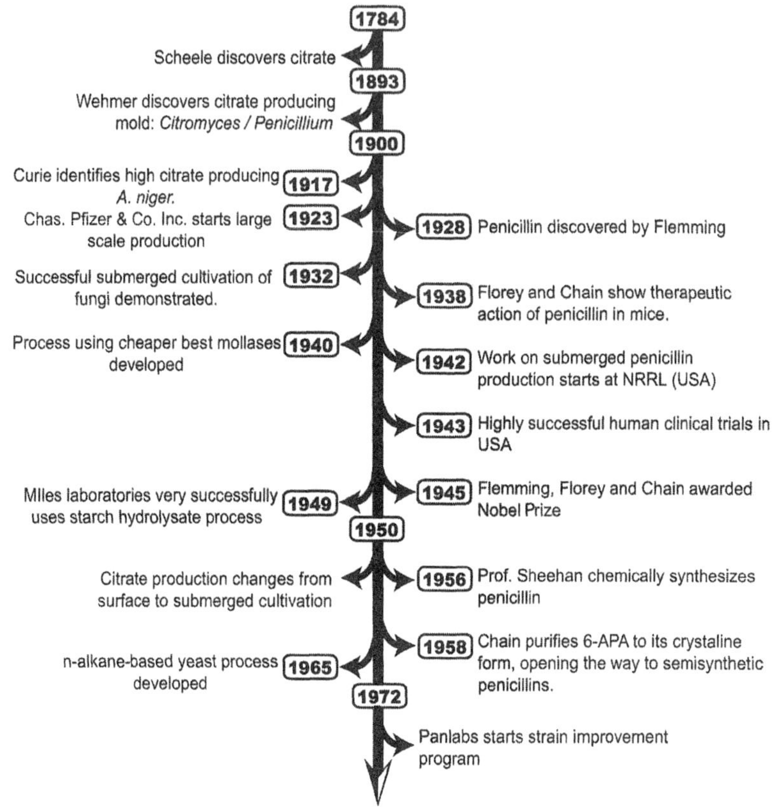

Figure 1. Timeline of key historical developments for citrate and penicillin production.

As early as 1893 there was the first hint at using a biotechnological process when Wehmer discovered a mold that produced citrate (Mattey, 1992). He called it *Citromyces*, which later became better known as *Penicillium*. In 1917 Curie found an *A. niger* capable of producing high concentrations of citrate in a sugar medium. This discovery quickly led the U.S. company, Chas. Pfizer & Co. Inc., to start producing citrate on a large scale using surface cultures. The process was soon improved by using beet molasses instead of sugar, thereby making the process more economically viable.

The fortuitous discovery of penicillin by Alexander Fleming in September 1928 opens a new chapter in the early history of fungal hyperproducers. In 1938, Howard Florey and Ernest Chain made a systematic survey of antibacterial substances known to be produced by microorganisms. They started with penicillin and published their results in *The Lancet* (Chain *et al.*, 1940), where they showed methods for production of penicillin and its therapeutic action in mice. The initial success of this study quickly led to a human trial being attempted. Their second paper described detailed methods for large scale production of penicillin (Abraham *et al.*, 1941), but the severity of World War II, made it very difficult to produce large quantities of penicillin in the United Kingdom.

Researchers Howard Florey and Norman G. Heatley therefore went to the United States. Here they were directed to the Northern Regional Research Laboratory (NRRL) where the chief of the fermentation division suggested using submerged cultivation. Corn steep liquor was used as a carbon source, which led to more penicillin G being produced, while in the U. K. mostly penicillin F was produced. Florey also tried to interest several large U. S. pharmaceutical companies in producing penicillin. At that time the companies generally thought that it would only be a matter of time before chemical synthesis could be applied to produce penicillin, and they were therefore reluctant to invest in production scales larger than pilot scale for submerged cultures (Nielsen, 1995). The extremely successful clinical trials that were conducted in 1943, led the U. S. War Production Board to encourage pharmaceutical companies to increase supply, and to invest in the fermentation route, and this eventually resulted in large-scale production of penicillin by fermentation. It was only in 1956 that John Sheehan, a professor at MIT, succeeded in establishing a chemical synthesis route towards penicillin. The chemical synthesis turned out to be very inefficient and expensive and never presented a real challenge to fermentation.

In connection with design of large-scale penicillin production there were several problems to be solved to achieve successful submerged cultivation. These included: sterile design, dirt- and oil- free equipment design, large-scale agitators and sterile aeration capabilities (Shuler and Kargi, 1992). Similar problems were encountered for citrate production, but in the case of citrate production sterile air was less of a problem, as the very low pH at which citrate fermentations are conducted makes contamination much less likely than for penicillin fermentations. Successful submerged penicillin production was rapidly achieved, which is generally attributed to the transfer of know-how from the submerged fumarate production process using *R. oryzae*. The *R. oryzae* process was the first submerged fermentation process using filamentous fungi and was therefore used as a model for scale-up and fermentation techniques for the penicillin production process (Roehr and Kubicek, 1996) Hurdles in submerged cultivation were overcome by a multidisciplinary team, which included biologists and chemical engineers, and can be seen as the birth of biochemical engineering as a discipline.

Shortly after, 40,000 L fermenters were implemented, followed by Pfizer completing a plant containing 14 x 7,000 gallons fermenters in just 6 months. Thus, the intense interest created in penicillin during World War II soon paid off and by the end of the war, the U.S. capability for patient treatment was 100,000 a year (Hobby, 1985). For their contributions Fleming, Florey and Chain received the Nobel Prize in Medicine in 1945 "for the discovery of penicillin and its curative effects in various infectious diseases."

Rapid progress was made in the optimisation of fermentation parameters during the following years leading to large increases in yields and productivities for both citrate and penicillin.

Along with the increases in capacity for penicillin production, new antibiotics were being developed. K. Kato of Japan in 1953 found that 6-APA, the nucleus of penicillin, was produced during fermentation and this component was purified to its crystalline form by Chain in his Italian laboratory in 1958, leading the way towards the production of semisynthetic penicillins (Abraham, 1983).

These advances in fermentation technology soon started to be applied to the production of citrate, and from 1950 the process changed over time from surface to submerged cultures. In 1965 various yeast-based citrate production processes were introduced. These processes used n-alkanes, since hydrocarbons were cheap at the time, but most have now changed to regular carbon sources (Barbesgaard *et al.*, 1992). Furthermore, today *A. niger* is the main cell factory used for citrate production.

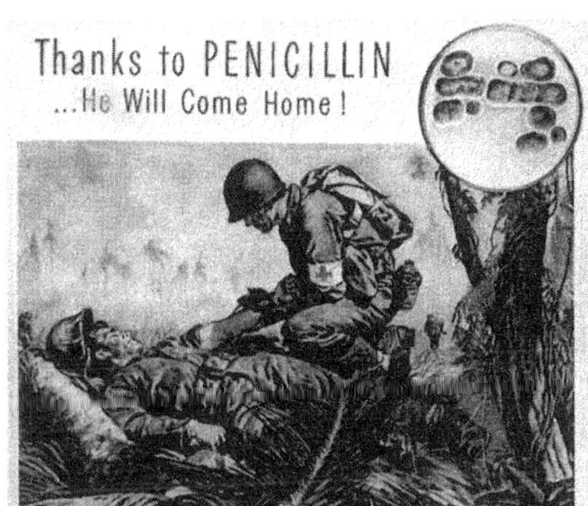

Figure 2. Historic poster from WWII.

Table 1. Products of interest and the fungi that produce them (Adapted from Ruijter, Kubicek, and Visser, 2002; Demain and Elander, 1999, with additional information from Roehr, Kubicek, and Kominek, 1992; Schuster *et al.*, 2002; Mattey, 1992; Elander, 2003).

Organism	Metabolite	Application
Aspergillus niger	Citrate	□ Food and beverage industries □ Acidifier, pH adjustment, flavour enhancer, reduces sweetness, antioxidant, preservative □ Pharmaceutical industry and cosmetics □ pH adjustment, anticoagulant, antioxidant, fast dissolution of active agent, preservative in stored blood, iron citrate as a source of iron □ Others □ Cleaning of metal surfaces, oil well treatment, retards concrete setting, hardening of adhesives, plastics industry, washing agents, household cleaners, removal of sulfur dioxide from waste gases, leather tanning, electroplating
	Gluconate	Food additive, therapeutic metal salts, dissolving of calcium deposits such as milkstone in dairy industry, metal cleaning
Aspergillus terreus	Itaconic acid	Plastics and paper industries
Rhizopus oryzae	Fumarate	Food additive, synthetic polymers
Penicillium chrysogenum	Penicillin / 6-APA	□ Penicillin G and V, which are either used directly or to produce the penicillin precursor 6-APA □ Semi-synthetic penicillins □ Broad spectrum penicillins: ampicillin, amoxycillin □ Isoxazolyl penicillins: oxacillin, cloxacillin, flucloxacillin □ Expanded spectrum ureido penicillins: azlocillin, mezlocillin, piperacillin □ Prolong pharmocokinetically improved penicillins: pivampicillin, bacampicillin □ Antipseudomonal penicillins: ticarcillin, indanyl piperacillin, ticarcillin and carbenicillin □ Penicillin-resistant penicillins: methicillin and dicloxacillin □ Others : epicillin, nafcillin and cyclacillin
	ad-7-ADCA	Used to produce the cephalosporin precursor 7-ADCA through enzymatic cleavage

Process improvements using classical strain improvement approaches to obtain higher producing strains soon started and this was the main approach to this aim up to the 1980s. Here it is of particular importance to mention the project undertaken by the American company Panlabs Inc., which led to large increases in penicillin yields and productivities between 1972 and the early 1980s (see example in the classical strain improvement section). DNA-mediated transformation started in the mid-1980s and quickly led to new disciplines such as genetic engineering and, later, metabolic engineering.

Citrate and penicillin production techniques are still being improved today, and it is a measure of the historical, as well as current, importance of these products that they still generate such interest. Figure 3 shows the increase in the world consumption of penicillin and citrate, as well as the distribution of citrate producers throughout the world and the effect of gross national product on penicillin consumption. Table 2 provides some additional economic data on the products.

Table 2. Citrate and Penicillin production statistics

	Citrate *A. niger'*	Penicillin* *P. chrysogenum*
World market [million $]	360	5000
Cost [U.S.$/kg]	0,4	< 8
Titer [g/L]	>110''	40-50
Yield [%]	>80'''	10-12

' Comparatively small percentage of citrate produced by other organisms.'' For the year 2000 ''' Mattey, 1992,* Elander, 2003.

The fast increase in citrate consumption along with its current and expanding global production capacity has placed citrate as one of the largest bulk commodity products produced by biotechnology worldwide. The sharp increasing trend for penicillin consumption is almost guaranteed to continue for the foreseeable future. This can especially be deduced from Figure 3 (D), which shows that annual per capita consumption of penicillin is closely linked to the gross national product (GNP). Coupled with the rapid increase in the GNP in India and China there will obviously be a further increase in the worldwide consumption of penicillin. It is interesting to note that the annual per capita intake of penicillin is close to being at Western levels even for relatively low GNP, which is explained by the fact that even 60 years after its introduction, penicillin is still one of the most cost-effective drugs.

PRODUCER STRAINS

Several producer strains are currently employed in industrial processes: citric acid and gluconic acid production by *A. niger*, penicillin by *P. chrysogenum*, itaconic acid production by *A. terreus*, malate and fumarate production by *Rhizopus oryzae*, as well as numerous enzyme production processes. Table 2 gives a list of products of interest and the organisms that produce them, and Figure 4 depicts microscopic images of these fungi.

Platform organisms

Even though there are several fungal producer strains used at the moment by industry for organic acid and antibiotic production, only *A. niger* and *P. chrysogenum* can be characterized as platform organisms for metabolite production. A platform organism is an industrial organism that is extensively used to produce several commodity products. The main advantages of these organisms are that they are well understood and have been used by industry for an extended period of time. This allows rapid transfer of new products from the research and development stage to production. Table 3 lists the advantages and disadvantages of three industrially used producer strains.

Health and safety

Both *A. niger* and *P. chrysogenum* are particularly suitable for food or drug production, as they have been certified as generally regarded as safe (GRAS) by the Food and Drug Administration (U.S.A) and the World Health Organisation. Many specific *A. niger* production processes have been certified as GRAS, among these are several enzymes: α-amylase, cellulase, amyloglucosidase, catalase, glucose oxidase, lipase and pectinase. Bulk chemicals produced by *A. niger* have also attained this status, such as citrate and gluconate. In the case of *P. chrysogenum*, production of penicillin and ad-7ADCA have also been recognised as GRAS.

As is the case for other fungi, it is important to avoid too much contact with the spore dust, but *A. niger* (Schuster *et al.*, 2002) and *P. chrysogenum* are not considered particularly dangerous. *A. niger* has only been found to cause human infection in people with a compromised immune system, and while Ochratoxin A is produced by *A. niger* in rare cases (Schuster *et al.*, 2002), it has not been cited as a problem for any of the current biotechnological applications.

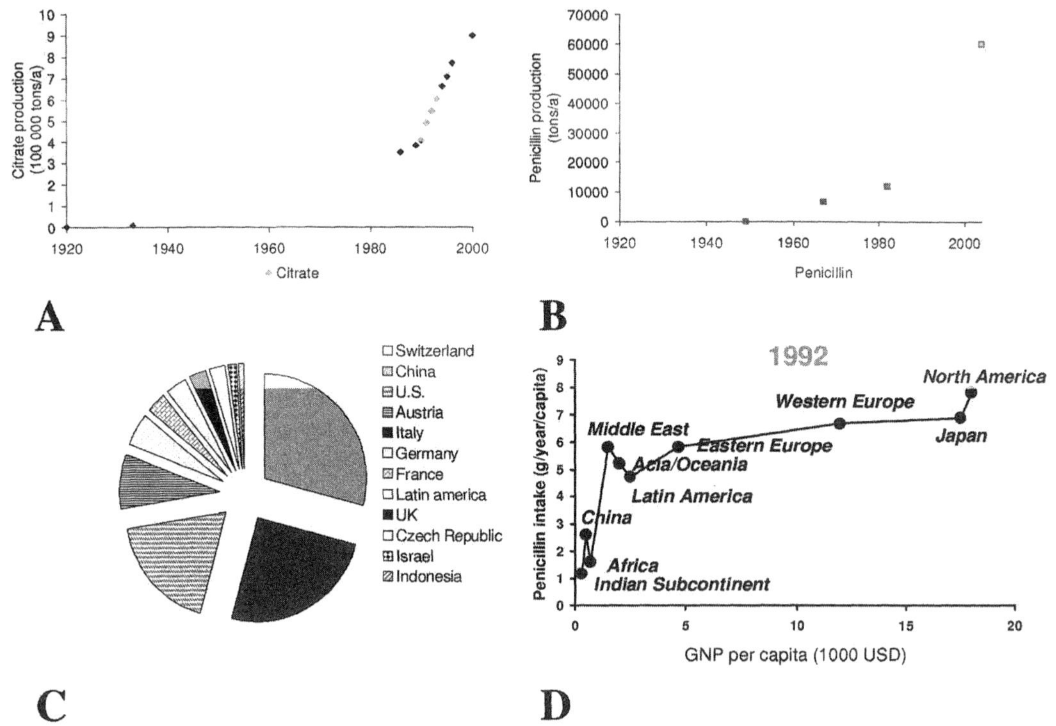

Figure 3. World market development of (A) citrate (Connor, 1998), (B) penicillin (Demain and Elander, 1999), (C) production capacity by country for 1996 (Connor, 1998), and (D) penicillin usage vs. gross national product for 1992.

The safety and ease of use of these platform organisms has led to their extensive application in industry and has made them the focus of intense study and process improvement projects.

Classical strain improvement

Classical strain improvement have been used since the early 1950s to increase yield, titer, productivity or general ease of use of *Aspergillus niger* and *Penicillium chrysogenum*, and has been responsible for virtually all the published superproducers to date. Mutagens such as short wavelength radiation (UV), X- and γ-ray ionizing radiation and chemical agents (base analogs: 5-chlorouracil, hydroxylamine; alkylating agents: N-methyl-N'-nitro N-nitrosoguanidine; intercalating agents, etc.) have been extensively used. The most popular agent of these, because of its very high mutant to survivor ratio and multiplicity of mutations, has been nitrosoguanidine (Parekh *et al.*, 2000). Most mutations occur at very low frequency (10^{-5}-10^{-10}/ generation, Parekh *et al.*, 2000) and mostly with negative impact on the characteristic of interest. The challenge is to isolate the true beneficial mutants, and therefore, well-designed screening and directed mutagenesis techniques are of vital importance.

Random mutation and selection

These were initially used because the lack of any detailed biochemical knowledge of the organisms prevented a more direct approach. This empirical method can be described as "hit and miss," requiring a brute force approach. A classical case would involve the screening of more than 100,000 mutants to obtain a positive result (Lein, 1986).

By applying rational selection techniques a tenfold improvement in hit rate can be

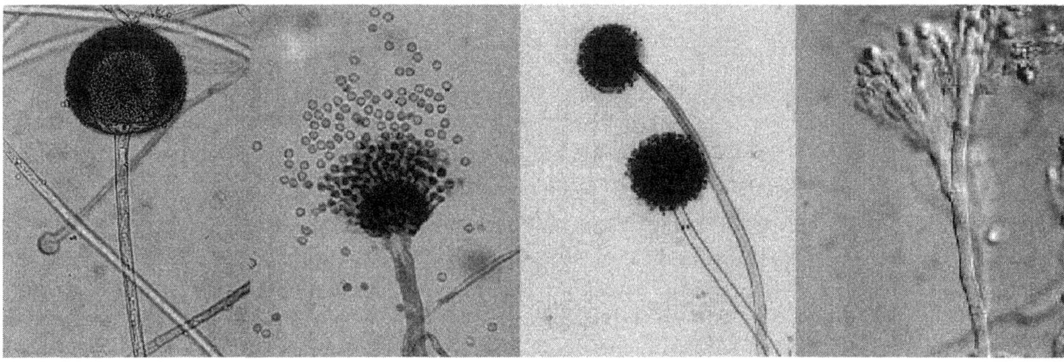

Figure 4. Microscopic images of the sporulating structures of *Rhizopus oryzae, Aspergillus terreus, Aspergillus niger* and *Penicillium chrysogenum* (from left to right). Pictures kindly provided by Dana Savicka of the Institute of Chemical Technology Prague, and Michel Cavalla.

Table 3. The advantages and disadvantages of producer strains used by industry

Organism	Advantages	Disadvantages
Aspergillus niger	• High acid tolerance (>200 g/L citrate) • High productivities possible (r_P> 1.4 g/L/h) • Can utilize complex substrates (biomass) • Extensively applied in industry • GRAS organism • 4 x coverage of genome publicly available, public sequencing project to 9 x coverage.	• Does not naturally produce other useful organic acids (other than citrate and gluconate) at high titers • Can be consumer problems with GM strains
Aspergillus terreus	• Used to produce itaconic acids at high titers and productivities (15,000 tons/a) • Itaconic process is considered GRAS • In process of being sequenced for a public assessable database	• Less industrially utilized than *A. niger* • Seen as an emerging antigen for causing aspergillosis, and resistant to amphotericin B, a crucial treatment for fungal infections (Steinbach *et al.*, 2004)
Penicillium chrysogenum	• High titers of β-lactams possible (>50 g/L) • High productivities • Superproducers can be used as platforms for new products (ad-7-ADCA) • Extensively applied in industry • GRAS organism	• Not yet sequenced • Only used for secondary metabolite production

achieved (Vournakis and Elander, 1983), which results in a direct efficiency increase for strain improvement programs. Rational selection techniques involve screening, not for the product of interest, but for a biochemical characteristic associated with it. Such techniques involve, for example, direct colony selection after a bioassay overlay for better penicillin producers (Vournakis and Elander, 1983), or citrate specific indicator (P-di-methylaminobenzaldehyde) assays for citrate hyperproducers (Mattey, 1992).

Resistance to metals

Selection of mutants resistant to metallic ions was a particularly innovative solution for antibiotic producing strains. Certain toxic heavy-metals complex with β-lactam antibiotics, and

EXAMPLE

The efforts by the U.S. company Panlabs Inc. to increase penicillin production is one of the best examples available in the literature on a classical strain improvement project. It combined several rounds of mutation (using dimethyl sulfate, ethyl methanesulfonate and UV-irradiation) and selection, and two different starting strains (P1 and P2, obtained from Toyo Jozo Co. for Penicillin V production and Nippon Kayaku Co. for penicillin G production, respectively), to create mutants with vastly higher penicillin production capabilities. Nitrosoguadine was mentioned earlier as the most popular mutagen used, however, it was purposefully not used in the Panlabs penicillin project. It was thought at the time that the high rate of mutagenesis would cause multiple mutations per mutant, which would lead to crippled strains after several rounds of mutagenesis. Two methodologies were employed in the mutant selection process: random selection of mutants was used for the P2 strains, while rational selection techniques were applied to the P1 strains. The rational selection techniques were designed around breaking down the biosynthetic control mechanisms or increased antibiotic secretion to prevent feedback inhibition on its own biosynthetic pathway (e.g., 2-aminoadipic acid, sodium sulfide, lysine, valine hydroxamate, etc.). Protoplast fusion was also employed later in the program, but only in the period after 1977.

Initial experiments were carried out in shake flask and large-scale fermenters, but as it was found that the shake flask experiments adequately predicted behaviour in large-scale fermentations, shake flasks were used for the remainder of the strain improvement process (see Figure 5). [(Lein, 1986), (Demain and Elander, 1999)].

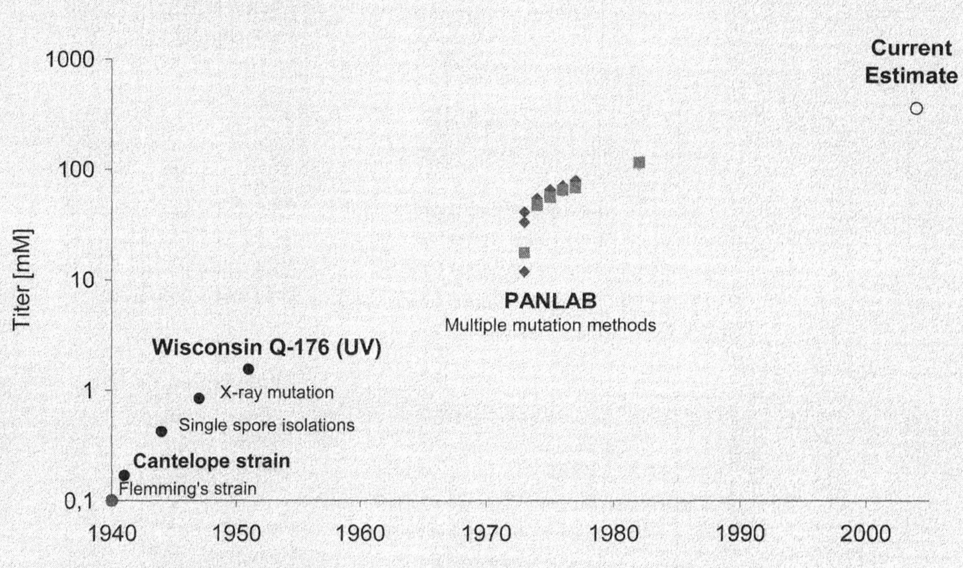

Figure 5. The development of penicillin titers with an indication of Panlabs' contribution to penicillin production improvement through classical strain improvement in the 1970s. Strain lineages P1 ◆ and P2 ■ are marked.

antibiotic production therefore detoxifies the medium. It was therefore possible to screen for faster growing colonies to find the best penicillin producers. This technique can also be used iteratively, with increased metal concentrations leading to increased production in a step-wise manner (Penalva *et al.*, 1998). Metallic ions are also of great importance to citrate production. It has long been known that the presence of even trace amounts of Mn^{2+} (more than 1 ppb, Mattey, 1992) will drastically decrease citrate production in submerged cultivation. Mn^{2+} has been reported to have multiple effects on *A. niger* physiology, e.g., morphology changes from pellet to filamentous, increased protein turnover, impaired DNA synthesis and altered composition of plasma membrane and cell walls (Karaffa and Kubicek, 2003). Traditionally, the feedstock has been de-ionised using ion-exchange pre-treatment, which adds to the labour and costs involved in using substrates such as molasses. Classical selection techniques have therefore been used to select for mutants less sensitive to Mn^{2+} [(Lesniak *et al.*, 2002); (Schreferl *et al.*, 1986); (Gupta and Chandra, 2002)].

Extension of substrate range

Obtaining mutants with the ability to grow or produce on new substrates can lead to reduced substrate costs and thereby better process economics. Ikram-ul-Haq *et al.* (2001) used chemical and UV-radiation to create mutants with enhanced abilities of citrate production on black strap molasses. The best mutant produced 86.1 g/L citrate in 168 h, compared to 31.1 g/L by the wild-type; it also had a faster specific growth rate. Starch and dextrose syrups have also been successfully used for citrate production (80% yield) by trace metal resistant mutants, without the necessity for ion-exchange pre-treatment (Lesniak *et al.*, 2002). Ikram-Ul Haq *et al.* (2003) selected similar mutants, which had enhanced productivity for direct citrate production from raw starch. Several studies (Leangon *et al.*, 1999) also looked at growth and citrate production characteristics of mutants on solid media. Some *A. niger* strains have also been adapted to grow on gluconic acid, an unwanted side product, which can decrease process productivities (Mattey, 1992).

Decreased by-product formation

Selecting mutants deficient in the responsible enzymes can be used to decrease unwanted by-product formation. This has the added advantage of increasing process productivity, while hopefully increasing yields. Unwanted gluconate and oxalate productions were prevented in this way by selecting for mutants deficient in glucose oxidase or oxaloacetate hydrolase, respectively (Ruijter *et al.*, 1999). A double mutant obtained by Ruijter *et al.*, (1999) also had the unexpected characteristic of being able to produce citrate at high pH, and in the presence Mn^{2+}. Side product formation also plays a role in penicillin production, where part of the side-chain precursors (phenylacetic and phenoxyacetic acids) fed during penicillin fermentations are metabolized to undesirable side products, such as 2-or 4-hydroxyphenylacetate, and are then incorporated into penicillin. Classical mutagenesis techniques where therefore employed to prevent these expensive side-chain precursors from being wasted (Penalva *et al.*, 1998).

Toxic precursors, toxic end products or metabolic antagonists

The use of toxic precursors, toxic end products or metabolic antagonists has been extensively employed to select for higher citrate or penicillin producers, e.g., S-2-aminoethyl-L-cysteine has been used to select for lysine feed-back insensitive strains of *P. chrysogenum* for β-lactam antibiotic production (Vournakis and Elander, 1983). Selection for greater glycolytic capacity, and subsequent faster growth rate and higher citrate production rate, was achieved using 2-deoxyglucose-resistance (Kirimura *et al.*, 1992). Methanol was known to impair protein synthesis in *A. niger* with the side-effect of increased citrate production Selection for cycloheximide sensitive strains of *A. niger* has yielded strains with impaired protein synthesis (Rugsaseel *et al.*, 1996). These strains accumulated similar amounts of citrate, without the addition of methanol, compared to the wild type with methanol added to the production media. Use of mutants that are resistant to analogues of metabolic intermediates, which over-produce these intermediates, have been applied to the selection of penicillin producer strains by using analogues of natural amino acid precursors of penicillin biosynthesis (Penalva *et al.*, 1998; Vournakis and Elander, 1983; Lein, 1986).

Other strategies

Selecting a penicillin-producing mutant with pellet morphology instead of the wild type filamentous morphology, obtained better mass transfer rates and subsequent better penicillin productivities (Penalva *et al.*, 1998). Similar studies have been done for *A. niger*, as pellet morphology is of particular importance for citrate production. Repeated rounds of mutation can adversily affect many cell functions. The successful selection for better sporulation and germination efficiencies as markers for cell damage has been achieved and led to higher production (Penalva *et al.*, 1998). Better producers have also been isolated through selection for reversion of auxotrophs of biosynthetic intermediates (Vournakis and Elander, 1983).

METABOLIC ENGINEERING

"The combination of analytical methods to quantify fluxes and their control with molecular biological techniques to implement suggested genetic modifications is the essence of metabolic engineering." Stephanopoulos, G. N., Aristidou A. A., Nielsen J., (1998)

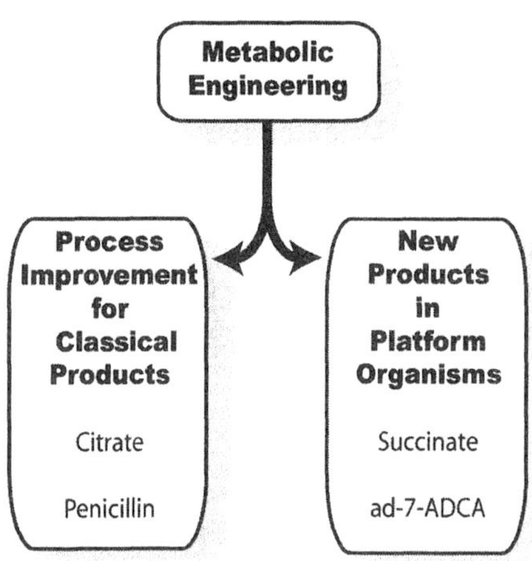

Figure 6. The two main branches of metabolic engineering.

Metabolic engineering can be divided into two main branches: improvement of current processes and new product development (See Figure 6). These can be seen as the next logical biotechnological step following on classical strain improvement and genetic engineering. Genetic engineering, which can be defined as the transformation of a cell by foreign DNA, supplied the necessary tools for the directed genetic changes, which is a central theme of metabolic engineering (Figure 7).

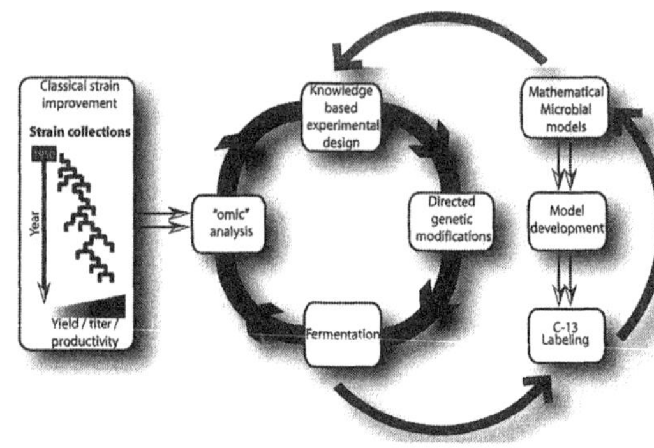

Figure 7. Process improvement through metabolic engineering is an iterative approach, where, using all available information, several rounds of directed strain improvements, and in-depth experimental analysis, leads to a strain with preconceived physiological attributes. The importance of classically improved strain lineages becomes apparent when this approach is applied to penicillin and citrate production. The information obtained from the in-depth study of these strain lineages can be directly applied to better design strategies for directed strain improvement.

Improving our fundamental understanding of hyperproducers

The application of metabolic engineering to improve penicillin production, as well as citrate production, is a real challenge as the current applied industrial strains are the result of more than 50 years of classical strain improvement programs. However, metabolic engineering has been very successfully applied to elucidate the mechanisms behind the improvements achieved through classical means. Here, especially the study of the SmithKline Beecham *Penicillium*-strain improvement series has been illuminating (Penalva *et al.*, 1998). This and similar studies provided both the genetic rationale behind the improvements (Christensen *et al.*, 1995; Newbert *et al.*, 1997), as well as biochemical understanding of the interplay between the individual pathway intermediates and production (Theilgaard *et al.*, 2001).

It was consistently found that increased productivity equalled increased biosynthetic cluster copy number (proportionally increased production can be seen up to a copy number of four (Thykaer and Nielsen, 2003)), while lower producing revertants displayed decreased copy numbers (Christensen *et al.*, 1995). However, it has been shown that not all increases in productivity can be linked to copy number alone, but concerted efforts to find specific point mutants in the promoter areas of the high producing strains were unsuccessful (Newbert *et al.*, 1997). Possibly transcription factors might play an important role here, and some of these have been identified in *A. nidulans* (Brakhage, 1997; Litzka *et al.*, 1999). The work of Kiel *et al.* (2005) offers another possible explanation for increased penicillin production in the absence of an increased number of tandem repeats. They have shown that increasing the microbody numbers in the cell directly leads to increased penicillin production, without any increase in penicillin biosynthetic pathway enzyme activities.

The availability of sequences of individual pathway genes allowed detailed studies on the effects of individual genes in the tandem repeats observed in the classically improved strains. Subsequent studies have revealed the importance of balanced expression or over-expression of the penicillin biosynthetic genes (Theilgaard *et al.*, 2001) and partly explained why classically improved hyperproducers practically always have the entire gene cluster in multiple copies. Unbalanced expression of the individual genes led to accumulation of toxic intermediates of the pathway, slowing the maximum specific growth rate and lowering penicillin production. Only when the entire biosynthetic gene cluster was introduced in multiple copies did the penicillin production increase (Theilgaard *et al.*, 2001).

Towards a systems approach

Unfortunately, researchers on citrate production have not always been as successful in translating knowledge gained from classically improved strains through metabolic engineering. In most cases these studies demonstrate one of the fundamental difficulties in metabolic engineering, namely the need for extensive and detailed information about the process, pathway or metabolic branch point one wishes to manipulate. Promper *et al.* (1993) investigated the role of complex I (NADH:ubiquinone oxidoreductase) in citrate accumulation. The study was undertaken following the observation that the mutant citrate-producing B60 strain lost complex I at the onset of citrate accumulation (Wallrath *et al.*, 1991). The NADH binding subunit of complex I was therefore deleted in strain nuo51, which then disrupts the assembly of a functional complex I, and compared to the strain B60 and the parent. As expected the intracellular citrate concentration increase over that of the parent strain (up to twenty-fold), but unexpectedly the total citrate produced was much reduced. The reason for this could not be proven, although the authors speculated that it was probably the lack of a citrate transporter, possibly mitochondrial, which led to strain nuo51 growing much slower than either the parent or B60, and that this lack also accounted for the low citrate production.

Another example was the improvement of the flux towards citrate production by over-expressing pyruvate kinase and phophofructokinase (Ruijter *et al.*, 1997). The rationale for this approach was mainly based on previous work, which suggested that hexokinase and phosphofructokinase may be important steps in flux control (Schreferl-Kunar *et al.*, 1989). It has also been found that high-producing strains, selected for their ability to grow faster in high sucrose concentrations with increased citrate yields, had twofold higher activities of these two enzymes (Schreferl-Kunar *et al.*, 1989). These conditions of high sucrose concentration have also been found to increase intracellular fructose 2,6-bisphosphate concentration (Kubi-cek-Pranz *et al.*, 1990), which is in turn a potent activator of phosphofructokinase. Although the approach was well thought out, and the necessary tools were available, an unexpected compensatory mechanism negated the genetic modifications. Over-expression of citrate synthase also had no effect on citrate production, most likely because the native citrate synthase had excess capacity (Ruijter *et al.*, 2000).

Modelling in metabolic engineering

The above mentioned study on improving citrate production through over-expression of key genes in the pathway clearly points to the fact that the use of directed genetic changes can be problematic, without systems knowledge. A major step towards applying a systems approach in metabolic engineering is arising in the form of using genome-scale microbial models (Patil *et al.*, 2004). Two major classes of models are of current interest: kinetic models and stoichiometric models.

Both of these modelling approaches offer directly applicable advantages for metabolic engineers. Kinetic models represent a dynamic view of the cellular metabolism, but its application is severely constrained. The main reason for this is the lack of enzymatic kinetic data, which are cumbersome to obtain, and the discrepancy between *in vivo* and *in vitro* enzyme activities. These problems have limited the use of kinetic models for large metabolic networks, but, interesting results have been obtained after extensive study of specific and relevant pathways, such as those for citrate (Guebel and Torres Darias, 2001; Torres *et al.*, 1996; Varez-Vasquez *et al.*, 2000) and penicillin synthesis (Conejeros and Vassiliadis, 2000). Varez-Vasquez *et al.* (2000) predicted a five-to-tenfold increase in citrate production rate through the over-expression of 13 genes or more. This is currently outside the possible practical range of genetic modifications and a reworking of this problem, using a macroscopic approach and different physiological parameters (Guebel and Torres Darias, 2001) suggested a less extensive manipulation of the strain. It was predicted that with over-expression of the glucose carrier alone, a significantly increased citrate production rate could be achieved. Unfortunately, Papagianni and Mattey (2004) have demonstrated that glucose import during citrate production is a passive transport process, and therefore over-expression of the glucose carrier is likely to have little effect, unless a heterologous active transporter is expressed. Torres *et al.* (1996) used biological systems theory coupled with constrained linear optimization, to show that at least 7 glycolytic enzymes needed to be over-expressed to achieve a significant increase in flux towards citrate. This could help explain why the single and double over-expressions of glycolytic genes (Ruijter *et al.*, 2000; Ruijter *et al.*, 1997) tried so far has had no (or little) effect.

Stoichiometric models describe the biochemical reactions in the cell as a set of algebraic equations, and can be used to simulate steady-state intracellular fluxes (Henriksen *et al.*, 1996; Patil *et al.*, 2004) (see Figure 8). The main advantages of stoichiometric models compared with kinetic models are: no kinetic information is needed, comprehensive models of cellular biochemistry are possible, and these models can be used to test pathway insertions or deletions, *in silico*, with relatively little effort (David *et al.*, 2003; Forster *et al.*, 2003). This last feature of stoichiometric models is particularly useful for metabolic engineering, as it allows a systems approach for planning directed genetic changes. A recent stoichiometric model of the central carbon metabolism by David *et al.*, (2003) is the most comprehensive effort to date for *A. niger* and includes more than 230 reactions.

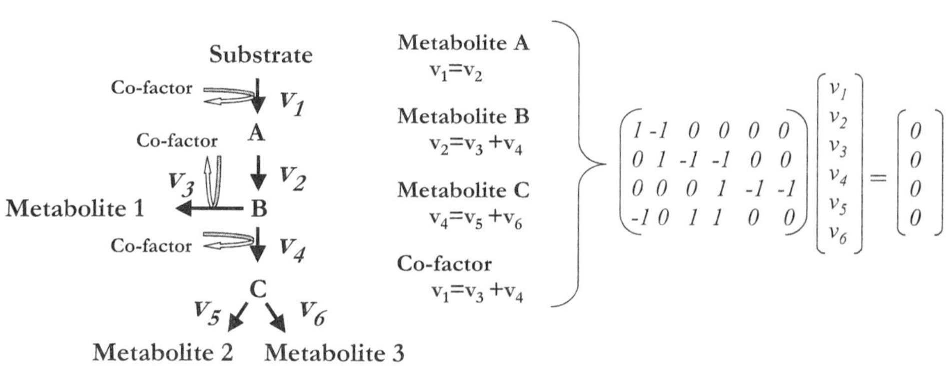

Figure 8. Example of the mathematical formula representing the stoichiometric model of a simple branched pathway.

The power of large-scale stoichiometric models have been illustrated through the work of Forster [(Forster *et al.*, 2002); (Forster *et al.*, 2003)] in *S. cerevisiae*, where it was shown to be possible to predict intracellular fluxes with high accuracy for anaerobic conditions, and predict extreme pathways and maximum theoretical yields for aerobic conditions.

Carbon-13 labeling

Combining stoichiometric models with ^{13}C-labeling enables quantitative knowledge of the intracellular fluxes (Christensen *et al.*, 2001; Christensen and Nielsen, 2000; Christensen and Nielsen, 1999), and can also lead to the elucidation of previously unknown pathways (Christensen and Nielsen, 2000). This tool plays an important role in elucidating the major physiological differences between high and low citrate producers, and has already been applied to this end in penicillin production (Christensen *et al.*, 2000). It has also extended the understanding of catabolism of co-substrates (glucose and phenoxyacetic acid (Christensen and Nielsen, 2002), and glucose and adipate (Thykaer *et al.*, 2002)) during penicillin production, and the effects of glucose concentration on glycolytic metabolism under citrate producing conditions (Peksel *et al.*, 2002).

Examples of strain improvement through metabolic engineering

The high level of secrecy surrounding industrial producer strains means that we can only focus on the work done by academia (see Table 4). To date, there have been only a few success stories for the application of metabolic engineering to citrate production in *A. niger*. The public resistance to GM products probably has a dampening effect on the application of metabolic engineering for citrate production.

The only direct improvement in citrate productivity was achieved through the increased rate of citrate accumulation in the early stages of fermentation on sucrose. Knowing that *A. niger* hexokinase is inhibited by trehalose-6-phosphate *in vitro* Panneman *et al.* (1998) later showed that it was a particularly strong inhibition) and reports that hexokinase is a flux controlling step (Panneman *et al.*, 1998), Arisan-Atac *et al.*, (1996) decided to study the *in vivo* relevance of this inhibition. Several *A. niger* strains were constructed, containing either an amplification or disruption of trehalose-6-phosphate synthase A (T6PSA)-encoding gene (*ggsA*). It was found that, although equal final concentrations were achieved, the disruptant strain accumulated citrate faster than the wild type strain, while the amplified strain led to slower accumulation. It therefore appears that decreased trehalose-6-phosphate leads to higher hexokinase activity, bearing out the *in vitro* results *in vivo*.

An excellent example of a novel metabolic engineering approach for increased penicillin production was achieved through the over-expression of a single gene, *Pc-Pex11*. Pc Pex11p is a peroxin involved in microbody abundance, which have been found to increase microbody abundance upon over-expression in, among others: yeast, *Trypanosoma brucei* and mammals (Kiel *et al.*, 2005). The rationale behind this strategy was the observation that some high-penicillin producing strains often had significantly increased microbody numbers, and it was thought that the increase in overall microbody surface area could lead to increased yields. Another fact that was taken into consideration was the importance of subcellular localization of the penicillin biosynthetic pathway enzymes (Müller *et al.*, 1992), which highlighted the apparent importance of microbodies in general for penicillin production. Over-expression of *Pc-Pex11* led to a 2.7-fold increase in penicillin production in a low producing strain (Wis54-1255) of *P. chrysogenum* (0.32 vs. 0.12 g/L after 96 h of shake flask fermentation), but significantly, it also doubled the production in DS04825 (2.03 g/L vs. 0.98 g/L after 96 h of shake flask fermentation), a *P. chrysogenum* strain which already had increased production when compared to Wis54-1255.

Decreased by-product formation is an important target for yield, titer and productivity improvements through metabolic engineering. Unlike the classical strain improvement approach, where mutation was employed as an imprecise tool for by-product reduction, metabolic engineering techniques allow for precision excision of offending genes and subsequent detailed analysis of the resulting strain. A good example of this is the cloning and deletion of the oxaloacetate hydrolyase (OAH) gene in *A. niger* by Pedersen (Pedersen *et al.*, 2000a,b). The deletion of OAH resulted in the elimination of the by-product, oxalate, which causes problems during the downstream purification of proteins. It was also shown through a carbon labelling study that such a deletion can be done without pleiotropic consequences (Pedersen *et al.*, 2000a).

The importance of morphology during citrate fermentations, and the negative influence of Mn^{2+} on the desired pelleted morphology, has been stressed earlier. Dai *et al.* (2004) used suppression subtractive hybridization to identify key genes involved in the negative effects of Mn^{2+} on pellet morphology in *A. niger*. Anti-sense expression of one of these genes (*Brsa-25*, a possible amino acid transporter) was then employed to allow pelleted growth in the presence of high Mn^{2+} concentrations.

Metabolic engineering offers the promise of extending the product range of the most frequently used platform organisms. The best example to date is the production of ad-7-ADCA by a penicillin-hyperproducing *P. chrysogenum*. The expandase gene (also known as deacetoxycephalosporin C synthase) from *Streptomyces clavuligerus* was inserted into this industrially used *P. chrysogenum* strain to expand the β-lactam ring (Crawford *et al.*, 1995). Adding adipic acid as side-chain precursor then allows ad-7-ADCA to be produced. This strain improvement led to the predominant production of ad-7ADCA in fed-batch fermentation. 7-ADCA, a cephalosporin with great market potential, can then be produced by enzymatic cleavage of the adipoyl side-chain from ad-7ADCA. The use of the platform organism *P. chrysogenum* allowed for the rapid transfer of the research to industrial application as many years of strain improvement for penicillin production could be exploited, and the process is currently being employed industrially by DSM (van de Sandt and de Vroom, 2000).

The power of large scale stoichiometric models have been illustrated through the work of Forster (Forster *et al.*, 2002; Forster *et al.*, 2003)in *S. cerevisiae*, where it was shown to be possible to predict intracellular fluxes with high accuracy for anaerobic conditions, and predict extreme pathways and maximum theoretical yields for aerobic conditions. The developments in genomics provide knowledge and techniques that can lead to better understanding of how cell factories operate. Beside genome sequencing, which by itself is potentially extremely valuable for metabolic engineering as the identification of genes and gene targeting is greatly facilitated, genomics encompass transcriptomics, proteomics, metabolomics, etc. Using the information gathered in these fields, and integrating it with mathematical models (Akesson *et al.*, 2004), it may generally lead to an improved systems approach to strain design.

Transcriptome

Aspergillus nidulans was the first filamentous fungus to have its genome sequenced. It was chosen because of its use as a model organism for genetics in filamentous fungi, and the long history of work in *A. nidulans* meant that all the tools necessary to take advantage of the genome sequence were readily available. The successful sequencing of *A. niger* was announced in a press release by DSM in 2001 and Affymetrix arrays have been created (see Table 5). A 4 x coverage of the genome sequence, provided by Integrated Genomics (Genencor), has been publicly released in 2005, and a new public sequencing project, using the Integrated Genomics sequence as starting point, has been initiated in 2005 by the Department of Energy in the USA to improve the coverage to 9 x (http://www.jgi.doe.gov/sequencing/DOEmicrobes2005.html). *A. terreus* is in the process of being sequenced (http://www.ebi.ac.uk/genomes/wgs. html) and *P. chrysogenum* is currently listed as a candidate for sequencing by the Fungal Genome Initiative, to date no set time table has been put forward. At the

Table 4. Examples of successful process enhancements through metabolic engineering

Reference	Species	Strategy	Result
		Productivity improvement	
Arisan-Atac *et al.*, 1996	*Aspergillus niger*	Trehalose-6-phosphatase synthase A deletion	**Citrate** Increased in high conc. sucrose
Theilgaard *et al.*, 2001	*Penicillium chrysogenum*	Complete penicillin biosynthetic cluster over-expression.	**Penicillin** 176% increase
Kiel *et al.*, 2005	*P. chrysogenum*	Overexpression of Pc-Pex11, leading to massive proliferation of microbody numbers.	**Penicillin** 100% increase
		Decreased by-product formation	
Pedersen *et al.*, 1999	*A. niger*	Targeted deletion of oxaloacetate hydrolyase gene.	**Oxalate** by-product eliminated
		Lower sensitivity to trace Mn^{2+}	
Dai *et al.*, 2004	*A. niger*	Target identification by suppression subtractive hybridisation, followed by antisense expression of Brsa-25.	**Morphology** and **Citrate** Less sensitive to Mn^{2+}
		Extension of product range	
Crawford *et al.*, 1995	*P. chrysogenum*	Insertion of bacterial expandase gene, and adding adipic acid to the medium.	**Ad-7-ADCA** industrial production

moment about 2,500 ESTs (Expressed Sequence Tags) are publicly available for *A. niger* (Hofmann *et al.*, 2003), and a quick search on the NCBI website (http://www.ncbi.nlm.nih.gov/entrez/) gives in the region of 200 and 100 partially or fully sequenced and annotated genes for *A. niger* and *P. chrysogenum*, respectively.

Although no global transcription analysis publications are available for *A. niger* or *P. chrysogenum* or other industrially important fungi, the possibilities for strain improvement through this approach have already been amply demonstrated. For example, a limited sized array was used to analyse *A. oryzae* growth on solid media and in liquid culture; from this array it was possible to identify several genes involved in morphology (Hofmann *et al.*, 2003). Another possible result of great value would be the identification of global signal transduction pathways and global regulators, which have been achieved in such diverse organisms as *E. coli* (Ma *et al.*, 2004a; Ma *et al.*, 2004b), *S. cerevisae* (Bro *et al.*, 2004) and *A. nidulans* (Brakhage, 1997; Litzka *et al.*, 1999b).

Table 5. Genome characteristics for *Aspergillus niger* and *A. nidulans* (adapted from Archer *et al.*, 2004)

	A. niger	***A. nidulans***
Genome size [Mb]	35.9	**30.1**
Predicted Genes	14097	**9967**
Genes with Pfam hits	5306	**4512**

Proteome

Proteomics has not yet been applied to any great extent to the elucidation or improvement of superproducers. The field will soon become more pertinent when the genomes of more superproducers are sequenced and annotated. Until then it is relevant at this stage to mention that proteomics has been applied to the determination of all cephalosporin protein targets in *E. coli* and the determination of proteins involved in reducing methicillin-resistance (Cordwell *et al.*, 2002), which could lead to eventual advances through metabolic engineering in antibiotic development.

Metabolome

"It has become clear that even a complete understanding of the state of the genes, messages, and proteins in a living system does not reveal its phenotype. Therefore, researchers have started to study the metabolome (or the metabolic complement of functional genomics)." (Villas-Boas *et al.*, 2004)

In order to define the metabolic state of an organism, increasing emphasis is being laid on the central role of mass spectrometry. An extensive review on this field and its application to phenotypical analysis is presented by Villas-Boas *et al.* (2004). Techniques, such as gas chromatography coupled with mass spectrometry have been used to determine the phenotype of otherwise null-mutants in yeast, by identifying most of the relevant metabolites in the central carbon metabolism (Villas-Boas *et al.*, 2005). Askenazi *et al.* (2003) used HPLC and electrospray MS to perform secondary metabolite profiling of the high and low producers of *A. terreus*, and was able to identify lovastatin and related monacolins, as well as a variety of (+)-geodin related compounds. This analysis coupled to gene expression data enabled the identification of novel components of the (+)-geodin biosynthetic pathway. A study on the applications of metabolite profiling of fungi by Smedsgaard and Nielsen (2005) demonstrated how direct infusion MS and HPLC with diode array detection could be used to identify novel compounds. It was also shown to be possible to differentiate between closely related fungal species and gain a deeper understanding of the phenotypic behaviour of fungal and yeast strains. Detailed phenotypical analysis plays a central role in most modern strain improvement programs, and advances in the field of metabolite profiling is therefore of great interest.

FUTURE PERSPECTIVES

As more fungal species are sequenced, and our abilities in proteomics and metabolomics are perfected, it will become possible to extend the product range of our platform organism beyond what has been done so far, and also improve on existing industrial processes.

For instance, the industrial penicillin production process still has the potential for significant yield improvements. The high relative cost of the substrate, glucose, makes yield improvements particularly attractive in penicillin production. Currently reported yields are in the range of 0.04-0.06 moles penicillin/mole glucose, while theoretically yields of up to 8-10 times higher are possible (depending on if the model employed for calculation includes the transsulfuration or direct sulfhydrylation pathway for L-cysteine formation with subsequent 6-oxopiperide-2-carboxylic acid production) (Nielsen, 1995). Obtaining improvements so close to the maximum theoretical yield is, however, unlikely in a real process as penicillin biosynthesis is indirectly coupled to other reactions in the cell (e.g., ATP dissipation during precursor metabolite production). Improvements in the range of four-to fivefold are considered possible (Nielsen, 1995), and therefore provide a strong economical driving force for continued investment in strain and process improvements.

The penicillin yield stands in stark contrast to citrate production process, which already has a yield of more than 80% (see Table 1). Here the future aim is therefore predominantly to improve process productivity. Productivities of 0.7 g/L/h-1 g/L/h have been reported (Mattey, 1992; Roehr *et al.*, 1992), and we presented an example where increased productivity was achieved for citrate production through the deletion of trehalose-6-phosphatase synthase A (Arisan-Atac *et al.*, 1996). Mathematical kinetic modelling predicts possible productivity increases in the range of 45% (if glucose transport into the cell can be doubled (Guebel and Torres-Darias, 2001) to over 500% (depending on the over-expression of 13 key enzymes and increases in enzyme concentrations (Torres *et al.*, 2000)). It is not currently practically possible to test these predictions by over-expressing so many enzymes in a balanced fashion, and many of the enzymes have not yet been identified, but in the future we can expect attempts to be made as technologies for controlled expression of genes improves.

The future importance of product range extension was also demonstrated by the transfer of the expandase gene into a hyperproducing *P. chrysogenum* strain, and the resulting high titer industrial production of ad-7ADCA. There are, however, still several pitfalls that need to be overcome before such a successful product transfer, or successful productivity increases, can be routinely repeated. For one, more complete mathematical models of the relevant hyperproducers need to be developed, which should be greatly facilitated by the public release of the *A. niger* genomic sequence. It will also be important to improve the integration of data from several "omic" techniques with large-scale microbial models, as this should in the future provide tools for a stronger systems approach to strain design.

An indication of the prominent role combinatory approaches will play in future successful strain improvement programs can be seen in the work of Askenazi *et al.* (2003). A combination of secondary metabolite profiling and transcriptional analysis techniques, coupled to statistical analysis allowed the development of a metabolic engineering strategy to obtain higher lovastatin production in *A. terreus*. Here the analysis of engineered lovastatin-producing strains, which produce greater or lesser amounts of lovastatin, enabled the identification of metabolite related genes. Particularly lovastatin-associated genes were identified, which were then used to develop a reporter-based strategy around the promoter sequences of these genes. This allowed for the rapid identification of higher producing mutants using the promoter from the lovastatin biosynthetic gene *lovF* fused to the bacterial phleomycin resistance gene (*ble*). Similar approaches could yield information on transcription factors and global regulators, which would be invaluable in finding genetic targets for productivity and titer improvement of hyperproducers.

The value of classical strain improvement methods will still remain, especially as some processes cannot currently use genetically modified organisms, but it is clear that modern methods and techniques will play an ever-greater role in the development and improvement of current and novel processes in the future.

REFERENCES

Abraham, E. P., Chain, E., Fletcher, C. M., Florey, H. W., Gardner, A. D., Heatley, N. G., and Jennings, M. A. (1941). Further observations on Penicillin. The Lancet 2:177.

Abraham, E. P. (1983). History of Beta-Lactam Antibiotics, *In* Antibiotics containing the beta-lactam structure. Demain and Solomon, Eds., Springer-Verlag, Berlin, Germany, 1983, 1.

Akesson, M., Forster, J., and Nielsen, J. (2004). Integration of gene expression data into genome-scale metabolic models. Metabolic Engineering 6:285.

Arisan-Atac, I., Wolschek, M. F., and Kubicek, C. P. (1996). Trehalose-6-Phosphate synthase A affects citrate accumulation by *Aspergillus niger* under conditions of high glycolytic flux. FEMS Microbiological Letters 140:77.

Askenazi, M., Driggers, E. M., Holtzmann, A. D., Norman, C. T., Iverson, S., Zimmer, P. D., Boers, M., Blomquist, R. P., Martinez, J. E., Monreal, W. A. *et al.* (2003). Integrating transcriptional and metabolite profiles to direct the engineering of lovastatin-producing fungal strains. Nature Biotechnology 21:150.

Barbesgaard, P., Heldt-Hansen, H. P., and Diderichsen, B. (1992). On the safety of *Aspergillus oryzae*: a review. Applied Microbiology and Biotechnology 36:569.

Brakhage, A. A. (1997). Molecular regulation of penicillin biosynthesis in *Aspergillus (Emericella) nidulans*. FEMS Microbiological Letters 148:1-10.

Bro, C., Regenberg, B., and Nielsen, J. (2004). Genome-wide transcriptional response of a *Saccharomyces cerevisiae* strain with an altered redox metabolism. Biotechnology and Bioengineering 85:269.

Chain, E., Florey, H. W., Gardner, A. D., Heatley, N. G., Jennings, M. A., Orr-Ewing, J., and Sanders, A. G. (1940). Penicillin as a chemotherapeutic agent. The Lancet 2:226.

Christensen, B., and Nielsen, J. (1999). Isotopomer analysis using GC-MS. Metabolic Engineering 1:282.

Christensen, B., Thykaer, J., and Nielsen, J. (2000). metabolic characterization of high- and low-yielding strains of *Penicillium chrysogenum*. Applied Microbiology and Biotechnology 54:212.

Christensen, B., and Nielsen, J. (2000). Metabolic network analysis of *Penicillium chrysogenum* using (13)C-Labeled glucose. Biotechnology and Bioengineering 68:652.

Christensen, B., Christiansen, T., Gombert, A. K., Thykaer, J., and Nielsen, J. (2001). Simple and robust method for estimation of the split between the Oxidative Pentose Phosphate pathway and the Embden-Meyerhof-Parnas pathway in microorganisms. Biotechnology and Bioengineering 74:517.

Christensen, B., and Nielsen, J. (2002). Reciprocal 13C-labeling: a method for investigating the catabolism of cosubstrates. Biotechnological Progress 18:163.

Christensen, L. H., Henriksen, C. M., Nielsen, J., Villadsen, J., and Egel-Mitani, M. (1995). Continuous cultivation of *Penicillium chrysogenum*. Growth on glucose and penicillin production. Journal of Biotechnology 42:95.

Conejeros, R., and Vassiliadis ,V. S. (2000). Dynamic Biochemical reaction process analysis and pathway modification predictions. Biotechnology and Bioengineering 68:285.

Connor, J. M. (1998). The Global citric acid conspiracy: Legal-economic lessons. Agribusiness 14:435.

Cordwell, S. J., Larsen, M. R., Cole, R. T., and Walsh, B. J. (2002). Comparative proteomics of *Staphylococcus aureus* and the response of methicillin-resistant and methicillin-sensitive strains to triton X-100. Microbiology 148:2765.

Crawford, L., Stepan, A. M., McAda, P. C., Rambosek, J. A., Conder, M. J., Vinci, V. A., and Reeves, C. D. (1995). Production of cephalosporin intermediates by feeding adipic acid to recombinant *Penicillium chrysogenum* strains expressing ring expansion activity. Biotechnology (N.Y.) 13:58.

Dai, Z., Mao, X., Magnuson, J. K., and Lasure, L. L. (2004). Identification of genes associated with morphology in *Aspergillus niger* by using suppression subtractive hybridization. Applied and Environmental Microbiology 70:2474.

David, H., Akesson, M., and Nielsen, J. (2003). Reconstruction of the central carbon metabolism of *Aspergillus niger*. European Journal of Biochemistry 270:4243.

Demain, A. L. and Elander, R. P. (1999). The Beta-Lactam antibiotics: Past, present, and future. Antonie van Leeuwenhoek 75:5-19.

Elander, R. P. (2003). Industrial production of Beta-lactam antibiotics. Applied Microbiology and Biotechnology 61:381-392.

Forster, J., Gombert, A. K., and Nielsen, J. (2002). A functional genomics approach using metabolomics and in silico pathway analysis. Biotechnology and Bioengineering 79:703.

Forster, J., Famili, I., Palsson, B. O., and Nielsen, J. (2003). Large-scale evaluation of in silico gene deletions in *Saccharomyces cerevisiae*. OMICS 7: 193.

Guebel, D. V., and Torres-Darias, N. V. (2001). Optimization of citric acid production by *Aspergillus niger* through a metabolic flux balance model, Electronic Journal of Biotechnology 4.

Gupta, S., and Chandra, B. S. (2002). Biochemical studies of citric acid production and accumulation by *Aspergillus niger* mutants. World Journal of Microbiology & Biotechnology 18:379.

Henriksen, C. M., Christensen, L. H., Nielsen, J., and Villadsen, J. (1996). Growth energetics and metabolic fluxes in continuous cultures of *Penicillium chrysogenum*. Journal of Biotechnology 45:149.

Hobby, G. L. (1985). Penicillin. Meeting the Challenge. Yale University Press, New Haven, 1985.

Hofmann, G., McIntyre, M., and Nielsen, J. (2003). Fungal genomics beyond *Saccharomyces cerevisiae*? Current Opinion in Biotechnology 14:226.

Ikram-ul-Haq, Khurshid, S., Ali, S., Ashraf, H., Qadeer, M.A., and Ibrahim Rajoka, M. (2001). Mutation of *Aspergillus niger* for hyperproduction of citric acid from black strap molasses. World Journal of Microbiology and Biotechnology 17:35.

Ikram-Ul Haq, A. S., and Iqbal, J. (2003). Direct production of citric acid from raw starch by *Aspergillus niger*. Process Biochemistry 38:921.

Karaffa, L., and Kubicek, C. P. (2003). *Aspergillus niger* citric acid accumulation: Do we understand this well working black box? Applied Microbiology and Biotechnology 61:189.

Kiel, J. A. K. W., Klei, I. J. van der, Berg, M. A. van den, Bovenberg, R. A. L., Veenhuis, M. (2005). Overproduction of a single protein, Pc-Pex11p, results in 2-fold enhanced penicillin production by *Penicillium chrysogenum*. Fungal Genetics and Biology 42:154.

Kirimura, K., Sarangbin, S., Rugsaseel, S., and Usami, S. (1992). Citric acid production by 2-deoxyglucose-resistant mutant strains of *Aspergillus niger*. Applied Microbiology and Biotechnology 36:573.

Kubicek-Pranz, E. M., Mozelt, M., Rohr, M., and Kubicek, C. P. (1990). Changes in the concentration of fructose 2,6-bisphosphate in *Aspergillus niger* during stimulation of acidogenesis by elevated sucrose concentration. Biochimica et Biophysica Acta 1033:250.

Leangon, S., Maddox, I. S., and Brooks, J. D. (1999). Influence of the glycolytic rate on the production of citric acid and oxalic acid by *Aspergillus niger* in solid state fermentation. World Journal of Microbiology and Biotechnology 15:493.

Lein, J. (1986). The Panlabs penicillin strain improvement program. *In* Overproduction of microbial metabolites. Strain improvement and process control strategies (Vanek, Z., and Hostalek, Z., eds.), Butterworths Publishers, Stoneham, Chapter 6.

Lesniak, W., Pietkiewicz, J., and Podgorski, W. (2002). Citric acid fermentation from starch and dextrose syrups by trace metal resistant mutant of *Aspergillus niger*. Biotechnology Letters 24:1065.

Litzka, O., Bergh, K., Brulle, J. van den, Steidl, S., and Brakhage, A. A. (1999). Transcriptional control of expression of fungal beta-lactam biosynthesis genes. Antonie Van Leeuwenhoek 75:95.

Ma, H. W., Buer, J., and Zeng, A. P. (2004a). Hierarchical structure and modules in the *Escherichia coli* transcriptional regulatory network revealed by a new top-down approach. BMC Bioinformatics 5:199.

Ma, H. W., Kumar, B., Ditges, U., Gunzer, F., Buer, J., and Zeng, A. P. (2004b). An extended transcriptional regulatory network of *Escherichia coli* and analysis of its hierarchical structure and network motifs. Nucleic Acids Research 32:6643.

Mattey, M. (1992). The Production of Organic Acids, Critical Reviews in Biotechnology 12:87.

Müller, W. M., Bovenberg, R. A. L., Groothuis, M. H., Kattevilder, R., Smaal, E. B., Voort, L. H. M. van der, and Verkleij, A. J. (1992). Involvement of microbodies in penicillin biosynthesis. Biochimica et Biophysica Acta 1116:210.

Newbert, R. W., Barton, B., Greaves, P., Harper, J., and Turner, G. (1997). Analysis of a commercially Improved *Penicillium chrysogenum* strain series: Involvement of recombinogenic regions in amplification and deletion of the penicillin biosynthesis gene cluster. Journal of Industrial Microbiology and Biotechnology 19:18.

Nielsen, J. (1995). Physiological Engineering Aspects of *Penicillin chrysogenum*. Doctor Technices, Technical University of Denmark, 1995.

Panneman, H., Ruijter, G. J., van den Broeck, H. C., and Visser, J. (1998). Cloning and biochemical characterisation of *Aspergillus niger* hexokinase—the enzyme is strongly inhibited by physiological concentrations of trehalose-6-phosphate. European Journal of Biochemistry 258:223.

Papagianni, M., and Mattey, M. (2004). Modeling the mechanisms of glucose transport through the cell membrane of *Aspergillus niger* in submerged citric acid fermentation processes. Biochemical Engineering Journal 20:7.

Parekh, S., Vinci, V. A., and Strobel, R. J. (2000). Improvement of microbial strains and fermentation processes. Applied Microbiology and Biotechnology 54:287.

Patil, K. R., Akesson, M., and Nielsen, J. (2004). Use of genome-scale microbial models for metabolic engineering. Current Opinion in Biotechnology 15:64.

Pedersen, H., Christensen, B., Hjort, C., and Nielsen, J. (2000a). Construction and characterization of an oxalic acid nonproducing strain of *Aspergillus niger*. Metabolic Engineering 2:34.

Pedersen, H., Hjort, C., and Nielsen, J. (2000b). Cloning and characterization of *Oah*, the gene encoding oxaloacetate hydrolase in *Aspergillus niger*. Molecular and General Genetics 263:281.

Peksel, A., Torres, N. V., Liu, J., Juneau, G., and Kubicek, C. P. (2002). 13C-NMR Analysis of glucose metabolism during citric acid production by *Aspergillus niger*. Applied Microbiology and Biotechnology 58:157.

Penalva, M. A., Rowlands, R. T., and Turner, G. (1998). The optimization of penicillin biosynthesis in fungi. Trends in Biotechnology 16:483.

Promper, C., Schneider, R., and Weiss, H. (1993). The role of the proton-pumping and alternative respiratory chain NADH:ubiquinone oxidoreductases in overflow catabolism of *Aspergillus niger*. European Journal of Biochemistry 216:223.

Roehr, M., Kubicek, C. P., and Kominek, J. (1992). Industrial acids and other small molecules. Biotechnology 23:91.

Roehr, M. and Kubicek, C. P. (1996). Further organic acids. *In* Biotechnology, Vol. 6, 2nd edition (Roehr, M., ed.), VCH Verlagsgesellschaft mbH, Weinheim, Germany, Chapter 11.

Rugsaseel, S., Kirimura, K., and Usami, S. (1996). Citric acid accumulation by cycloheximide-sensitive mutant strains of *Aspergillus niger*. Applied Microbiology and Biotechnology 45:28.

Ruijter, G. J., Panneman, H., and Visser, J. (1997). Overexpression of phosphofructokinase and pyruvate kinase in citric acid-producing *Aspergillus niger*. Biochimica et Biophysica Acta 1334:317.

Ruijter, G. J., Vondervoort van der, P. J., and Visser, J. (1999). Oxalic acid production by *Aspergillus niger*: an oxalate-non-producing mutant produces citric Acid at pH 5 and in the presence of manganese. Microbiology 145:2569.

Ruijter, G. J., Panneman, H., Xu, D., and Visser, J. (2000). Properties of *Aspergillus niger* citrate synthase and effects of *citA* overexpression on citric acid production. FEMS Microbiology Letters 184: 35.

Ruijter, G. J. G., Kubicek, C. P., and Visser, J. (2002). Production of Organic Acids by Fungi. *In* The Mycota X. Industrial Applications. (Osiewacz, H. D., ed.), Springer-Verlag, Berlin, Heidelberg, p213.

Sandt, E. J. A. X. van de, and Vroom, E. de (2000). Innovations in cephalosporin and penicillin production: Painting the antibiotics industry green. Chemica Oggi-ChemistryToday 18:72.

Schreferl, G., Kubicek, C. P., and Röhr, M. (1986). Inhibition of citric acid accumulation by manganese ions in *Aspergillus niger* mutants with reduced citrate controle of phosphofructokinase, Journal of Bacteriology165:1019–1022.

Schreferl-Kunar, G., Grotz, M., Roehr, M., and Kubicek, C. P. (1989). Increased citric acid production by mutants of *Aspergillus niger* with increased glycolytic capacity. FEMS Microbiology Letters 59:297.

Schuster, E., Dunn-Coleman, N., Frisvad, J. C., and Dijck, P. W. van (2002). On the safety of *Aspergillus niger* - a Review. Applied Microbiology and Biotechnology 59:426.

Shuler, M. L., and Kargi, F. (1992). Bioprocess Engineering: Basic Concepts. Prentice-Hall, New Jersey, 1992.

Smedsgaard, J., and Nielsen, J. (2005). Metabolite profiling of fungi and yeast: from phenotype to metabolome by MS and Informatics. Journal of Experimental Botany 56:273.

Steinbach, W. J., Benjamin, D. K. Jr., Kontoyiannis, D. P., Perfect, J. R., Lutsar, I., Marr, K. A., Lionakis, M. S., Torres, H. A., Jafri, H., and Walsh, T. J. (2004). Infections due to *Aspergillus terreus*: a multicenter retrospective analysis of 83 cases. Clinical Infectious Diseases 39:192.

Stephanopoulos, G. N., Aristidou, A. A., and Nielsen, J. (1998). Metabolic Engineering: Principles and Methodologies. Academic Press, San Diego, 1998.

Theilgaard, H., Berg, M. van den, Mulder, C., Bovenberg, R., and Nielsen, J. (2001). Quantitative analysis of *Penicillium chrysogenum* Wis54-1255 transformants overexpressing the penicillin biosynthetic genes. Biotechnology and Bioengineering 72:379.

Thykaer, J., Christensen, B., and Nielsen, J. (2002). Metabolic network analysis of an adipoyl-7-ADCA-producing strain of *Penicillium chrysogenum*: Elucidation of adipate degradation. Metabolic Engineering 4:151.

Thykaer, J., and Nielsen, J. (2003). Metabolic engineering of beta-lactam production. *Metabolic Engineering* 5:56.

Torres, N. V., Voit, E., and Gonzalez-Alcon, C. (1996). Optimization of nonlinear biotechnological processes with linear programming: Application to citric acid production by *Aspergillus niger*. Biotechnology and Bioengineering 49:247.

Varez-Vasquez, F., Gonzalez-Alcon, C., and Torres, N. V. (2000). Metabolism of citric acid production by *Aspergillus niger:* Model definition, steady-state analysis and constrained optimization of citric acid production rate. Biotechnology and Bioengineering 70:82.

Villas-Boas, S. G., Mas, S., Akesson, M., Smedsgaard, J., and Nielsen, J. (2004). Mass spectrometry in metabolome analysis. Mass Spectrometry Reviews 24:613-646.

Villas-Boas, S. G., Moxley, J. F., Akesson, M., Stephanopoulos, G., and Nielsen J. (2005). High-throughput metabolic state analysis: the missing link in integrated functional genomics of yeasts. Biochemical Journal 388:669-677.

Vournakis, J. N. and Elander, R. P. (1983). Genetic manipulation of antibiotic-producing microorganisms. Science 219:703.

Wallrath, J., Schmidt, M., and Weiss, H. (1991). Concomitant loss of respiratory chain NADH:ubiquinone reductase (Complex I) and citric acid accumulation of *Aspergillus niger*. Applied Microbiology and Biotechnology 36:76.

Chapter 10

Hyperproduction of enzymes by fungi

Han A. B. Wösten[1]*, Karin Scholtmeijer[2] and Ronald P. de Vries[1]
[1]*Microbiology, Institute of Biomembranes, University of Utrecht, Padualaan 8, 3584 CH Utrecht, The Netherlands;* [2]*BiOMade Technology, Nijenborgh 4, 9747 AG Groningen, The Netherlands.*

INTRODUCTION

Fungi play a crucial role in nature by degrading organic material. They secrete a wide variety and large amount of enzymes that cleave complex substrates into small products that can be taken up by the cell. The filamentous mode of growth of mycelial fungi contributes to the efficacy of degradation. By growing at their tips, hyphae can penetrate the substrate. Moreover, the hypha exposes a large surface that contacts the substrate, which may thereby facilitate uptake of nutrients. Yeast-like fungi do not effectively penetrate organic material. Their mode of growth only allows superficial colonization.

The capacity of fungi to secrete proteins is used by the industry to produce large amounts of homologous and heterologous proteins. Fungal enzymes are applied in the food and feed industry as well as in other industrial sectors, such as the paper and pulp industry. Fungi are also interesting hosts for producing therapeutic proteins of animal or human origin, especially since the first steps of glycosylation of proteins in fungi and animals are similar (Gerngross, 2004). Genetic engineering of the yeast *Pichia pastoris* is now one step away from completion to produce proteins with a glycosylation pattern identical to that found in humans (Hamilton *et al.*, 2003; Bobrowicz *et al.*, 2004).

In this review we will focus on the production of fungal enzymes that are used in the food and feed industry. The different fungal hosts that are used and the strategies to improve protein production are discussed.

FUNGAL ENZYMES USED BY THE FOOD AND FEED INDUSTRY

Many fungal enzymes are being used in the food and feed industry (Table 1) and this number is continuously increasing. Plant material is the major source for the food and feed sector. It mainly consists of the cell wall polysaccharides cellulose, hemicellulose and pectin (McNeill *et al.*, 1984). In addition, protein levels are relatively high. Therefore, application of plant polysaccharide degrading enzymes and proteases is most important and widespread. The enzymes needed for processing of the plant material depend on the nature of the plant source and the desired effect. In most industrial applications enzyme mixtures rather than pure enzymes are used. These mixtures are cheaper than purified enzymes because they require less downstream processing (de Vries and Visser, 2001; de Vries, 2003). Moreover, the desired effect (e.g., clarification of fruit juice) can often only be obtained by the combined action of a number of enzymes (Grassin and Fauquembergue, 1996).

Table 1. Fungal enzymes used in food and feed applications. More details can be found in the *Handbook of Food Enzymology* (Whitaker *et al.*, 2003).

Enzyme	Source	Application(s)	References
α-Amylase	*Aspergillus niger, A. oryzae, Rhizopus oryzae*	Preparation of starch syrup and dextrose; preparation of alcohol and beer.	Wong and Robertson, 2003
Arabinofuranosidase/ endoarabinase	*A. niger, A. aculeatus*	Preparation and clarification of fruit juices.	de Vries and Visser, 2003
Catalase	*A. niger*	Preservation of colour, texture, flavour, taste and aroma of frozen foods.	Diehl *et al.*, 1936
Cellulase	*A. niger, Trichoderma reesei, T. viride, A. aculeatus*	Brewing and baking; wine and juice production; isolation of starch and other polysaccharides; improvement of digestibility of feed.	Johnston, 2003; Tenkanen *et al.*, 2003
Feruloyl esterase	*A. niger, A. tubingensis, Neocallimastix, Penicillium funiculosum, A. oryzae*	Release of ferulic acid for vanillin production.	Williamson, 1998
Galactanase	*A. niger, A. aculeatus, A. tubingensis*	Extraction of fruit juices.	de Vries and Visser, 2003
Glucoamylase	*A. niger, A. oryzae, R. oryzae, R. niveus, R. delemar*	Saccharification of steamed rice and potato; preparation of glucose syrup.	Reilly, 2003
Glucose oxidase	*A. niger*	Removal of residual glucose or oxygen to increase shelf life; flavour and colour stability; reduction of alcohol percentage in wine, production of H_2O_2..	Frederick *et al.*, 1990; Kapat *et al.*, 2001; Ohlmeyer *et al.*, 1957; Pickering, 1998; Vermulapalli *et al.*, 1998; Malherbe *et al.*, 2003
Laccase	*Pycnoporus cinnabarinus, P. versicolor*	Biosensor for determination of polyphenols/catechols in tea; removal of phenols in apple juice and wine.	Ghindilis *et al.*, 1992; Piacquadio *et al.*, 1997; Brenna and Bianchi, 1994
B-Galactosidase/ Lactase	*A. niger, S. cerevisiae, Candida pseudotropicalis, K. lactis*	Hydrolysis of lactose.	Mahoney, 2003
Lipase	*A. niger, A. oryzae, hizomucor miehei, C. rugosa, C. lipolytica, R. delemar, R. oryzae, R. niveus, P. roqueforti,P. camemberti, Mucor javanicus*	Lipid hydrolysis; manufacture of cheese, cheese flavours and other dairy products; modification of lipids; development of flavours in processed food.	Wong, 2003
Mannanase	*A. niger, A. aculeatus, T. reesei, T. harzianum*	Improving animal feed; treatment of extracted coffee beans; reducing the viscosity of pineapple juice.	Stålbrand, 2003
Pectic esterase	*A. niger*	Fruit juice extraction and clarification; preparation of specific pectins.	Benen *et al.*, 2003
Phytase	*A. niger, A. oryzae, A. melleus, R. niveus, R. oryzae*	Degradation of phytate in animal feed; starch processing.	Misset, 2003

Polygalacturonase	*A. niger, K. lactis, S. cerevisiae*	Extraction and clarification of fruit juices; production of pulpy nectars; production of pectin preparations.	Benen and Visser, 2003
Proteolytic enzymes	*P. chrysogenum, A. niger, M. pusillus, Rhizomucor miehei, S. cerevisiae, Kluyveromyces lactis*	Softening of doughs; improvement of texture, elasticity and volume of bread; brewing; production of miso and tofu; flavour development in cheese; improving digestibility of animal feeds; preparation of soy bean milk; preparation of dehydrated soups; clarification of wine.	Whitaker, 2003
Xylanolytic enzymes	*A. niger, T. longibrachiatum, Disporotrichum dimorphosporum, T. reesei*	Release of xylose for conversion into ethanol; production of xylo-oligosaccharides (food-additives); improvement of digestibility of feed; preparation of baking products; clarification of fruit juices.	Biely, 2003

FUNGAL HOSTS FOR PROTEIN PRODUCTION

Hosts used in industry

Both filamentous fungi and yeasts are used in industry to produce enzymes.

In general, filamentous fungi secrete more protein than yeasts. Production levels of the former can be as high as 30 gr l^{-1}, while yeasts produce usually tenfold less. On the other hand, growth of yeasts in fermentors is easier and results in a less viscous medium.

Most of the enzymes produced in yeast have been expressed in *Saccharomyces*. This choice was largely based on the fact that this yeast has a long history in the brewing and baking industry and is thus considered a GRAS (generally regarded as safe) organism. Moreover, molecular biology, genetics and physiology of *Saccharomyces* are well established. In fact, *S. cerevisae* can be considered the best studied eukaryote. However, other yeasts like *P. pastoris, Yarrowia lipolytica* and *Kluyveromyces lactis* give higher production yields than *S. cerevisae* and therefore have attracted interest from the industry. Nowadays, a number of these so-called non-conventional yeasts can be genetically modified, and grown in large scale fermentors (Buckholz and Gleeson, 1991).

Several filamentous fungi are used in the industry including *Rhizomucor* (e.g., *R. miehei*), *Rhizopus* (e.g., *R. oryzae* and *R. niveus*), *Aspergillus* (e.g., *A. niger* and *A. oryzae*) and *Trichoderma reesei* (Pariza and Johnson, 2001). The latter two are used most widely. It should be noted that some strains of these organisms are known to produce mycotoxins. For instance, some *A. niger* isolates produce ochratoxin A (Blumenthal, 2004). The Environmental Protection Agency (1997a; 1997b) concluded that mycotoxin production in *A. niger* and *A. oryzae* can be avoided by controlling the fermentation conditions. Their fermentation products are therefore regarded as safe.

Novel hosts for protein production

It is unlikely that one single host is equipped to efficiently produce all enzymes of industrial interest. Therefore, new hosts should be introduced to increase the number of proteins that can be produced at an industrial scale. When developing a new host for production of (food) enzymes several requirements have to be met. Firstly, the fungal species should be amenable to genetic engineering. Secondly, the strain should be tested for the production of any toxicologically significant amount of mycotoxin such as aflatoxin B1, ochratoxin A, sterigmatocystin, T-2 toxin (a trichothecene toxin) or zearalenone (Blumenthal, 2004). Finally, fermentation technology should be developed and media optimized. Clearly, this is a very costly process and therefore new production platforms (hosts) are only sporadically introduced.

Fusarium graminearum A 3/5 (reclassified as *F. venenatum*) is the Quorn® mycoprotein fungus. This strain is an interesting host for protein production. Unlike many other species of *Fusarium* it produces no detectable amounts of mycotoxin and is therefore safe as a food source (Trinci, 1992). Moreover, its growth in fermentors has been well characterized (Trinci, 1992) and it can be genetically engineered (Royer *et al.*, 1995). Finally, the low background levels of proteases and other secreted proteins are advantageous for production of relatively pure enzymes. The production of an active tryptic protease of *Fusarium oxysporum* was only low when expressed in *A. oryzae*, possibly due to the absence of (a) protease(s) required for processing of the enzyme. However, the tryptic protease was successfully produced in *F. graminearum* A 3/5 (Royer *et al.*, 1995). Also a fungal lipase, a cellulase and a carboxypeptidase were produced in this strain (Royer *et al.*, 1995; Blinkovsky *et al.*, 1999).

So far only zygomycetes and ascomycetes are used by the industry. However, homobasidiomycetes offer great potential for production of industrial proteins as well. They secrete enzymes in their culture media with activities or amounts unsurpassed in other fungi. For instance, homobasidiomycetes produce various metalloenzymes such as laccase, and lignin and manganese peroxidase. Until now, expression of basidiomycete metalloenzymes in ascomycete production systems such as *Aspergillus* and *T. reesei* have met limited success (see Conesa *et al.*, 2001a). Therefore, it is an interesting option to develop basidiomycetes as hosts for large-scale protein production. *Schizophyllum commune* and *Pycnoporus cinnabarinus* have been studied as candidate basidiomycete production systems. They are amenable to genetic engineering (Schuren and Wessels, 1994; Scholtmeijer *et al.*, 2001; Alves *et al.*, 2004). Moreover, the *GPD* and *SC3* promoters of *S. commune* (Wessels *et al.*, 1987; Schuren and Wessels, 1994; Alves *et al.*, 2004) and the *lac1* laccase promoter of *P. cinnabarinus* (Alves *et al.*, 2004) can be used for high-level gene expression. Gram levels of laccase per liter were produced in *P. cinnabarinus* using the *GPD* or the *lac1* promoter. These production levels are at least tenfold higher than other fungal production systems. It is not yet known whether *S. commune* has the capacity to produce these levels of protein. However, it has been reported that wild-type strains secrete up to 60 mg L^{-1} of SC3 hydrophobin (Wösten *et al.*, 1999).

Strategies to improve protein production

Efficient production of a secreted protein requires high mRNA levels of the encoding gene as well as machinery to translate the mRNA and to fold and secrete the protein. The secretion pathway of proteins is complex and involves several cellular compartments. In *S. cerevisae* more than 300 genes are known to function in the secretory process. Ultrastructural data suggest that the processes in filamentous fungi and in yeast are essentially the same (Punt *et al.*, 1994; Archer and Peberdy, 1997; Gouka *et al.*, 1997). Secretion of proteins involves several processes. The mRNA is translated and the resulting protein translocated across the endoplasmic reticulum (ER). This is accompanied by the removal of the signal peptide from the newly synthesized protein. The protein is folded in the ER lumen, which involves the combined action of chaperones (e.g., binding protein (BiP) and calreticulin/calnexin/UDP-Glc transferase complex) and foldases (e.g., protein disulphide isomerase (PDI) and peptidyl prolyl *cis-trans* isomerases). The chaperones and foldases also prevent transport of uncorrectly folded proteins to the Golgi. Incorrectly folded proteins are recognized and are degraded by proteases in the cytoplasm. As in other eukaryotes, this process probably involves ubiquitin labeling of the protein (Coux *et al.*, 1996) and transport from the ER lumen to the cytoplasm (Wiertz *et al.*, 1996). Correctly folded proteins are transported from the ER to the Golgi equivalent (a recognizable Golgi apparatus is absent in most fungi), followed by transport to the plasmamembrane via vesicles. It is believed that the exocyst, consisting of a complex of 8 proteins, tethers secretory vesicles to the plasma membrane in preparation for membrane fusion (Guo *et al.*, 2000), with subsequent release of the proteins into the cell wall. The proteins may then be carried to the outside of the wall by the flow of wall constituents at the growing hyphal apex (Wessels, 1993) and released into the medium. The complexity of the system enabling the journey of secreted proteins through the secretory pathway is illustrated by recent data indicating that the exocyst and the translocon in the ER communicate (reviewed in Guo and Novick, 2004). This communication would maintain the balance between protein synthesis and secretion. For instance, expression of several ribosomal proteins is severely reduced when secretion is malfunctioning (Mizuta and Warner, 1994).

Increasing chymosin production in *A. niger*

The glucoamylase (*glaA*) promoter was used to express the bovine chymosin cDNA in *A. niger*. The signal sequence of glucoamylase or that of chymosin was used to direct the protein to the secretion pathway. The low production levels of about 15 mg L^{-1} were shown not to be caused by mRNA levels (Ward, 1989a,b) but by inefficient secretion and probable degradation (Ward *et al.*, 1990). Secretion of chymosin was increased tenfold by fusing the chymosin cDNA to the last codon of *glaA* of *A. niger* (Ward *et al.*, 1990). In this case the chymosin was cleaved autocatalytically from the fusion protein. By introducing the glucoamylase-chymosin fusion into an *A. niger* strain in which an aspartyl protease gene had been deleted (Berka *et al.*, 1990) production levels were increased to 250 mg L^{-1} (Dunn-Coleman *et al.*, 1991). This strain was subjected to mutagenesis and mutants were selected by a robotic screening procedure (Dunn-Coleman *et al.*, 1991). In this way mutants were obtained secreting more than 1 gr L^{-1} of chymosin.

In a number of cases protein production has been strongly improved (for example see the **text box**). However, increasing production yields is a slow process for most proteins. It is impossible to predict the bottleneck(s) in production but limitations may occur at the transcriptional and/or the (post)-translational level. Several strategies have been developed to overcome these limitations including: (1) the introduction of multiple copies of the gene encoding the secreted protein or its regulator(s), (2) targeting of constructs into positions in the genome that have a high transcriptional activity, (3) the use of strong promoters and efficient secretion signals, (4) optimization of codon usage, (5) introduction of introns in cDNA sequences, (6) gene fusion strategies in which the gene of interest is fused at the 3' end of a gene encoding a well-secreted homologous protein, (7) introduction of N-glycosylation sites, (8) increasing expression of genes of the unfolded protein response (UPR), (9) use of protease-deficient host strains, and (10) optimization of production medium and development of new fermentation technologies. These strategies will be discussed in the next sections (see also Figure 1).

Optimization of heterologous protein production at the transcriptional level

The most straightforward approach to enhance the transcriptional level of a gene is to increase its copy number. In *S. cerevisae* these copies can be maintained on self-replicating vectors. However, in case of most other fungi these copies have to integrate into the genome. Analysis of *A. niger* multicopy transformants showed a gene dosage dependent expression of glucoamylase up to about 20 copies (Verdoes *et al.*, 1993). Studies with the glucoamylase promoter fused to a reporter gene suggested that transcription in multi-copy transformants was limited due to titration of regulatory proteins (Verdoes *et al.*, 1994a). Limiting availability of regulatory proteins was also suggested to hamper expression of an endochitinase from a *cbh1* promoter in *T. reesei* (Margolles-Clark *et al.*, 1996). This was based on the observation that endogenous expression of *cbh1* was reduced when 10 copies of the *cbh1*::endochitinase construct were introduced.

Other data suggested that the presence of three copies of the *cbh1* promoter would already titrate regulatory proteins or transcription factors (Karhunen *et al.*, 1993). This kind of

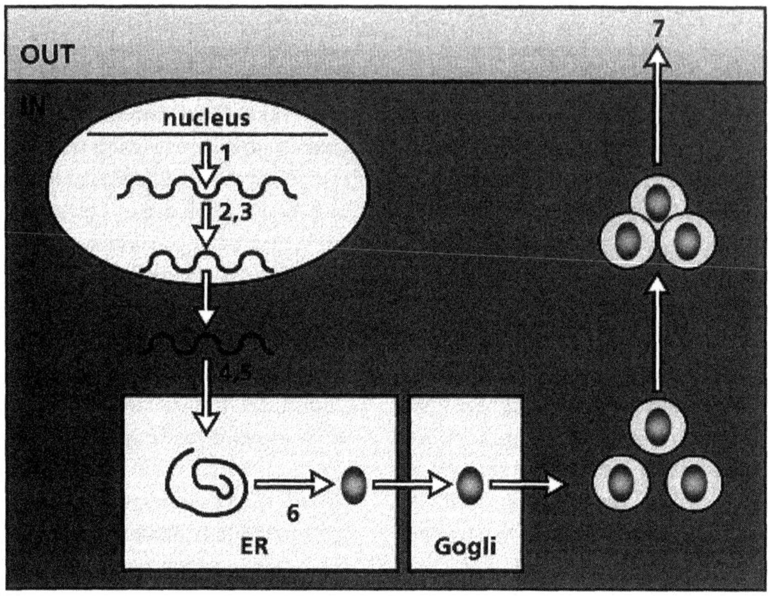

Figure 1: Protein production in fungi has been improved using a variety of strategies. Transcription of the gene encoding the secreted protein has been increased by introducing multiple copies in the host (**1**). A similar approach was followed for regulatory genes. Targetting of transgenes to highly active transcriptional regions and use of strong promoters have also been used to improve transcription. Premature termination of transcripts (**2**) has been overcome by removing polyadenylation signals within coding sequences, while stability of mRNA could be increased in some cases by introducing introns in cDNA sequences (**3**). Translation has been increased by optimising codon usage (**4**). Fusing the protein of interest to a well-secreted protein improved translocation across the ER membrane (**5**) as well as folding in the ER lumen (**6**). Folding was also facilitated in some cases by introduction of N-glycosylation sites, or by over-expressing the unfolded protein response pathway or its components such as foldases or chaperones. Improving protein production by targeting transport from ER to Golgi or from Golgi to the plasma membrane has not yet been reported. However, reduced activity of proteases in the medium has been reported to improve production levels of secreted proteins (**7**).

promoter competition has also been reported for the *GAL* system in *S. cerevisae*. In this case it was overcome by over-expression of the transcriptional activator GAL4 (Schulz *et al.*, 1987).

Although introduction of multiple copies should result in over-expression of a gene, in some cases expression is reduced or even absent. Upon introduction of more than one copy of the *SC3* gene in a wild-type strain of *S. commune*, both the endogenous and introduced *SC3* genes were silenced.

Silencing was shown to act at the transcriptional level accompanied by methylation of the DNA (Schuurs *et al.*, 1997). Similarly, *SC3* was silenced in *P. cinnabarinus* although in this case more copies had to be introduced to induce the phenomenon (AMCR Alves and HAB Wösten, unpublished). Silencing of genes due to the introduction of multiple copies has also been described in other fungi. The best studied examples are repeat-induced point mutation (RIP) in *Neurospora crassa*, methylation induced premeiotically (MIP) in *Ascobolus immerses* (see Irelan and Selker, 1996), and quelling in *N. crassa* (Romano and Macino, 1992). In MIP, DNA is silenced by a process of homology-based methylation. In case of RIP both the introduced and the endogenous gene are mutated, which is often accompanied by methylation of the DNA. Both RIP and MIP occur during the sexual stage of the fungus. In contrast, quelling, which is also known as RNA interference (RNAi), occurs in the vegetative mycelium. Little is known about gene silencing in industrial fungi. RIP has been reported in *A. nidulans* and *A. fumigatus* (see Galagan and Selker, 2004) and it cannot be excluded that homology-based gene silencing also occurs in *Aspergilli* that are used in the industry.

The site of integration affects expression of the introduced gene (Ward *et al.*, 1990; Verdoes *et al.*, 1993; 1994b). The effect of random integration can be overcome by targeting the expression vector to loci of known high transcriptional activity, like those of the *glaA* and *cbhI* genes in *A. niger* and *T. reesei*, respectively. Harkki *et al.* (1991) reported that in the best performing transformants the expression constructs for endoglucanase I (EGI) were integrated into the *cbhI* locus.

Limitation of heterologous protein production at the transcriptional level can also be caused by incorrect processing of pre-mRNA and/or a low mRNA stability. A truncated transcript was observed when wild-type α-galactosidase (*aglA*) of *Cyamopsis tetragonoloba* (guar, a beanlike plant) was expressed in *A. niger* (Gouka *et al.*, 1996). The truncated transcript was also observed in *A. nidulans* but not in *Hansenula polymorpha* (Fellinger *et al.*, 1991), *S. cerevisiae* (Verbakel, 1991) and *K. lactis* (Bergkamp *et al.*, 1992). A full-length *aglA* transcript was produced in *A. niger* by replacing an AT-rich sequence in the *aglA* gene with a more GC-rich sequence ($aglA_{syn}$) (Gouka *et al.*, 1996). Incorrect processing of heterologous genes resulting in truncated mRNAs also occurs in *P. pastoris* (Scorer *et al.*, 1993) and *S. commune* (Schuren and Wessels, 1998). Production of full length mRNAs of the *hygromycin B resistance* gene in the latter fungus was restored by increasing the GC content in an AT rich region (Scholtmeijer *et al.*, 2001). The results suggest that AT-rich sequences in the coding region of heterologous genes result in premature polyadenylation of the mRNA. The mechanisms for recognizing polyadenylation signals (or AU-rich stretches in transcripts) are unclear and may vary between organisms and genes to be expressed. Therefore, the occurrence of premature termination cannot be predicted but should be considered when expression of the target gene is absent.

The absence of introns in or near the coding sequence of a gene was shown to cause mRNA instability in *S. commune* (Lugones *et al.*, 1999; Scholtmeijer *et al.*, 2001). Intron dependent mRNA accumulation was also observed in the ascomycete *Podospora anserina* (Dequard-Chablat and Rötig, 1997) and the basidiomycete *Phanerochaete chrysosporium* (Ma *et al.*, 2001). For all examined genes, accumulation was low, if present at all, when a cDNA was expressed. However, accumulation increased dramatically upon introduction of gDNA or cDNA containing natural or artificial intron(s). The phenomenon of intron-dependent mRNA accumulation may be more widespread, but seems not of general occurence. In the industrial relevant fungi *A. niger* and *T. reesei,* at least some cDNAs are efficiently expressed (Roberts *et al.*, 1989; Aifa *et al.*, 1999; Conesa *et al.*, 2001a; Rose and van Zyl, 2002).

Optimization of heterologous protein production at the (post)-translational level

Fusion of the gene of interest behind a highly expressed homologous gene can resolve limitations at early stages in the secretion pathway. Using glucoamylase of *A. niger* or cellobiohydrolase (CBHI) of *T. reesei* as a carrier increased levels of secreted protein 5-1000 fold to 5-250 mg L^{-1} (Ward *et al.*, 1990; Contreras *et al.*, 1991; Roberts *et al.*, 1992; Broekhuijsen *et al.*, 1993; Jeenes *et al.*, 1993; Nyyssönen and Keränen, 1995; Ward *et al.*, 1995). The carrier improves translocation of the protein into the ER and allows proper folding, thereby protecting the heterologous protein from degradation. In most cases the fusion protein is cleaved at a later stage in the secretory pathway. Cleavage can occur by autocatalytic processing of the heterologous protein (Ward *et al.*, 1990), by an unknown protease (Roberts *et al.*, 1992; Baron *et al.*, 1992; Nyyssönen *et al.*, 1993; Nyyssönen and Keränen, 1995) or by a KEX2-like protease for which a recognition site is introduced in the fusion protein (Contreras *et al.*, 1991; Broekhuijsen *et al.*, 1993; Ward *et al.*, 1995).

Introduction of an N-glycosylation site in the protein of interest has been used as an alternative to improve folding of the protein in the ER (Sagt *et al.*, 2000). A hydrophobic cutinase aggregated in the ER of *S. cerevisae* in association with BiP (Sagt *et al.*, 1998). It was reasoned that this was due to an N-terminal hydrophobic stretch that was prone to aggregation after being translocated in the lumen of the ER. By introduction of an N-glycosylation site preceding this hydrophobic stretch secretion was improved tenfold (Sagt *et al.*, 2000).

The examples described above show that folding may be improved by genetic engineering of the protein of interest. However, upon over-expression of native or engineered proteins, foldases and chaperones may become limiting to ensure correct folding. When unfolded proteins accumulate in the ER, a signal transduction pathway is activated which is called the unfolded protein response (UPR). The UPR controls expression of a variety of genes including genes encoding ER-located chaperones and foldases such as BiP and PDI (Kaufman, 1999; Patil and Walter, 2001). These chaperones and foldases have been over-expressed individually to improve production of secreted proteins. In some cases this improved protein production (Robinson *et al.*, 1994; Harmsen *et al.*, 1996; Punt *et al.*, 1998). As an alternative to over-expression of individual chaperones or foldases, the UPR response was constitutively induced (Valkonen *et al.*, 2003). By over-expressing the UPR induced form of the transcription factor *hacA* in *A. niger*, production of mature bovine chymosin as well as a fungal laccase was improved. In the former case, improvement was up to sevenfold while in the latter case a 2.8-fold increase was observed. However, over-expression of the *ire1* gene did not increase production of a fungal laccase in *T. reesei* (Valkonen *et al.*, 2004). This gene encodes the most upstream element of the UPR, which functions in sensing the folding status of proteins in the ER.

Proteins that have been successfully synthesized can be degraded proteolytically (Gouka *et al.*, 1996; Suzuki *et al.*, 1989; Wingfield and Dickinson, 1993). In *S. cerevisiae*, the use of strains deficient in vacuolar proteases has led to increased levels of heterologous proteins (Suzuki *et al.*, 1989; Wingfield and Dickinson, 1993). Extracellular proteases can also affect production of proteins. These enzymes are responsible for the degradation of many heterologous proteins in *Aspergillus* (Berka *et al.*, 1990; Archer *et al.*, 1992; Broekhuijsen *et al.*, 1993), *S. commune* (van Wetter, 2000) and *P. cinnabarinus* (Alves *et al.*, 2004). Fungal strains deficient in extracellular proteases have been isolated by random mutagenesis (Mattern *et al.*, 1992; van den Hombergh *et al.*, 1995, 1997) or molecular genetic approaches (Berka, *et al.*, 1990; van den Hombergh *et al.*, 1997). The use of these protease-deficient strains has resulted in improvement of production levels (Berka *et al.*, 1991; Roberts *et al.*, 1992; Broekhuijsen *et al.*, 1993). Recently, an *Aspergillus* species (*A. vadensis*) has been described that is closely related to *A. niger* (de Vries *et al.*, 2005). This species is an interesting host for protein production since it hardly produces extracellular proteases and does not acidify the culture medium (de Vries *et al.*, 2004). In addition, unlike many of the reported proteolytic *Aspergillus* mutants, growth of *A. vadensis* is not reduced when grown on solid or liquid media.

Another way to improve production of heterologous proteins is medium development (Smith and Wood, 1991) and controlled, large-scale fermentation (reviewed by Greasham, 1991; Dunn-Coleman, 1992). For a number of fungal proteins (e.g., glucoamylase) the fermentation conditions have been optimized. The use of heterologous expression signals, like those of the glucoamylase gene, might allow high level production of proteins under already optimized culture conditions, as was shown for phytase (van Gorcom *et al.*, 1990) and aspartic protease (Ward and Kodama, 1991).

Fungi have evolved to grow on moist solid substrates. Since solid state fermentation resembles the natural habitat of fungi it is not surprising that under these conditions more proteins are secreted than in submerged fermentation technology. Catabolite repression is less pronounced in solid state fermentations (Nandakumar *et al.*, 1999; Solis-Pereira *et al.*, 1993). As a consequence, fungi grow better and enzymes are more efficiently produced (Favela-Torres *et al.*, 1998; Viniegra-Gonzales *et al.*, 2003). An additional advantage of solid state fermentation is a reduced protease activity (Solis-Pereira *et al.*, 1993; Maldonado and Strasser de Saad, 1998), lower water demand, higher end-concentration of products, and lower demands on sterility (see Hölker *et al.*, 2004). Despite the fact that enzyme production in solid state fermentation is estimated to be much cheaper (Tengerdy, 1996), it is not used on a large scale. This is due to engineering problems. During cultivation in up-scaled solid state fermentations gradients in temperature, pH, pO_2, moisture, and substrate concentration are being built up. Due to the low water activity, these gradients are difficult to control (Hölker *et al.*, 2004). When these problems would be solved technologically solid state fermentation would be a very good alternative for submerged fermentation.

CONCLUSIONS AND FUTURE PERSPECTIVES

A large number of fungal proteins are used in the food and feed industry. Fungi are potent production platforms for these proteins. A small increase in yield results in a large financial benefit. Therefore, it is attractive to improve production processes. This can be done by improving existing platforms, by introducing new hosts, or by developing new fermentation technologies. In the past production was improved by random mutagenesis. This strategy has proven very successful. With the advent of molecular biology strategies became more rational. Multiple copies of individual genes were introduced or were not combined with the use of strong promoters. Alternatively, regulators or proteins involved in the secretion process were over-expressed. A new strategy is to increase expression of the whole secretion system or parts thereof (e.g., the unfolded protein response). With the availability of the genome sequence of a number of fungal production platforms, it is now worthwhile to assess which genes have been affected during mutagenesis programs in the past. Expression of these genes could be further improved or additional mutations introduced. Studying the secretion pathway will also give new targets for strain improvement.

REFERENCES

Aifa, M. S., Sayadi, S., and Gargouri, A. (1999). Heterologous expression of lignin peroxidase of *Phanerochaete chrysosporium* in *Aspergillus niger*. Biotechnological Letters 21:849-853.

Alves, A. M. R. C., Scholtmeijer, K., Wessels, J. G. H., and Wösten H. A. B. (2004). Highly efficient production of laccase in the basidiomycete *Pycnoporus cinnabarinus*. Applied and Environmental Microbiology 70:6379-6384.

Archer, D. B., MacKenzie, D. A., Jeenes, D. J., and Roberts, I. N. (1992). Proteolytic degradation of heterologous proteins expressed in *Aspergillus niger*. Biotechnological Letters 14:357-362.

Archer, D. B., and Peberdy, J. F. (1997). The molecular biology of secreted enzyme production by fungi. Critical Reviews in Biotechnology 17:273-306.

Baron, M., Tiraby, G., Calmels, T., Parriche, M., and Durand, H. (1992). Efficient secretion of human lysozyme fused to the Sh *ble* phleomycin resistance protein by the fungus *Tolypocladium geodes*. Journal of Biotechnology 24:253-266.

Benen, J. A. E., and Visser, J. (2003). Polygalacturonases. *In* Handbook of Food Enzymology (Whitaker, J. R., Voragen, A. G. J., and Wong, D. W. S., eds.), Marcel Dekker, Inc., New York, U.S.A., pp. 857-866.

Benen, J. A. E., Alebeek, G. J. W. M. van, Voragen, A. G. J., and Visser, J. (2003). Pectic esterases. *In* Handbook of Food Enzymology (Whitaker, J. R., Voragen, A. G. J., and Wong, D. W. S., eds), Marcel Dekker, Inc., New York, U.S.A., pp. 849-856.

Bergkamp, R. J. M., Kool, I. M., Geerse, R. H., and Planta, R. J. (1992). Multiple copy integration of the α-galactosidase gene from *Cyamopsis tetragonoloba* into the ribosomal DNA of *Kluyveromyces lactis*. Current Genetics 21:365-370.

Berka, R. M., Ward, M., Wilson, L. J., Hayenga, K. J., Kodama, K. H., Carlomagno, L. P., and Thompson, S. A. (1990). Molecular cloning and deletion of the aspergillopepsin A gene from *Aspergillus awamori*. Gene 86:153-162.

Berka, R. M., Bayliss, F. T., Bloebaum, P., Cullen, D., Dunn-Coleman, N. S., Kodama, K. H., Hayenga, K. J., Hitzeman, R. A., Lamsa, M. A., Przetak M. M., Wilson, L. J., and Ward, M. (1991). *Aspergillus* var. *awamori* as a host for the expression of heterologous genes. *In* Applications of Enzyme Biotechnology (Kelly, J. W. and Baldwin, T. O., eds.), Plenum, New York, U.S.A., pp. 273-292.

Biely, P. (2003). Xylanolytic enzymes. *In* Handbook of Food Enzymology (Whitaker, J. R., Voragen, A. G. J., and Wong, D. W. S., eds.), Marcel Dekker, Inc., New York, U.S.A., pp. 879-915.

Blinkovsky, A. M., Byun, T., Brown, K. M., and Golightly, E. J. (1999). Purification, characterization and heterologous expression in *Fusarium venenatum* of a novel serine carboxypeptidase from

Aspergillus oryzae. Applied and Environmental Microbiology 65:3298-3303.

Blumenthal, C. Z. (2004). Production of toxic metabolites in *Aspergillus niger*, *Aspergillus oryzae* and *Trichoderma reesei*: justification of mycotoxin testing in food grade enzyme preparations derived from the three fungi. Regulatory Toxicology and Pharmacology 39:214-228.

Bobrowicz, P., Davidson, R. C., Li, H., Potgieter, T. I., Nett, J. H., Hamilton, S. R., Stadheim, T. A., Miele, R. G., Bobrowicz, B., Mitchell ,T., Rausch, S., Renfer, E., and Wildt, S. (2004). Engineering of an artificial glycosylation pathway blocked in core oligosaccharide assembly in the yeast *Pichia pastoris*: production of complex humanized glycoproteins with terminal galactose. Glycobiology 14:757-766.

Brenna, O., and Bianchi, E. (1994). Immobilized laccase for phenolic removal in must and wine. Biotechnological Letters 16:35-40.

Broekhuijsen, M. P., Mattern, I. E., Contreras, R., Kinghorn, J. R., and Hondel C. A. M. J. J. van den (1993). Secretion of heterologous proteins by *Aspergillus niger*: production of active human interleukin-6 in a protease deficient mutant by KEX-2 processing of a glucoamylase-HIL-6 fusion protein. Journal of Biotechnology 31:135-145.

Buckholz, R. G., and Gleeson, M. A. G. (1991). Yeast systems for the commercial production of heterologous proteins. Bio/Technology 9:1067-1072.

Conesa, A., Hondel, C. A. M. J. J. van den, and Punt P. J. (2001a). Studies on the production of fungal peroxidases in *Aspergillus niger*. Applied Environmental Microbiology 66:3016-3023.

Conesa, A., Punt, P. J., Luijk, N. van, and Hondel, C. A. M. J. J. van den (2001b). The secretion pathway in filamentous fungi: a biotechnological view. Fungal Genetics and Biology 33:155-171.

Contreras, R., Carrez, D., Kinghorn, J. R., Hondel, C. A. M. J. J. van den, and Fiers, W. (1991). Efficient KEX-like processing of a glucoamylase-interleukin-6 fusion protein by *Aspergillus nidulans* and secretion of mature interleukin-6. Bio/Technology 9:378-381.

Coux, O., Tanaka, K., and Goldberg, A. L. (1996). Structures and functions of the 20S and 26S proteasomes. Annual Reviews in Biochemistry 65:801-847.

Dequard-Chablat, M., and Rötig, A. (1997). Homologous and heterologous expression of a ribosomal protein gene in *Podospora anserina* requires an intron. Molecular and General Genetics 253:546-552.

Diehl, H. C., Campbell, H., and Berry, J. A. (1936). Freezing of Alderman peas. Food Research 1:61-71.

Dunn-Coleman, N. S., Bloebaum, P., Berka, R. M., Bodie, E., Robinson, N., Armstrong, G., Ward, M., Przetak, M., Carter, G. L., LaCost, R., Wilson, L. J., Kodama, K. H., Baliu, E. F., Bower, B., Lamsa, M., and Heinsohn, H. (1991). Commercial levels of chymosin production by *Aspergillus*. Bio/Technology 9:976-981.

Dunn-Coleman, N. S., Bodie, E. A., Carter, G. L., and Armstrong, G. L. (1992). Stability of recombinant strains under fermentation conditions. *In* Applied Molecular Genetics of Filamentous fungi (Kinghorn, J. R., and Turner G., eds.), Blackie, Glasgow, Scotland, pp. 152-174.

Environmental Protection Agency (EPA) (1997a). Final decision document: TSCA section 5(H) (4) exemption for *Aspergillus niger*. Attachment I. Item #: 3171.

Environmental Protection Agency (EPA) (1997b). Final decision document: TSCA section 5(H) (4) exemption for *Aspergillus oryzae*. Attachment I. Item #: 3173.

Favela-Torres, E., Cordova-Lopez, J., Garcia-Rivero, M., and Gutierrez-Rojas, M. (1998). Kinetics of growth of *Aspergillus niger* during submerged, agar surface and solid state fermentations. Proceedings in Biochemistry 33:103-107.

Fellinger, A. J., Verbakel, J. M. A., Veale, R. A., Sudbery, P. E., Bom, I. J., Overbeeke, N., and Verrips, C. T. (1991). Expression of the α-galactosidase from *Cyamopsis tetragonoloba* (Guar) by *Hansenula polymorpha*. Yeast 7:463-473.

Frederick, K. R., Tung, J., Emerick, R. S., Masiarz, F. R., Chamberlain, S. H ., Vasavada, A., Rosenberg, S., Chakraborty, S., Schopter, L. M., and Massey, V. (1990). Glucose oxidase from *Aspergillus niger*: Cloning, gene sequence, secretion from *Saccharomyces cerevisiae* and kinetic analysis of a yeast-derived enzyme. Journal of Biological Chemistry 265:3793-3802.

Galagan, J. E., and Selker, E. U. (2004). RIP: the evolutionary cost of genome defense. Trends in Genetics 20:417-420.

Gerngross, T. U. (2004). Advances in the production of human therapeutic proteins in yeasts and filamentous fungi. Nature Biotechnology 22:1409-1414.

Ghindilis, A. L., Gavrilova, V. P., and Yaropolov, A. I. (1992). Laccase-based biosensor for determination of polyphenols: determination of catechols in tea. Biosensors and Bioelectronics 7:127-131.

Gorcom, R. F. M. van, Hartingsveldt, W. van, Paridon, P. A. van, Veenstra, A. E., Luiten, R. G. M., and Selten, G. C. M. (1990). Cloning and expression of microbial phytase. European Patent Application (EPA) 0420358.

Gouka, R. J., Punt, P. J., Hessing, J. G. M., and Hondel, C. A. M. J. J. van den (1996). Analysis of heterologous protein production in defined recombinant *Aspergillus awamori* strains. Applied and Environmental Microbiology 62:1951-1957.

Gouka, R. J., Punt, P. J., and Hondel, C. A. M. J. J. van den (1997). Efficient production of secreted proteins by *Aspergillus*: progress, limitations and prospects. Applied and Environmental Microbiology 47:1-11.

Grassin, C., and Fauquembergue, P. (1996). Applications of pectinases in beverages. *In* Pectin and Pectinases, vol. 14 (Visser, J., and Voragen, A. G. J., eds), Elsevier Science, Amsterdam, The Netherlands, pp. 453-462.

Greasham, R. L. (1991). Growth kinetics and fermentation scale-up. *In* Biotechnology of Filamentous Fungi: Technology and Products (Finkelstein, D. B., and Ball, C., eds.), Butterworth-Heinemann Stoneham, MA, U.S.A, pp. 65-88.

Guo, W., Sacher, M., Barrowman, J., Ferro-Novick, S., Novick, P. (2000). Protein complexes in transport vesicle targeting. Trends in Cell Biology 10:251-255.

Guo, W., and Novick, P. (2004). The exocyst meets the translocon: a regulatory circuit for secretion and protein synthesis. Trends in Cell Biology 14:61-63.

Hamilton, S. R., Bobrowicz, P., Bobrowicz, B., Davidson, R. C., Li, H., Mitchell, T., Nett, J. H., Rausch, S., Stadheim, T. A., Wischnewski, H., Wildt, S., and Gerngross, T. U. (2003). Production of complex human glycoproteins in yeast. Science 301:1244-1246.

Harkki, A., Mäntylä, A., Penttilä, M., Muttilainen, S., Bühler, R., Suominen, P., Knowles, J., and Nevalainen, H. (1991). Genetic engineering of *Trichoderma* to produce strains with novel cellulase profiles. Enzyme and Microbial Technology 13:227-233.

Harmsen, M. M., Bruyne, M. I., Raué, H. A., and Maat, J. (1996). Overexpression of binding protein and disruption of the *PMRI* gene synergistically stimulate secretion of bovine prochymosin but not plant thaumatin in yeast. Applied Microbiology and Biotechnology 46:365-370.

Hölker, U., Höfer, M., and Lenz, J. (2004). Biotechnological advantages of laboratory-scale solid-state fermentation with fungi. Applied Microbiology and Biotechnology 64:175-186.

Hombergh, J. P. T. W. van den, Vondervoort, P. J. I. van de, Heijden, N. C. B. A. ven der, and Visser, J. (1995). New protease mutants in *Aspergillus niger* result in strongly reduced *in vitro* degradation of target proteins; genetical and biochemical characterization of seven complementation groups. Current Genetics 28:299-308.

Hombergh, J. P. T. W. van den, Vondervoort, P. J. I. van de, Fraissinet-Tachet, L., and Visser, J. (1997). *Aspergillus* as a host for heterologous protein production: the problem of proteases. TIBTECH 15:256-263.

Irelan, J. T., and Selker, E. U. (1996). Gene silencing in filamentous fungi: RIP, MIP and quelling. Journal of Genetics 75:313-324.

Jeenes, D. J., Marczinke, B., MacKenzie, D. A., and Archer, D. B. (1993). A truncated glucoamylase gene fusion for heterologous protein secretion from *Aspergillus niger*. FEMS Microbiology Letters 107:267-271.

Johnston, D. (2003). Methodologies for assaying the hydrolysis of cellulose by cellulases. *In* Handbook of Food Enzymology (Whitaker, J. R., Voragen, A. G. J., and Wong, D. W. S., eds.), Marcel Dekker, Inc., New York, U.S.A., pp. 761-769.

Kapat, A., Jung, J. K., and Park, Y. H. (2001). Enhancement of glucose oxidase production in batch cultivation of recombinant *Saccharomyces cerevisiae*: optimization of oxygen transfer conditions. Journal of Applied Microbiology 90:216-222.

Kaufman, R. J. (1999). Stress signaling from the lumen of the endoplasmic reticulum: coordination of gene transcriptional and translational controls. Genes and Development 13:1211–1233.

Karhunen, T., Mäntylä, A., Nevalainen, K. M. H., and Suominen, P. L. (1993). High frequency one step gene replacement in *Trichoderma reesei*. I. Endoglucanase I overproduction. Molecular and General Genetics 241:515-522.

Lugones, L. G., Scholtmeijer, K., Klootwijk, R., and Wessels, J. G. H. (1999). Introns are necessary for mRNA accumulation in *Schizophyllum commune*. Molecular Microbiology 32:681-700.

Ma, B., Mayfield, M. B., and Gold, M. H. (2001). The green fluorescent protein gene functions as a reporter of gene expression in *Phanerochaete chrysosporium* Applied and Environmental Microbiology 67:948-955.

Mahoney, R. R. (2003). β-Galactosidase. *In* Handbook of Food Enzymology (Whitaker, J. R., Voragen, A. G. J., and Wong, D. W. S., eds.), Marcel Dekker, Inc., New York, U.S.A., pp. 823-828.

Maldonado, M. C., and Strasser de Saad, A. M. (1998). Production of pectin esterase and polygalacturonase by *Aspergillus niger* in submerged and solid state systems. Journal of Industrial Microbiology and Biotechnology 22:622-626.

Malherbe, D. F., du Toit, M., Cordero Ottero, R. R., van Rensburg, P., and Pretorius, L. S. (2003). Expression of the *Aspergillus niger* glucose oxidase gene in *Saccharomyces cerevisiae* and its potential applications in wine production. Applied Microbiology and Biotechnology 61:502-511.

Margolles-Clark, E., Hayes, C. K., Harman, G. E., and Penttilä, M. (1996). Improved production of *Trichoderma harzianum* endochitinase by expression in the filamentous fungus *Trichoderma reesei*. Applied and Environmental Microbiology 62:2145-2151.

Mattern, I. E., Noort, J. M. van, Berg, P. van den, Archer, D. B., Roberts, I. N., and Hondel, C. A. M. J. J. van den (1992). Isolation and characterization of mutants of *Aspergillus niger* deficient in extracellular proteases. Molecular and General Genetics 234:332-336.

Misset, O. 2003. Phytase. *In* Handbook of Food Enzymology (Whitaker, J. R., Voragen, A. G. J., and Wong, D. W. S., eds.), Marcel Dekker, Inc., New York, U.S.A., pp. 687-706.

Mizuta, K., and Warner, J. R. (1994). Continued functioning of the secretory pathway is essential for ribosome synthesis. Molecular and Cellular Biology 14:2493-2502.

Nandakumar, M. P., Thakur, M. S., Raghavaro, K. S. M. S, and Ghildyal, N. P. (1999). Studies on catabolite repression in solid state fermentation for biosynthesis of fungal amylases. Letters in Applied Microbiology 29:380-384.

Nyyssönen, E., Penttilä, M., Harkki, A., Saloheimo, A., Knowles, J. K. C., and Keränen, S. (1993). Efficient production of antibody fragments by the filamentous fungus *Trichoderma reesei*. Bio/Technology 11:591-595.

Nyyssönen, E., and Keränen, S. (1995). Multiple roles of the cellulase CBHI in enhancing production of fusion antibodies by the filamentous fungus *Trichoderma reesei*. Current Genetics 28:71-79.

Ohlmeyer, D. W. (1957). Use of glucose oxidase to stabilize beer. Food Technology October, pp. 503-507.

Pariza, M. W., and Johnson, E. A. (2001). Evaluating the safety of microbial enzyme preparations used in food processing: update for a new century. Regulatory Toxicology and Pharmacology 33:173-186.

Patil, C., and Walter, P. (2001). Intracellular signaling from the endoplasmic reticulum to the nucleus: the unfolded protein response in yeast and mammals. Current Opinion in Cell Biology 13:349–355.

Piacquadio, P., DeStefano, G., Sammartino, M., and Sciancalepore, V. (1997). Phenols removed from apple juice by laccase immobilized on Cu^{2+}-chelate regenerable carrier. Biotechnology Techniques 11:515-517.

Pickering, G. (1998). The use of enzymes to stabilise colour and flavour in wine – an alternative to SO_2. Aust. Grapegrower and Winemaker September, pp. 101-103.

Punt, P. J., Veldhuizen, G., and van den Hondel, C. A. M. J. J. (1994). Protein targetting and secretion in filamentous fungi. Antonie van Leeuwenhoek 65:211-216.

Punt, P. J., van Gemeren, I. A., Drint-Kuijvenhoven, J., Hessing, J. G., Muijlwijk-Harteveld, G. M. van, Beijersbergen, A., Verrips, C. T., and Hondel, C. A. M. J. J. van den (1998). Analysis of the role of the gene bipA, encoding the major endoplasmic reticulum chaperone protein in the secretion of homologous and heterologous proteins in black Aspergilli. Applied and Microbial Biotechnology 50:447–454.

Reilly, J. (2003). Glucoamylase. *In* Handbook of Food Enzymology (Whitaker, J. R., Voragen, A. G. J., and Wong, D. W. S., eds.), Marcel Dekker, Inc., New York, U.S.A., pp. 727-738.

Roberts, I. N., Oliver, R. P., Punt, P. J., and Hondel, C. A. M. J. J. van den (1989). Expression of the *Escherichia coli* β-glucuronidase gene in industrial and phytopathogenic filamentous fungi. Current Genetics 15:177-180.

Roberts, I. N., Jeenes, D. J., MacKenzie, D. A., Wilkinson, A. P., Sumner, I.G., and Archer, D. B. (1992). Heterologous gene expression in *A. niger*: a glucoamylase-porcine pancreatic phospholipase A_2 fusion protein is secreted and processed to yield mature enzyme. Gene 122:155-161.

Robinson, A. S., Hines, V., and Wittrup, K. D. (1994). Protein disulfide isomerase overexpression increases secretion of foreign proteins in *Saccharomyces cerevisiae*. Biotechnology 12:381–384.

Romano, N., and Macino, G. (1992). Quelling: transient inactivation of gene expression in *Neurospora crassa* by transformation with homologous sequences. Molecular Microbiology 6:3343-3353.

Rose, S. H., and Zyl, W. H. van (2002). Constitutive expression of the *Trichoderma reesei* β-1,4-xylanase gene (*xyn2*) and the β-1,4-endoglucanase gene (*egl*) in *Aspergillus niger* in molasses and defined glucose media. Applied Microbiology Biotechnology 58:461-468.

Royer, J. C., Moyer, D. L., Reiwitch, S. G., Madden, M. S., Jensen, E. B., Brown, S. H., Yonker, C. C., Johnstone, J. A., Golightly, E. J., Yoder, W. T., and Shuster, J. R. (1995). *Fusarium graminearum* A 3/5 as a novel host for heterologous protein production. Biotechnology 13:1479-1483.

Sagt, C. M. J., Müller, W. H., Boonstra, J., Verkleij, A. J., and Verrips, C. T. (1998). Impaired secretion of a hydrophobic cutinase by *Saccharomyces cerevisae* correlates with an increased association with immunoglobulin heavy chain binding protein (BiP). Applied Environmental Microbiology 64:316-324.

Sagt, C. M. J., Kleizen, B., Verwaal, R., Jong, M. D. M. de, Müller, W. H., Smits, A., Visser, C., Boonstra, J., Verkleij, A. J., and Verrips, C. T. (2000). Introduction of an N-glysoylation site increases secretion of heterologous proteins in yeasts. Applied and Environmental Microbiology 66:4940-4944.

Scholtmeijer, K., Wösten, H. A. B., Springer, J., and Wessels, J. G. H. (2001). Effect of introns and AT-rich sequences on expression of the bacterial hygromycin B resistance gene in the basidiomycete *Schyzophyllum commune*. Applied and Environmental Microbiology 67:481-483.

Schultz, L. D., Hofmann, K. J., Mylin, L. M., Montgomery, D. L, Ellis, R. W., and Hopper, J. E. (1987). Regulated overproduction of the *Gal4* gene product greatly increases expression from galactose inducible promoters on multi-copy expression vectors in yeast. Gene 61:123-133.

Schuren, F. H. J., and Wessels, J. G. H. (1994). Highly-efficient transformation of the homobasidiomycete *Schizophyllum commune* to phleomycin resistance. Current Genetics 26:179-183.

Schuren, F. H. J., and Wessels, J. G. H. (1998). Expression of heterologous genes in *Schizophyllum commune* is often hampered by the formation of truncated transcripts. Current Genetics 33:151-156.

Schuurs, T. A., Shaeffer, E. A., and Wessels, J. G. H. (1997). Homology-dependent silencing of the *SC3* gene in *Schizophyllum commune*. Genetics 147:589-596.

Scorer, C. A., Buckholz, R. G., Clare, J. J., and Romanos, M. A. (1993). The intracellular production and secretion of HIV-1 envelope protein in the methylotrophic yeast *Pichia pastoris*. Gene 136:111-119.

Smith, D. C., and Wood, T. M. (1991). Xylanase production by *Aspergillus awamori*. Development of a medium and optimization of the fermentation parameters for the production of extracellular xylanase and β-xylosidase while maintaining low protease production. Biotechnology and Bioengineering 38:883-890.

Solis-Pereira, S., Favela-Torres, E., Viniegra-Gonzalez, G., and Gutierez-Rojas, M. (1993). Effect of different carbon sources on the synthesis of pectinases in *Aspergillus niger* in submerged and solid state fermentation. Applied and Microbiology Biotechnology 39:36-41.

Stålbrand, H. (2003). Enzymology of endo-1,4-β-mannanases. *In* Handbook of Food Enzymology (Whitaker, J. R., Voragen, A. G. J., and Wong, D. W. S., eds.), Marcel Dekker, Inc., New York, U.S.A., pp. 961-969.

Suzuki, K., Ichikawa, K., and Jigami, Y. (1989). Yeast mutants with enhanced ability to secrete human lysozyme: isolation and identification of a protease deficient mutant. Molecular and General Genetics 219:58-64.

Tengerdy, R. P. (1996). Cellulase production by solid state fermentation. Journal of Scientific Industrial Research 55:313-316.

Tenkanen, M., Niku-Paavola, M. J., Linder, M., and Viikari, L. (2003). Cellulases in food processing. *In* Handbook of Food Enzymology (Whitaker, J. R., Voragen, A. G. J., and Wong, D. W. S., eds.), Marcel Dekker, Inc., New York, U.S.A., pp. 771-789.

Trinci, A. P. J. (1992). Mycoprotein: a twenty-year overnight success story. Mycological Research 96:1-13.

Valkonen, M., Ward, M., Wang, H., Penttilä, M., and Saloheimo, M. (2003). Improvement of foreign-protein production in *Aspergillus niger* var. *awamori* by constitutive induction of the unfolded-protein response. Applied and Environmental Microbiology 69:6979-6986.

Valkonen, M., Penttilä, M., and Saloheimo, M. (2004). The *ire*1 and *ptc*2 genes involved in the unfolded protein response pathway in the filamentous fungus *Trichoderma reesei*. Molecular and General Genomics 272:443-451.

Verbakel, J. M. A. (1991). Heterologous gene expression in the yeast *Saccharomyces cerevisiae*. PhD Thesis, University of Utrecht, the Netherlands.

Verdoes, J. C., Punt, P. J., Schrickx, J. M., Verseveld, H. W. van, Stouthamer, A. H., and Hondel, C. A. M. J. J. van den (1993). Glucoamylase overexpression in *Aspergillus niger*: molecular genetic analysis of strains containing multiple copies of the *glaA* gene. Transgenic Research 2:84-92.

Verdoes, J. C., Diepeningen, A. D. van, Punt, P. J., Debets, A. M. J., Stouthamer, A. H., and Hondel, C. A. M. J. J. van den (1994a). Evaluation of molecular and genetic approaches to generate glucoamylase overproducing strains of *Aspergillus niger*. Journal Biotechnology 36:165-175.

Verdoes, J. C., Punt, P. J., Stouthamer, A. H., and Hondel, C. A. M. J. J. van den (1994b). The effect of multiple copies of the upstream region of the *Aspergillus niger* glucoamylase-encoding gene on expression. Gene 145:179-187.

Vermulapalli, V., Miller, K. A., and Hoseney, R. C. (1998). Glucose oxidase in breadmaking systems. Cereal Chemistry 75:439-442.

Viniegra-Gonzales, G., Favela-Torres, E., Aguilar, C. N., Romero-Gomez, S., Diaz-Godinez, G., and Augur C. (2003). Advantages of fungal enzyme production in solid state over liquid fermentation systems. Biochemical Engineering Journal 13:157-167.

Vries, R. P. de, and Visser, J. (2001). *Aspergillus* enzymes involved in degradation of plant cell wall polysaccharides. Microbiology and Molecular Biology Reviews 65:497-522.

Vries, R. P. de (2003). Regulation of *Aspergillus* genes encoding plant cell wall polysaccharide-degrading enzymes; relevance for industrial production. Applied Microbiology and Biotechnology 61:10-20.

Vries, R. P. de, and Visser, J. (2003). Enzymes releasing L-arabinose and D-galactose from the side chains of pectin. *In* Handbook of Food Enzymology (Whitaker, J. R., Voragen, A. G. J., and Wong, D. W. S., eds.), Marcel Dekker, Inc., New York, U.S.A., pp. 867-877.

Vries, R. P. de, Burgers, K., Vondervoort, P. J. I. van de, Frisvad, J. C., Samson, R. A., and Visser, J. (2004). A new black *Aspergillus* species, *A. vadensis*, is a promising host for homologous and heterologous protein production. Applied and Enviromental Microbiology 70:3954-3959.

Vries, R. P. de, Frisvad, J. C., Vondervoort, P. J. I. van de, Burgers, K., Kuijpers, A. F. A., Samson, R. A., and Visser, J. (2005). *Aspergillus vadensis*, a new species of the group of black aspergilli. Antonie van Leeuwenhoek 87:95-103.

Ward, M. (1989a). Production of calf chymosin by *Aspergillus awamori*. *In* Genetics and Molecular Biology of Industrial Microorganisms. (Hershberger, C. L., Queener, S. W., and Hageman, G., eds.), American Society for Microbiology, Washington DC, U.S.A., pp. 288-294.

Ward, M. (1989b). Heterologous gene expression in *Aspergillus*. *In* Proceedings of the EMBO-Alko Workshop on Molecular Biology of Filamentous Fungi (Nevalainen, H., and Penttilä, M., eds), Foundation for Biotechnical and Industrial Fermentation Research 6:119-128.

Ward, M., Wilson, L. J., Kodama, K. H., Rey, M. W., and Berka, R. M. (1990). Improved production of chymosin in *Aspergillus* by expression as a glucoamylase-chymosin fusion. Bio/Technology 8:435-440.

Ward, M., and Kodama, K. H. (1991). Introduction to fungal proteinases and expression in fungal systems. *In* Advances in experimental medicine and biology, Vol. 306. Structure and function of the aspartic proteinases: genetics, structures and mechanisms (Dunn, B. M., ed.). Plenum, New York, U.S.A., pp. 149-160.

Ward, P. P., Piddington, C. S., Cunningham, G. A., Zhou, X., Wyatt, R. D., and Conneely, O. M. (1995). A system for production of commercial quantities of human lactoferrin: a broad spectrum natural antibiotic. Bio/Technology 13:498-503.

Wessels, J. G. H., Mulder, G. H., and Springer, J. (1987). Expression of dikaryon specific and non-specific mRNAs of *Schizophyllum commune* in relation to environmental conditions and fruiting. Journal of General Microbiology 133:2557-2561.

Wessels, J. G. H. (1993). Cell wall growth, protein excretion and morphogenesis in fungi. Transley Review No 45. New Phytologist 123:397-413.

Wetter, M. A. van (2000). Functions of hydrophobins in *Schizophyllum commune*. PhD Thesis, University of Groningen, The Netherlands.

Whitaker, J. R. (2003). Proteolytic enzymes. *In* Handbook of Food Enzymology (Whitaker, J. R., Voragen, A. G. J., and Wong, D. W. S., eds.), Marcel Dekker, Inc., New York, U.S.A., pp. 993-1018.

Whitaker, J. R., Voragen, A. G. J., and Wong, D. W. S. (eds.) (2003). Handbook of Food Enzymology. Marcel Dekker, Inc., New York.

Wiertz, E. J. H. J., Tortorella, D., Bogyo, M., Yu, J., Mothes, W., Jones, T. R., Rapoport, T. A., and Ploegh, H. L. (1996). Ces61-mediated transfer of a membrane protein from the endoplasmatic reticulum to the proteosome for destruction. Nature 384:432-438.

Williamson, G., Kroon, P. A., and Faulds, C. B. (1998). Hairy plant polysaccharides: a close shave with microbial esterases. Microbiology 144:2011-2023.

Wingfield, J. M., and Dickinson, J. R. (1993). Increased activity of a model heterologous protein in *Saccharomyces cerevisiae* strains with reduced vacuolar proteinases. Applied Microbiology and Biotechnology 39:211-215.

Wong, D. W. S. (2003). Lipase. *In* Handbook of Food Enzymology (Whitaker, J. R., Voragen, A. G. J., and Wong, D. W. S., eds.), Marcel Dekker, Inc., New York, U.S.A., pp. 667-680.

Wong, D. W. S., and Robertson, G. H. (2003). α-Amylases. *In* Handbook of Food Enzymology (Whitaker, J. R., Voragen, A. G. J., and Wong, D. W. S., eds.), Marcel Dekker, Inc., New York, U.S.A., pp. 707-718.

Wösten, H. A. B., Wetter, M. A. van, Lugones, L. G., Mei, H. C. van der, Busscher, H. J., Wessels, J. G. H. (1999). How a fungus escapes the water to grow into the air. Current Biology 9:85-88.

Part 5

FUNGAL SPOILAGE: ECOLOGY, GROWTH AND DETECTION

A substantial amount of food in our daily life does not end in our stomach, but finds a way to the litter box, while it is spoiled by different species of fungi. A remarkable observation is that fungi are more or less specific for certain food products and that the number of dominant fungal species associated with a certain type of food is surprisingly low. Frisvad, Andersen and Samson provide an update of the knowledge on this ecological aspect of spoilage in Chapter 11 and discuss the specific fungal ("the mycological flora") and their association with food. Once fungi develop on food from, for example, germinating spores they colonise the food product. Rahardjo and Rinzema describe in Chapter 12 the growth of fungi in food-like media. They highlight the important role of gradients that are present in a food matrix and describe the different fluxes that can evolve during spoilage. In all this, the aerial mycelium plays a surprising role.

Is it possible to detect the presence of fungi in food far before they appear? In other words is it possible to detect very low amounts of fungi in the food matrix. Up till now, plating out of food samples is still an important facet of detection of food fungi. Geisen highlights in Chapter 13 the state of the art of detection by means of molecular tools. He addresses the question of the lower limit of detection of fungi in the food matrix and describes many different techniques that also report on the possible metabolic state of the fungus in such an environment. Fungi also produce specific volatiles in such a way that a device called "electronic nose" literally "smells" the spoilage in an accurate way. Karlshøj, Nielsen and Larsen describe the different modes of detection of volatiles in Chapter 14 and highlight the different patterns of volatiles fungi produce.

The last 2 chapters of this part describe spoilage fungi in action in two types of food products. In Chapter 15 Leong addresses the topic of grape infection by fungi, which may be reflected on the level of a mycotoxin in wine, namely ochratoxin A. This chapter deals with many different aspects of fungal spoilage ranging from post-harvest problems to mycotoxin detection, all related to the production chain from early grape formation to the bottle of wine on your table. In Chapter 16 Stark covers the fungi associated with cheese and sausages, which in a way display a mixture of wanted and unwanted fungi.

Chapter 11

Association of moulds to foods

Jens C. Frisvad[1], Birgitte Andersen[1] and Robert A. Samson[2]
[1]*Center for Microbial Biotechnology, Biocentrum-DTU, Technical University of Denmark, DK-2800 Kgs. Lyngby, Denmark;* [2]*CBS Fungal Biodiversity Centre, Utrecht, The Netherlands.*

INTRODUCTION

The traditional division of filamentous fungi has been between the food spoiling fungi, which were considered opportunistic fungi with no substrate preferences and animal or plant pathogenic fungi with tight associations to their hosts. Most food spoiling fungi have been regarded as saprophytic organisms thriving on any substrate they could encounter and they can indeed be isolated on many different laboratory media. They can grow and differentiate on minimal media containing only nitrate and sucrose as nitrogen and carbon source, as well as on complex media based on cereal, vegetable, fruit and meat, such as peptones, corn steep liquor, malt and yeast extract. (Raper and Thom, 1949; Smith, 1960; Samson *et al.*, 2004c). However, as early as 1949 Westerdijk suggested that certain Penicillia were associated to certain food substrates, such as *P. italicum* and *P. digitatum* to citrus fruits and *P. expansum* to pomaceous and stone fruits. These associations were regarded as the exceptions rather than the rule for filamentous fungi. Several authorities (Thom, 1930; Raper and Thom, 1949; Pitt, 1979; Ramírez, 1982) were of the opinion that the many *Penicillium* species they treated were saprophytic generalists rather than species associated to specific natural habitats or processed foods and feedstuff. Difficult taxonomy and lack of data may have obscured the less obvious associations between *Penicillium* species and habitats or food products (Frisvad, 1988; Frisvad and Filtenborg, 1988). Furthermore, using inadequate techniques and methods would often make it impossible to distinguish between simple surface contamination of a food product and true infection resulting from the fungal-substrate association. Fungi isolated from surface disinfected products are with some probability thriving on the product and therefore associated with it, whereas fungi growing from non-surface-disinfected products could be accidental contaminations from other materials, storage facilities or the air (King *et al.*, 1986; Frisvad and Samson, 1991; Samson *et al.*, 1992; Filtenborg *et al.*, 1996; Hocking *et al.*, 2006). This applies not only to *Penicillium*, but also to all other major food spoiling genera, such as *Aspergillus*, *Alternaria* and *Fusarium*.

The associated mycobiota of a food product can be defined as all the fungal species that are able to infect and actively grow on the product under harvest, storage or processing conditions. In connection with food mycology and safety, the fact that each fungal species found in a food product produces a species specific profile of extrolites is of particular importance. An extrolite can be a volatile or non-volatile secondary metabolite, an organic acid, an extracellular enzyme or other outwards directed biochemical compounds (Frisvad *et al.*, 1998, 2004; Frisvad and Samson, 2004; Larsen *et al.*, 2005). For example, only three of the approximately 90 food spoiling *Penicillium* species are able to produce penicillin (Samson *et al.*, 2004c). Among the extrolites, mycotoxins and other bioactive compounds, such as ochratoxin or patulin, are of direct health concern. The production of mycotoxins is highest and most diverse under optimal conditions in a laboratory and mycotoxins are only produced in the food products during storage or processing when conditions change to the advantage of the fungi. Therefore, knowledge of production of mycotoxins and other extrolites by individual fungal species under controlled conditions in the laboratory is important in order to relate production to specific food products. Furthermore, extrolites that once were regarded as non-toxic or at best of little importance in known mycotoxicoses have now shown to be highly toxic if they are inhaled rather than ingested. For example, brevianamide A, mycophenolic acid and roquefortine C have been shown to be cytotoxic and inflammatory in mouse lungs (Rand *et al.*, 2005). This could be of consequence for employees in food factories, which may inhale mycotoxins via fungal spored in the air. Another reason for determining the mycobiota is that secondary metabolites which are now considered non-toxic may later turn out to be mycotoxins, and retrospective analysis may then help explain certain mycotoxicosis, where no known mycotoxins could be found. A third reason for examining foods for the mycobiota is that some mycotoxins may act synergistically. Knowing the mycobiota on a particular food product, and thereby which mycotoxins are theoretically possible to encounter in the product, can give an indication on which mycotoxins the product should analyze for. However, a negative result for a mycotoxin analysis in a food product may give a false sense of security, as it is now known that mycotoxins produced in a crop plant may be masked by glucosylation or by reaction with β-D-glucans by the crop plant (Berthiller *et al.*, 2005; Yiannikouris *et al.*, 2006). These masked mycotoxins are a good reason for finding the mycobiota of foods before a mycotoxin analysis, to ensure that all mycotoxins and their derivatives are recovered in an analysis.

This chapter gives an overview of the associated mycobiota of different raw materials and food products together with the mycotoxins and other bioactive compounds to consider in a chemical food product analysis. A detailed encyclopaedic list on the occurrence of mycotoxins in different foods, spices and other edible substrata has been given by Weidenbörner (2001a), but this list is only backed up by a few references.

GROWTH FACTORS

The main factors that determine fungal growth on a food product are water activity (aw), pH, temperature, oxygen and other microorganisms. Filamentous fungi tolerate low pH and low aw better than prokaryotes and are therefore seen more often spoiling acidic foods with low aw, because they have less prokaryotic competitors at these conditions (Scott, 1957; Corry, 1987). At neutral pH and high aw the prokaryotes have the advantage and filamentous fungi are most often out-competed by bacteria. In a few cases the filamentous fungi may be able to compete with bacteria due to their production of antibiotic extrolites, but in other cases filamentous fungi may co-exist with lactic acid bacteria and acidophilic yeasts.

Often species will occur in a succession. In dry stored foods, species such as *Aspergillus penicillioides*, *Asp. restrictus* and *Eurotium echinulatum*, *E. herbariorum*, *E. rubrum*, *E. chevalieri*, *E. amstelodami* and *Wallemia* spp. may pave the way for the more toxigenic species of *Aspergillus* and *Penicillium*. In other foods the field mycobiota, mostly species of *Alternaria*, *Stemphylium*, *Ulocladium*, *Drechslera*, and *Fusarium* may initially grow and after some drying toxigenic *Penicillium* and *Aspergillus* species will take over.

At low temperature the fungi have the advantage over the bacteria and many fungi are adapted to spoil refrigerated foods. A series of terverticillate Penicillia are able to grow well at 5, 10 and 15°C and several of these species have a temperature optimum between 15 and 20°C (Pitt, 1979; Frisvad and Samson, 2004). Few of these species are able to grow well at 37°C, even though species such as *Penicillium chrysogenum* and *P. aethiopicum* may grow at this temperature (Frisvad and Samson, 2004). Most Aspergilli and many *Penicillium* species from subgenus *Furcatum* are able to grow at 37°C, and these species are common in subtropical and tropical climates (Pitt, 1979; Pitt and Hocking, 1997). *Cladosporium* and *Fusarium* species are generally able to grow well at low temperatures, but some of these and most *Alternaria* species also grow well at high temperatures (Pitt and Hocking, 1997).

Temperature, aw and pH have been shown to interact strongly and each fungal species has its own set of optimal growth factors (Andersen and Frisvad, 2002).

The presence of other microorganisms can also restrict growth and mycotoxin formation. The absence of moulds on fresh meat is explained by the rapid growth of bacteria. Some moulds can also hinder the development for other moulds. For example, *Asp. flavus* may form only a little amount of aflatoxin in the presence of other moulds (Ashworth *et al.*, 1965). The pH and composition of the substrate have minor influence on the actual growth of species that are discussed in this chapter, but these factors can greatly influence the formation of specific mycotoxins (Kinderlerer and Hatton, 1990). The formation of aflatoxins, for example, is highly stimulated by the presence of certain amino acids, fatty acids and the element zinc (Venkitasubramanian, 1977). Being tolerant to lactic acid these fungi may also be resistant to acidic preservatives such as acetic, propionic, benzoic, and sorbic acid. However, these factors are not the only determinants for the mycobiota of acidic foods. For example *Penicillium roqueforti* and *P. paneum* are common on silage (containing lactic acid bacteria), while *P. roqueforti* and *P. carneum* are common on rye bread (containing preservatives) (Boysen *et al.*, 1996; Lund *et al.*, 1996; Sumarah *et al.*, 2006; Nielsen *et al.*, 2006). Maize silage may contain mycotoxins from the species thriving under lactic acid bacterium generated silage conditions, such as patulin, mycophenolic acid, citrinin, roquefortine C and marcfortins produced by *Byssochlamys nivea, Monascus ruber, Paecilomyces variotii, Penicillium roqueforti* and *P. paneum*, but silage may also contain mycotoxins produced on the maize plants before they were processed, including aflatoxins, fumonisins, deoxynivalenol and zearalenone produced by *Aspergillus flavus, Fusarium verticillioides, F. proliferatum* and other fungi (El-Shanawany *et al.*, 2005; Garon *et al.*, 2006). Even though silage is only used as feedstuff, this example shows that the same kind of plant material may have two completely different mycobiotas from different stages in the process. Concerning plant products, the mycobiota is strongly dependent on the stage of processing. Raw material has a typical field mycobiota, stored plant material will have a different storage mycobiota, dried plant materials will have a third mycobiota composition and finally heat processed plant products may contain heat-resistant fungi only (Pitt and Hocking, 1997; Filtenborg *et al.*, 2004).

Oxygen is usually necessary for the growth of fungi, but certain species can also grow under anaerobic conditions with the formation of ethanol and organic acids. Oxygen also influences production of mycotoxins. The production of patulin and penicillic acid decrease sharply at low oxygen concentrations, while fungal growth is not noticeably influenced (Northolt, 1979). The production of aflatoxins is greatly restricted at an oxygen concentration of less than 1% (Landers *et al.*, 1967).

Finally, the factor time must be mentioned. The time necessary for the germination of spores or conidia increases to weeks under unfavourable conditions, while that under favourable conditions can be approximately one day (Northolt, 1979). In establishing safe limits of a_w and temperature for storage of foodstuffs, it is necessary to consider that the absence of fungal growth after one month does not always mean that a product can be stored safely for a much longer time.

FIELD AND STORAGE CONDITIONS

Raw material, semi-manufactured and finished food products can be contaminated with spores and mycelium fragments from the environment. Contamination can occur at different stages of production: during growth, ripening and harvest of the crops in the field, during processing in the factory and during storage of the final products. The presence of large numbers of fungal propagule in products that are not visibly mouldy can either point to a general contamination of the environment or to the processing of mouldy raw material. During processing, the fungi may be inactivated and may loose viability.

Fungal growth only occurs under favourable conditions. The conditions for each species vary and adaptability determines which species will dominate. The precise cause why a particular species dominates in a product is often not known, but is certainly correlated with the species, characteristics and the properties of the product. The predominance of one species can be due to heavy contamination from ecological niches where the mould has developed. The frequent occurrence of *Penicillium expansum* on apples is probably due to growth of the mould on rotten matter in orchards (Börner, 1963), from where it infects the apples. Salads are made from ingredients that are sometimes kept for long periods under refrigeration, favouring the dominance of the psychrotolerant species of *Cladosporium* and *Penicillium*. The occurrence of *Eurotium herbariorum* on grains can be explained by its xerophilic characteristics. Also many types of bakery products have a low water activity (Northolt *et al.*, 1980b), thus explaining the predominance of the xerophilic species *E. herbariorum*.

During ripening of cheese in storage, the moisture level and water activity decrease, which influence the composition of the mycobiota: *Penicillium commune* remains, while *P. brevicompactum* is replaced by *Asp. versicolor* and the xerophilic *E. herbariorum*.

ASSOCIATED MYCOBIOTA ON RAW MATERIALS AND PROCESSED FOODS

The components of the food product are not only determining the mycotoxin formation, but also the range of species, which are able to grow and thus spoil the individual food types. Normally less than ten and often one to three species are responsible for a particular spoilage (Frisvad and Filtenborg, 1988, 1993). As far as fungi in foods are concerned, the discovery of unique food/fungi associations is fairly new (Frisvad and Filtenborg, 1988), and is due to the development of new mycological methods and taxonomy of food-borne moulds especially the genera *Penicillium, Aspergillus* and *Fusarium* and to a certain extent *Alternaria*. The former dominating role of morphology in mould identification has been aided by the combined use of extrolite profiles (Frisvad and Filtenborg, 1983, 1990), isozyme profiles, physiological (Frisvad, 1981) and ecological characteristics, DNA patterns and morphology (Samson and Pitt, 1990; Frisvad and Samson, 2004).

This significant development in methodology and taxonomy of food-borne moulds within the past 20 years means that the major part of the publications on the mycobiota of foods up till now should be "translated" according to those changes. Some fungi from culture collections have been re-examined and their new identity has been published (Marasas *et al.*, 1984; Frisvad, 1989), but much of the earlier work needs to be repeated because most isolates have not been kept. This problem has been taken into consideration here, so reference is mainly given to publications from the last decade, which is in agreement with the new concepts agreed by the international working groups. Examples will be given to illustrate the confusion and disagreement, which arise when former methods and taxonomy are being used to determine which species are responsible for the spoilage of each type of foods. Some species such as *A. niger*, *A. ochraceus* and *Alt. alternata* have often been reported as *sensu lato*, so they may represent other related species in some cases (Samson *et al.*, 2004c; Simmons, 1999b; Andersen *et al.*, 2002).

When describing the associated mycobiota of foods, it is important to differentiate between species which are infesting the food and species which are actually able to infect the food due to growth. Only the last species belong to the associated mycobiota. Fungal propagules (conidia, ascospores, mycelial fragments) may be present for several reasons, either as part of the "normal" airspora of outdoor or indoor air, or as part of the "normal" mycobiota (formerly mycoflora), which is not necessarily damaging the food, rather the opposite. An example of the latter situation is the occurrence of several yeast species and *Rhizopus oligosporus* in tempe (Nout, 2004) and the use of yeasts and fungal starter cultures for meat and cheese (Spotti *et al.*, 1994). Growth of such species may prevent growth of serious fungal spoilers and mycotoxin producers, by interaction.

Table 1. Most common associated fungal species

Crop	Product	Fungal species
Beans and peas	Black beans, cowpeas	*Alternaria alternata, Aspergillus flavus, Asp. ochraceus, Asp. parasiticus, Fusarium proliferatum, Penicillium citrinum*
Cereal	Maize	*Asp. flavus, Asp. niger, Asp. ochraceus, F. graminearum, F. proliferatum, F. verticillioides, P. citrinum*
	Rice	*Asp. flavus, Asp. niger, P. citrinum*
	Rye bread	*Eurotium repens, Eur. rubrum, P. carneum, P. paneum, P. roqueforti*
	Sorghum	*Alt. alternata, Asp. flavus, F. verticillioides, F. semitectum, P. citrinum*
	Wheat bread	*Asp. flavus, Eur. repens, Eur. rubrum*
	Wheat, rye, barley, oat	*Alt. tenuissima* and *infectoria* sp.-grps., *Asp. flavus, Asp. parasiticus, F. avenaceum, F. culmorum, F. graminearum, P. aurantiogriseum, P. cyclopium, P. freii, P. melanoconidium, P. polonicum, P. verrucosum*
Cheese	Hard cheese	*Asp. versicolor, P. commune, P. discolor, P. nalgiovense, P. solitum*
Coffee	Coffee - monsoon	*Asp. candidus, Asp. niger, Asp. tamarii*
	Coffee - traditional	*Asp. carbonarius, Asp. steynii, Asp. westerdijkiae, P. citrinum*
Fruit	Citrus	*Alt. tangelonis, Alt. tenuissima* sp.-grp., *Alt. turkisafria, P. digitatum, P. italicum*
	Dried fruits	*Asp. carbonarius, Asp. flavus, Asp. niger, Asp. ochraceus, Xeromyces bisporus, Wallemia sebi*
	Fruit juice	*Byssochlamys nivea, B. spectabilis* (= *Paecilomyces variotii*), *Eupenicillium* spp., *Neosartorya* spp,. *Talaromyces* spp.
	Grapes	*Asp. carbonarius, Asp. niger, Asp. tubingensis, P. expansum*
	Pomaceous and stone	*Alt. arborescens* sp.-grp., *Alt. tenuissima* sp.-grp., *F. lateritium, P. crustosum, P. expansum, P. solitum*
Meat	Sausages	*P. nalgiovense, P. nordicum, P. olsonii, P. chrysogenum, Eurotium* spp.
Nuts	Almonds, hazelnuts, pistachio, walnuts	*Alt. arborescens* sp.-grp., *Asp. flavus, Asp. niger, Asp. tamari, F. acuminatum, F. avenaceum, F. semitectum, P. crustosum, P. discolor*
Oil crop	Olives	*Alt. alternata, Asp. versicolor, P. citrinum, P. expansum*
	Peanuts	*Asp. flavus, Asp. niger*
	Sunflower	*Alt. alternata, Asp. flavus, Asp. niger, Asp. parasiticus, F. verticilloides, F. semitectum*
Vegetables	Ginger	*P. brevicompactum*
	Onion and garlic	*P. allii, P. glabrum, Petromyces alliaceus*
	Pepper - bell	*Alt. alternata*
	Pepper - black	*Asp. flavus, Asp. parasiticus, Asp. tamarii*
	Potatoes	*Alt. alternata, Alt. solani, F. coeruleum., F. sambucinum*
	Tomatoes	*Alt. alternata, Alt. subtropica, Alt. tenuissima* sp.-grp., *P. expansum, P. olsonii, P. tularense, Stemphylium eturmiunum, St. solani*
	Yams	*Botryosphaeria rhodina, F. verticillioides, P. sclerotigenum*
	Yam chips	*Asp. flavus, Asp. niger*

Crop plants themselves may be damaged by numerous plant pathogenic fungi, like smuts (*Tilletia* spp. and *Ustilago* spp.), powdery and downy mildew (*Sphaerotheca* spp., *Microsphaera* spp., *Peronospora* spp. and others), but this Chapter will focus on the crop part that is used for food or feed.

Accurate identification of fungal species from foods is important, but not always easy. In many cases only rather outdated keys are available, at least for large genera as *Alternaria,*

Fusarium, Penicillium and *Aspergillus*. Keys for the most common fungi in foods are available in Samson *et al.* (2004c), Pitt and Hocking (1997) and elsewhere, but keys based on polyphasic approaches to taxonomy will probably be more common in the future. Keys based only on morphology, such as those in Raper and Fennell (1965) may lead to incorrect identifications. For example Raper and Fennell (1965) provided two alternative keys for the *Aspergillus niger* group (*Aspergillus* section *Nigri*). The data provided by Samson *et al.* (2004a) may help in accurate identification of this group, as both morphological, extrolite and molecular data are available for identification. Data provided in a polyphasic approach to the *Aspergillus ochraceus* group (*Aspergillus* section *Circumdati*) (Frisvad *et al.*, 2004b) may be helpful for identification of these Aspergilli.

For example it is possible that the most important toxigenic species growing in green coffee beans is *Aspergillus westerdijkiae* rather than *Asp. ochraceus*. This may be of some importance because the former species is a much more consistent and efficient producer of ochratoxin A than the latter species (Frisvad *et al.*, 2004b). Different types of data and keys are provided for the terverticillate Penicillia in Frisvad and Samson (2004), including the possibility to identify the species based on β-tubulin sequences (Samson *et al.*, 2004b). However, more keys for food-borne fungi, based on different kinds of features, are needed. For example in *Alternaria* a series of papers have been published on new species (Simmons, 1986, 1990, 1992, 1993, 1994, 1995, 1999a,b; Simmons and Roberts, 1993), but there is still no overall key published to these species.

Simple chemical methods may be used for detecting secondary metabolites, including mycotoxins, in filamentous fungi. For example TLC methods have been used based on the direct analysis of agar plugs taken from common identification media, such as yeast extracts sucrose (YES) agar (Filtenborg and Frisvad, 1980; Filtenborg *et al.*, 1983). Agar plug based methods now include HPLC and direct inlet electrospray mass spectrometry (reviewed in Nielsen et al., 2004 and Larsen et al., 2005).

CEREALS AND CEREAL PRODUCTS

Wheat, rye, barley and oat in the field

Fusarium, Alternaria, Cladosporium and *Claviceps* are very common on grain crops in the field and can reduce the grain quality by their growth due to discolouration (Abdel-Kader *et al.*, 1979; Luke and Barnett, 1979; Chong and Sheridan, 1982; Mazen *et al.*, 1984; Pelhate and Agosin, 1985; Müller and Schwadorf, 1993; Pitt *et al.*, 1994; Andersen *et al.*, 1996; Berleth *et al.*, 1998; Ackermann, 1998; González *et al.*, 1998; Castella *et al.*, 1999a,b; Enikuomehin, 2005; Medina *et al.*, 2006). *Fusarium* ear diseases of cereals (also called head blight or scab) is caused primarily by *Fusarium culmorum* and *F. graminearum* (Clear *et al.*, 2002a,b; González *et al.*, 1996; Wiese, 1987). Both species can produce deoxynivalenol (and related trichothecenes), zearalenone and several other biological active metabolites in the grains (Gareis *et al.*, 1989). While the Fusaria will be eliminated during food processing, a significant carry-over of their mycotoxins will be possible as they are resistant to cleaning of grains, milling, brewing, baking and other food processes. *Fusarium* mycotoxins, including trichothecenes, zearalenone and fumonisins are common in cereal products (see for example Medina *et al.*, 2006 and Schollenberger *et al.*, 2006). Another important species is *F. avenaceum*. This species can produce moniliformin, antibiotic Y, enniatins and fusarin C, whereas earlier reports on production of trichothecenes and zearalenone have been insufficiently verified (Thrane, 1989). Reports on natural occurrence of metabolites from growth of *F. avenaceum* are very scarce. *Fusarium* infections take place by airborne conidia on the heads or by systemic infection. So far no highly resistant wheat or barley cultivars have been developed. For prevention of *Fusarium* diseases crop rotation is advised, as chemical treatment of seeds or application of fungicides to emerged heads is either not 100% effective or profitable nor is it desirable from an environmental point of view.

Alternaria together with *Cladosporium* can cause discolouration of the grains (black or sooty heads) by their abundant presence on the grain during rainy growth seasons. The most common *Alternaria* found on grain belong to the *Alt. infectoria* species-group (Andersen *et al.*, 1996; Kosiak *et al.*, 2004), but neither these *Alternaria* species nor any *Cladosporium* species are known to be able to produce mycotoxins. The most common, toxigenic *Alternaria* spp. belong to the *Alt. tenuissima* species-group and are able to produce alternariol, alternariol monomethyl ether, altertoxin I and tenuazonic acid (Andersen *et al.*, 2002). Alternariols and other *Alternaria* toxins have been detected infrequently in grains (Andrews, 1986; Champ *et al.*, 1991; Chełkowski and Visconti, 1992; Medina *et al.*, 2006).

Ergot, *Claviceps purpurea,* occurs mainly on rye, but certain wheat lines have known to been infected too. The sclerotia, replacing grains, are the visible damage, but in addition *C. purpurea* produces a series of alkaloids toxic towards humans. These alkaloids have been detected in rye and wheat (grains and flour) (Möller *et al.*, 1993; Scott *et al.*, 1992). Crop rotation and good farming practice is the only way to prevent ergot formation.

Wheat, rye, barley and oat in storage

In temperate and mild subtropical climate the dominating storage moulds are species of *Penicillium* and *Aspergillus* (Cylland *et al.*, 1981, Ylimaki, 1981; Hill and Lacey, 1984; Ackermann, 1988; Kunwar, 1989; Adisa, 1994; Frisvad, 1995; Weidenbörner and Kunz, 1995; Andersen *et al.*, 1996; Hasan, 1999). Data on occurrence of these fungi in Canada can be found in Mills *et al.*, (1995). Data based primarily on barley, but also samples of rye and wheat (Scudamore *et al.*, 1993) strongly indicate that a restricted number of *Penicillium* species are of paramount importance in stored cereals. Seventy samples containing ochratoxin A from spoiled barley in Denmark were colonized by the Penicillia in the series *Viridicata, Verrucosa* and *Corymbifera,* including *P. aurantiogriseum, P. cyclopium, P. freii, P. melanoconidium, P. polonicum, P. viridicatum, P. verrucosum* and *P. hordei* (Lund and Frisvad, 1994, 2003; Frisvad and Samson, 2004). The taxonomy of the series *Viridicata*, earlier named *P. verrucosum* var. *cyclopium* and var. *verrucosum* (Samson *et al.*, 1976; Ramírez, 1982), has been revised by Lund and Frisvad (1994) and Frisvad and Samson (2004).

These data are generally in agreement with those of Scudamore *et al.* (1993) and Mills *et al.* (1995), despite the fact that wheat was the commodity examined in those studies. Several toxin-producing Aspergilli have been reported to dominate on cereals, especially *Asp. candidus, Asp. flavus, Asp. niger, Asp. versicolor,* and *Asp. penicillioides* and *Eurotium* spp. at lower water activities (Lacey *et al.*, 1991; Sauer *et al.*, 1992). The toxicology of the latter species has not been thoroughly examined. The most common extrolite produced by *Eurotium* species is echinulin, which causes feed refusal in swine (Vesonder *et al.*, 1988).

Ochratoxin A, citrinin, xanthomegnin, viomellein and vioxanthin have all been found in barley, rye, and/or wheat (Hald *et al.*, 1983; Scudamore *et al.*, 1986, Frisvad, 1995). Several other possibly toxic secondary metabolites are produced by species in series *Viridicata* (Lund and Frisvad, 1994), such as verrucosidin, penicillic acid, cyclopenin, viridicatol, pseurotins, viridic acid, brevianamide A, nephrotoxic glycopeptides, anacine, auranthine, aurantiamine, terrestric acid, puberulonic acid, verrucofortine, puberuline, roquefortine C, meleagrin, oxaline, viridamine and aspterric acid (Frisvad and Lund, 1993), but present day analyses do not cover these metabolites. As mentioned above, the *Penicillium* species most common in cereals are *P. aurantiogriseum, P. cyclopium, P. freii, P. hordei, P. melanoconidium, P. polonicum* and *P. viridicatum*. Furthermore the citrinin and ochratoxin A producing *P. verrucosum* is very common in cereals in temperature climates (Lund and Frisvad, 2003), but much less common in Spain and other Mediterranean countries (Medina *et al.*, 2006). The latter authors found that 30 out of 40 strains of *Aspergillus* section *Nigri* produced ochratoxin (26 of these were *Asp. carbonarius*), while only three out of 20 strains of *Aspergillus* section *Circumdati* produced ochratoxin. Thus in warmer climates *Aspergillus* section *Nigri* and *Circumdati* are probably more important sources of ochratoxin A in barley than *P. verrucosum*. On the other hand the Aspergilli producing ochratoxin A produced relatively low amounts. Medina *et al.*

(2006) examined Spanish malting barley and found that all isolates of *Fusarium verticillioides* and 15 out of 18 isolates of *F. proliferatum* produced fumonisins, 26 out of 90 strains of *Alternaria* produced alternariols, 30 out of 34 strains of *F. graminearum* produced trichothecenes and 10 out of 50 isolates of *Aspergillus flavus* or *Asp. parasiticus* produced aflatoxins.

Other mycotoxins from Penicillia growing in stored cereals in subtropical or tropical climate could include viridicatumtoxin (*P. aethiopicum*), citrinin (*P. citrinum*), cyclopiazonic acid, patulin and roquefortine C (*P. griseofulvum*) and secalonic acid D (*P. oxalicum*) (Frisvad and Filtenborg, 1989), but this has not yet been examined. The importance of mycotoxins produced in cereals by Aspergilli, like the aflatoxins and cyclopiazonic acid, has been pointed out by Pier and Richard (1992), but the taxonomy of these species seems to be less complicated. Other species in *Aspergillus* section *Flavi* may be more common than originally thought as *Asp. nomius* and *Asp. parasiticus* are occasionally identified as *Asp. flavus*.

Mixed feeds may be contaminated with all the mycotoxins listed above, including ochratoxin A produced by black and ochre Aspergilli (Accensi *et al.*, 2004).

Bread (rye and wheat)

The most important species spoiling rye bread with no preservatives added are *Penicillium roqueforti, P. paneum, P. carneum, P. corylophilum, Eurotium repens* and *E. rubrum* (Lund *et al.*, 1996; Spicher, 1985). *Penicillium paneum* and *P. carneum* are newly described species based on significant differences in mycotoxin, DNA and morphological characteristics (Boysen *et al.*, 1996). Isolates belonging to these species have earlier been identified as *P. roqueforti* or *P. roqueforti* var. *carneum* (Frisvad and Filtenborg, 1989). The cultures used for blue-cheese production all belongs to *P. roqueforti*. If preservatives like sorbic acid and propionic acid are added, the spoilage mycobiota would be completely dominated by *P. roqueforti* and the new species *P. paneum* and *P. carneum* (Spicher, 1985; Lund *et al.*, 1996). Species of less importance to the quality of rye bread are: *P. commune, P. solitum, Asp. flavus, Asp. niger, P. decumbens, Paecilomyces variotii, Monascus ruber* and *Mucor* spp. Some of these species may become important for a very short period of time and some are always there but only causing few infections.

Only a few mycotoxins have been detected in rye bread (Reiss, 1972, 1977; Dich *et al.*, 1979; Scott *et al.*, 1992): Aflatoxins, citrinin, ergot alkaloids and patulin. Furthermore the inclusion of citrinin on the list may not be relevant, as the toxin is not produced by the important spoiling species, but has been detected after artificial inoculation. As potential mycotoxins in rye bread, in accordance with the above mentioned list of important spoiling species, the following can be mentioned: isofumigaclavine A and B, roquefortin C, patulin, penitrem A, and penicillic acid, mycotoxins from *P. roqueforti* and *P. carneum* and *P. paneum*. However this list is by far exhaustive, since these species have been found to produce several secondary metabolites which are toxic in certain biological test systems. The list of mycotoxins should also include the toxins formed in the cereals used for rye bread, as mentioned elsewhere in this chapter. This "carry over" is important as it is well known that mycotoxins in contrary to the moulds often are not inactivated during the baking process (Charmley and Prelusky, 1994).

The spoilage is due to species tolerating a lowered water activity (around 0.95) and by the presence of organic acids like acetic acid and propionic acid, which are formed during the fermentation or have been added as preservatives (Frisvad *et al.*, 1992). The infection takes place after the baking process, which obviously kills all fungal propagules and is due to airborne conidia originating from growth of the spoiling species on product wastes in a few specific places in the plant (Lund *et al.*, 1996).

Beer (barley)

The most likely mycotoxin to occur in beer is ochratoxin A (Payen *et al.*, 1983; El-Dessouki, 1992; Guldborg, 1997; Gareis, 1999; Nakajima *et al.*, 1999; Medina *et al.*, 2005b) because it is produced by most isolates of *P. verrucosum* which is common in barley (Lund and Frisvad, 2003). Trichothecenes and zearalenone (Shim *et al.*, 1997; Molto *et al.*, 2000) and fumonisins have been reported from beer (Torres *et al.*, 1998). Beer products have not been examined for other mycotoxins that are produced by Penicillia associated to barley, and some of the potential mycotoxins include citrinin, penicillic acid, verrucosidin and xanthomegnin (Frisvad and Lund, 1993). Sorghum based beers have been shown to be contaminated with aflatoxins, fumonisins and zearalenone (Nkwe *et al.*, 2005). Fumonisins have also been found in Xhosa maize beer (Shephard *et al.*, 2005). *Aspergillus clavatus* may be a major problem in certain situations in malting houses (Flannigan, 1986). This species produces several mycotoxins, including patulin and cytochalasin E (Cole and Cox, 1981).

Maize

The mycobiota associated with maize comprises *Fusarium, Alternaria, Penicillium* and *Aspergillus* species (Pelhate, 1979, 1981; Mastrodisalgado and Carvalho, 1980; McLean and Berjak, 1987; Gbodi *et al.*, 1986; Siradhana *et al.*, 1978, Baird *et al.*, 1995; Broggi *et al.*, 2002). Weidenbörner (2001a) gives an overview of the mycotoxins found in maize. The most important toxigenic species in maize are the fumonisin producers, *F. verticillioides* and *F. proliferatum* (Ono *et al.*, 1999, 2002, 2006; Wilson *et al.*, 2006; Abbas *et al.*, 2006; Samapundo *et al.*, 2005; Farnochi *et al.*, 2005; Scaff and Scussell, 2004; Ghiasian *et al.*, 2004; González *et al.*, 1995, 2002; Almeida *et al.*, 2002a,b; De Farias *et al.*, 2000; Orsi *et al.*, 2000; Castella *et al.*, 1999a,b; Julian *et al.*, 1995; Castro *et al.*, 1995; Adisa, 1994), the trichothecene producers, *F. graminearum* and other spp. (El-Maghraby *et al.*, 1995); the aflatoxin producing *Asp. flavus* (Razzaghi-Abyaneh *et al.*, 2006; Arrus *et al.*, 2005; Da Silva *et al.*, 2004; Machinski *et al.*, 2001; Aziz and Shahin, 1997; Adebajo *et al.*, 1994; Adebajo and Idowu, 1994; Asevedo *et al.*, 1994; Sinha and Sinha, 1992, Doupnik, 1972), the ochratoxin producers *Asp. niger* and *Asp. ochraceus* (Jurjević *et al.*, 1999; Magnoli *et al.*, 2006), the citrinin producer *P. citrinum* (Broggi *et al.*, 2002) and *P. oxalicum*, which produce secalonic acid D (Steyn, 1970; Ehrlich and Lee, 1984; Ehrlich *et al.*, 1985). *Penicillium funiculosum* has also been found in maize (Broggi *et al.*, 2002), but no mycotoxins are known from this species. Of all these mycotoxins, the aflatoxins and fumonisins appear to be the most important. *Fusarium moniliforme* (now *F. verticillioides*) and probably fumonisins may be implicated in esophagal cancer in Transkei (Rheeder *et al.*, 1992; Marasas *et al.*, 1981).

Rice

The mycobiota of freshly harvested rice grain from Italy have shown to be infected with *Alternaria alternata* and *Fusarium equiseti* (Fisher and Petrini, 1992). Stored rice from Argentina and Paraguay is dominated by *Penicillium citrinum, P. islandicum, Aspergillus niger, Asp. flavus* and *F. semitectum* (Tonon *et al.*, 1997). *Penicillium citrinum, Asp. candidus, F. proliferatum* and *P. verrucosum* have been isolated from different rice samples (Pitt *et al.*, 1994; Park *et al.*, 2005a,b,c). *Penicillium islandicum* producing luteoskyrin has been isolated from Japanese rice, together with *Alternaria* spp., *Eurotium* spp., and the potentially toxigenic *Asp. versicolor* and *Asp. ochraceus* (Sakai *et al.*, 2005). Aflatoxins have been found in parboiled rice, with *Asp. parasiticus* and *Asp. flavus* being the most important producers (Lau and Sheridan, 1975; El-Almeida *et al.*, 1991; Reshma and Ahmad, 1998) and trichothecenes, zearalenone and sterigmatocystin have also been found in rice (El-Maghraby, 1996). *Fusarium tricinctum* and *Cylindrocarpon candidum*, which are able to produce trichothecenes, were among the toxigenic species recovered (El-Maghraby, 1996). Problems with yellow rice toxicosis, caused by *P. citrinum, P. islandicum* and *P. citreonigrum* (Enomoto and Ueno, 1974; Ueno, 1974) may still be relevant, as Jayaraman and Kalyanasundarum (1990, 1994a,b) have shown that the mycobiota of rice may establish itself on the parboiled rice, even though this mycobiota has been eliminated by boiling.

Sorghum

Freshly harvested sorghum from Argentina showed *Fusarium* and *Alternaria* as the dominating genera with *F. moniliforme, F. semitectum* and *Alt. alternata* as the most prevalent species, whereas *Penicillium citrinum, P. funiculosum,*

Aspergillus flavus and *Asp. niger* were found less frequently (González *et al.*, 1997, Saubois *et al.*, 1999; Dejene *et al.*, 2004). Other reports show sorghum infected with *Asp. flavus* and *F. verticilloides* and *F. proliferatum* and many of the isolates examined produced aflatoxins and fumonisins, respectively (Da Silva *et al.*, 2004). Analyses of stored sorghum samples from Agentina showed the presence of *Fusarium, Alternaria, Penicillium* and *Aspergillus* as well as mycotoxins such as zearalenone, alternariols and aflatoxins (González *et al.*, 1997). A disease outbreak caused by consumption of mouldy sorghum and maize containing fumonisins has been reported (Sashidhar *et al.*, 1992; Bhat *et al.*, 1997) and deoxynivalenol contamination in sorghum has also caused mycotoxicoses (Ramakrishna *et al.*, 1989).

Pearl millet

Pearl millet was reported to be infected with *Aspergillus flavus, Fusarium moniliforme* (= *F. verticillioides*), *F. chlamydosporum* and *F. semitectum* (Ganbobo and Dostaler, 1990). This resulted in formation of mycotoxins such as aflatoxin B_1, fumonisins, moniliformin and beauvericin (Wilson *et al.*, 2006).

Amaranth, anise, fennel and blackgram seeds

Stored amaranth seeds may contain several mycotoxigenic moulds, including *Aspergillus flavus, Asp. parasiticus. Asp. tamarii*, and *Asp. niger* (Adebanjo and Ikotun, 1994; Bresler *et al.*, 1995). Aflatoxin may also be produced in blackgram seeds (Ahmad, 1993). Fennel and anise seeds also contain potentially mycotoxigenic species (Moharram *et al.*, 1989).

OIL CROPS AND OIL PRODUCTS

Olives

Damaged olives are susceptible to *Alternaria alternata* infection and *Alternaria* mycotoxins, such as alternariols, altertoxin I and tenuazonic acid, have been found in olives before pressing, but not in olive oil (Logrieco *et al.*, 2003). Heperkan *et al.* (2006) found *Penicillium crustosum, P. roqueforti, P. viridicatum, P. citrinum, P. brevicompactum, P. solitum* and *Aspergillus versicolor* in olives, but they also found citrinin in the olives, so the report of the citrinin-producing *P. citrinum, P. expansum* and *P. verrucosum* (Sahin *et al.*, 1999) in olives may explain the production of this nephrotoxin.

Sunflower seeds

Alternaria and *Fusarium* are the main genera infecting the internal seed of sunflower and *Alt. alternata* can be found in between 15 and 60% of surface disinfected seeds, while *F. moniliforme, F. semitectum* and *F. solani* constitute up to 3% of infected seeds (Zad, 1978; Dawar and Ghaffar, 1991; Nawaz *et al.*, 1997; Bhutta, 1998; Begum *et al.*, 2003; Logrieco *et al.*, 2003). Alternariol and alternariol monomethyl ether have be detected in seeds from infected sunflower heads in quantities up to 1.8 and 0.13 mg kg^{-1}, respectively (Torres *et al.*, 1993; Logrieco *et al.*, 2003; Pozzi *et al.*, 2005). As the sunflower seeds are de-scaled and stored, the mycobiota will change in favour of *Aspergillus* and other storage fungi. The most common toxigenic *Aspergillus* species in sunflower seeds are *Asp. flavus, Asp. niger, Asp. parasiticus* and *Asp. tamarii* (Raut, 1975; Jimenez *et al.*, 1991; Logrieco *et al.*, 2003) and high levels of aflatoxins (230 ng kg^{-1}) were found in Tunisian seeds (Logrieco *et al.*, 2003).

Pumpkin, pine, cotton, melon and fenugreek seeds also may contain mycotoxins if not properly handled, as several mycotoxigenic moulds have been found (Gupta and Agrawal, 1976; Hashmi and Thrane, 1990; Mazen *et al.*, 1990; Ekundayo and Idze, 1990; Weidenbörner, 2001b,c; Hasan, 2001; El-Nagerabi, 2002). Aflatoxins may also occur in *Foeniculum vulgare* seeds (Rani and Singh, 1989).

Rape seeds

A Spanish study showed that *Alternaria alternate, Penicillium* spp. and *Apergillus flavus* were dominant on oilseed rape, but only one sample contained mycotoxins and then only aflatoxin (Vinas *et al.*, 1994). Another study of British oilseed rape products showed that 33% of the samples contained one or more of *Alternaria* mycotoxins (Nawaz *et al.*, 1997).

Peanuts

Peanuts are very often degraded by *Aspergillus flavus* and *Asp. niger* (Diener, 1960; Doupnik, 1969; Joffe, 1969a,b; Barnes *et al.*, 1970; Hanlin, 1973; Moubasher *et al.*, 1979; Waliyar and Roquebert, 1979; Waliyar and Zambettakis, 1979; Lisker *et al.*, 1993; Pitt *et al.*, 1993; Weidenbörner and Kunz, 1994; Ranzani and Fonseca, 1995; Ismail, 2001; Mphande *et al.*, 2004). The growth of *Asp. flavus* and aflatoxin B_1 production has been documented since aflatoxin was discovered and was first found in peanuts sold for human consumption in 1967 (Taber and Schroeder, 1967; Lopez and Crawford, 1967; Mehan and McDonald, 1984; El-Maghraby and El-Maraghy, 1987, 1988; Mehan *et al.*, 1991; Lisker *et al.*, 1993; Martinsmaciel *et al.*, 1996) and is probably one of the most severe mycotoxins problems of all. *Aspergillus flavus* can act as an endophyte in peanuts or invade the peanut fruits (Hanlin, 1970; McDonald, 1970; Diener *et al.*, 1987; Pitt *et al.*, 1991a; Calori-Domingues *et al.*, 1996; Horn and Dorner, 1998; Horn, 2003, 2005; Bankole *et al.*, 2005).

Coconut

Several fungi are important in coconut, including *Asp. niger, Asp. flavus, P. crustosum* and *P. chrysogenum* (Kinderlerer, 1984a,b; Kinderlerer and Hatton, 1991; Ismail, 2001). Other potentially toxigenic fungi from coconut included *Asp. ochraceus, Paecilomyces variotii, Penicillium crustosum* and *P. islandicum*. Aflatoxin and ochratoxin have been found in coconut (Zohri and Saber, 1993).

Oils and margarines

The mycobiota of oils, margarines and similar products (mayonnaise and dressing) is quite restricted to species of *Cladosporium, Aspergillus* and *Penicillium* (Kurtzman *et al.*, 1971; Hocking, 1994; Okpokwasili and Molokwu, 1996). Species such as *P. expansum, P. chrysogenum, P. corylophilum* and *P. glabrum* were frequent spoilers of margarine (Hocking, 1994)

NUTS

The mycobiota of nuts is generally dominated by *Fusarium, Alternaria* and *Cladosporium* in the field (Belisario *et al.*, 2002) and by *Aspergillus, Penicillium* and *Trichoderma* in storage (Abdel-Hafez and Saber, 1993; Abdel-Gawad and Zohri, 1993; Bayman *et al.*, 2002a; Weidenbörner *et al.*, 1995). However certain species of *Alternaria, Aspergillus, Fusarium* and *Penicillium* are particularly common on nuts, especially at low water activities. On hazelnuts, pistachio and walnuts *Alt. alternata* and *Alt. arborescens* species-group are the most common *Alternaria* species (Belisario *et al.*, 2002, 2004; Pryor and Michailides, 2002), while *F. acuminatum, F. avenaceum* and *F. semitectum* were the most common *Fusarium* species in Italian walnuts (Belisario *et al.*, 2002). In storage *Asp. niger* complex strains were dominating on pistachios, almonds and walnuts, *Asp. flavus, Asp. tamarii* and *Penicillium* spp. were dominant on brazil nuts (Mojtahedi *et al.*, 1979; Bayman *et al.*, 2002a). *Penicillium* species were also dominant on walnuts in many cases and often negatively correlated with occurrence of *Asp. niger* and *Rhizopus* (Bayman *et al.*, 2002a). The Penicillia often occurring on pecans and walnuts are *P. discolor, P. expansum* and *P. crustosum* (Wells and Payne, 1976; Frisvad, 1989; Frisvad *et al.*, 1996). Brazil nuts are predominantly contaminated with *Asp. flavus*, giving problems with aflatoxins, cyclopiazonic acid and 3-nitropropionic acid (Arrus *et al.*, 2005). Hazelnuts may also be contaminated with aflatoxin (Simsek *et al.*, 2002). Almonds are also prone to deterioration and aflatoxin production by *Asp. flavus* (King and Schade, 1986; Purcell *et al.*, 1980; Fuller *et al.*, 1977) as are pistachio nuts (Sommer *et al.*, 1986; Heperkan *et al.*, 1994; Doster and Michailides, 1995, 1999). Chestnuts are often infected by *Asp. flavus, Asp. niger, P. discolor* and *P. crustosum* (Wells and Payne, 1979; Overy *et al.*, 2003). *Emericella nidulans, Asp. ochraceus, Asp. melleus,* and *Asp. fumigatus* are less frequently found on nuts (Bayman *et al.*, 2002a).

In general aflatoxins may be produced on nut seeds (Abdel-Gawad and Zohri, 1993). Emodin is a less common mycotoxin which has been detected in chestnuts and in salted seeds (Wells *et al.*, 1975; Hasan, 1998), but aflatoxin and cyclopiazonic acid are the most important mycotoxins.

BEANS AND PEAS

Beans and peas are often infested with the same filamentous fungi as cereals, including *Alternaria, Aspergillus, Fusarium* and *Penicillium* series *Viridicata*, (Mislivec *et al.*, 1975; Åkerstrand and Josefsson, 1979; Abdel-Hafez, 1984; Sanchis *et al.*, 1988; Naumova, 1988; Mantle and McHugh, 1993; Pitt *et al.*, 1994; Tseng *et al.*, 1995a,b; Sinha *et al.*, 1999; Roy *et al.*, 2001; Castillo *et al.*, 2004). In freshly harvested black beans (*Phaseolus vulgaris*) from Argentina, *Alt. alternata* was the most prevalent species with 23% infection, while *Aspergillus* and *Penicillium* constituted 7 and 6%, respectively (Castillo *et al.*, 2004). A Brasilian study showed that *Asp. flavus, Asp. parasiticus, Asp. ochraceus* and *P. citrinum* were common on stored black and coloured cultivars of beans (Costa and Scussel, 2002). Ochratoxin A has been found in beans, but only those infected with *Penicillium* and/or *Aspergillus* species (Milanez and Sabino, 1989; Costa and Scussel, 2002; Domijan *et al.*, 2005). Kritzinger *et al.* (2003) found natural occurrence of fumonisins in cowpeas (*Vigna unguiculata*) together with the producer, *F. proliferatum*. They also found other *Fusarium* species, including *F. equiseti, F. chlamydosporum, F. graminearum, F. sambucinum, F. semitectum* and *F. subglutinans*. Cowpeas are in general quite well protected from fungal attack (Onesirosan, 1982a,b; Beuchat, 1984; Halls and Harman, 1991). Lentils also contain aflatoxins and maybe other toxins from potentially mycotoxic moulds (Abdel-Hafez, 1988; El-Nagerabi and Elshafie, 2001; Abd-Allah and Hashem, 2006).

Mycotoxigenic moulds and aflatoxins and other mycotoxins are occasionally found in peas and soya beans (Miller and Roy, 1982; El-Kady and Youssef, 1993; Paster *et al.*, 1993; Mills and Woods, 1994; Roy and Ratnayake, 1997; Weidenbörner, 1997; El-Nagerabi *et al.*, 2000; Schollenberger *et al.*, 2006). Aflatoxin has also been found in pigeon peas (Bankole *et al.*, 1995) and chickpeas (Ahmad and Singh, 1991).

FRUIT AND FRUIT PRODUCTS

Pomaceous and stone fruits

Pomaceous and stone fruits can be degraded by a number of pathogenic species including *Monilia laxa, M. fructigena* and *Rhizopus stolonifer*. For example the "box rot" of dried French prunes, which is soft, sticky, macerated areas on the fruit and slippage of the skin under slight pressure due to the activity of pectinolytic enzymes produced by these fungi (Sholberg and Ogawa, 1983). However these fungi are probably not mycotoxin producers. *Alternaria alternata, Alt. tenuissima* species-group, *Fusarium lateritium, Penicillium expansum, P. crustosum* and *P. solitum* were reported as toxigenic organisms able to produce rot in apples (Raper and Thom, 1949; Samson *et al.*, 1976; Frisvad, 1981; Pitt *et al.*, 1991b; Vinas *et al.*, 1992; Reuveni *et al.*, 2002; Serdani *et al.*, 2002; Andersen and Thrane, 2006). *Alternaria arborescens* species-group and *P. expansum* are commonly isolated from cherry (Roberts *et al.*, 2000; Andersen and Thrane, 2006). In Ya Li pear fruit *Alt. yaliinficiens* had been reported as a severe fruit spoiler (Roberts, 2005).

Penicillium expansum is known for its production of patulin and citrinin and these mycotoxins have been found in mouldy fruits (Harwig *et al.*, 1973; Ciegler *et al.*, 1977; McKinley and Carlton, 1991; Vinas *et al.*, 1993). Other mycotoxins produced by *P. expansum*, such as roquefortine C and chaetoglobosin C (Frisvad and Filtenborg, 1989), or by *P. crustosum*, such as terrestric acid, roquefortine C and penitrem A, have not yet been reported from rotting pomaceous fruits. Naturally occurring mycotoxins, such as patulin, chaetoglobosins, communisin B and roquefortine C, have been detected in windfall apples and in an apple still on the tree (Andersen *et al.*, 2004).

When artificially inoculated into apples, *Alt. alternata* produced alternariols in both the rotten and sound part of apples (Stinson *et al.*, 1980, 1981; Ozcelik *et al.*, 1990; Robiglio and Lopez, 1995). *Alternaria* species from the *Alt.*

CHAPTER 1, Figure 1: *Colletotrichum gloeosporioides* symptoms in avocado (with permission of L. Coates).

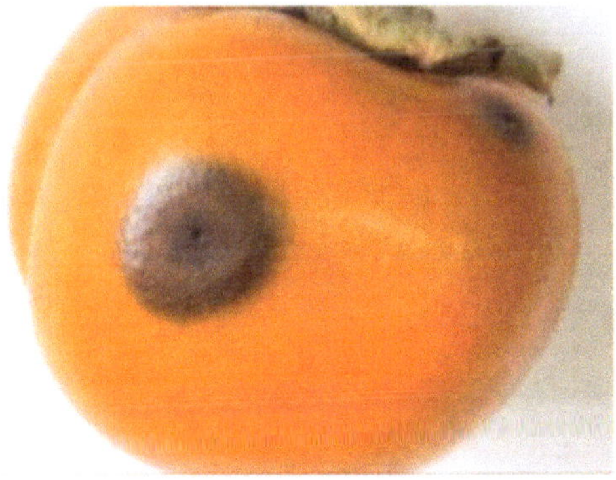

CHAPTER 1, Figure 3: *Alternaria alternata* symptoms in persimmon fruits.

CHAPTER 6, Figure 1: Formation of fruit bodies of the fungus *Talaromyces macrosporus*; numerous yellow coloured ascomata (fruiting bodies) can be observed after 7 (left) and 14 days (right) after inoculation of the fungus.

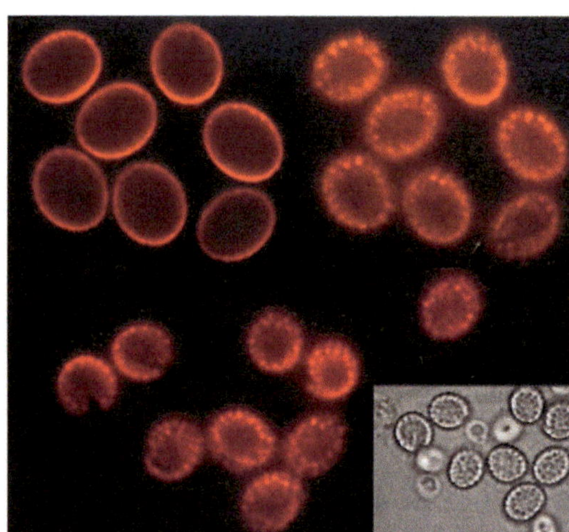

CHAPTER 6, Figure 5: Ascospores of *T. macrosporus* exhibit a strong autofluorescence in a broad range of excitation wavelengths. Top panel shows two focal planes through the spores, illustrating the spikes on the surface of the cell wall. The lower panel shows prosilited ascospores with ruptured cell walls where ejection has taken place (for an image with transmitted light, see the inset).

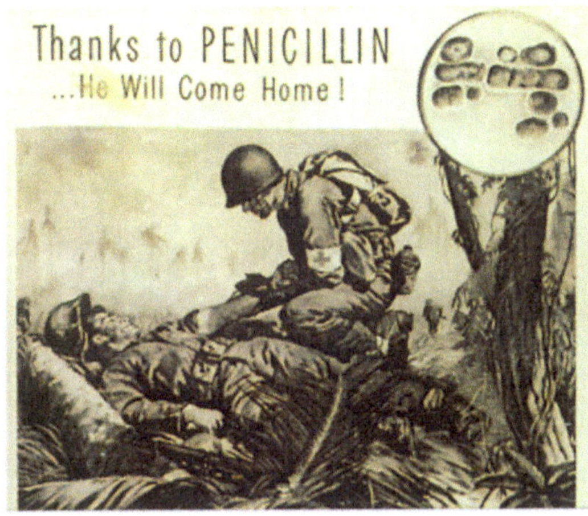

CHAPTER 9, Figure 2: Historic poster from WWII.

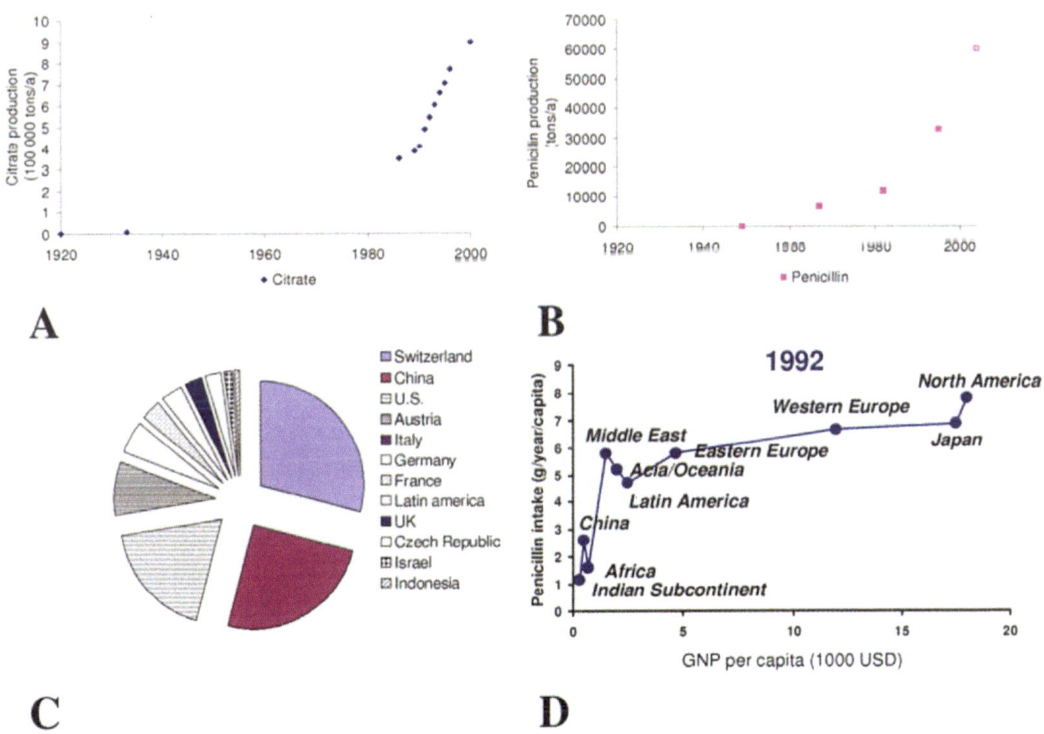

CHAPTER 9, Figure 3: World market development of (A) citrate (Connor, 1998), (B) penicillin (Demain and Elander, 1999), (C) production capacity by country for 1996 (Connor, 1998), and (D) penicillin usage vs. gross national product for 1992.

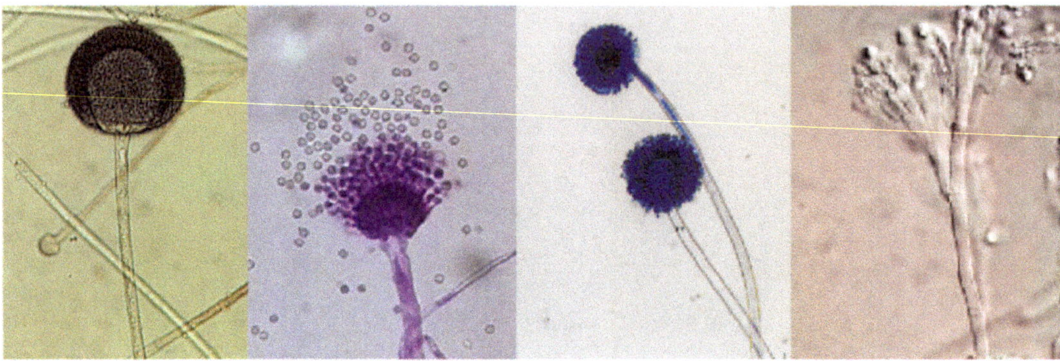

CHAPTER 9, Figure 4: Microscopic images of the sporulating structures of *Rhizopus oryzae*, *Aspergillus terreus*, *Aspergillus niger* and *Penicillium chrysogenum* (from left to right). Pictures kindly provided by Dana Savicka of the Institute of Chemical Technology Prague, and Michel Cavalla.

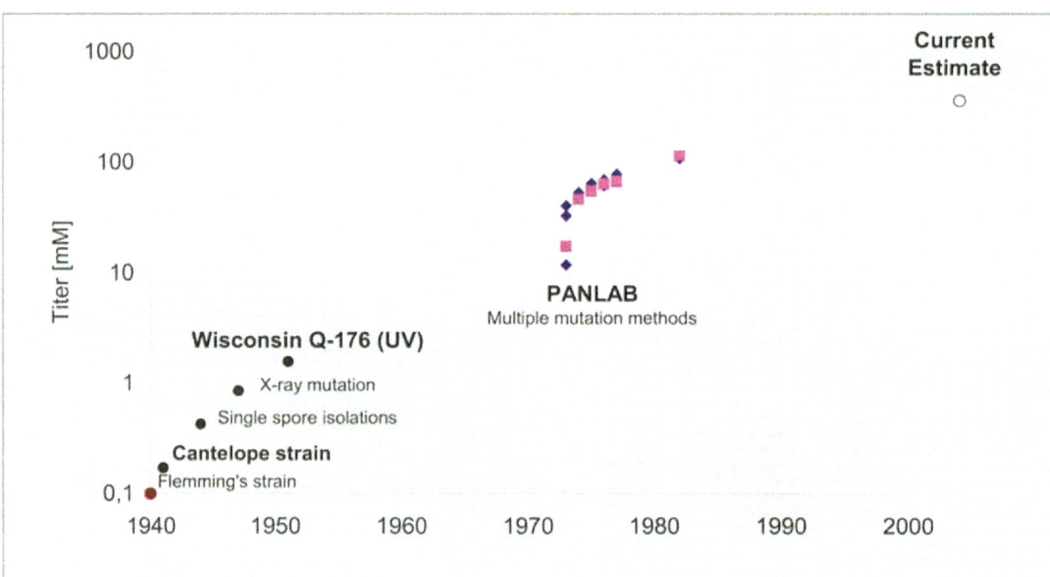

CHAPTER 9, Figure 5: The development of penicillin titers with an indication of Panlabs. contribution to penicillin production improvement through classical strain improvement in the 1970s. Strain lineages P1 u and P2 are marked.

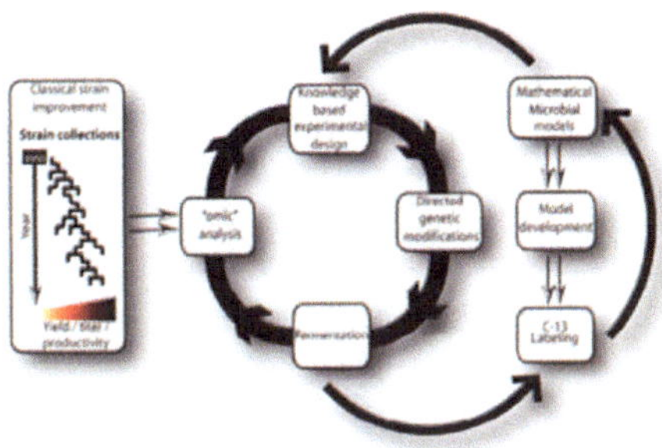

CHAPTER 9, Figure 7: Process improvement through metabolic engineering is an iterative approach, where, using all available information, several rounds of directed strain improvements, and in-depth experimental analysis, leads to a strain with preconceived physiological attributes. The importance of classically improved strain lineages becomes apparent when this approach is applied to penicillin and citrate production. The information obtained from the in-depth study of these strain lineages can be directly applied to better design strategies for directed strain improvement.

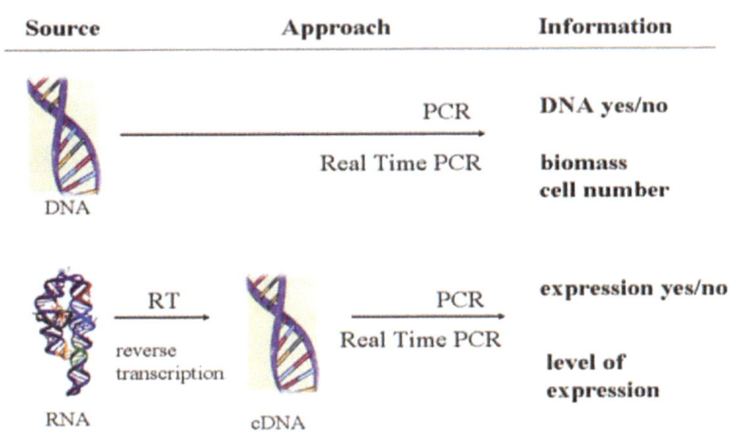

CHAPTER 13, Figure 2: Level of information obtained by PCR/Real Time PCR, respectively.

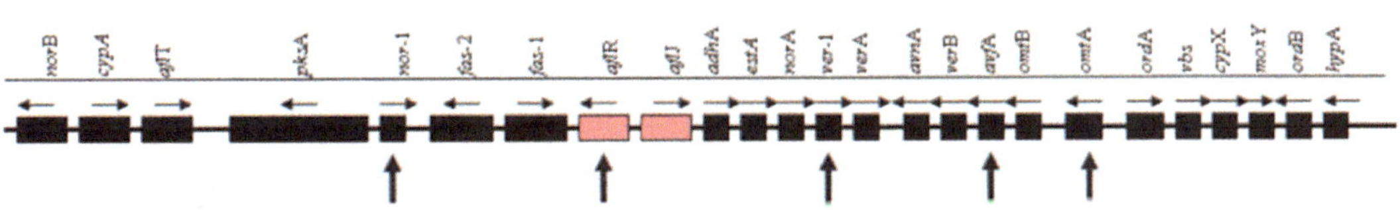

CHAPTER 13, Figure 5: Scheme of the complete aflatoxin biosynthesis cluster according to Yu *et al.* (2004). The genes used as target sequences in diagnostic PCR are indicated as vertical arrows.

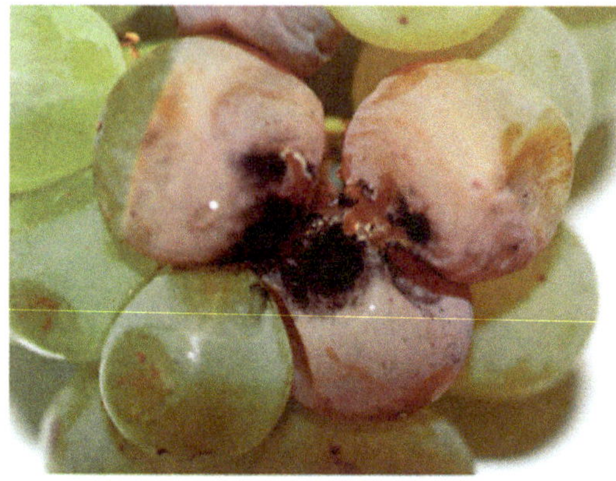

CHAPTER 15, Figure 2: *Aspergillus carbonarius* infection of Semillon berries.

CHAPTER 16, Figure 1: Roquefort and Camembert cheeses.

CHAPTER 16, Figure 3: *Penicillium discolor.* Sporulating structures and conidia.

CHAPTER 17, Figure 1: Fungal fermented foods (a: sufu; b: men; c: tempe; d: oncom; e: soy sauce; f: Camembert; g: blue-veined cheese).

CHAPTER 19, Figure 1: A dense crop of mushrooms (top) is produced by the following scheme of cultivation of *A. bisporus* according to the Dutch system (bottom).

CHAPTER 19, Figure 2: Mushroom infected with *Verticillium fungicola* (dry bubble disease).

CHAPTER 19, Figure 3: *Boletus edulis* in a sixty-year old wooded bank of *Quercus robur* in The Netherlands.

tenuissima species-groups may be involved in core rot of apples (Serdani *et al.*, 1998, 2002) and *Alternaria* mycotoxins may accumulate in apples (Robiglio and Lopez, 1995).

Fruit juice

Naturally occurring alternariols have been found in Spanish apple juice concentrates (Delgado and Gómez-Cordovés, 1998) and in Danish apple pulp (Andersen *et al.*, 2004). Communesin B, chaetoglobosin A and C and expansolide have all been detected in artificially contaminated black currant and cherry juices (Larsen *et al.*, 1998a) and chaetoglobosins and communesins have been found in naturally contaminated apple fruits (Andersen *et al.*, 2004).

Fruit juices can be contaminated during production and storage at the plant as well as in the home of the consumer. This kind of fungal growth is often directly observable, and therefore such mouldy products will rarely be consumed. Contamination of the fruits before juice production can be a severe problem, resulting in carry-over of mycotoxins, especially patulin (see above), but the most important problem is contamination of pasteurized juice, where heat-resistant fungi may grow in the juice and produce mycotoxins (Beuchat and Rice, 1979). The mycotoxin that is most relevant is patulin, which can be produced by *Byssochlamys nivea* (Kis *et al.*, 1969; Scurti *et al.*, 1973; Splittstoesser and Splittstoesser, 1977). Another heat-resistant patulin producing fungus is *Eupenicillium lapidosum* (Williams *et al.*, 1941; Myrchink, 1967). There have been no reports since 1967 on toxigenic *Eupenicillium* species being a problem in fruit products but *Talaromyces, Neosartorya* and *Eurotium* species are also often found in pasteurized juices and other fruit-based products (Splittstoesser and Splittstoesser, 1977; Beuchat, 1986; King and Halbrook, 1987; Scott and Bernard, 1987; Splittstoesser *et al.*, 1989; Udagawa, 1991; Jesenská *et al.*, 1991, 1993; Spotti *et al.*, 1992; Enigl *et al.*, 1993; Quintavalla and Spotti, 1993; Tournas, 1994; Suresh *et al.*, 1996; Udagawa, 2000; Yaguchi *et al.*, 2005). *Neosartorya fisheri* has been shown to produce the very toxic fumitremorgins and verruculogen under different laboratory conditions (Nielsen *et al.*, 1988, 1989a,b).

Canned fruits like apricots and peaches sometimes suffer from textural changes due to heat-resistant fungal enzymes produced in the raw fruits (Harris and Dennis, 1980) or to the enzymatic activity of surviving heat resistant fungi like *Byssochlamys fulva* (Rice *et al.*, 1979).

Mango and papaya

Mango and papaya fruits in the orchards will contain many toxigenic species of *Aspergillus* in the rhizospere soil (Dube *et al.*, 1980; Singh, 1972). Two of the most important species are *Asp. niger* and *Asp. carbonarius*, producing toxic naphtho-γ-pyrones (Ghosal *et al.*, 1979), but given the production of ochratoxin A reported from these black Aspergilli, this toxin is a far more important mycotoxin to check for in mango and papaya based products (Abdel-Sater *et al.*, 2001).

Citrus fruits

Citrus trees (lemons, oranges, mandarins, tangerine, cumquats, tangelo, limes, pomelos, grapefruits) are susceptible to many different plant pathogenic *Alternaria* species attacking leaves (Simmons, 1999a; Peever *et al.*, 2002) depending on citrus cultivar. Few of these *Alternaria* species have also been isolated from citrus fruit lesions: *Alt. citri* and *Alt. tenuissima* species-group from oranges (Simmons, 1990). *Alternaria tangelonis* and *Alt. turkisafria* have on occasion been isolated from lesions of tangelo (Simmons, 1999a). The two latter species have been shown to produce tenuazonic acid, tentoxin, altertoxin I, alternariols and altersetin in pure culture (Andersen *et al.*, 2005). In storage three spoilage Penicillia are of paramount importance, *Penicillium digitatum, P. italicum* and *P. ulaiense* (Birkinshaw *et al.*, 1931; Raper and Thom, 1949; Westerdijk, 1949; Holmes *et al.*, 1994) and so is a series of *Alternaria* species in the field (Simmons, 1999a). According to Holmes *et al.* (1994), *P. ulaiense* only appears when the other two pathogens are inhibited by fungicides and *P. ulaiense* is much more related to *P. italicum* than *P. digitatum*. *Fusarium poae* has also been reported on decayed citrus fruits in Georgia and Russia (Booth, 1971).

Germination of *P. digitatum* conidia on citrus, which are non-climatic fruits, is stimulated by certain combinations of the volatiles surrounding wounded fruit, notably limonene, α-pinene, sabinene, ß-myrcene, acetaldehyde, ethanol, ethylene and CO_2. Ethylene did not stimulate the germination of *P. digitatum* conidia in the non-climacteric fruit (Eckert and Ratnayake, 1994), whereas ethylene invited fungal attack in the climacteric tomatoes, avocadoes and bananas (Flaishman and Kolattukudy, 1994). Other components in oranges, such as simple sugars and organic acids also stimulate conidium germination in *P. digitatum* (Pelser and Eckert, 1977). Furthermore, a few fruits spoiled by fungi can cause reduced shelf life of the wound fruits due to accelerated ripening or senescence triggered by the releasing of the gas ethylene (Rippon, 1980). This shows that the associated mycobiota can tolerate and is sometimes even stimulated by the acids and other protecting volatile and non-volatile phytoalexins of citrus fruits in combination with the ability to produce pectinases and other citrus skin degrading enzymes. The fungal activities result in serious weight loss, shrinkage and softening of the citrus fruits (Ben-Yehoshua *et al.*, 1987).

Alternaria mycotoxins have been found in mandarins (Logrieco *et al.*, 1990) and they can also be produced in lemons and oranges (Stinson *et al.*, 1981). The mycotoxins found include tenuazonic acid and alternariols. No mycotoxins from *P. italicum* and *P. digitatum* have been found in citrus fruits, yet these fungi produce compounds that are toxic to bacteria, plants, brine shrimps and chick embryos (Faid and Tantaoui-Elaraki, 1989; Tantaoui-Elaraki *et al.*, 1994). One *P. italicum* isolate that was toxic in laboratory animals (Kriek and Werner, 1981) was found to produce the mycotoxin 5,6-dihydro-4-methoxy-2H-pyran-2-one (Gorst-Allman *et al.*, 1982) related to verrucolone (arabenoic acid) (Isaac *et al.*, 1991; Larsen *et al.*, 1998b). Several other *P. italicum* secondary metabolites have been identified, but not tested for toxicity (Arai *et al.*, 1989). *P. digitatum* has been found to produce tryptoquialanins and tryptoquialanons (Ariza *et al.*, 2002; Frisvad and Samson, 2004).

Grapes

Grapes on the vine are easily attacked by filamentous fungi like *Botrytis cinerea*, but traditional field fungi, such as *Alternaria* and *Fusarium* are rarely reported. There have, however, been reports on *Alt. alternata* rot in table grapes in cold-stores (Swart and Holz, 1991). The most important fungi on grape are the black Aspergilli (Gupta, 1956, Nair, 1985; Snowdon, 1990), as many of these may produce ochratoxin A in the grape juice that may also end up in wines (Cabañes *et al.*, 2002). There is general agreement that *Asp. carbonarius* is the main producer of ochratoxin A in grapes (Heenan *et al.*, 1998; Da Rocha Rosa *et al.*, 2002; Sage *et al.*, 2002, 2004; Battilani and Pietri, 2002; Abarca *et al.*, 2003; Battilani *et al.*, 2003; Magnoli *et al.*, 2003; Serra *et al.*, 2003, 2005; Leong *et al.*, 2004; Bellí *et al.*, 2004, 2005; Bau *et al.*, 2005; Valero *et al.*, 2005), but a certain rather low percentage of strains of *Asp. niger* and *Asp. tubingensis* may also produce ochratoxin A (Medina *et al.*, 2005a; Perrone *et al.*, 2006). Few members of *Aspergillus* section *Circumdati* have been found on grapes, and some of these strains were reported to produce ochratoxin A too (Pardo *et al.*, 2005b; Bellí *et al.*, 2004).

The well-known producer of ochratoxin A, *Penicillium verrucosum* has never been found in grapes, but other *Penicillium* species have been found such as *P. expansum* producing citrinin and patulin in pure culture. Only one strain out of 51 of *P. expansum* could produce citrinin in grape juice, whereas 33 strains out of 51 could produce patulin (Abrunhosa *et al.*, 2001). Out of 379 strains of *Penicillium* growing on "passito" grapes, three were claimed to produce ochratoxin A (Torelli *et al.*, 2006), but the identification of these three Penicillia was equivocal. To screen for ochratoxin A, the Penicillia were incubated at 37 °C on coconut cream agar; however, most Penicillia are unable to grow at 37 °C, let alone produce ochratoxin at that temperature. The results are therefore doubtful. The most important ochratoxin A producer in wine by far is thus *Asp. carbonarius*, followed by *Asp. niger* and in rare cases *Asp. ochraceus*. The water and temperature relationships of the important ochratoxin A producing species,

Asp. carbonarius, has been examined by Mitchell *et al.* (2004).

Wine (grape)

Wine has been shown to contain ochratoxin A in several investigations (see for example Zimmerli and Dick, 1996; Burdaspal and Legarda, 1999; Otteneder and Majerus, 2000; Majerus *et al.*, 2001; López de Cerain *et al.*, 2002; Stefaniki *et al.*, 2003; Ng *et al.*, 2004). Few filamentous fungi are able to grow in wine, so the ochratoxin A contamination is usually caused by ochratoxin A production in the grape fruits (see above). In few instances wine has been contaminated, but this has only been recorded in homemade wines. In one case, *Penicillium crateriforme* was growing in homemade rhubarb wine, which resulted in the production of large amounts of rubratoxin, which again caused disease in one boy that needed to have a liver transplantation, because of rubratoxin intoxication (Richer *et al.*, 1997).

Dried fruits

The most common toxigenic fungi associated to dried fruits are Aspergilli and their perfect states: *Eurotium* spp., *Aspergillus flavus, Asp. niger, Asp. carbonarius, Asp. ochraceus* and *Petromyces alliaceus* (Pitt *et al.*, 1993; Zohri and Abdel-Gawad, 1993; Iamanaka *et al.*, 2005). *Aspergillus ochraceus, Asp. melleus* and *Petromyces alliaceus* are common on figs and isolates of the latter species could produce ochratoxin A in pure culture (Doster *et al.*, 1996; Bayman *et al.*, 2002b). Dried grapes (raisins, sultanas, etc.) may contain significant amounts of ochratoxin A (MacDonald *et al.*, 1999) and the ochratoxin A producers include *Asp. carbonarius, Asp. niger* and to a smaller extent *Asp. ochraceus* (Abarca *et al.*, 2003; Iamanaka *et al.*, 2005; Romero *et al.*, 2005). In raisins, strains of *Asp. flavus* produced cyclopiazonic acid, while strains of *Penicillium citrinum* produced citrinin and *Alternaria alternata* produced alternariols and tenuazonic acid in pure culture (Romero *et al.*, 2005).

BEVERAGES

Coffee

Green coffee beans and coffee cherries may be a host for many different fungi (Arocho *et al.*, 2005), but the most important toxigenic ones include *Aspergillus* spp. and *Penicillium* spp. (Pardo *et al.*, 2005a; Martins *et al.*, 2003; Batista *et al.*, 2003; Mislivec *et al.*, 1983; Betancourt and Frank, 1983; Levi *et al.*, 1974). Monsooned coffee had a different mycobiota than traditional coffee (Tharappan and Ahmad, 2006), consisting of *Aspergillus niger, Asp. tamarii, Asp. candidus, Penicillium* spp. and *Absidia heterospora*. Insects play a role in the spreading of different fungal species. For example the coffee berry borer carries the species *Penicillium coffeae* (Perez *et al.*, 2003).

Among the Penicillia, *P. citrinum* is probably the most important toxigenic species (Suárez-Quiroz *et al.*, 2004), but citrinin has not been found naturally occurring in coffee. Other Penicillia recovered were *P. glabrum, P. minioluteum,* and *P. brevicompactum* (Frank, 2001; Suárez-Quiroz *et al.*, 2004; Batista *et al.*, 2003), but few recognized mycotoxins are produced by these species. Ochratoxin A is the most important mycotoxin in coffee (Levi *et al.*, 1974). *Aspergillus ochraceus* and related species appear to be the prominent ochratoxin A producers in coffee (Urbano *et al.*, 2001; Ahmad and Magan, 2002; Mislivec *et al.*, 1983), but *Asp. carbonarius* is more and more frequently recovered (Joosten *et al.*, 2001; Taniwaki *et al.*, 2003). *Aspergillus niger, Asp. lacticoffeatus,* and *Asp. sclerotioniger* are less common ochratoxin A producing black *Aspergillus* species recovered from green coffee beans (Taniwaki *et al.*, 2003; Samson *et al.*, 2004a). Concerning *Asp. niger* and *Asp. tubingensis,* these fungi are very common in coffee, but few of the strains from coffee are able to produce ochratoxin A, and if they do, they produce ochratoxin in low amounts (Taniwaki *et al.*, 2003; Batista *et al.*, 2003; Urbano *et al.*, 2001; Joosten *et al.*, 2001). Even though *Asp. flavus* is occasionally recovered, and some of the strains produce aflatoxins (Batista *et al.*, 2003), this fungus is not genuinely associated with coffee. The most prominent group of toxigenic fungi associated with coffee are species from *Aspergillus* section *Circumdati* (formerly the *Aspergillus ochraceus* group). Nearly all species in this section have been found in coffee (Batista *et al.*, 2003), but the correct identification of species in this group is difficult, and should be based on a polyphasic approach (Frisvad *et al.*, 2004b). Even though *Asp. ochraceus* is often listed as the most common toxigenic fungus on green coffee beans (Batista *et al.*, 2003; Taniwaki *et al.*, 2003; Urbano *et al.*, 2001; Mislivec *et al.*, 1983), it may be *Asp. westerdijkiae* that is most common (Frisvad *et al.*, 2004b; Vega *et al.*, 2006). The best producers of ochratoxin A in green coffee beans are *Asp. westerdijkiae* and *Asp. steynii*, while *Asp. ochraceus* is a weak and also inconsistent producer of ochratoxin. For example, the strains often used in laboratory studies on ochratoxin A production in coffee or cereals are not *Asp. ochraceus*, but *Asp. steynii* (D 2306 in Mantle and Chow, 2000) or *Asp. westerdijkiae* (Blank *et al.*, 1998; Lee and Magan, 1999; Pardo *et al.*, 2004).

Ochratoxin A is by far the most important mycotoxin in coffee (von der Stegen *et al.*, 1997, 2001; Otteneder and Majerus, 2001; Joosten *et al.*, 2001; Leoni *et al.*, 2000; Romani *et al.*, 2000; Blanc *et al.*, 1998; Patel *et al.*, 1997; Nakajima *et al.*, 1997; Pittet *et al.*, 1996; Viani, 1996; Studer-Rohr *et al.*, 1995; Micco *et al.*, 1989). Roasting of the green coffee beans will reduce the amount of ochratoxin A, but some of the toxin will remain (Tsubouchi *et al.*, 1987; von der Stegen *et al.*, 2001; Suárez-Quiroz *et al.*, 2005). Fungal growth in coffee after it is being brewed is probably not relevant, since the used coffee is discarded.

Tea

Tea and herb tea may also contain mycotoxins, such as aflatoxin and fumonisins, but there are few data on mould or mycotoxin problems with tea (Halt and Klapec, 2005; Suga *et al.*, 2004; Omurtak and Yazicioglu, 2004; Martins *et al.*, 2001a,b; Elshafie *et al.*, 1999; Halt, 1998; Efuntoye, 1996; Abdel-Hafez and El-Maghraby, 1992; Mahmoud *et al.*, 1992; Halweg and Podsiadlo, 1991/1992; Cloete and Kotze, 1990). Some of the common fungi on tea are *Aspergillus flavus* and *Aspergillus* section *Nigri* (Martins *et al.*, 2001b). Aflatoxigenic fungi, such as *Asp. pseudotamarii*, has been found in tea field soils (Ito *et al.*, 1998), but the production of aflatoxin in tea by this species has not been studied.

Cocoa

Ochratoxin A has been found in cocoa beans and in chocolate (Amezqueta *et al.*, 2005; Brera *et al.*, 2005; Bonvehi, 2004; Tafuri *et al.*, 2004). Toxigenic fungi of particular importance for cocoa are *Aspergillus* sections *Circumdati* and *Nigri*, but *Penicillium* species are also found.

VEGETABLES

Many vegetables, including cale, broccoli, cauliflower, etc., are not particularly prone to fungal attack, as these are often infected and spoiled by bacteria, resulting in a low fungal count (Tournas, 2005). On the other hand, certain dried vegetables may be infected with *Aspergillus flavus, Asp. niger, Asp. fumigatus, Rhizopus oryzae, Penicillium oxalicum, Rhizomucor pusillus* and *Fusarium equiseti* (Adebanjo and Shopeju, 1993), so there is potential for mycotoxin production in such products. Aflatoxin may be produced in cold storage chillies for example (Kiran *et al.*, 2005).

Pepper

Damaged or overriped pepper fruits (*Capsicum annuum*) are commonly infected by *Alternaria alternata* and spoilage can continue during low temperature storage (Logrieco *et al.*, 2003). Alternariols, altenuene and tenuazonic acid have been detected in mouldy pepper fruit from Italy (Bottalico and Logrieco, 1998).

Black Pepper (*Piper nigrum*) and other pepper types have been reported to contain aflatoxins (Pal and Kundu, 1972; Scott and Kennedy, 1973; Suzuki *et al.*, 1973; Seenappa and Kempton, 1980; Garrido *et al.*, 1988, 1992; Martinez-Magana *et al.*, 1989; Aziz and Yussef, 1991). Black pepper is dominated by *Aspergillus, Eurotium, Emericella, Rhizopus, Penicillium, Curvularia, Cladosporium* and *Paecilomyces* species and the toxigenic species *Aspergillus flavus, Asp. parasiticus* and *Asp. tamarii* are common, but *Asp. ochraceus* is not very common in pepper (Christensen *et al.*, 1967; Flannigan and Hui, 1976; Moreau and Moreau, 1978; Banerjee *et al.*, 1993; Freira *et al.*, 2000; Gatti *et al.*, 2003). Other spices often also contain toxigenic fungi (Llewellyn *et al.*, 1981; Martinez *et al.*, 1988; Hashmi and Ghaffar, 1991; Hashmi and Thrane, 1990; Takahashi, 1993; Shrivastava and Jain, 1992; McKee, 1995; Weidernbörner, 2001a; El-Nagerabi, 2002), but they may be more of a contamination risk concerning the foods they are added to, than actually containing mycotoxins themselves.

Tomatoes

Tomato plants are susceptible to different plant pathogenic *Alternaria* species attacking leaves (Simmons, 2000). However, few *Alternaria* species have on occasion also been isolated from tomato fruit lesions: *Alt. subtropica, Alt. tomato* and *Alt. tomatophila* (Simmons, 2000), *Alt. solani* and *Alt. alternata* (Weir *et al.*, 1998; Morris *et al.*, 2000) and *Alt. tenuissima* species-group (Andersen and Frisvad, 2004). Undamaged tomatoes are quite resistant to fungal spoilage, but *Penicillium tularense, P. expansum* and *Stemphylium eturmiunum* and *S. lycopersici* have been reported on healthy tomatoes (Andersen and Frisvad, 2004). *Alternaria arborescens* (= *Alt. alternata* f. sp. *lycopersici*) is a stem canker pathogen of tomato (Simmons, 1999b; Morris, *et al.*, 2000) and has to the knowledge of the authors never been isolated from tomato fruit, but since this species is able to produce AAL toxin (Bottalico and Logrieco, 1998) it should be taken into account in the chemical analyses. AAL toxin has not been reported in other *Alternaria* species than *Alt. arborescens*.

Several mycotoxins have been found to occur naturally in tomatoes and tomato products. Tenuazonic acid was detected in tomatoes for American catsup production (Stack *et al.*, 1985), alternariols and tenuazonic acid have been found in Italian tomatoes (Ozcelik *et al.*, 1990; Bottalico and Logrieco, 1998), cyclopiazonic acid and/or tenuazonic acid was detected in Brazilian tomato puree and pulp (Da Motta and Soares, 2001) and several samples of mouldy Danish tomatoes contained aternariols, tentoxin, tenuazonic acid, infectopyrone, macrosporin, stemphol, paxillin, and/or janthitrems (Andersen and Frisvad, 2004).

Penicillium olsonii has often been found on damaged tomatoes and may grow directly on commercial tomatoes, but no mycotoxins have been reported from this species. *Penicillium olsonii* has been shown to produce verrucolone (= arabenoic acid), bis(2-ethyl-hexyl) phthalate and 2-(4-hydroxyphenyl)-2-oxo-acetaldehyde oxime, which are all bioactive compounds (Amade *et al.*, 1994; Frisvad *et al.*, 2004a).

Potato tubers

Dry rot of potatoes is mainly caused by *Fusarium sambucinum* and *F. coeruleum*. The taxonomy of *F. sambucinum* has been revised recently (Nirenberg, 1995) and does now contain *F. sulphureum* which often has been mentioned in connection with potatoes. The other species frequently reported in relation to dry rot of potatoes is *F. coeruleum* (Gerlach and Nirenberg, 1982). *F. crookwellense* is also frequently isolated from damaged potato tubers; however, its role as a primary pathogen is unclear. Several other fungi occur on potatoes, but few of these are known mycotoxin producers (Dashwood *et al.*, 1993).

The *Fusarium* dry rot is normally so pronounced that the tubers are not suitable for consumption. However, as the full extent of the damage is not always visible from the outside, it may be possible that partly rotted potatoes could pass on to further processing in the food industry. In addition to the physical damage of potatoes, mycotoxins may also be produced in the tubers. Diacetoxyscirpenol and related trichothecenes have been detected in tubers artificially inoculated with *F. sambucinum* (Desjardins and Plattner, 1989). El-Banna *et al.* (1984) reports deoxynivalenol production in tubers inoculated with *F. sambucinum* and *F. coeruleum*; however, deoxynivalenol production has never been verified from these species. This could be explained by the fact that *F. sambucinum* sometimes resembles *F. cerealis*, and that this species is known to produce deoxynivalenol. Production of unidentified toxins in tubers infected with *F. sambucinum* and *F. coeruleum* has been demonstrated in brine shrimp test (Siegfried and Langerfeld, 1978). Co-infection with bacteria (*Erwinia carotovora*) had no significant influence on the dry rot.

Two *Alternaria* species have been reported repeatedly from potato tubers and other *Solanaceae: Alt. alternata* (Weir *et al.*, 1998) and *Alt. solani,* which cause early blight on potato plants (Simmons, 2000), but also are able to infect damaged tubers during harvest. Seeds of potato may also contribute to the mycobiota of the potato tubers (Somani *et al.*, 1986).

The pathogens are present in soil and tubers and the infection takes place by damage of the periderm. High soil humidity raises the infection rate whereas crop rotation will lower it. The harvest should be done with caution to minimize physical damage of the tubers and the storage kept dry and cooled. Efficient ventilation is important to keep the tubers free from dry rot.

Yam tubers

Several species of *Aspergillus, Penicillium, Botryosphaeria, Cladosporium, Fusarium* and *Rhizopus* are able to spoil yams (Yamamoto *et al.*, 1955; Okigbo and Ikediugwu, 2000; Okigbo, 2003; Bankole and Mabekoje, 2004). *Penicillium sclerotigenum* has repeatedly been isolated from yam tubers, but never from any other product. It has been isolated from a *Dioscorea* sp. in Japan, on *D. cayenensis* in Jamaica and in blue yams flour from the Philippines. This fungus produces patulin, griseofulvin, roquefortine C and gregatins (Frisvad and Filtenborg, 1989, 1993; Frisvad and Samson, 2004), but yam products have to our knowledge never been examined for those toxins. Other Penicillia found included *P. oxalicum* and *P. citrinum* together with *Aspergillus niger, Asp. flavus, Asp. tamarii, Asp. ochraceus* and *Asp. fumigatus* were among the most common Aspergilli present on dried yam chips (Bankole and Mabekoje, 2004). *Botryosphaeria rhodina, Fusarium verticillioides* and *Penicillium sclerotigenum* were among the most common on the yam tubers (Okigbo, 2003). Most samples of yams have low levels of aflatoxins and few samples had high levels of aflatoxin B_1 (Bankole and Mabekoje, 2004).

Garlic and onions

Apart from *Botrytis aclada,* few species are able to spoil garlic and onions (Abdel-Sater and Eraky, 2002). *Penicillium allii* is a widespread spoiler of garlic (Vincent and Pitt, 1989; Frisvad and Filtenborg, 1989) while the closely related *P. albocoremium, P. radicicola* and *P. tulipae* are more common on other onions (Overy and Frisvad, 2003; Overy *et al.*, 2005a;b). *Petromyces alliaceus, Aspergillus niger* and *Penicillium glabrum* are cited as producers of rots in onions (Raper and Fennell, 1965; Raper and Thom, 1949; Hayden and Maude, 1994; Hayden *et al.*, 1994). *Penicillium glabrum* appears to grow only in the outer layers of onions. *Alternaria porri* may cause damage on onion leaves, but is rarely found in the onions themselves (Fokkema and Lorbeer, 1974). Onion seeds may harbour toxigenic species that may eventually grow in the onions (Hayden and Maude, 1994; El-Nagerabi and Abdalla, 2004).

Petromyces alliaceus is a very efficient producer of ochratoxin A (Hesseltine *et al.*, 1972; Ciegler, 1972; El-Shayeb *et al.*, 1992), but onions have not been analyzed for natural occurrence of this toxin. *Penicillium allii* produces the roquefortine C, meleagrin and the viridicatin chemosynthetic family (Frisvad and Filtenborg, 1989; Overy *et al.*, 2005a;b). *Penicillium glabrum* produces the nephrotoxin citromycetin (Domsch *et al.*, 1993).

Allinin and other antimicrobial compounds produced by the onions strongly selects for the associated mycobiota (Overy *et al.*, 2005a) and can be added to other foods to prevent fungal growth (Abdel-Hafez and El-Said, 1997).

Ginger

Several fungi may grow on ginger, but *Penicillium brevicompactum* is quite commonly found on mouldy ginger and *P. brevicompactum* is able to produce mycophenolic acid on this root (Overy and Frisvad, 2004). Mycophenolic acid is not acutely toxic, but being a very strong immunosuppressant, it may pave the way for bacterial diseases.

MILK AND MEAT PRODUCTS

Cheese (hard and soft)

The most important spoilage species of hard, semi-hard and soft cheese without preservatives are *Penicillium commune, P. palitans* and *P. nalgiovense* (Lund *et al.*, 1995). Species of less importance are: *P. verrucosum, P. nordicum, P. solitum, P. roqueforti, Scopulariopsis brevicaulis* and *Aspergillus versicolor*. It has been shown that important isolates from cheese which have been identified as *P. verrucosum var. cyclopium, P. aurantiogriseum, P. cyclopium* and *P. puberulum* (Northolt *et al.*, 1980a; Aran and Eke, 1987, Erdogan *et al.*, 2003; Korukluoglu *et al.*, 2005) could be re-identified as *P. commune* (Frisvad and Filtenborg, 1993; Lund *et al.*, 1995). Another species, *P. discolor*, has been isolated from natamycin treated hard cheeses (Frisvad *et al.*, 1996).

The most important mycotoxin found in cheese is sterigmatocystin (Northolt *et al.*, 1980a; Abd Alla *et al.*, 1996), but cyclopiazonic acid, rugulovasine A and B and ochratoxin A should also be taken into account due to their toxic potential. Aflatoxin can be produced by *Asp. flavus* under special conditions (Sinigaglia *et al.*, 2004). Besides it should be noted that *P. nalgiovense* is a potential producer of penicillin (Andersen and Frisvad, 1994; Färber and Geisen, 1994, 2000).

Fungal growth also spoils cheese by production of off-flavours. If sorbates are used as preservatives, resistant species are able to metabolize these compounds under formation of a plastic-like or "kerosene" off flavour caused by the metabolites trans-1,3-pentadiene or trans-piperylene (Sensidoni *et al.*, 1994).

Meat products (fermented and dried)

The associated mycobiota in naturally fermented sausages is *Penicillium* species: *P. nalgiovense, P. olsonii, P. chrysogenum, P. nordicum, P. solitum, P. polonicum, P. commune, P. oxalicum, P. expansum, P. miczynskii* and *P. simplicissimum* (Ciegler *et al.*, 1972; Andersen, 1995a; Lopez-Diaz *et al.*, 2001; Tabuc *et al.*, 2004). Dominating species of *Aspergillus* and *Scopulariopsis* have also been reported, including *Eurotium* spp., *Asp. versicolor* and *Asp. niger* (Grazia *et al.*, 1986).

In the beginning of the fermentation process yeasts are dominating the surface mycobiota, but after a few weeks the above mentioned naturally occurring moulds take over, *P. nalgiovense* being dominating. This species in some cases is added as a starter culture. It has been claimed that several biotypes of *P. nalgiovense* exist (Fink-Gremmels and Leistner, 1990). However, it has been shown that all tested isolates belonged to one species, *P. nalgiovense*, even though the colour of the isolates ranged from white to dark blue green (Andersen, 1995b).

The *Penicillium* species in the associated mycobiota are known to produce several mycotoxins and antibiotics. Some of these mycotoxins have been detected in fermented sausages after mould inoculation in pure cultures: citreoviridin, citrinin, cyclopiazonic acid, isofumigaclavin A, ochratoxin A, patulin, roquefortin C and rugulovasin A (Fink-Gremmels and Leistner, 1990; Bailly *et al.*, 2005). Other mycotoxins like viomellein and xanthomegnin are produced by the associated mycobiota and all tested isolates of *P. nalgiovense* were shown to produce penicillin (Andersen and Frisvad, 1994; Färber and Geisen, 1994). Penicillin production in sausages is possible (Laich *et al.*, 1999).

CONCLUSIONS

Different foods will have different associated mycobiotas, both in a qualitative and quantitative sense. Certain filamentous fungi are specifically associated to a narrow range of plants or parts of plants, as *Penicillium italicum, P. ulaiense* and *P. digitatum* are associated to citrus fruits. Other fungi, such as *P. expansum* have a broader host range, primarily associated to pome and stone fruits, but also to nuts. Yet other fungi, such as *Aspergillus niger, P. brevicompactum* and *P. citrinum* appear to be generalists. Even within these generalists, there are clear quantitative differences. For example, *Asp. niger* and *Asp. carbonarius* are more common than *Asp. ochraceus* and *Asp. westerdijkiae* on grapes, while the reverse is the case for green coffee beans. Even on processed foods, where the chemical defense mechanisms characteristic of living plants and animals are removed, a specific associated mycobiota may be obvious. Soft and hard cheeses, for example, harbour a specific mycobiota dominated by *Asp. versicolor*, *P. commune* and *P. palitans*, while rye bread is often spoiled by *P. roqueforti*, *Monascus ruber* and *Paecilomyces variotii*. At more extreme conditions of pH, water activity and temperature, fungi that tolerate these conditions may grow on all kinds of foods and feedstuffs, maybe because of less competition by other fungi at these rather extreme conditions. *P. roqueforti* probably co-evolved with lactic acid bacteria, but is often found both in habitats with lactic acid bacteria and in acid preserved foods. *Eurotium*, *Wallemia* and *Xeromyces* species can be found in all kinds of habitats with low water activity, whether this low water activity is caused by adding salt or sugar or whether the substrate is cakes, salty fish, salami or leather. Still, however, there is a certain quantitative preference, meaning that certain species of *Eurotium* and *Wallemia* are more common on sugar preserved foods, while others are more prevalent on salt-preserved foods.

Knowledge on the associated mycobiota on foods will help prevent mycotoxin production, because fungi produce a species specific profile of mycotoxins and other bioactive extrolites and therefore mycotoxin prevention can be based on detailed knowledge on fungal species and their physiology and specific mycotoxin production. An example is green coffee, where ochratoxin A is an important mycotoxin, while aflatoxin is not. This is caused by the high tolerance of *Aspergillus* section *Circumdati* and *Nigri* species to caffeine and caffeic acid and the low tolerance of *Aspergillus* section *Flavi* to these compounds, but probably also to other environmental factors.

A more accurate database on which fungal species belong to which types of foods and other substrata would be very helpful in preventing fungal growth and mycotoxin formation and spoilage and allergenic problems caused by off-flavour formation, discolouration, formation of allergenic parts of fungal thalli, enzyme production, etc. Also a better knowledge on the factors that govern the association of specific fungi to specific foods is needed. These factors may include the abiotic environment, nutritional factors, secreted enzymes, secondary metabolites, adhesins, hydrophobins, insect, mite, rodent or bird vectors, etc., and often the relative importance of these factors is unknown.

Acknowledgements
BA and JCF would like to thank the Danish Council for Technology and Production Science for financial support to the project Excretomics and proteomics of mycotoxigenic foodborne fungi (no. 23-03-0137).

REFERENCES

Abarca, M. L., Accensi, F., Bragulat, M. R., Castellá, G., and Cabañes, F. J. (2003). *A. carbonarius* as the main source of OTA contamination in dried vine fruits from Spanish market. Journal of Food Protection 66:504-506.

Abbas, H. K., Cartwright, R. D., Xie, W. P., and Shier, W. T. (2006). Aflatoxin and fumonisin contamination in corn (Maize, *Zea mays*) hybrids in Arkansas. Crop Protection 25:1-9.

Abd Alla, E. A. M., Metwaly, M. M., Mehriz, A. M. and Abu Sree, Y. H. (1996). Sterigmatocystin incidence, fate and production by *Aspergillus versicolor* in Ras cheese. Nahrung 40:310-313.

Abd-Allah, E. F., and Hashem, A. (2006). Seed mycoflora of *Lens esculenta* and their biocontrol by chitosan. Phytoparasitica 34:213-218.

Abdel-Gawad, K. M., and Zohri, A. A. (1993). Fungal flora and mycotoxins of 6 kinds of nut seeds for human consumption in Saudi-Arabia. Mycopathologia 124:55-64.

Abdel-Hafez, S. I. I. (1984). Mycoflora of bean, broad bean, lentil, lupine and pea seeds in Saudi-Arabia. Mycopathologia 88:45-49.

Abdel-Hafez, A. I. I. (1988). Mycoflora of broad bean, chick pea and lentil seeds in Egypt. Cryptogamie Mycologie 9:35-343.

Abdel-Hafez, A. I. I., and El-Maghraby, A. O. O. (1992). Fungal flora and aflatoxin associated with cocoa, roasted coffee and tea powders. Cryptogamie Mycologie 3:31-45.

Abdel-Hafez, A. I. I., and El-Said, A. H. M. (1997). Effect of garlic, onion and sodium benzoate on the mycoflora of pepper, cinnamon and rosemary in Egypt. International Biodeterioration and Biodegradation 39:67-77.

Abdel-Hafez, A. I. I., and Saber, S. M. (1993). Mycoflora and mycotoxins of hazelnut (*Corylus avellanea* L.) and walnut (*Juglans regia* L.) seeds in Egypt. Zentralblatt für Microbiologie 148:137-147.

Abdel-Kader, M. I. A., Moubasher, A. H., and Abdel-Hafez, S. I. I. (1979). Survey of the mycoflora of barley grains in Egypt. Mycopathologia 69:143-147.

Abdel-Sater, M. A., and Eraky, S. A. (2002). Bulbs mycoflora and their relation to three stored product mites. Mycopathologia 153:33-39.

Abdel-Sater, M. A., Zohri, A. A., and Ismail, M. A. (2001). Natural contamination of some Egyptian fruit juices and beverages by mycoflora and mycotoxins. Journal of Food Science and Technology Mysore 38:407-411.

Abrunhosa, L., Paterson, R. R. M., Kozakiewicz, Z., Lima, N., and Venencio, A. (2001). Mycotoxin production from fungi isolated from grape. Letters in Applied Microbiology 32:240-242.

Accensi, F. Abarca, M. L., and Cabañes, F. J. (2004). Occurrence of *Aspergillus* species in mixed ffeds and component raw materials and their ability to produce ochratoxin A. Food Microbiology 21:623-627.

Ackermann, A. (1988). Mycoflora of South African barley and malt. Journal of the American Society of Brewing Chemists 56:169-176.

Adebajo, L. O., and Idowu, A. A. (1994). Mycoflora and aflatoxins in a West-African corn-groundnut based convenience food. Mycopathologia 126:21-26.

Adebajo, L. O., Idowu, A. A., and Adesanya, O. O. (1994). Mycoflora, and mycotoxins production in Nigerian corn and corn-based snacks. Mycopathologia 126:183-192.

Adebanjo, A., and Ikotun, T. (1994). Effect of harvest seasons on the incidence of seed-borne mycoflora of 3 *Amaranthus* cultivars. Mycopathologia 128:25-32.

Adebanjo, A., and Shopeju, E. (1993). Sources and mycoflora associated with some sundried vegetables in storage. International Biodeterioration and Biodegradation 31:255-263.

Adisa, A. (1994). Mycoflora of postharvest maize and wheat grains and the implication of their contamination by molds. Nahrung 38:318-326.

Ahmad, R., and Magan, N. (2002). Microfloral contamination and hydrolytic enzyme differences between monsooned and non-monsooned coffees. Letters in Applied Microbiology 34:279-282.

Ahmad, S. K. (1993). Mycoflora changes and aflatoxin production in stored blackgram seeds. Journal of Stored Food Research 29:33-36.

Ahmad, S. K., and Singh, P. L. (1991). Mycofloral changes and aflatoxin contamination in stored chickpea seeds. Food Additives and Contaminants 8:723-730.

Åkerstrand, K., and Josefsson, E. (1979). Fungi and mycotoxins in beans and peas. Vår Főda 31:405-414.

Almeida, A. P., Fonseca, H., Fancelli, A. L., Direito, G. M., Ortega, E. N. M., and Corrêa, B. (2002a). Mycoflora and fumonisin contamination in Brazilian corn from sowing to harvest. Journal of Agricultural and Food Chemistry 50:3877-3882.

Almeida, A. P., Corrêa, B. Mallozzi, M. A. B., Sawazaki, E., and Soares, L. M. V. (2002b). Mycoflora and aflatoxin/fumonisin production by fungal isolates from freshly harvested corn hybrids. Brazilian Journal of Microbiology 31:321-326.

Amade, P., Mallea, M. and Bouaicha, N. (1994). Isolation, structural identification and biological activity of two metabolites produced by *Penicillium olsonii* Bainier and Sartory. Journal of Antibiotics 47:201-207.

Amézqueta, S., González-Peñas, E., Murillo M., and López de Cerain, A. L. (2005). Occurrence of ochratoxin A in cocoa beans: Effect of shelling. Food Additives and Contaminants 22:590-596.

Andersen, B., and Frisvad, J. C. (2002). Characterization of *Alternaria* and *Penicillium* species from similar substrata based on growth at different temperature, pH and water activity. Systematic and Applied Microbiology 25:162-172.

Andersen, B., and Frisvad, J. C. (2004). Natural occurrence of fungi and fungal metabolites in moldy tomatoes. Journal of Agricultural and Food Chemistry 52:7507-7513.

Andersen, B., and Thrane, U. (2006). Food-borne fungi in fruit and cereals and their production of mycotoxins. Advances in Experimental Medicine and Biology 571:137-152.

Andersen, B., Hansen, M. E., and Smedsgaard, J. (2005). Automated and unbiased image analyses as tools in phenotypic classification of small-spored *Alternaria* spp. Phytopathology 95:1021-1029.

Andersen, B., Krøger, E., and Roberts, R. G. (2002). Chemical and morphological segregation of *Alternaria arborescens*, *A. infectoria* and *A. tenuissima* species-groups. Mycological Research 106:170-182.

Andersen, B., Smedsgaard, J., and Frisvad, J. C. (2004). *Penicillium expansum*: Consistent production of patulin, chaetoglobosins, and other secondary metabolites in culture and their natural occurrence in fruit products. Journal of Agricultural and Food Chemistry 52:2421-2428.

Andersen, B., Thrane, U., Svendsen, A., and Rasmussen, I. A. (1996). Associated field mycobiota on malt barley. Canadian Journal of Botany 74:854-858.

Andersen, S. J. (1995a). Compositional changes in surface mycoflora during ripening of naturally fermented sausages. Journal of Food Protection 58:426-429.

Andersen, S. J. (1995b). Taxonomy of *Penicillium nalgiovense* isolates from mould-fermented sausages. Antonie van Leeuwenhoek 68:165-171.

Andersen, S. J., and Frisvad, J. C. (1994). Penicillin production by *Penicillium nalgiovense*. Letters in Applied Microbiology 19:486-488.

Andrews, S. (1986). Dilution plating versus direct plating of various cereal samples. *In* Methods for the mycological examination of foods (King, A. D., Pitt, J. I., Beuchat, L. R., and Corry, A. J. E .L., eds.), Plenum Press, New York, NY, U.S.A., pp. 40-45.

Arai, K., Miyajima, H., Mushiroda, T., and Yamamoto, Y. (1989). Metabolites of *Penicillium italicum* Wehmer: Isolation and structures of new metabolites including naturally occurring 4-ylidene-acyltetronic acids, italicinic acid and italicic acid. Chemical and Pharmaceutical Bulletin 37:3229-3235.

Aran, N., and Eke, D. (1987). Mould mycoflora of Kasar cheese at the stage of consumption. Food Microbiology 4:101-104.

Ariza, M. R., Larsen, T. O., Petersen, B. O., Duss, J. Ø., and Barrero, A. F. (2002). Biochemistry of *Penicillium digitatum* on citrus fruits and synthetic media. Journal of Agricultural and Food Chemistry 50:6361-6365.

Arocho, L. A., Rodriguez, R. D. P., and Betancourt, C. (2005). Pathogenic role of the mycoflora associated with coffee fruit spots. Journal of the Agricultural University of Puerto Rico 89:85-96.

Arrus, K., Blank, G., Abramson, D, Clear, R., and Holley, R. A. (2005). Aflatoxin production by *Aspergillus flavus* in Brazil nuts. Journal of Stored Products Research 41:513-527.

Asevedo, I. G., Gambale, W., Corrêa, B., Paula, C. R., Almeida, R. M. A., and Souza, V. M. (1994). Mycoflora and aflatoxigenic species of *Aspergillus* spp isolated from stored maize. Revista de Microbiologia 25:46-50.

Ashworth, L. J., Schroeder, H. W., and Langley, B. C. (1965). Aflatoxin environmental factors governing occurrence in Spanish peanuts. Science 148:1228-1229.

Aziz, N. H., and Shahin, A. A. M. (1997). Influence of other fungi on aflatoxin production by *Aspergillus flavus* on maize kernels. Journal of Food Safety 17:113-123.

Aziz, N. H., and Youssef, Y. A. (1991). Occurrence of aflatoxins and aflatoxin-producing moulds in fesh and processed meat in Egypt. Food Additives and Contaminants 8:321-331.

Bailly, J. D., Tabuc, C., Querin, A., and Guerre, P. (2005). Production and stability of patulin, ochratoxin A, citrinin, and cyclopiazonic acid on dry cured ham. Journal of Food Protection 68:1516-1520.

Baird, R. E., Huber, D. M., and Mullinix, B. G. (1995). The mycobiota from seeds of shrunken-2 (sh2) sweet corn. Mycopathologia 132:147-154.

Banerjee, A., Matthews, R. P., Prakash, H. S., and Shetty, H. S. (1993). Mycobiota and toxigenic *Aspergillus flavus* associated with cardamom and pepper. Mycological Research 97:1403-1406.

Bankole, S. A., Eseige, D. A., and Enikuomehin, O. A. (1995). Mycoflora and aflatoxin production in pigeon pea stored in jute sacks and iron bins. Mycopathologia 132:155-160.

Bankole, S. A., and Mabekoje, O. O. (2004). Mycoflora and occurrence of aflatoxin B_1 in dried yam chips from markets in Ogun and Oyo States, Nigeria. Mycopathologia 157:111-115.

Bankole, S. A., Ogunsanwo, B. M., and Eseigbe, A. D. (2005). Aflatoxins in Nigerian dry-roasted groundnuts. Food Chemistry 89:503-506.

Barnes, G. L., Nelson, G. L., and Manbeck, H. B. (1970). Effects of drying, storage gases, and temperature on development of mycoflora and aflatoxins in stored high-moisture peanuts. Phytopathology 60:581.

Batista, L. R., Chalfoun, S. M., Prado, G., Schwan, R. F., and Wheals, A. E. (2003). Toxigenic fungi associated with processed (geen) coffee beans (*Coffea Arabica* L.). International Journal of Food Microbiology 85:293-300.

Battilani, P., and Pietri, A. (2002). Ochratoxin A in grapes and wine. European Journal of Plant Pathology 108:639-643.

Battilani, P., Pietri, A., Bertuzzi, T., Languasco, L., Giorni, P., and Kozakiewicz, Z. (2003). Occurrence of ochratoxin A-producing fungi in grapes grown in Italy. Journal of Food Protection 66:633-636.

Bau, M., Bragulat, M. R., Abarca, M. L., Minguez, S., and Cabañes, F. J. (2005). Ochratoxigenic species from Spanish wine grapes. International Journal of Food Microbiology 98:125-130.

Bayman, P., Baker, J. L., and Mahoney, N. E. (2002a). *Aspergillus* on tree nuts: incidence and associations. Mycopathologia 155:161-169.

Bayman, P., Baker, J. L., Doster, M. A., Michailides, T. J., and Mahoney, N. E. (2002b). Ochratoxin production by the *Aspergillus ochraceus* group and *Aspergillus alliaceus*. Applied and Environmental Microbiology 68:2326-2329.

Begum, M., Rai, V. R., and Lokesh, S. (2003). Effect of mycoflora on the physico-chemical characteristics of oil obtained from the infected sunflower, toria and sesame seeds. Journal of Food Science and Technology Mysore 40:626-628.

Belisario, A., Maccaroni, M., and Corazza, L. (2002). Occurrence and ethiology of brown apical necrosis on Persian (English) walnut fruit. Plant Disease 86:599-602.

Belisario, A., Maccaroni, M., Coramusi, A., and Corazza, L. (2004). First report of *Alternaria* species groups involved in disease complex of hazelnut and walnut fruit. Plant Disease 88:426.

Bellí, N., Pardo, E., Marín, S., Farre, G., Ramos, A. J., and Sanchis, V. (2004). Occurrence of ochratoxin A and toxigenic potential of fungal isolates from Spanish grapes. Journal of the Science of Food and Agriculture 84:541-546.

Bellí, N., Mitchell, D., Marín, S., Alegre, I., Ramos, A. J., Magan, N., and Sanchis, V. (2005). Ochratoxin A-producing fungi in Spanish grapes and their relationship with meteorological conditions. European Journal of Plant Pathology 113:233-239.

Ben-Yehoshua, A, Barak, E., and Shapiro, B. (1987). Postharvest curing at high temperatures reduces decay of individually sealed lemons, pomelos and other citrus fruit. Journal of the American Society of Horticultural Science 112:658-663.

Berleth, M., Backes, F., and Kramer, J. (1998). Mould spectrum and mycotoxins (deoxynivelanol and ochratoxin A) in grain samples from ecological and integrated cultivated sites. Agribiological Research 51:369-376.

Berthiller, F., Dall'Asta, C., Schuhmacher, R., Lemmens, M., Adam, G., and Krska, R. (2005). Masked mycotoxins: determination of a deoxynivalenol glucoside in artificially and naturally contaminated wheat by liquei chromatography-tandem mass spectrometry. Journal of Agricultural and Food Chemistry 53:3421-3425.

Betancourt, L. E., and Frank, H. K. (1983). Bedingungen des mikrobiellen Verderbs von grunem Kaffee. Deutsche Lebensmittel-Rundschau 79:366-369.

Beuchat, L. R. (1984). Survival of *Aspergillus flavus* conidiospores and other fungi on cowpeas during long-term storage under various environmental conditions. Journal of Stored Product Research 20:119-123.

Beuchat, L. R. (1986). Extraordinary heat resistance of *Talaromyces flavus* and *Neosartorya fischeri* ascospores in fruit products. Journal of Food Science 51:1506-1510.

Beuchat, L. R., and Rice, S. L. (1979). *Byssochlamys* spp. and their importance in processed fruits. Advances in Food Research 25:237-288.

Bhat, R. V., Shetty, P. H., Rao, P. A., and Rao, V. S. (1997). A foodborne disease outbreak due to the consumption of moldy sorghum and maize containing fumonisin mycotoxins. Journal of Toxicology. Clinical Toxicology 35:249-255.

Bhutta, A. R. (1998). Biological studies on some fungi associated with sunflower in Pakistan. PhD thesis, Sindh Agriculture University, Tandojam, Pakistan.

Birkinshaw, J. H., Victor, J. H., and Raistrick, H. (1931). Studies on the biochemistry of microorganisms. Part XVIII. Biochemical characteristics of species of *Penicillium* responsible for the rot of citrus fruits. Transactions of the Royal Society (London) B220:355-367.

Blanc, M., Pittet, A., Munoz-Box, R., and Viani, R. (1998). Behaviour of ochratoxin A during green coffee roasting and soluble coffee manufacture. Journal of Agricultural and Food Chemistry 46:673-675.

Blank, G., Nwoko, U., Frohlich, A., and Marquardt, R. (1998). Ochratoxin A production in relation to growth morphology of *Aspergillus alutaceus*. Lebensmittel Wissenshaft und Technologie 31:210-214.

Booth, C. (1971). The genus Fusarium. Commonwealth Mycological Institute, Kew, Surrey, U.K.

Dörner, H. (1963). Untersuchungen über die Bildung antiphytotischer und antimikrobieller Substanzen durch Mikroorganismen im Boden und ihre mögliche Bedeutung für die Bodenmüdigkeit beim Apfels *Pirus malus* L. Phytopathologisch Zeitschrift 49:1-28.

Bonvehi, J. S. (2004). Occurrence of ochratoxin A in coxoa products and chocolate. Journal Agricultural and Food Chemistry 52:6347-6352.

Bottalico, A., and Logrieco, A. (1998). Toxigenic *Alternaria* species of economic importance. *In* Mycotoxins in Agriculture and food safety (Sinha, K. K., and Bhatnagar, D., eds), Marcel Dekker, Inc., New York, USA, pp. 65-108.

Boysen, M., Skouboe, P., Frisvad, J. C., and Rossen, L. (1996). Reclassification of the *Penicillium roqueforti* group into three species on the basis of molecular genetic and biochemical profiles. Microbiology (UK) 142:541-549.

Brera, C., Grossi, S., and Miraglia, M. (2005). Interlaboratory study for ochratoxin A determination in cocoa powder samples. Journal of Liquid Chromatography and Related Technologies 28:35-61.

Bresler, G., Brizzio, S. B., and Vaamonde, G. (1995). Mycotoxin-producing potential of fungi isolated from Amaranth seeds in Argentina. International Journal of Food Microbiology 25:101-108.

Broggi, L. E., Gonzáles, H. H. L., Resnik, S., and Pacin, A. M. (2002). Mycoflora distribution of dry-milled fractions of corn in Argentina. Cereal Chemistry 79:741-744.

Burdaspal, P. A., and Legarda, T. M. (1999). Ocratoxina A en vinos, mostos y zumos de uva elaborados en España y en otros países europeos. Alimentaria 36:107-113.

Cabañes, F. J., Accensi, F., Bragulat, M. R., Abarca, M. L., Castellá, G., Minguez, S. and Pons, A. (2002). What is the source of ochratoxin A in wine? International Journal Food Microbiology 79:213-215.

Calori-Domingues, M. A., Fonseca, H. and Ranzani, M. R. T. D. (1996). Effect of propionic acid on fungal growth and aflatoxin production in moist inshell groundnuts. Revista de Microbiologia 27:71-77.

Castella, G., Bragulat, M. R., and Cabañes, F. J. (1999a). Surveillance of fumonisins in maize-based feeds and cereals from Spain. Journal of Agricultural and Food Chemistry 47:4707-4710.

Castella, G., Bragulat, M. R., and Cabañes, F. J. (1999b). Fumonisin production by *Fusarium* species isolated from cereals and feeds in Spain. Journal of Food Protection 62: 811-813.

Castillo, M. D., González, H. H. L., Martínez, E. J., Pacin, A. M., and Resnik, S. (2004). Mycoflora and potential for mycotoxin production of freshly harvested black bean from the Argentinean main production area. Mycopathologia 158:107-112.

Castro, M. F. P. M., Soares, L. M. V., and Furlani, R. P. Z. (1995). Mycoflora, aflatoxigenic species and mycotoxins in freshly harvested corn (*Zea mays* L): A preliminary study. Revista de Microbiologia 26:289-295.

Champ, B. R., Highley, C., Hocking, A. D., and Pitt, J. I. (1991). Fungi and mycotoxins in stored products. Australian Centre for International Agricultural Research, Canberra.

Charmley, L. L., and Prelusky, D. B. (1994). Decontamination of *Fusarium* mycotoxins. *In* Mycotoxins in grain. Compounds other than aflatoxin (Miller, J. D., and Trenholm, H. L., eds.), Eagan Press, St. Paul, U.S.A., pp. 421-435.

Chelkowski, J., and Visconti, A. (eds) (1992). *Alternaria* Biology, plant diseases and metabolites. Elsevier, Amsterdam.

Chong, L. M., and Sheridan, J. E. (1982). Mycoflora of barley (*Hordeum vulgare* L) seed in New Zealand. New Zealand Journal of Botany 20:187-189.

Christensen, C. M., Fanse, H. A., Nelson, G. H., Bates, F. and Mirocha, C. J. (1967). Mycoflora of black and red pepper. Applied Microbiology 15:622-626.

Ciegler, A. (1972). Bioproduction of ochratoxin A and penicillic acid by members of the *Aspergillus ochraceus* group. Canadian Journal of Microbiology 18:631-636.

Ciegler, A., Mintzlaff, H.-J., Machnik, W., and Leistner, L. (1972). Untersuchungen über das Toxinbildungs-vermögen von Rohwürsten isolieter Schimmelpilze der Gattung *Penicillium*. Fleischwirtschaft 52:1311-1318.

Ciegler, A. Vesonder, R. F., and Jackson, L. K. (1977). Production and biological activity of patulin and citrinin from *Penicillium expansum*. Applied and Environmental Microbiology 33:1004-1006.

Clear, R. M., Patrick, S. K., and Gaba, D. (2002a). Prevalence of fungi and fusariotoxins on barley seed from Western Canada, 1995-1997. Canadian Journal of Plant Pathology 22:44-50.

Clear, R. M., Patrick, S. K., and Gaba, D. (2002b). Prevalence of fungi and fusariotoxins on oat seed from Western Canada, 1995-1997. Canadian Journal of Plant Pathology 22:310-314.

Cloete, T. E., and Kotza, J. M. (1990). Microbial aspects of tea manufacture in South Africa. Acta Horticultura 275:691-698.

Cole, R. J., and Cox, R. H. (1981). Handbook of toxic fungal metabolites. Academic Press, New York.

Corry, J. E. L. (1987). Relationships of water activity to fungal growth. *In* Food and beverage mycology (Beuchat, L. R., ed.), AVI, New York, U.S.A, pp. 51-100.

Costa, L. L. F., and Scussel, V. M. (2002). Toxigenic fungi in beans (*Phaseolus vulgaris* L.) classes black and color cultivated in the state of Santa Caterina, Brazil. Brazilian Journal of Microbiology 33:138-144.

Da Motta, S., and Soares, L. M. V. (2001). Survey of Brazilian tomato products for alternariol, alternariol monomethyl ether, tenuazonic acid and cyclopiazonic acid. Food Additives and Contaminants 18:630-634.

Da Rocha Rosa, C. A., Palacios, V., Combina, M., Fraga, M. E., De Oliveira Rekson, A., Magnoli, C. E., and Dalcero, A. M. (2002). Potential ochratoxin A producers from wine grapes in Argentina and Brazil. Food Additives and Contaminants 19:408-414.

Da Silva, J. B., Dilkin, P., Fonseca, H., and Corrêa, B. (2004). Production of aflatoxins by *Aspergillus flavus* and of fumonisins by *Fusarium* species isolated from Brazilian sorghum. Brazilian Journal of Microbiology 35:182-186.

Dashwood, E. P., Fox, R. A. and Duncan, J. M. (1993). Effect of substrate and plant maturity on the incidence of potato roots by pathogenic and non-pathogenic fungi. Mycological Research 97:733-745.

Dawar, S., and Ghaffar, A. (1991). Detection of the seed-borne mycoflora of sunflower. Pakistan Journal of Botany 23:173-178.

Dejene, M., Yuen, J., and Sigvald, R. (2004). The impact of storage methods on storage environment and sorghum grain quality. Seed Science and Technology 32:511-529.

Delgado, T., and Gómez-Cordovés, C. (1998). Natural occurrence of alternariol and alternariol methyl ether in Spanish apple juice concentrates. Journal of Chromatography A 815:93-97.

De Farias, A. X., Robbs, C. F., Bittencourt, A. M., Andersen, P. M., and Corrêa, T. B. S. (2000). Endogenous *Aspergillus* spp. contamination of post-harvest corn in Parana State, Brazil. Pesquisa Agropecuaria Brasiliera 35:617-621.

Desjardins, A. E., and Plattner, R. D. (1989). Trichothecene toxin production by strains of *Gibberella pulicaris* (*Fusarium sambucinum*) in liquid culture and in potato tubers. Journal of Agricultural and Food Chemistry 37:388-392.

Dich, D., Åkerstrand, K., Andersson, A., Josefsson, E., and Jansson, E. (1979). Konserveringsmedels förekomst och inverkan på mögel- og mycotoxinbildning i matbröd. Vår Föda 31:385-403.

Diener, U. L. (1960). The mycoflora of peanuts in storage. Phytopathology 50:220-223.

Diener, U. L., Cole, R. J., Sanders, T. H., Payne, G. A., Lee, L. S., and Klich, M. A. (1987). Epidemiology of aflatoxin formation by *Aspergillus flavus*. Annual Reviews in Phytopathology 25:249-270.

Domijan, A.-M., Peraica, M., Žlender, V., Cvetković, B., Jurjević, Ž., Topolovec-Pintarić, S., and Ivić, D. (2005). Seed-borne fungi and ochratoxin A contamination of dry beans (*Phaseolus vulgaris* L.) in the republic of Croatia. Food and Chemical Toxicology 43:427-432.

Domsch, K. H., Gams, W., and Anderson, T.-H. (1993). Compendium of soil fungi. Academic Press, London.

Doster, M. A., and Michailides, T. J. (1995). The relationship between date of hull splitting and decay of pistachio nuts by *Aspergillus* species. Plant Disease 79:766-769.

Doster, M. A., and Michailides, T. J. (1999). Relationship between shell discoloration of pistachio nuts and incidence of fungal decay and insect infestation. Plant Disease 82:669-673.

Doster, M. A., Michailides, T. J., and Morgan, D. P. (1996). *Aspergillus* species and mycotoxins in figs from California orchards. Plant Disease 80:484-489.

Doupnik, B. (1969). Aflatoxins in farmers stock peanuts. Peanut quality mycoflora and climatological conditions as influencing factors. Journal of the American Oil Chemist's Society 46:A121.

Doupnik, B. (1972). Mycoflora and aflatoxin contamination of *Helminthosporium maydis* damaged corn. Phytopathology 62:802.

Dube, V. P., Charaya, M. U., and Modi, P. (1980). Ecological and in vitro studies on the soil mycoflora of mango orchards. Proceedings of the Indian Acadamy of Sciences- Plant Sciences 89:151-160.

Eckert, J. W., and Ratnayake, M. (1994). Role of volatile compounds from wounded oranges in induction of germination of *Penicillium digitatum* conidia. Phytopathology 84:746-750.

Efuntoye, M. O. (1996). Fungi associated with herbal drug plant during storage. Mycopathologia 136:115-118.

Ehrlich, K. C., Ciegler, A. Klich, M., and Lee, L. S. (1985). Fungal competition and mycotoxin production in corn. Experientia 41:691-693.

Ehrlich, K. C., and Lee, L. S. (1984). Mycotoxins in grain dust: method for analysis of aflatoxins, ochratoxin A, zearalenone, vomitoxin, and secalonic acid D. Journal of the Association of Official Analytical Chemists 67:963-967.

Ekundayo, C. A., and Idze, E. (1990). Mycoflora and nutritional value of shelled melon seeds (*Citrulus vulgaris* Schrad) in Nigeria. Plant Foods and Human Nutrition 40:215-222.

El-Almeida, R. M. A., Gambale, W., Corrêa, B., Paula, C. R., and Deasevedo, I. G. (1991). Mycoflora and aflatoxigenic species of *Aspergillus* spp isolated from rice. Revista de Microbiologia 22:161-163.

El-Banna, A. A., Scott, P. M., Lau, P.-Y., Sakuma, T., Platt, H. W., and Campbell, V. (1984). Formation of trichothecenes by *Fusarium solani* var. *coeruleum* and *Fusarium sambucinum* in potatoes. Applied and Environmental Microbiology 47:1169-1171.

El-Dessouki, S. (1992). Ochratoxin A in Bier. Deutsche Lebensmittel-Rundschau 88:354-355.

El-Kady, I. A., and Youssef, M. S. (1993). Survey of mycoflora and mycotoxins in Egyptian soybean seeds. Journal of Basic Microbiology 33:371-378.

El-Maghraby, O. M. O. (1996). Mycotoxins and mycoflora of rice in Egypt with special reference to trichothecenes production and control. Journal of Natural Toxins 5:49-59.

El-Maghraby, O. M. O., and El-Maraghy, S. S. M. (1987). Mycoflora and mycotoxins of peanuts (*Arachis hypogea* L.) seeds in Egypt. I. Sugar fungi and natural occurrence of mycotoxins. Mycopathologia 98:165-170.

El-Maghraby, O. M. O., and El-Maraghy, S. S. M. (1988). Mycoflora and mycotoxins of peanuts (*Arachis hypogea* L.) seeds in Egypt. III. Cellulose-decomposing and mycotoxin-producing fungi. Mycopathologia 104:19-24.

El-Maghraby, O. M. O., El-Kady, I. A., and Soliman, S. (1995). Mycoflora and *Fusarium* toxins of 3 types of corn grains in Egypt with special reference to production of trichothecene-toxins. Microbiological Research 150:225-232.

El-Nagerabi, S. A. F. (2002). Determination of seed-borne fungi and detection of aflatoxins in Sudanese fenugreek seeds. Phytoparasitica 30:61-66.

El-Nagerabi, S. A. F., and Abdalla, R. M. O. (2004). Survey of seedborne fungi of Sudanese cultivars of onion, with new records. Phytoparasitica 32:413-416.

El-Nagerabi, S. A. F., and Elshafie, A. E. (2001). Incidence of seed-borne fungi and aflatoxins in Sudanese lentil seeds. Mycopathologia 149:151-156.

El-Nagerabi, S. A. F., Elshafie, A. E., and Abdalla, A. H. (2000). Composition of mycoflora and aflatoxins in pea seeds from the Sudan. Kuwait Journal of Science and Engineering 27:109-121.

El-Shanawany, A. A., Mostafa, M. E., and Barakat, A. (2005). Fungal populations and mycotoxins in silage in Assiut and Sohag governorates in Egypt, with a special reference to characteristic Aspergilli toxins. Mycopathologia 159:281-289.

Elshafie, A. E., Al-Lawatia, T., and Al-Bahry, S. (1999). Fungi associated with black tea and tea quality in the Sultana of Oman. Mycopathologia 145:83-93.

El-Shayeb, N. M. A., Mabrouk, S. S., and Abd-El-Fattah, A. M. M. (1992). Production of ochratoxins by some Egyptian *Aspergillus* strains. Zentralblatt für Mikrobiologie 147:86-91.

Enigl, D. C., King, A. D., and Török, T. (1993). *Talaromyces trachyspermus*, a heat-resistant mold isolated from fruits juice. Journal of Food Protection 56:1039-1042.

Enikuomehin, O. A. (2005). Seed abnormalities and associated mycoflora of rainfed wheat (*Triticum aestivum* L.) in South Western Nigeria. African Journal of Biotechnology 4:672-675.

Enomoto, M., and Ueno, I. (1974). *Penicillium islandicum* (toxic yellowed rice) – luteoskyrin – islanditoxin æ cyclochlorotine. *In* Mycotoxins (Purchase, I. F. H., ed.), Elsevier, Amsterdam, The Netherlands, pp. 303-326.

Erdogan, A., Gurses, M., and Sert. S. (2003). Isolation of moulds capable of producing mycotoxins from blue mouldy Tulum cheeses produced in Turkey. International Journal of Food Microbiology 85:83-85.

Faid, M., and Tantaoui-Elaraki, A. (1989). Production of toxic metabolites by *Penicillium italicum* and *P. digitatum* isolated from citrus fruits. Journal of Food Protection 52:194-197.

Färber, P., and Geisen, R. (1994). Antagonistic activity of the food relevant filamentous fungus *Penicillium nalgiovense* is due to the production of penicillin. Applied and Environmental Microbiology 60:3401-3404.

Färber, P., and Geisen, R. (2000). Karyotype of *Penicillium nalgiovense* and assignment of the penicillin biosynthetic genes to chromosome IV. International Journal of Food Microbiology 58:59-63.

Farnochi, M. C., Torres, A. M., Magan, N., and Chulze, S. N. (2005). Effect of antioxidants and competing mycoflora on *Fusarium verticilloides* and *F. proliferatum* populations and fumonisin production on maize grain. Journal of Stored Product Research 41:211-219.

Filtenborg, O., and Frisvad, J. C. (1980). A simple screening method for toxigenic fungi in pure cultures. Lebensmittel-Wissenshaft und Technologie 13:128-130.

Filtenborg, O., Frisvad, J. C., and Svendsen, J. A. (1983). Simple screening method for moulds producing intracellular mycotoxins in pure cultures. Applied and Environmental Microbiology 45:581-585.

Filtenborg, O., Frisvad, J. C., and Thrane, U. (1996). Moulds in food spoilage. International Journal of Food Microbiology 33:85-102.

Filtenborg, O., Frisvad, J. C., and Samson, R. A. (2004). Specific association of fungi to foods and influence of physical environmental factors. *In* Introduction to food- and airborne fungi 7th edition (Samson, R. A., Hoekstra, E. H., and Frisvad, J. C., eds.), Centraalbureau voor Schimmelcultures, Utrecht, The Netherlands, pp. 306-320.

Fink-Gremmels, J, and Leistner L. (1990). Toxicological evaluation of moulds. Food Biotechnology 4:579-584.

Fisher, P. J., and Petrini, O. (1992). Fungal saprobes and pathogens as endophytes of rice (*Oryza sativa* L). New Phytologist 120:137-143.

Flaishman, M. A., and Kolattukudy, P. E. (1994). Timing of fungal invasion using host's ripening hormone as a signal. Proceedings of the National Academy of Sciences USA 91:6579-6583.

Flannigan, B. (1986). *Aspergillus clavatus* – an allergenic, toxigenic deteriogen of cereals and cereal products. International Biodeterioration and Biodegradation 22:79-89.

Flannigan, B., and Hui, S. C. (1976). The occurrence of aflatoxin-producing strains of *Aspergillus fla-*

vus in the mould floras of ground spices. Journal of Applied Bacteriology 41:411-418.

Fokkema, N. J., and Lorbeer, J. W. (1974). Interactions between *Alternaria porri* and saprophytic mycoflora of onion leaves. Phytopathology 64:1128-1133.

Frank, J. M. (2001). Development of critical control points for preventing ochratoxin A accumulation in coffee. *In* Mycotoxins and phycotocins in perspective at the turn of the millennium (Koe, W. J. de, Samson, R. A., van Egmond, H. P., Gilbert, J., and Sabino, M., eds.), IUPAC and AOAC International, Wageningen, The Netherlands, pp. 289-298.

Freira, F. C. O., Kozakiewicz, Z., and Paterson, R. R. M. (2000). Mycoflora and mycotoxins in Brazilian black pepper, white pepper and Brazil nuts. Mycopathologia 149:13-19.

Frisvad, J. C. (1981). Physiological criteria and mycotoxin production as aids in identification of common asymmetric penicillia. Applied and Environmental Microbiology 41:568-579.

Frisvad, J. C. (1988). Fungal species and their specific production of mycotoxins. *In* Introduction to Food-borne fungi, 3rd ed., (Samson, R. A., and Reenen-Hoekstra, E. H. van, eds.), Centraalbureau voor Schimmelcultures, Baarn, The Netherlands, pp. 239-249.

Frisvad, J. C. (1989). The connection between penicillia and aspergilli and mycotoxins with special emphasis on misidentified isolates. Archives of Environmental Contamination and Toxicology 18, 452-467.

Frisvad, J. C. (1995). Mycotoxins and mycotoxigenic fungi in storage. *In* Stored grain ecosystems (Jayas, D. S., White, N. D. G., and Muir, W. E., eds.), Marcel Dekker, New York, U.S.A., pp. 251-288.

Frisvad, J. C., and Filtenborg, O. (1983). Classification of terverticillate penicillia based on profiles of mycotoxins and other secondary metabolites. Applied and Environmental Microbiology 46:1301-1310.

Frisvad, J. C., and Filtenborg, O. (1988). Specific mycotoxin producing *Penicillium* and *Aspergillus* mycoflora of different foods. Proceedings of the Japanese Association of Mycotoxicologists Suppl. 1:163-166.

Frisvad, J. C., and Filtenborg, O. (1989). Terverticillate Penicillia: chemotaxonomy and mycotoxin production. Mycologia 81:836-861.

Frisvad, J. C., and Filtenborg, O. (1990). Secondary metabolites as consistent criteria in Penicillium taxonomy and a synoptic key to *Penicillium* subgenus *Penicillium*. *In* Modern concepts in *Penicillium* and *Aspergillus* classification (Samson, R. A., and Pitt, J. I., eds.), Plenum Press, New York, U.S.A., pp. 373-384.

Frisvad, J. C., and Filtenborg, O. (1993). Mycotoxin production by *Penicillium* and *Aspergillus* species associated with different foods and other substrates. *In* Occurrence and significance of mycotoxins (Scudamore, K. A., ed.), Central Science Laboratory, Slough, U.K., pp. 138-145.

Frisvad, J. C., Filtenborg, O., Lund F., and Thrane, U. (1992). New selective media for the detection of toxigenic fungi in cereal products, meat and cheese. *In* Modern methods in food mycology (Samson, R. A., Hocking, A. D., Pitt, J. I., and King, A. D., eds.), Elsevier, Amsterdam, U.S.A., pp. 275-284.

Frisvad, J. C., Frank, J. M., Houbraken, J. A. M. P., Kuijpers, A. F. A., and Samson, R. A. (2004b). New ochratoxin producing species of *Aspergillus* section *Circumdati*. Studies in Mycology 50:23-43.

Frisvad, J. C., and Lund, F. (1993). Toxin and secondary metabolite production by Penicilium species growing in stored cereals. *In* Occurrence and significance of mycotoxins (Scudamore, K. A., ed.), Central Science Laboratory, Slough, U.K., pp. 146-171.

Frisvad, J. C., and Samson, R. A. (1991). Filamentous fungi in foods and feeds: ecology, spoilage and mycotoxin production. *In* Handbook of Applied Mycology. Vol. 3. Foods and Feeds (Arora, D. K., Mukerji, K. G., and Marth, E. H., eds.), New York. Marcel Dekker, U.S.A., pp. 31-68.

Frisvad, J. C., and Samson, R. A. (2004). Polyphasic taxonomy of *Penicillium* subgenus *Penicillium*. A guide to identification of food and air-borne terverticillate Penicillia and their mycotoxins. Studies in Mycology 49:1-173.

Frisvad, J. C., Samson, R. A., Rassing, B. R., and Horst, M. I. van der (1996). *Penicillium discolor*, a new species from cheese, nuts and vegetables. Antonie van Leeuwenhoek 72:119-126.

Frisvad, J. C., Smedsgaard, J., Larsen, T. O., and Samson, R. A. (2004a). Mycotoxins, drugs and other extrolites produced by species in *Penicillium* subgenus *Penicillium*. Studies in Mycology 49:201-241.

Frisvad, J. C., and Thrane, U. (1995). Mycotoxin production by food-borne fungi. *In* Introduction to food-borne fungi 4th edition (Samson, R. A., Hoekstra, E. S., Frisvad, J. C., and Filtenborg, O., eds.), Centraalbureau voor Schimmelcultures, Baarn, The Netherlands, pp. 251-260.

Frisvad, J. C., Thrane, U., and Filtenborg, O. (1998). Role and use of secondary metabolites in fungal taxonomy. *In* Chemical fungal taxonomy (Frisvad, J. C., Bridge, P. D., and Arora, D. K., eds.) Marcel Dekker, New York, U.S.A., pp. 289-319.

Fuller, G., Spooner, W. W., King, A. D., Schade, J., and MacKey, B. (1977). Survey of aflatoxins in California tree nuts. Journal of the American Oil Chemists Society 54:231A-234A.

Ganbobo, M. S. and Dostaler, D. (1990). Inventory and incidence of mycoflora associated with pearl-millet seeds in Niger. Seed Science and Technology 18:567-576.

Gareis, M. (1999). Contamination of German malting barley and of malt produced from it with the mycotoxins ochratoxin A and B. Archiv für Lebensmittelhygiene 50:83-87.

Gareis, M., Bauer, J., Enders, C., and Gedek, B. (1989). Contamination of cereals and feed with *Fusarium* mycotoxins in European countries. *In* Fusarium: Mycotoxins, taxonomy, and pathogenicity (Chełkowski, J., ed.), Elsevier Science Publishers B.V. Amsterdam, The Netherlands, pp. 441-472.

Garon, D., Richard, E., Sage, L., Bouchart, V., Pottier, D., and Lebailly, P. (2006). Mycoflora and multitoxin detection in corn silage: experimental study. Journal of Agricultural and Food Chemistry 54:3479-3484.

Garrido, M. D., Jordano, R., Martinez, P., Jodral, M., and Pozo, R. (1988). Fungal contamination of commercial spices. Alimentaria 189:81 + 83-84.

Garrido, M. D., Jodral, M., and Pozo, R. (1992). Mold flora and aflatoxin-producing strains of *Aspergillus flavus* in spices and herbs. Journal of Food Protection 55:451-452.

Gatti, M. J., Fraga, M. E., Magnoli, C., Dalcero, A. M., and da Rocha Rosa, C. A. (2003). Mycological survey for potential aflatoxin and ochratoxin producers and their toxicological properties in harvested Brazilian black pepper. Food Additives and Contaminants 20:1120-1126.

Gbodi, T. A., Nwude, N., Aliu, Y. O., and Ikediobi, C. O. (1986). The mycoflora and some mycotoxins found in maize (zea-mays) in the plateau state of Nigeria. Veterinary and Human Toxicology 28:1-5.

Gerlach, W., and Nirenberg, H. (1982). The genus *Fusarium* - a pictorial atlas. Mitt. Biol. Bundesanst. Land- Forstw. Berlin-Dahlem 209:1-406.

Ghiasian, S. A., Kord-Bacheh, P., Rezayat, S. M., Maghsood, A. H., and Taherkhani, H. (2004). Mycoflora of Iranian maize harvested in the main production areas in 2000. Mycopathologia 158:113-121.

Ghosal, S., Biswas K., and Chakrabarti D. K. (1979) Toxic naphtho-γ-pyrones from *Aspergillus niger*. Journal of Agricultural and Food Chemistry 27:1347-1351.

González, H. H. L., Martinez, E. J., and Resnik, S. L. (1997). Fungi associated with sorghum grain from Agentina. Mycopathologia 139:35-41.

González, H. H. L., Martinez, E. J., Pacin, A., and Resnik, S. L. (1998). Relationship between *Fusarium graminearum* and *Alternaria alternata* contamination and deoxynivalenol occurrence on Argentinian durum whate. Mycopathologia 144:97-102.

González, H. H. L., Resnik, S. L., Boca, R. T., and Marasas, W. F. O. (1995). Mycoflora of Argentinean corn harvested in the main production area in 1990. Mycopathologia 130:29-36.

González, H. H. L., Pacin, A., Resnik, S. L. and Martinez, E. J. (1996). Deoxynivalenol and contaminant mycoflora in freshly harvested Argentinean wheat in 1993. Mycopathologia 135:129-134.

González, H. H. L., Resnik, S., and Pacin, A. M. (2002). Mycoflora of freshly harvested flint corn from northwestern Provinces of Argentina. Mycopathologia 155:207-211.

Gorst-Allman, C. P., Maes, C. M. T. P., Steyn, P. S., and Rabie, C. J. (1982). 5,6-dihydro-4-methoxy-2H-pyran-2-one, a new mycotoxin from *Penicillium italicum*. Suid-Afrikaans Tydskrift voor Chemie 35:102-103.

Grazia, L., Romano, P., Bagni, A., Roggiani, D., and Guglielmi, G. (1986). The role of moulds in the ripening process of salami. Food Microbiology 3:19-25.

Guldborg, M. (1997). Ochratoxin A in Danish beer. Brygmesteren 54:16-17.

Gupta, M. N., and Agrawal, V. K. (1976). Mycoflora associated with melons (*Cucumis melo* L) seeds. Current Science 45:393.

Gupta, S. L. (1956). Occurrence of *Aspergillus carbonarius* (Bainier) Thom causing grape rot in India. Science and Culture 22:167-168.

Gylland, H., Kjellén, K., Haikara, A., and Sigsgaard, P. (1981). Evaluation of fungal contamination of barley and malt. Journal of the Institute of Brewing 87:248-251.

Hald, B., Christensen, D. H., and Krogh, P. (1983). Natural occurrence of the mycotoxin viomellein in barley and the associated quinone-producing penicillia. Applied and Environmental Microbiology 46:1311-1317.

Halls, J. S., and Harman, G. E. (1991). Protection of stored legume seeds against attack by storage fungi and weevils – mechanism of action of lipoidal and soil seed treatments. Crop Protection 10:375-380.

Halt, M. (1998). Moulds and mycotoxins in herb tea and medicinal plants. European Journal of Epidemiology 14:269-274.

Halt, M., and Klapec, T. (2005). Microbial populations in medicinal and aromatic plants and herbal teas from Croatia. Italian Journal of Food Science 17:349-354.

Halweg, H., and Podsiadlo, B. (1991/1992). Mikoflora herbaty. Acta Mycologica 27:115-120.

Hanlin, R. T. (1970). Invasion of peanut fruits by *Aspergillus flavus* and other fungi. Mycopathologia et Mycologia Applicata 40:341-348.

Hanlin, R. T. (1973). The distribution of peanut fungi in Southeastern United States. Mycopathological Mycological Application 49:227-241.

Harris, J. E., and Dennis, C. (1980). Heat stability of endo-polygalacturonases of mucoraceous spoilage fungi in relation to canned fruits. Journal of the Science of Food and Agriculture 31:1164-1172.

Harwig, J., Chen, Y.-K., Kennedy, B. P. C., and Scott, P. M. (1973). Occurrence of patulin-producing strains of *Penicillium expansum* in natural rots of apple in Canada. Canadian Institute of Food Science and Technology Journal 6:22-25.

Hasan, H. A. H. (1998). Studies on toxigenic fungi in roasted foodstuff (salted seed) and halotolerant activity of emodin-producing *Aspergillus wentii*. Folia Microbiologica 43:383-391.

Hasan, H. A. H. (1999). Mycoflora and changes of safflower, wheat and faba bean seed quality during the storage. Rostlinna Vyroba 45:85-91.

Hasan, H. A. H. (2001). Role of pencycuron in aflatoxin production and cotton seed production. Journal of Natural Toxins 10:127-136.

Hashmi, M. H., and Ghaffar, A. (1991). Seed-borne mycoflora of *Coriandrum sativum* L. Pakistan Journal of Botany 23:165-172.

Hashmi, M. H., and Thrane, U. (1990). Mycotoxins and other secondary metabolites in species of *Fusarium*, isolated from seeds of capsicum, coriander and fenugreek. Pakistan Journal of Botany 22:106-116.

Hayden, N. J., and Maude, R. B. (1994). The effect of heat on the growth and recovery of *Aspergillus* spp. from the mycoflora of onion seeds. Plant Pathology 43:627-630.

Hayden, N. J., Maude, R. B., and Proctor, F. J. (1994). Studies on the biology of black mould (*Aspergillus niger*) on temperate and tropical onions. 1. A comparison of sources of the disease in temperate and tropical field crops. Plant Pathology 43:562-569.

Heenan, C. N., Shaw, K. J., and Pitt, J. I. (1998). Ochratoxin A production by *Aspergillus carbonarius* and *A. niger* isolates and detection using coconut cream agar. Journal of Food Mycology 1:67-72.

Heperkan, D. V., Aran, V., and Ayfer, M. (1994). Mycoflora and aflatoxin contamination in shelled pistacio nuts. Journal of the Science of Food and Agriculture 66:273-278.

Heperkan, D., Meric, B. E., Sismanoglu, G. Dalkiliç, G., and Güler, F. K. (2006). Mycobiota, mycotoxigenic fungi, and citrinin production in black olives. Advances in Experimental Medicine and Biology 571:203-210.

Hesseltine, C.W., Vandegraft, E. E., Fennell, D. I., Smith, M. L., and Shotwell, O. L. (1972). Aspergilli as ochratoxin producers. Mycologia 64:539-550.

Hill, R. A., and Lacey, J. (1984). *Penicillium* species associated with barley grain in the U.K. Transactions of the Britisch Mycological Society 82:297-303.

Hocking, A. D. (1994). Fungal spoilage of high-fat foods. Food Australia 46:30-33.

Hocking, A. D., Pitt, J. I., Samson, R. A., and Thrane, U. (eds) (2006). Advances in Experimental Medicine and Biology 571, Springer, New York, U.S.A.

Holmes, G. J., Eckert, J. W., and Pitt, J. I. (1994). Revised description of *Penicillium ulaiense* and its role as a pathogen of citrus fruit. Phytopathology 84:719-727.

Horn, B. W. (2003). Ecology and population biology of aflatoxigenic fungi in soil. Journal of Toxicology-Toxin Reviews 22:351-379.

Horn, B. W. (2005). Colonization of wounded peanut seeds by soil fungi: selectivity for species from *Aspergillus* section *Flavi*. Mycologia 97:202-217.

Horn, B. W., and Dorner, J. W. (1998). Soil populations of *Aspergillus* species from section *Flavi* along a transect through peanut-growing regions of the United States. Mycologia 90:767-776.

Iamanaka, B. T., Taniwaki, M. H., Menezes, H. C., Vicente, E. and Fungaro, M. H. P. (2005). Incidence of toxigenic fungi and ochratoxin A in dried fruits soil in Brazil. Food Additives and Contaminants 22:1258-1263.

Isaac, B., Ayer, S. W., and Stonard, R. J. (1991). Arabenoic acid, a natural product herbicide of fungal origin. Journal of Antibiotics 44:793-794.

Ismail, M. A. (2001). Deterioration and spoilage of peanuts and dessicated coconuts from two sub-Saharan tropical East-African countries due to the associated mycobiota and their degradative enzymes. Mycopathologia 150:67-84.

Ito, Y., Peterson, S. W. ,and Goto, T. (1998). Properties of *Aspergillus tamarii, A. caelatus* and related species in tea field soils in Japan. Mycopathologia 144:169-175.

Jayaraman, P., and Kalyanasundaram, I. (1990). Natural occurrence of toxigenic fungi and mycotoxins in rice bran. Mycopathologia 110:81-85.

Jayaraman, P., and Kalyanasundaram, I. (1994a). Changes in storage mycoflora of parboiled rice through different stages of processing. Journal of Food Science and Technology Mysore 31:219-224.

Jayaraman, P., and Kalyanasundaram, I. (1994b). Changes in moisture-content, mycoflora and aflatoxin content of rice bran during storage. Mycopathologia 126:115-120.

Jesenská, Z., Piecková, E., and Bernát, D. (1993). Heat resistance of fungi from soil. International Journal of Food Microbiology 19:187-192.

Jesenská, Z., Piecková, E., and Sepitková, J. (1991). Thermoresistant propagules of *Neosartorya fischeri*: some ecological considerations. Journal of Food Protection 54:582-584.

Jimenez, M., Mateo, R., Querol, A., Huerta, T., and Hernandez, E. (1991). Mycotoxins and mycotoxigenic moulds in nuts and sunflower seeds for human consumption. Mycopathologia 115:122-128.

Joffe, A. Z. (1969a). Relationships between *Aspergillus flavus*, *A. niger* and some other fungi in the mycoflora of groundnut kernels. Plant and Soil 31:57-64.

Joffe, A. Z. (1969b). The mycoflora of fresh and stored groundnut kernels in Israel. Mycopathologia et Mycologia Applicata 68:39-46.

Joosten, H. M. L. J., Goetz, J., Pittet, A., Schellenberg, M., and Bucheli, P. (2001). Production of ochratoxin A by *Aspergillus carbonarius* on coffee cherries. International Journal of Food Microbiology 65:39-44.

Julian, A. M., Wareing, P. W., Phillips, S. I., Medlock, V. F. P., MacDonald, M. V., and Delrio L. E. (1995). Fungal contamination and selected mycotoxins in preharvest and postharvest maize in Honduras. Mycopathologia 129:5-16.

Jurjević, Z., Solfrizzo, M., Cvetković, B., Avvantagiato, G., and Visconti, A. (1999). Ochratoxin A and fumonisins (B_1 and B_2) in maize from Balkan nephropathy endemic and non-endemic areas of Croatia. Mycotoxin Research 15:67-80.

Kinderlerer, J. (1984a). Fungi in desiccated coconut. Food Microbiology 1:205-207.

Kinderlerer, J. (1984b). Spoilage in desiccated coconut resulting from growth of xerophilic fungi. Food Microbiology 1:23-28.

Kinderlerer, J., and Hatton, P. V. (1990). Fungal metabolites of sorbic acid. Food Additives and Contaminants 7:657-669.

Kinderlerer, J., and Hatton, P. V. (1991). The effect of temperature, water activity and sorbic acid on ketone rancidity produced by *Penicillium crustosum* Thom in coconut and palm kernel oils. Journal of Applied Microbiology 70:502-506.

King, A. D., and Halbrook, W. U. (1987). Ascospore heat resistance and control measures for *Talaromyces flavus* isolated from fruit juice concentrate. Journal of Food Science 52:1252-1254.

King, A. D., and Schade, J. E. (1986). Influence of almond harvest, processing and storage on fungal population and flora. Journal of Food Science 51:202-215.

King, A. D., Jr., Pitt, J. I., Beuchat, L. R., and Corry, J. E. L., eds. (1986). Methods for the mycological examination of foods. Plenum Press, New York, U.S.A.

Kiran, D. R., Narayana, K. J. P., and Vijayalakshmi, M. (2005). Aflatoxin B_1 production in chillies (*Capsicum annuum* L.) kept in cold stores. African Journal Biotechnology 4:791-795.

Kis, Z., Furger, P., and Sigg, H. P. (1969). Über die Isolierung von Pyrenophenol. Experientia 25:123.

Korukluoglu, M., Yigit, A., and Sahan, Y. (2005). Mycoflora of some cheese samples in Bursa, Turkey. Indian Veterinary Journal 82:340-341.

Kosiak, B., Torp, M., Skjerve, E., and Andersen, B. (2004). *Alternaria* and *Fusarium* in Norwegian grains of reduced quality – a matched pair sample study. International Journal of Food Microbiology 93:51-62.

Kriek, N. P. J., and Werner, F. C. (1981). Toxicity of *Penicillium italicum* to laboratory animals. Food and Cosmetics Toxicology 19:311-315.

Kritzinger, Q., Aveling, T. A. S., Marasas, W. F. O., Rheeder, J. P., Westhuizen, L. van der, and Shephard, G. S. (2003). Mycoflora and fumonisin mycotoxins associated with cowpea (*Vigna unguiculata* (L.) Walp) seeds. Journal of Agricultural and Food Chemistry 51:2188-2192.

Kunwar, I. K. (1989). Mycoflora associated with stored wheat and its milling fractions in India. Procreddings of the Indian Academy of Sciences – Plant Sciences 99:437-443.

Kurtzman, C. P., Rogers, R., and Hesseltine, C. W. (1971). Microbiological spoilage of mayonnaise and salad dressings. Applied Microbiology 21:870-874.

Lacey, J., Ramakrishna N., Hamer A., Magan N., and Marfleet, I. C. (1991). Grain fungi. *In* Handbook of applied mycology Vol. 3. Foods and feeds (Arora, D. K., Mukerji, K. G., and Marth, E. H., eds.), Marcel Dekker, New York, U.S.A., pp. 121-177.

Laich, F., Fierro, F. Cordoza, R. E., and Martin, J. F. (1999). Organization of the gene cluster for biosynthesis of penicillin in *Penicillium nalgiovense* and antibiotic production in cured dry sausages. Applied Environmental Microbiology 65:1236-1240.

Landers, K. E., Davis, N. D. and Diener, U. L. (1967). Influence of atmospheric gases on aflatoxins production by *Aspergillus flavus* in peanuts. Phytopathology 57:1086-1090.

Larsen, T. O., Frisvad, J. C., Ravn, G., and Skaaning, T. (1998a). Mycotoxin production by *Penicillium expansum* on blackcurrant and cherry juice. Food Additives and Contaminants 15:671-675.

Larsen, T. O., Frisvad, J. C., and Christophersen, C. (1998b). Arabenoic acid (verrucolone), a major chemical indicator of *Penicillium verrucosum*. Biochemical Systematics and Ecology 26:463-465.

Larsen, T. O., Smedsgaard, J., Nielsen, K. F., Hansen, M. E., and Frisvad, J. C. (2005). Phenotypic taxonomy and metabolite profiling in microbial drug discovery. Natural Products Reports 22:672-695.

Lau, K. H., and Sheridan, J. E. (1975). Mycoflora of rice (*Oryza sativa* L) seed in New Zealand. New Zealand Journal of Agricultural Research 18:237-250.

Lee, H. B., and Magan, N. (1999). Environment factors influence in vitro interspecific interactions between *A. ochraceus* and other maize spoilage fungi, growth and ochratoxin production. Mycopathologia 146:43-47.

Leong, S. L., Hocking A. D., and Pitt, J. I. (2004). Occurrence of fruit rot fungi (*Aspergillus* section *Nigri*) on some drying varieties of irigated grapes. Australian Journal of Grape Wine Research 10:83-88.

Leoni, L. A. B., Soares, L. M. V., and Oliveira, P. L. C. (2000). Ochratoxin A in Brazilian roasted and instant coffees. Food Additives and Contaminants 17:867-870.

Levi, C. P., Trenk, H. L. and Mohr, H. K. (1974). Study of the occurrence of ochratoxin A in green coffee beans. Journal of the Association of Official Analytical Chemists 57:866-870.

Lisker, N., Michaeli, R., and Frank, Z. R. (1993). Mycotoxigenic potential of *Aspergillus flavus* strains isolated from groundnut growing in Israel. Mycopathologia 122:177-183.

Llewellyn, E. G., Dixon, E. C. X., and Eadie, T. (1981). Aflatoxin formation on whole and ground cumin and anise seeds. Journal of the American Oil Chemists Society 58:985A-988A.

Logrieco, A., Bottalico, A., Mulé, M., Moretti, A., and Perrone, G. (2003). Epidemiology of toxigenic fungi and their associated mycotoxins for some Mediterranean crops. European Journal of Plant Pathology 109:645-667.

Logrieco, A., Visconti, A., and Bottalico, A. (1990). Mandarin fruit rot caused by *Alternaria alternata* and associated mycotoxins. Plant Disease 74:415-417.

Lopez, A., and Crawford, M. A. (1967). Aflatoxin content of groundnuts sold for human consumption in Uganda. Lancet 2:1351.

López de Cerain, A., González-Peñas, E. Jiménez, A. M., and Bello, J. (2002). Contribution to the study of ochratoxin A in Spanish wines. Food Additives and Contaminants 19:1058-1064.

Lopez-Diaz, T. M., Santon, J. A., Prieto, M., Garcia-Lopez, M. L., and Otero, A. (1995). Mycoflora of a traditional Spanish blue cheese. Netherlands Milk and Dairy Journal 49:191-199.

Lopez-Diaz, T. M., Santos, J. A., Garcia-Lopez, M. L., and Otero, A. (2001). Surface mycoflora of Spanish fermented meat sausage and toxigenicity of *Penicillium* isolates. International Journal of Food Microbiology 68:69-74.

Luke, H. H., and Barnett, R. D. (1979). Influence of harvest conditions on the germinability and mycoflora of rye seed. Seed Science Technology 7:431-437.

Lund, F., Filtenborg, O., and Frisvad, J. C. (1995). Associated mycoflora of cheese. Food Microbiology 12:173-180.

Lund, F., Filtenborg, O., Westall, S., and Frisvad, J. C. (1996). Associated mycoflora of rye bread. Letters in Applied Microbiology 23:213-217.

Lund, F. and Frisvad, J. C. (1994). Chemotaxonomy of *Penicillium aurantiogriseum* and related species. Mycological Research 98:481-492.

Lund, F. and Frisvad, J. C. (2003). *Penicillium verrucosum* in cereals indicates production of ochratoxin A. Journal of Applied Microbiology 95:1117-1123.

MacDonald, S., Wilson, P., Barnes, K., Damant, A., Massey, R., Mortby, E., and Shepherd, M. J. (1999). Ochratoxin A in dried vine fruit, method development and survey. Food Additives and Contaminants 16:253-260.

Machinski, M., Soares, L-M. V., Sawazaki, E., Bolonhezi, D., Castro, J. L., and Bortoletto, N. (2001). Aflatoxins, ochratoxin A and zearalenone in Brazilian corn cultivars. Journal of the Science of Food and Agriculture 81:1001-1007.

Magnoli, C., Hallak, C., Astoreca, A., Ponsone, L., Chiacchiera, S., and Dalcero, A. M. (2006). Occurrence of ochratoxin A-producing fungi in commercial corn kernels in Argentina. Mycopathologia 161:53-58.

Magnoli, C., Violante, M., Combina, M., Palacio, G., and Dalcero, A. (2003). Mycoflora and ochratoxin-producing strains of *Aspergilllus* section *Nigri* in wine grapes in Argentina. Letters in Applied Microbiology 37:179-184.

Mahmoud, M. I., El-Bazza, Z. E., and Mahamed, Z. G. (1992). Aflatoxin production at different relative humidities on gamma-irradiated herbs used as Egyptian drinks. Egyptian Journal of Pharmaceutical Science 33:21-30.

Majerus, P., Bresch, H., and Otteneder, H. (2001). Ochratoxin A in wines, fruit juices and seasonings. Archiv für Lebensmittelhygiene 51:95-97.

Mantle, P. G., and Chow, A. M. (2000). Ochratoxin A formation in *Aspergillus ochraceus* with particular reference to spoilage of coffee. International Journal of Food Microbiology 56:105-109.

Mantle, P. G., and McHugh, K. M. (1993). Nephrotoxic fungi in foods from nephropathy households in Bulgaria. Mycological Research 97:205-212.

Marasas, W. F. O., Wehner, F. C., van Rensburg, S. J., and van Schalkwyk, D. J. (1981). Mycoflora of corn produced in human esophageal cancer areas in Transkei, Southern Africa. Phytopathology 71:792-796.

Marasas, W. F. O., Nelson, P. E., and Toussoun, T. A. (1984). Toxigenic *Fusarium* species. Identity and mycotoxicology. The Pennsylvania State University Press, University Park, PN, U.S.A.

Martinez, P., Jodral, M. Garrido, M. D., Jordano, R., and Pozo, R. (1988). Identification of toxigenic *Aspergilllus flavus* in commercial spices. Alimentaria 189:85-86.

Martinez-Magana, P., Jodral-Villarejo, M., and Pozo-Lora, R. (1989). Mycoflora and *Aspergillus flavus* in pepper on sale in Spain. Microbiologie Aliments Nutrition 7:311-314.

Martins, M. L., Martins, H. M., and Bernardo, F. (2001a). Fumonisins B_1 and B_2 in black tea and medicinal plants. Journal of Food Protection 64:1268-1270.

Martins, M. L., Martins, M. L., Dias, M. I., and Bernardo, F. (2001b). Evaluation of microbiological quality of medicinal plants used in natural infusions. International Journal of Food Microbiology 68:149-153.

Martins, M. L., Martins, H. M., and Gimeno, A. (2003). Incidence of microflora and of ochratoxin A in green coffee beans (*Coffea arabica*). Food Additives and Contaminants 20:1127-1131.

Martinsmaciel, E. R., Machinski, M., Pereira, S. R. C., Takahachi, G., Kemmelmeier, C., and Nishiyama, P. (1996). Incidence of aflatoxins and *Aspergillus flavus* in peanuts consumed in Maringa City, Brazilian Archives of Biology and Technology 39:807-813.

Mastrodisalgado, J., and Carvalho, P. C. T. D. (1980). Toxigenic fungi associated to grains. 1. Survey of mycoflora associated to corn, wheat and rice. Revista de Microbiologia 1160-63.

Mazen, M. B., Abdel-Hafez, S. I. I., and Shaban, G. M. M. (1984). Survey of the mycoflora of Egyptian wheat grains and their lemmae and paleae. Mycopathologia 85:155-159.

Mazen, M. B., El-Kady, I. A., and Saber, S. M. (1990). Survey of the mycoflora and mycotoxins of cotton seeds and cotton seed products in Egypt. Mycopathologia 110:133-138.

McDonald, D. (1970). Fungal infection of groundnut fruit before harvest. Transactions of the British Mycological Society 54:453-460.

McKee, L. H. (1995). Microbial contamination of spices and herbs – a review. Lebensmittel-Wissenshaft und Technologie 28:1-11.

McKinley, E. R., and Carlton, W. W. (1991). Patulin. *In* Mycotoxins and phytoalexins (Sharma, R. P., and Salunkhe, D. K., eds.), CRC Press, Boca Raton, FL, U.S.A., pp. 191-236.

McLean, M., and Berjak, P. (1987). Maize grains and their associated mycoflora – a microecological consideration. Seed Science and Technology 15:831-850.

Medina, Á., Mateo, R., López-Ocaña, L., Valle-Algarra, F. M., and Jiménez, M. (2005a). Study of Spanish grape mycobiota and ochratoxin A production by isolates of *Aspergillus tubingensis* and other members of *Aspergillus* section *Nigri*. Applied and Environmental Microbiology 71:4696-4702.

Medina, Á., Jiménez. M., Valle-Algarra, F. M., Gimeno-Adelantado, J. V., and Mateo, R. (2005b). Determination of ochratoxin A in beer marketed in Spain by liquid chromatography with fluorescence detection using lead hydroxyacetate as a clean-up agent. Journal of Chromatography A 1083:7-13.

Medina, Á., Valle-Algarra, F. M., Mateo, R., Gimeno-Adelantado, J. V., Mateo, F., and Jiménez. M. (2006). Survey of the mycobiota of Spanish malting barley and evaluation of the mycotoxin producing potential of species of *Alternaria*, *Aspergillus* and *Fusarium*. International Journal of Food Microbiology 108:196-203.

Mehan, V. K., and McDonald, M. (1984). Mycotoxin-producing fungi in groundnuts – potential for mycotoxin contamination. Oliagineux 39:25-29.

Mehan, V. K., Mayee, C. D., Jayanthi, S., and McDonald, M. (1991). Preharvest seed infection by *Aspergillus flavus* group fungi and subsequent aflatoxin contamination in groundnuts in relation to soil types. Plant and Soil 136:239-248.

Micco, C., Grossi, M., Miraglia, M., and Brera, C. (1989). A study of the contamination by ochratoxin A of green and roasted coffee beans. Food Additives and Contaminants 6:333-339.

Milanez, T. V., and Sabino, M. (1989). Ochratoxin A in beans and its stability after cooking. Revista do Instituto A. Lutz, Brasil 49:131-135.

Miller, W. A., and Roy, K. (1982). Mycoflora of soybean leaves, pods, and seeds in missisippi. Canadian Journal of Botany 60:2716-2723.

Mills, J. T., and Woods, S. M. (1994). Factors affecting storage life of farm-stored field peas (*Pisum sativum* L) and white beans (*Phaseolus vulgaris* L). Journal of Stored Products Research 30:215-226.

Mills, J. T., Seifert, K. A., Frisvad, J. C., and Abramson, D. (1995). Nephrotoxigenic *Penicillium* species occurring on farm-stored cereal grains in western Canada. Mycopathologia 130:23-28.

Mislivec, P. B., Dieter, C. T., and Bruce, V. R. (1975). Mycotoxin-producing potential of mold flora of dried beans. Applied Microbiology 29:522-526

Mislivec, P. B., Bruce, V. R., and Gibson, R. (1983). Incidence of toxigenic and other molds in green coffee beans. Journal of Food Protection 46:969-973.

Mitchell, D., Parra, R., Aldred, D., and Magan, N. (2004). Water and temperature relations of growth and ochratoxin A production by *Aspergillus carbonarius* strains from grapes in Europe and Israel. Journal of Applied Microbiology 97:439-445.

Moharram, A. M., Abdel-Malek, A. Y., and Abdel-Hafez, A. L. I. (1989). Mycoflora of anise and fennel seeds in Egypt. Journal of Basic Microbiology 29:427-435.

Mojtahedi, H., Rabie, C. J., Lubben A., Steyn, M., and Danesh, D. (1979). Toxic Aspergilli from pistacio nuts. Mycopathologia 115:122-128.

Möller, T., Brostedt, S., and Johansson, M. (1993). Ergot alkaloids in Swedish cereals. Vår Föda 45:308-311.

Molto, G., Samara, M. M., Resnik, S., Martinez, E. J., and Pacin, A. (2000). Occurrence of trichothecenes in Argentinean beer: a preliminary exposure assessment. Food Additives and Contaminants 17:809-813.

Moreau, C., and Moreau, M. (1978). La contamination des épices, ses conséquence dans les industries alimentaires. Industries Alimentaires et Agricoles 95:497-502.

Morris, P. F., Connolly, M. S., and St Clair, D. A. (2000). Genetic diversity of *Alternaria alternata* isolated from tomato in California assessed using RAPDs. Mycological Research 104:286-292.

Moubasher, A. H., El-Hissy, E. I., Abdel-Hafez, S. I., and Hassan, S. K. M. (1979). The mycoflora of peanuts in Egypt. Mycopathologia 98:165-170.

Mphande, F. A., Siame, F. A., and Taylor, J. E. (2004). Fungi, aflatoxins, and cyclopiazonic acid associated with peanut retailing in Boyswana. Journal of Food Protection 67:96-102

Müller, H. M., and Schwadorf, K. (1993). A survey of the natural occurrence of *Fusarium* toxins in wheat grown in a Southwestern area of Germany. Mycopathologia 121:115-121.

Myrchink, T. G. (1967). Production of patulin by a group of fungi *Penicillium lapidosum* Raper and Fennell. Antibiotiki 12:762-766.

Nair, N. G. (1985). Fungi associated with bunch rot of grapes in the Hunter Valley. Australian Journal of Agriculture Research 36:435-442.

Nakajima, M., Tsubouchi, H. Miyabe, M., and Ueno, Y. (1997). Survey of aflatoxin B1 and ochratoxin A in commercial green coffee beans by high-performance liquid chromatography linked with immunoaffinity chromatography. Food and Agriculture Immunology 9:77-83.

Nakajima, M., Tsubouchi, H., and Miaybe, M. (1999). A survey of ochratoxin A and aflatoxins in domestic and imported beer in Japan by immunoaffinity and liquid chromatography. Journal of the Association of Official Analytical Chemists 82:897-902.

Naumova, E. S. (1988). The specific composition of fungi on the soy-beans from Voronezh region. Mikologia Fitopatologia 22:217-223.

Nawaz, S., Scudamore, K. A., and Rainbird, B. C. (1997). Mycotoxins in ingredients of animal feeding stuff: I. Determination of *Alternaria* mycotoxins in oilseed rape and sunflower seed meal. Food Additives and Contaminants 14:249-262.

Ng, W., Mankotia, M., Pantazopoulos, P., Neil, R. J., and Scott, P. M. (2004). Ochratoxin A in wine and grape juice sold in Canada. Food Additives and Contaminants 21:971-981.

Nielsen, K. F., Smedsgaard, J., Larsen, T. O., Lund, F., Thrane, U., and Frisvad, J. C. (2004). Chemical identification of fungi - metabolite profiling and metabolomics. *In* Fungal Biotechnology in Agricultural, Food and Environmental Applications (Arora, D. K., ed.), Marcel Dekker, New York, U.S.A., pp. 19-35.

Nielsen, K. F., Sumarah, M. W., Frisvad, J. C., and Miller, J. D. (2006). Production of metabolites from the *Penicillium roqueforti* complex. Journal of Agricultural and Food Chemistry 54:3756-3763.

Nielsen, P. V., Beuchat, L. R., and Frisvad, J. C. (1988). Growth and fumitremorgin production by *Neosartorya fischeri* as affected by temperature, light, and water activity. Applied and Environmental Microbiology 54:1504-1510.

Nielsen, P. V., Beuchat, L. R. and Frisvad, J. C. (1989a). Influence of atmospheric oxygen-content on growth and fumitremorgin production by a heat resistant mould. *Neosartorya fischeri*. Journal of Food Science 54:679-682.

Nielsen, P. V., Beuchat, L. R., and Frisvad, J. C. (1989b). Growth and fumitremorgin production by *Neosartorya fischeri* as affected by food preservatives and organic acids. Journal of Applied Bacteriology 66:197-207

Nirenberg, H. (1995). Morphological differentiation of *Fusarium sambucinum* Fuckel sensu lato, *F. torulosum* (Berk. and Curt.) Nirenberg comb. nov. and *F. venenatum* sp. nov. Mycopathologia 129:131-141.

Nkwe, D. O., Taylor, J. E., and Siame, B. A. (2005). Fungi, aflatoxins, fumonosin B1 and zearalenone contaminating sorghum-based traditional malt, wort and beer in Botswana. Mycopathologia 160:177-186.

Northolt, M. D. (1979). The effect of water activity and temperature on the production of some mycotoxins. Ph. D. Dissertation University of Wageningen.

Northolt, M. D., Egmond, H. P. van, Soentoro, P., and Deijll, E. (1980a). Fungal growth and the presence of sterigmatocystin in hard cheese. Journal of the Association of Official Analytical Chemists 63:115-119.

Northolt, M. D., Eikelenboom, C., Hartog, B. J., Nooitgedacht, A. J., and Pateer, P. M. (1980b). Onderzoek naar de biologische gesteldheid en houdbaarheid van droog gebak en vruchtengebak. English Summary. De Ware(n) Chemicus 10:116-124.

Nout, M. J. R. (2004). Useful role of fungi in food processing. *In* Introduction to food- and airborne fungi 7th edition (Samson, R. A., Hoekstra, E. S., and Frisvad, J. C., eds.*)*, Centraalbureau voor Schimmelcultures, Utrecht, The Netherlands, pp. 364-374.

Okigbo, R. N. (2003). Mycoflora of tuber surface of white yam (*Dioscorea rotundata* Poir) and postharvest control of pathogens with *Bacillus subtilis*. Mycopathologia 156:81-85.

Okigbo, R. N., and Ikediugwu, F. E. O. (2000). Studies on the biological control of postharvest rot in yams (*Dioscorea* spp.) using *Trichoderma viride*. Journal of Phytopathology 148:351-355.

Okpokwasili, G. C., and Molokwu, C. N. (1996). Yeast and mould contamination of vegetable oils. Bioresources and Technology 57:245-249.

Omurtak, G. Z., and Yazicioglu, D. (2004). Determination of fuminisins B_1 and B_2 in herbal tea and medicinal plants in Turkey by high-performance liquid chromatography. Journal of Food Protection 67:1782-1786.

Onesirosan, P. T. (1982a). Effect of moisture content and temperature on the invasion of cowpeas by storage fungi. Seed Science and Technology 10:619-629.

Onesirosan, P. T. (1982b). Effect of seed coat type on the susceptibility of cowpeas to invasion by storage fungi. Seed Science and Technology 10:631-637.

Ono, E. Y. S., Sugiura, Y., Homechin, M., Kamogae, M., Vizzoni, E., Ueno, Y., and Hiroka, E. Y. (1999). Effect of climatic conditions on natural mycoflora and fumonisins in freshly harvested corn of the State of Parana, Brazil. Mycopathologia 147:139-148.

Ono, E. Y. S., Sasaki, E. Y., Hashomoto, E. H., Hara, L. N., Corrêa, B., Itano, E. N., Sugiura, T., Ueno, Y., and Hirooka, E. Y. (2002). Post-harvest storage of corn; effect of beginning moisture content on mycoflora and fumonisin contamination. Food Additives and Contaminants 19:1081-1090.

Ono, E. Y. S., Biazon, L., da Silva, M., Vizoni, E. Sugiura, Y., Ueno, Y., and Hirroka, E. Y. (2006). Fumonisins in corn: Correlation with *Fusarium* sp count, damaged kernels, protein and lipid content. Brazilian Archives of Biology and Technology 49:63-71.

Orsi, R. B., Corrêa, B., Possi, C. R., Schammass, E. A., Nogueira, J. R., Dias, S. M. C., and Malozzi, M. A. B. (2000). Mycoflora and occurrence of fumonisins in freshly harvested and stored hybrid maize. Journal of Stored Products Research 36:75-87.

Otteneder, H., and Majerus, P. (2000). Occurrence of ochratoxin A (OTA) in wines: influence of type of wine and its geographical origin. Food Additives and Contaminants 17:793-798.

Otteneder, H., and Majerus, P. (2001). Ochratoxin A (OTA) in coffee: nation-wide evaluation of data collected by German Food Control 1995-1999. Food Additives and Contaminants 18:431-435.

Overy, D. P., and Frisvad, J. C. (2003). New *Penicillium* species associated with bulbs and root vegetables. Systematic and Applied Microbiology 26:631-639.

Overy, D. P., and Frisvad, J. C. (2004). Mycotoxin production and postharvest storage rot of ginger (*Zingiber officinale*) by *Penicillium brevicompactum*. Journal of Food Protection 68:607-609.

Overy, D., Seifert, K., Savard, M. E., and Frisvad, J. C. (2003). Spoilage fungi and their mycotoxins in commercially marketed chestnuts. International Journal of Food Microbiology 88:69-77.

Overy, D. P., Frisvad, J. C., Steinmeier, U., and Thrane, U. (2005a). Clarification of the agents causing blue mold storage rot upon various flower and vegetable bulbs: implications for mycotoxin contamination. Postharvest Biology and Technology 35, 217-221.

Overy, D. P., Valdez, J. G., and Frisvad, J. C. (2005b). Revisions to the *Penicillium* ser. *Corymbifera*: agents responsible for the blue mould storage rot of various flower and vegetable bulbs. Canadian Journal of Botany 83:1422-1433.

Ozcelik, S., Ozcelik, N., and Beuchat, L. R. (1990). Toxin production by *Alternaria alternata* in tomatoes and apples stored under various conditions and quantitation of the toxins by high-performance liquid chromatography. International Journal of Food Microbiology 11:187-194.

Pal, N., and Kundu, A. K. (1972). Studies on *Aspergillus* spp. from Indian spices in relation to aflatoxin production. Science and Culture 38:252-254.

Pardo, E., Marín, S., Sanchis, V., and Ramos, A. J. (2004). Prediction of fungal growth and ochratoxin A production by *Aspergillus ochraceus* on irradiated barley grain as influenced by temperature and water activity. International Journal of Food Microbiology 95:79-88.

Pardo, E., Marín, S., Sanchis, V., and Ramos, A. J. (2005a). Effect of water activity and temperature on mycelial growth and ochratoxin A production by isolates of *Aspergillus ochraceus* on irradiated green coffee beans. Journal of Food Protection 68:133-138.

Pardo, E., Marín, S., Sanchis, V., and Ramos, A. J. (2005b). Impact of relative humidity and temperature on visible fungal growth and OTA production of ochratoxigenic *Aspergillus ochraceus* on grapes. Food Microbiology 22: 383-389.

Park, J. W., Choi, S. Y., Hwang, H. J., and Kim, Y. B. (2005a). Fungal mycoflora and mycotoxins in Korean polished rice destined for humans. International Journal of Food Microbiology 103:305-314.

Park, J. W., Chung, S. H., Lee, C., and Kim, Y. B. (2005b). Fate of ochratoxin A during cooking of naturally contaminated rice. Journal of Food Protection 68:2107-2111.

Park, J. W., Lee, Y. C., and Kim, Y. B. (2005c). Fate of aflatoxin B_1 during the cooking of Korean polished rice. Journal of Food Protection 68:1431-1434.

Paster, N., Droby, S., Chalutz, E., Menasherov, M., Mitzan, R., and Wilson, C. L. (1993). Evaluation of the potential of the yeast *Pichia guillermondii* as a biocontrol agent against *Aspergillus flavus* and fungi of stored soya beans. Mycological Research 97:1201-1206.

Patel, S., Hazel, C. M., Winterton, A. G. M., and Gleadle, A. E. (1997). Survey of ochratoxin A in UK retail coffees. Food Additives and Contaminants 14:217-222.

Payen, J., Girard, T., Gaillardin, M., and Lafont, P. (1983). Presence of ochratoxin A in beers. Microbiologie Aliments Nutrition 1:143-146.

Peever, T. L., Ibanez, A. Akimitsu, K., and Timmer, L. W. (2002). Worldwide phylogeography of citrus brown spot pathogen, *Alternaria alternata*. Phytopathology 92:794-802.

Pelhate, J. (1979). Mycoflora of corn seeds. 1. Pre-harvesting contamination. Revue de Mycologie 43:109-129.

Pelhate, J. (1981). Mycoflora of corn grains. 2. Keeping of corn cobs. Cryptamic Mycology 2:61-84.

Pelhate, J., and Agosin, E. (1985). Natural wheat straw mycoflora. Cryptogamic Mycology 6:1-19.

Pelser, P., Du, T., and Eckert, J. W. (1977). Constituents of orange juice that stimulate the germination of conidia of *Penicillium digitatum*. Phytopathology 67:747-754.

Perez, J., Infante, F., Vega, F. E., Holguin, F., Macias, J., Valle, J., Nieto, G., Peterson, S. W., Kurtzman, C. P., and O'Donnell, K. (2003). Mycobiota associated with the coffee berry borer (*Hypothenemus hampei*) in Mexico. Mycological Research 107:879-887.

Perrone, G., Mulé, G., Susca, A., Battilani, P., Pietri, A., and Logrieco, A. (2006). Ochratoxin A production and amplified length polymorphism analysis of *Aspergillus carbonarius*, *Aspergillus tubingensis*, and *Aspergillus niger* strains isolated from grapes in Italy. Applied and Environmental Microbiology 72:680-685.

Pier, A. C., and Richard, J. L. (1992). Mycoses and mycotoxicoses of animals caused by Aspergilli. *In* Aspergillus, Biology and Industrial applications (Bennett, J. W., and Klich, M. A., eds.), Butterworth-Heinemann, Boston, MS, U.S.A., pp. 233-248.

Pitt, J. I. (1979). The genus *Penicillium* and its teleomorphic states *Eupenicillium* and *Talaromyces*. Academic Press, New York.

Pitt, J. I., Dyer, S. K., and McCammon, S. (1991a). Systemic invasion of developing peanuts plants by *Aspergillus flavus*. Letters in Applied Microbiology 13:16-20.

Pitt, J. I., Spotts, R. A., Holmes, R. J., and Cruickshank, R. H. (1991b). *Penicillum solitum* revived, and its role as a pathogen of pomaceous fruit. Phytopathology 81:1108-1112.

Pitt, J. I., and Hocking, A. D. (1997). Fungi and food spoilage, 2nd edition. Blackie Academic and Professional, London, U.K.

Pitt, J. I., Hocking, A. D., Bhudhasamai, K., Miscamble, B. F., Wheeler, K. A., and Tanboonek, P. (1993). The normal mycoflora of commodities from Thailand. 1. Nuts and oilseeds. International Journal of Food Microbiology 20:211-226.

Pitt, J. I., Hocking, A. D., Bhudhasamai, K., Miscamble, B. F., Wheeler, K. A., and Tanboonek, P. (1994). The normal mycoflora of commodities from Thailand. 2. Beans, rice, small grains and other commodities. International Journal of Food Microbiology 23:35-53.

Pittet, A., Tornare, D., Huggett, A., and Viani, R. (1996). Liquid chromatographic determinaton of ochratoxin A in pure and adulterated soluble coffee using an immunoaffinity column clean up procedure. Journal of Agricultural and Food Chemistry 44:3564-3569.

Pozzi, C. R., Braghini, R., Arcaro, J. R. P., Zorzete, P., Israel, A. L. M., Pozar, I. C., Denucci, S., and Corrêa, B. (2005). Mycoflora and occurrence of alternariol and alternariol monomethyl ether in Brazilian sunflower from sowing to harvest. Journal of Agricultural Food Chemistry 53:5824-5828.

Pryor, B. M., and Michailides, T. J. (2002). Morphological, pathologic and molecular characterization of *Alternaria* isolates associated with Alternaria late blight of pistachio. Phytopathology 92:406-416.

Purcell, S. L., Phillips, D. J., and MacKey, B. E. (1980). Distribution of *Aspergillus flavus* and other fungi in several almond-growing areas of California. Phytopathology 70: 926-929.

Quintavalla, S., and Spotti, E. (1993). Heat resistance of *Talaromyces flavus, Neosartorya fischeri* and *Byssochlamys nivea* isolated from fresh fruits. Microbiologie Aliments Nutrition 11:335-341.

Ramakrishna, Y., Bhat, R. V., and Ravindranath, V. (1989). Production of deoxynivalenol by *Fusarium* isolates from samples of wheat associated with a human mycotoxicosis outbreak and from sorghum cultivars. Applied and Environmental Microbiology 55:2619-2620.

Ramírez, C. (1982). Manual and atlas of the Penicillia. Elsevier Biomedical, Amsterdam, The Netherlands.

Rand, T. G., Giles, S., Flemming, J., Miller, J. D., and Puniani, E. (2005). Inflammatory and cytotoxic response in mouse lungs exposed to purified toxins from building isolated *Penicillium brevicompactum* Dierckx and *P. chrysogenum* Thom. Toxicological Science 87:213-222.

Rani, N., and Singh, S. (1989). Natural occurrence of mycoflora and aflatoxins in *Foeniculum vulgare* Mill seeds. National Academy of Sciences Letters India 12:411-413.

Ranzani, M. R. T. C., and Fonseca, H. (1995). Mycological evaluation of chemically treated unshelled peanuts. Food Additives and Contaminants 12:343-346.

Raper, K. B., and Fennell, D. I. (1965). The genus *Aspergillus*. Williams and Wilkins, Baltimore.

Raper, K. B., and Thom, C. (1949). Manual of the Penicillia. Williams and Wilkins, Baltimore.

Raut, J. G. (1975). Fungi associated to sunflower seeds. Current Sciences 44:789-790.

Razzaghi-Abyaneh, M., Shams-Ghahfarokhi, M., Allameh, A., Kazerron-Shiri, A., Ranjbar-Bahadori, S., Mirzahoseini, H., and Rezaee, M. B. (2006). A survey on distribution of *Aspergillus* section *Flavi* in corn field soils in Iran: Population patterns based on aflatoxins, cyclopiazonic acid and sclerotia production. Mycopathologia 161:183-192.

Reiss, J. (1972). Mykotoxine in Nahrungsmitteln. II. Nachweis von Patulin in spontan verschimmeltem Brot und Gebäck. Naturwissenschaften 59:37.

Reiss, J. (1977). Mycotoxins in foodstuffs. X. Production of citrinin by *Penicillium chrysogenum* in bread. Food and Cosmetics Toxicology 15:303-307.

Reshma, S. V., and Ahmad, R. (1998). Natural incidence of aflatoxins in parboiled rice during various states of processing. Journal of Food Science and Technology Mysore 35:451-454.

Reuveni, M., Sheglov, D., Ben-Arie, R., and Prusky, D. (2002). Sensitivity of red delicious apple fruit at various phonologic stages to infection by *Alternaria alternata* and moldy-core rot. Journal of Plant Pathology 108:421-427.

Rheeder, J. P., Marasas, W. F. O., Thiel, P. G., Sydenham, E. W., Shephard, G. S., and Schalkwyk, D. J. van (1992). *Fusarium moniliforme* and fumonisins in corn in relation to human esophageal cancer in Transkei. Phytopathology 82:353-357.

Rice, S. L., Beuchat, L. R. and Heaton, E. K. J. (1979). Changes in the composition and texture of canned peach halves infected with *Byssochlamys fulva*. Journal of Food Science 42:1562-1565.

Richer, L., Sigalet, D., Kneteman, N., Shapiro, J., Jones, A., Scott, R. B., Ashbourne, R., Sigler, L., Frisvad, J. C., and Smith, L. (1997). Fulminant hepatic failure following ingestion of moldy homemade rhubarb wine. Gastroenterology 112:A1366.

Rippon, L. E. (1980). Wastage of postharvest fruit and its control. CSIRO Food Research Quarterly 40:1-12.

Roberts, R. G. (2005). *Alternaria yaliinficiens* sp. nov. on Ya Li pear fruit: From interception to identification. Plant Disease 89:134-145.

Roberts, R. G., Reymond, S. T., and Andersen, B. (2000). RAPD fragment pattern analysis and morphological segregation of small-spored *Alternaria* species and species groups. Mycological Research 104:151-160.

Robiglio, A. L., and Lopez, S. E. (1995). Mycotoxin production by *Alternaria alternata* strains isolated from red delicious apples in Argentina. International Journal of Food Microbiology 24:413-417.

Romani, S., Sacchetti, G. López, C. C., Pinnavia, G. G., and Rosa, M. D. (2000). Screening on the occurrence of ochratoxin A in green coffee beans of different origins and types. Journal of Agricultural and Food Chemistry 48:3616-3619.

Romero, S. M., Comerio, R. M., Larumbe, G., Ritieni, A., Vaamonde, G., and Pinto, V. F. (2005). Toxigenic fungi isolated from dried vine fruits in Argentina. International Journal of Food Microbiology 104:43-49.

Roy, K. W., and Ratnayake, S. (1997). Frequency of occurrence of *Fusarium pallidoroseum*, effect on seeds and seedlings, and associations with other fungi in soybean seeds and pods. Canadian Journal of Plant Pathology 19:188-192.

Roy, K. W., Baird, R. E., and Abney, T. S. (2001). A review of soybean (*Glycine max*) seed, pod, and flower mycofloras in North America, with methods and a key for identification of selected fungi. Mycopathologia 150:15-27.

Sage, L., Krivobok, S., Delbos, E., Seigle-Murandi, F. and Creppy, E. E. (2002). Fungal flora and ochratoxin A production in grapes and musts from France. Journal of Agricultural and Food Chemistry 50:1306-1311.

Sage, L., Garon, D., and Seigle-Murandi, F. (2004). Fungal microflora and ochratoxin A risk in French vineyards. Journal of Agricultural and Food Chemistry 52:5764-5768.

Sahin, I., Basoglu, F., Korukluoglu, M., and Gőcmen, D. (1999). Salamura siyah zeytinlerde rastlanan kürfler ve mikotoksin riskleri. Kükem Dergisi 22:1-8.

Sakai, A., Tanaka, H., Konishi, Y., Hanazawa, R., Ota, T., Nakahara, Y., Sekiguchi, S., Oshida, E., Takino, M., Ichinoe, M., Yoshikawa, K., Yoshizawa, T., and Takatori, K. (2005). Mycological examination of domestic unpolished rice and mycotoxin production by isolated *Penicillium islandicum*. Journal of the Food Hygiene Society of Japan 46:205-212.

Samapundo, S., Devliehgere, F., de Meulenaer, B., and Debevere, J. (2005). Effect of water activity and temperature on growth and the relationship between fumonisin production and the radial growth of *Fusarium verticillioides* and *Fusarium proliferatum* on corn. Journal of Food Protection 68:1054-1059.

Samson, R. A., and Pitt, J. I. (eds.) (1990). Modern concepts in *Penicillium* and *Aspergillus* systematics. Plenum Press, New York, U.S.A.

Samson, R. A., Hocking, A. D., Pitt, J. I., and King, A. D. (eds.) (1992). Modern methods in food mycology. Elsevier, Amsterdam, The Netherlands.

Samson, R. A., Stolk, A. C., and Hadlok, R. (1976). Revision of the subsection Fasciculata of *Penicillium* and some allied species. Studies in Mycology 11:1-47.

Samson, R. A., Houbraken, J. A. M. P., Kuijpers, A. F. A., Frank, J. M., and Frisvad, J. C. (2004a). New ochratoxin A or sclerotium producing species in *Aspergillus* section *Nigri*. Studies in Mycology 50:45-61.

Samson, R. A., Seifert, K. A., Kuijpers, A. F. A., Houbraken, J. A. M. P., and Frisvad, J. C. (2004b). Phylogenetic analysis of *Penicillium* subgenus *Penicillium* using partial β-tubulin sequencing. Studies in Mycology 49:175-200.

Samson, R. A., Hoekstra, E. H., and Frisvad, J. C. (eds.). (2004c). Introduction to food- and airborne fungi. Centraalbureau voor Schimmelcultures, Utrecht, The Netherlands.

Samson, R. A., Stolk, A. C., and Hadlok, R. (1976). Revision of the subsection Fasciculata of *Penicillium* and some allied species. Studies in Mycology 11:1-47.

Sanchis, V., Scott, P. M., and Farber, J. M. (1988). Mycotoxin-producing potential of fungi isolated from red kidney beans. Mycopathologia 104:157-162.

Sashidhar, R. B., Ramakrishna, Y., and Bhat, R. V. (1992). Molds and mycotoxins in sorghum stored in traditional containers in India. Journal of Stored Products Research 28:257-260.

Saubois, A., Piontelli Laforet, E., Nepote M. C., and Wagner M. L. (1999). Mycological evaluation of a sorghum grain of Agentina, with emphasis on the characterization of *Fusarium* species. Food Microbiology 16:435-445.

Sauer, D. B., Meronouk, R. A., and Christensen, C. M. (1992). Microflora. *In* Storage of cereal grains and their products (Sauer, D. B., ed.), American Association of Cereal Chemists, St. Paul, MN, U.S.A., pp. 313-340.

Scaff, R. M. C., and Scussell, V. M. (2004). Fumonisins B_1 and B_2 in corn-based products commercialized in the state of Santa Catarina – Southern Brazil. Brazilian Archives of Biology and Technology 47:911-919.

Schollenberger, M., Müller, H.-M., Rüfle, M., Suchy, S., Plank, S., and Drochner, W. (2006). Natural occurrence of 16 *Fusarium* toxins in grains and feedstuffs of plant origin from Germany. Mycopathologia 161:43-52.

Scott, W. J. (1957). Water relations of food spoilage microorganisms. Advances in Food Research 7:83-127

Scott, V. N., and Bernard D. T. (1987). Heat-resistance of *Talaromyces flavus* and *Neosartorya fischeri* isolated from commercial fruit juices. Journal of Food Protection 50:18-20.

Scott, P. M., and Kennedy, B. P. C. (1973). Analysis and survey of ground black, white and capsicum peppers for aflatoxins. Journal of the Association of Official Analytical Chemistry 56:1452-1457.

Scott, P. M., Lombaert, G. A., Pellaers, P., Bacler, S., and Lappi, J. (1992). Ergot alkaloids in grain foods sold in Canada. Journal of the Association of Official Analytical Chemistry 75:773-779.

Scudamore, K. A., Atkin, P. M., and Buckle, A. E. (1986). Natural occurrence of the naphthoquinone mycotoxins xanrhomegnin and viomellein, and vioxanthin in cereals and animal feedstuffs. Journal of Stored Products Research 22:81-84.

Scudamore, K. A., Clarke, J. H., and Hetmanski, M. T. (1993). Isolation of *Penicillium* strains producing ochratoxin A, citrinin, xanthomegnin, viomellein, and vioxanthin from cereal grains. Letters in Applied Microbiology 17:82-87.

Scurti, J. C., Codignola, A., Nobile, G., and Caputo, O. (1973). Un ceppo di *Byssochlamys nivea* Westl., isolato da insilato di mais integrale, producenta patulina. Allionia 19:39-42.

Seenappa, M., and Kempton, A. G. (1980). Application of a minicolumn detection method for screening spices for aflatoxin. Journal of Environmental Science 15:219-231.

Sensidoni, A., Rondinini, G., Peressini, D., Maifreni, M., and Bortolomeazzi, R. (1994). Presence of an off-flavour associated with the use of sorbates in cheese and margarine. Italian Journal of Food Science 6:237-242.

Serdani, M., Crous, P. W., Holz, G., and Petrini, O. (1998). Endophytic fungi associated with core rot of apples in South Africa, with specific reference to *Alternaria* species. Sydowia 50:257-271.

Serdani, M., Kang, J.-C, Andersen, B., and Crous P. W. (2002). characterisation of *Alternaria* species-groups associated with core rot of apples in South Africa. Mycological Research 106: 561-569.

Serra, R., Abrunhosa, L., Kozakiewicz, Z., and Venâncio, A. (2003). Black *Aspergillus* species as ochratoxin A producers in Portugese wine grapes. International Journal of Food Microbiology 88:63-68.

Serra, R., Braga, A., and Venâncio, A. (2005). Mycotoxin-producing and other fungi isolated from grapes for wine production, with particular emphasis on ochratoxin A. Research in Microbiology 156:515-521.

Shephard, G. S., Westhuizen, L. van der, Gatyeni, P. M., Somdyala, N. I. M., Burger, H. M., and Marasas, W. F. O. (2005). Fumonisin mycotoxins in traditional Xhosa maize beer in South Africa. Journal of Agricultural and Food Chemistry 53:9634-9637.

Shim, W. B., Kim, J. C., Seo, J. A., and Lee, Y. W. (1997). Natural occurrence of trichothecenes and zearalenone in Korean and imported beers. Food Additives and Contaminants 14:1-5.

Sholberg, P. L., and Ogawa, J. M. (1983). Relation of postharvest decay fungi to the slip-skin maceration disorder of dried Frensh prunes. Phytopathology 73:708-713.

Shrivastava, A., and Jain, P. C. (1992). Seed mycoflora of some spices. Journal of Food Science and Technology Mysore 29:228-230.

Siegfried, R., and Langerfeld, E. (1978). Vorläufige Untersuchungen über die Produktion von Toxinen durch Fäuleerreger bei Kartoffeln. Potato Research 21:335-339.

Simmons, E. G. (1986). *Alternaria* themes and variations (22-26). Mycotaxon 25:287-308.

Simmons, E. G. (1990). *Alternaria* themes and variations (27-53). Mycotaxon 37:79-119.

Simmons, E. G. (1992). *Alternaria* taxonomy: current status, viewpoints, challenge. *In* Alternaria. Biology, plant diseases and metabolites (Chełkowski, J., and Visconti, A., eds.), Elsevier, Amsterdam, The Netherlands, pp. 1-36.

Simmons, E. G. (1993). *Alternaria* themes and variations (63-72). Mycotaxon 48:91-107.

Simmons, E. G. (1994). *Alternaria* themes and variations (106-111). Mycotaxon 50:409-427.

Simmons, E. G. (1995). *Alternaria* themes and variations (112-144). Mycotaxon 55:55-163.

Simmons, E. G. (1999a). *Alternaria* themes and variations (226-235). Classification of citrus pathogens. Mycotaxon 70:263-323.

Simmons, E. G. (1999b). *Alternaria* themes and variations (236-243). Host-specific toxin producers. Mycotaxon 70:325-369.

Simmons, E. G. (2000). *Alternaria* themes and variations (244-286). Species on Solanaceae. Mycotaxon 75:1-115.

Simmons, E. G., and Roberts, R. G. (1993). *Alternaria* themes and variations (73). Mycotaxon 48:109-140

Simsek, O., Arici, M., and Demir, C. (2002). Mycoflora of hazelnut (*Corylus avellana* L.) and aflatoxin content in hazelnut kernels artificially infected with *Aspergillus parasiticus*. Nahrung 46:194-196.

Sinha, K. K., and Sinha, A. K. (1992). Impact of stored grain pests on seed deterioration and aflatoxin contamination in maize. Journal of Stored Products Research 28:211-219.

Sinha, A., Singh, S. K., and Qaisar, J. (1999). Seed mycoflora of French bean and its control by means of fungicides. Tropenlandwirt 100:59-67.

Singh, A. B. (1972). Effect of virus-infection on rhizosphere mycoflora of papaya plants. Plant and Soil 36:205.

Sinigaglia, M., Albenzio, M. Corbo, M. R., and Ciccarone, C. (2004). Comparison of mycoflora associated to Canestrato Pugliese cheese produced according to protocols. Sciences des Alimentes 24:159-172.

Siradhana, B. S., Dange, S. R. S., Rathore, R. S., and Singh, S. D. (1978). Seed mycoflora of maize with reference to mycotoxins. Current Science 47:59-60.

Smith, G. (1960). An introduction to industrial mycology. 5th edition. Edward Arnold, London.

Snowdon, A. L. (1990). A colour atlas of post-harvest diseases and disorders of fruits and vegetables. 1. General introduction and fruits. Wolfe Scientific, London.

Somani, A. K., Lange, L., and Mathur, S. B. (1986). Preliminary studies on the mycoflora of true seed of potato. Seed Science and Technology 14:213-216.

Sommer, N. F., Buchanan, J. R., and Fortlage, R. J. (1986). Relation of early splitting and tattering of pistachio nuts to aflatoxin in the orchard. Phytopathology 76:692-694.

Spicher, G. (1985). Die erreger der Schimmelbildung bei Backwaren. IV. Weitere Untersuchungen über die auf verpackte Schnittbroten auftretenden Schimmelpilze. Deutsche Lebensmittel Rundschau 81:16-20.

Splittstoesser, D. F., and Splittstoesser, C. M. (1977). Ascospores of *Byssochlamys fulva* compared with those of a heat resistant *Aspergillus*. Journal of Food Science 42:685-688.

Splittstoesser, D. F., Lammers, J. M., Downing, D. L., and Churey, J. J. (1989). Heat resistance of *Eurotium herbariorum,* a xerophilic mould. Journal of Food Science 42:685-688.

Spotti, E., Mutti, P., and Scalare, F. (1994). *P. nalgiovense, P. gladioli, P. candidum,* and *A. candidus*: possibility of their use as starter cultures. Industria Conserve 69:237-241.

Spotti, E., Quintavalla, S., and Mutti, P. (1992). Contaminazione da spore fungine termoresistenti di frutta, pomodoro e loro derivati. Industria Conserve 67:421-425.

Stack, M. I., Mislivec, P. B., Roach, J. A., and Pohland, A. F. (1985). Liquid chromatographic determination of tenuazonic acid and alternariol methyl ester in tomatoes and tomato products. Journal of the Association of Official Chemists 68:640-642.

Stefaniki, I., Foufa, E., Tsatsou-Dritsa, A., and Dais, P. (2003). Ochratoxin A concentrations in Greek domestic wines and dried vine fruits. Food Additives and Contaminants 20:74-83.

Stegen von der, G., Jorissen, U., Pittet, A., Saccon, M., Steiner, W., Vicenzi, M., Zapp, J., and Sclatter, C. (1997). Screening of European coffee final products for occurrence of ochratoxin A (OTA). Food Additives and Contaminants 14:211-216.

Stegen, G. H. D. von der, Essens, P. J. M., and Lijn, J. von der, (2001). Effect of roasting conditions on reduction of ochratoxin A in coffee. Journal of Agricultural and Food Chemistry 49:4713-4715.

Steyn, P. S. (1970). The isolation, structure and absolute configuration of secalonic acid D, the toxic metabolite of *Penicillium oxalicum*. Tetrahedron 26:51-57.

Stinson, E. E., Bills, D. D., Osman, S. F., Siciliano, J., Ceponis, N. J., and Heisler, E. G. (1980). Mycotoxin production by *Alternaria* species grown on apples, tomatoes, and blueberries. Journal of Agricultural and Food Chemistry 28:960-963.

Stinson, E., Osman, E., Heisler, D. F., Siciliano, D. G., and Bills, D. D. (1981). Mycotoxin production in whole tomatoes, apples, oranges and lemons. Journal of Agricultural and Food Chemistry 29:790-792.

Studer-Rohr, I., Dietrich, D. R., Schlatter, J., and Schlatter, C. (1995). The occurrence of ochratoxin A in coffee. Food Chemistry and Toxicology 33:341-355.

Suárez-Quiroz, M., Gonzáles-Rios, O., Barel, M., Guyot, B., Schorr-Galindo, S., and Guiraud, J.-P. (2004). Study of ochratoxin A-producing strains in coffee processing. International Journal of Food Science and Technology 39:501-507.

Suárez-Quiroz, M., De Louise B., Gonzáles-Rios, O., Barel, M., Guyot, B., Schorr-Galindo, S., and Guiraud, J.-P. (2005). The impact of roasting on the ochratoxin A content of coffee. International Journal of Food Science and Technology 40:605-611.

Suga, K., Mochizuki, N., Harayama, K., and Yamashita, H. (2004). Analysis of trichothecenes in barley tea and beer by liquid chromatography/tandem mass spectrometry. Journal of the Food Hygiene Society Japan 45:307-312.

Sumarah, M. W., Miller, J. D., and Blackwell, B. A. (2006). Isolation and metabolite production by *Penicillium roqueforti, P. paneum* and *P. crustosum*. Mycopathologia 159:571-577.

Suresh, E. R., Ethiraj, S., and Jayraman, H. L. (1996). Heat resistance of *Neosartorya fischeri* isolated from grapes. Journal of Food Science and Technology Mysore 33:76-78.

Suzuki, J. I., Dainius, B., and Kilbuck, J. H. (1973). A modified method for aflatoxin determination in spices. Journal of Food Science 38:949-950.

Swart, A. E., and Holz, G. (1991). *Alternaria alternata* rot of cold-stored table grapes in the Cape Province of South Africa. Phytophylactica 23:217-222.

Taber, R. A., and Schroeder, H. W. (1967). Aflatoxin-producing potential of isolates of the *Aspergillus flavus-oryzae* group from peanuts (*Arachis hypogea*). Applied Microbiology 15:140-144.

Tabuc, C., Bailly, J. D., Bailly, S., Querin, A., and Guerre, P. (2004). Toxigenic potential of fungal mycoflora isolated from dry cured meat products: preliminary study. Revue de Medecine Veterinaire 155:287-291.

Tafuri, A., Ferracane, R., and Ritieni, A. (2004). Ochratoxin A in Italian marketed cocoa products. Food Chemistry 88:487-494.

Takahashi, T. (1993). Aflatoxin contamination in nutmeg: analysis of interfering TLC spots. Journal of Food Science 58:197-198.

Taniwaki, M. H., Pitt, J. I., Teixeira, A. A., and Iimanaka, B. T. (2003). The source of ochratoxin A in Brazilian coffee and its formation in relation to processing methods. International Journal of Food Microbiology 82:173-179.

Tantaoui-Elaraki, A., Lemrani, M., and Gonzalez-Vila, F. J. (1994). Tentative identification of metabolites in toxic extracts from *Penicillium digitatum* (Pers ex. Fr.) Sacch. and *P. italicum* Wehmer cultures. Microbiologie Aliments Nutrition 12:225-230.

Tharappan, B., and Ahmad, R. (2006). Fungal colonization and biochemical changes in coffee beans undergoing monsooning. Food Chemistry 94:247-252.

Thom, C. (1930). *The Penicillia*. Williams and Wilkins, Baltimore, USA.

Thrane, U. (1989). *Fusarium* species and their specific profiles of secondary metabolites. *In* Fusarium: Mycotoxins, taxonomy and pathogenicity (Chełkowski, J., ed.), Elsevier, Amsterdam, The Netherlands, pp. 199-225.

Tonon, S. A., Marucci, R. S., Jerke, G., and García, A. (1997). Mycoflora of paddy and milled rice produced in the region of Northeastern Argentina and Southern Paraguay. International Journal of Food Microbiology 37:231-235.

Torelli, E., Firrao, G., Locci, R., and Gobbi, E. (2006). Ochratoxin A-producing strains of *Penicillium* spp. isolated from grapes used for the production of "passito" wines. International Journal of Food Microbiology 106:307-312.

Torres, A., Chulze, S., Varsavasky, E., and Rodriguez. (1993). *Alternaria* metabolites in sunflower seeds. Mycopathologia 121:17-20.

Torres, M. R., Sanchis, V., and Ramos, A. J. (1998). Occurrence of fumonisins in Spanish beers analyzed by enzyme-linked immunosorbent assay method. International Journal of Food Microbiology 39:139-143.

Tournas, V. (1994). Heat-resistant fungi of importance to the food and beverage industry. Critical Reviews in Microbiology 20:243-263.

Tournas, V. H. (2005). Moulds and yeast in fresh and minimally processed vegetables and sprouts. International Journal of Food Microbiology 99:71-77.

Tseng, T. C., Tu, J. C., and Tzean, S. S. (1995a). Mycoflora and mycotoxins in dry bean (*Phaseolus vulgaris*) produced in Taiwan and in Ontario, Canada. Botanical Bulletin of Academia Sinica 36:229-234.

Tseng, T. C., Tu, J. C., and Soo, L. C. (1995b). Comparison of the profiles of seedborne fungi and the occurrence of aflatoxins in mould-damaged beans and soybeans. Microbios 84:105-116.

Tsubouchi, H., Yamamoto, K., Hisada, K., Sakabe, Y., and Udagawa, S. (1987). Effect of roasting on ochratoxin A level in green coffee beans inoculated with *Aspergillus ochraceus*. Mycopathologia 97:111-115.

Udagawa, S. (1991). Contamination and spoilage problems of foods by heat-resistant moulds. Japanese Journal of Food Microbiology 8:121-130.

Udagawa, S. (2000). Heat resistant moulds, a specific topic on food and beverage mycology. Mycotoxins 50:3-11.

Ueno, Y. (1974). Citreoviridin from *Penicillium citreoviride* Biourge. *In* Mycotoxins (Purchase, I. F. H., ed.), Elsevier, Amsterdam, The Netherlands, pp. 283-302.

Urbano, G. R., Taniwaki, M. H., Leitao, M. F., and Vicentini, M. C. (2001). Occurrence of ochratoxin A-producing fungi in raw Brazilian coffee. Journal of Food Protection 64:1226-1230.

Valero, A., Marín, S., Ramos, A. J., and Sanchis, V. (2005). Ochratoxin A-producing species in grapes and sundried grapes and their relation to ecophysiological factors. Letters in Applied Microbiology 41:196-201.

Vega, F. E., Posada, F., Gianfagna, T. J., Chaves, F. C. and Peterson, S. W. (2006). An insect parasitoid carrying an ochratoxin producing fungus. Naturwissenschaften 93:297-299.

Venkitasubramanian, T. A. (1977). Biosynthesis of aflatoxin and its control. *In* Mycotoxins in human and animal health (Rodricks, J. V., Hesseltine, C. W., and Mehlman, M. A., eds.), Pathotox, Park Forest South, Illinois, U.S.A., pp. 83-98.

Vesonder, R. F., Lambert, R., Wicklow, D. T., and Biehl, M. L. (1988). *Eurotium* spp. and echinulin in feed refusal by swine. Applied and Environmental Microbiology 54:830-831.

Viani, R. (1996). Fate of ochratoxin A (OTA) during processing of coffee. Food Additives and Contaminants 13(Suppl.):29-33.

Vinas, I., Bonet, J., and Sanchis, V. (1992). Incidence and mycotoxin production by *Alternaria tenuis* in decayed apples. Letters in Applied Microbiology 14:284-287.

Vinas, I., Dadon, J., and Sanchis, V. (1993). Citrinin-producing capacity of *Penicillium expansum* strains from apple packinghouses of Leirda (Spain). International Journal of Food Microbiology 19:153-156.

Vinas, I., Palma, J., Garza, S., Sibilia, A. Sanchis, V., and Visconto, A. (1994). Natural occurrence of aflatoxin and *Alternaria* mycotoxins in oilseed rape from Catalonia (Spain): Incidence of toxigenic strains. Mycopathologia 128:175-179.

Vincent, M. A., and Pitt, J. I. (1989). *Penicillium allii*, a new species from Egyptian garlic. Mycologia 81:300-303.

Waliyar, F., and Roquebert, M. F. (1979). Mycoflora of peanuts cultivated in Senegal. Revue de Mycologie 43:169-186.

Waliyar, F., and Zambettakis, C. (1979). Study of the mycoflora if groundnut seeds and pods in Senegal. Oleagineux 34:191-198.

Weidenbörner, M. (1997). The mycoflora of soybean seeds in relation to the used plating medium DG18 or MSA. Nahrung 41:239-240.

Weidenbörner, M. (2001a). Encyclopedia of food mycotoxins. Springer Verlag, Berlin, Germany.

Weidenbörner, M. (2001b). Pumpkin seeds – the mycobiota and potential mycotoxins. European Journal of Food Research and Technology 212:279-281.

Weidenbörner, M. (2001c). Pine nuts: the mycobiota and potential mycotoxins. Canadian Journal of Microbiology 47:460-463.

Weidenbörner, M., and Kunz, B. (1994). Contamination of different muesli components by fungi. Mycological Research 98:583-586.

Weidenbörner, M., and Kunz, B. (1995). Mycoflora of cereal flakes – a research note. Journal of Food Protection 58:809-812.

Weidenbörner, M., Wieczorek, C., and Kunz, B. (1995). Mold spectra of various foods in relation to plating medium. Journal of Food Protection 58:661-665.

Weir, T. L., Huff, D. R., Christ, B. J., and Romaine, C. P. (1998). RAPD-PCR analysis of genetic variation among isolates of *Alternaria solani* and *Alternaria alternata* from tomato and potato. Mycologia 90:813-821.

Wells, J. M., and Payne, J. A. (1976). Toxigenic species of *Penicillium*, *Fusarium*, and *Aspergillus* from weevil-damaged pecans. Canadian Journal of Microbiology 22:281-285.

Wells, J. M., and Payne, J. A. (1979). Mycoflora and market quality of chestnuts treated with hot water for control of the chestnut weevil. Phytopathology 68:533.

Wells, J. M., Cole, R. J., and Kirksey, J. (1975). Emodin, a toxic metabolite of *Aspergillus wentii* isolated from weevil-damaged chestnuts. Applied Microbiology 30:26-28.

Westerdijk, J. (1949). The concept "association" in mycology. Antonie van Leeuwenhoek 15:187-189.

Wiese, M. V. (1987). Compendium of wheat diseases. Second edition. APS Press, St. Paul, MN, U.S.A.

Williams, C. C., Cameron, E. J., and Williams, O. B. (1941). A facultatively anaerobic mould of unusual heat resistance. Food Research 6:69-73.

Wilson, J. P., Jurjević, Z., Hanna, W. W., Wilson, D. M., Potter, T. L., and Coy, A. E. (2006). Host-specific variation in infection by toxigenic fungi and contamination by mycotoxins in pearl millet and corn. Mycopathologia 161:101-107.

Yaguchi, T., Udagawa, S., and Nishimura, K. (2005). *Geosmithia argillacea* is the anamorph of *Talaromyces eburneus* as a heat resistant fungus. Cryptogamie Mycologie 26:133-141.

Yamamoto, W., Yoshitani, K., and Maeda, M. (1955). Studies on the *Penicillium* and *Fusarium* rots of Chinese yam and their control. Scientific Reports of the Hyogo University of Agriculture, Agricultural Biology Series 2,1:69-79.

Yiannikouris, A., André, G., Poughon, L., François, J., Dussap, C.-G., Jeminet, G., Bertin, G., and Jouany, J.-P. (2006). Chemical and conformational study of the interactions involved in mycotoxin complexation with β-D-glucans. Biomacromolecules 7:1147-1155.

Ylimaki, A. (1981). The mycoflora of cereal seeds and some feedstuffs. Annales Agriculturae Fenniae 20:74-88.

Zad, J. (1978). Note on the mycoflora of sunflower seeds in Iran. Seed Science and Technology 6:953-956.

Zimmerli, B., and Dick, R. (1996). Ochratoxin A in table wine and grape-juice: occurrence and risk assessment. Food Additives and Contaminants 13:655-668.

Zohri, A. A., and Abdel-Gawad, K. M. (1993). Survey of mycoflora and mycotoxins of some dried fruits in Egypt. Journal of Basic Microbiology 33:279-288

Zohri, A. A., and Saber, S. M. (1993). Filamentous fungi and mycotoxin detected in coconut. Zentralblatt für Mikrobiologie 148:325-332.

Chapter 12

Transport phenomena in fungal colonisation on a food matrix

Yovita S.P. Rahardjo[1] and Arjen Rinzema[2]
[1]*DSM Food Specialties, P.O. Box 1, 2600 MA Delft;* [2]*Wageningen University, Food and Bioprocess Engineering Group, P.O. Box 8129, 6700 EV Wageningen, The Netherlands.*

INTRODUCTION

Food is correlated with fungi in two opposite ways. In a positive way, we think of traditional and exotic fermented foods from different parts of the world. The variety of those foods is wide, ranging from tempe and soy sauce in Asia to different types of cheese in Europe and the Middle East (Campbell-Platt, 1994). For the case of these fermented foods, we use the ability of fungi to degrade complex polymeric substrates to smaller compounds that are easily taken up by men. Some fungi can also consecutively convert the degraded compounds into other type of desirable metabolic products that determine the typical composition, taste, odour, consistency and colour of fermented foods. In a negative way, we think of unwanted spoilage or rotting of foods. A negative view on fungi includes exactly the same principles of digestion of complex substrates and production of other products. In both views the benefit of the fungi is served namely, the optimal survival by living in, growing on, and colonising a food matrix.

The process of how fungi grow on and colonise a solid food matrix is described as solid-substrate or solid-state fermentation (SSF). A growing interest in SSF has been shown by a significant increase in numbers of publications on this topic in the past 20 years. Sufficient knowledge about the process of how fungi interact with a substrate matrix is absolutely needed to manipulate SSF processes, in order to: (1) improve the desired fermented (food) products yield and quality, and (2) prevent food spoilage. However, the knowledge of basic phenomena and their influence on metabolic responses in SSF is still relatively limited including such subjects as kinetics of enzyme production, release and transport, and kinetics of substrates conversion.

Those processes are determined by transport phenomena of substrates, metabolites and hydrolytic enzymes. Knowledge of these transport phenomena is crucial to a further understanding of the kinetics in SSF. In this chapter, transport phenomena during fungal colonisation on a substrate matrix are described. Firstly, the occurrence of those phenomena and the complex consequences in SSF are explained. Secondly, some examples from published modelling and experimental work using diffusive transport phenomena are shown. Due to the complex features of SSF and based on diffusion as the transport mechanism, many authors have produced mathematical models to study and understand SSF, e.g., to predict biomass growth. Finally, possible transport phenomena other than diffusion are discussed.

FUNGAL BIOMASS AND SUBSTRATE LAYERS

The "system" of fungal colonisation on or/and in a substrate matrix consists of three entities: fungal biomass, substrate fragments, and water (Figure 1).

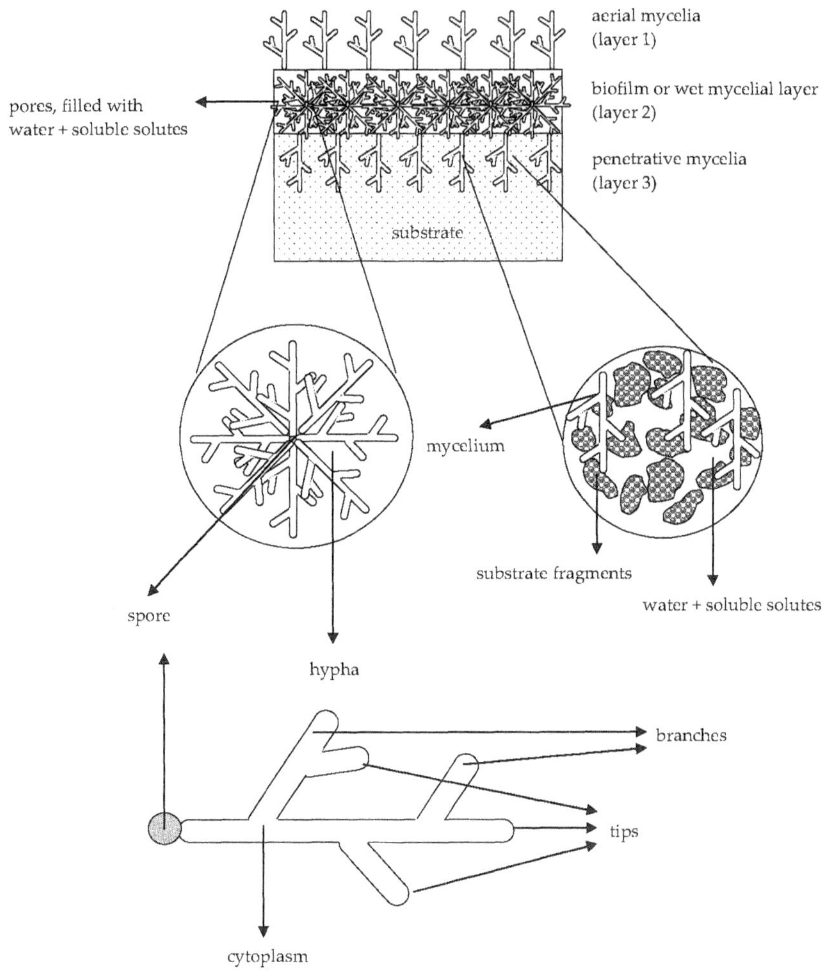

Figure 1. "System" of (filamentous) fungal colonisation on a substrate matrix and mode of growth of filamentous fungi, development of a spore to fungal biomass.

Fungal growth starts with spore germination. Under favourable conditions spores will swell and germinate. The germinated spore continues elongating at the tip and soon new branches are formed along the tubular hyphae.

It is believed that this mode of growth allows filamentous fungi to colonise the surface of and penetrate into the substrate matrix in search for nutrients. The substrate matrix itself contains mainly polymeric substrate fragments and water. At a later stage, the continuously elongating and branching hyphae form a porous three-dimensional net, which is known as mycelium (Figure 1). The interactions of fungal biomass with the substrate matrix are heterogeneous. Young mycelia, depending on where the spores were located, grow on the surface of the substrate matrix (layer 2) and from there between the substrate fragments, i.e., inside (layer 3) or into the air (layer 1).

As the hyphae elongate and branch, they secrete enzymes that are needed to degrade the polymeric substrates into smaller and more digestible molecules. The degradation of the polymeric substrates also reduces the firmness of the substrate matrix. This in combination with the turgor pressure inside the hyphae (i.e., osmotic uptake of water (Wessels, 1999)), which is present in most fungi (Harold, 2002; Harold *et al.*, 1995), allows the tips of elongating hyphae to penetrate into the substrate matrix (Mendgen *et al.*, 1996). This is an advantage because penetrative mycelia can directly access the substrate. For aerobic fungi, as long as oxygen is available for the penetrating tips, growth into the substrate can be as rapid as hyphal elongation on the substrate surface.

As the mycelia on the substrate surface continue growing, layer 1 becomes so dense that its pores get filled with water and it transforms into layer 2 and the packing density and/or thickness of the previous layer 2 increases to such an extent that oxygen is depleted in its lower part and in deeper layers of the substrate. Because of the water-filled pores, the mycelial layer is then regarded as a biofilm layer or a thin layer of water filled with growing biomass (Oostra *et al.*, 2001; Rahardjo *et al.*, 2002). At the same time, a new layer 1 is formed on top of layer 2. The pores of layer 1 are filled with gas and therefore the fungal biomass in this layer is called the aerial mycelium.

TRANSPORT PHENOMENA

As the fungal biomass grows, it consumes substrates and secretes metabolites and enzymes. Both consumption of substrates and production of metabolites can be regarded as conversion reactions that cause concentration gradients of substrates and metabolites. Because of these gradients, transport of substrates and metabolites occurs. These gradients affect the metabolic activity of the fungi; for example, gradients in the concentrations of inducers (e.g., oligosaccharides) or repressors (e.g., glucose), and oxygen may affect enzyme production. The transport and the conversion reactions occur simultaneously and therefore have to be (mathematically) treated at the same time.

Coupled reaction and diffusion phenomena have been used to describe and model many biological systems. Some examples are microbial biofilms for off-gas treatment (Ottengraf and van den Oever, 1983), immobilised cells in bioreactors (De Gooijer *et al.*, 1991; Wijffels *et al.*, 1991; Hunik *et al.*, 1994; Beuling *et al.*, 1998), cultivation of cartilage (Malda *et al.*, 2004), and many others. These phenomena are originally based on the approach used by Thiele (1939). He described coupled reaction and diffusion phenomena in a catalyst particle. Using Fick's law and a mass balance over a thin slice z and surface area A of the layer that is to be described (Figure 2):

$$A\left(dz\frac{\partial C}{\partial t}\right) = -A\left(-D\frac{\partial C}{\partial z}\bigg|_{z} + D\frac{\partial C}{\partial z}\bigg|_{z+dz} + dz.r\right)$$

the basic equation of all reaction diffusion models can be derived:

$$\frac{\partial C}{\partial t} = D\frac{\partial^2 C}{\partial z^2} + r$$

$\frac{\partial C}{\partial t}$ is the substance concentration change over time, D is the diffusion coefficient, $\frac{\partial^2 C}{\partial z^2}$ is the substance concentration change over place, r is the substance reaction, which can be consumption or production.

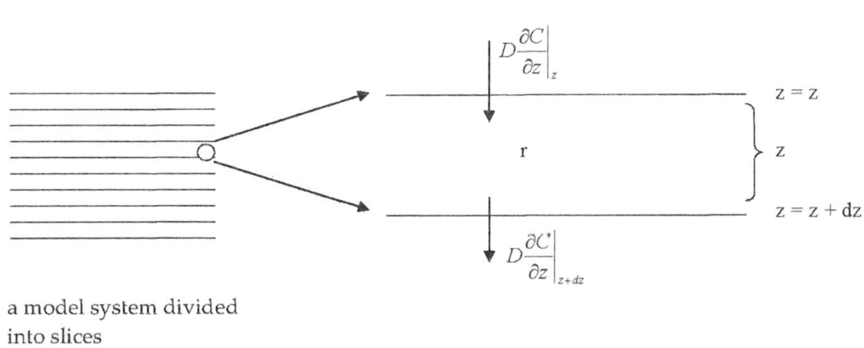

Figure 2. A mass balance over a thin slice z and surface area A, with Fick's law for the diffusion of substrates entering and leaving the thin slice z and substrates reaction r (which can be consumption or formation) within the slice z.

Coupled reaction and diffusion phenomena in the fungal biomass and the substrate matrix that have been used to describe SSF are presented in Figure 3. All descriptions were for over-culture system and only the studies of Georgiou and Shuler (1985), and Molin *et al.* (1993) were modelled according to the fungal colony (over-culture cultivation is obtained when the spores are homogenously spread over the substrate surface, while colony cultivation is obtained from point inoculation). Figure 3 summarises and illustrates the studies discussed below (except the studies of Georgiou and Shuler, 1985 and Molin *et al.*, 1993). Despite their simplicity, most models are able to predict (part of) the dynamic state of fungal growth and the most common substrates and products, such as glucose, oxygen, water and enzymes. Results from those modelling works as well as few experimental validations done to test the model are also presented. Since the scope of this chapter is the description of transport phenomena in fungal colonisation in a substrate matrix, critical mathematical evaluation of each model is not discussed here and can be found in a separate publication (Rahardjo *et al.*, 2005d).

Fungal biofilms

For a proper sustainment of the fungal biofilm layer (layer 2 in Figure 3), substrates have to be supplied from the substrate matrix and oxygen from the air. At the same time, metabolites and enzymes have to be transported from this layer into the substrate matrix. Nandakumar *et al.* (1994 and 1996), Oostra *et al.* (2001), and Rahardjo *et al.* (2002) applied a simple reaction diffusion model for oxygen to fungal biofilms on a solid substrate (substrate layer in Figure 1).

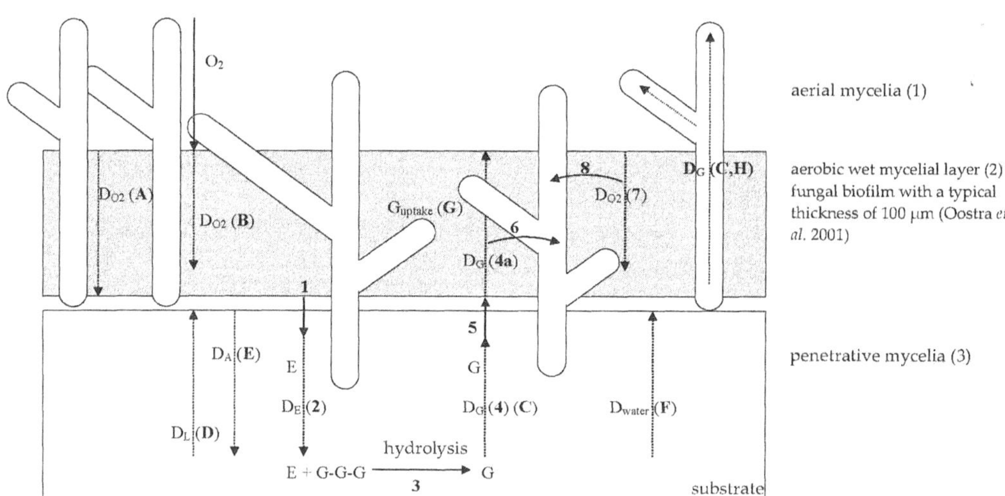

Figure 3. Reaction diffusion phenomena studied by several authors. **A**: for oxygen (Nandakumar *et al.*, 1994; 1996), biochemical reaction takes place in a very narrow zone; **B**: for oxygen (Oostra *et al.*, 2001 and Rahardjo *et al.*, 2002), a typical thickness of an aerobic (fungal) biofilm of 100 μm and no oxygen diffusion limitation in aerial mycelia; **C**: for glucose (Edelstein and Segel, 1983), for glucose and glucoamylase (Mitchell *et al.*, 1991): (1) biomass releases glucoamylase (E), (2) glucoamylase is transported by diffusion, (3) hydrolysis of starch by glucoamylase into free glucose, (4) glucose is transported by diffusion, (5) glucose is taken up by biomass; and Rajagopalan *et al.*, 1995; 1997: steps 1 – 5 are the same as Mitchell's, (6) glucose is converted to biomass with a yield coefficient, (7) diffusion of oxygen, (8) oxygen is converted to biomass with a yield coefficient; **D**: for lactate (Aldarf *et al.*, 2004; 2005); **E**: for ammonium (Aldarf *et al.*, 2004; 2005); **F**: for water (Nagel *et al.*, 2002); **G**: for glucose through hyphal membrane of growing tips (Wayman and Mattey, 1999); **H**: for glucose within hyphae (Edelstein and Segel, 1983; Olsson and Jennings, 1991).

Nandakumar *et al.* (1994 and 1996) used the model to calculate the substrate particle degradation in SSF. The degradation of the substrate particle was derived from the space occupied by the growing biomass
The biomass growth was assumed to take place only at the biomass substrate interface, and is dependent on the diffusing oxygen (without being consumed) through the fungal biofilm and constant supply of carbon source. They cultivated *Aspergillus niger* (Nandakumar *et al.*, 1994) and *Bacillus coagulans* (Nandakumar *et al.*, 1996) on wheat bran and measured the changes in the size of the particles during fermentation. The particle diameters calculated from the model were in agreement for cultures of *A. niger* (Nandakumar *et al.*, 1994), except towards the end of the fermentation. In case of cultures of *B. coagulans*, the model could predict the substrate diameters only in the case of small particles (Nandakumar *et al.*, 1996).

Oostra *et al.* (2001) used the model to predict oxygen uptake and demonstrate that intraparticle oxygen diffusion limitation does exist in the cultivation of *Rhizopus oligosporus* on a defined agar medium. They assumed that the diffusion of oxygen only occurred in vertical direction of the fungal biofilm, that transport and reaction properties were constant throughout the fungal biofilm, that oxygen was consumed with a certain yield coefficient, and that the biofilm was in a pseudo-steady state ($\frac{\partial C_{O2}}{\partial t} = 0$). For the validation of the model, Oostra *et al.* (2001) measured the oxygen concentration gradient during the cultivation using microelectrodes. The measured oxygen concentration profiles were in agreement with the model. Rahardjo *et al.* (2002) used the same model to calculate oxygen uptake in the cultivation of *Aspergillus oryzae* on a wheat-flour solid substrate. Interestingly, from the experimental validation, they obtained solid evidence that aerial mycelia of *A. oryzae* contributed up to 75% of the oxygen uptake rate at 50 hours and over 90% after 70 hours (Figure 4). Furthermore, they also showed that the model could predict the oxygen uptake of the fungal biofilm (layer 2) (Figure 4).

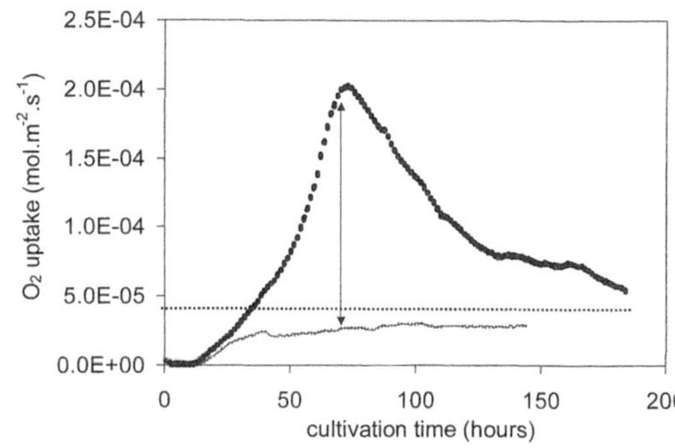

Figure 4. (Adapted from Rahardjo et al., 2002): Oxygen uptake rate of over-culture of *A. oryzae* in time (●●●), *A. oryzae* without aerial mycelia in time (solid line), and oxygen uptake rate predicted by the model of Oostra et al. (2001) (dashed line, 3.8×10^{-5} $mol.m^{-2}.s^{-1}$), ↔ or the area above the solid line is the oxygen uptake by aerial mycelia only.

Substrate matrix

A substrate matrix consists of carbon source, which can be glucose or starch-based medium, water and other nutrients. Several authors have used reaction diffusion phenomena for glucose and/or other nutrients, water and hydrolytic enzymes in a solid substrate matrix to study and eventually predict fungal biomass growth.

Georgiou and Shuler (1985) proposed a reaction-diffusion model for glucose and nitrate in the solid-substrate layer during SSF. They used these models to describe the growth of a fungal colony on the surface of a flat agar substrate. They applied Monod kinetics to describe biomass growth with either glucose or nitrate or both as the limiting substrate. The results of the simulation suggested that mass-transfer limitation in the substrate matrix is a very important factor for the biomass growth. Unfortunately, the study lacked an experimental validation to test the model. Molin *et al.* (1993) provided a reaction-diffusion model for glucose to calculate the fungal biomass production during SSF. Since they used the results from the same experiments for both the determination of the parameter and the validation of the model, it is rather difficult to judge the predictive value of their model.

Edelstein and Segel (1983) used reaction-diffusion phenomena for nutrients in a model-medium and within the hyphae (in cytoplasm) in their model for growth and morphogenesis (other examples on diffusion within hyphae are given in a later section). For the growth of fungal biomass, they described the mycelium as a continuum, a function of a tip extension rate and a production rate of new tips. Their study was pure theoretical and no experimental validation was done.

In a study on the ripening of Camembert cheese, Aldarf *et al.* (2004) used reaction diffusion phenomena for lactate and ammonium in a solid model media during the cultivation of *Geotrichum candidum*. It was shown that reaction-diffusion phenomena for lactate (as the substrate) and ammonium (as the metabolite) indeed exist during the cultivation and the concentration gradients both of lactate and ammonium match with the prediction of the reaction-diffusion model. They applied a logistic model for the biomass growth and did a separate experiment to measure the diffusion coefficient of lactate and ammonium in the associated media. In a subsequent study, the same authors (Aldarf *et al.*, 2005) indicated that although the diffusion of substrates occurred in a solid model media during the cultivation of *Geotrichum candidum* and *Penicillium camemberti*, there was no substrate diffusion limitation of growth. Instead, oxygen limitation was suggested to cause the linear growth phase and inhibitory effects of pH was shown to cause the growth phase deceleration.

Nagel *et al.* (2002) succeeded in measuring glucose and water gradients during cultivation of *A. oryzae* on a substrate matrix using nuclear magnetic resonance (NMR). They were also the first to apply a reaction diffusion model for water in the substrate matrix toward the fungal biomass. The fungal biomass on the substrate surface contains a significant amount of water (up to 2 kg.kg DM^{-1}, Nagel *et al.*, 2001), and the water comes mainly from the substrate matrix. The model simulation on the moisture gradients in the solid substrate fitted to the experimental measurements using NMR. Furthermore, using the model they could also predict the decrease in fungal growth rate due to water loss from the culture when evaporation takes place. Because the accumulation of glucose and amino acids in the substrate matrix could also affect the water transport, Nagel *et al.* (2002) concluded that transport of other solutes should be included to improve their description.

The models of Georgiou and Shuler (1985), Molin *et al.* (1993), Edelstein and Segel (1983), and Aldarf *et al.* (2004 and 2005) are based on the assumption that the carbon source in the substrate matrix is readily available for the fungi, which can be unrealistic. Perhaps for some studies using defined media, this assumption can be used. For most SSF applications, the carbon source is present as a polymer such as starch, cellulose, pectin or lignin. The potential of fungi to produce enzymes in order to hydrolyse complex polymeric substances is one of the most important features of SSF. Mitchell *et al.* (1991) were the first to indicate that diffusion of a hydrolytic enzyme (glucoamylase) is the rate-limiting step during cultivation of *R. oligosporus* on a kappa-carrageenan matrix containing cassava starch (Figure 3). They used reaction diffusion models for glucoamylase (step 2) and glucose (step 4) in a solid matrix to predict the biomass growth. It was assumed that penetrative mycelia are absent and therefore glucoamylase is produced only in the fungal biomass on the substrate surface (step 1) and have to diffuse downward into the substrate matrix. In their model, they also included the hydrolysis rate of starch (step 3) to calculate the glucose generation rate. Glucose generated in the substrate matrix had to diffuse toward the mycelium on the substrate surface, where the fungus consumed it at a certain rate (step 5). All glucose reaching the surface is converted into biomass with yield coefficient of glucose on biomass $Y_{G/X}$. In this model biomass growth was calculated with Monod kinetics. Taken together, the diffusion of glucoamylase determines the glucose generation rate, which consecutively determines the glucose transport rate to the surface of the substrate matrix.

Unfortunately, Mitchell *et al.* (1991) had to adjust the diffusion coefficient of glucoamylase determined from the experimental validation to fit the predicted biomass production and glucoamylase concentration profiles. It is however not yet possible to have a reliable measurement of glucoamylase diffusivity nor of glucoamylase concentration profiles in moistened solid substrate. An indirect proof of their conclusion is the study of Nagel *et al.* (2002) who observed during cultivation of *A. oryzae* that glucose was absent deep in the matrix. This can be the result of the slow diffusion of glucoamylase indicated by Mitchell *et al.* (1991).

Fungal biofilms and substrate matrix

Extending the work of Mitchell *et al.* (1991), Rajagopalan *et al.* (1995 and 1997) predicted that oxygen diffusion limitation was more severe than glucose diffusion limitation. From the simulation of glucoamylase concentration profiles in the substrate matrix, they showed that glucoamylase remained in the exterior part of the particle and therefore glucoamylase took more time to diffuse to the core of larger particles. This suggests that for larger substrate particles, i.e., thicker substrate matrix, enzyme diffusion is more growth limiting than oxygen diffusion. Like Mitchell *et al.* (1991), Rajagopalan *et al.* (1995 and 1997) used reaction diffusion models for glucose and glucoamylase in a solid matrix to predict the biomass growth (Figure 3, step 1–step 5 is as Mitchell's description). However, they extended the model system of Mitchell *et al.* (1991) to include diffusion for glucose (step 4a) and oxygen (step 7) in the fungal biofilm, and glucose (step 6) and oxygen (step 7) conversion into biomass with yield coefficients $Y_{G/X}$ and $Y_{O/X}$, respectively. They assumed that fungal biomass is a layer of water with constant biomass concentration, in which oxygen and glucose were transported by diffusion. The fungal biofilm expanded in time due to growth of the fungus, but the fungus did not penetrate the substrate matrix or form aerial mycelia in this model. Rajagopalan *et al.* (1997) showed that their model was able to predict the general trend in particle size reduction from experimental data collected by Nandakumar *et al.* (1996). Rajagopalan *et al.* (1997) assumed that the reduction in particle volume was proportional to the amounts of glucose consumed by the fungal biomass and accumulated in the fungal biofilm.

Through the hyphal cell wall

A fungal hypha can be regarded as a long tubular living body deliminated by a mostly rigid cell wall as a "skeletal" element. The elongation of a hypha is basically a continuous insertion of new material into the plastic wall mixture at the apex and at the same time a conversion of the plastic material into more rigid lateral wall. Wayman and Mattey (2000) showed that diffusion phenomena could explain the specific rate of glucose uptake observed during cultivation of *A. niger* in liquid culture. They treated the fungal biomass as a single very long cylinder of hyphal diameter having a volume of all the branched mycelium. Furthermore, they assumed that only a certain percentage (18% in this case) of the total biomass, in the form of growing tips, was particularly active. Recent work (Vinck *et al.*, 2005) has indicated that even differentiation occurs between hyphal tips in their ability to produce glycoamylase and that only part of the hyphae was active.

Although this study was done in liquid culture, it can be expected that the diffusion mechanism of glucose uptake by the fungal biomass is also valid in SSF. The substrate matrix contains water and in all models of diffusion phenomena in a substrate matrix discussed above, it is assumed that the diffusion phenomena took place in "water." They used a permeability coefficient instead of diffusion coefficient to describe the glucose flux through the hyphal cell wall. It would be very worthwhile to obtain the percentage of the active biomass and the permeability of the cell wall.

Within hyphae

Olsson and Jennings (1991) experimentally showed that diffusion was the mechanism of glucose transport in the hyphae. The direction of this diffusion transport was from the tips region, where a new source of nutrient was available, towards the inoculation point. In their work, they studied the translocation of label added as [^{14}C] glucose and [^{32}P] orthophosphate to cultures of *Rhizopus nigricans* grown on opposing gradients of glucose and other nutrients in glass fiber filters. Nopharatana *et al.* (1998) used this concept, however in the opposite direction, to describe the growth of aerial mycelia on the surface of a substrate matrix. Glucose was assumed to diffuse through the hyphae from the substrate matrix to the mycelial tips in the air and be the only limiting nutrient. The consumption of glucose was due to growth and maintenance. The biomass production rate was calculated from the number of tips at a particular position and time and the tip elongation rate, which was assumed to depend on the concentration of glucose at some distance behind the tip. In their later work Nopharatana *et al.* (2003) used confocal scanning laser microscopy to measure the spatial biomass distribution of aerial mycelia, with a maximum density of 39.54 mg dry wt cm^{-3}, in SSF and compared the results to their previous modelling work (Nopharatana *et al.*, 2003). From this preliminary experimental validation, it was clear that an important assumption made in their modelling work namely the maximal aerial mycelium density was invalid (Nopharatana *et al.*, 1998).

Is diffusion the only transport phenomenon in fungal colonisation on a food matrix?

The experimental validation of the oxygen concentration gradient in a fungal biofilm layer growing on a flat substrate surface by Oostra *et al.* (2001), for *Rhizopus oligosporus*, and Rahardjo *et al.* (2002), for *Aspergillus oryzae*, showed that (reaction) diffusion phenomena are relevant for the fungal biofilm layer in SSF. Hitherto, diffusion is regarded as the most common transport phenomenon during fungal colonisation on and in a food matrix. However, the actual processes in SSF are more complex and there is evidence that diffusion does not account for oxygen transport alone. This evidence is based on preliminary experimental works without direct validation or on the base of theoretical considerations. Figure 5 shows a conceptual illustration of other transport phenomena in fungal growth on a substrate matrix (Figure 5 is complementary to Figure 3). Clearly, more research is needed to elucidate these transport phenomena.

Elongating and penetrating hyphae

It is generally accepted that the growth of filamentous fungi is basically an elongation of the tips of hyphae and generation of new tips along the hyphal body. Chang and Trevithick (1974) suggested that the growing apical wall is more porous than the rigid mature wall and therefore allows the rapid secretion of proteins, such as hydrolytic enzymes.

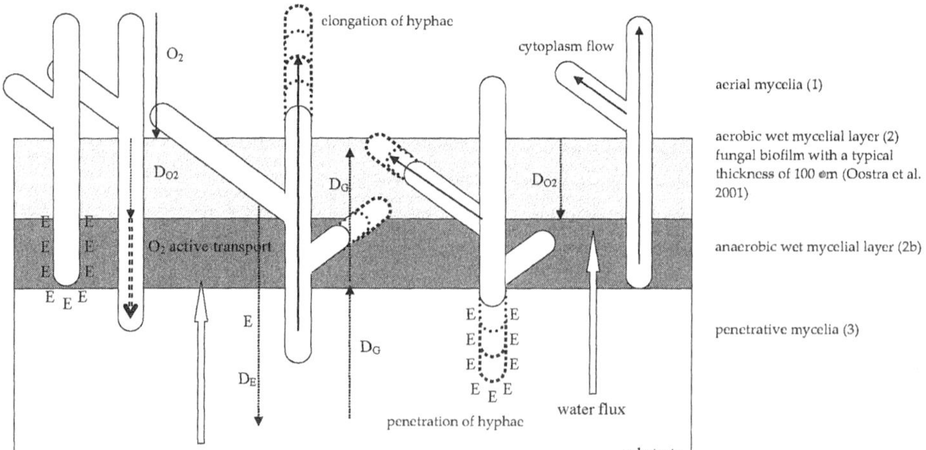

Figure 5. Other transport phenomena that might be significant in fungal colonisation in and/or on a substrate matrix. D is the diffusion transport phenomena explained in Figure 3.

Testing this theory, Wösten *et al.* (1991) experimentally showed that glucoamylase secretion is located at the tips of growing hyphae. The growing tips of penetrative mycelia can be regarded as a moving source of enzymes (Varzakas, 1998), which accelerates the transport of hydrolytic enzymes into the substrate matrix beyond passive diffusion. Intricately, recent findings indicate that not all hyphal tips at the explorative mycelium of *A. niger* produce glycoamylase to the same extent, but that lower and higher producing hyphae can be discerned (Vinck *et al.*, 2005). This type of reasoning is not usual in most process engineering studies that treat the fungus as a unicellular organism.

Using a simple calculation, Rahardjo *et al.* (2005d) showed that transport of enzymes by penetration of growing tips is faster than by diffusion (Figure 6, adapted from Rahardjo *et al.*, 2005d). They compared the time needed for diffusion of glucoamylase to that for penetration of new tips. For their calculation, they used diffusion coefficient values from literature and extension rate values derived from unpublished experiments. Only for very small particles (with diameter smaller than 10^{-4} m), in combination with a relatively high diffusion coefficient, diffusion would be faster than the penetration of growing tips.

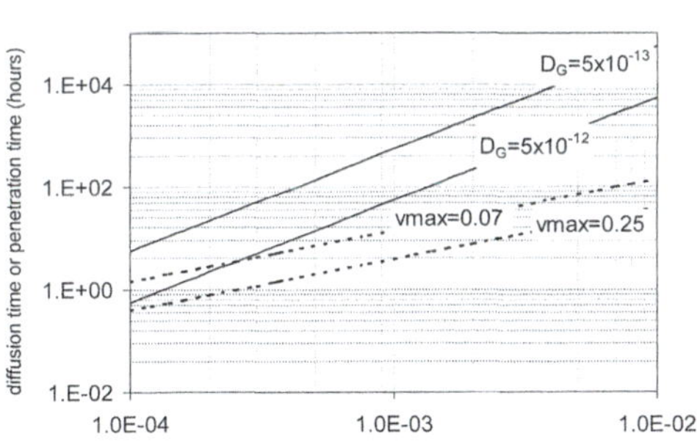

Figure 6. (Adapted from Rahardjo *et al.*, 2005d): Theoretical diffusion time (solid line) and penetrating time (dashed-line) of glucoamylase for different sizes of substrate particles, calculated with two different D values (Mitchell *et al.* (1991) used a D value of 7 x 10^{-13} $m^2.s^{-1}$) and two different penetration rate values, 0.25 $mm.h^{-1}$ for *A. oryzae* in wheat flour (Rahardjo, unpublished results), and 0.07 $mm.h^{-1}$ for *C. minitans* in starch-rich agar (Oostra, unpublished results).

If the porous (Chang and Trevithick, 1974) or plastic (Wessels, 1999) growing apical wall allows protein translocation to the medium, it also could support uptake of glucose and other nutrients as is suggested by Wayman and Mattey (2000, see above). Thus, the penetrating tips would then act as a moving substrate-collector and accelerate the overall uptake of substrates correlated with an increase in the number of fungal tips as a result of branching.

However, we need to consider the significance of the elongating tips (apical cells) or penetrative hyphae in comparison to the total biomass (e.g., 18% for cultivation in liquid culture was assumed by Wayman and Mattey (2000)). This is especially important in relation to the availability of oxygen and the ability of fungi to deal with low amounts of oxygen. As oxygen is depleted already at depths over 100 μm (Oostra *et al.*, 2001; Rahardjo *et al.*, 2002) in solid culture, the presence of penetrative hyphae of aerobic fungi would be negligible. Rahardjo *et al.* (2005a) observed that hyphal extension rate of *A. oryzae* grown at very low oxygen concentration (0.25% v/v) was comparable to the rate value obtained at 21% (v/v). Their observation on constant extension rates was valid for parent hyphae (with an extension rate of a single hypha) and young mycelium (with an average extension rate of several hyphae). However, they found that low oxygen concentrations reduced the branching frequency. This was shown by the reduction in hyphal growth unit (a ratio of total hyphal length to number of tips).

Water flux

Generally speaking, there is no free-moving water in SSF system. However, as the hydrolysed substrates are mostly soluble solutes, water is of high significance for fungal growth and metabolism. As has been explained previously, Nagel *et al.* (2002) were the first who realised that water movement from the substrate matrix to the fungal mats could play an important role in SSF. In their calculation, there will be a significant water loss from the system in case of water evaporation at the gas-solid interface (meant for cooling of the system) in combination with water uptake by the fungal biomass. This means that there will be a significant transfer of water from the centre of the substrate towards the fungal biomass. This flux of water can be an additional means of carbon transport to the fungal biomass. However, it can be unfavourable for oxygen and enzyme transport into the biofilm and substrate (Figure 5), if oxygen and enzyme are transported by diffusion only. From their modelling works, Mitchell *et al.* (1991) and Rajagopalan *et al.* (1997) concluded that glucoamylase remained at the periphery of the substrate particle because of the slow diffusion. This might be due to the water flux towards the fungal biofilm, which might be the main cause for the slow movement of glucoamylase and it is therefore important to incorporate the counter-current (convective) flow of water in description for the transport of enzymes and other substrates.

Enzyme transport

In addition to the conclusion of Mitchell *et al.* (1991) and Rajagopalan *et al.* (1997) that glucoamylase remained at the periphery of the substrate particle, Rahardjo *et al.* (2004 and 2005b) observed that most of α-amylase produced in SSF remains in the fungal biomass layer and does not move into the substrate. Other than the counter-current flow of water upward (see previous paragraph), the apparent immobility of the enzymes may also be due to binding to an extracellular polysaccharide matrix, or entrapment in an impermeable outer wall component (Zlotnik *et al.*, 1984), or cross-linking to the lateral wall (Schreuder *et al.*, 1993). Proteins can become trapped in the cell wall as is proposed by the bulk-flow hypothesis for protein translocation through the wall in a growing hyphal apex by Wessels (1988). From their experiments, Wösten *et al.* (1991) also indicated that the walls of *A. niger* seem to contain glucoamylase and some of this enzyme seemed to leak from older non-growing culture. In the case for aerial mycelia, the apparent immobility of the enzymes are most probably due to the fact that extracellular enzymes produced simply have no water to move in. While individual hyphae can differentially express proteins (Vinck *et al.*, 2005) aerial hyphae may not produce any enzymes while growing into the air. However, quick degradation of newly found substrata after an aerial "episode" could give an advantage in case of enzymes that are secreted but present on the outer boundary of the hypha. Reliable measurements of, for example, local enzyme concentration would provide information on the apparent transport phenomena and useful sets of data for model validation purposes.

Cytoplasmic streaming, most notably in aerial mycelia

Elongating hyphae, including aerial hyphae, need a flow of cytoplasm with concomitant glucose transport. The dominant role of diffusion in hyphae was established in colony-growth experiments (Olsson and Jennings, 1991), where the glucose has to be transported against the cytoplasmic flow from the periphery of the colony towards the centre of the colony. The diffusion of glucose in aerial hyphae might be insignificant compared to the transport of glucose by the flow of cytoplasm. Alternatively, glucose might be transported as glycogen granules that can be readily present inside fungal hyphae (Dijksterhuis *et al.*, 1991). The cytoplasmic glucose flow might explain the formation of inconceivably long aerial hyphae (of at least 4.5 mm (Rahardjo *et al.*, 2002)). More studies on the physiology and cellular biology of fungi might clarify whether there is an active transport of glucose or other derived carbon sources.

Recently, aerial mycelia were shown to be very significant for the oxygen uptake, biomass production and α-amylase production of *A. oryzae* cultivated on wheat-flour disks (Rahardjo *et al.*, 2002 and 2005b). They also found that the amount of α-amylase produced by *A. oryzae* is proportional to the amount of oxygen consumed (Rahardjo *et al.*, 2005b and 2005c). This suggests that aerial mycelia can accelerate enzyme production, but there is no proof that they produce these enzymes themselves. It is clear that a large part of the oxygen uptake occurred in the aerial mycelia layer (Rahardjo *et al.*, 2002). However, it is not known whether oxygen is fully consumed by the aerial mycelia or the fungus possesses an active transport for oxygen to supply the anaerobic part of fungal biofilm or penetrative biomass deep in the substrate matrix. Te Biesebeke *et al.* (2006) experimentally showed overproduction of *Aspergillus* haemoglobin domains and improvement of biomass and enzymes (amylase, protease and glucoamylase) production in cultures of transformed *A. oryzae* in SSF. They also observed that the oxygen uptake of the transformants was significantly higher than that of the untransformed wild-type strain. They obtained the transformed *A. oryzae* by isolating DNA-fragments coding for haemoglobin-domains from *A. oryzae* and *A. niger*, and introducing it in wild-type *A. oryzae*. The exact mechanism of this possible active oxygen transport is however not known from the current study. It would be interesting to understand how this active oxygen transport counteracts cytoplasmic flow and what the consequences are on enzyme production. It would also be interesting to find out the physiological role of aerial mycelia, apart from a sporulation purpose or simply a means to search for nutrients.

CONCLUDING REMARKS

Diffusion is the only transport phenomenon that has been rigorously validated in fungal colonisation on a food matrix as well as in the hyphae. As the significance of other transport phenomena mentioned above is not yet known, it would be worthwhile to experimentally verify and measure them. For example, measurements of glucose concentration in a fungal biofilm (Beuling *et al.*, 1998) could be combined with the measurements of water concentration using NMR to study the effect of the upward water flux on the glucose distribution in the culture. Measurements of glucose concentration in aerial hyphae could hopefully provide preliminary information about the cytoplasmic flow. Measurements of the apparent transport of glucose or other carbon forms and oxygen (which might be bound to haemoglobin) within the hyphae would provide not only knowledge on fungal physiology, but also a solid basis for modelling works. Advanced molecular techniques, such as labelling might also be helpful to follow the exact distribution of certain compounds. Imaging techniques, such as confocal scanning laser microscopy (CSLM), which allows three-dimensional visualisation of mycelia, might be a good alternative to study the elongation of hyphae in more detail.

REFERENCES

Aldarf, M., Fourcade, F., Amrane, A., and Prigent, Y. (2004). Diffusion of lactate and ammonium in relation to growth of *Geotrichum candidum* at the surface of solid media. Biotechnology and Bioengineering 87:69-80.

Aldarf, M., Fourcade, F., and Amrane, A. (2005). Solid-state culture of *Geotrichum candidum* and *Penicillium camembertii* on a glutamate and lactate based medium. Enzyme and Microbial Technology 35:159-167.

Beuling, E. E., Dusschoten, D. van, Lens, P., Heuvel, J. C. van den, As, H. van, and Ottengraf, S. P. P. (1998). Characterization of the diffusive properties of biofilms using pulsed field gradient-nuclear magnetic resonance. Biotechnology and Bioengineering 60:283-291.

Biesebeke, R te., Boussier, A., van Biezen, N., Braaksma, M., Hondel, C. A. M J J van den, Vos, W. M. de, and Punt, P. J. (2006). Expression of Aspergillus hemoglobin domain activities in *Aspergillus oryzae* grown on solid substrates improves growth rate and enzyme production. Biotechnological Journal 1:822-827.

Campbell-Platt, G. (1994). Fermented foods – a world perspective. Food Research International 27:253-257.

Chang, P. L. Y., and Trevithick, J. R. (1974). How important is secretion of exoenzymes through apical cell walls of fungi. Archives of Microbiology 101:281-293.

De Gooijer, C. D., Wijffels, R. H., and Tramper, J. (1991). Growth and substrate consumption of *Nitrobacter agilis* cells immobilized in carrageenan: Part 1. Dynamic modelling. Biotechnology and Bioengineering 38:224-231

Dijksterhuis, J., Veenhuis, M., Wyss, U., and Harder, W. (1991). Colonization and digestion of nematodes by the endoparasitic fungus *Drechmeria coniospora*. Mycological Research 95: 873-878.

Edelstein, L., and Segel, L. A. (1983). Growth and metabolism in mycelial fungi. Journal of Theoretical Biology 104:187-210.

Georgiou, G., and Shuler, M. L. (1985). A computer model for the growth and differentiation of a fungal colony on solid substrate. Biotechnology and Bioengineering 28:405-416.

Harold, F. M. (2002). Force and compliance: rethinking morphogenesis in walled cells. Fungal Genetics and Biology 37:271-282.

Harold, R. L., Money, N. P., and Harold, F. M. (1995). Growth and morphogenesis in *Saprolegnia ferax*: is turgor required? Protoplasma 191:105-114.

Hunik, J. H., Bos, C. G., van den Hoogen, M. P., De Gooijer, C. D., and Tramper, J. (1994). Co-immobilized *Nitrosomonas europaea* and *Nitrobacter agilis* cells: validation of a dynamic model for simultaneous substrate conversion and growth in kappa-carrageenan gel beads. Biotechnology and Bioengineering 43:1153-1163.

Malda, J., Rouwkema, J., Martens, D. E., le Comte, E. P., Kooy, F. K., Tramper, J., van Blitterswijk, C. A., and Riesle, J. (2004). Oxygen gradients in tissue-engineered Pegt/Pbt cartilaginous constructs: Measurement and modelling. Biotechnology and Bioengineering 86:9-18.

Mendgen, K., Hahn, M., and Deising, H. (1996). Morphogenesis and mechanisms of penetration by plant pathogenic fungi. Annual Review of Phytopathology 34:367-386.

Mitchell, D. A., Do, D. D., and Greenfield, P. F. (1991). A semimechanistic mathematical model for growth of *Rhizopus oligosporus* in a model solid-state fermentation system. Biotechnology and Bioengineering 38:353-362.

Molin, P., Gervais, P., and Lemiere, J. P. (1993). A computer model based on reaction-diffusion equations for the growth of filamentous fungi on solid substrate. Biotechnological Progress 9:385-393.

Nagel, F. J. I., Tramper, J., Bakker, M. S. N., and Rinzema, A. (2001). Model for on-line moisture-content control during solid-state fermentation. Biotechnology and Bioengineering 72:231-243.

Nagel, F. J., As, H. van, Tramper, J., and Rinzema, A. (2002). Water and glucose gradients in the substrate measured with NMR imaging during solid-state fermentation with *Aspergillus oryzae*. Biotechnology and Bioengineering 79:653-663.

Nandakumar, M. P., Thakur, M. S., Raghavarao, K. S. M. S., and Ghildyal, N. P. (1994). Mechanism of solid particle degradation by *Aspergillus niger* in solid state fermentation. Process Biochemistry 29:545-551.

Nandakumar, M. P., Thakur, M. S., Raghavarao, K. S. M. S., and Ghildyal, N. P. (1996). Substrate particle size reduction by *Bacillus coagulans* in solid-state fermentation. Enzyme and Microbial Technology 18:121-125.

Nopharatana, M., Howes, T., and Mitchell, D. (1998). Modelling fungal growth on surfaces. Biotechnological Research 12:313-318.

Nopharatana, M., Howes, T., and Mitchell, D. A. (2003). Use of confocal scanning laser microscopy to measure the concentrations of aerial and penetrative hyphae during growth of *Rhizopus oligosporus* on a solid surface. Biotechnological and Bioengineering 84:71-76.

Ottengraf, S. P. P., and van den Oever, A. H. C. (1983). Kinetics of organic compound removal from waste gases with a biological filter. Biotechnology and Bioengineering 25:3089–3102.

Olsson, S., and Jennings, D. H. (1991). Evidence for diffusion being the mechanism of translocation in the hyphae of three molds. Experimental Mycology 15:302-309.

Oostra, J., le Comte, E. P., van den Heuvel, J. C., Tramper, J., and Rinzema, A. (2001). Intraparticle oxygen diffusion limitation in solid-state fermentation. Biotechnology and Bioengineering 75:13-24.

Rahardjo, Y. S. P., Weber, F. J., Comte, E. P. le, Tramper, J., and Rinzema, A. (2002). Contribution of aerial hyphae of *Aspergillus oryzae* to respiration in a model solid-state fermentation system. Biotechnology and Bioengineering 8:539-544.

Rahardjo, Y. S. P., Korona, D., Haemers, S., Weber, F. J., Tramper, J., and Rinzema, A. (2004). Limitations of membrane cultures as a model solid-state fermentation system. Letters in Applied Microbiology 39:504-508.

Rahardjo, Y. S. P., Sie, S., Weber, F. J., Tramper, J., and Rinzema, A. (2005a). Effect of low oxygen concentrations on growth and α-amylase production of *Aspergillus oryzae* in model solid-state fermentation systems. Biomolecular Engineering 21:163-172.

Rahardjo, Y. S. P., Weber, F. J., Haemers, S., Tramper, J., and Rinzema, A. (2005b). Aerial mycelia of *Aspergillus oryzae* accelerate α-amylase production in a model solid-state fermentation system. Enzyme and Microbial Technology 36:900-902.

Rahardjo, Y. S. P., Jolink, F., Haemers, S., Tramper, J., and Rinzema A. (2005c). Significance of bed porosity, bran and specific surface area in solid-state cultivation of *Aspergillus oryzae*. Biomolecular Engineering 22:133-139.

Rahardjo, Y. S. P., Tramper, J., and Rinzema, A. (2005d). Modelling conversion and transport phenomena in solid-state fermentation: A review and perspectives. Biotechnology Advances 24:161-179.

Rajagopalan, S, and Modak, J. M. (1995). Evaluation of relative growth limitation due to depletion of glucose and oxygen during fungal growth on a spherical solid particle. Chemical Engineering Science 50:803-811.

Rajagopalan, S., Rockstraw, D. A., and Munson-McGee, S. H. (1997). Modeling substrate particle degradation by *Bacillus coagulans* biofilm. Bioresources Technology 61:175-183.

Schreuder, M. P., Brekelmans, S., van den Ende, H., and Klis, F. M. (1993). Targeting of a heterologous protein to the cell wall of *Saccharomyces cerevisiae*. Yeast 9:399-409.

Thiele, E. W. (1939). Relation between catalytic activity and size of particle. Industrial Engineering and Chemistry 31:916-920.

Varzakas, T. (1998). *Rhizopus oligosporus* mycelial penetration and enzyme diffusion in soya bean tempe. Process Biochemistry 33.741-747.

Vinck, A., Terlou, M., Pastman, W. R., Martens, E. P., Ram, A. F., Hondel, C. A. M. J. J. van den, and Wösten, H. A. B. (2005). Hyphal differentiation in the exploring mycelium of *Aspergillus niger*. Molecular Microbiology 58:693-699.

Wayman, F. M., and Mattey, M. (2000). Simple diffusion is the primary mechanism for glucose uptake during the production phase of the *Aspergillus niger* citric acid process. Biotechnology and Bioengineering 67:451-456.

Wessels, J. G. H. (1999). Fungi in their own right. Fungal Genetics and Biology 27:134-145.

Wessels, J. G. H. (1988). A steady-state model for apical wall growth in fungi. Acta Botanica Neerlandica 37:3-16.

Wijffels, R. H., De Gooijer, C. D., Kortekaas, S., and Tramper, J. (1991). Growth and substrate consumption of *Nitrobacter agilis* cells immobilized in carrageenan: Part 2. Model evaluation. Biotechnology and Bioengineering 38:232-240.

Wösten, H. A. B., Moukha, S. M., Sietsma, J. H., and Wessels, J. G. H. (1991). Localization of growth and secretion of proteins in *Aspergillus niger*. Journal of General Microbiology 137:2017-2023.

Zlotnik, H., Fernandes, M. P., Bowers, B., and Cabib, E. (1984). *Saccharomyces cerevisiae* mannoproteins form an external cell wall layer that determines wall porosity. Journal of Bacteriology 159:1018-1026.

Chapter 13

Molecular detection and monitoring

Rolf Geisen
Federal Research Centre for Nutrition and Food, Location Karlsruhe, Haid-und-Neu-Str. 9, 76131 Karlsruhe, Germany.

INTRODUCTION

Fungi play an important role as spoilage or toxigenic microorganisms mainly for plant-derived foods. It is estimated that about 25% of the annual production of plants for human and animal nutrition is spoiled due to the growth of fungi (Smith *et al.*, 1994). Besides their spoiling activity many of the food-relevant fungi are able to produce mycotoxins (Samson *et al.*, 2004), secondary metabolites with a variety of toxic activities. About 300 different mycotoxins are known; however, only a few play a role for human health, as they can occur in measurable amounts in food and feed. Table 1 shows the most important mycotoxins occurring in food and the main producing fungi. For all of these fungi molecular detection systems have been described.

Most molecular detection systems developed for food-relevant fungi are targeted against important mycotoxin-producing fungi and only a few have been described for spoilage fungi. The fungal contamination of a food sample is usually assessed by classical mycological means, such as the plate count technique (Samson *et al.*, 2004). This technique has the great advantage that the spectrum of fungal species and their numbers present in a food sample can be determined. This method is very important for analysing the ecological change of fungal communities during production, storage or ripening of a food product. A prerequisite for this type of analysis, however, is trained personnel with taxonomical skills. A major disadvantage of this methodology is its time consumption. Usually up to 5 days (or more) are needed for correct identification of the fungal genera. This time-consuming process was originally the driving force for the development of molecular detection methods. Now, new applications of these methods have been evolved which are pointing towards correct taxonomical identification or parallel identification of various important mycotoxigenic fungi by multiplex PCR (Polymerase Chain Reaction). Even quantification of fungal biomass or cell numbers and determination of growth kinetics or physiological activity (e.g., mycotoxin production) of the fungal cell can be the subject of this type of analysis. In terms of mycotoxigenic fungi especially correct taxonomical identification is an important aspect in conjunction with molecular PCR techniques. Fungi may be misidentified by morphological methods, especially when identified by a non-expert (Frisvad *et al.*, 2004). Diagnostic PCR methods can be very valuable tools in identifying fungi and have been described to identify morphologically non-distinguishable strains of Penicillia, Aspergilli and Fusaria.

A prerequisite for a functional diagnostic PCR system is the availability of unique target sequences. In the era of genomics, where whole genomes become accessible from various fungal species, proposed primer sequences can be checked for possible cross-hybridisation by means of genomic databases.

Table 1. Selection of described molecular detection systems of food related fungi

Species	Targeted sequence	PCR product	Primer sequences	Assay	Reference
Aflatoxin-producing fungi					
Aspergillus flavus *A. parasiticus*	*ver*-1 gene aflatoxin biosynthesis	895	atgtcggataatcaccgtttagat cgaaaagcgccaccatccaccccaatg	PCR	Shapira *et al.*, 1996
	*afl*R gene aflatoxin biosynthesis	1032	tatctccccccgggcatctcccgg ccgtcagacagccactggacacgg		
	omt-1 gene aflatoxin biosynthesis	1024	ggcccggttccttggctcctaagc cgccccagtgagacccttcctcg		
A. flavus *A. parasiticus*	*nor*-1 gene aflatoxin biosynthesis	400	accgctacgccggcactctcggcacgt tggccgccagcttcgacactccg	multiplex PCR	Geisen, 1996
	ver-1 gene aflatoxin biosynthesis	537	gccgcaggccgcggagaaagtggtg gggatatactcccgcgacacagcc		
	omt-1 gene aflatoxin biosynthesis	797	gtggacggacctagtccgacatcacgt cggcgccacgcactgggttgggg		
A. flavus *A. parasiticus*	*avf*A gene aflatoxin biosynthesis	950	atggtcacatacgccctcctcggg gcctcgcattctctcggcgaccgaa	multiplex PCR	Yang *et al.*, 2004
	omt-1 gene aflatoxin biosynthesis	797	gtggacggacctagtccgacatcacgt cggcgccacgcactgggttgggg		
	ver-1 gene aflatoxin biosynthesis	452	gccgcaggcgcggagaaaggtggtc cgcagtcaatggccatgcagcg		
A. flavus *A. parasiticus*	*nor*-1 gene aflatoxin biosynthesis	400	accgctacgccggcactctcggcacgt tggccgccagcttcgacactccg	multiplex PCR	Chen *et al.*, 2002
	ver-1 gene aflatoxin biosynthesis	538	gccgcaggccgcggagaaagtggtg gggatatactcccgcgacacagcc		
	omt-1 gene aflatoxin biosynthesis	1025	gtggacggacctagtccgacatcacgt cggcgccacgcactgggttgggg		
	apa-2 gene aflatoxin biosynthesis	1032	tatctccccccgggcatctcccgg ccgtcagacagccactggacacgg		
A. flavus *A. parasiticus*	*nor*-1 gene aflatoxin biosynthesis	66	gtccaagcaacaggccaagt tcgtgcatgttggtgatggt tgtcttgatcggcgcccg	Real Time PCR RT Real Time PCR	Mayer *et al.*, 2003 a and b
A. flavus *A. parasiticus*	*afl*R gene aflatoxin biosynthesis	630	cgcgctcccagtccccttgatt cttgttccccgagatgacca	RT PCR	Sweeney *et al.*, 2000
Trichothecene-producing fungi					
Fusarium sp.	*tri*5 gene trichothecene biosynthesis	658	gctgctcatcactttgctcag ctgatctggtcacgctcatc	PCR	Niessen and Vogel, 1998

F. graminearum	*gao* gene galactose oxidase	900	agggacaataagtgcagac actgtgcactgtcgcaagtg	PCR	Niessen and Vogel, 1997
Gibberella zeae	*tri*7 gene trichothecene biosynthesis	161 NIV 173–327 DON	ggctttacgactcctcaacaatgg agagccctgcgaaag(ct)actggtgc	PCR	Lee *et al.*, 2001
F. graminearum	RAPD fragment anonymous	332	gcagggtttgaatccgagac agaatggagctaccaacggc	PCR	Schilling *et al.*, 1996
F. culmorum	RAPD fragment anonymous	472	gatgccagaccaagacgaag gatgccagacgcactaagat		
F. avenaceum	ITS region ribosomal DNA	272	ccagaggacccaaactctaa accgcagaagcagagccaat		
F. moniliforme (verticillioides)	RAPD fragment anonymous	561	tttacgaggcggcgatgggt ggccgtttacctggcttctt	PCR	Möller *et al.*, 1999
F. subglutinans	RAPD fragment anonymous	445	ggccactcaagaggcgaaag gtcagaccagagcaatgggc		
F. culmorum	RAPD fragment anonymous	570	atggtgaactcgtcgtggc cccttcttacgccaatctcg	PCR	Nicholson *et al.*, 1998
F. graminearum	RAPD fragment anonymous	280	acagatgacaagattcaggcaca ttctttgacatctgttcaaccca		
F. graminearum	RAPD fragment anonymous	300	ctccggatatgttgcgtcaa ggtaggtatccgacatggcaa	PCR	Doohan *et al.*, 1998
F. poae	RAPD fragment anonymous	250	caagcaaacaggctcttcacc tgttcccacctcagtgacaggtt		
F. avenaceum	RAPD fragment anonymous	920	caagcattgtcgccactctc gtttggctctaccgggactg		
Fusarium sp.	*tri*5 gene trichothecene biosynthesis	260	cagatggagaactggatggt gcacaagtgccacgtgac	quantitative competitive PCR	Edwards *et al.*, 2001
Fusarium sp.	*tri*5 gene trichothecene biosynthesis	658	gctgctcatcactttgctcag ctgatctggtcacgctcatc	Real Time PCR Light Cycler	Schnerr *et al.*, 2001
			Fumonisin-producing fungi		
Fusarium sp.	ITS region ribosomal DNA	431	aactcccaaacccctgtgaacata tttaacggcgtggccgc	multiplex PCR	Bluhm *et al.*, 2002
fumonisin-producing *Fusarium* sp.	*fum*5 fumonisin biosynthesis	845	gtcgagttgttgaccactgcg cgtatcgtcagcatgatgtagc		
trichotecene-producing *Fusarium* sp.	*tri*6 trichothecene biosynthesis	596	ctctttgatcgtgttgcgtc cttgtgtatccgcctatagtgatc		

F. moniliforme (verticillioides)	DNA fragment obtained by differential screening	1800	cttggtcatgggccagtcaagac cacagtcacatagcattgctagcc	PCR	Murillo *et al.*, 1998
			Ochratoxin-producing fungi		
A. ochraceus	AFLP fragment anonymous	260	ataccaccgggtctaatgca tgccgacagaccgagtggatt	PCR	Schmidt *et al.*, 2003
A. ochraceus	AFLP fragment anonymous	260	ataccaccgggtctaatgca tgccgacagaccgagtggatt	Real Time PCR Light Cycler	Schmidt *et al.*, 2004b
A. carbonarius	AFLP fragment anonymous	189	gaattcaccacacatcatagc ttaactaggatttggcattgaac	PCR	Schmidt *et al.*, 2004a
	AFLP fragment anonymous	351	gaattcacggtgctcgaccc ttaactgctggcggaagaggc		
A. carbonarius	RAPD fragment anonymous	809	aggctaatgttgataacggatgat gctgtcagtattggaccttagag	PCR	Fungaro *et al.*, 2004
P. nordicum	*otapks*PN gene ochratoxin biosynthesis	490	tacggccatcttgagcaacggcactgc atgcctttctgggtcagta	PCR	Bogs *et al.*, 2005
P. nordicum *P. verrucosum*	*nps*PN gene ochratoxin biosynthesis	750	agtcttcgctgggtgcttcc cagcacttttccctccatctatcc	PCR	
P. nordicum	*otapks*PN gene ochratoxin biosynthesis	-	cacggtttggaacaccacaat tgaagatctcccccgcct cgtaccaatccccatccagggctc	RT Real Time PCR Taq Man	Geisen *et al.*, 2004
			Patulin-producing fungi		
P. expansum *P. brevicompactum*	*idh* gene patulin biosynthesis	600	caatgtgtcgtactgtgccc accttcagtcgctgttcctc	PCR	Paterson *et al.*, 2000
P. expansum	polygalacturonase gene	404	atcggctgcggattgaaag agtcacgggtttggaggga	PCR	Marek *et al.*, 2003
			Alternaria toxin-producing fungi		
Alternaria sp.	ITS region ribosomal DNA	212	attgcaatcagcgtcagtaac caagcaaagcttgagggtaca	PCR	Zur *et al.*, 1999
A. radicina	RAPD fragment anonymous	900	gggcgttatgcgagatcagg gtatttgtaggaatttccag	PCR	Pryor and Gilbertson, 2001
A. alternata	ITS region ribosomal DNA	340	tgcaatcagcgtcagtaacaaat atggatgctagacctttgctgat	PCR	Konstantinova *et al.*, 2002
A. dauci	ITS region ribosomal DNA	345	gcaatcagcgtcagtaacaaca cgcaaggggagacaaaaa		
A. radicina	ITS region ribosomal DNA	251	aatcagcgtcagtaaacaaacg agaggctttgtggatgctg		

Ergot alkaloid-producing fungi					
Claviceps purpurea	mating type genes	250	ccaagccggtatcagtgaatgc cgacctgttgtcgaacaaaggt	PCR	Yokoyama *et al.*, 2004
Spoilage fungi					
P. roqueforti *P. carneum*	ITS region ribosomal DNA	300	ctgtctgaagaatgcagtctgagaac ccatacgctcgaggaccggac	PCR	Pedersen *et al.*, 1997
Penicillium subgenus *Penicillium*	ITS region ribosomal DNA	336	aaatataaattatttaaaactttc ctggataaaaatttgggttg		

During the time of writing this chapter the sequences of several fungal genomes are available and research is being done with additional species (http://www.broad.mit.edu/annotation/fungi/fgi/; http://www.genome.gov/ 11008243). Among the sequenced fungal species are *Neurospora crassa, Aspergillus nidulans, Fusarium graminearum, Aspergillus flavus, Ustilago maydis, Magnaporte grisea* and others; however, from this selection only *F. graminearum* and *A. flavus* are important mycotoxin-producing fungi. Another important aspect is the availability of protocols to isolate DNA from a food sample in a PCR-required purity. It has long been known that components of the food samples can interfere with the PCR reaction (Rossen *et al.*, 1992). In order to avoid pre-enrichment steps, which extend the duration of the detection method, it must be ensured by optimized protocols that pure fungal DNA can be isolated from food.

Theoretically, a diagnostic PCR reaction is very sensitive and can detect down to 1 or 10 molecules. However, practically there are different constraints leading to a sensitivity which is much less and roughly comparable with conventional plate count method.

These and more aspects will be discussed in this chapter. Also, examples for particular PCR systems will be given for most food-relevant fungi. However, this chapter does not claim to give a complete overview and to cover all possible aspects and examples. Recently, a very interesting and comprehensive review about this topic has been published (Edwards *et al.*, 2002).

FUNDAMENTALS OF MOLECULAR DETECTION, QUANTIFICATION AND MONITORING METHODS

The principle of PCR as a diagnostic tool is the detection of a DNA sequence, which is unique for the fungus of interest. If this specific DNA sequence can be detected by PCR, it can be concluded that the sample is contaminated by the fungus. With conventional PCR, only the mere presence of fungi can be detected. Because of its simplicity and speed this approach is very well suited to get a first survey of the presence of a fungal species in food. However, nothing can be said about the cell number or spore number in a sample, or about mycotoxin production. A derivative of conventional PCR with higher specificity, the so-called PCR-ELISA method has been described for detection of *Fusarium moniliforme* (*F. verticillioides*) and other fumonisin-producing fungi (Grimm and Geisen, 1998). Here, the specificity of the reaction is highly increased compared to PCR by using an internal capture probe which binds to the PCR product. This capture probe is biotinylated and binds to streptavidin coated microtiter plate wells. The DNA hybrid can be detected by antibodies against Streptavidin by a colour reaction. With Real Time PCR or competitive PCR the copy number of target genes can be determined. Both methods can therefore be used for fungal quantification.

Real Time PCR is a method which enables the direct online determination of the generated PCR product during the reaction by an increase of the fluorescence of the reaction mixture. This increase in fluorescence is either achieved by using Sybr Green, which intercalates into the nascent PCR product or by applying the TaqMan® or Light Cycler® principle.

In competitive PCR systems the fungal DNA is quantified in the presence of an internal standard. This is a second artificial target sequence of known concentration. The PCR reaction in which both, the target sequence and the artificial target sequence, show the same signal intensity indicate the concentration of the target sequence. This approach has been used in food mycology to quantify trichothecene producing *Fusarium* species in wheat (Edwards *et al.*, 2001; Nicholson *et al.*, 2002) or in environmental mycology to quantify *Verticillium chlamydosporum* in soil (Mauchline *et al.*, 2002). The drawback of this method is the necessity of an artificial target with the same primer binding sites as the sample target to ensure the same hybridization kinetics for both types of targets. If the concentration of the standard target is carefully chosen, the influence upon the sensitivity of the method is minimal. Once this target is available, this is a reliable method for quantification.

As with other methods, quantification of the fungal biomass is not an easy task and several considerations have to be made. If one target gene is present in the genome, which is the case for many genes and most of the mycotoxinbiosynthetic genes, the number of copies reflects the genome number, which is proportional to the fungal biomass or cell number. In case of bacterial quantification by Real Time PCR, there is approximately a 1 to 1 ratio between genome number and cell number. This is not the case with filamentous fungi, which consist of different cell types including vegetative spores, ascospores and hypha. Spores can be uninucleate or multinucleate depending on the fungal species. Many food-related mycotoxigenic fungi possess uninucleate spores, but hyphal cells are multinucleate. Many nuclei can occur in a single fungal cell. Figure 1 shows various cells in hyphae of *Penicillium nalgiovense* compared to its uninucleate spores. This observation complicates the interpretation of the data. Nevertheless a correlation between the Real Time PCR data and fungal growth occurs.

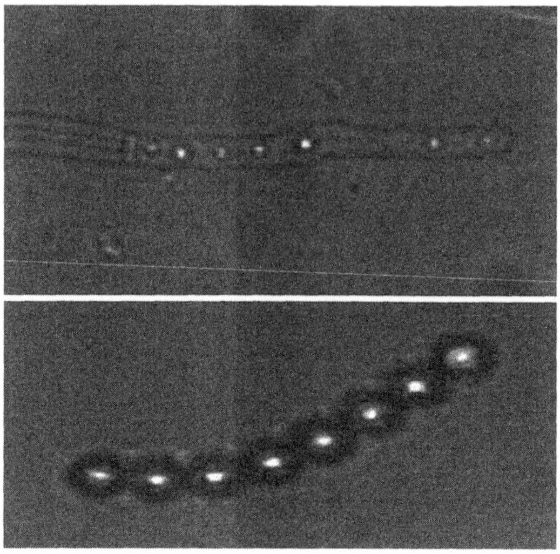

Figure 1. Dapi stained nuclei of a growing hyphal tip (top) and of a chain of conidia of *P. nalgiovense*. The multikaryotic nature of the filamentous cells and the unikaryotic nature of the conidia are clearly visible.

Generally, the copy numbers determined by Real Time PCR are higher than the numbers of viable cells determined by the plate count method (Mayer *et al.*, 2003a). This can be explained by the detection of dead cells or released DNA and by the multinucleate status of the fungal cells. With the plate count method, only the number of single-cell spores is quantified exactly. Mycelial fragments can consist of many cells which however will give rise to 1 colony only. With molecular detection methods, if the DNA isolation procedure is efficient, all nuclei of these mycelial fragments will contribute to the copy number determined by Real Time PCR or competitive PCR. Haugland *et al.* (1999) compared the results obtained by Real Time PCR for the quantification of spores of *Stachybotrys chartarum* with the results obtained by counting the spores in a microscopic counting chamber. They found a very good correlation between the spore number and the copy number of the target sequence. If these relationships are kept in mind, quantitative Real Time PCR is a rapid and reliable tool for quantification of fungal cells and for the establishment of growth kinetics (Mayer *et al.*, 2003a).

Reverse Transcriptase PCR (RT PCR) or Reverse Transcriptase Real Time PCR (RT Real

Time PCR) are approaches to detect the level of mRNA in a cell instead of DNA. This approach therefore does not aim at detecting and quantifying fungal biomass, but at monitoring gene expression. Depending on the target sequence, information on the regulation of gene activity is obtained. For this approach mRNA is isolated and has to be reverse-transcribed into cDNA. This cDNA can be quantified by Real Time PCR or detected by conventional PCR. This system is very useful to analyse the activity of mycotoxin biosynthetic genes in relation to food safety. It can be used to identify molecular critical control points (MCCPs) which are environmental conditions that allow the activation of mycotoxin biosynthetic genes. Several monitoring systems have already been described for various mycotoxins and are discussed below. In-depth information to the technological background of the various methods is not given here, but can be found in various general reviews.

TARGET SEQUENCES

Unique target sequences are vital for the development of molecular detection methods. Nowadays, many genes of the biosynthetic pathways of the most important mycotoxins are known. Nearly all complete pathways are known in the case of aflatoxin (Yu *et al.*, 2004), trichothecenes (Brown *et al.*, 2001) or fumonisins (Proctor *et al.*, 2003).

For other mycotoxins the information is less complete; however, knowledge about single key enzymes of ochratoxin A biosynthesis (O'Callaghan *et al.*, 2003, Karolewiez and Geisen, 2005), patulin biosynthesis (Beck *et al.*, 1990) or PR toxin biosynthesis exists (Proctor and Hohn, 1993). For all fungi producing these mycotoxins, molecular detection methods are available based on single or multiple genes of biosynthetic pathways. The specificity of these methods depends strongly on the choice of the respective target gene and on the taxonomic relations between species. For example if a polyketide synthetase (*pks*) gene (in case of aflatoxin, ochratoxin A or fumonisin synthesis) is selected, cross-reactions with other species may occur as this gene is a very common gene for secondary metabolite pathways. Fungi may contain several *pks* genes per genome. *A. fumigatus* carries up to 14 putative polyketide synthase genes (Varga *et al.*, 2003). Cytochrome oxidases or non-ribosomal peptide synthetases are also common enzymes which are often involved in various secondary metabolite biosynthesis pathways and further, silent genes in taxonomically related species can occur, as is the case for the aflatoxin biosynthetic genes in *A. oryzae* and *A. sojae* (Chang *et al.*, 1995, Kusumoto *et al.*, 1998). These facts have to be kept in mind, when choosing the appropriate target sequence.

Other identification systems are based on ribosomal or ITS sequences, despite the fact that the variability is not very high between related taxa.

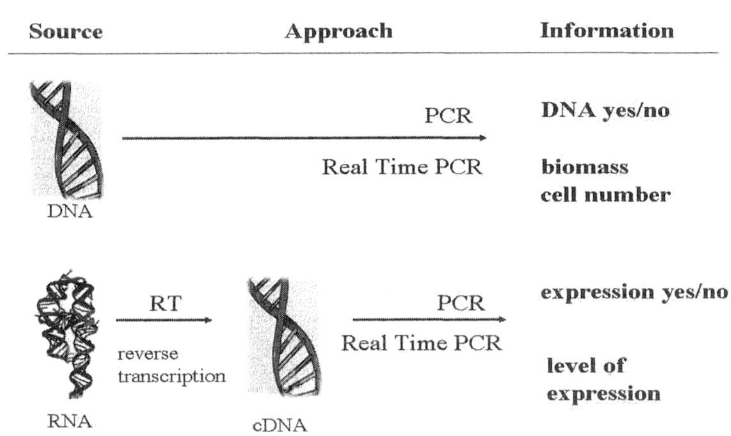

Figure 2. Level of information obtained by PCR/Real Time PCR, respectively.

For this reason it was not possible to differentiate between *Fusarium culmorum*, *F. graminearum* and *F. crookwellense* based on the ITS region of the rDNA (Bateman *et al.*, 1996). This was confirmed by studying the 28S rDNA (Edel *et al.*, 1996) or other attempts (Schilling *et al.*, 1996; Möller *et al.*, 1999; Nicholson *et al.*, 1998; Marek *et al.*, 2003). Pedersen *et al.* (1997) reported that the rDNA sequences within the genus *Penicillium* subgenus *Penicillium* are too conserved for differentiation between individual species. Despite these difficulties, Kulik *et al.* (2004) was able to use a polymorphism in the ITS2 region of *F. sporotrichioides* to specifically detect and differentiate this fungus from other mycotoxin-producing Fusaria. They were able to identify specific primers based on at least 4 nucleotide heterogeneity in the ITS2 region of the ribosomal rDNA. These authors could achieve a positive reaction with 11 of 12 *F. sporotrichioides* strains. They conclude and could confirm by morphological analysis that the negative strain has obviously been misidentified.

Many copies of rDNA are arranged in the genome in a tandem-like fashion (O´Donnell, 1992) and may occur in 40–240 copies per genome, depending on the species analysed (Griffin, 1994). This higher copy number compared to other sequences like for example secondary metabolite biosynthetic genes, increases the sensitivity of the reaction considerably. The copies of the rDNA genes in turn are separated by itself by so-called IGS sequences (intergenic spacer) which are random sequences between each transcription unit of the rDNA region. These IGS regions are variable too and have been used for detection and identification purposes (Mirete *et al.*, 2003). A scheme of the genetic organisation of that genomic region is depicted in Figure 3.

Further candidates are the unique SCAR sequences (sequence characterized amplified regions). Here, anonymous markers are identified based on RAPD (Schilling *et al.*, 1996) or AFLP fragments (Schmidt *et al.*, 2003). Based on the sequence of specific bands for mycotoxigenic species, primers can be developed. If it turns out that these primers are specific for the respective mycotoxin-producing species they can serve as an "anonymous" marker and can be used for diagnostic purposes.

Finally, miscellaneous genes unique for certain species are candidates for diagnostic PCR methods. Niessen and Vogel (1997) used a gene for a galactose oxidase (*gao*) to specifically detect the mycotoxigenic fungus *F. graminearum*. According to these authors, this gene is only present in a few fungal species, namely *F. graminearum*, *F. fujikuroi* and *Beltraniella portoricensis*. Perrone *et al.* (2004) used a variable sequence of the calmodulin gene to detect and differentiate members of the black Aspergilli, namely *A. japonicus* and the ochratoxigenic *A. carbonarius*.

Sensitivity

Theoretically the PCR method is a very sensitive method. In the scientific literature statements do exist that down to one molecule can be detected. This goal is usually not achieved under practical conditions. Firstly, the target molecule has to be present in the reaction vessel, which is not a trivial task, if not impossible to be achieved, when a large food sample has to be analysed. Secondly the DNA preparation has to be pure enough to ensure that no PCR inhibition occurs.

The sampling scheme during an assessment of the quality of the food is very important. The distribution of fungal contamination is usually heterogeneous throughout the food sample.

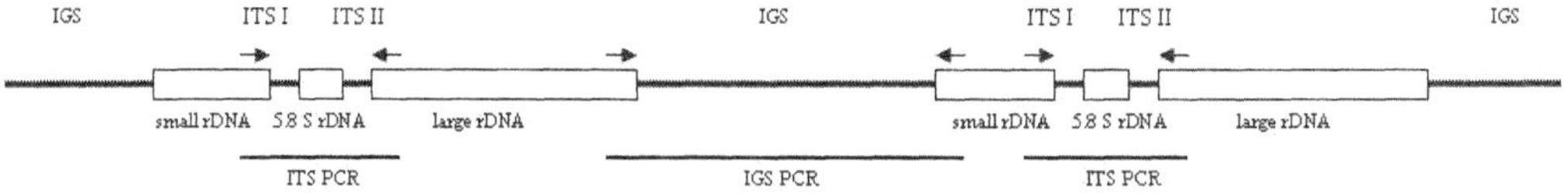

Figure 3. Schematic presentation of the organisation of fungal rDNA genes and indication of possible target sequences.

The fungal colony may grow only in a certain part of the food matrix or only on certain kernels. So, in case of a low contamination, it is important to analyse a large sample to increase the possibility to identify undesired fungi. For the analytical assessment of foods for the presence of mycotoxins sampling schemes have been described (Whitaker, 2000). Similar principles should be followed for the sampling of probes to be analysed by molecular methods. However molecular methods usually deal with small amounts of the sample material. For a PCR reaction a sample amount of 100 mg up to 1 g is used for DNA isolation. The DNA is subsequently solubilized in buffer solution and only 2 to 5 μl are used for the PCR. If in 1 g sample material 100 copies of the target sequence are present, which are all isolated by the DNA preparation procedure and redissolved in 100 μl buffer, a concentration of 1 template copy/μl will be realised. This would mean that per reaction 2 to 5 copies of the template are present. This could be detected if the whole PCR reaction runs optimal. However, under practical circumstances a vast amount of sample DNA (mainly plant DNA) is also present, which has a negative influence on the performance of the specific PCR reaction (Rossen *et al.*, 1992; Färber *et al.*, 1997). In addition, only a part of the original 100 copies will be isolated from the food depending on the optimization of the isolation procedure. Impurities from the food sample which are copurified with the DNA, like proteins, lipids or carbohydrates have a negative effect on the activity of the PCR enzyme (Rossen *et al.*, 1992).

Taken together all the aspects discussed above lead to a maximum sensitivity of 1000 to 10,000 copies per gram of sample material (in optimal cases perhaps down to 100 copies per gram). This is in agreement with many results described in the literature. Bluhm *et al.* (2002) for example reported a sensitivity of their PCR reaction to detect *F. graminearum* in corn meal of 8 x 10^4 CFU/g. If an enrichment step is involved prior to the PCR reaction, a lower number of target sequences can be detected at the expense of time. Shapira *et al.* (1996) used an enrichment step to detect aflatoxigenic fungi in wheat and identified 10^2 spores/gram after a 24 hour incubation period in potato dextrose broth. Böhm *et al.* (1999) detected 8.9 pg *P. citricola* DNA per microgram of plant material without an amplification step, but did not calculate the cell number or biomass of *P. citricola*. Haugland *et al.* (1999) described the detection of 23 spores of *S. chartarum* in air samples by TaqMan PCR. Aflatoxinogenic moulds growing on figs have been detected by PCR by Färber *et al.* (1997). These authors also determined the sensitivity of the method. They could detect 25 pg of purified *A. flavus* DNA in a sample of 0.5 gram, corresponding to 0.7 x 10^4 genome equivalents. However, when fig DNA was present in the reaction, the sensitivity decreased tenfold, resulting in a minimum of 250 pg of *A. flavus* DNA per reaction. Bluhm *et al.* (2002) described a multiplex PCR for the simultaneous detection of trichothecene or fumonisin producing *Fusarium* species. They found a sensitivity of 10 to 100 pg of genomic DNA for the primers specific for trichothecene production and between 0.1 and 1 ng for fumonisin producers. The sensitivity of the trichothecene-specific reaction is in agreement with the results from Schnerr *et al.* (2001) who reported a sensitivity of about 50 pg based on the *tri5* gene as target sequence. However, these values have been determined with purified fungal DNA and not under practical conditions in a food matrix. Perrone *et al.* (2004) reported a sensitivity of 12.5 pg of genomic DNA or a calculated copy number of 260 *A. carbonarius* genomes determined in laboratory media. Schilling *et al.* (1996) reported a sensitivity of 50 pg of DNA of *F. culmorum* in their PCR assay based on SCAR sequences. With a similar assay Möller *et al.* (1999) demonstrated a sensitivity of 100–200 genomes per reaction to detect *F. moniliforme* or *F. subglutinans*. The reaction still worked when the *Fusarium* DNA was diluted with maize DNA by a factor of 10.000. This system could be used to identify contamination by *F. moniliforme* or *F. subglutinans* in visually sound kernels.

It is clear that the quality of DNA is of utmost importance for the reliablity of the PCR results. For each specific food matrix, a DNA preparation protocol has to be optimized. Very often a simple dilution step is enough to counteract against the inhibition, as the inhibitor is diluted below its activity threshold. Lantz *et al.* (1994a) has optimized a method for the isolation of DNA from soft cheeses. The method is based upon an aqueous two-phase system containing polyethylene glycol and dextran. By applying this system the inhibitory factors are distributed to the polyethylene glycol phase, whereas the template DNA concentrates in the dextran phase. The sensitivity of the system could be improved by a factor of 1000 by using this purification method. Maher *et al.* (2001) used an magnetic bead capture method for improving the sensitivity of a PCR method to detect *Pneumocystis carinii* from an airborne environment. They could increase the number of positive reactions from 0/12 to 28/30 by applying this purification method. A rapid DNA isolation procedure based on sonification and extraction was described by Knoll *et al.* (2002). These authors achieved a sensitivity to detect *F. graminearum* in cereals of one infected kernel (0.04 g) in a 40 g sample. An overview about the topic of DNA purification and possibilities to improve the results was given by Lantz *et al.* (1994b).

Controls and regulations

If a diagnostic PCR is used to assess the safety of a food sample (Rossen *et al.*, 1992), false negative reactions resulting from PCR inhibition have to be distinguished from true negative reactions. For this purpose an external control reaction can be performed. In this approach, a known target sequence with appropriate primer pairs (the same as for the sample sequence) is added to an otherwise identical sample. If in this reaction, after electrophoresis, a band with the expected molecular weight appears it can be concluded that the analysed sample does not contain inhibitors of the reaction. However, the use of an internal control, where a known DNA template is added to each reaction is most reliable. The template is designed in such a way, that the same primers bind to both control template and target. This prevents differences in hybridization kinetics. The control target is genetically engineered resulting in a PCR product of different length than the sample product. This ensures easy detection in an agarose gel. The possible results of that setup are given in Table 2.

This approach definitely demonstrates the functionality of the reaction and clearly distinguishes between false and true negatives and gives unambiguous results. Hoorfar *et al.* (2004) demonstrates the importance of an internal standard for a diagnostic PCR reaction to give reliable results.

Table 2. Possible outcome of a PCR reaction containing an internal control

Sample	Internal positive control	Negative control	Result
+	+	-	positive
+	+	+	contamination
-	+	-	negative
-	-	-	inhibition

The validation of PCR methods for food safety must be performed in interlaboratory studies with different independent laboratories which all analyse the same samples under standardized conditions. The results have to be compared among the laboratories and with the current standard detection method (in this case the plate count technique). To ensure standardizations during the performance of routine diagnostic PCR methods, national and international agencies like DIN, AFNOR, CEN or ISO are currently dealing with standard PCR protocols, which define the conditions that have to be met before routinely using PCR to assess the safety of a food sample (www.din.de: ISO/FDIS 21571, DIN 10134, DIN 10135, DIN 58967-60, DIN 58969-61). These protocols are currently described for bacteria, but can be adapted to the detection of fungi in food.

Recently a European Union research project called "Food PCR" (www.pcr.dk) has been established which deals with questions of standardization and validation of PCR systems used for the analysis of food commodities.

Differentiation between mycotoxin producing and non-producing strains

The trial to differentiate between mycotoxin-producing and non-producing strains by means of PCR-methods based on DNA as the target molecule is done in a number of studies (Criseo *et al.*, 2001; Chen *et al.*, 2002; Fungaro *et al.*, 2004). This approach is based on the assumption that the genetic difference, which is responsible for the mycotoxin negative phenotype, can be detected by PCR. However in most cases the mutation or the variation leading to a non-mycotoxigenic genotype will not be detectable by PCR because it might not be located in the target sequence. In cases when strains show a strong rearrangement of the mycotoxin biosynthetic genes (which would be detectable by PCR), they obviously belong to different subpopulations or species, as is the case for *A. flavus* and *A. oryzae* or *A. parasiticus* and *A. sojae*. All four species carry homologues of the aflatoxin biosynthetic genes, but the genes in *A. oryzae* and *A. sojae* are inactive and only partly homologous (Watson *et al.*, 1995; Klich *et al.*, 1995; Klich et al., 2000; Kusumoto *et al.*, 1998). This indicates that positive or negative PCR results may occur dependent on the analysed strain; however, the result is not directly correlated to mycotoxin production.

Criseo *et al.* (2001) attempted to differentiate aflatoxin producing from non-producing *A. flavus* strains by quadruplex PCR. They found varying results with non-producing strains, ranging from one, two, three or even four bands (the same as the producing strains). They came to the conclusion that the PCR pattern is not sufficient for this purpose. With *A. flavus/parasiticus* Chen *et al.* (2002) found no clear correlation between the completeness of the quadruple pattern of a multiplex PCR and the ability of the strains to produce aflatoxin. All strains which could produce aflatoxin exhibited the complete banding pattern. However 7 out of 15 strains with a complete banding pattern were non-aflatoxigenic. A number of the analysed non-aflatoxigenic *A. oryzae* and *A. sojae* strains proved to be positive in all 4 bands, indicating the presence of homologous but silent genes. All these data indicate that differentiation between toxigenic and non-toxigenic strains of related fungi by (multiplex) PCR is not an easy task.

A better way to address this question is the monitoring of the activity of mycotoxin biosynthetic genes (see below) by measuring mRNA levels of key enzymes of the biosynthetic pathways. This method is closer to the phenotypic "front" than the DNA-based approach, as it gives information about the expression of the genes. Scherm *et al.* (2005) demonstrated in an elegant analysis, that not all analysed 9 aflatoxin biosynthetic genes are equally expressed during aflatoxin biosynthesis. They identified 3 genes, namely the *nor*-1, the *omt*B and the *omt*A genes, which showed reliable expression under production conditions and their expression was suggested as decision points between aflatoxin and non-aflatoxin producing Aspergilli. Klich *et al.* (1997) analysed various strains of *A. parasiticus* able or not able to produce aflatoxin and non-toxic *A. sojae* strains. All aflatoxin-producing strains showed expression of all analysed genes; however, the *A. parasiticus* strain which was unable to produce the toxin, as well as one *A. sojae* isolate did not show any expression. Two other *A. sojae* isolates showed expression of only part of genes. The finding of Ehrlich *et al.* (2003) may also differentiate between highly producing or low producing and non-producing strains. These authors describe a sequence variability within the *afl*R gene and the intergenic region of the *afl*R/*afl*J genes, the two regulatory genes of the aflatoxin biosynthetic pathway. The authors could identify sequence differences which separate highly producing strains or lineages (*A. parasiticus* and *A. flavus* S_B) from low or non-consistently producing ones (*A. nomius, A. flavus* L.). If primers for a diagnostic PCR are based on such well-analysed regions, shown to correspond in their variability to secondary metabolite production, it might be possible to develop PCR systems which can give an indication to the mycotoxin producing capacity of a given strain.

Aspects of gene monitoring by PCR

It is a well-known fact that the window for growth influenced by environmental parameters like temperature, pH, water activity or presence of nutrients is much broader than the window for mycotoxin biosynthesis (Häggblom, 1982; Skrinjar and Dimic, 1992; Lee and Magan, 2000; Cuero *et al.*, 2003; Mitchell *et al.*, 2004). The reason for this difference is the tight regulation of secondary metabolite genes like mycotoxin biosynthetic genes. For aflatoxin biosynthetic genes Feng and Leonhard (1998) demonstrated a better expression when nitrate, instead of ammonium, was present in the medium. Expression of the *afl*R gene was highly influenced by the presence of the nitrate source (Ehrlich and Cotty, 2002). Contrastingly, *P. nordicum* showed a higher expression of the ochratoxin A polyketide synthase gene (*otapks*PN) and higher production of ochratoxin A, after growth on medium containing NH_4^+ compared to NO_2^- (Geisen, 2004). In *A. parasiticus*, the external pH has an effect on expression of aflatoxin biosynthetic genes (Keller *et al.*, 1997).

Induction of expression of mycotoxigenic genes occurs a certain time before the first mycotoxin can be detected (Xu *et al.*, 2000; Mayer *et al.*, 2003b). Xu *et al.* (2000) analysed the expression of the *nor*-1 gene of the aflatoxin biosynthetic pathway in *A. flavus* on peanut pods. Expression of the gene was 12 h before the first aflatoxin could be detected. This is similar for trichodiene synthase (*tri*5), a key enzyme in the biosynthesis of trichothecenes (Hohn and Beremand, 1989). So, expression of the *tri*5 gene could be a good indicator for subsequent trichothecene biosynthesis. Mayer *et al.* (2003b) demonstrated expression of the *nor*-1 gene even 48 hours before the first aflatoxin was detected. High expression of the *otapks*PN gene (ochratoxin A polyketide synthase) in *P. nordicum* occurred when ochratoxin A itself was only barely detectable by analytical methods (Geisen *et al.*, 2004).

Thus, monitoring of the mRNA production from key enzymes of mycotoxin biosynthetic pathways enables the exact measurement of environmental parameters which allow expression of the mycotoxin biosynthetic genes BEFORE the mycotoxin can be measured by analytical methods like TLC or HPLC.

It was demonstrated in an expression analysis of the ochratoxin A polyketide synthase gene (*otapks*PN) of *P. nordicum*, that the level of transcription of this gene is strongly dependent on environmental parameters and that this level corresponds to ochratoxin A production (Geisen, 2004). Doohan *et al.* (1999) developed a quantitative RT PCR assay to study the expression of the *tri*5 gene of *Fusarium* species in relation to trichothecene production. They found a direct relationship between expression of that gene and DON production. This system was used to study the influence of fungicides upon induction of the biosynthesis genes. The application of several fungicides leads to an increase of expression, explaining the well-known observation that fungicides may increase the production of mycotoxins.

Only when growth and mycotoxin production are strictly correlated, which is dependent on environmental parameters present in the food matrix, the amount of biomass or DNA is a measure for mycotoxin production. Lund and Frisvad demonstrated that the occurrence of *P. verrucosum* in wheat indicates the presence of ochratoxin A. They could show that more than 7% (number of infected kernels) infestation of wheat by *P. verrucosum* lead to detectable amounts of ochratoxin A in the samples. They also found a correlation between the rate of infection and the ochratoxin A produced. Schnerr *et al.* (2002) correlated the amount of *Fusarium* spp. DNA to the amount of deoxynivalenol (DON) produced. They found a linear relation between the DNA concentration, determined by Real Time PCR and the detectable amount of DON (see also Edwards *et al.*, 2001).

All these examples show the current molecular tools enable the exact measurement of the induction process and the exact determination of environmental conditions which enable gene activation. In view of a quality control concept, like the HACCP (Hazard Analysis Critical Control Point) concept the combination of the parameters which allow the activation of the mycotoxin biosynthetic genes can be regarded as Molecular Critical Control Point (MCCPs) which enables control of mycotoxin production by changing the environmental parameters in a way that they inhibit induction of these genes. Because of the capacity of these molecular methods to directly detect the induction of toxin genes before the first secondary metabolite can be detected, they can be regarded as online monitoring systems.

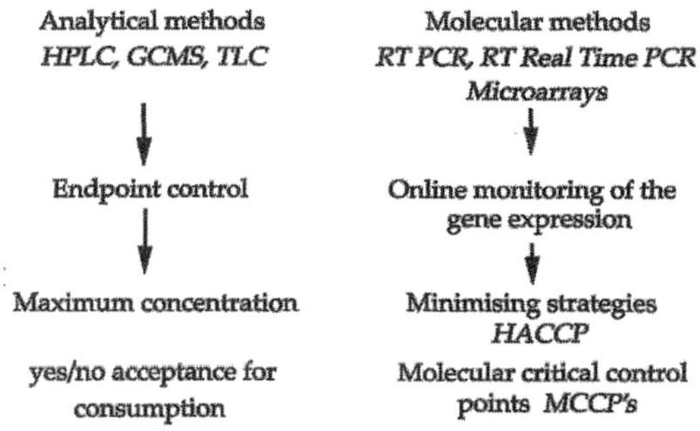

Figure 4. Differences in the outcome between analytical and molecular methods.

If this information about expression is complete, the data can be used to model and predict mycotoxin production under given conditions. In contrast to this, the determination of the mycotoxins in foods by analytical methods like TLC, HPLC, LC-MS or others is an endpoint control. Figure 4 gives a comparison between the information obtained by both approaches.

Detection of aflatoxin-producing fungi

The biosynthetic pathway for the production of aflatoxin by *A. flavus* and *A. parasiticus* has well been elucidated (Brown *et al.*, 1999; Yu *et al.*, 2004). The sequence of the whole gene cluster (consisting of 25 genes) has been completed recently (Yu *et al.*, 2004). Each of them can serve as potential target sequence for selected primers. The target regions which have been developed for aflatoxin-producing fungi are depicted in Figure 5. They are more or less located in the middle of the gene cluster.

The aflatoxin biosynthetic genes are very similar in *A. flavus* and *A. parasiticus* (Yu *et al.*, 1995). So, molecular methods could detect both species at once. Not much is known about the genetic background of aflatoxin production in *A. nomius*, *A. tamarii* and other potential aflatoxin-producing species like *A. pseudotamarii* (Ito *et al.*, 2001), *A. bombycis* (Peterson *et al.*, 2001) or *A. ochraceoroseus* (Klich *et al.*, 2000). However, these species are less important under food safety aspects.

The biosynthetic cluster for the production of sterigmatocystin, a metabolite produced as a precursor during aflatoxin production, but excreted as an end-product by *A. nidulans* and other fungi, is also very well characterized (Brown *et al.*, 1996). The sterigmatocystin biosynthesis genes of *A. nidulans* are homologues to the aflatoxin biosynthetic genes of *A. flavus*/*A. parasiticus* (Brown *et al.*, 1996), however with distinct differences at the nucleotide level. The *afl*R gene for example has a similarity of 33% to the *afl*R gene of *A. flavus*/*A. parasiticus*, respectively (Yu *et al.*, 1995). This situation implies that under most circumstances primers for the aflatoxin biosynthetic fungi will not cross-hybridize to sterigmatocystin-producing fungi. In addition according to Pohland (1993) no significant contamination of foods by sterigmatocystin could be identified, even after intensive analysis.

Several diagnostic PCR systems have been described for aflatoxin producing *A. flavus* and *A. parasiticus* strains. Two similar systems have been described independently (Shapira *et al.*, 1996; Geisen, 1996). Shapira *et al.* (1996) used three separate PCR reactions to determine the presence of aflatoxinogenic Aspergilli in corn and used the genes *apa*2 (now renamed as *afl*R), *ver*-1 and *omt*-1 (now *omt*A). They could easily identify *A. parasiticus*, but had difficulties to detect *A. flavus* DNA with only faint PCR product bands indicating micro-heterogeneities at the binding sites. In the other system also 3 genes are targeted in a multiplex PCR reaction (Geisen, 1996). In this reaction the *nor*-1, *ver*-1 and *omt*A gene sequences have been used.

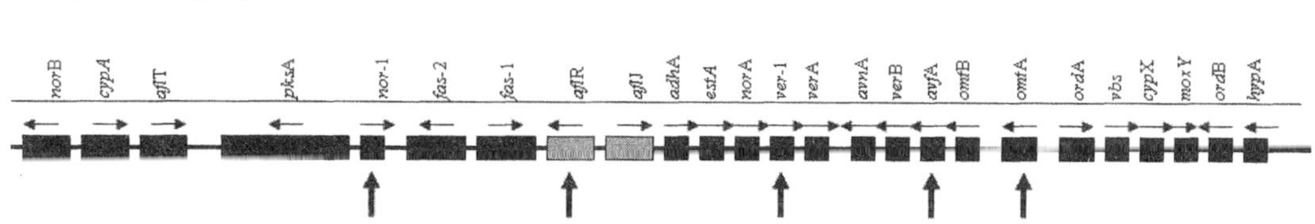

Figure 5. Scheme of the complete aflatoxin biosynthesis cluster according to Yu *et al.* (2004). The genes used as target sequences in diagnostic PCR are indicated as vertical arrows.

Differences in the detection of *A. parasiticus* or *A. flavus* were not observed here and the reaction has been used later to detect aflatoxinogenic fungi in figs (Färber *et al.*, 1997). The system clearly differentiates between infected and non-infected figs. Zachová *et al.*, (2003) targeted the *afl*R gene and the *ver*-1 gene in two separate reactions in 50 feed samples. Only two samples showed fungal growth on AFPA medium, but were negative in the PCR reaction. These strains did not produce detectible quantities of aflatoxin. Yang *et al.* (2004) used a multiplex PCR against *avf*A, *omt*A and *ver*-1 in Korean fermented foods and grains. In parallel, the presence of aflatoxin was tested by ELISA. These authors found three positive samples within 32 food samples. The ELISA test was negative, but enrichment of the positive samples in medium resulted in positive ELISA tests. This illustrates the sensitivity of the PCR method. Chen *et al.* (2002) detected aflatoxinogenic *A. parasiticus* strains in infected kernels by using a multiplex PCR targeting four biosynthetic genes (*afl*R, *nor*-1, *ver*-1 and *omt*A). In addition, a Real Time PCR system for the quantification of the fungal contamination based on the *nor-1 gene* was developed (Mayer *et al.*, 2003a). Biomass and/or cfu and Real Time PCR results clearly correlated. The system has been used to quantify the amount of aflatoxinogenic fungi in pepper, paprika and maize. The most congruent results have been obtained with pepper.

Two expression monitoring systems based either on conventional RT PCR (Sweeney *et al.*, 2000) or RT Real Time PCR (Mayer *et al.*, 2003b) have been described. Sweeney *et al.* (2000) targeted the *afl*R and the *ord*1 (*ord*A) genes. During growth on YES medium, which is optimal for aflatoxin production, both genes were highly induced. Under restrictive conditions, during growth in YEP medium, no specific mRNA was detected. The ß-tubulin gene, as a housekeeping gene, was expressed under both conditions. The phenotypic detection of aflatoxin by thin layer chromatography (TLC) paralleled the gene monitoring results. Mayer *et al.* (2003b) used a RT Real Time PCR system to monitor and quantify the induction of the *nor*-1 gene in wheat. A correlation between induction of the gene and production of aflatoxin could be demonstrated even 48 h before aflatoxin was phenotypically detectable.

Detection of toxic and pathogenic Fusarium species

The genes needed for trichothecene biosynthesis are located at three different positions in the genome, which is unique for this secondary metabolite. The core gene cluster consists of 12 genes (*tri*3 – *tri*14, Brown *et al.*, 2003). The second locus consists of a single gene (*tri*101) and the third locus encodes one to two biosynthetic genes (*tri*1; *tri*16). Species that produce the B type trichothecenes (DON, NIV) like *F. graminearum* and species which produce the A type trichothecenes (T2- and HT2 toxin, NEO and DAS) like *F. sporotrichioides* are similar, but characteristic differences occur. According to Brown *et al.* (2003) the cluster of *F. sporotrichioides* is an ancestor of the cluster of *F. graminearum*. This makes the development of specific primer pairs for all trichothecene producing species possible, or even for differentiation of the produced mycotoxin (see below).

Not all of the known genes for trichothecene biosynthesis are currently unambiguously assigned to biosynthetic functions in the pathway. Some of the assigned genes are indicated in Figure 6, which shows the main cluster of the trichothecene biosynthesis genes which contain the *tri*5 gene, the gene for the trichodiene synthase. This is a key enzyme in trichothecene biosynthesis and most of the methods are directed against this gene.

Niessen and Vogel (1998) developed a group specific PCR based assay to detect *Fusarium* species, which are potentially able to produce trichothecenes. They could identify the specific *tri*5 PCR product from *Fusarium* belonging to different sections like *Discolor, Sporotrichiella, Arthrosporiella, Gibbosum* and *Dlaminia* and the PCR products showed a high degree of sequence homology. The results of the PCR correlated well with the capabilities of the species to produce the toxin. The system was also useful to detect potential trichothecene producing species in cereals and malts.

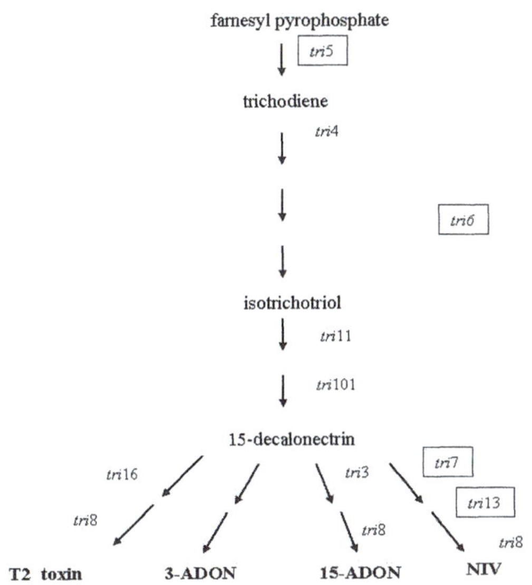

Figure 6. Proposed trichothecene biosynthetic pathway with assigned genes. The target genes used for diagnostic PCR are framed.

The same authors developed a PCR assay specific for *F. graminearum* based on a metabolic gene (galactose oxidase, gao) not involved in trichothecene production (Niessen and Vogel, 1997).

Lee *et al.* (2001) developed a PCR system for characterization and differentiation of DON and NIV producing phenotypes of *Gibberella zea*. They sequenced the gene clusters of these phenotypes and most of the genes were highly conserved, except the *tri7* gene. In case of the DON phenotype, the gene was altered because of several mutations and several copies of an 11 bp repeat. In case of the NIV phenotype this gene was active and did not carry the repeated structure. It has been formerly shown by Ichinoue *et al.* (1983) at the phenotypical level, that the *G. zeae* species can be divided into two chemotypes, one producing DON and acetyl-DON, the other NIV and 4ANIV. The NIV phenotype seems mainly to occur in several countries of Africa, Asia and Europe, whereas the DON phenotype has been reported in North America. The authors developed specific primers for the DNA region which distinguishes both chemotypes. The primers gave different PCR products. The PCR product was the same for all NIV chemotypes analysed; however, it varied considerably for the DON phenotypes. This was due to a high variability of the number of the 11 bp repeated sequence. Besides *tri7* the gene *tri13* is also responsible for the difference in secondary metabolite production between NIV and DON chemotypes. Jennings *et al.* (2004) used the same approach to analyse the occurrence of DON and NIV producing phenotypes of *F. graminearum*. They could demonstrate with this method that obviously the DON producing phenotypes increased during the recent years in fields of the U.K. By use of the variability of the 11 bp region they could subgroup the DON producing phenotypes. These authors also used a primer set directed against the *tri3* gene, which differentiates whether the DON phenotype was able to produce 3-ADON or 15-ADON. From 76 analysed strains 72 proved to be 15-ADON and 4 strains 3-ADON producers.

These results demonstrate the potential of diagnostic PCR methods to differentiate between highly related but distinguishable genotypes, if the molecular background is known.

PCR approaches to detect toxic or pathogenic Fusarium species, which were not based on toxin biosynthetic genes have been described by Schilling *et al.* (1996). In this work the anonymous marker approach has been followed. With this approach specific primers for *F. culmorum, F. avenaceum* and *F. graminearum* were constructed. Variabilities in the ITS regions were used. Stems of rye that were infected with *F. culmorum* or *F. graminearum* gave positive signals. The same research group developed primers specific for the maize pathogens *F. moniliforme* and *F. subglutinans* (Möller *et al.*, 1999). A method based on SCAR primers for *Fusarium culmorum* and *Fusarium graminearum* have been described by Nicholson *et al.* (1998). They could confirm that colonization of barley and wheat was more extensive with trichothecene-producing isolates compared to non-producing isolates. The functionality of both primer sets could be demonstrated in a multiplex reaction to detect both pathogens at once. The primers could also be used to quantify the pathogens by competitive PCR. The obtained results showed a correlation between disease symptoms and inoculum sizes. PCR was also used to quantify the infection of crops by stem-based pathogenes, including *Microdochium nivale, Rhizoctonia cerealis* and *Fusarium* spp. (Nicholson *et al.*, 2002). Doohan *et al.* (1998) used species-specific PCR to analyse the occurrence of *F. culmorum, F. poae, F. avenaceum, F. graminearum, M. nivale* var. *majus* and *M. nivale* var. *nivale* on different parts of crop plants. They found a good correlation between visual disease assessment and PCR results. A competitive quantification system based on the *tri*5 gene was described by Edwards *et al.* (2001). The system works at least with 6 trichothecene-producing species. They used this system to analyse the influence of fungicides upon the occurrence of trichothecene producers in cereals. According to the results obtained, it was hypothesized that fungicides may alter the occurrence of trichothecene-producing *Fusarium* species. A Real Time PCR system for the quantification of trichothecene producing *Fusarium* species, also based on the *tri*5 gene has been described by Schnerr *et al.* (2001). The method was used to analyse the contamination of various wheat samples for the presence of *Fusarium tri*5 DNA. The Real Time PCR data correlated well with mycological data obtained by conventional methods.

A multiplex PCR system for the differential detection of trichothecene and fumonisin-producing Fusaria has been described by Bluhm *et al.* (2002). Primers for genus-specific detection were designed according to differences in the ITS regions and group-specific primers were generated either from the *tri*6 gene for trichothecene-producing fungi or from the *fum*5 gene for fumonisin-producing fungi. It was possible to use this system for detection and differentiation of *F. graminearum* and *F. verticillioides* in corn meal at a concentration of $>10^5$ CFU/gram. Murillo *et al.* (1998) described a method to detect *F. moniliforme* a pathogen and fumonisin-producing species from maize. To obtain specific primers a gene bank of *F. moniliforme* was probed with maize DNA to identify clones with no cross-hybridization with the host DNA. One of these clones was sequenced and specific primers could be generated according to this sequence. The specificity was proven against a range of other *Fusarium* species.

For most of the relevant trichothecene- and fumonisin-producing species, PCR systems are now available and can be used for research or routine purposes.

Detection of ochratoxin A-producing fungi

Ochratoxin A is a mycotoxin which is produced by *Aspergillus* and *Penicillium* species. Mainly *A. ochraceus, A. carbonarius* and *A. niger* are responsible for the presence of ochratoxin A in products like coffee (Mantle, 1998) or wines (Pietri *et al.*, 2001). *P. verrucosum* is nearly exclusively responsible for ochratoxin A in cereals or cereal products and *P. nordicum* is an ochratoxin A-producing species which mainly grows on fermented foods (Lund and Frisvad, 2003). In contrast to the aflatoxins or trichothecenes, much less is known about the ochratoxin A biosynthetic pathway as source for potential specific target sequences. Recently a publication about the cloning of a part of the ochratoxin A polyketide synthase from *A. ochraceus* has been released (O'Callaghan *et al.*, 2003). However, until now no diagnostic PCR system for ochratoxin A-producing Aspergilli based on this sequence has been published. However, several diagnostic PCR systems based on anonymous markers or gene sequences directed against ochratoxigenic Aspergilli are available. Perrone *et al.* (2004) described a PCR system based upon sequence variabilities within the calmodulin gene. Because of these variabilities they could develop primers specific for *A. carbonarius* or *A. japonicus* and none of the other Aspergilli Section *Nigri* showed cross-reactivity. With the AFLP technique it was possible to isolate a fragment which occurred only in *A. carbonarius* (Schmidt *et al.*, 2004a). From the sequence of this fragment specific primers were made and used to detect *A. carbonarius* in green coffee. Non-infected coffee gave negative PCR results, whereas samples consisting of a mixture of 99% non-infected and 1% infected coffee yielded a positive signal. Only at a concentration of 0.1% infected coffee the primers failed to give a signal. All analysed naturally infected samples proved to be positive for the presence of *A. carbonarius*, whereas non-infected coffee samples gave no signal. Schmidt *et al.* (2003) also developed primers for the second important ochratoxinogenic *Aspergillus* species, *A. ochraceus.* The primers were very specific and used to develop a Real Time PCR system for detection and quantification of *A. ochraceus* in green coffee samples (Schmidt *et al.*, 2004b). Many of them were weakly positive, but some of them gave relatively high Real Time PCR data, indicating a high contamination and a correlation between the PCR data and ochratoxin A in the sample was established. Fungaro *et al.* (2004) developed a method effective for *A. carbonarius* in coffee with a RAPD marker as basis. For *Penicillium nordicum,* part of the gene cluster responsible for ochratoxin A production has been elucidated (Karolewiez and Geisen, 2005). It covers two putative genes, namely the ochratoxin A polyketide synthase (*otapks*PN) and a non-ribosomal peptide synthetase (*otanps*PN). Recently Schmidt-Heydt has identified two additional open reading frames from *P. nordicum* with homology to a chloroperoxidase (*otachl*PN) and a putative ochratoxin A transporter gene (*otatra*PN, personal communication). Figure 7 summarizes the current knowledge about ochratoxin A biosynthesis genes in Penicillia.

Interestingly the polyketide synthase of *P. nordicum* has only limited homology to that of *P. verrucosum.* The gene of the non-ribosomal peptide synthetase in contrast has high homology in both *Penicillium* species. No homology could be found by gene comparison or by hybridization techniques (Karolewiez and Geisen, 2005) with ochratoxinogenic Aspergilli. Based on this situation specific primers for *P. nordicum* or *P. nordicum* and *P. verrucosum* were developed. The system has been used to detect the presence of *P. nordicum* on cured meats (Bogs *et al.*, 2005). About 11% of the isolated Penicillia from this habitat showed a positive reaction with the diagnostic PCR system. All strains with a positive reaction later showed ochratoxin A production. Further, a clear correlation between expression of *otapsksPN* and ochratoxin A production was found (Geisen *et al.*, 2004).

These examples show that the genetic background of the biosynthesis of ochratoxin A is beginning to be unravelled. Despite the incompleteness of the data, several diagnostic PCR systems for all important ochratoxin A-producing species have been described and are available.

Detection of miscellaneous toxic, pathogenic or spoilage moulds by molecular methods

Molecular detection methods for a wide range of food-related fungi are reported. Patulin-producing fungi have been detected with a method based on the iso-epoxy dehydrogenase gene of the patulin metabolic pathway (Paterson *et al.*, 2000).

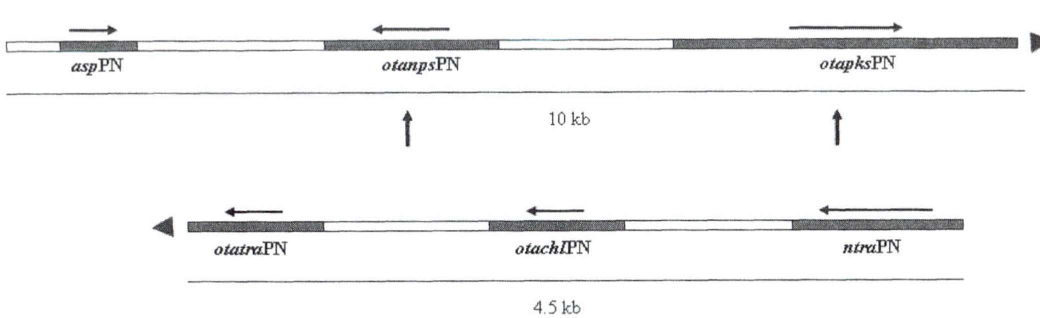

Figure 7. Current knowledge about the ochratoxin biosynthetic genes in *P. nordicum* and *P. verrucosum* (Karolewiez and Geisen, 2005; Geisen *et al.*, 2006). The genes used for diagnostic purposes are indicated by vertical arrows. The newly identified ochratoxin biosynthetic genes *otatra*PN and *otachl*PN have still to be assigned to the cluster. *asp*PN = alkaline serin protease (obviously not involved in ochratoxin A biosynthesis and representing the 5′ end of the cluster), present in various Penicillia, *otanps*PN = non-ribosomal peptide synthetase, only present in *P. nordicum* and *P. verrucosum*, *otapks*PN = ochratoxin A polyketide synthase, only present in *P. nordicum*, *otatra*PN, putative ochratoxin A transport protein, present in *P. nordicum* and *P. verrucosum*; *otachl*PN, putative halogenating enzyme, present in *P. nordicum* and *P. verrucosum*.

P. expansum, a known patulin producer, gave positive results, but cross-reactions with *P. brevicompactum* occurred. In a further analysis Paterson *et al.* (2003) analysed several potential patulin producers and demonstrated the gene in *P. brevicompactum, P. paxilli* and *P. roquefortii.* One *P. brevicompactum* strain was able to produce patulin. Another system able to detect *P. expansum* as the "blue mould rot" of various fruits like apples, cherries, nectarines and peaches was described by Marek *et al.* (2003) with polygalacturonase as a target.

Alternaria sp. is a contaminant of fruits and vegetables and may be able to produce toxic secondary metabolites like alternariol, alternariol methyl ether, altenuene or tenuazonic acid. Zur *et al.* (1999) detected *Alternaria alternata* with a much higher sensitivity with a PCR method based on ITS (intergenic spacer) regions, than with the usually used Howard mould count. The same assay specific for *A. alternata* and *A. solani* detected these fungi in cereals (Zur *et al.*, 2002). None of these samples was positive for the presence of the toxins with after HPLC analysis, but the PCR assay revealed the presence of *A. alternata,* but not *A. solani.* A subsequent mycological analysis of the samples confirmed the PCR results. This result of course raises the question about the correspondence of PCR data based on DNA as target and of mycotoxin production data as discussed in a former paragraph.

Alternaria was detected in carrots by two PCR methods (Pryor and Gilbertson, 2001; Konstantinova *et al.*, 2002). The first system was based on the sequence of a specific RAPD product; the second was directed against ribosomal sequences.

Pedersen *et al.* (1997) detected specific spoilage moulds of bread namely, *P. roqueforti* and *P. carneum* based on the ribosomal gene sequences. One primer pair is group specific for *Penicillium* Subgenus *Penicillium.* The other primer set identified the two species *P. roqueforti* and *P. carneum.* The primer sets were applied to two types of soft cheese, one fermented with *P. camenberti,* the other with *P. roqueforti.* The subgenus specific primers were able to amplify the specific PCR fragment with both cheeses, whereas the specific primer pair produced only a fragment with the *P. roqueforti* fermented cheese.

Claviceps purpurea, the causative agent of the occurrence of ergot alkaloids in cereals was recently detected by molecular methods (Yokoyama *et al.*, 2004).

PCR as taxonomic tool

In addition to diagnostic applications described above, PCR methods can be used to objectively differentiate between morphologically similar or cryptic species. Yoder and Christianson (1998) described primer pairs for differentiation between different *Fusarium* species. They used this molecular taxonomic tool especially to reassign some formerly classified strains of *F. sambucinum* to *F. crookwellense, F. torulosum* or *F. venenatum.* Accensi *et al.* (1999) described a PCR method to differentiate between species of the *A. niger* aggregate. The taxonomical situation of the *A. niger* aggregate is still under debate (Abarca *et al.*, 2004). In the work of Accensi *et al.* (1999) a RFLP (restriction fragment length polymorphism) of a PCR product has been described. The authors amplified the ribosomal region consisting of the ITSI–5.8S–ITSII region and digested the PCR products with the restriction enzyme *Rsa*I. With this method they were able to differentiate between the morphologically similar species *A. tubingensis* and *A. niger.* This system has been used to analyse a set of strains from the *A. niger* aggregate. According to the PCR RFLP, these strains could clearly be classified as *A. niger* or *A. tubingensis.* Some of the analysed strains were able to produce ochratoxin A. All of these producing strains belonged to *A. niger* (Accensi *et al.*, 2001).

F. langsethiae have recently been described as a new species with high morphological similarity to *F. sporotrichioides* (Torp and Nirenberg, 2004). Wilson *et al.* (2004) were able to generate two primer pairs which unambiguously differentiate between both species. High similarity of the *tri5* sequence between *F. sporotrichioides* and *F. langsethia* was reported by Niessen *et al.* (2004). By comparing the variability of an intron sequence within the *tri5* gene, the authors were able to develop species specific primer pairs by combining the conventional *tri5* forward primer (exon targeted) with the variable intron *tri5* primer. With this approach it was possible to develop primer pairs for *F. poae, F. kyushuense, F. sporotrichioides* and *F. langsethiae*. All primer pairs proved to be species specific, except the primer pair for *F. langsethiae* which cross-reacted with *F. sporotrichioides*.

Very recently a differentiating PCR system for the two morphologically related species *P. verrucosum* and *P. nordicum* has been described (Bogs *et al.*, 2005).

These examples show the potential of PCR systems to differentiate between morphologically similar species. These PCR methods are useful tools for the objective assessment of food safety by their ability to characterize mycotoxin-producing fungal species, otherwise hard to characterize.

CONCLUSIONS AND PROSPECTS

According to the current literature for most important toxic and pathogenic food-relevant filamentous fungi, diagnostic PCR methods have been described. Many of them have been proven to be robust and reliable and are ready to be applied at the practical level. Most of them can be used to analyse the food sample directly without additional preenrichment steps. The sensitivity of the molecular approaches is high and in particular cases exceeds that of comparable methods. In fact conventional PCR has become a routine method for the detection of mycotoxinogenic fungi in food samples. As the results can be obtained much faster, molecular methods are very well suited as screening methods to check if particular mycotoxin producing fungi are present in a food sample or as confirmation method to objectively identify a fungal species. The molecular detection methods in this way ideally complement classical methods like the cultural technique or the determination of mycotoxins by analytical approaches.

Besides conventional diagnostic PCR, more sophisticated systems like Real Time PCR or Reverse Transcriptase PCR or even microarrays have been described for food-relevant fungi or are under development. These systems deliver other levels of information depending on whether the target molecule is DNA or RNA.

In fact if RNA instead of DNA is used as a target molecule, the activity of mycotoxin biosynthetic genes in a food matrix can be monitored and the obtained information can be used to predict mycotoxin production. If gene expression analysis by RT Real Time PCR is analysed in a systematic manner, in the future detailed expression profiles (for example in relation to environmental parameter like water activity, temperature, pH for a given product) for mycotoxin biosynthetic fungi can be generated. It is reasonable to assume that these expression profiles will enable the prediction if mycotoxin biosynthesis is possible under a set of given conditions.

A model for the biosynthesis of a given mycotoxin in a given food substrate will enable the food producer to control the process even more than possible today. This approach can be further improved by using microarray technology to monitor the whole pathway, groups of genes or even the whole genome instead of only single key genes. First promising results with a microarray for ochratoxin A biosynthesis have been achieved. Recently Nicolaisen *et al.* (2005) published a microarray for identification and differentiation of trichothecene-producing and non-producing *Fusarium* species.

All these new technologies will lead to the capacity to exactly predict and control the mycological status of foods even more than possible today.

REFERENCES

Abarca, M. L., Accensi, F., Cano, J., and Cabanes, F. J. (2004). Taxonomy and significance of black aspergilli. Antonie van Leeuwenhoek 86:33-49.

Accensi, F., Cano, J., Figuera, L., Abarca, M. L., and Cabanes, F. J. (1999). New PCR method to differentiate species in the *Aspergillus niger* aggregate. FEMS Microbiology Letters 180:191-196.

Accensi, F., Abarca, M. L., Cano, J., Figuera, L., and Cabanes, F. J. (2001). Distribution of ochratoxin A producing strains in the *A. niger* aggregate. Antonie van Leeuwenhoek 79:365-370.

Bateman, G. L., Kwasna, H., and Ward, E. (1996). Relationships among *Fusarium* spp. estimated by comparing restriction fragment length polymorphisms in polymerase chain reaction-amplified nuclear rDNA. Canadian Journal of Microbiology 42:1232-1240.

Beck, J., Ripka, S., Siegner, A., Schiltz, E., Schweizer, E. (1990). The multifunctional 6-methylsalicylic acid synthase gene of *Penicillium patulum*. European Journal of Biochemistry 192:487-498.

Bluhm, B. H., Flaherty, J. E., Cousin, M. A., and Woloshuk, C. P. (2002). Multiplex polymerase chain reaction assay for the differential detection of trichothecene- and fumonisin-producing species of *Fusarium* in corn meal. Journal of Food Protection 67:536-543.

Bogs, C., Battilani, P., and Geisen, R. (2005). Development of a molecular detection and differentiation system for ochratoxin A producing *Penicillium* species and its application to analyse the occurrence of *P. nordicum* in cured meats. International Journal of Food Microbiology 107:39-47.

Böhm, J., Hahn, A., Schubert, R., Bahnweg, G., Adler, N., Nechwatal, J., Oehlmann, R., and Oßwald, W. (1999). Real-time Quantitative PCR: DNA determination in isolated spores of the mycorrhizal fungus *Glomus mosseae* and monitoring of *Phytophthora infestans* and *Phytophthora citricola* in their respective host plants. Journal of Phytopathology 147:409-416.

Brown, D. W., Yu, J., Kelkar, H. S., Fernández, M., Nesbit, T. C., Keller, N. P., Adams, T. H., and Leonhard, T. J. (1996). Twenty-five coregulated transcripts define a sterigmatocystin gene cluster in *Aspergillus nidulans*. Proceedings of the National Academy of Sciences U.S.A. 93:1418-1422.

Brown, M. P., Brown-Jenco, C. S., and Payne, G. A. (1999). Genetic and Molecular Analysis of aflatoxin biosynthesis. Fungal Genetics and Biology 26:81-98.

Brown, D. W., McCormick, S. P., Alexander, N. J., Proctor, R. H., and Desjardins, A. E. (2001). A genetic and biochemical approach to study trichothecene diversity in *Fusarium sporotrichioides* and *Fusarium graminearum*. Fungal Genetics and Biology 32:121-133.

Brown, D. W., Proctor, R. H., Dyer, R. B., and Plattner, R. D. (2003). Characterization of a *Fusarium* 2-gene cluster involved in trichothecene C-8 modification. Journal of Agricultural and Food Chemistry 51:7936-7944.

Chang, P. K., Bhatnagar, D., Cleveland, T. E., and Bennett, J. W. (1995). Sequence variability in homologs of the aflatoxin pathway gene *aflR* distinguishes species in *Aspergillus* section *Flavi*. Applied and Environmental Microbiology 61:40-43.

Chen, R.-S., Tsay, J.-G., Huang, Y.-F., and Chiou, R. Y. Y. (2002). Polymerase chain reaction-mediated characterization of molds belonging to the *Aspergillus flavus* group and detection of *Aspergillus parasiticus* in peanut kernels by a multiplex polymerase chain reaction. Journal of Food Protection 65:840-844.

Criseo, G., Bagnara, A., and Bisignano, G. (2001). Differentiation of aflatoxin-producing and non-producing strains of *Aspergillus flavus* group. Letters in Applied Microbiology 33:291-295.

Cuero, R., Ouellet, T., Yu, J., and Mogongwa, N. (2003). Metal ion enhancement of fungal growth, gene expression and aflatoxin synthesis in *Aspergillus flavus*: RT-PCR characterization. Journal of Applied Microbiology 94:953-961.

Doohan, F. M., Parry, D. W., Jenkinson, P., and Nicholson, P. (1998). The use of species-specific PCR-based assays to analyse *Fusarium* ear blight of wheat. Plant Pathology 47:197-205.

Doohan, F. M., Weston, G., Rezanoor, H. N., Parry, D. W., and Nicholson, P. (1999). Development and use of a reverse transcription-PCR assay to study expression of *Tri5* by *Fusarium* species in vitro and In planta. Applied and Environmental Microbiology 65:3850-3854.

Edel, V., Steinberg, C., Gautheron, N., and Alabouvette, C. (1996). Evaluation of restriction analysis of polymerase chain reaction (PCR)-amplified ribosomal DNA for the identification of *Fusarium* species. Mycological Research 101:179-187.

Edwards, S. G., Pirgozliev, S. R., Hare, M. C., and Jenkinson, P. (2001). Quantification of trichothecene-producing *Fusarium* Species in harvested grain by competitive PCR to determine efficacies of fungicides against Fusarium Head Blight of Winter Wheat. Applied Environmental Microbiology 67:1575-1580.

Edwards, S. G., O'Callaghan, J., and Dobson, A. D. W. (2002). PCR-based detection and quantification of mycotoxigenic fungi. Mycological Research 106:1005-1025.

Ehrlich, K. C., and Cotty, P. J. (2002). Variability in nitrogen regulation of aflatoxin production by *Aspergillus flavus* strains. Applied Microbiology and Biotechnology 60:174-178.

Ehrlich, K. C., Montalbano, B. G., and Cotty, P. J. (2003). Sequence comparison of *afl* R from different *Aspergillus* species provides evidence for variability in regulation of aflatoxin production. Fungal Genetics and Biology 38:63-74.

Färber, P., Geisen, R., and Holzapfel, W. H. (1997). Detection of aflatoxinogenic fungi in figs by a PCR reaction. International Journal of Food Microbiology 36:215-220.

Feng, G. H., and Leonhard, T. J. (1998). Culture conditions control expression of the genes for aflatoxin and sterigmatocystin biosynthesis in *Aspergillus parasiticus* and *A. nidulans*. Applied and Environmental Microbiology 64:2275-2277.

Frisvad, J. C., Smedsgaard, J., Larsen, T. O., and Samson, R. A. (2004). Mycotoxins, drugs and other extrolites produced by species in *Penicillium* subgenus *Penicillium* pp. 201-242.

Fungaro, M. H. P., Vissotto, P. C., Sartori, D., Vilas-Boas, L. A., Furlaneto, M. C., and Taniwaki, M. H. (2004). A molecular method for detection of *Aspergillus carbonarius* in coffee beans. Curr Microbiology 49:123-127.

Geisen, R. (1996). Multiplex polymerase chain reaction for the detection of potential aflatoxin and sterigmatocystin producing fungi. Systematic and Applied Microbiology 19:388-392.

Geisen, R., Mayer, Z., Karolewiez, A., and Färber, P. (2004). Development of a Real Time PCR system for detection of *Penicillium nordicum* and for monitoring ochratoxin A production in foods by targeting the ochratoxin polyketide synthase gene. Systematic and Applied Microbiology 27:501-507.

Geisen, R. (2004). Molecular monitoring of environmental conditions influencing the induction of ochratoxin A biosynthetic genes in *Penicillium nordicum*. Molecular Nutrition and Food Research 48:532-540.

Geisen, R., Schmidt-Heydt, M. and Karolewiez, A. (2006). A gene cluster of ochratoxin A biosynthetic genes in *Penicillium*. Mycotoxin Research, 22:134-141.

Griffin, D. H. (1994). Fungal Physiology. Wiley-Liss., New York.

Grimm, C., and Geisen, R. (1998). A PCR-ELISA for the detection of potential fumonisin producing *Fusarium* species. Letters in Applied Microbiology 26, 456-462.

Häggblom, P. (1982). Production of ochratoxin A in barley by *Aspergillus ochraceus* and *Penicillium viridicatum*: effect of fungal growth, time, temperature and inoculum size. Applied and Environmental Microbiology 43:1205-1207.

Haugland, R. A., Vesper, S. J., and Wymer, L. J. (1999). Quantitative measurement of *Stachybotrys chartarum* conidia using real time detection of PCR products with the TaqMan TM fluorogenic probe system. Molecular and Cellular Probes 13: 329-340.

Hohn, T. M., and Beremand, M. N. (1989). Regulation of trichodiene synthase in *Fusarium sporotrichoides* and *Gibberella pulicaris (Fusarium sambucinum)*. Applied and Environmental Microbiology 55:1500-1503.

Hoorfar, J., Malorny, N., Wagner, M., De Medici, D., Abdulmawjood, A., and Fach, P. (2004). Diagnostic PCR: making internal amplification control mandatory. Journal of Applied Microbiology 96:221-222.

Ichinoue, M., Kurata, H., Sugiura, Y., and Ueno, Y. (1983). Chemotaxonomy of *Gibberella zeae* with special reference to production of trichothecenes and zearalenone. Applied and Environmental Microbiology 46:1364-1369.

Itoh, Y., Peterson, S. W., Wicklow, D. T., and Goto, T. (2001). *Aspergillus pseudotamarii*, a new aflatoxin producing species in *Aspergillus* section *Flavi*. Mycological Research 105:233-239.

Jennings, P., Coates, M. E., Walsh, K., Turner, J. A., and Nicholson, P. (2004). Determination of deoxynivalenol- and nivalenol-producing chemotypes of *Fusarium graminearum* isolated from wheat crops in England and Wales. Plant Pathology 53:643-652.

Karolewiez, A., and Geisen, R. (2005). Cloning a part of the ochratoxin A biosynthetic gene cluster of *Penicillium nordicum* and characterization of the ochratoxin polyketide synthase gene. Systematic and Applied Microbiology 28:588-595.

Keller, N. P., Nesbitt, C., Sarr, B., Phillips, T. D., and Burow, G. B. (1997). pH regulation of sterigmatocystin and aflatoxin biosynthesis in *Aspergillus* spp. Phytopathology 87:643-648.

Klich, M., Yu, J., Chang, P. K., Mullaney, E. J., Bhatnagar, D., and Cleveland, T. E. (1995). Hybridization of genes involved in aflatoxin biosynthesis to DNA of aflatoxinogenic and non-aflatoxinogenic Aspergilli. Applied and Microbiological Biotechnology 44:439-443.

Klich, M., Montalbano, B., and Ehrlich, K. (1997). Northern analysis of aflatoxin biosynthesis genes in *Aspergillus parasiticus* and *Aspergillus sojae*. Applied Microbiology and Biotechnology 47:246-249.

Klich, M., Mullaney, E. J., Daly, C. B., and Cary, J. W. (2000). Molecular and physiological aspects of aflatoxin and sterigmatocystin biosynthesis by *Aspergillus tamarii* and *A. ochraceoroseus*. Applied Microbiology and Biotechnology 53:605-609.

Knoll, S., Mulfinger, S., Niessen, L., and Vogel, R. F. (2002). Rapid preparation of *Fusarium* DNA from cereals for diagnostic PCR using sonification and an extraction kit. Plant Pathology 51:728-734.

Konstantinova, P., Bonants, P. J. M., van Gent-Pelzer, M. P. E., van der Zouwen, P., and van den Bulk,

R. (2002). Development of specific primers for detection and identification of *Alternaria* spp. in carrot material by PCR and comparison with blotter and plating assays. Mycological Research 106:23-33.

Kulik, T., Fordonski, G., Pszczólkowska, A., Plodzien, K., and Lapinski, M. (2004). Development of PCR assay based on ITS2 rDNA polymorphism for the detection and differentation of *Fusarium sporotrichioides*. FEMS Microbiology Letters 239:181-186.

Kusumoto, K. I., Yabe, K., Nogata, Y., and Ohta, H. (1998). *Aspergillus oryzea* with and without a homolog of aflatoxin biosynthetic gene *ver-1*. Applied Microbiology and Biotechnology 50:98-104.

Lantz, P. G., Tjernfeld, F., Borch, E., Hahn-Hägerdal, B., and Radström, P. (1994a). Enhanced sensitivity in PCR detection of *Listeria monocytogenes* in soft cheese through use of an aquous two-phase system as a sample preparation method. Applied and Environmental Microbiology 60:3416-3418.

Lantz, P. G., Hahn-Hägerdal, B., and Radström, P. (1994b). Sample preparation methods in PCR-based detection of food pathogens. Trends in Food Science and Technology 5:384-389.

Lee, H. B., and Magan, N. (2000). Impact of environment and interspecific interactions between spoilage fungi and *Aspergillus ochraceus* on growth and ochratoxin production in maize grain. International Journal of Food Microbiology 61:11-16.

Lee, T., Oh, D. W., Kim, H. S., Lee, J., Kim, Y. H., Yun, S. H., and Lee, Y. W. (2001). Identification of Deoxynivalenol- and Nivalenol-Producing Chemotypes of *Gibberella zeae* by Using PCR. Applied and Environmental Microbiology 67:2966-2972.

Lund, F., and Frisvad, J. C. (2003). *Penicillium verrucosum* in wheat and barley indicates presence of ochratoxin A. Journal of Applied Microbiology 95:1117-1123.

Maher, N., Dillon, H. K., Vermund, S. H., and Unnasch, T. R. (2001). Magnetic bead capture eliminates PCR inhibitors in samples collected from the airborne environment, permitting detection of *Pneumocystis carinii* DNA. Applied and Environmental Microbiology 67:449-452.

Mantle, P. G. (1998). Ochratoxin A in coffee. Journal of Food Mycology 2:63-65.

Marek, P., Annamalai, T., and Venkitanarayanan, K. (2003). Detection of *Penicillium expansum* by polymerase chain reaction. International Journal of Food Microbiology 89:139-144.

Mauchline, T. H., Kerry, B. R., and Hirsch, P. R. (2002). Quantification in soil and the rhizosphere of the nematophagous Fungus *Verticillium chlamydosporium* by competitive PCR and comparison with selective plating. Applied and Environmental Microbiology 68:1846-1853.

Mayer, Z., Bagnara, A., Färber, P., and Geisen, R. (2003a). Quantification of the copy number of *nor-1*, a gene of the aflaxtoxin biosynthetic pathway by real-time PCR, and its correlation to the cfu of *Aspergillus flavus* in foods. International Journal Food Microbiology 82:143-151.

Mayer, Z., Färber, P., and Geisen, R. (2003b). Monitoring the production of aflatoxin B_1 in wheat by measuring the concentration of *nor-1* mRNA. Applied and Environmental Microbiology 69:1154-1158.

Mirete, S., Patino, B., Vazquez, C., Jiménez, M., Hinojo, M. J., Soldevilla, C., and González-Jaén, M. T. (2003). Fumonisin production by *Gibberella fujikuroi* strains from *Pinus* species. International Journal of Food Microbiology 89:213-221.

Mitchell, D., Parra, R., Aldred, D., and Magan, N. (2004). Water and temperature relations of growth and ochratoxin A production by *Aspergillus carbonarius* strains from grapes in Europe and Israel. Journal of Applied Microbiology 97:445.

Möller, E. M., Chelkowski, J., and Geiger, H. H. (1999). Species-specific PCR assays for the fungal pathogenes *Fusarium moniliforme* and *Fusarium subglutinans* and their application to diagnose maize ear rot disease. Journal of Phytopathology 147:497-508.

Murillo, I., Cavallarin, L., and San Segundo, B. (1998). The development of a rapid PCR assay for detection of *Fusarium moniliforme*. European Journal of Plant Pathology 104:301-311.

Nicolaisen, M., Justesen, A., Thrane, U., Skouboe, P., and Holmström, K. (2005). An oligonucleotide microarray for the identification and differentiation of trichothecene-producing and non-producing *Fusarium* species occurring in cereal grain. Journal of Microbiological Methods 62:57-69.

Nicholson, P., Simpson, D. R., Weston, G., Rezanoor, H. N., Lees, A. K., Parry, D. W., and Joyce, D. (1998). Detection and quantification of *Fusarium culmorum* and *Fusarium graminearum* in cereals using PCR assays. Physiological and Molecular Plant Pathology 53:17-37.

Nicholson, P., Turner, A. S., Edwards, S. G., Bateman, G. L., Morgan, L. W., Parry, D. W., Marshall, J., and Nuttall, M. (2002). Development of stem-base pathogens on different cultivars of winter wheat determined by quantitative PCR. European Journal of Plant Pathology 108:163-177.

Niessen, M. L. and Vogel, R. F. (1997). Specific identification of *Fusarium graminearum* by PCR with *gaoA* targeted primers. Systematic and Applied Microbiology 20:111-113.

Niessen, M. L., and Vogel, R. F. (1998). Group Specific PCR-Detection of potential trichothecene-producing Fusarium-species in pure cultures and cereal samples. Systematic and Applied Microbiology 21:618-631.

Niessen, L., Schmidt, H., and Vogel, R. F. (2004). The use of *tri5* gene sequences for PCR detection and taxonomy of trichothecene-producing species in the *Fusarium* section *Sporotrichiella*. International Journal of Food Microbiology 95:305-319.

O'Callaghan, J., Caddick, M. X., and Dobson, A. D. W. (2003). A polyketide synthase gene required for ochratoxin A biosynthesis in *Aspergillus ochraceus*. Microbiology 149:3485-3491.

O'Donnell, K. (1992). Ribosomal DNA internal transcribed spacers are highly divergent in the phytopathogenic ascomycete *Fusarium sambucinum (Gibberella pulicaris)*. Current Genetics 22:213-220.

Paterson, R. R. M., Archer, S., Kozakiewicz, Z., Lea, A., Locke, T., and O'Grady, E. (2000). A gene probe for the patulin metabolic pathway with potential for use in patulin and novel disease control. Biocontrol Science and Technology 10:509-512.

Paterson, R. R. M., Kozakiewicz, Z., Locke, T., Brayford, D., and Jones, S. C. B. (2003). Novel use of the isoepoxydon dehydrogenase gene probe of the patulin metabolic pathway and chromatography to test penicillia isolated from apple production systems for the potential to contaminate apple juice with patulin. Food Microbiology 20:359-364.

Perrone, G., Susca, G., and Mule, G. (2004). PCR assay for identification of *Aspergillus carbonarius* and *Aspergillus japonicus*. European Journal of Plant Pathology 110:641-649.

Pedersen, H. L., Skouboe, P., Boysen, M., Soule, J., and Rossen, L. (1997). Detection of *Penicillium* species in complex food samples using the polymerase chain reaction. International Journal of Food Microbiology 35:169-177.

Peterson, S. W., Itoh, Y., Horn, B. W., and Goto, T. (2001). *Aspergillus bombycis*, a new aflatoxigenic species and genetic variation in its sibling species, *A. nomius*. Mycologia 93:689-703.

Pietri, A., Bertuzzi, T., Pallaroni, L., and Piva, G. (2001). Occurence of ochratoxin A in Italian wines. Food Additives and Contaminants 18:647-654.

Pohland, A. E. (1993). Mycotoxins in review. Food Additives and Contaminants 10:17-28.

Proctor, R. H., and Hohn, T. M. (1993). Aristolochen Synthase. Isolation, characterization, and bacterial expression of a sesquiterpenoid biosynthetic gene *(ARI1)* from *Penicillium roqueforti*. Journal of Biological Chemistry 268:4543-4548.

Proctor, R. H., Brown, D. W., Plattner, R. D., and Desjardins, A. E. (2003). Co-expression of 15 contiguous genes delineates a fumonisin biosynthetic gene cluster in *Gibberella moniliformis*. Fungal Genetics and Biology 38:237-249.

Pryor, B. M., and Gilbertson, R. L. (2001). A PCR-based Assay for Detection of *Alternaria radicina* on Carrot Seed. Plant Disease 85:18-23.

Rossen, L., Norskov, P., Holmstrom, K., and Rasmussen, O. F. (1992). Inhibition of PCR by components of food samples microbial diagnostic assays and DNA-extraction solutions. International Journal of Food Microbiology 17:37-45.

Samson, R. A., Hoekstra, E. H., and Frisvad, J. C. (Eds.). (2004). Introduction to food- and airborne fungi. Centraalbureau voor Schimmelcultures, Utrecht, The Netherlands.

Scherm, B., Palomba, M., Serra, D., Marcello, A., and Migheli, Q. (2005). Detection of transcripts of the aflatoxin genes *afl*D, *afl*O, and *afl*P by reverse-transcription-polymerase chain reaction allows differentiation of aflatoxin-producing and non-producing isolates of *Aspergillus flavus* and *Aspergillus parasiticus*. International Journal of Food Microbiology 98:201-210.

Schilling, A. G., Möller, E. M., and Geiger, H. H. (1996). Polymerase chain reaction-based assays for species-specific detection of *Fusarium culmorum, F. graminearum* and *F. avenaceum*. Phytopathology 86:515-522.

Schmidt, H., Ehrmann, M., Vogel, R. F., Taniwaki, M. H., and Niessen, L. (2003). Molecular typing of *Aspergillus ochraceus* and construction of species specific SCAR-primers based on AFLP. Systematic and Applied Microbiology 26:138-146.

Schmidt, H., Taniwaki, M. H., and Niessen, L. (2004a). Utilization of AFLP markers for PCR-based identification of *Aspergillus carbonarius* and indication of its presence in green coffee samples. Journal of Applied Microbiology 97:899-909.

Schmidt, H., Bannier, M., Vogel, R. F., and Niessen, L. (2004b). Detection and quantification of *Aspergillus ochraceus* in green coffee by PCR. Letters in Applied Microbiology 38:464-469.

Schnerr, H., Niessen, M. L., and Vogel, R. F. (2001). Real Time detection of the *tri5* gene in *Fusarium* species by LightCycler™-PCR using SYBR Green I for continuous fluorescence monitoring. International Journal of Food Microbiology 71:53-61.

Schnerr, H., Vogel, R. F., and Niessen, L. (2002). Correlation between DNA of trichothecene-producing *Fusarium species* and deoxynivalenol concentrations in wheat-samples. Letters in Applied Microbiology 35:121-125.

Shapira, R., Paster, N., Eyal, O., Menasherov, M., Mett, A., and Salomon, R. (1996). Detection of aflatoxinogenic molds in grains by PCR. Applied and Environmental Microbiology 62:3270-3273.

Skrinjar, M., and Dimic, G. (1992). Ochratoxigenicity of *Aspergillus ochraceus* group and *Penicillium verrucosum* var. *cyclopium* strains on various media. Acta Microbiologica Hungarica 39:257-261.

Smith, J. E., Lewis, C. W., Anderson, J. G., and Solomons, G. L. (1994). Mycotoxins in Human Nutrition and Health. Publication EUR 16048 EN, DGXII.

Sweeney, M. J., Pamies, P., and Dobson, A. D. W. (2000). The use of reverse transcription-polymerase chain reaction (RT-PCR) for monitoring aflatoxin production in *Aspergillus parasiticus* 439. International Journal of Food Microbiology 56:97-103.

Torp, M., and Nirenberg, H. I. (2004). *Fusarium langsethia* sp. no. on cereals in Europe. International Journal of Food Microbiology 95:247-256.

Varga, J., Rigó, K., Kocsube, S., Farkas, B., and Pal, K. (2003). Diversity of polyketide synthase gene sequences in Aspergillus species. Research in Microbiology 154:593-600.

Watson, A. J., Archer, D. B., Seal, S., and Linz, J. (1995). Homologues of aflatoxin biosynthetic genes in *Aspergillus oryzea*. FGN 42A, 82.

Whitaker, T. B. (2000). Sampling Techniques. *In* Mycotoxin Protocols (Trucksess, M. W., and Pohland, A. E., eds), Humana Press, Ottawa, Canada, pp. 11-14,

Wilson, A., Simpson, D., Chandler, E., Jennings, P., and Nicholson, P. (2004). Development of PCR assays for the detection and differentiation of *Fusarium sporotrichioides* and *Fusarium langsethiae*. FEMS Microbiology Letters 233:69-76.

Xu, H., Annis, S., Linz, J., and Trail, F. (2000). Infection and colonization of peanut pods by *Aspergillus parasiticus* and the expression of the aflatoxin biosynthetic gene, *nor-1*, in infection hyphae. Physiological and Molecular Plant Pathology 56:185-196.

Yang, Z. Y., Shim, W. B., Kim, J. H., Park, S. J., Kang, S. J., Nam, B. S., and Cigan, A. L. (2004). Detection of aflatoxin-producing molds in Korean fermented foods and grains by multiplex PCR. Journal of Food Protection 67:2622-2626.

Yoder, W. T. and Christianson, L. M. (1998). Species-specific primers resolve members of Fusarium Section Fusarium. Genetics and Biology 23:68-80.

Yokoyama, E., Yamagishi, K., and Hara, A. (2004). Development of a PCR-based mating-type assay for *Clavicipitaceae*. FEMS Microbiology Letters 237:205-212.

Yu, J., Chang, P. K., Cary, J. W., Wright, M., Bhatnagar, D., Cleveland, T. E., Payne, G. A., and Linz, J. (1995). Comparative mapping of aflatoxin pathway gene clusters in *Aspergillus parasiticus* and *Aspergillus flavus*. Applied and Environmental Microbiology 61:2365-2371.

Yu, J. H., Butchko, R. A. E., Fernández, M., Keller, N. P., Leonhard, T. J., and Adams, T. H. (1996). Conservation of structure and function of the aflatoxin regulatory gene *aflR* from *Aspergillus nidulans* and *A. flavus*. Current Genetics 29:549-555.

Yu, J., Bhatnagar, D., and Cleveland, T. E. (2004). Completed sequence of aflatoxin pathway gene cluster in *Aspergillus parasiticus*. FEBS Letters 564:126-130.

Zachová, I., Vytrasová, J., Pejchalová, M., Cervenka, L., and Tavcar-Kalcher, G. (2003). Detection of aflatoxinogenic fungi in feeds using the PCR method. Folia Microbiologica 48:817-821.

Zur, G., Hallerman, E. M., Sharf, R., and Kashi, Y. (1999). Development of a polymerase chain reaction-based assay for the detection of *Alternaria* fungal contamination in food products. Journal of Food Protection 62:1191-1197.

Zur, G., Shimoni, E., Hallerman, E. M., and Kashi, Y. (2002). Detection of *Alternaria* fungal contamination in cereal grains by a polymerase chain reaction-based assay. Journal of Food Protection 65:1433-1440.

Chapter 14

Fungal volatiles: Biomarkers of good and bad food quality

Kristian Karlshøj, Per Væggemose Nielsen, and Thomas Ostenfeld Larsen
Center for Microbial Biotechnology, BioCentrum-DTU, Technical University of Denmark, Kgs. Lyngby, Denmark.

INTRODUCTION

Most people have experienced the smell of fungal spoiled foods such as bread. The typical reaction is "this really smells bad" and most people will associate this with a mouldy smell. It has been shown that the odour thresholds of some off-flavour-related fungal volatile compounds are very low. The odour threshold for the earthy smelling terpene alcohol geosmin (Figure 1) in water is between 0.0082-0.018 ppb and the musty smelling 2-methyl-isoborneol (Figure 1) has an odour threshold of 0.1 ppb in water (Medsker *et al.*, 1969).

Traditionally, specific fungal species have been used as starter cultures in certain fermented foods such as blue and white mould cheeses (*Penicillium roqueforti*; and *Penicillium camemberti* and *Geotrichum candidum*, respectively) and soy sauce (*Aspergillus oryzae* and *Aspergillus sojae*). It is also well known that various mycotoxin-producing fungi occur as contaminants in foods and feed stuffs, which is a considerable problem in terms of food and feed quality and safety and hence economically, as in 1985 FAO (the Food and Agriculture Organization) estimated that 25% of world crops (Pitt and Hocking, 1985) and as much as 50% of crops in developing countries, are contaminated with mycotoxins (Waller and Brayford, 1990; Pohland, 1993). It is therefore important that it is ensured that starter cultures are pure.

Fungal detection is not only of importance in terms of food safety. For centuries, truffles, a most valuable fungal commodity, have been found by use of pigs, which can smell the truffles odorous compounds very well and find them covered under soil. In the field of medicine, it is also important to be able to detect fungal infections, such as aspergillosis, as soon as possible.

Food quality has, traditionally, often been assessed by sensory panel evaluation, for instance in quality control of cereals (Börjesson *et al.*, 1996). Sensory panel analysis is a very laborious process as it requires a panel of sensory judges which is very expensive and time consuming to train to a proficient level. Even a well-trained sensory judge will give an at least partially subjective score in sensory panel analysis. There are further restrictions in using sensory panel analysis, as potentially toxic samples cannot be analyzed in this manner.

Of the traditional analytical methods, GC-MS analysis is time consuming, but somewhat less expensive than sensory panel analysis or HPLC-DAD and LC-MS analysis. Data analysis

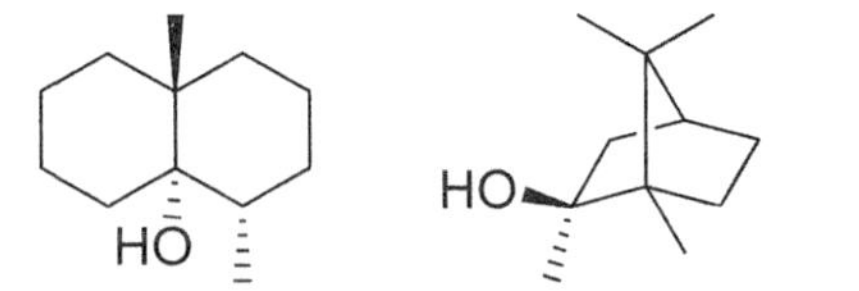

Geosmin 2-Methylisoborneol

Figure 1. Chemical structure of the mouldy smelling volatile fungal biomarkers geosmin and 2-methylisoborneol, two volatile compounds with extremely low odour-threshold values.

with all these techniques is complicated and requires skill and experience. HPLC-DAD and LC-MS analysis are expensive, both in terms of equipment, running costs and time consumption (Guernion *et al.*, 2001).

In recent years, however, the electronic nose (e-nose) has been tested successfully for quality control of foods and feed stuffs. It has so far not been thoroughly tested whether it is possible to link e-nose analysis with mycotoxin content/production, or to establish whether it is possible to predict mycotoxin production during production or storage. This is desirable as e-nose measurements are faster, cheaper and easily automated in industrial processes (Sim *et al.*, 2003). Potentially, e-nose prediction models can be constructed, and with these data analysis could principally be performed by unskilled personnel.

Following is a brief discussion of fungal biochemistry, with emphasis on volatile metabolites and mycotoxins as well as their application as biomarkers, the traditional chemical analysis methods for mycotoxins (HPLC and LC-MS) as well as for volatile organic compound (VOC) analysis (GC-MS) and finally, an in-depth description of an electronic nose system and its potential applications for mycotoxin prediction. A couple of cases will be shown for illustrative purposes.

FUNGAL BIOCHEMISTRY

Fungal presence can be detected in many ways. One such way concerns the production of the sterol ergosterol, a plasma membrane component unique for fungi, and production of cell walls containing chitin. Fungi are also well known to produce a broad variety of extracellular enzymes, which are utilized in degradation of nutrient macromolecules. Some metabolites, like ergosterol, are produced by almost all fungi, but most known secondary metabolites, such as mycotoxins and volatile terpenes, have been shown to be more restricted in their distribution, for instance being produced only by 1 to 15 species within genus *Penicillium* (Larsen and Frisvad, 1995a).

VOLATILE METABOLITES

Among the volatile metabolites produced by fungi are alcohols, aliphatic C8 compounds, alkanes, alkenes, esters, ketones, lactones, pyrazines and terpenes, an overview of the volatile metabolite pathways is shown in Figure 2.

The alcohols produced can be put into three categories according to their synthesis pathway. The first group comprises of primary alcohols which are produced in two reductive steps from fatty acid-CoA esters (Luckner, 1990). The second group, the fusel alcohols, is a product of the Ehrlich pathway. In this pathway amino acids (such as leucine, isoleucine and valine) are deaminated and the resulting β-keto acid decarboxylated, the resulting aldehyde is reduced to the alcohol by the alcohol dehydrogenase enzyme (Figure 3) (Suomalainen, 1971; Berry, 1988).

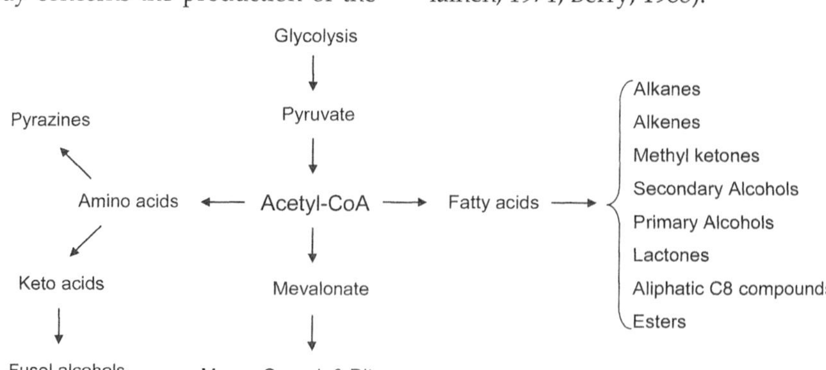

Figure 2. Overview of the biosynthesis of important fungal volatile metabolites (adapted from Börjesson, 1993; Larsen, 1994; Jelén and Wasowicz, 1998).

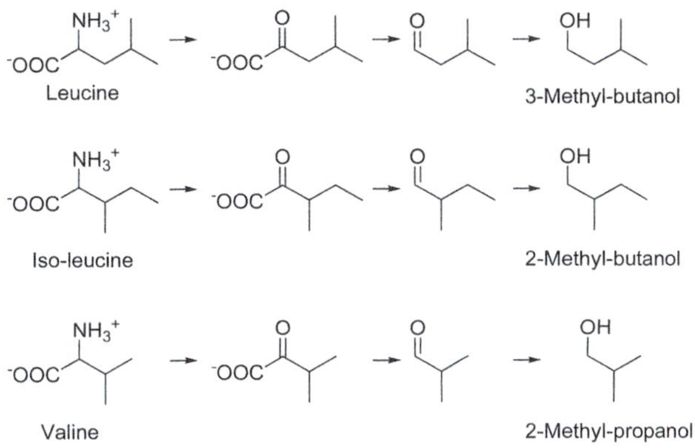

Figure 3. Fusel alcohol pathways. The amino acid is deaminated, then decarboxylated and the aldehyde reduced to the resulting alcohol (Suomalainen, 1971; Gurney, 1997).

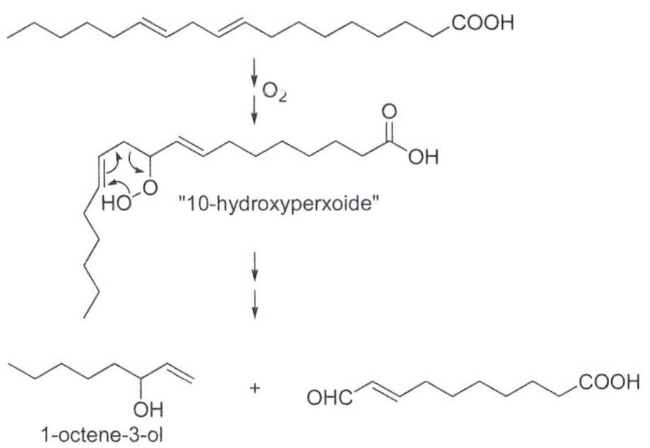

Figure 4. 1-Octen-3-ol synthesis. Linoleic acid is oxidized into a 10-hydroxyperoxide which in turn is cleaved into 1-octen-3-ol and a ten-carbon fragment (Wurzenberger and Grosch, 1982; Wurzenberger and Grosch, 1984).

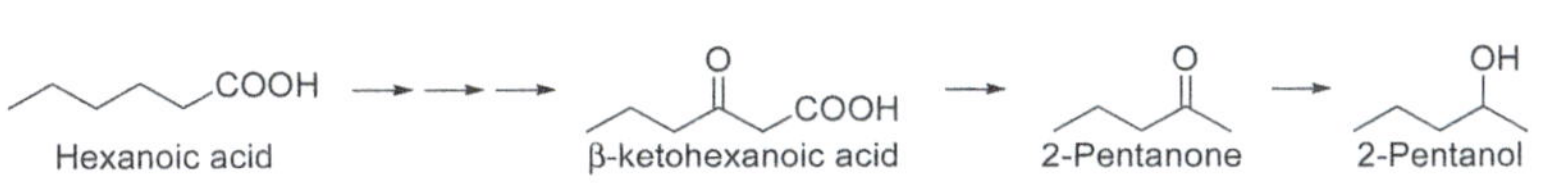

Figure 5. Methyl ketone synthesis example. The fatty acid is β-oxidatized by the usual pathway in the fatty acid metabolism. The β-keto acid is then decarboxylated to form the methyl ketone. The methyl ketone can be further reduced to the resulting secondary alcohol (Luckner, 1990).

Secondary alcohols, which comprise the third group, are formed by reduction of methyl ketones (see below and Figure 5) (Hawke, 1966; Kinsella and Hwang, 1976; Kinderlerer, 1989).

Aliphatic C8 compounds are produced by lipoxygenation; for instance 1-octen-3-ol is produced by oxidation of linoleic acid into a 10-hydroxyperoxide which is then cleaved into 1-octen-3-ol and a ten-carbon fragment (Figure 4) (Wurzenberger and Grosch, 1982; Wuzenberger and Grosch, 1984).

Synthesis of alkanes and alkenes is done by decarboxylation of the corresponding fatty acids. The likely pathway for this is an α-oxidation with a β-keto acid intermediate. Further unsaturation of the alkene can be achieved by hydroxylation and dehydrogenation of the alkene (Luckner, 1990).

The acid moiety of esters produced by fungi can be formed by three possible pathways: by activation of monocarboxylic acids; from an intermediate of the long chain monocarboxylic acid synthesis or from oxidative decarboxylation of β-keto acids (Kempler, 1983). Methyl ketones are synthesized during fatty acid catabolism. The β-oxidation pathway is followed until β-ketoacyl-CoA has been formed, β-ketoacyl-CoA is then either both deacylated and decarboxylated to form a methyl ketone one carbon shorter than the fatty acid or the fatty acid is further β-oxidized (Figure 5) (Hawke, 1966; Kinsella and Hwang, 1976; Kinderlerer, 1989). Lactones are formed from γ-keto and δ-keto acids synthesized from fatty acids (Kempler, 1983). Synthesis of tetramethylpyrazines has been suggested to be a condensation reaction between acetoin and ammonia. Production of 2-methoxy-3-isopropylpyrazine has been proposed to be produced from valine and ethanedione whereas 2-methoxy-3-isopropyl-5-methyl pyrazine is suggested to be formed from valine and pyruvaldehyde (Kempler, 1983; Leete *et al.*, 1991).

Among the most diverse volatile metabolites produced by fungi are terpenes and terpene alcohols. Terpenes are comprised by with isoprene units. The synthesis of terpenes starts with acetoacetyl-CoA formation from two units of acetyl-CoA. Acetoacetyl-CoA and acetyl-CoA are synthesized into β-hydroxy-β-methyl glutaryl-CoA, which is converted into mevalonate by reduction of the carbonyl group into a primary alcohol. Mevalonate is decarboxylated yielding isopentenyl pyrophosphate. Isopentenyl pyrophosphate is the polymerized into the following terpene precursors: geranyl phosphate (monoterpene precursor), farnesyl pyrophosphate (sesquiterpenes precursor) and geranylgeranyl phosphate (diterpene precursor) (Figure 6) (Lynen, 1959; Richards and Hendrickson, 1964; Herbert, 1989; Luckner, 1990). Other volatiles produced by fungi are compounds such as dimethyldisulphide (Figure 7, (Demarigny *et al.*, 2000)) and other sulphur containing volatiles, which are produced by degradation of methionine.

Knowledge of the nutrient content of a food product is therefore of importance when predicting which type of volatiles will be relevant for determination of food quality; e.g., for fat rich foods, monitoring for alcohols, alkanes, alkenes and methyl ketones will be relevant for detection of possible spoilage.

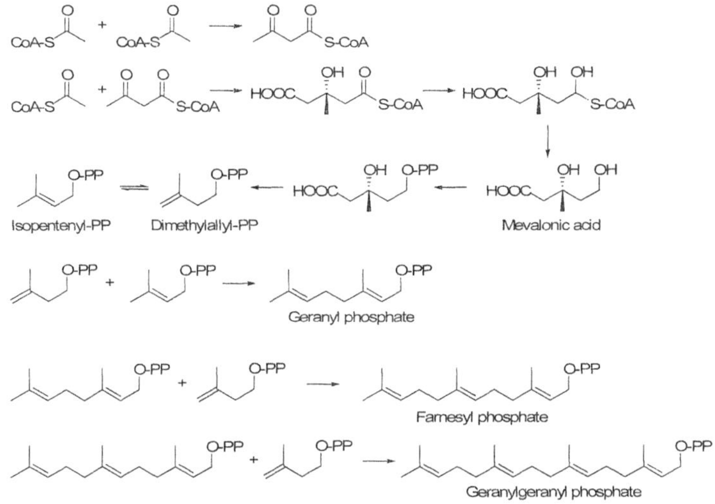

Figure 6. Synthesis pathway of terpene precursors. Acetoacetyl-CoA is formed from two units of acetyl-CoA. β-hydroxy-β-methyl glutaryl-CoA is synthesized from acetoacetyl-CoA and acetyl-CoA and then converted into mevalonate by reduction of the carbonyl group into a primary alcohol. Mevalonate is decarboxylated yielding isopentenyl pyrophosphate which is then polymerized into the following terpene precursors: geranyl phosphate (monoterpene precursor), farnesyl pyrophosphate (sesquiterpenes precursor) and geranylgeranyl phosphate (diterpene precursor) (Herbert, 1989).

Table 1. Most commonly encountered associated spoilage funga of *Aspergillus* and *Penicillium* for selected foods (adapted from Filtenborg *et al.*, 1996; Samson *et al.*, 2002).

Foods	Fungal Species
Bread, rye	*P. roqueforti, P. paneum*
Cheese	*P. commune, P. nalgiovense, P. atramentosum, P. nordicum, A. versicolor*
Fruits; pomaceous and stone	*P. expansum, P. crustosum, P. solitum*
Grain, stored	*P. cyclopium, P. freii, P. hordei, P. melanoconidium, P. polonicum, P. verrucosum, P. aurantiogrise-um, A. flavus, A. niger, A. candidus*
Nuts	*P. commune, P. crustosum, P. discolor, A. flavus*
Fermented sausages	*P. nalgiovense, P. olsonii, P. chrysogenum, P. nordicum*

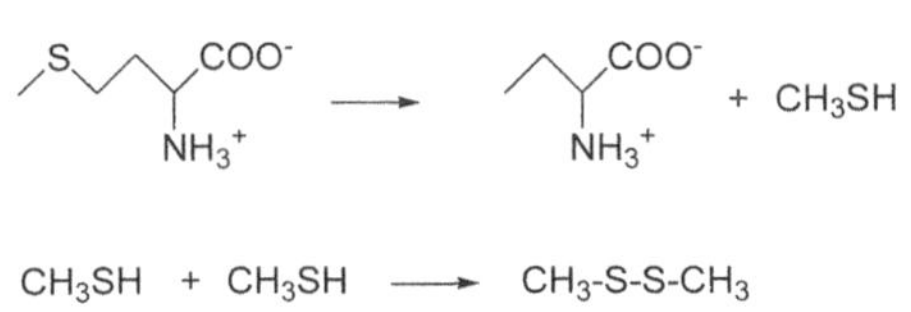

Figure 7. Dimethyldisulphide biosynthesis. Methanethiol is synthesized by via γ-demethiolase by degradation of metheonine; dimethyldisulphide is then produced by polymerization of two units of methanethiol (Demarigny *et al.*, 2000).

NON-VOLATILE SECONDARY METABOLITES

The number of non-volatile secondary metabolites produced by fungi is extremely high (Samson *et al.*, 2002). It has been shown that each fungal species has specific affinity for food and feed habitats thus leading to specific species occupying the different ecologic niches. The fungi most often seen in a given habitat are referred to as that habitat's *associated funga* (Filtenborg *et al.*, 1996).

Species from genera *Aspergillus*, *Penicillium* and *Fusarium* are among the most frequent food contaminants. Typical contaminants, with typical habitat and key mycotoxins produced shown in brackets, are *Aspergillus flavus* (nuts and cereals; aflatoxin and cyclopaldic acid), *Aspergillus terreus* (silage; patulin and citrinin), *Penicillium carneum* (silage; patulin and roquefortin C), *Penicillium commune* (cheese; cyclopaldic acid), *Penicillium expansum* (pomaceous fruit; patulin, roquefortine C and citrinin), *Penicillium nordicum* (meat products and cheese; ochratoxin A), *Penicillium paneum* (silage; patulin and roquefortin C), *Penicillium verrucosum* (cereal and cheese; ochratoxin A and citrinin), *Fusarium graminearum* (cereals; trichothecenes), *Fusarium poae* (cereals; trichothecenes), *Fusarium sambucinum* (cereals and potatoes; trichothecenes) and *Fusarium sporotrichioides* (cereals; trichothecenes) (Samson *et al.*, 2002), Table 1 grants an overview of the associated funga of genera *Aspergillus* and *Penicillium* for selected food stuffs. Detection of fungal spoilage is therefore of utmost importance.

Like the volatile metabolites, fungal secondary metabolites derive from pathways linked to key compounds formed in primary metabolism (Herbert, 1989). The plasma membrane sterol ergosterol is produced via the terpene pathway, and thus originates from acetyl-CoA converted into mevalonate (Herbert, 1989).

Many mycotoxins, such as the aflatoxins, citrinin, ochratoxin A, patulin, penicillic acid, are polyketides, and thus originate from the polyketide pathway which begins with acetyl-CoA. Alkaloid mycotoxins, such as roquefortine C and chaetoglobosin A originate from α-amino acids (lysine, ornithine, phenylalanine, tyrosine and tryptophan) and penicillins, which are β-lactams, being modified tripeptides, thus also originate from amino acids (valine, serine and α-aminoadipic acid) (Herbert, 1989). Trichothecenes are sesquiterpenes and thus originate from the terpene pathway which starts from acetyl-CoA converted into mevalonate.

As an example an overview of mycotoxins produced by fungi associated with pomaceous and stone fruits is found in Table 2. The toxicity of these mycotoxins is listed in Table 3.

Structures of selected mycotoxins are shown in Figure 8. Further information on mycotoxin production is found in Frisvad, Thrane and Samson's Chapter 8, *Mycotoxin producers*.

Table 2. Mycotoxins, capable of evoking acute or chronic diseases in vertebrate animals, produced by pomaceous and stone fruit associated fungi of genus *Penicillium*.

Species	Mycotoxins
P. crustosum	Penitrem A – F; roquefortine C; terrestric acid
P. expansum	Chaetoglobosin C; citrinin; communesins; patulin; roquefortine C
P. solitum	–

Table 3. Toxicity type of mycotoxins listed in Table 2.

Mycotoxin	Toxic activity
Chaetoglobosins	Cytotoxic
Citrinin	Nephrotoxic
Communesins	Cytotoxic
Patulin	Carcinogenic, cytotoxic, generally toxic
Penitrems	Acutely toxic, tremorgenic
Roquefortine C	Neurotoxic
Terrestric acid	Cardiotoxic

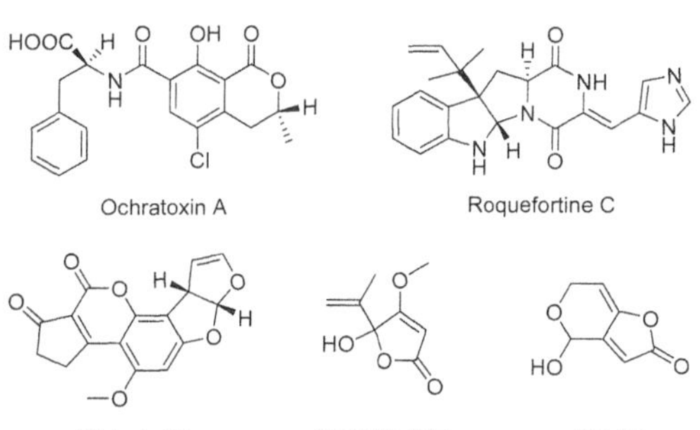

Figure 8. Chemical structure of selected mycotoxins.

BIOMARKERS AND ASSOCIATED FUNGA

In industrial food quality control, sensory panel analysis has traditionally been used. The trained panelists use terms such as musty, acidic or putrid to describe off-flavours of the spoiled product. Unfortunately even the best trained sensory panel will yield subjective scores. Off-flavours can also be studied by analytical chemistry, for instance by use of GC-MS. In such an analysis the level of geosmin, 2-methyl-isoborneol and other off-flavour related compounds, can be determined and chemometrics can be used to classify samples according to whether they have off-flavours of any kind. It is important to remember that sample treatment can influence the volatile metabolite profile since for instance the compound 1-octen-3-ol will be formed in large amounts when fungal mycelium is destroyed (Karahadian *et al.*, 1985).

To make analysis easier it can be worthwhile to focus on compounds generally encountered in spoiled food. Two such compounds which indicate fungal spoilage are ergosterol and chitin, which thus are *biomarkers* for fungal growth/biomass and thus fungal spoilage. Chemical analysis, usually by HPLC, to determine the level of ergosterol has been proven to be a reliable measure for estimation of fungal CFU (colony-forming unit) in a given sample (Cahagnier *et al.*, 1983).

By applying knowledge of the associated funga of a specific product and knowledge of volatile compounds produced by these fungi, it is possible to determine the fungal species in a sample. A list of associated spoilage fungi, and their habitat, of genus *Aspergillus* and *Penicillium* is shown in Table 1. For instance, if analysis for geosmin and 2-methyl-isoborneol is done, it is possible to distinguish between *P. expansum* (geosmin producer), *P. solitum* (2-methyl-isoborneol producer) and *P. crustosum* (produces both geosmin and 2-methyl-isoborneol), three *Penicillium* species associated with apple spoilage (for production of volatile compounds by species, see Table 4). These results were obtained on synthetic media, but they indicate a possibility for differentiation of spoilage fungi in foods as well.

This distinction of course relies on the given media stimulating the production of these volatile compounds. Thus both volatile and non-volatile metabolites can be used as biomarkers. A biomarker can be more-or-less specific indicating fungal spoilage by pointing at a selected group, such as the terverticillate Penicillia series *viridicata* or series *verrucosa*, or even an individual species.

Table 4. Volatiles produced by the most commonly encountered food spoilage species in Table 1, adapted from (Börjesson *et al.*, 1992; Zeringue, Jr. *et al.*, 1993; Larsen and Frisvad, 1995a; Fischer *et al.*, 1999; Karlshøj and Larsen, 2005).

Species	Volatiles metabolites
A. candidus	3-methylfuran, 2-methyl-1-propanol, 1-penten-3-ol, 2-methyl-1-butanol, thujopsene, ethyl hexanoate, 1-octen-3-ol ethyl ester, 2,3,5-trimethylfuran, anisole, 3-octanone, 3-cyclohepten-1-one, 3-methyl-1-butanol, 1-octen-3-ol, 3-methyl-1-heptene, 1,3,6-octatriene and one unidentified monoterpene
A. flavus	3-methylfuran, 2-methyl-1-propanol, 1-penten-3-ol, octadiene, limonene, thujopsene, 3-methyl-1-butanol, 3-octanone, 3-octanol, 1-octen-3-ol, 1-octanol, *cis*-2-octen-1-ol, α-gurjunene, *trans*-caryophyllene, epi-bucyclosesqui-phellandrene, eremophilene, β-cubebene, valencene, epizonaren, γ-selinene, γ-cadinene, cadinene, δ-cadinene, α-muurolene, aristolen, α-copaene
A. niger	2-methyl-borane, 2-methyl-bornene, α-pinene, 3-methyl-1-butanol, 3-octanone, 3-octanol, 1-octen-3-ol, 2-octen-1-ol, 1-octanol
A. versicolor	3-methylfuran, 2-methyl-1-propanol, 1-penten-3-ol, 2-methyl-1-butanol, octadiene, limonene, thujopsene, anisole, 1-(3-methylphenyl)-ethanone, 6-methyl-2-heptanone, χ-curcumene, α-muurolene, myrcene, 3-methyl-1-butanol, 1-octen-3-ol
P. atramentosum	Ethyl acetate, methyl isobutanoate, ethyl isobutanoate, isobutyl acetate, ethyl 2-methyl-butanoate, ethyl isopentanoate, isobutyl isobutanoate, isobutyl 2-methyl butanoate, butyl isopentanoate
P. aurantiogriseum	2-methyl-1-propanol, 3-methyl-1-butanol, 1-ethyl-cyclopentene, 1,3-octadiene (two isomers), 3-heptanone, 3-octanone, γ-elemene, and two unidentified sesquiterpenes
P. chrysogenum	1-heptene, 1,3-octadiene (two isomers), 3-heptanone, 1-nonene, 1,3-nonadiene, 1-octen-3-ol, 3-octanone, 3-octanol
P. commune	Ethyl acetate, 2-methyl-1-propanol, 3-methyl-1-butanol, 3-hexanone, 1,3-octadiene (two isomers), 3-heptanone, styrene, 1-octen-3-ol, 3-octanone, 3-octanol, 2-methyl-isoborneol, β-caryophyllene, (+)-aristolochene, and seven unidentified sesquiterpenes
P. crustosum	Ethyl acetate, 2-methyl-1-propanol, ethyl propanoate, 3-methyl-1-butanol, dimethyldisulphide, ethyl isobutanoate, 1,3,5-cycloheptatriene, isobutyl acetate, ethyl butanoate, ethyl-2-methyl-butanoate, ethyl isopentanoate, Isopentyl acetate, styrene, ethyl pentanoate, 3-octanone, ethyl hexanoate, ethyl octanoate, 2-methyl-isoborneol, geosmin and one unidentified monoterpene
P. cyclopium	One unidentified sesquiterpene
P. discolor	2-methyl-3-butene-2-ol, 2-methyl-1-propanol, 3-methyl-1-butanol, isobutyl acetate, 3-octanone, 2-methyl-isoborneol, geosmin, four unidentified monoterpenes and three unidentified sesquiterpenes
P. expansum	Ethanol, 2-methyl-1-propanol, 3-methyl-1-butanol, ethyl acetate, β-pinane, 1-methyoxy-3-methyl-benzene, zingiberene, α-bergamotene, β-bisabolene, geosmin and one unidentified sesquiterpene
P. freii	2-methyl-1-propanol, 3-methyl-1-butanol, 1,3-octadiene (two isomers), 3-octanone
P. hordei	2-methyl-1-propanol, 3-methyl-1-butanol, 1,3,6-octatriene, 3-heptanone, β-phellandrene, 3-octanone, limonene, 1,8-cineol, geosmin, γ-elemene and three unidentified sesquiterpenes
P. melanoconidium	2-methyl-1-propanol, 3-methyl-1-butanol, 1-ethyl cyclopentene
P. nalgiovense	Ethyl acetate, 2-methyl-1-propanol, 3-octanone, RI1404
P. nordicum	Acetone, 2-butanone, 2-methyl-1-propanol, 3-methyl-butanol, 2-pentanone, 2-methyl-isoborneol
P. olsonii	2-butanone, 2-methyl-1-propanol, 2-methyl-butanol, 2-heptanone, limonene, 2-nonanone
P. paneum	Acetone, 2-methyl-1-propanol, 3-methyl-1-butanol, 2-pentanone, β-elemene, β-caryophyllene, (+)-aristolochene, eremophilene, α-selinene, 14 unidentified sesquiterpenes and two unidentified diterpenes

P. polonicum	Ethyl acetate, 3-octanone, 2-methyl-isoborneol, γ-elemene, β-farnesene and three unidentified sesquiterpenes
P. roqueforti	Acetone, 2-methyl-1-propanol, 3-methyl-1-butanol, 2-methyl-butanol, isobutyl acetate, 1-octene, 2-pentanone, 3-octanone, β-myrcene, p-cymene, limonene, linalool, β-patchoulene, β-elemene, diepi-α-cedrene, β-caryophyllene, patchoulene isomer, (+)-aristolochene, RI1528, eremophilene, α-selinene, valencene, β-bisabolene, himachalene, 17 unidentified sesquiterpenes and one unidentified diterpene
P. solitum	2-methyl-1-propanol, 3-methyl-1-butanol, ethyl 2-methyl-butanoate, isobutyl 2-methyl-butanoate, 2-methyl-butyl 2-methyl-butanoate (two enantiomers), 2-methyl-isoborneol, β-elemene, (+)-aristolochene and one unidentified monoterpene
P. verrucosum	2-butanone, 2-methyl-1-propanol, 2-pentanone, 3-pentene-2-one, 3-methyl-1-butanol, 3-octanone, 2-methyl-isoborneol

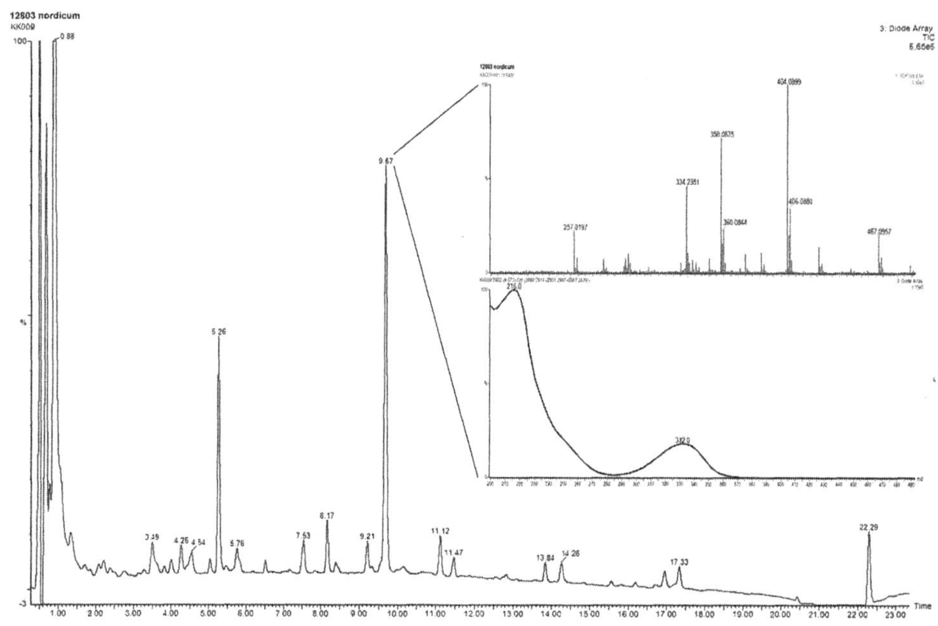

Figure 9. Positive identification of ochratoxin A by LC-DAD-MS analysis. In a LC-MS chromatogram mycotoxins in the sample can be identified by comparison of the UV spectra and the mass given by the mass spectra to literature. In this case the mycotoxin ochratoxin A is identified, the mycotoxin has a mono-isotopic mass of 403.0823, and thus a $[M+H]^+$ adduct of 404.0902, the detected mass is 404.0899, less than 1 ppm deviation from the calculated mass.

Mycotoxins form another group of compounds which are excellent as biomarkers, in the example of apple spoilage Penicillia, detection of the mycotoxin patulin will indicate that the apple spoilage has been done by *P. expansum* (Table 2, mycotoxins produced by food spoilage species from Table 1).

TRADITIONAL CHEMICAL ANALYSIS METHODS

Sampling

One of the big hurdles when an attempt is made to discover fungal growth in huge stocks of food is that fungal growth often occurs as a very local infestation, a so-called hot spot. Thus accurate sampling can be extremely difficult, particularly if dealing with spot tests, such as a grain from a silo. When dealing with volatile metabolite sampling, the problem lies in making sure that the correct sampling technique is applied, so that relevant compounds can be collected for analysis.

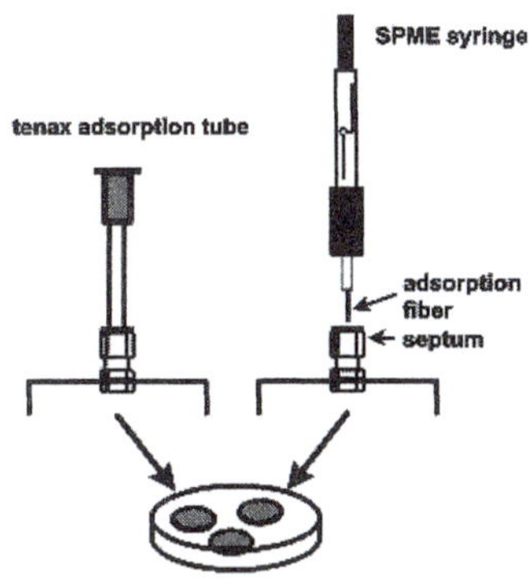

Figure 10. Collection set-up of volatile metabolites from a 3-point inoculated fungal culture on Tenax tube (left) and SPME fiber (right) (Nielsen *et al.*, 2003).

How to make correct spot sampling and the sampling problem in general, is a science in itself, and going into further details is beyond the scope of this chapter. The spot sampling problem can; however, be avoided if head space analysis by GC-MS or electronic nose is performed. This analysis type is very suitable for large samples, such a grain in a silo.

HPLC

One of the traditional ways of analyzing non-volatile compounds is by HPLC. In order to attain good results, it is of utmost importance to have a good scheme of sample preparation. It is important to choose the right extraction procedure related to what type of compound has to be detected, i.e., a non-polar extraction method should be used to capture non-polar compounds and so forth. It is also desirable to minimize the amount of matrix within the sample, both to increase HPLC performance and sensitivity (Nielsen *et al.*, 2003). The most commonly used sample preparation technique is solid phase extraction (SPE). In a typical HPLC set-up screening for mycotoxins, C18 columns are used, thus the equipment is run in the reverse phase mode.

This is ideal for separation of non-polar and semi-polar compounds. The typical detector system is a diode array detector (DAD) coupled to a mass spectrometer (MS). The combination of UV absorbance, retention index and accurate mass can be used to identify compounds by comparison to findings in literature (Figure 9) (Nielsen *et al.*, 2003).

Gas Chromatography and GC-MS

Gas Chromatography (GC) is used for the analysis of volatile and semi-volatile compounds such as sugars, lipids, amino acids, sterols and trichothecenes. If analysis of semi-volatile compounds is desired, the compounds need to be derivatised prior to analysis in order to improve their volatility. In this chapter focus will be kept on GC as a tool for analysis of volatiles collected by headspace sampling. The first step in headspace analysis is sample collection and preparation. Typically samples are collected in Tenax TA (2,6-diphenylene-oxide polymer resin) adsorption tubes (Figure 10) or onto solid phase microextraction (SPME) fibers (Nielsen *et al.*, 2003). For both methods, the material in the tube or on the fiber adsorbs volatile components from the headspace. Where Tenax TA adsorption most often takes place overnight, sample collection on SPME fibers is done in less than an hour as the equilibrium between gas phase and SPME fiber occurs rapidly.

Sampling time depends on the level of volatiles present in the headspace, thus prior knowledge of the headspace concentration makes a good estimation of sampling time possible.

Adsorption of VOCs on Tenax TA material is dependent on the affinity of the compounds. SPME fibres consist of a fused silica core coated by a polymer adsorbent. The coating material on the fibre can be varied in polarity to suit the type of volatiles desired. Typical SPME fiber coatings are polydimethylsiloxane (for volatile compounds) and Carbowax/divinylbenzene (for alcohols and polar compounds) (Hamm *et al.*, 2003); coating thickness is usually in the range of 30 to 100 μm. To achieve good separation during gas chromatography, good injection is required (Skoog *et al.*, 1996). For SPME injection is done by thermal desorption directly in the GC injection port, thus SPME relies on rapid desorption to achieve good injection.

Tenax TA adsorption tubes are usually thermally desorbed, over a course up to half an hour, into a cold-trap. Once desorption of the Tenax adsorption tube is completed the cold trap is rapidly heated to release the volatile compounds which are then immediately introduced in the GC injector.

Since the capacity of capillary columns is limited, it is important that the system is not overloaded. Therefore, most injector systems are of the split/splitless type (Wilson and Walker, 1994; Skoog *et al.*, 1996). This is important in case samples are collected from a concentrated source, for instance very close to the source of contamination or even immediately over a fungal culture. The split/splitless injector works by ensuring a constant flow onto the column, which is required to get reliable chromatography, while part of the sample is injected onto the column, the majority of the sample will be ejected from the system through the split/vent, thus preventing the need for diluting samples (Grob, 1993). Since the amount of volatile compounds bound to an SPME fiber will be substantially less than what is bound in a Tenax tube, GC-MS analysis from SPME samples typically start in splitless mode to get proper sample application on the GC column. The columns typically used in GC-MS systems are fused silica columns. The stationary phase on these columns can vary in polarity and film thickness and is chosen according to the polarity and volatility of the sample to be analyzed. Column polarity commonly varies between the non-polar DB-5 columns [(5%-Phenyl)-methylpolysil-oxane] and low/ medium polar DB-1701 columns [(14%-Cyanopropyl-phenyl)-methylpolysil-oxane] (Skoog *et al.*, 1996). For analytical purposes the column dimensions are typically between 0.18 mm and 0.32 mm internal diameter (capillary column) by 30 m length. When selecting column film diameter, it is a choice between high separation power (thin column film) and higher capacity (thick column film). The thick film columns are better suited for separation of highly volatile compounds (Grob, 1993; Skoog *et al.*, 1996). For instance for non-polar compounds such as mono- and sesquiterpenes, a non-polar column, such as a DB-5 column, is preferable for better separation of the compounds (König *et al.*, 1999). GC columns are placed inside an oven for control of temperature. During a typical GC run the oven temperature is increased from 30 to 270 °C over a period between 30 to 60 minutes. By increasing the column temperature the volatile compounds in the sample will further be split by affinity to the stationary phase as a compound with high affinity to the stationary phase requires a higher temperature to leave the stationary phase than a compound with less affinity for the stationary phase.

The most common detector on any GC system is a flame ionization detector (FID). This detector utilizes a hydrogen/air flame for detection primarily of carbon containing compounds (Skoog *et al.*, 1996). It is a highly robust detector, with a sensitivity level of approximately 10^{-13} g/s (Wilson and Walker, 1994; Skoog *et al.*, 1996), but it does not yield any structural information, and on a single column GC instrument it cannot be used for compound identification. If identification is desired, using only an FID detector, the sample must be analyzed on two columns of differing type, and a standard of the compound must be analyzed as well for comparison of retention time/retention index. When more information about the volatile compounds is required, such as mass, for identification of the compound, mass spectrometers are used. Mass spectrometers are often used in combination with an FID. The typical mass spectrometer used for GC is a quadropole instrument. Mass spectral analysis provides structural information through the fragmentation pattern, the mass spectrum, formed by electron impact (EI) ionization this pattern can be searched in a database library for compound identification. Sensitivity in MS can be improved by a factor of around 50 from ng to pg level by scanning for few selected characteristic ions, i.e., selected ion recording (SIR/SIM). Unfortunately MS cannot yield information on isomers and some compounds (Ramaswami *et al.*, 1988) unless coupled to a GC system equipped with chiral columns as well as usage of chiral standards.

Case I: Identification of Penicillia and Detection of Mycotoxin Production by Volatile Metabolite Profiling and Identification

It is believed that volatile metabolites play an important role in the chemical interactions between fungi and their surrounding organisms. Recently, it has been shown that volatile metabolites produced by *Penicillium paneum* inhibit mycelial growth of different species of fungi belonging to a variety of genera (Chitarra *et al.*, 2004) and it has also been shown that 1-octen-3-ol inhibits germination of spores from *P. paneum*, as well as induction of microcycle conidiation, showing that this compound is acting like a fungal hormone during fungal development in *P. paneum* (Chitarra *et al.*, 2004; Chitarra *et al.*, 2005).

It has been shown that the volatile metabolite profile is usually species specific within genus *Penicillium* (Larsen and Frisvad, 1995a). Fungal volatile metabolites include alcohols, ketones, esters, hydrocarbons such as small alkenes and mono- and sesquiterpenes, of which the terpenes were shown to be most relevant for species identification (Larsen and Frisvad, 1995b). It is important to remember that the production of volatile metabolites is highly media specific, for instance the production of ketones and secondary alcohols derive from lipid degradation. When analyzing various series such as series *viridicata*, series *camemberti* and series *verrucosa*, within genus *Penicillium*, it has been shown to be difficult to differentiate the fungi to species level (Larsen *et al.*, 2001). In those cases, the differentiation will be on a series level and the volatile compounds will be series specific instead of species specific. In some cases differentiation of species proves easier as with the very closely related species *Penicillium carneum*, *P. paneum* and *P. roqueforti* can be differentiated through volatile metabolite profiling and identification of, particularly, terpenes (Karlshøj and Larsen, 2005) (Figure 11). For *P. roqueforti*, for instance, the major sesquiterpene compounds produced are β-elemene, selenine, patchuline (Larsen and Frisvad, 1995b) as well as (+)-aristolochene (Demyttenaere *et al.*, 2001), whereas *P. carneum* produces far less terpenes but large amounts of 3-methyl-1-butanol.

It has been shown, using the knowledge that it is possible to differentiate species from genus *Penicillium* on basis of terpene profile, that it is possible to detect a fungal contamination at a ratio of 1000:1 of a *P. roqueforti* culture with *P. commune* within three days by analysis of volatile metabolites (Larsen, 1997), at a stage where it was very difficult to detect the contamination by morphological studies of the mixed culture.

This was achieved by combining SPME with SIR MS analysis mainly of ions characteristic to sesquiterpenes specific to the cheese associated fungi, such as 2-methyl-isoborneol and β-caryophyllene for *P. commune* and limonene, β-elemene and β-caryophyllene for *P. roqueforti*, providing a method for starter culture cross-contamination checking.

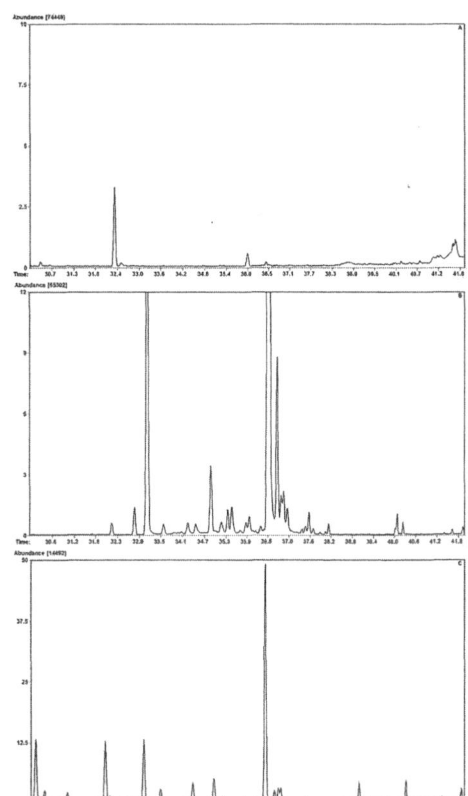

Figure 11. Chromatograms of the RI interval 1340-1800 for: *P. carneum* (top), *P. roqueforti* (middle) and *P. paneum* (bottom). The abundance scale is in percentage of the abundance given in the top left corner of each chromatogram. The difference in volatile metabolite profile between the species is evident.

A series of studies have shown correlations between production of specific volatile metabolites and mycotoxin production. Release of specific sesquiterpenes unique to mycotoxenogenic *Aspergillus flavus* isolates, among them α-gurjunene, *trans*-caryophyllene and γ-cadinene, was shown to be correlated to aflatoxin biosynthesis, and the decline of aflatoxin biosynthesis was also correlated to the disappearance of the specific sesquiterpenes from the headspace (Zeringue, Jr. *et al.*, 1993). Production of terpenes, among them trichodiene, has been shown to be biomarkers for production of trichothecenes in *Fusarium* species (Jelén *et al.*, 1995; Pasanen *et al.*, 1996; Demyttenaere *et al.*, 2004) and it has been shown that it is possible to distinguish between different toxinogenic *Fusarium* species through sesquiterpene profiling (Demyttenaere *et al.*, 2004). A good review on fungal volatile metabolites can be found in Jelén and Wasowicz (1998).

ELECTRONIC NOSE ANALYSIS

How the Electronic Nose Works

The electronic nose can be compared to the mammalian olfactory system where gasses stimulate receptors. The stimulated receptors send nerve signals to the olfactory cortex where they are analyzed and interpreted. It has been shown that the receptors generally are quite non-selective, though a few compounds like geosmin have a very low threshold value, thus a given receptor responds to many compounds and that many receptors respond to a given compound. This gives rise to patterns of responses for the olfactory cortex to analyze and interpret (Pearce, 1997). Like its counterpart, the electronic nose consists of a number of non-specific receptors, its sensors, whose signal patterns are analyzed, by either a neural network or chemometrics for interpretation. Like the mammalian olfactory system, the electronic nose relies on an array of receptors, a sensor array.

As seen in the following section, there are several different approaches to sensor design in terms of how the compounds are detected and thus what kind of signal pattern is generated for analysis.

Electronic nose technology has many applications. It has been used for screening for toxic gases, volatile organic compounds and food-related compounds. Especially within the field of food technology, electronic noses have been applied in quality control (Maul *et al.*, 2000; Werlein, 2001; Rye and Mercer, 2003; Berna *et al.*, 2004; Vinaixa *et al.*, 2004; García-González *et al.*, 2004; Balasubramanian *et al.*, 2004; Trihaas *et al.*, 2005a), process control (Zondervan *et al.*, 1999), maturity monitoring (Brezmes *et al.*, 2005; Trihaas and Nielsen, 2005; Marrazzo *et al.*, 2005; Trihaas *et al.*, 2005b), etc. This has been done on both raw materials and manufactured products.

Until now the potential of the electronic nose to replace methods like GC-MS and HPLC / LC-MS for indirect mycotoxin estimation has not been explored. Obviously, for this potential to be realistic the e-nose has to be capable of separation of fungal species among a given associated funga when analyzing samples from a particular habitat. When comparing e-nose analysis with GC-MS analysis, some of the key differences are that the e-nose sensors may not yield a sensitive response to some of the compounds which can be found as key compounds, by GC-MS analysis, for a given food contamination problem (Schaller *et al.*, 1998; Kohl, 2001). On the other hand, e-nose analysis has a potential for automation through construction of prediction models by chemometrics or neural network analysis. In addition, e-nose analysis is a very rapid method and thus suitable for high throughput screening.

Electronic Nose Sensors

The different sensor types used in electronic noses can be divided into four groups.

1. *Conductivity sensors*: Metal oxide semiconductors (MOS), intrinsically conductive polymer chemiresistors (ICP) and conductive polymer composite chemiresistors (CP)
2. *Electrostatic potential sensors*: Metal oxide semiconductor field effect transistors (MOSFET) and gas sensitive field effect transistor sensors (GASFET)

3. *Acoustic resonance sensors*: Thickness-shear mode / quartz crystal microbalance / bulk acoustic wave (TSM / QCM / BAW) and surface acoustic wave (SAW)
4. *Optical vapour sensors*: Polymer-deposited optical sensors (DPO) and self-encoded bead (SEB)

Conductivity Sensors

Metal Oxide Semiconductor Sensors

Metal oxide semiconductors (MOS) have been used commercially as gas alarms since the 1960s (Schaller *et al.*, 1998). It took more than 20 years for the first cross-reactive MOS sensor array to be demonstrated by Persaud and Dodd (1982). The sensors usually consist of a cylindrical ceramic former, which contains a heating element. The ceramic former is coated with a film of semiconductor material (Bartlett and Gardner, 1992; Strike *et al.*, 1999; Gardner and Bartlett, 1999).

There are two types of semiconductors used: negative electron type (n-type) or positive hole (p-type). For p-type conductors, the density of holes in the valence band exceeds electron density in the conduction band the opposite is the case for n-type semiconductors. Electrical conduction in p-type semiconductors is mostly due to the movement of positive holes whereas electrical conductivity in n-type semiconductors is mostly due to the movement of electrons. N-type semiconductors, which usually consist of zinc oxide, tin dioxide, iron (III) oxide or titanium dioxide, respond mainly to reducing gases, while p-type semiconductors, which normally are oxides of nickel or cobalt, mainly respond to oxidizing compounds (Mielle, 1996). The semiconductor film can be coated as either a thick film (10-300 μm) or a thin film (6-1000 nm). Thin films, though harder to manufacture reproducibly, offer faster responses as well as higher sensitivity. Most often, commercial MOS sensors are of the thick film type (Schaller *et al.*, 1998).

Sensor selectivity can be changed by various means. The semiconductor film can be doped with catalytic metals, (usually platinum or palladium), the operating temperature can be shifted (in the range of 50-400 °C) and the particle size in the semiconductor films polycrystalline structure can be changed (Watson and Yates, 1985; Morrison, 1987; Mielle, 1996; Albert *et al.*, 2000; Strike *et al.*, 1999).

Both reactions will lead to a change in the semiconductor material and thus to the measured change in conductivity.

Overall, the MOS sensors are less selective than for instance CP, BAW, SAW and MOSFET sensors (Mielle, 1996). MOS sensors are typically operated at high temperatures to increase reactivity and decrease the sensitivity to water (Albert *et al.*, 2000).

Sensitivity to water, and therefore high operating temperature, along with a very high sensitivity to ethanol, poisoning by sulphur containing compounds as well as weak acids and slow baseline recovery when subjected to high molecular weight compounds are all drawbacks to this sensor type (Mielle, 1996; Schaller *et al.*, 1998).

Conducting Polymer Sensors

Conducting polymer sensors have been applied as electronic nose sensors since the 1980s (Pelosi and Persaud, 1988; Bartlett *et al.*, 1989). They are comprised of a substrate, fiberglass or silicone, coated by a conducting organic polymer between the two electrodes, which are usually gold-plated (Amrani *et al.*, 1995; Mielle, 1996). The polymer used is typically polypyrrole, polyaniline or polythiopene (Figure 13). The polymers act as the sensing unit of the sensor. The polymers in intrinsically conductive polymer chemiresistor sensors (ICP) are linear backboned comprised of repeating conjugated organic monomers that act as one-dimensional conductors as electrons will travel mostly through the conjugated linear backbone. In conductive polymer composite chemiresistor sensors (CP) polymers, carbon black or polypyrrole is used as the conductor, while non-conducting organic polymers serve as the insulating substrate (Strike *et al.*, 1999).

To change the selectivity of the conductive polymer sensor changes in the polymerization conditions as well as the counter ion in use can be made. Furthermore, the oxidation state of the polymer can be changed after deposition on the sensor (Schaller *et al.*, 1998). It has been

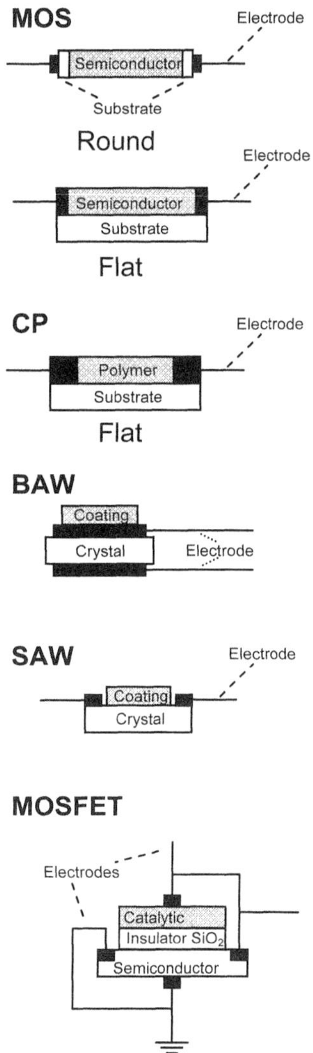

Figure 12. Schematic diagrams of five of the most common sensor types. MOS, Metal oxide semiconductor; CP, Conducting polymer; BAW, Bulk acoustic wave; SAW, Surface acoustic wave; MOSFET, Metal oxide semiconductor field effect transistor (Schaller *et al.*, 1998; Albert *et al.*, 2000).

shown that use of chiral material for the polymer in carbon black polymer composites (CP sensors) leads to the ability to differentiate between enantiomer compounds (Severin *et al.*, 1998). It is also possible to imbed biomaterials such as antibodies, enzymes or cells in the polymer if so desired.

When a CP or ICP sensor is exposed to volatile compounds some of these will be adsorbed in the polymer causing swelling and altering the electron flow in the polymer. This will cause the measured change in conductivity (Shiers, 1995; Albert *et al.*, 2000).

Conducting polymer sensors are operated at low temperatures (below 50 °C), which causes extreme sensitivity to moisture (Shiers, 1995). It has proven difficult to manufacture conductive polymer sensors reproducibly, due to the polymerization step that is hard to control (Mielle, 1996; Partridge *et al.*, 1996).

Electrostatic Potential Sensors

Metal oxide semiconductor field effect transistor sensors (MOSFET) were firstly reportedly used by Lundström *et al.* in 1975 (see also; Lundström *et al.*, 1990; Lundström *et al.*, 1993). The MOSFET sensor is constructed of three layers: the top layer, the so-called gate, comprises of metal, underneath this is an insulating layer of an oxide, usually SiO_2, and in the bottom the semi-conducting substrate, often a p-type silicon with n-type channels on both sides of the metal gate (Schaller *et al.*, 1998; Albert *et al.*, 2000). In MOSFET sensors the metal gate traditionally consists of aluminum, whereas catalytic metals, such as palladium, platinum and iridium, are used, as the only difference in construction, in gas sensitive field effect transistor sensors (GASFET).

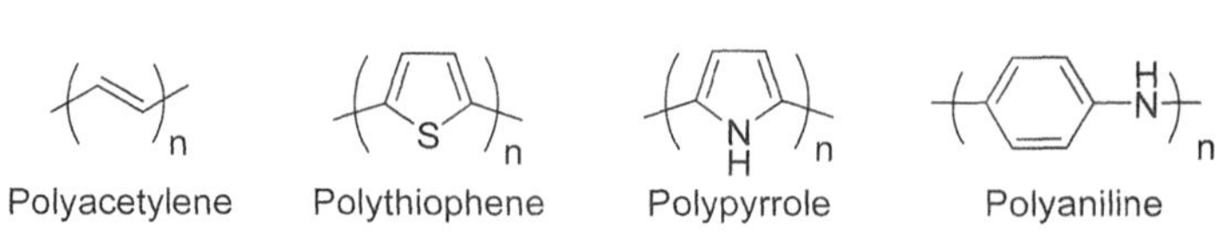

Figure 13. Structure of polymer backbone in insulating form for CP sensors (Albert *et al.*, 2000).

Selectivity and sensitivity of MOSFET sensors can be changed by alterations in operating temperature, the type of metal used in the metal gate (i.e., change between catalytic and non-catalytic metals) and the microstructure of the metal in the metal gate (Lundström *et al.*, 1975; Lundström *et al.*, 1990; Albert *et al.*, 2000). Palladium is preferable for hydrogen sensing and platinum and iridium for sensing polar compounds (Lundström *et al.*, 1992; Albert *et al.*, 2000).

The metal gate can be applied as either a continuous (thick) film (100-400 nm), or a discontinuous (thin) film (3-30 nm) (Müller and Lange, 1986; Sundgren *et al.*, 1990; Winquist *et al.*, 1992; Schaller *et al.*, 1998). The thick film sensors primarily respond to compounds, which can be dehydrogenated, whereas the thin film sensors also respond to compounds such as carbon monoxide and ammonia (Lundström *et al.*, 1975; Lundström *et al.*, 1990; Spetz *et al.*, 1992; Schaller *et al.*, 1998). The dehydrogenation takes place on the metal gate, the hydrogen can adsorb to the gate and diffuse to the SiO_2 / metal interface were it forms a dipole layer. This changes the electrostatic potential of the MOSFET sensor (Bergveld, 1985; Lundström *et al.*, 1992). Apart from adsorptions and reactions on the metal gate and hydrogen diffusion to the metal-insulator interface, thin film sensors also have their electrostatic potential changed on the insulator surface due to reactions of polar compounds on the metal oxide surface (Lundström *et al.*, 1975; Lundström *et al.*, 1990).

MOSFET sensors are silicon-based and thus operate at temperatures below 250 °C (Strike *et al.*, 1999). Typical operating temperatures for MOSFET sensors are in the range of 50 to 200 °C (Lundström *et al.*, 1975; Lundström *et al.*, 1990). These sensors, like MOS sensors, exhibit a fairly low sensitivity to moisture and furthermore are quite robust. On the down side, high manufacturing expertise is required for good sensor quality and reproducibility (Schaller *et al.*, 1998).

Acoustic Resonance Sensors

King (1964) introduced bulk acoustic wave sensors (BAW) also referred to as thickness-shear mode sensors (TSM), and by the name quartz crystal microbalance (QCM) (King, 1964). Wohltjen and Dessy introduced surface acoustic wave sensors (SAW) in 1979 (Wohltjen and Dessy, 1979a; Wohltjen and Dessy, 1979b; Wohltjen and Dessy, 1979c). Martin *et al.* reported the first use of SAW sensors in the 1980s (Martin *et al.*, 1983; Martin *et al.*, 1984; Martin *et al.*, 1985).

Both BAW and SAW sensors consist of crystal discs, usually made of either quartz, lithium niobate or lithium thantalate coated with for instance chromatographic stationary phases, polymer films or other non-volatile compounds that adsorb vapors (Guilbault and Jordan, 1988; Nieuwenhuizen and Nederlof, 1992; Holmberg, 1997; Strike *et al.*, 1999). In a BAW sensor, the electrodes are positioned on top of and below the crystal, with coating on top of the electrodes. In a SAW sensor, the electrodes are both positioned on top of the crystal with the coating in between the electrodes (Albert *et al.*, 2000). The coating on BAW sensors is quite thin (1 μm to 10 nm) and SAW sensors are constructed on such a minute scale as to be compatible with planar integrated circuits fabrication technology (Caliendo and Verona, 1992; Wünsche *et al.*, 1995; Mielle, 1996).

In order to change the selectivity of the sensor, the coating used can be changed, and choices in coating are nearly limitless (Mielle, 1996; Hodgins, 1997; Strike *et al.*, 1999).

Both BAW and SAW sensors, being vibrating crystals, exhibit resonance vibrations when an alternating current is applied to them, even at room temperature. This vibration is also dependant on the mass of the crystal. Thus when volatile compounds are adsorbed on the coating, the mass of the sensor increases, which causes the resonance frequency of the sensor to change. This change is measured (Hodgins, 1997; Strike *et al.*, 1999; Albert *et al.*, 2000). BAW sensors generate three-dimensional waves through the crystal, perpendicular to the surface of the crystal and are operated at frequencies between 10-30 MHz. In SAW sensors the waves generated are two-dimensional, only penetrating approximately one wavelength into the crystal. The SAW sensors are operated at frequencies between 100 MHz to 1 GHz (Nieuwenhuizen and Nederlof, 1992; Holmberg, 1997; Albert *et al.*, 2000). Due to its minute size, SAW sensors are very robust. They are also more sensitive than BAW sensors though both are less sensitive than the other sensor types. Because of their operating frequencies SAW sensors are much noisier than BAW sensors (Mielle, 1996; Hodgins, 1997). Unfortunately the coating technology is as yet not fully controlled, thus leading to poor batch-to-batch reproducibility. The acoustic sensors exhibit a high sensitivity to temperature and humidity fluctuations (Mielle, 1996; Doleman *et al.*, 1998).

Optical Vapour Sensors

Polymer-deposited optical sensors (PDO) are comprised of an optical fibre on which an indicator is immobilized on the tip. The immobilized indicator is coated with a polymer. In self-encoded bead (SEB) sensors thousands of tiny beads (3.2 μm), of polymer or ceramic materials, are immobilized in acid-etched wells on the tip of the optical fibre. The beads consist of a ceramic or polymer material containing a vapour sensing dye. One of the dyes typically used in either sensor type is Nile Red. If a change in specificity is required the polymer, and for SEB also ceramic, material can be changed (Dickinson *et al.*, 1997; Walt *et al.*, 1998).

When vapour is adsorbed in the polymer, the polarity of the surroundings of the vapour-sensing dye changes and the dye changes colour. In general, the more polar the vapour adsorbed, the more red-shifted the absorption and/or emission spectra of the dye will be. In PDO sensors many data can be collected simultaneously. Among those are changes in intensity, fluorescence lifetime and spectral shape (Albert *et al.*, 2000). SEB sensors can be constructed in very small arrays, which will give rise to short response times and increased sensitivity. Five seconds is usually ample time for proper response and recovery of the sensor in experiments (Albert *et al.*, 2000).

The most commonly used sensor types are MOS and CP sensors. Due to the high power consumption of MOS systems, almost all portable e-nose systems apply CP sensors.

Electronic Nose Signal Analysis

As with any other analysis method, proper data analysis of electronic nose signals is crucial. It is therefore important to know what you wish to achieve with your analysis in order to choose a proper data analysis technique. Apart from choice of technique, there are details to keep in mind when performing data analysis. Even the most successful analysis is worthless unless proper data analysis is performed.

Chemometrics

Use of chemometrics is the traditional data analysis tool for e-nose data. There is a broad variety of chemometric tools available and the most common ones are briefly described in the following section.

Principal Component Analysis

Principal component analysis (PCA) is frequently used to get an overview of the data obtained. One of the objectives of PCA is to reduce the dimensionality of data in order to make data analysis easier. Reduction of dimensionality also reduces the level of noise in the data, as the noise is omitted from the PCA model. PCA will also change the coordinate system of the data, moving the centre point to where the average data are (Esbensen, 2001). In PCA, data are described by principal components (PCs). The first PC is always chosen so that it explains the maximal variance, which is the same as minimizing the summed square transverse distances from the data points to the PC (least squares approach), the second PC is always orthogonal to the first PC and is chosen so it describes the second highest variance. The third PC must be orthogonal to the first two and so on (Esbensen, 2001).

The maximal number of PCs in any PCA is no higher than the number of components minus one or the number of variables, whichever of the two is lower. In most cases, PCA models consist of few PCs (typically one to four). Thus in PCA, dividing data into structure and noise can be achieved.

In the PCA model the loadings give a relation between the real variables, for instance e-nose sensors, and the PCs. Similarly the scores show the relation between the samples/objects in the PC system. Scores and loadings are most often viewed visually in scores and loadings plots, and to see which loadings correlate positively or negatively with given objects, a bi-plot can be made (Figure 15).

The closer an object is to a specific loading in the bi-plot, the higher the positive correlation, if the object is on the opposite side of the ordinate or abscissa compared to the loading, the two are negatively correlated (Esbensen, 2001).

To check whether the PCA is modelling the data accurately, object residuals can be viewed. The lower these residuals are, the better the model represents the data. To check model robustness, it is important to validate the model. This is typically done by leave one out (LOO) validation, also known as full cross-validation. In this validation method PCA models are made with one object removed from the model. Thus if there are 20 objects, 20 different models with 19 objects are made and compared.

Preprocessing is as important to data analysis as sample preparation is to data collection. If data are analyzed in which the scale of certain data is very different from each other, variables can be made more comparable so that no variable is allowed to dominate because of its range. This is typically done by multiplying the variable by 1/SDev (one divided by the variable's standard deviation). This is not always the best solution if empirical variance is more-or-less comparable, as there is a risk of overemphasizing noise. This is overcome by using 1/(h + SDev) thus using an offset.

Finally it is important, when analyzing the results, to look for outliers. If the outlier is a known erroneous measurement, it should be removed.

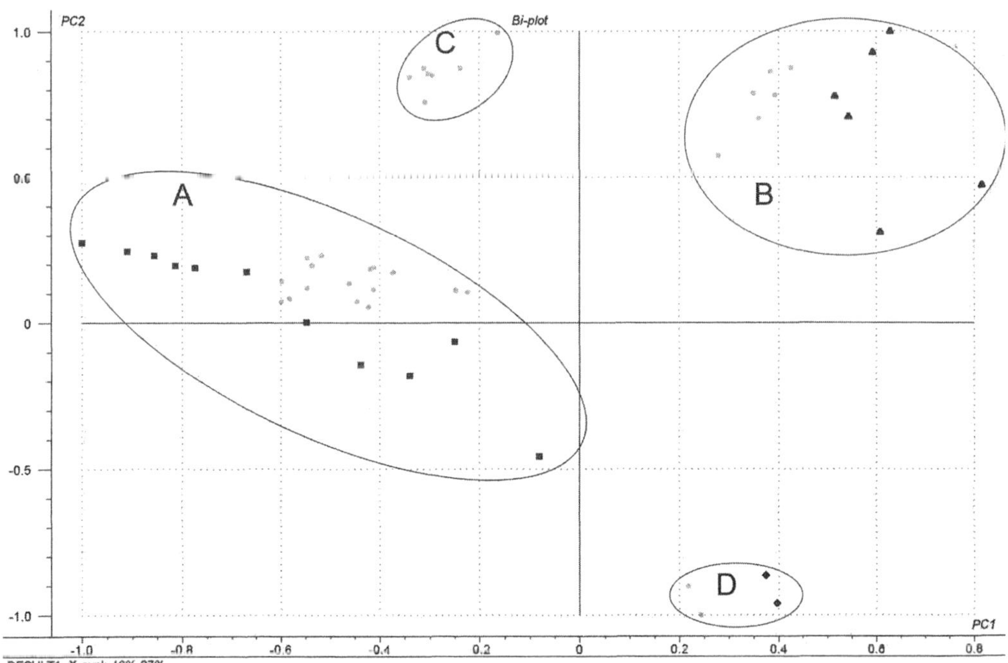

Figure 15. A bi-plot of PCA objects (dark grey squares, triangles and diamonds) and loadings (light grey circles). Loadings are positively correlated to objects which are on the same side of the ordinate or abscissa and negatively correlated to objects on the opposite side. Positive correlations are also seen between closely located loadings and objects. In this bi-plot, loadings in ellipse C are positively correlated to all objects in ellipse B and all objects above the abscissa in ellipse A. All loadings in ellipse B are positively correlated to objects in ellipse B. Loadings in ellipse D are positively correlated to objects in the same ellipse, but negatively correlated to all other objects.

The influence plot can be used to check whether a sample has a high residual (poorly fitted by the model), a high leverage (high influence on the model) or both (Esbensen, 2001).

Principal Component Regression and Partial Least Squares Regression

These chemometric methods are used when prediction is desired. It is used to predict the value of Y, the dependent variable, (for instance the level of a given volatile component) from an X matrix, the independent variable, of measurements (for instance e-nose measurements) by regression. Construction of the prediction model is divided into calibration and validation. In the calibration step known X and Y values, measured in the way that is going to be used in future measurements, and which are representative for the future X measurements are used to construct the prediction model. Validation is then performed with a (or several) test set to ensure the prediction ability of the model. The test sets are important, and the use of multiple test sets for validation is advised. It is important that test sets used differ from one another (Esbensen, 2001). In principal component regression (PCR) the prediction model is made by multi-linear regression of Y on a PCA model of X.

An approximated view of partial least squares (PLS) regression is that the PLS prediction model is constructed by multi-linear regression of a PCA model of Y on a PCA model of X.

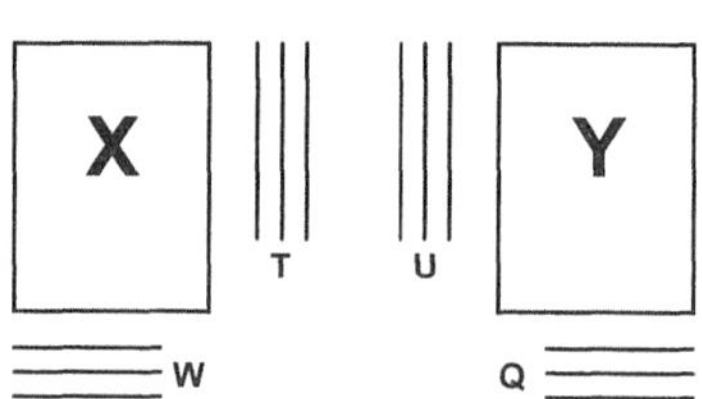

Figure 16. PLS prediction model is construction. The PCA done on X is based on a starting point score vector obtained from Y ($\mathbf{u}_1$) thus obtaining the "loading-weight" vector **w** which in turn is used to calculate the vector **t**. The PCA done on Y is based on **t** being the starting point score vector instead of **vector** u yielding the vector **q**. Multi-linear regression of the PCA model of Y on the PCA model of X is then performed yielding the PLS model (Esbensen, 2001).

This is not completely the case, as the PCA done on X is based on a starting point score vector obtained from Y and the later PCA done on Y is based on a starting point score vector obtained from the PCA on X (Figure 16). This is done to reduce the influence of large X values that do not correlate with Y (Esbensen, 2001).

Soft Independent Modelling of Class Analogy

For classification purposes, soft independent modelling of class analogy (SIMCA) is used. The classification model is usually based on individual PCA models of the classes. Construction of the classification model is a two-step process consisting of training and classification. The model is trained by construction of the PCA models consisting of each class. If the classes are known, the classification is a supervised classification (Esbensen, 2001).

During the actual classification, the models constructed in training are used to predict whether a new, unknown sample, belongs to any (several) of the classes described by the SIMCA model (Esbensen, 2001).

Neural Networks

Neural networks (NNs), or more accurately artificial neural networks, consist of an interconnected group of artificial neurons through which information is processed through a mathematical or computational model. This means that NNs are non-linear statistical data modelling tools with which for instance patterns in data can be found (Gurney, 1997). When using NNs for data analysis the NN has to learn how to interpret data. The typical way of teaching the NN to interpret data, for classification or prediction modelling, is to do supervised learning. This is done by feeding the NN data pairs of input data (x) and output data (y) with which the NN will seek to find the function that matches the examples. This is typically done by trying to minimize the mean-squared error between the NNs output value and the target output value (y) (Gurney, 1997). When using the learning algorithm it is important to make sure that the NN does not over-fit the training data and thus fails to find the true statistical process which generates the data.

This is particularly a problem when using small training sets (Gurney, 1997).

Case II: Discriminating between Food Spoilage Fungi by Electronic Nose Technology

Several studies have focused on food and feed quality control by electronic nose analysis. It has been shown that it is possible to classify grain samples, by use of a MOSFET and MOS sensor system coupled with neural network analysis, into three off-flavour groups and a normal group (75% correct sample classification) as well as into two simple categories (90% correct sample classification), namely good or bad with higher percentage probability than the classification done by two grain inspectors (Börjesson *et al.*, 1996). A more recent study has shown that it is possible to use a CP sensor e-nose to classify beef according to whether it is unspoiled (<10^6 cfu/g) or spoiled (>10^6 cfu/g) stored at 4 and 10 °C. The best classification was obtained on samples stored at 10°C where a classification accuracy of over 96% could be reached for individual data sets, and classification accuracy of approximately 70% when data sets were combined (Balasubramanian *et al.*, 2004).

Differentiation or discrimination of bread spoilage organisms, grown on milled wheat agar, has been achieved on four *Eurotium* species, *Wallemia sebi* and one unknown *Penicillium* species using a 14 sensor CP system and PCA and discriminant function analysis (Keshri *et al.*, 1998). Even before visible growth, 93% separation was achieved. This illustrates the potential of using electronic nose analysis for early spoilage detection in food stuffs.

Skimmed milk spoilage by *Pseudomonas fluorescens, Bacillus cereus, Candida pseudotropicalis* and *Kluyveromyces lactis* was investigated by electronic nose analysis with a 14 CP sensor system. After 60-minute incubation *Candida pseudotropicalis* could be distinguished from unspoiled milk and *Kluyveromyces lactis* was differentiated from control samples after 5 hours incubation. By discriminant function analysis it was possible to make a prediction model as to which organism was responsible for spoilage and thus correctly reclassify all samples (Magan *et al.*, 2001).

In a very recent study, distinction between deoxynivalenol (DON) levels, from unknown source(s), in durum wheat (*Triticum durum*) samples was shown to be possible using a MOS sensor e-nose, classifying the samples in three groups, blank, medium and high DON levels (Cheli *et al.*, 2005).

CONCLUSIONS

It is clear that volatile metabolites produced by fungi can be used as biomarkers for food quality. As previously stated, spot analysis of volatile metabolites can in many cases be more efficient than spot analysis of non-volatile metabolites such as mycotoxins (e.g., grain in a grain silo, coffee in coffee bags, etc.). Analysis of volatile metabolites can be used to make spot sampling of, e.g., mycotoxin content more efficient since analysis of volatile metabolites present in the sample can give an overall estimate of the presence of spoilage organisms. Since it is not necessarily of interest to know exactly which compounds are present in the headspace of a given food sample, the e-nose seems to be a very promising tool for initial food quality screening. With the e-nose rapid sampling can be performed and with proper modelling it should be possible to determine whether a given sample analyzed is "good," "bad" or "unknown." The "bad" samples can immediately be discarded, whereas the "good" samples need no further testing and only the "unknown" samples need to be tested through traditional methods, thus improving speed, safety and efficiency of food quality spot sampling by using volatile metabolites as predictors of good and bad food quality. A final remark is that it would be even better if it becomes possible to predict mycotoxin content in samples by electronic nose analysis.

Acknowledgements

This project was supported by the Food Technology, Safety and Quality research program, Ministry of Food, Agriculture and Fisheries, Denmark, grant no. FSK03-DTU-4.

REFERENCES

Albert, K. J., Lewis, N. S., Schauer, C. L., Sotzing, G. A., Stitzel, S. E., Vaid, T. P., and Walt, D. R. (2000). Cross-reactive chemical sensor arrays. Chemical Reviews 100:2595-2626.

Amrani, M. E. H., Persaud, K. C., and Payne, P. A. (1995). High-frequency measurements of conducting polymers: development of a new technique for sensing volatile chemicals. Measurement Science and Technology 6:1500-1507.

Balasubramanian, S., Panigrahi, S., Logue, C. M., Marchello, M., Doetkott, C., Gu, H., Sherwood, J., and Nolan, L. (2004). Spoilage identification of beef using an electronic nose system. Transactions of the American Society of Agricultural Engineers 47:1625-1633.

Bartlett, P. N., Archer, P. B. M., and Ling-Chung, S. K. (1989). Conducting polymer gas sensors Part I: Fabrication and characterisation. Sensors and Actuators 19:125-140.

Bartlett, P. N., and Gardner, J. W. (1992). Odour sensors for an electronic nose *In* Sensors and sensory systems for an electronic nose (Gardner, J. W. and Bartlett, P. N., ed.), Proceedings of the NATO advanced research workshop, Reykjavik 1991. Kluwer, Dordrecht, The Netherlands, pp. 31-51.

Bergveld, P. (1985). Impact of MOSFET-based sensors. Sensors and Actuators 8:109-127.

Berna, A. Z., Lammertyn, J., Saevels, S., Natale, C. D., and Nicolai, B. M. (2004). Electronic nose systems to study shelf life and cultivar effect on tomato aroma profile. Sensors and Actuators B: Chemical 97:324-333.

Berry, D. R. (1988). Physiology of industrial fungi. Blackwell Scientific Publications, Oxford, UK, pp. 1-285.

Börjesson, T. (1993). Volatile fungal metabolites as indicators of mould growth in stored cereals, Doctoral thesis at the Department of Microbiology, Uppsala Genetic Center, Swedish University of Agricultural Sciences, ISBN: 91-576-4706-2.

Börjesson, T., Eklöv, T., Jonsson, A., Sundgren, H., and Schnürer, J. (1996). Electronic nose for odor classification of grains. Cereal Chemistry 73:457-461.

Börjesson, T., Stöllman, U., and Schnürer, J. (1992). Volatile metabolites produced by six fungal species compared with other indicators of fungal growth on cereal grains. Applied and Environmental Microbiology 58:2599-2605.

Brezmes, J., López Fructuoso, M. L., Llobet, E., Vilanova, X., Recasens, I., Orts, J., Saiz, G., and Correig, X. (2005). Evaluation of an electronic nose to assess fruit ripeness. IEEE Sensors Journal 5:97-108.

Cahagnier, B., Richard-Molard, D., and Poisson, J. (1983). Évolution de la teneur en ergostérol des grains au cours de la conservation, une possibilité d'évaluation quantitative et rapide de leur mycoflore. Sciences des Aliments 3:219-244.

Caliendo, C., and Verona, E. (1992). Surface acoustic wave (SAW) gas sensor *In* Gas Sensors (Sberveglieri, G., ed.), Kluwer Academic Publishers, Dordrecht, The Netherlands, pp. 281-306.

Cheli, F., Dell'Orto, V., Pinotti, L., Campagnoli, A., and Tognon, G. (2005). Implementation of the electronic nose for the identification of mycotoxins in durum wheat (*Triticum durum*). Veterinary Research Communications 29:391-393.

Chitarra, G. S., Abee, T., Rombouts, F. M., Posthumus, M. A., and Dijksterhuis, J. (2004). Germination of *Penicillium paneum* conidia is regulated by 1-octen-3-ol, a volatile self-inhibitor. Applied and Environmental Microbiology 70:2823-2829.

Chitarra, G. S., Abee, T., Rombouts, F. M., and Dijksterhuis, J. (2005). 1-Octen-3-ol inhibits conidia germination of *Penicillium paneum* despite of mild effects on membrane permeability, respiration, intracellular pH, and changes the protein composition. FEMS Microbiology Ecology 54:67-75.

Demarigny, Y., Berger, C., Desmasures, N., Gueguen, M., and Spinnler, H. E. (2000). Flavour sulphides are produced from methionine by two different pathways by *Geotrichum candidum*. Journal of Dairy Research 67:371-380.

Demarne, V. and Sanjinés, R. (1992). Thin film semiconducting metal oxide gas sensors *In* Gas Sensors (Sberveglieri, G., ed.), Kluwer Academic Publishers, Dordrecht, The Netherlands, pp. 89-116.

Demyttenaere, J. C. R., Adams, A., Belleghem, K. van, De Kimpe, N., König, W. A., and Tkachev, A. V. (2001). De novo production of (+)-aristolochene by sporulated surface cultures of *Penicillium roqueforti*. Phytochemistry 59:597-602.

Demyttenaere, J. C. R., Moriña, R. M., De Kimpe, N., and Sandra, P. (2004). Use of headspace solid-phase microextraction and headspace sorptive extraction for the detection of the volatile metabolites produced by toxigenic Fusarium species. Journal of Chromatography A 1027:147-154.

Dickinson, T. A., Walt, D. R., White, J., and Kauer, J. S. (1997). Generating sensor diversity through combinatorial polymer synthesis. Analytical Chemistry 69:3413-3418.

Doleman, B. J., Severin, E. J., and Lewis, N. S. (1998). Trends in odor intensity for human and electronic noses: relative roles of odorant vapor pressure vs. molecularly specific odorant binding. Proceedings of the National Academy of Sciences of the USA 95:5442-5447.

Esbensen, K. H. (2001). Multivariate data analysis -In practice. CAMO AS, Oslo, Norway pp. 1-598.

Filtenborg, O., Frisvad, J. C., and Thrane, U. (1996). Moulds in food spoilage. International Journal of Food Microbiology 33:85-102.

Fischer, G., Schwalbe, R., Möller, M., Ostrowski, R., and Dott, W. (1999). Species-specific production of microbial volatile organic compounds (MVOC) by airborne fungi from a compost facility. Chemosphere 39:795-810.

García-González, D. L., Barie, N., Rapp, M., and Aparicio, R. (2004). Analysis of virgin olive oil volatiles by a novel electronic nose based on a miniaturized SAW sensor array coupled with SPME enhanced headspace enrichment. Journal of Agricultural and Food Chemistry 52:7475-7479.

Gardner, J. W. and Bartlett, P. N. (1999). Electronic nose principles and applications. Oxford University Press, Oxford, UK, pp. 1-264.

Grob, K. (1993). Split and splitless injection in capillary gas chromatography. Hüthig, Heidelberg, Germany, pp. 1-547.

Guernion, N., Ratcliffe, N. M., Spencer-Phillips, P. T. N., and Howe, R. (2001). Identifying bacteria in human urine: Current practice and the potential for rapid, near-patient diagnosis by sensing volatile organic compounds. Clinical Chemistry and Laboratory Medicine 39:893-906.

Guilbault, G. and Jordan, J. M. (1988). Analytical uses of piezoelectric crystal: a review. Critical Reviews in Analytical Chemistry 19:1-28.

Gurney, K. (1997). An introduction to neural networks. University College London, London, U.K., pp. 1-248.

Hamm, S., Lesellier, E., Bleton, J., and Tchapla, A. (2003). Optimization of headspace solid phase microextraction for gas chromatography/mass spectrometry analysis of widely different volatility and polarity terpenoids in olibanum. Journal of Chromatography A 1018:73-83.

Hawke, J. C. (1966). Reviews of the progress of dairy science. Section D. Dairy chemistry. The formation and metabolism of methyl ketones and related compounds. Journal of Dairy Research 33:225-243.

Herbert, R. B. (1989). The biosynthesis of secondary metabolites. St Edmundsbury Press Ltd, Bury St Edmunds, UK, pp. 1-231.

Hodgins, D. (1997). Techniques for analyzing food aroma. Marcel Dekker Inc., New York, U.S.A., pp. 331-371.

Holmberg, M. (1997). Data evaluation for an electronic nose. Doctoral thesis at the Department of Physics and Measurement Technology, Linköping University, ISBN: 91-7871-913-5.

Jelén, H. H., Mirocha, C. J., Wasowicz, E., and Kaminski, E. (1995). Production of volatile sesquiterpenes by *Fusarium sambucinum* strains with different abilities to synthesize trichothecenes. Applied and Environmental Microbiology 61:3815-3820.

Jelén, H. H., and Wasowicz, E. (1998). Volatile fungal metabolites and their relation to the spoilage of agricultural commodities. Food Reviews International 14:391-426.

Karahadian, C., Josephson, D. B., and Lindsay, R. C. (1985). Contribution of *Penicillium* sp. to the flavour of Brie and Camembert cheese. Journal of Dairy Science 68:1865-1877.

Karlshøj, K., and Larsen, T. O. (2005). Differentiation of species from the *Penicillium roqueforti* group by volatile metabolite profiling. Journal of Agricultural and Food Chemistry 53:708-715.

Kempler, G. M. (1983). Production of flavour compounds by microorganisms. Advances in Applied Microbiology 29:29-51.

Keshri, G., Magan, N., and Voysey, P. (1998). Use of an electronic nose for the early detection and differentiation between spoilage fungi. Letters in Applied Microbiology 27:261-264.

Kinderlerer, J. (1989). Volatile metabolites of filamentous fungi and their role in food flavour *In* Journal of Applied Bacteriology Symposium Supplement pp. 133S-144S.

King, W. H. (1964). Piezoelectric sorption detector. Analytical Chemistry 36:1735-1739.

Kinsella, J. E., and Hwang, D. H. (1976). Enzymes of *Penicillium roqueforti* involved in the biosynthesis of cheese flavour. Critical Reviews in Food Science and Nutrition 8:191-228.

Kohl, D. (2001). Function and applications of gas sensors. Journal of Physics D: Applied Physics 34:R125-R149.

König, W. A., Bölow, N., and Saritas, Y. (1999). Identification of sesquiterpene hydrocarbons by gas phase analytical methods. Flavour and Fragrance Journal 14:367-378.

Larsen, T. O. (1994). Chemosystematics of species in genus Penicillium based on profiles of volatile metabolites. Doctoral thesis at the The Mycology Group, Department of Biotechnology, Technical University of Denmark, ISBN: 87-88584-03-8.

Larsen, T. O. (1997). Identification of cheese-associated fungi using selected ion monitoring of volatile terpenes. Letters in Applied Microbiology 24:463-466.

Larsen, T. O., and Frisvad, J. C. (1995a). Characterization of volatile metabolites from 47 *Penicillium*-taxa. Mycological Research 99:1153-1166.

Larsen, T. O., and Frisvad, J. C. (1995b). Chemosystematics of *Penicillium* based on profiles of volatile metabolites. Mycological Research 99:1167-1174.

Larsen, T. O., Svendsen, A., and Smedsgaard, J. (2001). Biochemical characterization of ochratoxin A-producing strains of the genus *Penicillium*. Applied and Environmental Microbiology 67:3630-3635.

Leete, E., Bjorklund, J. A., Reineccius, G. A., and Cheng, T. B. (1991). Biosynthesis of 3-isopropyl-2-methoxypyrazine and other alkylpyrazines: widely distributed flavour compounds. *In* Bioformation of Flavours (Patterson, R. L. S., Charlwood, B. V., MacLeod, G., and Williams, A. A., eds.), The Royal Society of Chemistry, Cambridge, U.K., pp. 75-95.

Luckner, M. (1990). Secondary metabolism in microorganisms, plants and animals. Springer-Verlag, Berlin, Germany, pp. 1-549.

Lundström, I., Hedborg, E., Spetz, A., Sundgren, H., and Winquist , F. (1992). Electronic noses based on field effect structures *In* Sensors and sensory systems for an electronic nose (Gardner, J. W. and Bartlett, P. N., ed.), Proceedings of the NATO advanced research workshop, Reykjavik 1991, Kluwer, Dordrecht, The Netherlands, pp. 303-319.

Lundström, I., Shivaraman, S., Svensson, C., and Lundkvist, L. (1975). A hydrogen-sensitive MOS field-effect transistor. Applied Physics Letters 26:55-57.

Lundström, I., Spetz, A., Winquist, F., Ackelid, U., and Sundgren, H. (1990). Catalytic metals and field-effect devices. A useful combination. Sensors and Actuators B: Chemical B1:15-20.

Lundström, I., Svensson, C., Spetz, A., Sundgren, H., and Winquist, F. (1993). From hydrogen sensors to olfactory images - twenty years with catalytic field-effect devices. Sensors and Actuators, B: Chemical B13:16-23.

Lynen, F. (1959). New aspects of acetate incorporation into isoprenoid precursors *In* Biosynthesis of terpenes and sterols (Wolstenholme, G. E. W., and O'Conner, M., eds.), J. and A. Churchill Ltd., London, U.K., pp. 95-116.

Magan, N., Pavlou, A., and Chrysanthakis, I. (2001). Milk-sense: a volatile sensing system recognises spoilage bacteria and yeasts in milk. Sensors and Actuators B: Chemical 72:28-34.

Marrazzo, W. N., Heinemann, P. H., Crassweller, R. E., and LeBlanc, E. (2005). Electronic nose chemical sensor feasibility study for the differentiation of apple cultivars. Transactions of the American Society of Agricultural Engineers 48:1995-2002.

Martin, S. J., Schwartz, S. K., Gunshor, R. L., and Pierret, R. F. (1983). Surface acoustic wave resonators on a ZnO-on-Si layered medium. Journal of Applied Physics 54:561-569.

Martin, S. J., Schweizer, K. S., Schwartz, S. K., and Gunshor, R. L. (1984). Vapor sensing by means of a ZnO-on-Si surface acoustic wave resonator. *In* Proceeding IEEE Ultrasonics Symposium, Dallas, USA, pp. 207-212.

Martin, S. J., Schweizer, K. S., Ricco, A. J., and Zipperian, T. E. (1985). *In* Gas sensing with surface acoustic wave devices. Transducers '85: 1985, International Conference on Solid-State Sensors and Actuators - Digest of Technical Papers, pp. 71-73.

Maul, F., Sargent, S. A., Sims, C. A., Baldwin, E. A., Balaban, M. O., and Huber, D. J. (2000). Tomato flavour and aroma quality as affected by storage temperature. Journal of Food Science 65:1228-1237.

Medsker, L., Jenkins, L. D., Thomas, J. F., and Kock, C. (1969). Odorous compounds in natural waters. 2-Exo-hydroxy-2-methyl-boran, the major compound produced by several actinomycetes. Environmental Science and Technology 3:476-477.

Mielle, P. (1996). "Electronic noses": Towards objective instrumental characterization of food aroma. Trends in Food Science and Technology 7:432-438.

Morrison, S. R. (1987). Selectivity in semiconductor gas sensors. Sensors and Actuators 12:425-440.

Müller, R., and Lange, E. (1986). Multidimensional sensor for gas analysis. Sensors and Actuators 9:39-48.

Nielsen, K. F., Smedsgaard, J., Larsen, T. O., Lund, F., Thrane, U., and Frisvad, J. C. (2003). Chemical identification of fungi: Metabolite profiling and metabolomics. *In* Fungal Biotechnology in Agricultural, Food, and Environmental Applications (Arora, D. K., ed.), Marcel Dekker, Inc., New York, U.S.A., pp. 19-35.

Nieuwenhuizen, M. S., and Nederlof, A. J. (1992). Silicon based surface acoustic wave gas sensors *In* Sensors and sensory systems for an electronic nose (Gardner, J. W., and Bartlett, P. N., ed.), Proceedings of the NATO advanced research workshop, Reykjavik 1991, Kluwer, Dordrecht, The Netherlands, pp. 131-145.

Partridge, A. C., Harris, P., and Andrews, M. K. (1996). High sensitivity conducting polymer sensors. Analyst 121:1349-1353.

Pasanen, A. L., Lappalainen, S., and Pasanen, P. (1996). Volatile organic metabolites associated with some toxic fungi and their mycotoxins. Analyst 121:1949-1954.

Pearce, T. C. (1997). Computational parallels between the biological olfactory pathway and its analogue 'The Electronic Nose': : Part I. Biological olfaction. Biosystems 41:43-67.

Pelosi, P. and Persaud, K. (1988). Sensors and sensory systems for advanced robots. Springer-Verlag, Berlin, Germany, pp. 361-381.

Persaud, K., and Dodd, G. (1982). Analysis of discrimination mechanisms in the mammalian olfactory system using a model nose. Nature 299:352-355.

Pitt, J. I., and Hocking, A. D. (1985). *Fungi and food spoilage*. Academic Press, Sydney, Australia, pp. 1-413.

Pohland, A. E. (1993). Mycotoxins in review. Food Additives and Contaminants 10:17-28.

Ramaswami, S. K., Briscese, P., Gargiullo, R. J., and von Geldern, T. (1988). Sesquiterpene hydrocarbons: from mass confusion to orderly line-up. *In* Flavours and Fragrances: A World Perspective. Proceedings of the 10th International Congress of Essential Oils, Fragrances and Flavours, Washington, DC, U.S.A., 16-20 November 1986, pp. 951-980.

Richards, J. H., and Hendrickson, J. B. (1964). The biosynthesis of steroids, terpenes and acetogenins. W.A. Benjamin, Inc., New York, USA, pp. 1-416.

Rye, G. G., and Mercer, D. G. (2003). Changes in headspace volatile attributes of apple cider resulting from thermal processing and storage. Food Research International 36:167-174.

Samson, R. A., Hoekstra, E. S., Frisvad, J. C., and Filtenborg, O. (2002). Introduction to food- and airborne fungi. Centraalbureau voor Schimmelcultures, Utrecht, The Netherlands, pp. 1-389.

Schaller, E., Bosset, J. O., and Escher, F. (1998). "Electronic Noses" and their application to food. Food Science and Technology 31:305-316.

Severin, E. J., Sanner, R. D., Doleman, B. J., and Lewis, N. S. (1998). Differential detection of enantiomeric gaseous analytes using carbon black-chiral polymer composite, chemically sensitive resistors. Analytical Chemistry 70:1440-1443.

Shiers, V. P. (1995). Electronic nose technology - evaluations and developments for the food industry *In* Food Ingredients Europe: Conference Proceedings, pp. 198-200.

Shimizu, Y., and Egashira, M. (1999). Basic aspects and challenges of semiconductor gas sensors. MRS Bulletin 24:18-24.

Sim, C. O., Ahmad, M. N., Ismail, Z., Othman, A. R., Noor, N. A. M., and Zaihidee, E. M. (2003). Chemometric classification of herb - *Orthosiphon stamineus* according to its geographical origin using virtual chemical sensor based upon fast GC. Sensors 458-471.

Skoog, D. A., West, D. M., and Holler, F. J. (1996). Fundamentals of analytical chemistry international edition. Saunders College Publishing, New York, U.S.A., pp. 1-870.

Spetz, A., Winquist, F., Sundgren, H., and Lundström, I. (1992). Field effect gas sensors *In* Gas Sensors (Sberveglieri, G., ed.), Kluwer Academic Publishers, Dordrecht, The Netherlands, pp. 219-279.

Strike, D. J., Meijerink, M. G. H., and Koudelka-Hep, M. (1999). Electronic noses - A mini-review. Fresenius' Journal of Analytical Chemistry 364:499-505.

Sundgren, H., Lundström, I., Winquist, F., Lukkari, I., Carlsson, R., and Wold, S. (1990). Evaluation of a multiple gas mixture with a simple MOSFET gas sensor array and pattern recognition. Sensors and Actuators, B: Chemical B2:115-123.

Suomalainen, H. (1971). Yeast and its effect on the flavour of alcoholic beverages. Journal of the Institute of Brewing 77:164-177.

Trihaas, J., and Nielsen, P. V. (2005). Electronic nose technology in quality assessment: monitoring the ripening process of Danish blue cheese. Journal of Food Science 70:44-49.

Trihaas, J., Nielsen, P. V., and van den Tempel, T. (2005a). Electronic nose technology in quality assessment. Predicting the volatile composition of Danish blue cheese. Journal of Food Science 70:392-400.

Trihaas, J., Vognsen, L., and Nielsen, P. V. (2005b). Electronic nose: New tool in modelling the ripening of Danish blue cheese. International Dairy Journal 15:679-691.

Vinaixa, M., Marin, S., Brezmes, J., Llobet, E., Vilanova, X., Correig, X., Ramos, A., and Sanchis, V. (2004). Early detection of fungal growth in bakery products by use of an electronic nose based on mass spectrometry. Journal of Agricultural and Food Chemistry 52:6068-6074.

Waller, J. M., and Brayford, D. (1990). *Fusarium* diseases in the tropics. Tropical Pest Management 36:181-194.

Walt, D. R., Dickinson, T., White, J., Kauer, J., Johnson, S., Engelhardt, H., Sutter, J., and Jurs, P. (1998). Optical sensor arrays for odor recognition. Biosensors and Bioelectronics 13:697-699.

Watson, J. and Yates, R. A. (1985). A solid-state gas sensor. Electronic Engineering 57:47-57.

Werlein, H. D. (2001). Discrimination of chocolates and packaging materials by an electronic nose. European Food Research and Technology 212:529-533.

Wilson, K. and Walker, J. (1994). Principles and techniques of practical biochemistry. Cambridge University Press, Cambridge, U.K., pp. 1-586.

Winquist, F., Sundgren, H., Hedborg, E., Spetz, A., and Lundström, I. (1992). Visual images of gas mixtures produced with field-effect structures. Sensors and Actuators, B: Chemical B6:157-161.

Wohltjen, H. and Dessy, R. (1979a). Surface acoustic wave probe for chemical analysis I. Introduction and instrument description. Analytical Chemistry 51:1458-1465.

Wohltjen, H. and Dessy, R. (1979b). Surface acoustic wave probe for chemical analysis II. Gas chromatography detector. Analytical Chemistry 51:1465-1470.

Wohltjen, H. and Dessy, R. (1979c). Surface acoustic wave probe for chemical analysis III. Thermomechanical polymer analyzer. Analytical Chemistry 51:1470-1478.

Wünsche, L. F., Vuilleumier, C., Keller, U., Byfield, M. P., May, I. P., and Kearney, M. J. (1995). Scent characterisation: from human perception to electronic nose *In* Proceedings of the 13th International Congress of Flavours, Fragrances and Essential Oils, 15-19 October 1995, Istanbul, Turkey, pp. 295-313.

Wurzenberger, M. and Grosch, W. (1982). The enzymatic oxidative breakdown of linoleic acid in mushrooms (*Psalliota bispora*). Zeitschrift für Lebensmittel-Untersuchung und -Forschung 175:186-190.

Wurzenberger, M. and Grosch, W. (1984). Origin of the oxygen in the products of the enzymatic cleavage reaction of linoleic acid to 1-octen-3-ol and 10-oxo-*trans*-8-decenoic acid in mushrooms (*Psalliota bispora*). Biochemica et Biophysica Acta 794:18-24.

Zeringue, H. J., Jr., Bhatnagar, D., and Cleveland, T. E. (1993). $C_{15}H_{24}$ volatile compounds unique to aflatoxigenic strains of *Aspergillus flavus*. Applied and Environmental Microbiology 59:2264-2270.

Zondervan, C., Muresan, S., Jonge, H. G., Velzen, E. U. T., Wilkinson, C., Nijhuis, H. H., and Leguijt, T. (1999). Controlling Maillard reactions in the heating process of blockmilk using an electronic nose. Journal of Agricultural and Food Chemistry 47:4746-4749.

Chapter 15

Wine and fungi – implications of vineyard infections

Su-lin L. Leong
Food Science Australia, P.O. Box 52, North Ryde NSW 1670; School of Agriculture, Food and Wine, University of Adelaide; Cooperative Research Centre for Viticulture, Australia.

INTRODUCTION

The making of wine is a fungal biotransformation; in addition, fungi have the potential to affect this process at every stage from the grape vine to consumption. Several fungi are pathogenic to grape vines, infecting the roots, trunk, canes, leaves and berries (Pearson and Goheen, 1988). Fungal diseases may affect crop yield and quality, which, in turn, may affect wine at its most fundamental level – the decision whether or not it is made! Fungi which infect berries (Hewitt, 1988; Emmett *et al.*, 1992) include the mildew pathogens *Erysiphe necator (Uncinula necator)* and *Plasmopara viticola, Alternaria* spp., *Aspergillus* spp., *Botrytis cinerea, Cladosporium* spp., *Penicillium* spp. and *Rhizopus* spp. These infections are well-recognised and either cause damage sporadically in seasons in which conditions favour disease, or are controlled through vineyard management techniques, such as fungicidal sprays. Changes in the berry due to fungal infection may in turn alter the properties of the wine.

Once vinification commences, filamentous fungi are less of a consideration, although their prior activity may affect the wine. Yeast strains are selected to confer desirable flavours and aromas to the wine during fermentation; conversely, undesirable strains and species may kill the desirable yeasts, hinder fermentation or produce off-flavours (Kunkee and Bisson, 1993; Fleet, 2003; Loureiro and Malfeito-Ferreira, 2003). Post-fermentation, yeasts may continue to produce off-flavours, referment residual sugars, grow on the surface as a film, or cause cloudiness and sediment formation. These problems are generally well managed in wineries; however, *Dekkera* spp. (anamorphs in *Brettanomyces*) are spoilage yeasts which appear to be of growing concern to the industry. After bottling, filamentous fungi may again affect wine through their involvement in "cork taint" (Lee and Simpson, 1993).

A recent mycological food safety challenge to confront the wine industry is the production of ochratoxin A (OA), an isocoumarin derivative linked to phenylalanine, by black *Aspergillus* spp. in grapes, and its subsequent passage into wine. Aspergillus bunch rots caused by black *Aspergillus* spp. sporadically occur in vineyards situated in warm to temperate regions (Snowdon, 1990). These fungi typically invade berries via insect punctures or splits in the berry skin. Abarca *et al.* (1994) reported production of OA by two strains of *A. niger* from feed, and in the following year, Horie (1995) reported OA production by a second species, *A. carbonarius*, which has long been known to cause grape rots (Gupta, 1956). OA, a nephrotoxin, was detected in wine only recently (Majerus and Otteneder, 1996; Zimmerli and Dick, 1996), and Heenan *et al.* (1998) isolated ochratoxigenic black *Aspergillus* spp. during dried grape processing, suggesting that such species may indeed be the source of OA in grapes. This chapter summarises current research on the occurrence of OA in wine, its production in grapes, fate during vinification, and the potential contribution of DNA-based studies to this problem. The effects of some other fungal infections on wine quality are also discussed.

OCHRATOXIN A IN WINE

The incidence and degree of OA contamination in wine is now fairly well documented (Figure 1; tabulated in Otteneder and Majerus (2000), Bellí *et al.* (2002), Stefanaki *et al.* (2003) and Blesa *et al.* (2006)). OA was detected more frequently in wines from Mediterranean countries and northern Africa, following a trend for increased prevalence in wines from southern (warmer) regions compared with northern regions, and in red compared to white wine (Majerus and Otteneder, 1996; Zimmerli and Dick, 1996; Ospital *et al.*, 1998; Otteneder and Majerus, 2000; Markarki *et al.*, 2001; Pietri *et al.*, 2001). Occurrence of OA in wines from the so-called "new world" (USA, Canada, South America, South Africa, Australia, New Zealand) was low. In Australia and South Africa, no obvious differences were observed between OA in red and white wines, and wines from warmer areas did not show increased contamination rates (Stander and Steyn, 2002; Hocking *et al.*, 2003; Leong *et al.*, 2006a). Less expensive wines, such as those sold in plastic-lined cardboard boxes, were more frequently contaminated than bottled wines in Australia, Italy and South Africa (Tateo *et al.*, 2000; Tateo and Bononi, 2001; Stander and Steyn, 2002; Hocking *et al.*, 2003; Tateo and Bononi, 2003). OA contamination was also more frequent in certain dessert or fortified wines such as Moscatel, Marsala and Malaga (Zimmerli and Dick, 1996; Burdaspal and Legarda, 1999; Stander and Steyn, 2002; Bellí *et al.*, 2004a; Blesa *et al.*, 2004), whereas its occurrence in ports and sherries was not greater than in still table wines (Festas *et al.*, 2000; Ratola *et al.*, 2004).

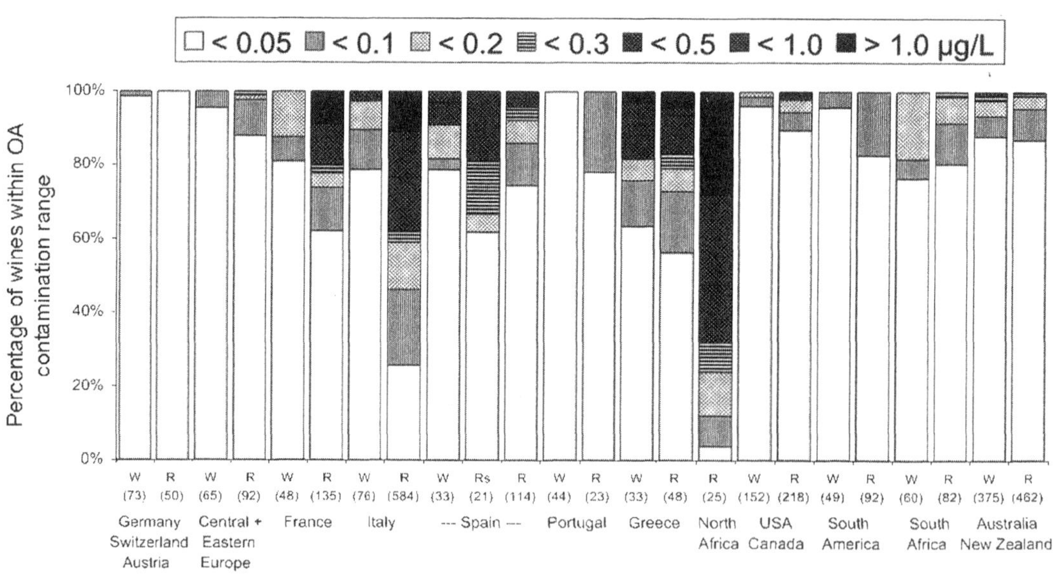

Figure 1. Incidence and degree of OA contamination in wines produced in viticultural regions worldwide. White wines (W), red wines (R), rosé (Rs); the cumulative number of wines tested for each region from various surveys is given in parentheses. Dessert and fortified wines are not included. Data presented here were calculated from Majerus and Otteneder (1996), Zimmerli and Dick (1996), MAFF (1997), Ospital *et al.* (1998), MAFF (1999), Tateo *et al.* (1999), Visconti *et al.* (1999), Castellari *et al.* (2000), Festas *et al.* (2000), Tateo *et al.* (2000), Filali *et al.* (2001), Markarki *et al.* (2001), Pietri *et al.* (2001), Soleas *et al.* (2001), Tateo and Bononi (2001), Eder *et al.* (2002), Stander and Steyn (2002), Hocking *et al.* (2003), Micheli *et al.* (2003), Siantar *et al.* (2003), Soufleros *et al.* (2003), Tateo and Bononi (2003), Blesa *et al.* (2004), Ng *et al.* (2004), Ratola *et al.* (2004), Rosa *et al.* (2004), Czerwiecki *et al.* (2005) and Leong *et al.* (2006a).

OA has also been detected in grape juices and wine vinegars [reviewed by Bellí *et al.* (2002); see also MAFF (1999), Majerus *et al.* (2000), Markarki *et al.* (2001), Roset (2003), Bellí *et al.* (2004a), Ng *et al.* (2004), Rosa *et al.* (2004), and Czerwiecki *et al.* (2005)]. Quantification of OA in wine is performed by a variety of methods [reviewed by Bellí *et al.* (2002), see also Saez *et al.* (2004), Visconti and De Girolamo (2005)]; currently, sample purification with immuno-affinity columns followed by quantification by HPLC is most widely used.

OCHRATOXIGENIC FUNGI IN VINEYARDS

Black *Aspergillus* spp. which produce ochratoxins have frequently been isolated from grapes in France (Sage *et al.*, 2002; Sage *et al.*, 2004; Bejaoui *et al.*, 2006), Greece (Tjamos *et al.*, 2004; Tjamos *et al.*, 2006), Italy (Battilani *et al.*, 2003b), Portugal (Serra *et al.*, 2003; Serra *et al.*, 2005), Spain (Cabañes *et al.*, 2002; Abarca *et al.*, 2003; Bellí *et al.*, 2004c; Bau *et al.*, 2005), Israel (Guzev *et al.*, 2006), South America (Da Rocha Rosa *et al.*, 2002; Magnoli *et al.*, 2003; Magnoli *et al.*, 2004) and Australia (Heenan *et al.*, 1998; Leong *et al.*, 2004). *A. niger* was typically isolated more frequently than *A. carbonarius* or uniseriate species such as *A. aculeatus* or *A. japonicus*. Most studies reported that *A. carbonarius* isolates were frequently toxigenic (up to 100% of isolates), whereas a relatively small proportion of *A. niger* isolates produced OA [reviewed by Abarca *et al.* (2004)]. Reports of toxin production by a few *A. aculeatus* or *A. japonicus* isolates are yet to be confirmed. Toxigenic isolates of *A. ochraceus* have only occasionally been isolated from grapes (Da Rocha Rosa *et al.*, 2002; Battilani *et al.*, 2003b; Serra *et al.*, 2003; Bellí *et al.*, 2004c; Bau *et al.*, 2005). It can safely be concluded that *A. carbonarius* is the primary source of OA contamination of grapes.

The occurrence of black *Aspergillus* spp. on grapes significantly correlated with increased temperature, and, to a lesser extent, with increased humidity and rainfall, based on data from Spanish viticultural regions over three seasons (Bellí *et al.*, 2005a). Italian data likewise showed a positive correlation between black *Aspergillus* spp. and temperature, but a negative correlation with rainfall (Battilani *et al.*, 2006c). Black aspergilli were isolated more frequently from warmer regions with a Mediterranean climate than from temperate regions in France (Sage *et al.*, 2004), Italy (Battilani *et al.*, 2003a), Portugal (Serra *et al.*, 2003; Serra *et al.*, 2005) and Spain (Bellí *et al.*, 2005a). Bellí *et al.* (2005a) also reported increased isolation of black aspergilli during the warmest of three seasons. Serra *et al.* (2003) observed that black aspergilli were more frequently isolated from a hot, dry region than from a temperate, humid region, suggesting that the effect of temperature is stronger than that of humidity. Roset (2003) noted that OA in grape juice correlated with increased pre-harvest temperature, rainfall, proximity to the coast, and later date of harvest; similar trends regarding temperature and rainfall were reported by Battilani *et al.* (2003a).

Figure 2. *Aspergillus carbonarius* infection of Semillon berries.

OCHRATOXIN A PRODUCTION IN GRAPES

Soil directly beneath vines was a common source of *A. carbonarius* in Australian vineyards, where incidence was affected by soil type, moisture, tillage and mulching (Leong *et al.*, 2006a). *A. carbonarius* was also isolated from air close to vines, thus it is likely that spores from soil are deposited on berries by wind. Populations of black aspergilli in soil and on bunches fluctuate with time, and several authors have reported an increase in black aspergilli on grapes from berry set until harvest (Nair, 1985; Battilani *et al.*, 2003a; Serra *et al.*, 2003; Bellí *et al.*, 2004c; Bau *et al.*, 2005; Bellí *et al.*, 2005a; Serra *et al.*, 2005), although this trend was not consistent for all black *Aspergillus* spp. over three years in Italian vineyards (Battilani *et al.*, 2006c). Leong *et al.* (2006f) postulated that black aspergilli were seldom isolated early during the season because the surface of green berries, and exposure to UV light, provided a hostile environment for the survival of *A. carbonarius* spores. After veraison, however, berries were susceptible to infection with *A. carbonarius* (Figure 2). Drying of grapes during the production of liqueurs appeared to increase the population of *A. carbonarius* relative to *A. niger*, thus increasing the risk of OA contamination of such wines (Gómez *et al.*, 2006).

Battilani *et al.* (2004) reported differences among grape varieties in susceptibility to infection and OA production *in vitro*; furthermore, OA production correlated with severity of infection for certain varieties but not others. These trends were not necessarily reflected by infection and OA contamination in vineyards; differences among varieties were often associated with seasonal variations in climate and time of ripening (Battilani *et al.*, 2006c; Leong *et al.*, 2006a).

The presence of toxigenic isolates on grapes did not always correlate with OA in those samples (Battilani *et al.*, 2006c; Bellí *et al.*, 2005a), as toxigenic isolates survived on the berry surface without causing infection (Leong *et al.*, 2006f). Damage to grapes, both in nature and *in vitro* increased the severity of infection with black *Aspergillus* spp. (Leong *et al.*, 2004; Leong *et al.*, 2006a) and also OA contamination (Battilani *et al.*, 2004; Leong *et al.*, 2007). More puzzling is the suggestion that OA is produced in berries displaying no visible symptoms of infection with black *Aspergillus* spp. (P. Battilani; N. Bellí, pers. comm.) and the report of OA in grapes from which toxigenic fungi were not isolated from berries that had been surface sterilised (Serra *et al.*, 2006b). As OA is an excreted secondary metabolite, it may remain in berries even after fungal death.

The effects of water activity and temperature on growth and toxin production by *A. carbonarius* and *A. niger* have been studied using a synthetic grape juice medium (SGM) designed to simulate the berry composition at early veraison. On this medium, the optimum temperature for growth of *A. carbonarius* (30 °C) was lower than that for *A. niger* (approximately 35 °C) (Battilani *et al.*, 2003c; Bellí *et al.*, 2004b; Leong *et al.*, 2004; Mitchell *et al.*, 2004; Bellí *et al.*, 2005b). For both species, the optimum water activity for growth was approximately 0.97–0.99 (Bellí *et al.*, 2004b; Mitchell *et al.*, 2004; Bellí *et al.*, 2005b); however, the optimum for *A. niger* was closer to the upper limit of this range (Leong *et al.*, 2006c). *A. niger* was also more tolerant of water activities below 0.95 (Bellí *et al.*, 2004b; Leong *et al.*, 2006c). The optimum temperature for toxin production for both species, typically reported as 15–20 °C, was lower than that for growth. Reports of optimum water activity for toxin production differed among authors but were within the range 0.95–0.995 (Battilani *et al.*, 2003c; Bellí *et al.*, 2004b; Mitchell *et al.*, 2004; Bellí *et al.*, 2005b; Leong *et al.*, 2006c). Reports of greater OA production at temperatures below 25 °C were supported by data on OA in grapes inoculated with *A. carbonarius in vitro* (Battilani *et al.*, 2004); however, these authors also noted more growth at 20 °C that at 25 °C, which is contrary to the reports of maximum growth on SGM at approximately 30 °C discussed above. Despite an optimal temperature for OA production by *A. carbonarius* around 20 °C, diurnal temperature fluctuations (28 °C/20 °C) did not increase OA yield on SGM over that obtained at 28°C. However, alternating photoperiods increased growth rate (Bellí *et al.*, 2006).

Black *Aspergillus* spp. generally degrade OA after it is produced (Bellí *et al.*, 2004d; Esteban *et al.*, 2004; Leong *et al.*, 2006c), and this capability is not restricted to toxigenic isolates (Abrunhosa *et al.*, 2002; Abrunhosa *et al.*, 2003). Degradation is thought to begin by cleavage of the molecule into the isocoumarin portion, ochratoxin α, and phenylalanine, and then to other undetermined products (Varga *et al.*, 2000b, Abrunhosa *et al.*, 2002). Hypothetically, degradation of OA could be beneficial to the fungus as a source of organic nitrogen to support further growth.

Many of the studies on the growth of black *Aspergillus* spp. and OA production in grapes have been based on the understanding that Aspergillus bunch rots occur post-veraison (Emmett *et al.*, 1992), and thus ochratoxigenic fungi as well as OA contamination of berries would increase as berries mature (Lataste *et al.*, 2004). However, the potential for OA production by *A. carbonarius* inoculated onto green, pea-sized berries has been demonstrated by Battilani *et al.* (2001) (0.1 μg/kg on intact berries, 25 μg/kg on damaged berries) and Serra *et al.* (2006b) (5244 μg/kg on homogenised grape berries). Serra *et al.* (2006b) went on to suggest that OA production was greater, specifically, on immature green berries, than on berries at early veraison and harvest, even though immature berries contain very little sugar and the pH below 3 should be somewhat restrictive for toxin production by a majority of strains (Esteban *et al.*, 2005). Furthermore, in certain vineyards, more toxin was detected in immature green berries than in berries at early veraison and harvest, but toxigenic fungi were not isolated from immature berries. This raises the following questions:

- Does a type of latent or slow infection occur early during the season, leading to OA production in immature berries?
- Does infection develop as a mycelial network loosely attached to the berry surface, causing no visible damage?
- Or, as is more commonly believed, are the black aspergilli exclusively opportunistic fungi, surviving on the surface of berries until damage to the skin allows invasion of the berry pulp?

Additional research is required to clarify the timing and nature of black *Aspergillus* infection and OA production.

OCHRATOXIN A DURING WINEMAKING

OA production in grapes ceases at the commencement of processing, typically sterilisation with sulfur dioxide in industrial juice and wine production (Roset, 2003); Fernandes *et al.* (2007) also demonstrated that OA is not produced during vinification. Hence, the concentration of OA in the final product is a function of the initial concentration in the grapes and any reduction during processing.

In the production of white wine, grapes are crushed, then pressed to remove the skins and seeds. The juice may be treated with a pectinase to enhance precipitation of grape solids prior to the commencement of fermentation. In the production of red wines, grapes are crushed then fermented in the presence of skins and seeds to extract colour and tannins. This mixture is later pressed to remove the skins and seeds. Both white and red wines undergo successive clarification stages to remove precipitated yeasts and other solids. Malolactic fermentation, in which malic acid is converted into lactic acid by lactic acid bacteria, may also occur after fermentation. Some OA is removed at each of these stages.

The greatest reduction in OA concentration (40–92%) typically occurs when grapes are pressed, due to binding of the toxin to skins and seeds (Fernandes *et al.*, 2003; Leong *et al.*, 2006b; Leong *et al.*, 2006g; Fernandes *et al.*, 2007). Bejaoui *et al.* (2004) noted decreases in OA during fermentation, which were affected by choice of yeast strain. They postulated that these decreases occurred due to binding of OA to yeast cells, rather than degradation by the yeasts, as no degradation products were observed. The fate of radio-labelled OA during fermentation supports this hypothesis (Lataste *et al.*, 2004). Such binding of OA to grape solids and yeast cells during clarification of juice or wine resulted in further reductions in OA (yielding 2–37% of the concentration initially present in grapes; Roset, 2003; Fernandes *et al.*, 2003; Leong *et al.*, 2006b; Leong *et al.*, 2006g; Fernandes *et al.*, 2007). Malolactic fermentation of wine by lactic acid bacteria was reported to reduce the OA concentration by 56% (Grazioli *et al.*, 2006). The extent of OA reduction varied according to the strain of lactic bacteria (Silva *et al.*, 2003). Certain strains appeared to degrade OA during malolactic fermentation (M.D. Fumi, unpublished data), whereas, for other strains, degradation products were not detected. Rather, the bacterial biomass bound OA and precipitated the toxin from the wine (Fernandes *et al.*, 2003; Fernandes *et al.*, 2007). The cumulative effects of treatments during vinification resulted in a final OA concentration in wine that was 4–13% of the concentration in the grapes initially (Leong *et al.*, 2006b; Leong *et al.*, 2006g; Fernandes *et al.*, 2007). OA appeared to be fairly stable in some finished wines over a year (Lopez de Cerain *et al.*, 2002), whereas in other wines, decreases in OA concentration of up to 29% were reported over a similar storage period (Garcia Moruno *et al.*, 2005; Grazioli *et al.*, 2006; Leong *et al.*, 2006e).

MINIMISATION OF OCHRATOXIN A IN WINE

Black *Aspergillus* infection and OA production in vineyards may be minimised by reducing the presence of these fungi in soil, such as through management of moisture, tillage and mulching (Leong *et al.*, 2006a). Lataste *et al.* (2004) and Cozzi *et al.* (2006) demonstrated the importance of minimising berry damage, such as that caused by insects. Sprays containing a mixture of cyprodinil and fludioxonil were effective in reducing *Aspergillus* rots and OA production when applied at veraison and later (Lataste *et al.*, 2004; Tjamos *et al.*, 2004; Battilani *et al.*, 2006c); however, application so soon before harvest is not permitted in countries such as Australia (Bell and Daniel, 2004). Biocontrol of Aspergillus rot through application of epiphytic yeasts onto grapes has shown some efficacy (Zahavi *et al.*, 2000; Bleve *et al.*, 2006).

Tools to assess the risk of OA contamination before harvest include a model based on seasonal temperature and rainfall to describe areas in Southern Europe and Israel likely to display increased incidence of black *Aspergillus* spp. on grapes (Battilani *et al.*, 2006a). A rapid immunogenic method (lateral flow device) is available to determine what proportion of such isolates produce OA when grown on standard mycological media (Danks *et al.*, 2003). Assessment of OA contamination of grapes (and wine) may be facilitated by a number of novel antibody-based methods, both qualitative and quantitative, currently under development [reviewed by Visconti and De Girolamo (2005)]. Selection of bunches throughout the vineyard according to a statistically validated protocol will ensure that the extent of OA contamination in those bunches is representative of the entire vineyard plot (Battilani *et al.*, 2006b). Such data can inform decisions regarding spraying of the crop, harvest date, and style of vinification. Cool storage of harvested grapes and sanitary wineries minimise the risk of OA contamination postharvest (Gambuti *et al.*, 2005).

Management of the solid-liquid separation stages are the key to reduction of OA during winemaking. Avoiding pressing of grapes at high pressures may limit extraction of OA into the wine (Gambuti *et al.*, 2005). Yeast strains with mannoproteins that adsorb more toxin may be selected for fermentation, or even added at some stage as a preparation of dead cells purely for the purpose of binding OA (Caridi, 2006). Such binding is enhanced by heat-treatment of yeast cells and in acidic conditions that cause protonation of the OA molecule (Bejaoui *et al.*, 2004). During malolactic fermentation, selection of lactic acid bacterial strains capable of degrading OA may confer additional reductions in concentration of up to 80% (Silva *et al.*, 2003; M.D. Fumi, unpublished data). Fining agents that contain carbon remove OA most effectively, although these are often detrimental to wine quality (Dumeau and Trione, 2000; Castellari *et al.*, 2001; Silva *et al.*, 2003, Gambuti *et al.*, 2005). Proteinaceous agents such as egg albumin and gelatin also bind OA; however, their efficacy, and that of other agents such as bentonite and yeast hulls (cell walls), may be dependent on other wine components present (Leong *et al.*, 2006e). Multiple methods exist for screening of finished wines for OA, and some methods are amenable to automation (Brera *et al.*, 2003).

One aspect of winemaking yet to be studied is the disposal of OA-contaminated solids, whether grape skins and seeds or precipitated proteins and yeasts. Discarded grape stems, skins and seeds may have some commercial value as sources of tannins or as substrate for fermentation and subsequent distillation to produce the "grape alcohol" used in fortified wines. OA is fairly heat stable (Roset, 2003; Arici *et al.*, 2004), and risk of carry-over into these extracts is yet to be assessed.

MOLECULAR ANALYSES OF OCHRATOXIGENIC BLACK *ASPERGILLUS* SPECIES

No consideration of a new mycological challenge in the area of winemaking would be complete without assessing the potential contribution of DNA-based techniques. Black *Aspergillus* spp. have been examined by a suite of molecular techniques, usually for the purpose of speciation and with a particular focus on members of the morphologically indistinguishable *A. niger* aggregate [reviewed by Varga *et al.* (2000a), Abarca *et al.* (2004), and Varga *et al.* (2004a)]. Toxigenicity occurs quite rarely among *A. niger* isolates, and, to date, has been demonstrated only in a proportion of strains having type N profile based on RFLP analysis of 5.8S rDNA (Accensi *et al.*, 2001). In contrast, strains having type T profile are generally non-toxigenic. This N / T type separation was also observed in strains isolated from grapes and analysed by amplified fragment length polymorphisms (AFLP) (Perrone *et al.*, 2006). Among *A. carbonarius* strains, certain non-toxigenic isolates have been characterised as a new species, *A. ibericus,* based on genetic analysis as well as morphology (Serra *et al.*, 2006a). Further studies of relatedness among toxigenic and non-toxigenic isolates of *A. niger* and *A. carbonarius* [reviewed by Niessen *et al.* (2005)] suggest that finding a molecular marker common only to toxigenic strains would be difficult.

Preliminary studies suggested that molecular relatedness of strains did not correspond with substrate or country of isolation for a set of 68 isolates of *A. niger* and *A. carbonarius,* including some isolates from vineyards in Europe and Australia (Esteban, A., Leong, S. and Tran-Dinh, N., unpublished data). This supports the belief that black *Aspergillus* spp. are cosmopolitan in their distribution over a range of substrates and locations (Klich and Pitt, 1988). This belief has two implications for strategies to reduce OA contamination in vineyards. First, as physiological differences in growth and toxin production among isolates have been reported (Mitchell *et al.*, 2004), strategies must be effective for a reasonably diverse range of isolates potentially encountered within a single region. Second, differences among isolates are unlikely to hinder the application of strategies developed in one region to another region, as inter-regional diversity is not necessarily greater than intra-regional diversity. Rather, strategies may need to vary from region to region due to differences in climate, grape variety and vineyard management.

In addition to enhancing the understanding of strain relatedness, molecular techniques may be applied in the viticultural setting for the rapid detection of toxigenic species in grapes. PCR-based techniques are more rapid than isolation, identification and screening for toxin production on traditional culture media; furthermore, they overcome the need for expertise in identification of isolates based on morphological characteristics. Sensitive and specific PCR assays for the detection of *A. carbonarius* have been developed [reviewed by Niessen *et al.* (2005)], including a real time PCR assay to quantify this species in crushed grapes (Mulè *et al.*, 2006). However, the current limitations of molecular methods as a screening tool in this setting should be highlighted [reviewed by Edwards *et al.* (2002) and Varga *et al.* (2004b)].

First, the presence of potentially toxigenic isolates on grapes does not necessarily indicate the presence of toxin. These fungi are ubiquitous in vineyards and are often isolated from berries as surface contaminants without producing OA (discussed in "Ochratoxin A production in grapes"). Thus, it is not surprising that Mulè *et al.* (2006) detected low levels of *A. carbonarius* DNA in four OA-free grape samples. Yet, for 11 other grape samples containing up to 0.25 μg/kg OA, those authors noted a correlation between amount of *A. carbonarius* DNA and OA in grapes (R^2=0.92). It is unknown whether this correlation holds for more severely contaminated grapes.

Second, species-specific PCR, such as that developed for *A. carbonarius* above, detects both toxigenic and non-toxigenic isolates. It would be desirable to exclusively amplify toxigenic isolates of both *A. carbonarius* and *A. niger* in the same assay. One strategy is to target genes directly involved in toxin synthesis and to demonstrate a correlation between amount of fungal DNA and amount of toxin (Schnerr *et al.*, 2002). Genes involved in OA synthesis in black *Aspergillus* spp. have not been identified, although some have been characterised for *A. ochraceus* (O'Callaghan *et al.*, 2003) and *P. nordicum* (Färber and Geisen, 2004). Once such genes are identified among black *Aspergillus* spp., it is likely that homologues of genes for toxin synthesis will be found also in non-toxigenic strains; furthermore, subtle differences in gene sequences among related species may hinder development of a single test for both *A. carbonarius* and *A. niger* [reviewed by Edwards *et al.* (2002)].

Third, upregulation of OA synthesis genes in various environmental conditions (temperature, water activity, pH) does not necessarily lead to increased OA production, thus weakening the predictive value of mRNA quantification (Geisen, 2004).

For these reasons, current molecular techniques offer little guidance for identifying and monitoring critical control points for the minimisation of OA contamination of wine. Perhaps development of a rapid and robust PCR-based identification system for toxigenic isolates of *A. carbonarius* and *A. niger* may have some application for the early identification of these strains in Aspergillus bunch rots. Detection of toxigenic species such as *A. carbonarius* in grapes at intake could be helpful in identifying loads that require further testing to establish the presence and concentration of OA. However, the need for such tests would be negated if a rapid detection method for the toxin itself and applicable to grapes were widely available. Molecular analyses of black *Aspergillus* spp. still have much to contribute in understanding the regulation of toxin production in conditions encountered by fungi in vineyards. DNA probes in combination with confocal microscopy may also be useful in elucidating the nature of infection of grapes.

EFFECTS OF SOME OTHER FUNGI ON WINE QUALITY

OA is currently the only mycological food safety issue in wine. The effects of other fungi can be classified as mycological spoilage, and include hindering vinification, or adversely altering the stability, taste, smell and colour of wine. The effects of infection of grapes by *B. cinerea* on wine quality have been studied more extensively than those of other moulds (Ribereau-Gayon *et al.*, 1980; Donèche, 1993). Negative effects include the production of laccase, which, in the presence of oxygen, causes rapid browning of wines; soluble polysaccharides hinder clarification of wine. Positive effects of *B. cinerea*, such as concentrating grape sugars through berry dehydration, are exploited in the production of sweet, botrytized wines such as Sauternes in France, Tokay in Hungary, and Auslese in Germany. The differences between "noble rot" in the production of botrytized wines and "grey mould," one of the primary bunch rots affecting grapes, were examined by Donèche (1993) and are the subject of continuing research (Geny *et al.*, 2003). The primary factors resulting in noble rot are mentioned here. Noble rot will occur only on grapes that have reached maturity intact, and alternating conditions of humidity and dryness are also required at this stage. In humid conditions, *B. cinerea* germinates on the berry surface, penetrates via stomatal openings, and grows on and just below the cuticle. Infection is limited by phytoalexins, part of the grapevine defense mechanism, and also by dry conditions and degradation of berry cells by fungal enzymes in the layer between pulp and cuticle. The latter cause dehydration of the berry which, in turn, limits further hyphal penetration. In contrast, growth occurs throughout the entire berry in the case of grey mould; this may occur much earlier in the season, and is particularly likely when excessive moisture or rain causes berries to split. For berries affected by noble rot, extended periods of high humidity may allow the development of grey mould.

The effect of other fungal infections on wine quality is an area of continuing interest for the wine industry. Data are required linking specific infection levels with measurable changes in wine parameters (chemical and sensory), such as reported by Stummer *et al.* (2003) for powdery mildew infection. Similar research on wine made from grapes affected by "bitter rot" (*Greenaria uvicola*) is being conducted (M. Meunier, pers. comm, National Grape and Wine Industry Centre, Wagga Wagga, NSW, Australia). It should be noted that fungal infection often alters the acidity and sugar content of grapes; slight changes in acidity and alcohol in turn modify perceptions of astringency (Gawel, 1998) and bitterness (Noble, 1998) in wine.

The topic of grapevine defence mechanisms and wine brings together viticulture, plant pathology, molecular biology and oenology. Pathogenesis-related (PR) proteins and phytoalexins confer on grapevines resistance to fungal pathogens. These are produced both constitutively at well-defined stages of berry development and induced in response to fungal infection (Jacobs *et al.*, 1999; Tattersall *et al.*, 2001; Jeandet *et al.*, 2002). Increasing production of these compounds in grapevines through genetic manipulation would increase resistance to fungal infection, reducing the need for costly fungicides, the long-term environmental impacts of which are not known. However, a conundrum arises. PR proteins may protect the vine against fungal infection, but are highly resistant molecules which pass through vinification unchanged, and cause the development of haze in white wines upon storage; methods for their removal often detrimentally alter wine quality (Ferreira *et al.*, 2004). Girbau *et al.* (2004) demonstrated that powdery mildew infection increased the amount of PR proteins in wine made from infected grapes. In contrast, infection by *B. cinerea* decreased the amount of PR proteins in grapes, probably due to production of a protease. Thus, whereas one fungal infection increases haze-forming proteins, another may present the solution. It is possible that the *B. cinerea* protease may display greater activity than commercial proteases derived from *Aspergillus* spp. (E.J. Waters, pers. comm.).

Increased levels of phytoalexins in grapevines may confer resistance to fungal pathogens, and, in addition, the phytoalexin, resveratrol, may have positive effects on human health (Jeandet *et al.*, 2002). It would seem that these compounds are ideal candidates for increased production through genetic manipulation. Indeed, infection by *A. carbonarius* stimulated production of certain phytoalexins in berries, and, in turn, these compounds incorporated into culture media restricted growth of this species (Bavaresco *et al.*, 2003). Whereas growth was restricted by phytoalexins in those media, OA production was stimulated! These complex interactions between grapevines, fungi and vinification present opportunities for further investigation.

THE NEXT CHALLENGES

Much progress has been made over the past 10 years regarding OA contamination of wine, in terms of identifying the source of contamination, conditions associated with the development of Aspergillus rots and toxin production, and the binding of OA to solids during vinification [reviewed by Battilani *et al.* (2006d), Blesa *et al.* (2006), Chulze *et al.* (2006), Leong *et al.* (2006a), and Varga and Kozakiewicz (2006)]. Strategies are being formulated for reducing incidence of toxigenic *Aspergillus* spp. in vineyards, predicting regions at risk for OA contamination and controlling rots using fungicidal sprays. Research to clarify the nature of infection, toxin production and degradation with regard to berry maturity is still required, in the light of conflicting data. Identifying the genes for OA production by black *Aspergillus* spp. and their regulation at a molecular level provides further opportunities to study this group of fungi. Extensive surveys of wines produced worldwide have demonstrated that OA is seldom present at levels above the 2 μg/L limit introduced by the European Union (European Commission, 2005), thus the initial degree of alarm from a food safety perspective was perhaps unwarranted. Much of this fundamental research on OA in wine can also be applied to the minimisation of OA produced by the same fungi in dried vine fruits, in which OA is concentrated during the drying process (MacDonald *et al.*, 1999; Abarca *et al.*, 2003;

Möller and Nyberg, 2003; Stefanaki *et al.*, 2003; Lombaert *et al.*, 2004; Magnoli *et al.*, 2004).

Further mycological food safety concerns in wine are unlikely, in the absence of discovering hitherto unknown mycotoxin production by fungi isolated from grapes, or identifying a new toxic fungal metabolite. Wine quality is likely to be the driver for further mycological research. The wine export market in 2004 was worth over U.S.$ 20 billion, and consumers increasingly demand wines of consistent quality at a lower price. In the fiercely competitive international wine market, "old" problems of fungal infection and yeast spoilage which may have been managed haphazardly or tolerated in the past will require new solutions, preferably those with minimal chemical intervention. The move towards organic viticulture and oenology will certainly require a deeper understanding of fungal communities on grapes and during vinification. Even simple processes such as fermentation will continue to be scrutinized, so that the outcome in terms of wine quality is defined for any given parameters. This concept will be extended to grapes – the outcomes of vinification of grapes infected at differing degrees of severity and with a variety of fungi will be rigorously defined and allowable limits of infection introduced.

Acknowledgements

The author wishes to thank Ailsa Hocking and John Pitt of Food Science Australia, Eileen Scott of the University of Adelaide, and Elizabeth Waters of the Australian Wine Research Institute for their critical reading of the manuscript. Research into ochratoxigenic fungi and OA in wine in Australia is supported by Australian grape growers and winemakers through their investment body, the Grape and Wine Research and Development Corporation, with matching funds from the Australian government and by the Commonwealth Cooperative Research Centres Program. Work has been undertaken at Food Science Australia and the University of Adelaide, and at the Department of Primary Industries, Victoria, as part of the research program of the Cooperative Research Centre for Viticulture.

REFERENCES

Abarca, M. L., Bragulat, M. R., Castellá, G., and Cabañes, F. J. (1994). Ochratoxin A production by strains of *Aspergillus niger* var. *niger*. Applied and Environmental Microbiology 60:2650-2652.

Abarca, M. L., Accensi, F., Bragulat, M. R., Castellà, G., and Cabañes, F. J. (2003). *Aspergillus carbonarius* as the main source of ochratoxin A contamination in dried vine fruits from the Spanish market. Journal of Food Protection 66:504-506.

Abarca, M. L., Accensi, F., Cano, J., and Cabañes, F. J. (2004). Taxonomy and significance of black aspergilli. Antonie van Leeuwenhoek 86:33-49.

Abrunhosa, L., Serra, R., and Venâncio, A. (2002). Biodegradation of ochratoxin A by fungi isolated from grapes. Journal of Agricultural and Food Chemistry 50:7493-7496.

Abrunhosa, L., Serra, R., Lima, N., and Venâncio, A. (2003). Ochratoxin A risk assessment in Portuguese wines: a one-year case study. Bulletin de l'O.I.V. 76:618-634.

Accensi, F., Abarca, M. L., Cano, J., Figuera, L., and Cabañes, F. J. (2001). Distribution of ochratoxin A producing strains in the *A. niger* aggregate. Antonie van Leeuwenhoek 79:365-370.

Arici, M., Gümüs, T., and Karah, F. (2004). The fate of ochratoxin A during the Pekmez production from mouldy grapes. Food Control 15:597-600.

Battilani, P., Giorni, P., Bertuzzi, T., and Pietri, A. (2001). Preliminary results on ochratoxin A production by fungi invading grape berries. *In* 11 Congress of Mediterranean Phytopathological Union and 3rd Congress of Sociedade Portuguesa de Fitopatologia, Evora, Portugal, pp. 120-122.

Battilani, P., Giorni, P., and Pietri, A. (2003a). Epidemiology of toxin-producing fungi and ochratoxin A occurrence in grape. European Journal of Plant Pathology 109:715-722.

Battilani, P., Pietri, A., Bertuzzi, T., Languasco, L., Giorni, P., and Kozakiewicz, Z. (2003b). Occurrence of ochratoxin A-producing fungi in grapes grown in Italy. Journal of Food Protection 66: 633-636.

Battilani, P., Pietri, A., Giorni, P., Bertuzzi, T., and Barbano, C. (2003c). Growth and ochratoxin A production of *Aspergillus* section *Nigri* isolates from Italian grapes. Aspects of Applied Biology 68:175-180.

Battilani, P., Logrieco, A., Giorni, P., Cozzi, G., Bertuzzi, T., and Pietri, A. (2004). Ochratoxin A production by *Aspergillus carbonarius* on some grape varieties grown in Italy. Journal of the Science of Food and Agriculture 84:1736-1740.

Battilani, P., Barbano, C., Marín, S., Sanchis, V., Kozakiewicz, Z., and Magan, N. (2006a). Mapping of *Aspergillus* section *Nigri* in Southern Europe and Israel based on geostatistical analysis. International Journal of Food Microbiology 111 Suppl.1:S72-82.

Battilani, P., Barbano, C., Rossi, V., Bertuzzi, T., and Pietri, A. (2006b). Spatial distribution of ochratoxin A in vineyard and sampling design to assess contamination. Journal of Food Protection 69:884-890.

Battilani, P., Giorni, P., Bertuzzi, T., Formenti, S., and Pietri, A. (2006c). Black Aspergilli and ochratoxin A in grapes in Italy. International Journal of Food Microbiology 111 Suppl.1:S53-60.

Battilani, P., Magan, N., and Logrieco, A. (2006d). European research on ochratoxin A in grapes and wine. International Journal of Food Microbiology 111 Suppl.1:S2-4.

Bau, M., Bragulat, M. R., Abarca, M. L., Minguez, S., and Cabañes, F. J. (2005). Ochratoxigenic species from Spanish wine grapes. International Journal of Food Microbiology 98:125-130.

Bavaresco, L., Vezzulli, S., Battilani, P., Giorni, P., Pietri, A., and Bertuzzi, T. (2003). Effect of ochratoxin-A producing *Aspergilli* on stilbenic phytoalexin synthesis in grapes. Journal of Agricultural and Food Chemistry 51:6151-6157.

Bejaoui, H., Mathieu, F., Taillandier, P., and Lebrihi, A. (2004). Ochratoxin A removal in synthetic and natural grape juices by selected oenological *Saccharomyces* strains. Journal of Applied Microbiology 97:1038-1044

Bejaoui, H., Mathieu, F., Taillandier, P., and Lebrihi, A. (2006). Black Aspergilli and ochratoxin A production in French vineyards. International Journal of Food Microbiology 111 Suppl.1:S46-52.

Bell, S-J. and Daniel, C. (2004). Agrochemicals registered for use in Australian viticulture 2004/2005. The Australian Wine Research Institute, Glen Osmond, SA, Australia.

Bellí, N., Marín, S., Sanchis, V., and Ramos, A. J. (2002). Ochratoxin A (OTA) in wines, musts and grape juices: occurrence, regulations and methods of analysis. Food Science and Technology International 8:325-335.

Bellí, N., Marín, S., Duaigües, A., Ramos A. J., and Sanchis, V. (2004a). Ochratoxin A in wines, musts and grapes juices from Spain. Journal of the Science of Food and Agriculture 84:591-594.

Bellí, N., Marín, S., Sanchis, V., and Ramos, A. J. (2004b). Influence of water activity and temperature on growth of isolates of *Aspergillus* section *Nigri* obtained from grapes. International Journal of Food Microbiology 96:19-27.

Bellí, N., Pardo, E., Marín, S., Farré, G., Ramos, A. J., and Sanchis, V. (2004c). Occurrence of ochratoxin A and toxigenic potential of fungal isolates from Spanish grapes. Journal of the Science of Food and Agriculture 84:541-546.

Bellí, N., Ramos, A. J., Sanchis, V., and Marín, S. (2004d). Incubation time and water activity effects on ochratoxin A production by *Aspergillus* section *Nigri* strains isolated from grapes. Letters in Applied Microbiology 38:72-77.

Bellí, N., Mitchell, D., Marín, S., Alegre, I., Ramos, A. J., Magan, N., and Sanchis, V. (2005a). Ochratoxin A-producing fungi in Spanish wine grapes and their relationship with meteorological conditions. European Journal of Plant Pathology, 113:233-239.

Bellí, N., Ramos, A. J., Coronas, I., Sanchis, V., and Marín, S. (2005b). *Aspergillus carbonarius* growth and ochratoxin A production on a synthetic grape medium in relation to environmental factors. Journal of Applied Microbiology 98:839-844.

Bellí, N., Ramos, A. J., Sanchis, V., and Marín, S. (2006). Effect of photoperiod and day-night temperatures simulating field conditions on growth and ochratoxin A production of *Aspergillus carbonarius* strains isolated from grapes. Food Microbiology 23:622-627.

Blesa, J., Soriano, J. M., Moltó, J. C., and Mañes, J. (2004). Concentration of ochratoxin A in wines from supermarkets and stores of Valencian community (Spain). Journal of Chromatography A 1054:397-401.

Blesa, J., Soriano, J. M., Moltó, J. C., and Mañes, J. (2006). Factors affecting the presence of ochratoxin A in wines. Critical Reviews in Food Science and Nutrition 46:473-478.

Bleve, G., Grieco, F., Cozzi, G., Logrieco, A., and Visconti, A. (2006). Isolation of epiphytic yeasts with potential for biocontrol of *Aspergillus carbonarius* and *A. niger* on grape. International Journal of Food Microbiology 108:204-209.

Brera, C., Grossi, S., De Santis, B., and Miraglia, M. (2003). Automated HPLC method for the determination of ochratoxin A in wine samples. Journal of Liquid Chromatography & Related Technologies 26:119-133.

Burdaspal, P. A. and Legarda, T. M. (1999). Ocratoxina a en vinos, mostos y zumos de uva elaborados en España y en otros países europeos. Alimentaria 299:107-113.

Cabañes, F. J., Accensi, F., Bragulat M. R., Abarca, M. L., Castellá G., Minguez, S., and Pons, A. (2002). What is the source of ochratoxin A in wine? International Journal of Food Microbiology 79:213-215.

Caridi, A. (2006). Enological functions of parietal yeast mannoproteins. Antonie van Leeuwenhoek 89:417-422.

Castellari, M., Fabbri, S., Fabiani, A., Amati, A., and Galassi, S. (2000). Comparison of different immunoaffinity clean-up procedures for high-performance liquid chromatographic analysis of ochratoxin A in wines. Journal of Chromatography A 888:129-136.

Castellari, M., Versari, A., Fabiani, A., Parpinello, G. P., and Galassi, S. (2001). Removal of ochratoxin A in red wines by means of adsorption treatments with commercial fining agents. Journal of Agricultural and Food Chemistry 49:3917-3921.

Chulze, S. N., Magnoli, C., and Dalcero, A. (2006). Occurrence of ochratoxin A in wine and ochratoxigenic mycoflora in grapes and dried vine fruits in South America. International Journal of Food Microbiology 111 Suppl.1:S5-9.

Cozzi, G., Pascale, M., Perrone, G., Visconti, A., and Logrieco, A. (2006). Effect of *Lobesia botrana* damages on black aspergilli rot and ochratoxin A content in grapes. International Journal of Food Microbiology 111 Suppl.1:S88-92.

Czerwiecki, L., Wilczynska, G., and Kwiecien, A. (2005). Ochratoxin A: an improvement clean-up and HPLC method used to investigate wine and grape juice on the Polish market. Food Additives and Contaminants 22:158-162.

Da Rocha Rosa, C. A., Palacios, V., Combina, M., Fraga, M. E., De Oliveira Rekson, A., Magnoli, C. E., and Dalcero, A. M. (2002). Potential ochratoxin A producers from wine grapes in Argentina and Brazil. Food Additives and Contaminants 19:408-414.

Danks, C., Ostoja-Starzewska, S., Flint, J., and Banks, J. N. (2003). The development of a lateral flow device for the discrimination of OTA producing and non-producing fungi. Aspects of Applied Biology 68:21-28.

Donèche, B. J. (1993). Botrytized wines. *In* Wine Microbiology and Biotechnology (Fleet, G. H., ed.), Harwood Academic Publishers, Chur, Switzerland, pp. 327-351.

Dumeau, F. and Trione, D. (2000). Trattamenti e "ochratossina A" nei vini. Vignevini 9:79-81.

Eder, R., Parr, E., Edinger, W., and Lew, H. (2002). Untersuchungen über den Gehalt an Ochratoxin A (OTA) in Weinen, insbesondere Prädikatsweinen aus Österreich. Mitteilungen Klosternburg 52:125-131.

Edwards, S. G., O'Callaghan, J., and Dobson, A. D. W. (2002). PCR-based detection and quantification of mycotoxigenic fungi. Mycological Research 106:1005-1025.

Emmett, R. W., Harris, A. R., Taylor, R. H., McGechan, J. K. (1992). Grape diseases and vineyard protection. *In* Viticulture. Volume 2 (Coombe, B. G., and Dry, P. R., eds.), Practices. Winetitles, Adelaide, Australia, pp. 232-278.

Esteban, A., Abarca, M. L., Bragulat, M. R., and Cabañes, F. J. (2004). Effects of temperature and incubation time on production of ochratoxin A by black aspergilli. Research in Microbiology 155:861-866.

Esteban, A., Abarca, M. L., Bragulat, M. R., and Cabañes, F. J. (2005). Influence of pH and incubation time on ochratoxin A production by *Aspergillus carbonarius* in culture media. Journal of Food Protection 68:1435-1440.

European Commission (2005). Commission Regulation (EC) No 123/2005 of 26 January 2005 amending Regulation (EC) No 466/2001 as regards ochratoxin A. Official Journal of the European Union L25:3-5.

Färber, P. and Geisen, R. (2004). Analysis of differentially-expressed ochratoxin A biosynthesis genes of *Penicillium nordicum*. European Journal of Plant Pathology 110:661-669.

Fernandes, A., Venâncio, A., Moura, F., Garrido, J., and Cerdeira, A. (2003). Fate of ochratoxin A during a vinification trial. Aspects of Applied Biology 68:73-80.

Fernandes, A., Ratola, N., Cerdeira, A., Alves, A., and Venâncio, A. (2007). Change in ochratoxin A concentration during winemaking. American Journal of Enology and Viticulture, 58: in press.

Ferreira, R. B., Monteiro, S. S., Piçarra-Pereira, M. A., and Teixeira, A. R. (2004). Engineering grapevine for increased resistance to fungal pathogens without compromising wine quality. Trends in Biotechnology 22:168-173.

Festas, I., Herbert, P., Santos, L., Cabral, M., Barros, P., and Alves, A. (2000). Ochratoxin A in some Portuguese wines: method validation and screening in port wine and Vinho Verde. American Journal of Enology and Viticulture 51:150-154.

Filali, A., Ouammi, L., Betbeder, A-M., Baudrimont, I., Soulaymani, R., Benayada, A., Creppy, E. E. (2001). Ochratoxin A in beverages from Morocco: a preliminary survey. Food Additives and Contaminants 18:565-568.

Fleet, G. H. (2003). Yeast interactions and wine flavour. International Journal of Food Microbiology 86:11-22.

Gambuti, A., Strollo, D., Genovese, A., Ugliano, M., Ritieni, A., and Moio, L. (2005). Influence of enological practices on ochratoxin A concentration in wine. American Journal of Enology and Viticulture 56:155-162.

Gawel, R. (1998). Red wine astringency: a review. Australian Journal of Grape and Wine Research 4:74-95.

Garcia Moruno, E., Sanlorenzo, C., Boccaccino, B., and Di Stefano, R. (2005). Treatment with yeast to reduce the concentration of ochratoxin A in red wine. American Journal of Enology and Viticulture 56:73-76.

Geisen, R. (2004). Molecular monitoring of environmental conditions influencing the induction of ochratoxin A biosynthesis genes in *Penicillium nordicum*. Molecular Nutrition and Food Research 48:532-540.

Geny, L., Darrieumerlou, A., and Donèche, B. (2003). Conjugated polyamines and hydroxycinnamic acids in grape berries during *Botrytis cinerea* disease development: differences between 'noble rot' and 'grey mould'. Australian Journal of Grape and Wine Research 9:102-106.

Girbau, T., Stummer, B. E., Pocock, K. F., Baldock, G. A., Scott, E. S., and Waters, E. J. (2004). The effect of *Uncinula necator* (powdery mildew) and *Botrytis cinerea* infection of grapes on levels of haze-forming pathogenesis-related proteins in grape juice and wine. Australian Journal of Grape and Wine Research 10:125-133.

Gómez, C., Bragulat, M. R., Abarca, M. L., Mínguez, S., and Cabañes, F. J. (2006). Ochratoxin A-producing fungi from grapes intended for liqueur wine production. Food Microbiology 23:541-545.

Grazioli, B., Fumi, M. D., and Silva, A. (2006). The role of processing on ochratoxin A content in Italian must and wine: a study on naturally contaminated grapes. International Journal of Food Microbiology 111 Suppl.1:S93-96.

Gupta, S. L. (1956). Occurrence of *Aspergillus carbonarius* (Bainier) Thom causing grape rot in India. Science and Culture 22:167-168.

Guzev, L., Danshin, A., Ziv, S., and Lichter, A. (2006). Occurrence of ochratoxin A producing fungi in wine and table grapes in Israel. International Journal of Food Microbiology 111 Suppl.1:S67-71.

Heenan, C. N., Shaw, K. J., and Pitt, J. I. (1998). Ochratoxin A production by *Aspergillus carbonarius* and *A. niger* and detection using coconut cream agar. Journal of Food Mycology 1:67-72.

Hewitt, W. B. (1988). Berry rots and raisin moulds. *In* Compendium of Grape Diseases (Pearson, R. G, and Goheen, A. C., eds), The American Phytopathological Society, Minnesota, pp. 26-28.

Hocking, A. D., Varelis, P., Pitt, J. I., Cameron, S., and Leong, S. (2003). Occurrence of ochratoxin A in Australian wine. Australian Journal of Grape and Wine Research 9:72-78.

Horie, Y. (1995). Productivity of ochratoxin A of *Aspergillus carbonarius* in *Aspergillus* section *Nigri*. Nippon Kingakukai Kaiho 36:73-76.

Jacobs, A. K., Dry, I. B., and Robinson, S. P. (1999). Induction of different pathogenesis-related cDNAs in grapevine infected with powdery mildew and treated with ethephon. Plant Pathology 48:325-336.

Jeandet, P., Douillet-Breuil, A-C., Bessis, R., Debord, S., Sbaghi, M., and Adrian, M. (2002). Phytoalexins from the Vitaceae: biosynthesis, phytoalexin gene expression in transgenic plants, antifungal activity, and metabolism. Journal of Agricultural and Food Chemistry 50:2731-2741.

Klich, M. A. and Pitt, J. I. (1988). A Laboratory Guide to Common *Aspergillus* species and their Teleomorphs. Commonwealth Scientific and Industrial Research Organisation, Division of Food Processing, North Ryde, NSW, Australia.

Kunkee, R. and Bisson, R. (1993). Winemaking yeasts. *In* The Yeasts (Rose, A. H., and Harrison, J. S., eds), Academic Press, London, UK, pp. 69-127.

Lataste, C., Guéri, L., Solanet, D., Berger, J. L., Cottereau, P., and Molot, B. (2004). Évolution de la contamination en ochratoxine A: du vignoble Français au vin. Progrès Agricole et Viticole 121: 57-64.

Lee, T. H., and Simpson, R. F. (1993). Microbiology and chemistry of cork taints in wine. *In* Wine Microbiology and Biotechnology (Fleet, G. H., ed.), Harwood Academic Publishers, Chur, Switzerland, pp. 353-372.

Leong, S., Hocking, A. D., and Pitt, J. I. (2004). Occurrence of fruit rot fungi (*Aspergillus* section *Nigri*) on some drying varieties of irrigated grapes. Australian Journal of Grape and Wine Research 10:83-88.

Leong, S. L., Hocking, A. D., Pitt, J. I., Kazi, B. A., Emmett, R. W., and Scott, E. S. (2006a). Australian research on ochratoxigenic fungi and ochratoxin A. International Journal of Food Microbiology 111 Suppl.1:S10-17.

Leong, S. L., Hocking, A. D., Pitt, J. I., Kazi, B. A., Emmett, R. W., and Scott, E. S. (2006b). Black *Aspergillus* species in Australian vineyards: from soil to ochratoxin A in wine. *In* Advances in Food Mycology (Hocking, A. D., Pitt, J. I., Samson, R. A., and Thrane, U., eds.), Springer, New York, U.S.A., pp. 153 171.

Leong, S. L., Hocking A. D., and Scott, E. S. (2006c). Effect of temperature and water activity on growth and ochratoxin A production by Australian *Aspergillus carbonarius* and *A. niger* isolates on a simulated grape juice medium. International Journal of Food Microbiology 110:209-216.

Leong, S. L., Hocking A. D., and Scott, E. S. (2006e). The effect of juice clarification, static or rotary fermentation and fining on ochratoxin A in wine. Australian Journal of Grape and Wine Research 12:245-251.

Leong, S. L., Hocking, A. D., and Scott, E. S. (2006f). Survival and growth of *Aspergillus carbonarius* on wine grapes before harvest. International Journal of Food Microbiology 111 Suppl.1:S83-87.

Leong, S. L., Hocking, A. D., Varelis, P., Giannikopoulos, G., and Scott, E. S. (2006g). Fate of ochratoxin A during vinification of Semillon and Shiraz grapes. Journal of Agricultural and Food Chemistry 54:6460-6464.

Leong, S. L., Hocking A. D., and Scott, E. S. (2007). *Aspergillus* species producing ochratoxin A: isolation from vineyard soils and infection of Semillon bunches in Australia. Journal of Applied Microbiology 102:124-133.

Lombaert, G. A., Pellaers, P., Neumann, G., Kitchen, D., Huzel, V., Trelka, R., Kotello, S., and Scott, P. M. (2004). Ochratoxin A in dried vine fruits on the Canadian retail market. Food Additives and Contaminants 21:578-585.

Lopez de Cerain, A., González-Peñas, E., Jiménez, A. M., and Bello, J. (2002). Contribution to the study of ochratoxin A in Spanish wines. Food Additives and Contaminants 19:1058-1064.

Loureiro, V. and Malfeito-Ferreira, M. (2003). Spoilage yeasts in the wine industry. International Journal of Food Microbiology 86:23-50.

MacDonald, S., Wilson, P., Barnes, K., Damant, A., Massey, R., Mortby, E., and Shepherd, M. J. (1999). Ochratoxin A in dried vine fruit: method development and survey. Food Additives and Contaminants 16:253-260.

MAFF (Ministry of Agriculture, Fisheries and Food) (1997). Survey of aflatoxins and ochratoxin A in cereals and retail products. Ministry of Agriculture, Fisheries and Food, London, U.K.

MAFF (Ministry of Agriculture, Fisheries and Food) (1999). 1998 Survey of retail products for ochratoxin A. Ministry of Agriculture, Fisheries and Food, London, U.K.

Magnoli, C. E., Violante, M., Combina, M., Palacio, G., and Dalcero, A. M. (2003). Mycoflora and ochratoxin-producing strains of *Aspergillus* section *Nigri* in wine grapes in Argentina. Letters in Applied Microbiology 37:179-184.

Magnoli, C., Astoreca, A., Ponsone, L., Combina, M., Palacio, G., Rosa, C. A. R., and Dalcero, A. M. (2004). Survey of mycoflora and ochratoxin A in dried vine fruits from Argentina markets. Letters in Applied Microbiology 39:326-331.

Majerus, P. and Otteneder, H. (1996). Nachweis und Vorkommen von Ochratoxin A in Wein und Traubsaft. Deutsche Lebensmittel-Rundschau 92:388-390.

Majerus, P., Bresch, H., and Otteneder, H. (2000). Ochratoxin in wines, fruit juices and seasonings. Archiv für Lebensmittelhygiene 51:81-128.

Markarki, P., Delpont-Binet, C., Grosso, F., and Dragacci, S. (2001). Determination of ochratoxin A in red wine and vinegar by immunoaffinity high-pressure liquid chromatography. Journal of Food Protection 64:533-537.

Micheli, A., Negro, C., Tommasi, L., Minoia, E., and De Leo, P. (2003). Determination of polyphenols, resveratrol, antioxidant activity and ochratoxin A in wines obtained from organic farming in Southern Apulia (Italy). Bulletin de l'O.I.V. 76:976-997.

Mitchell, D., Parra, R., Aldred, D., and Magan, N. (2004). Water and temperature relations of growth and ochratoxin A production by *Aspergillus carbonarius* strains from grapes in Europe and Israel. Journal of Applied Microbiology 97:439-445.

Möller, T. E. and Nyberg, M. (2003). Ochratoxin A in raisins and currants: basic extraction procedure used in two small marketing surveys of the occurrence and control of the heterogeneity of the toxins in samples. Food Additives and Contaminants 20:1072-1076.

Mulè, G., Susca, A., Logrieco, A., Stea, G., and Visconti, A. (2006). Development of a quantitative real-time PCR assay for the detection of *Aspergillus carbonarius* in grapes. International Journal of Food Microbiology 111 Suppl.1:S28-34.

Nair, N. G. (1985). Fungi associated with bunch rot of grapes in the Hunter Valley. Australian Journal of Agricultural Research 36:435-442.

Ng, W., Mankotia, M., Pantazopoulos, P., Neil, R. J., and Scott, P. M. (2004). Ochratoxin in wine and grape juice sold in Canada. Food Additives and Contaminants 21:971-981.

Niessen, L., Schmidt, H., Mühlencoert, E., Färber, P., Karolewiez, A., and Geisen, R. (2005). Advances in the molecular diagnosis of ochratoxin A-producing fungi. Food Additives and Contaminants 22:324-334.

Noble, A. C. (1998). Why do wines taste bitter and feel astringent? *In* Chemistry of Wine Flavour (Ebeler, S. E., and Waterhouse, A. L., eds), ACS Symposium Series 714, American Chemical Society, Washington, DC, USA, pp. 156-165.

O'Callaghan, J., Caddick, M. X., and Dobson, A. D. W. (2003). A polyketide synthase gene required for ochratoxin A biosynthesis in *Aspergillus ochraceus*. Microbiology 149:3485-3491.

Ospital, M., Cazabeil, J-M., Betbeder, A-M., Tricard, C., Creppy, E., and Medina, B. (1998). L'Ochratoxine A dans les vins. Revue Française d'Oenologie 169:16-18.

Otteneder, H. and Majerus, P. (2000). Occurrence of ochratoxin A (OTA) in wines: influence of the type of wine and its geographical origin. Food Additives and Contaminants 17:793-798.

Pearson, R. G. and Goheen, A. C. (1988). Compendium of Grape Diseases. The American Phytopathological Society, Minnesota.

Perrone, G., Susca, A., Epifani, F., and Mulè, G. (2006). AFLP characterization of Southern Europe population of *Aspergillus* sect. *Nigri* from grapes. International Journal of Food Microbiology 111 Suppl.1:S22-27.

Pietri, A., Bertuzzi, T., Pallaroni, L., and Piva, G. (2001). Occurrence of ochratoxin A in Italian wines. Food Additives and Contaminants 18:647-654.

Ratola, N., Martins, L., and Alves, A. (2004). Ochratoxin A in wines–assessing global uncertainty associated with results. Analytica Chimica Acta 513:319-324.

Ribereau-Gayon, J., Ribereau-Gayon, P., and Seguin, G. (1980). *Botrytis cinerea* in enology. *In* The Biology of Botrytis (Coley-Smith, J. R., Verhoeff, K., and Jarvis, W. R., eds), Academic Press, New York, USA, pp. 251-274.

Rosa, C. A. R., Magnoli, C. E., Fraga, M. E., Dalcero, A. M., and Santana, D. M. N. (2004). Occurrence of ochratoxin A in wine and grape juice marketed in Rio de Janeiro, Brazil. Food Additives and Contaminants 21:358-364.

Roset, M. (2003). Survey on ochratoxin A in grape juice. Fruit Processing 13:167-172.

Saez, J. M., Medina, A., Gimeno-Adelantado, J. V., Mateo, R., and Jimenez, M. (2004). Comparison of different sample treatments for the analysis of ochratoxin A in must, wine and beer by liquid chromatography. Journal of Chromatography A 1029:125-133.

Sage, L., Krivobok, S., Delbos, É., Seigle-Murandi, F., and Creppy, E. E. (2002). Fungal flora and ochratoxin A production in grapes and musts from France. Journal of Agricultural and Food Chemistry 50:1306-1311.

Sage, L., Garon, D., and Seigle-Murandi, F. (2004). Fungal microflora and ochratoxin A risk in French vineyards. Journal of Agricultural and Food Chemistry 52:5764-5768.

Schnerr, H., Vogel, R. F., and Niessen, L. (2002). Correlation between DNA of trichothecene-producing *Fusarium* species and deoxynivalenol concentrations in wheat-samples. Letters in Applied Microbiology 35:121-125.

Serra, R., Abrunhosa, L., Kozakiewicz, Z., and Venâncio, A. (2003). Black *Aspergillus* species as ochratoxin A producers in Portuguese wine grapes. International Journal of Food Microbiology 88:63-68.

Serra, R., Braga, A., and Venâncio, A. (2005). Mycotoxin-producing and other fungi isolated from grapes for wine production, with particular emphasis on ochratoxin A. Research in Microbiology 156:515-521.

Serra, R., Cabañes, F. J., Perrone, G., Castellà, G., Venâncio, A., Mulè, G., and Kozakiewicz, Z. (2006a). *Aspergillus ibericus*: a new species of section Nigri isolated from grapes. Mycologia 98:295-306.

Serra, R., Mendonça, A., and Venâncio, A. (2006b). Ochratoxin A occurrence and production in Portuguese wine grapes at various stages of maturation. International Journal of Food Microbiology, 111 Suppl.1:S35-39.

Siantar, D. P., Halverson, C. A., Kirmiz, C., Peterson, G. F., Hill, N. R., and Dugar, S. M. (2003). Ochratoxin A in wine: survey by antibody- and polymeric-based SPE columns using HPLC/fluorescence detection. American Journal of Enology and Viticulture 54:170-177.

Silva, A., Galli, R., Grazioli, B., and Fumi, M. D. (2003). Metodi di riduzione di residui di ocratossina A nei vini. Industrie delle Bevande 32:467-472.

Snowdon, A. L. (1990). A Colour Atlas of Postharvest Diseases and Disorders of Fruit and Vegetables. I. General introduction and fruits. Wolfe Scientific, London.

Soleas, G. J., Yan, J., and Goldberg, D. M. (2001). Assay of ochratoxin A in wine and beer by high-pressure liquid chromatography photodiode array and gas chromatography mass selective detection. Journal of Agricultural and Food Chemistry 49:2733-2740.

Soufleros, E. H., Tricard, C., and Bouloumpasi, E. C. (2003). Occurrence of ochratoxin A in Greek wines. Journal of the Science of Food and Agriculture 83:173-179.

Stander, M. A. and Steyn, P. S. (2002). Survey of ochratoxin A in South African wines. South African Journal of Enology and Viticulture 23:9-13.

Stefanaki, I., Foufa, E., Tsatsou-Dritsa, A., and Dais, P. (2003). Ochratoxin A concentrations in Greek domestic wines and dried vine fruits. Food Additives and Contaminants 20:74-83.

Stummer, B. E., Francis, I. L., Markides, A. J., and Scott, E. S. (2003). The effect of powdery mildew infection of grape berries on juice and wine composition and on sensory properties of Chardonnay wines. Australian Journal of Grape and Wine Research 9:28-39.

Tateo, F., and Bononi, M. (2001). Survey of ochratoxin A in wines. First data concerning a sampling of bottled red wines. Bulletin de l'O.I.V. 74: 772-780.

Tateo, F., and Bononi, M. (2003). Survey on ochratoxin A in wine. More data concerning bottled red wines. Bulletin de l'O.I.V. 76:766-778.

Tateo, F., Bononi, M., Fuso-Nerini, A., Lubian, E., and Martello, S. (1999). Ricerca e determinazione dell'ocratossina a nei vini. Industrie delle Bevande 28:592-596.

Tateo, F., Bononi, M., and Lubian, E. (2000). Survey on Ochratoxin A in wines. Data concerning the market of table wines in brik. Bulletin de l'O.I.V. 73:772-783.

Tattersall, D. B., Pocock, K. F., Hayasaka, Y., Adams, K., van Heeswijck, R., Waters, E. J., and Høj, P. B. (2001). Pathogenesis related proteins - their accumulation in grapes during berry growth and their involvement in white wine heat instability. *In* Molecular Biology & Biotechnology of the Grapevine. (Roubelakis-Angelakis, K. A., ed.), Kluwer Academic Publishers, Dordrecht, Netherlands, pp. 183-201.

Tjamos, S. E., Antoniou, P. P., Kazantzidou, A., Antonopoulos, D. F., Papageorgiou, I., and Tjamos, E. C. (2004). *Aspergillus niger* and *Aspergillus carbonarius* in Corinth raisin and wine-producing vineyards in Greece: population composition, ochratoxin A production and chemical control. Journal of Phytopathology 152:250-255.

Tjamos, S. E., Antoniou, P. P., and Tjamos, E. C. (2006). *Aspergillus* spp., distribution, population composition and ochratoxin A production in wine-producing vineyards in Greece. International Journal of Food Microbiology 111 Suppl. 1:S61-66.

Varga, J., Kevei, F., Hamari, Z., Tóth, B., Téren, J., Croft, J. H., and Kozakiewicz, Z. (2000a). Genotypic and phenotypic variability among black Aspergilli. *In* Integration of Modern Taxonomic Methods for *Penicillium* and *Aspergillus* Classification. (Samson, R. A., and Pitt, J. I., eds.), Harwood Academic Publishers, Amsterdam, The Netherlands, pp. 397-411.

Varga, J., Rigó, K., and Téren, J. (2000b). Degradation of achratoxin A by *Aspergillus* species. International Journal of Food Microbiology 59:1-7.

Varga, J., Juhász, Á., Kevei, F., and Kozakiewicz, Z. (2004a). Molecular diversity of agriculturally important *Aspergillus* species. European Journal of Plant Pathology 110:627-640.

Varga, J., Tóth, B., and Mesterházy, Á. (2004b). Molecular detection of mycotoxin-producing fungi. Cereal Research Communications 32:193-200.

Varga, J. and Kozakiewicz, Z. (2006). Ochratoxin A in grapes and grape-derived products. Trends in Food Science and Technology 17:72-81.

Visconti, A. and De Girolamo, A. (2005). Fitness for purpose - ochratoxin A analytical developments. Food Additives and Contaminants Supplement 1:37-44.

Visconti, A., Pascale, M., and Centonze, G. (1999). Determination of ochratoxin A in wine by means of immunoaffinity column clean-up and high-performance liquid chromatography. Journal of Chromatography A 864:89-101.

Zahavi, T., Cohen, L., Weiss, B., Schena, L., Daus, A., Kaplunov, T., Zutkhi, J., Ben-Arie, R., and Droby, S. (2000). Biological control of Botrytis, Aspergillus and Rhizopus rots on table and wine grapes in Israel. Postharvest Biology and Technology 20:115-124.

Zimmerli, B. and Dick, R. (1996). Ochratoxin A in table wine and grape-juice: occurrence and risk assessment. Food Additives and Contaminants 13:655-668.

Chapter 16

Cheese and fermented sausages

Jacques Stark
DSM Food Specialties, P.O. Box 1, 2600 MA Delft, The Netherlands.

INTRODUCTION

Cheese was first made about 8,000 years ago in the Middle East, most likely in Mesopotamia, today's Iraq. Herdsmen filled up their milk in bags made of animal stomachs, these stomachs still contained the coagulating enzyme now known as chymosin. After some time the milk was transformed into the thick curd and the watery whey, while lactic acid bacteria metabolised the milk sugar lactose into lactic acid.The manufacture of cheese is one of the oldest and most natural examples of food preservation. It is an effective way of preventing bacterial spoilage and preserving the nutritious components of the milk. Two classical methods are combined: souring with lactic acid bacteria and transforming milk into a semi-solid mass by separating the whey from the curd, the water activity is further reduced by adding dry salt or dipping in brine.

In addition to cheese, the history of fermented sausages and their preservation by fermentation, drying and salting probably also dates back to more than 2,000 years.

As in the case for cheese, the two factors primarily responsible for preservation are both low pH and low water activity. The fermentation process of dry fermented sausages such as salami, cervelat and pepperoni involves the conversion of added sugars to lactic acid, thus lowering the pH.

Dry salting originated from the Anglo-Saxon cultures. In Cheddar cheese or cured meat production processes, salt is mixed with the milled curd or meat particles. Traditionally, for the brine-salting process as applied in the production of many cheese types concentrated brine from seawater was used. In today's cheese production of for example Gouda cheese, blocks of curd are put into metal or plastic cheese moulds, pressed and put into the brine. In some meat products the brine is injected. After brining a further reduction of the moisture content is achieved during the ripening process.

The processing steps described above contribute to the protection against bacterial spoilage and the growth of pathogenic bacteria. However, the surface of cheese and fermented sausages can be considered as a good substrate for fungal growth. Moreover, the products are mostly ripened in open air at a high relative humidity, conditions ideal for mould growth.

Although for specific cheeses and sausages both yeasts and moulds are important for the ripening process, in most cases their development has to be prevented. Apart from the economic losses due to spoilage, the presence of fungi can be of considerable concern for human health by their formation of mycotoxins, i.e., fungal metabolites that are toxic.

As prevention of mould growth, especially during ripening and storage, is essential in the production of cheese and fermented sausages, preservatives such as natamycin or sorbate are applied. Modern production processes with ripening under higher humidity as well as the development of new products with less salt, enhance the risk of fungal spoilage and require more intensive preservation measures.

In this chapter, first mould-ripened cheeses and sausages will be described. The ripening conditions of these products are quite similar to the conditions of cheeses and sausages where fungi cause spoilage. Secondly contamination of cheeses and sausages is addressed, including which fungi may occur on these products and ways to prevent this. In addition the topic of mycotoxins will be discussed.

MOULD-RIPENED CHEESES AND SAUSAGES

Mould-ripened cheeses are divided into two major types: those varieties ripened by moulds growing inside the structure of the cheese and those ripened using moulds growing on the surface.

The best known of internally mould-ripened cheeses are Roquefort, Gorgonzola, Stilton and Danish blue, all ripened using different strains of *Penicillium roqueforti* which forms blue veins within the cheese. The strains used for the production of Gorgonzola are greener in colour and have more proteolytic activity, while for Stilton a highly lipolytic strain of *P. roqueforti* is used. Blue cheeses are traditionally made from high fat milk and ripened in caves (Roquefort, Gorgonzola) or cellars (Stilton) at low temperature and high humidity.

To allow the moulds to grow in the interior of the cheese, aerobic conditions are required. Therefore, open channels in the structure of the cheese are created by spiking. Also gas producing *Leuconostoc* and yeasts contribute to a more open curd. Finally the use of non-homogenized milk with low acidity gives a less dense and more crumbly-textured curd (Nichol, 2000).

Traditional white cheeses such as Camembert and Brie are produced using the white mould *Penicillium camemberti,* which gives the cheeses their characteristic white rind. It seems to be that till around the end of the nineteenth century also most surface-ripened cheeses were blue-green of colour due to growth of *P. roqueforti*.

Besides the mould-ripened soft cheeses and blue-veined cheeses, there are a small number of semi-hard mould-ripened cheese types. Tome de Savoie and Saint-Nectaire are examples of such cheeses of which the surface is covered with mixed populations which may contain *Penicillium, Mucor, Cladosporium, Geotrichum, Epicoccum* and *Sporotrichum* species. Other examples are Taleggio and Robiola from Italy and Gammelost from Norway (Beresford *et al.*, 2001).

For the production of dry-fermented sausages with a white/cream coloured appearance mostly a culture containing spores of *Penicillium nalgiovense* is applied. In some traditional production processes a spontaneously appearing home flora develops on the surface of the sausages. The fungal mycelium penetrates into the sausages and contributes to the characteristic flavour development through metabolic activities, e.g., by proteolytic and lipolytic activities and by degrading lactic acid resulting in an increase of the pH. The surface mould protects the surface against light and undesirable microorganisms preventing the development of off-flavours.

ROQUEFORT AND CAMEMBERT

The two most famous mould-ripened cheeses are Roquefort and Camembert (Figure 1). The history of both traditional French cheeses starts with a legend. The myth says that Roquefort cheese was discovered when a young shepherd eating his lunch in a cave saw a beautiful girl in the distance.

Figure 1. Roquefort and Camembert cheeses.

He followed her leaving behind a piece of cheese and bread. When he failed to catch her, he returned to the cave and saw that the cheese was covered with blue mould. Roquefort is mentioned in literature as far back as 79 A.D. In 1411 Charles VI of France gave sole rights to the ageing of Roquefort cheese to the village of Roquefort-sur-Soulzon, and French law still dictates that only those cheeses ripened in the natural caves of this village are allowed to use the name Roquefort.

According to the Camembert legend, in 1790 after the French revolution a priest on the run was sheltered in the village of Camembert by a woman called Marie Harel. In return for the refuge on her farm, this priest from the town of Brie gave her a secret recipe. Marie Harel invented the Camembert cheese by combining the process of Brie and the recipe for Livarot, one of the oldest Norman cheeses. In 1855 a descendant of Marie Harel presented Napoleon III with a sample of this cheese. He liked it and ever since then it became known as Camembert cheese. In 1890 the characteristic round wooden box was invented and Camembert quickly became one of the most popular French cheeses conquering foreign markets as well.

The original Roquefort is produced from raw sheep milk, which after filtering is prepared with rennet. After curdling the whey is drained off and the curds are cut into small pieces and placed into cheese moulds. *Penicillium roqueforti* spores are sprayed onto the curd or added to the milk just before setting. After one week the cheeses are taken to the caves, where they are salted and pierced about 40 times from top to bottom to encourage the growth of the mould. The final step is the ripening process in the caves for 3 to 6 months.

The Roquefort mould originated from the walls of the limestone caves where the cheese was ripened. Cracks in the walls of these caves provide for natural air circulation, creating an environment with the optimal temperature and humidity for the ripening of Roquefort cheese. Traditionally the cheese makers extracted the mould by leaving bread in the caves for 6 to 8 weeks. The interior of the bread was then dried to produce a powder. Nowadays the traditional mould cultures are still produced within the caves using barley and rye loaves as growth substrate.

In the traditional recipe of Camembert, the milk is heated to a maximum temperature of 37°C, whereas in the industrial production pasteurised milk and starter cultures are used. Rennet is added to promote curdling. The curdled milk is ladled into cheese moulds. After being drained, the cheeses are removed from the cheese moulds and *P. camemberti* conidia are either sprayed over the surface or added to the milk or brine. Nowadays in industrial production *P. candidum* more and more replaces the traditional strains. To promote optimal growth conditions, the cheeses are matured at 10 to 15 °C and 85% relative humidity. After 3 to 4 weeks of ripening, the cheese is ready for consumption.

The blue and white moulds not only give the cheeses their characteristic appearance, they also metabolise lactic acid and lactate resulting in a higher pH under which conditions their proteolytic and lipolytic enzymes have an optimal activity leading to the development of flavour compounds and cheese texture. Although *P. roqueforti* and *P. camemberti* play a major role in the ripening process, many other microorganisms are present. This secondary flora contributes to the specific taste of traditional cheeses.

During ripening of blue cheeses, the yeast *Debaryomyces hansenii* is the predominant species occurring at levels of 10^6 to 10^8 CFU/g. During ripening the atmospheric conditions are defined by increasing carbon dioxide levels (25%) and decreasing oxygen levels (0.3%). It was found that under these conditions *D. hansenii* stimulates the growth of *P. roqueforti*, it was suggested to use this yeast as starter culture in blue cheese. The contaminant *G. candidum* has shown a growth potential similar to *P. roqueforti*, which may lead to inhibition of growth and sporulation of *P. roqueforti*, especially in the centre of the cheese, affecting the quality of the product. *G. candidum* was also found to inhibit the growth of *D. hansenii* (Van den Tempel and Nielsen, 2000).

On the surface of traditional Camembert first a layer of lactose-fermenting yeasts develops. At that stage *Geotrichum candidum* starts to appear. The yeast population counts up to 10^8 cells/g. Also here *D. hansenii* is the predominant species, but also *Kluyveromyces lactis* and *Saccharomyces cerevisiae* are present. Only after one week of ripening *Penicillium camemberti* is observed, within two to three weeks it covers the entire surface of the cheese. During the ripening a bacterial flora with *Brevibacterium linens* and *Hafnia alvei* as predominant species develops on the surface of the cheese, while inside the cheese lactic acid bacteria are clearly dominant. When milk is pasteurised and starter cultures are applied, the microflora of the cheese is less diverse, and its taste and aroma are more flat (Gripon, 1987).

FUNGAL CONTAMINATION IN CHEESE AND SAUSAGES INDUSTRY

Cheese and dry fermented sausages are excellent substrates for moulds; also the environmental conditions during production and ripening are favourable for fungi. The conditions are in a way quite similar to those in the Roquefort caves or the cellars where sausages are hanged: a high humidity, relatively low temperatures, air-circulation and ripening in open air with many cheeses or sausages stored together. Under such conditions fungal growth can easily occur.

In the production process of Dutch semihard cheeses several sections can be recognized. The more humid locations are the production, pressing, brining and plasticization sites, while in the ripening rooms the conditions are dryer. After brining the cheeses are coated and stored for about 15 days in the short-term storage. Finally the cheeses are transported to warehouses for further ripening. Each section of the cheese production process has a typical fungal flora, which may differ per factory; one can even say that in each production location a characteristic home flora is present. The total fungal count measured in the air, on the equipment and on the cheese often differs greatly between and within factories. Contamination levels are determined by the environmental conditions, the factory layout, general hygienic conditions, cleaning and disinfections protocols, relative humidity and use of preservatives.

Hoekstra *et al.* (1998) examined the composition of the fungal flora in four Dutch cheese factories and three warehouses by air and surface sampling. The aim of the study was to make a qualitative and quantitative enumeration of the mycoflora in the cheese factory environment. Air samples were taken in the following main sections of the cheese factories: production, pressing, brining, plasticization and short-term storage. In the warehouses air samples were taken in the long-term storage sites.

The values obtained in the cheese factories showed great variation. The highest counts were measured in the most humid locations of the cheese factory: the production, pressing, plasticization and brining sites. In two factories counts of 3360 and 9000 CFU/m^3 were measured, the counts in the two other factories were considerably lower (1000–1500 CFU/m^3). The highest counts among the warehouses were measured in two ripening rooms of one factory (2500–3000 CFU/m^3), in the other ripening rooms the counts showed considerably lower levels of mostly less then 500 CFU/m^3.

In the more humid areas of the factories the composition of the mycoflora differed from that in the ripening rooms where xerophilic species such as *Aspergillus penicillioides, A. versicolor, Eurotium* spp. and *Wallemia sebi* were encountered more frequently.

It was remarkable to observe that in four factories where the same cheese types were produced (Gouda) a completely different home flora could be detected. In one factory *Penicillium corylophilum* was the most frequently isolated species in the production, brining and coating sections; in the other factories this species was hardly present. *Penicillium brevicompactum* was the predominant species in two factories. The fourth factory was known for its ripening conditions at a low relative humidity, leading to lower contamination levels and a more xerophilic mycoflora of *Aspergillus, Wallemia* and *Cladosporium* species.

The brines were dominated by the salt tolerant yeasts *Debaryomyces hansenii* and

Trichosporum inkin, other species encountered were *Cryptococcus laurentii*, *Candida versatilis*, *Candida apicola* and *Rhodotorula aurantiaca*.

Contamination in Dutch cheese industry is mostly caused by *Penicillium discolor*, a species first described by Frisvad *et al.* (1997). *P. discolor* is less sensitive to the fungicide natamycin, which is commonly applied in cheese industry (Stark and Tan, 2003). Hoekstra *et al.* (1998) encountered *P. discolor* in each factory and warehouse while spoilage of the cheese was not observed, which proves that under good hygienic and processing conditions this species will not cause spoilage problems. However when present in high amounts the risk is higher and measures should be taken.

The yeast flora in a South African Gouda cheese factory was identified by Welthagen *et al.* (1998). They also found *D. hansenii* as the most frequently isolated yeasts species. It could be isolated from the brine, the environment and the cheese.

Moulds from Jarlsberg and Norvegia cheese have been identified by Kure *et al.* (2001). These Norwegian semi-hard cheeses are wrapped in a vacuum film before ripening. A total of 118 visible mould samples were taken from Jarlsberg produced in two factories. On cheeses from one factory the yeast-like fungi *G. candidum* represented 51% of the total isolates, while on cheeses from the other factory this species was not or hardly detected, here *P. commune* (32%) and *P. roqueforti* (28%) were the predominant species. In the same study 134 isolates from Norwegian cheese produced in two factories were taken. On these cheeses *P. commune* (30%/14%), *P. palitans* (18%/27%) and *P. solitum* (10%/19%) were the most frequently isolated species.

In an additional study air, equipment, plastic film, brine and milk were sampled from four Norwegian cheese factories (Kure *et al.*, 2004). *P. brevicompactum* was the most frequently isolated species from three factories, while *G. candidum* was the most frequently isolated species from the fourth. Air was found to be the major source of cheese contaminants such as *P. commune* and *P. palitans*.

Moulds from various cheese types from different countries were isolated by Lund *et al.* (1995). Samples were taken before and after packaging. *Penicillium* species were dominant (88%); *P. commune* represented 42% of all isolates and was before and after packaging clearly the predominant species. *P. nalgiovense* represented 26% of the total flora, this species was not found in Dutch cheese industry (Hoekstra et al., 1998; De Boer *et al.*, 1977).

From a factory with a high mould infection Lund et al. (1995) took air, swab and cheese samples from surface ripened semi-soft cheeses, *P. nalgiovense* was the dominant species in air (99%), smear (99%) and on cheese (76%). In addition *P. commune* (20%) was frequently found on cheese. *A. versicolor* was reported as predominant species in the air of two ripening houses (83%/74%), but could only incidentally be isolated from cheese.

In Australian Cheddar cheese factories, *Cladosporium cladosporioides* and *C. herbarum* have been identified as main spoilage organisms, of 195 isolates from moulded cheese, 44% were *Cladosporium* species (Hocking and Faedo, 1992). In the retail sector *P. commune* and *P. roqueforti* were the most common species (Hocking, 1994).

An Argentinean Cheddar cheese factory with contamination problems showed *Phoma glomerata* to be the most important spoilage organism (Basílico *et al.*, 2001). From 40 vacuum-packed cheeses 94 isolates were obtained, *P. glomerata* was isolated from all cheese samples and made up 64% of the isolates.

Table 1. Frequently isolated fungal species in the cheese industry

	Cheese factory environment		
Species	**Production**	**Storage**	**Brine**
Penicillium			
commune	+		
brevicompactum	+		
corylophilum	+		
nalgiovense	+		
roqueforti	+		
Aspergillus			
penicillioides		+	
versicolor		+	
Cladosporium	+	+	
Debaryomyces			
hansenii			+

Whereas this mould was dominant on cheese, it was almost absent in the air and on the surfaces of the cheese factory environment.

P. glomerata was not known as spoilage organism of food products. It remains unclear how this species could become predominant in this particular Cheddar cheese factory, most likely it frequently occurred in the outdoor environment of the factory.

The red mould *Sporendonema casei* may cause red-orange coloured spots on especially Provolone cheese. The associated surface mycoflora of Provolone cheese consists of moulds of the genera *Aspergillus, Penicillium* and *Sp. casei*. The red mould may develop after 30 days of ripening and is together with *Aspergillus candidus* the predominant species at the end of the ripening period (Galli and Zambrini, 1978). Growth of *Sp. casei* and other spoilage moulds can be prevented using natamycin (Galli *et al.*, 1978).

Packaged cheeses such as Cheddar are susceptible for spoilage by *P. commune* and *P. roqueforti*. These species are able to grow at low temperatures, reduced water activity and low oxygen concentrations, while they are resistant to high free fatty acid concentrations.

The most frequently isolated spoilage species from the cheese factory environment are summarized in Table 1. When no preservatives are applied most of these and many other fungal species may develop on cheese. On cheeses treated with the fungicide natamycin, only under less hygienic circumstances or a too high humidity, *Penicillium discolor* (Frisvad *et al.*, 1997) may develop.

During the ripening of dry fermented sausages several species of *Penicillium* are able to develop. *P. nalgiovense*, also applied as starter culture for mould-ripened sausages, is the most common spoilage mould isolated from dry fermented sausages. Other *Penicillium* species frequently found are *P. brevicompactum, P. chrysogenum, P. nordicum, P. olsonii* and *P. solitum*. Also other moulds such as *Aspergillus, Scopulariopsis* and *Mucor* species may develop (Filtenborg *et al.*, 2002; Leistner and Ayres, 1969; Stark, 2003).

The most important spoilage yeasts isolated from dry fermented sausages are *D. hansenii* and *Candida* species. Buzzini and Haznedari (1995) examined the composition of the yeast flora on Italian fermented sausages; *D. hansenii* represented 50% of the isolates and was the predominant species, *D. vanrijiae* represented 12% of the isolates, *Candida* and *Rhodotorula* spp. each 11-12%. In another study 82% of the yeasts isolated from Italian salami were *D. hansenii* (Grazia *et al.*, 1989). Also on Greek dry salami 66% of the population were species of *Debaromyces*, of which 48% was *D. hansenii*. The remaining flora consisted mainly of *Candida* spp. (Metaxopoulos *et al.*, 1996).

SOURCES OF CONTAMINATION

Nowadays both the large companies and small producers such as farmers and butchers are aware of good manufacturing practise, HACCP and cleaning and disinfection protocols. Adequate hygiene conditions in the production and ripening facilities are common practise. Nevertheless fungal spoilage still occurs and sometimes even seems to increase in number and persistency.

Processing equipment, shelves and air are the main sources of fungal contamination of cheeses and sausages; raw materials such as milk or meat are almost never a source of fungal contamination. Continuous exposure to equipment surfaces and air in an environment with optimal temperature and humidity conditions promotes the development of fungi.

Brining baths may contain up to 10^6 salt tolerant yeasts/ml, leading to a contamination of 10^6 yeasts/dm^2 of cheese surface. Moulds may develop on the side of the brining bath. If cheeses are put into the brine, the brine level rises and the moulds come into the brine leading to concentrations of up to 10^3 CFU/ml of brine.

After brining, one side of the cheese is treated with a coating containing natamycin, while the other side is unprotected for one or two days (Gouda type of cheese). Just after production the cheese is more susceptible to fungal growth. A high yeasts contamination may use up the natamycin too fast, leading to insufficient protection against spoilage moulds during later stages of preparation.

The unprotected side of the cheese is in close contact with the wooden shelf in which moulds are always present. Disinfection of wooden shelves is difficult. In practise cleaning and disinfection procedures consisting of a heat treatment of a few seconds at 70° C are insufficient to eliminate the fungal mycelium present inside the wood (Persoon and Van Rijn, 1993). From experience by cheese makers it is known that especially the first 3 to 6 months new wooden shelves can be a severe source of contamination. It is not known why; perhaps the moulds will develop easier in new wood because more nutrients are available. Plastic and stainless steel shelves are on the market but most producers prefer the wooden shelves.

In industrial ripening rooms for cheeses and sausages, often closed air-conditioning systems without microbial filters are applied, as filters would reduce the capacity of the air-conditioning system too much, i.e., indoor air is pumped around. When moulds develop on a certain spot in the factory their spores or mycelium fragments can be taken up by the air-conditioning system and released through the whole factory. In a few warehouses we noticed more spoilage on cheeses ripened near air-conditioning outlets; sometimes also mould growth within the air-conditioning system was observed.

It is hard to say in general which levels of air-borne fungi may lead to spoilage problems. However, it is recommended to monitor at least monthly and take measures in case of increased viable counts in the air.

In modern cheese warehouses mould spots are simply not noticed anymore; turning of the cheeses and cleaning of the shelves is fully automated. In such warehouses improved air-conditioning systems maintain higher humidity levels. Less space between cheeses and also sausages during ripening influences the air transport in a negative way, also leading to a higher relative humidity in the microenvironment and wet products. In general it is recommended to adapt the capacity of the storage rooms to increased production levels.

Damaged or moulded returned products should be stored in a separated room which is not in contact with the production or ripening area, also not via the air-conditioning system. In the sausages industry, mould-ripened sausages should not be hung near or in the same area as regular sausages.

Large industrial cheeses with increased internal weight are also more sensitive to mould growth. An example is the 16-kg Gouda cheese, compared with the regular 10-kg cheese; the weight-surface ratio is higher, leading to cheeses with a higher humidity.

Consumers appreciate low salt and low fat products or products with fewer preservatives; however, such products are more sensitive to spoilage and might require adapted production and ripening conditions to prevent mould growth.

In case of mould outbreaks identification of the source, extra cleaning and disinfection measures, a temporary decrease of the relative humidity of the air and treatment with a coating or casing containing higher levels of natamycin are important measures to solve the problems.

MYCOTOXINS

Mycotoxins are toxic metabolites produced by certain filamentous fungi. As low concentrations of mycotoxins in food cause chronic effects when ingested by man or animals, this is a concern for the long-term health of the consumer.

Mycotoxin production largely depends on nutrient availability and environmental conditions such as temperature and humidity. Most mycotoxins are extremely persistent and survive storage and processing conditions even when heated to high temperatures.

Mycotoxins can be excreted by moulds growing on cheese or sausages and penetrate into the product, which is reviewed by Scott (1989). Sterigmatocystin excreted by *Aspergillus versicolor* on naturally infected Gouda and Edam cheeses was only found in the outer 2-mm layer of the cheese, while on Cheddar cheese ochratoxin A and citrinin produced by

Penicillium species were present up to a distance of 8 cm. In several studies on migration of aflatoxins in different cheese types, aflatoxins were detected in the first layer of 1 to 1.5 cm of the cheese surface, in the layers below no aflatoxins were detected. It can be concluded that the penetration depth depends on the type of cheese and the type of mycotoxin.

The most widely used official analytical method to detect mycotoxins is HPLC, often in combination with immunoaffinity column-based isolation and concentration of the toxin from complex matrices. In addition, commercial immunological test kits are available for detecting several mycotoxins such as aflatoxins and ochratoxin A. Since the precision of these rapid tests may not be as good as traditional laboratory methods, confirmation is mostly required. Most commercial tests kits are based on antibodies. In the case of ELISA (enzyme-linked immunosorbent assay) a sample homogenate containing toxin is either directly quantified using a standard micro-titre plate, tube or membrane-based format ELISA. Improved detection methods made authorities to establish stricter limits for mycotoxins, although for cheese and dry sausages in most countries there is no legislation yet.

The most important mycotoxins associated with cheese are sterigmatocystine (*Aspergillus versicolor*), aflatoxins (*A. parasiticus, A. flavus*), ochratoxin A (*A. ochraceus, Penicillium verrucosum*) and patulin (*P. patulum*, now known as *P. griseofulvum*) (Filtenborg *et al*., 1996; Northolt *et al*., 1980). Cows fed with feed contaminated with aflatoxin B_1 will metabolise this toxin to aflatoxin M_1, which is then excreted in the milk, survives the pasteurisation and may end up in cheese too.

Nielsen *et al.* (2005) stated that there are many false reports about mycotoxin production from the *Penicillium roqueforti* strains applied in blue cheese production. This is partly because two closely related species, *P. carneum* and *P. paneum*, are often identified as *P. roqueforti*. These very similar species have completely different profiles of mycotoxins and thus different toxicological responses. Nielsen also stated that scientific evidence for the toxicity of roquefortine C, which is consistently produced in blue cheeses such as Roquefort, is lacking.

López-Diaz *et al.* (1996) examined 10 blue cheeses and 12 samples of Manchego cheese for the presence of mycotoxins, also 24 *Penicillium* and *Aspergillus* strains were isolated and assessed for their mycotoxigenicity. Roquefortine was found in one sample of blue cheese. Seven out of nine *P. roqueforti* strains were able to produce roquefortine. Although there is no evidence that roquefortines are formed at significant levels in cheese, production of this metabolite by *P. roqueforti* on blue cheeses occurs. The mycotoxin mycophenolic acid could be detected in four Manchego cheeses.

Penicillium and *Aspergillus* species were isolated from Cheddar and Swiss cheese (Bullerman, 1976, 1977). Several isolates were capable of producing patulin, penicillic acid, ochratoxin A and aflatoxins.

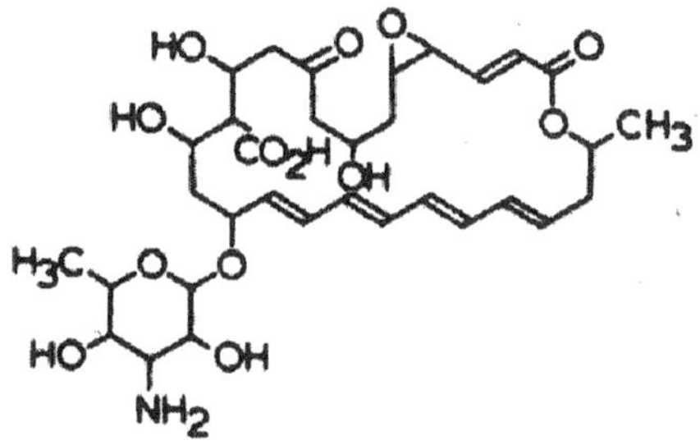

Figure 2. Structure of natamycine.

Aflatoxin B_1 was found in German Tilsit and Edelpilzkäse, while aflatoxin G_1 was detected in Tilsit, Gouda, Emmentaler and Romadur cheese (Kiermeier and Bohm, 1971).

In fermented sausages *Penicillium* species are known to produce mycotoxins. Ochratoxin A, patulin, citrinin, citreoviridin, cyclopiazonic acid, isofumigaclavin A, roquefortine C and rugulovasine A have been detected after inoculation in pure cultures of species isolated from fermented sausages (Fink-Gremmels and Leistner, 1990).

Growth of potentially toxinogenic moulds is of particular concern in the case of ripening under uncontrolled conditions and when unknown moulds develop. This may occur in case of traditional ripened cheeses such as Canestro Pugliese cheese produced from raw milk (Sinigaglia *et al.*, 2005). In this study mycotoxigenic moulds such as *Fusarium moniliforme*, *Aspergillus niger* and *A. flavus* were isolated. Optimization of the hygienic conditions and knowledge about the flora required for ripening of these traditional cheeses is needed to prevent mycotoxin formation.

As elimination of mycotoxins in milk, cheese or fermented sausages is not technologically and commercially feasible yet, prevention of growth of mycotoxin-producing moulds is of main importance. In several studies it has been demonstrated that all mycotoxin-producing moulds isolated from cheese and fermented sausages are extremely sensitive towards natamycin (Bullerman, 1977; Ray and Bullerman, 1982; Kiermeier and Zierer, 1975). Therefore a surface treatment of these products with natamycin is recommended.

NATAMYCIN AND SORBATES

In most countries natamycin and/or sorbates are the only approved antifungal agents for the surface treatment of cheese and dry sausages.

Although sorbates effectively inhibit the growth of many spoilage organisms occurring on these products, compared with natamycin, sorbates have several disadvantages. Relatively high concentrations of sorbate are needed to prevent mould growth and some common spoilage moulds have developed considerable resistance towards sorbate. While natamycin remains on the surface, sorbate migrates from the surface to the interior of the product, thus reducing its effectiveness. The use of sorbate may also affect the flavour and colour of the product in a negative way. Finally it inhibits the bacterial starter cultures (De Ruig and Van den Berg, 1985; Frank, 1989; Stiebing, 2001).

Natamycin (Figure 2) was discovered in the DSM research laboratories in Delft (Struyck *et al.*, 1957-1958). It is a polyene macrolide antimycotic produced by fermentation using *Streptomyces natalensis*, a filamentous bacterium originally found in a soil sample from the state of Natal, South Africa. Already for almost 50 years natamycin is successfully used to prevent fungal growth on the surface of cheese and dry sausages. It is permitted in many countries for these applications, in some countries wider use is permitted (Stark, 2003). The main advantages of natamycin are summarized in Table 2.

Although natamycin has been used for decades in the cheese and sausages industry, development of significant resistance has never been observed. De Boer *et al.* (1977) isolated moulds from cheese factories and tried to induce natamycin tolerance by serial transfers onto media with increasing natamycin concentrations, without success. Also attempts to induce resistance in the relatively natamycin tolerant mould *P. discolor* were not successful (Unpublished results DSM Food Specialties, Delft). Only ergosterol-free mutants produced in the laboratory are resistant to natamycin; however, such mutants cannot survive in nature (Ziogas *et al.*, 1983; Hamilton-Miller, 1974).

The mechanism of action of natamycin is binding to the ergosterol in the fungal cell membrane disrupting the ergosterol functions, leading to cell death. Since there is no separation between fungistatic and fungicidal concentrations, development of resistance is unlikely.

Table 2. Main advantages of natamycin

- □ Broad spectrum activity against moulds and yeasts
- □ Prevention of mycotoxin formation
- □ No development of resistance
- □ No effect on bacterial starter or surface ripening flora
- □ Effective at low concentrations
- □ Prolonged working time through slow release
- □ Remains on the surface
- □ Easy to apply via the coating or casing; by dipping or spraying
- □ No negative effect on the quality of cheese or sausages
- □ No colour, odour or taste
- □ Long history of safe use
- □ Permitted in most countries
- □ Chemically stable

Natamycin appears to have an all-or-none effect, perhaps explained by the single-hit theory, which suggests that even in diluted solutions natamycin forms micelles.

Figure 3. *Penicillium discolor.* Sporulating structures and conidia.

If a fungal cell comes into contact with such a micelle, the local concentration of natamycin is always high enough and the mould will not survive. Of course cells which do not come in contact with natamycin cannot develop resistance.

Due to its low water solubility of 30 to 50 ppm, natamycin will mainly be present in the solid state that guarantees a prolonged working time by slow release. The active dissolved natamycin slowly dissolves from the crystals and diffuses over the surface, compensating natamycin eliminated by interaction with fungal cells, hydrolysis or light.

On cheese, natamycin can be applied via the plastic cheese coating, mostly cheeses are treated several times with a coating containing 100 to 750 ppm of natamycin. Where dry sausages are concerned, casings can be soaked in a natamycin suspension of 1000 ppm. A coating or casing treatment guarantees a homogeneous distribution of natamycin over the surface. Cheeses or dry sausages can also be dipped in an aqueous suspension of natamycin, usually containing 1000 to 3000 ppm. The natamycin suspension can also be sprayed onto the surface of the product.

For dipping and spraying applications it is recommended to use Delvocid®-Dip for cheeses and Premi®Nat for sausages. These natamycin formulations contain xanthan, which optimise the distribution of natamycin over the surface and prevent too low concentrations at the top of the product (De Haan *et al.*, 1998).

A precondition for the efficacy of each preservative is a good hygienic production process, then 30 ppm of active natamycin is sufficient to prevent fungal growth on cheeses and sausages (see Table 3).

In many scientific publications the successful application of natamycin to prevent fungal growth on almost every type of cheese has been described: Gouda (Lück and Cheeseman, 1978; Engel *et al.*, 1983; De Ruig and Van den Berg, 1985; Daamen and Van den Berg, 1985), Edam (Engel *et al.*, 1983), Emmental (Fluckiger, 1973), Cheddar (Lück and Cheeseman, 1978; Sachdeva *et al.*, 1994); shredded cheese (Suloff *et al.*, 2003); cottage cheese (Woolf and Bender, 1991); Tilsiter (Engel *et al.*, 1983), Italian cheeses such as Caciotta (Neviani *et al.*, 1981) and Fontina, Tallegio, Montasio, Asiago, Provolone, Pecorino and Romano (Lodi *et al.*, 1989); Swedish hard cheeses (Mattsson, 1977); blue cheese

Table 3. Sensitivity to natamycin of fungi isolated from cheeses or dry sausages

Microorganism	MIC* (ppm)
Moulds	
Aspergillus penicillioides, A. flavus, A. parasiticus	10-20
Aspergillus versicolor	5-10
Cladosporium cladosporioides	< 5
Eurotium herborariorum	< 10
Geotrichum candidum	< 10
Penicillium discolor	20-30
Penicillium glabrum	5-10
Penicillium commune, P. chrysogenum, P. nalgiovense, P. verrucosum, P. brevicompactum, P. roqueforti, P. camemberti, P. corylophilum, P. solitum	< 5
Phoma glomerata	< 5
Wallemia sebi	< 10
Yeasts	
Candida zeylandoides	< 5
Debaryomyces hansenii	< 5

*MIC = Minimal Inhibitory Concentration.
Sources: DSM Food Specialties Research Laboratories Delft, The Netherlands; Centraalbureau voor Schimmelcultures, Utrecht, The Netherlands. The MIC was determined as described by Stark (2003).

(Morris and Castberg, 1980); Indian cheeses (Verma *et al.*, 1988; Pugazhenthi *et al.*, 1999).

Several researchers studied the inhibition of moulds on the surface of dry fermented sausages: Italian dry salami and Mortadella (Holley, 1981 and 1986; Cattaneo *et al.*, 1978; Baldini *et al.*, 1979), German Rohwurst (Stiebing *et al.*, 2001; Anonymous, 2001) and Dutch sausages (Moerman, 1972). Natamycin was effective in inhibiting the development of moulds without affecting the quality of the sausages.

REFERENCES

Anonymous (2001). Fit gegen Schimmel und Hefen, Oberflächebehandlung von Hartwurst. Zeitschrift für die Lebensmittelwirtschaft 49 (5):14 16. (German)

Baldini, P., Palmia, F., Raczynski, R. G., and Campanini, M. (1979). Impiego della pimaricina nella prevenzione della crescita delle muffe sui prodotti di salumeria Italiani. Industria Conserve54:305- 307. (Italian)

Basílico, J. C., deBasílico, M. Z., Chiericatti, C., and Vinderola, C. G. (2001). Characterization and control of tread mould in cheese. Letters in Applied Microbiology 32:419-423.

Beresford, T. P., Fitzsimons, N. A., Brennan, N. L., and Cogan, T. M. (2001). Recent advances in cheese microbiology. International Dairy Journal 11:259-274.

Boer, E. de, and Stolk-Horsthuis, M. (1977). Sensitivity to natamycin (pimaricin) of fungi isolated in cheese warehouses. Journal of Food Protection 40:533-536.

Bullerman, L. B. (1976). Examination of Swiss cheese for incidence of mycotoxin producing molds. Journal of Food Science 41:26-28.

Bullerman, L. B. (1977). Incidence and control of mycotoxin producing molds in domestic and imported cheeses. Annales de la Nutrition et de l'Alimentation 31:435-446.

Buzzini, P., and Haznedari, S. (1995). Caratterizzazione di lieviti isolati da insaccati fermentati prodotti in Umbria. Valutazione preliminare della loro attività proteolitica e lipolytica. Industrie Alimentari 34:620-625. (Italian)

Cattaneo, P., D' Aubert, S., and Righetti, A. (1978). Attività antifungina della pimaricina in salumi crudi stagionati. Industrie Alimentari 17:658-664. (Italian)

Daamen, C. B. G., and Berg, G. van den (1985). Prevention of mould growth on cheese by means of natamycin. Voedingsmiddelentechnologie 18: 26-29. (Dutch)

Engel, G., Hertel, K., and Teuber, M. (1983). Nachweis und Abbau von Natamycin (Pimaricin) auf der Käseoberfläche. Milchwissenschaft 38:145-147. (German)

Filtenborg, O., Frisvad, J. C., and Samson, R. A. (2002). Specific association of fungi to foods and influence of physical environmental factors. *In* Introduction to food- and airborne fungi, 6th Edition. (Samson, R. A. *et al.*, eds.), Centraal Bureau voor Schimmelcultures, Utrecht, The Netherlands, pp. 306-320.

Filtenborg, O., Frisvad, J. C., and Thrane, U. (1996). Moulds in food spoilage. International Journal of Food Microbiology 33:85-102.

Fink-Gremmels, J., and Leistner, L. (1990). Toxicological evaluation of moulds. Food Biotechnology 4:579-584.

Fluckiger, E. (1973). Prevention of mould growth on Emmental cheese. Alimenta 12:103-106.

Frank, J. (1989). Use of natamycin to prevent growth of mold contaminants on cheese. Milchforschung Milchpraxis 31:15-16.

Frisvad, J. C., Samson, R. A., Rassing, B. R., Horst, M. I. van der, Rijn, F. T. J. van, and Stark, J. (1997). *Penicillium discolor*, a new species from cheese, nuts and vegetables. Antonie van Leeuwenhoek 72:119-126.

Galli, A., and Zambrini, A. (1978). Surface microflora of Provolone cheese. Industria del Latte 14:3-12. (Italian)

Galli, A., Zambrini, A., and Suess, L. (1978). Trials with antiparasitic and antifungal agents on Provolone cheese. Industria del Latte 14:23-36 (Italian)

Grazia, L., Suzzi, L., Romano, P., and Giudici, P. (1989). The yeasts of meat products. Yeast S495-499.

Gripon, J. C. (1987). Mould-ripened cheeses. *In* Cheese: Chemistry, Physics and Microbiology, V2 Major cheese groups (Fox, P. F., ed.), Elsevier Applied Science, Crown House, England, pp. 121-150.

Haan, B. R. de, Stark, J., and Bozzetti, V. (1998). Antifungal composition. European Patent 0 867 124.

Hamilton-Miller, J. M. T. (1974). Fungal sterols and the mode of action of the polyene antibiotics. Advances in Applied Microbiology 17:109-134.

Hocking, A. D. (1994). Fungal spoilage of high fat foods. Food Australia 46:30-33.

Hocking, A. D., and Faedo, M. (1992). Fungi causing tread mould spoilage of vacuum packaged

Cheddar cheese during maturation. International Journal of Food Microbiology 16:123-130.

Hoekstra, E. S., Horst, M. I. van der, Samson, R. A., Stark, J., and Rijn, F. T. J. van (1998). Survey of the fungal flora in Dutch cheese factories and warehouses. Journal of Food Mycology 1:13-22.

Holley, R. A. (1981). Prevention of surface mold growth on Italian dry sausage by natamycin and potassium sorbate. Applied and Environmental Microbiology 41:422-429.

Holley, R. A. (1986). Effect of sorbate and pimaricin on surface mold and ripening of Italian dry salami. Lebensmittel-Wissenshaft und Technologie 19:59-65.

Kiermeier, F., and Bohm S. (1971). On aflatoxin formation in milk and milk products. V. Application of the chick embryo test for the affirmation of thin-layer chromatographic determination of aflatoxin in cheese. Zeitschrift Lebensmittel Untersuchung and Forschung 147:61-64. (German)

Kiermeier, F., and Zierer, E. (1975). Zur Wirkung von Pimaricin auf Schimmelpilze und deren Aflatoxinbildung bei Käsen. Zeitschrift Lebensmittel Untersuchung und Forschung 157:253-262. (German)

Kure, C. F., Wasteson, Y., Brendehaug, J., and Skaar, I. (2001). Mould contaminants on Jarlsberg and Norvegia cheese blocks from four factories. International Journal of Food Microbiology 70:21-27.

Kure, C. F., Skaar, I., and Brendehaug, J. (2004). Mould contamination in the production of semi-hard cheese. International Journal of Food Microbiology 93:41-49.

Leistner, L., and Ayres, J. C. (1969). Molds and meat. Fleischwirtschaft 48:62-65.

Lodi, R., Todesco, R., and Bozzetti, V. (1989). Nouvelles applications de la natamycine sur des fromages typiques Italiens. Microbiologie-Aliments-Nutrition 7:81-84. (French)

López-Díaz, T. M., Román-Blanco, C., García-Arias, M. T., García-Fernández, M. C., and García-López, M. L. (1996). Mycotoxins in two Spanish cheese varieties. International Journal of Food Microbiology 30:391-395.

Lück, H., and Cheesman, C. E. (1978). Mould growth on cheese as influenced by pimaricin or sorbate treatments. South African Journal of Dairy Technology 10:143-146.

Lund, F., Filterborg, O., and Frisvad, J. C. (1995). Associated mycoflora of cheese. Food Microbiology 12:173-180.

Mattsson, N. (1977). Mould control on cheese. Experience with potassium sorbate and pimaricin for surface treatment of hard cheese. Svenska Mejeritidningen 69:14-15. (Swedish)

Metaxopoulos, J., Stavropoulos, S., Kakouri, A., and Samelis, J. (1996). Yeasts isolated from traditional Greek dry salami. Italian Journal of Food Science 8:25-32.

Moerman, P. C. (1972). Schimmelwering op vleeswaren door pimaricine. Voedingsmiddelentechnologie 3:261-264. (Dutch)

Morris, H. A., and Castberg, H. B. (1980). Control of surface growth on blue cheese using pimaricin. Cultures Dairy Products Journal 15:21-23.

Neviani, E., Emaldi, G. C., and Carini, S. (1981). L'impiego di pimaricina come antifungo sulle croste dei formaggi: technologia e microflora di superficie. Latte 6:335-343. (Italian)

Nichol, A. W. (2000). Cheese/Mould-ripened varieties. *In* Encyclopedia of food microbiology (Robinson, R. K.), Academic Press, pp. 387-393.

Nielsen, K. F., Dalsgaard, P. W., Smedsgaard, J., and Ostenfeld Larsen, T. (2005). Andrastins A-D *Penicillium roquefortim*etabolites consistently produced in blue-mold-ripened cheese. Journal of Agricultural and Food Chemistry 53:2908-2913.

Northolt, M. D., Van Egmond, H. P., Soentoro, P., and Deijll, E. (1980). Fungal growth and the presence of sterigmatocystin in hard cheese. Journal-Association of Official Analytical Chemists 63:115-119.

Persoon, N. L. M., and Rijn, F. T. J. van (1993). Integrale schimmelbestrijding in de kaasindustrie. Voedingsmiddelentechnologie 26:11-13. (Dutch).

Pugazhenthi, T. R., Dhanalakshmi, B., Narasimhan, R., Shibu, A. V., and Madhan, S. (1999). Effect of anti-mycotic agents on *Penicillium citrinum* in cheese. Indian Veterinary Journal 76:537-539.

Ray, L. L., and Bullerman, L. B. (1982). Preventing growth of potentially toxic molds using antifungal agents. Journal of Food Protection 45:953-963.

Ruig, W. G. de, and Berg, G. van den (1985). Influence of the fungicides sorbate and natamycin in cheese coatings on the quality of the cheese. Netherlands Milk and Dairy Journal 39:165-172.

Sachdeva, S., Sing, S., Tiwari, B. D., and Ram, J. (1994). Effect of processing variables on the quality and shelf-life of processed cheese from buffalo milk Cheddar cheese. The Australian Journal of Dairy Technology 49:75-78.

Scott, P. M. (1989). Mycotoxigenic fungal contaminants of cheese and other dairy products. *In* Mycotoxins in dairy products (Egmond, H. P. van, ed.), Elsevier Applied Science, pp. 193-259.

Sinigaglia, M., Albenzio, M., Corbo, M. R., and Ciccarone, C. (2005). Comparison of mycoflora associated to Canestrato Pugliese cheese produced according to three protocols. Sciences des Aliments 24:159-172.

Stark, J. (2003). Natamycin, an effective fungicide for food and beverages. *In* Natural antimicrobials for the minimal processing of foods (Roller, S., ed.), Woodhead Publishing Limited, pp. 82-97.

Stark, J., and Tan, H. S. (2003). Natamycin. *In* Food Preservatives. Second Edition (Russel. N. J., and Gould, G. W., eds.), Kluwer Academic / Plenum Publishers, Dordrecht, The Netherlands, pp. 179-195.

Stiebing, A., Oberhaus, T., and Baumgart, J. (2001). Natamycin-Verhinderung des Schimmelpilz-wachstums bei Rohwurst. Fleischwirtschaft 81 (8):97-100. (German)

Struyck, A. P., Hoette, I., Drost, G., Waisvisz, J. M., Van Eek, T., and Hoogerheide, J. C. (1957-1958). Pimaricin, a new antifungal antibiotic. Antibiotics Annual 1957-1958:878-885.

Suloff, E. C., Marcy, J. E., Hackney, C. R., Sumner, S. S., and Bishop, J. R. (2003). Comparative study of a semisynthetic derivate of natamycin and the parent antibiotic on the spoilage of shredded Cheddar cheese. Journal of Food Protection 66:1499-1502.

Tempel, T van den, and Nielsen, M. S. (2000). Effects of atmospheric conditions, NaCl and pH on growth and interactions between moulds and yeasts related to blue cheese production. International Journal of Food Microbiology 2000:193-199.

Verma, H. S., Yadav, J. S., and Neelakantan, S. (1988). Preservative effect of selected antifungal agents on butter and cheese. Asian Journal of Dairy Research 7:34-38.

Welthagen, J. J., and Viljoen, B. C. (1998). Yeast profile in Gouda cheese during processing and ripening. International Journal of Food Microbiology 41:185-194.

Woolf, H. D., and Bender J. (1991). Extending the shelf-life of cottage cheese: identification of spoilage flora and their control using food grade preservatives. Cultured Dairy Journal 26:8-12.

Ziogas, B. N., Sisler, H. D., and Lusby, W. R. (1983). Sterol content and other characteristics of pimaricin-resistant mutants of *Aspergillus nidulans*. Pesticide Biochemistry and Physiology 20:320-329.

Part 6

FUNGI AS FOOD

Fungi do spoil enormous amounts of food, but they also serve as food. It even might be the case that long ago, the emergence of a new class of food product was the result of spoilage by fungi. Tempe consists out of cooked soybeans that are colonised by the fungus *Rhizopus oligosporus* and maybe started once as spoiled soybeans. However, the mass of beans and hyphae has a surprising good quality and taste, and no toxic compounds are found in the product. Further, its high protein content makes it an important food commodity in countries where people cannot afford meat on the menu. In Chapter 17 Nout gives a summary of many different food products that are based on the principle of "neatly-spoiled food" and in fact also blue cheeses belong to this group. He also addresses the different stages of preparation from a mycological point of view and discusses the production of secondary metabolites by the relevant fungi. In Chapter 18, Thrane discusses the possibilities of fungal protein as a meat-replacing food product. In the Western world the meat consumption is enormous and the wellbeing of the meat-delivering animals is a matter of hot debate for many people. Alternatives for meat, having a similar appearance and mouth-feeling, might by a very promising way of dealing with consumer demands in the future. In fact, these alternatives can already be bought in the supermarket for some time. In this chapter the safety on *Fusarium venenatum*-based products is discussed.

At last Chapter 19 describes the hallmark of Basidiomycetous and some Ascomycetous fungi, the fruiting body, as macroscopic fungal food products. The chapter addresses the techniques used in the intensive production of enormous densities of fruit bodies of *Agaricus bisporus*. It also summarizes the whole area of edible fruit bodies and it becomes clear that every new species has its own rules for proper production. Further, the chapter deals with pests and pathogen of edible fruit bodies in which the most devastating are fungal in nature.

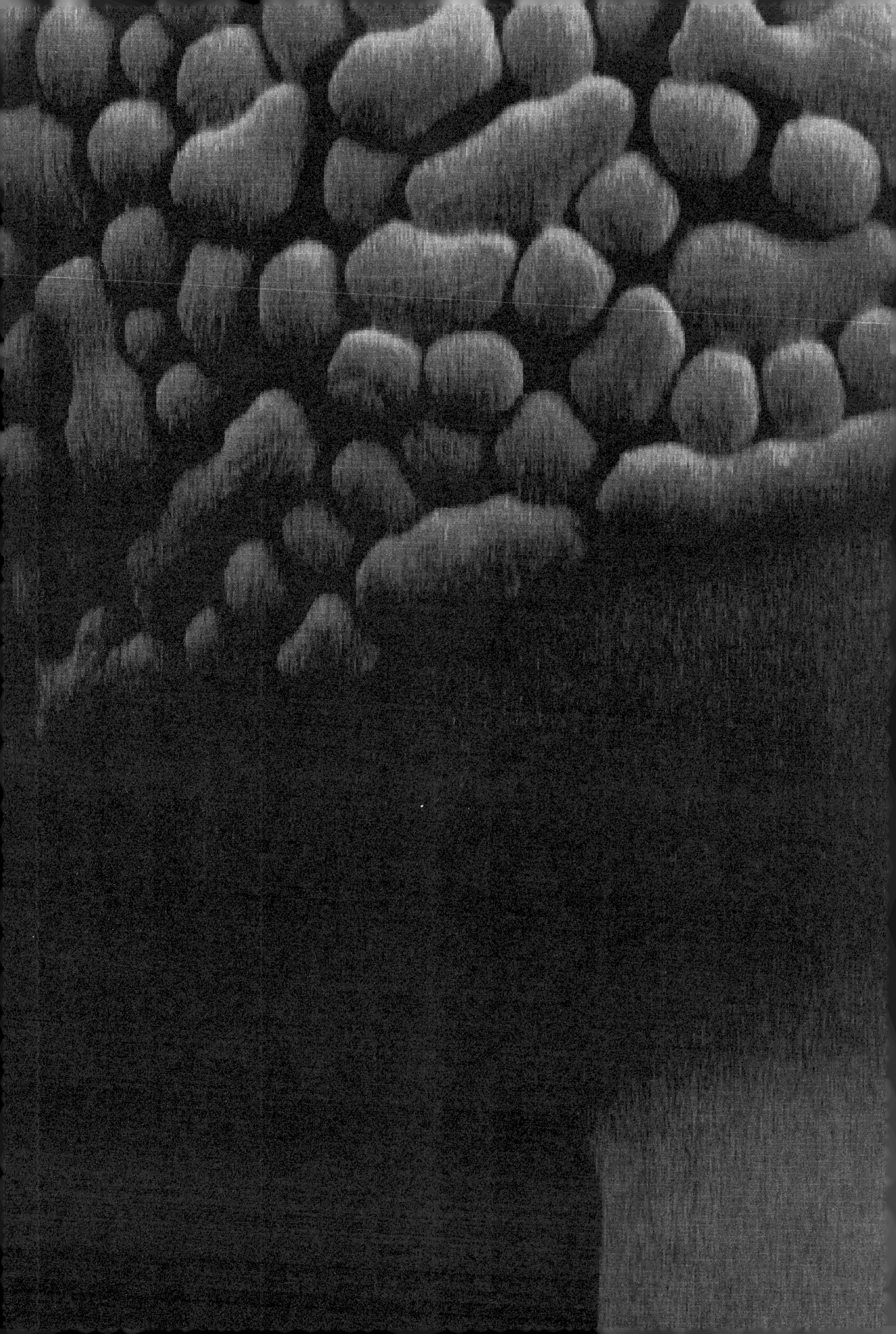

Chapter 17

The colonizing fungus as a food provider

Rob M. J. Nout
Laboratory of Food Microbiology, Wageningen University, P.O. Box 8129, 6700 EV Wageningen, The Netherlands.

INTRODUCTION

Filamentous fungi colonize food ingredients, penetrating into them, releasing a variety of enzymes, raising metabolites and reaction products and increasing their biomass. In many cases, this sequence of events is considered as spoilage (Pitt and Hocking, 1985) because of off-odours, unwanted discolourations, taste defects, and toxicity. Interestingly, however, mankind experienced that in certain situations, colonizing filamentous fungi (moulds in short) bring about desirable changes in foods, and termed them fermentations (Fukushima, 1985). Considering the chiefly aerobic metabolism of moulds, the usage of the term fermentations - indicating the anaerobic mechanism of energy generation - is incorrect by definition. Nevertheless the term fungal food fermentation is widely accepted in the sense of bioprocesses resulting in improvements of quality.

Whereas the origin of fungal food fermentation is the Orient, some specific processes have developed in Europe. Increasing international travel and trade have been the vehicle for the worldwide distribution of some of the derived products.

In principle, the birth of fungal food fermentations must have been from "spoilage"; from an ecological point-of-view it may thus be expected that the concept of "spoilage associations" (Dalgaard *et al.*, 2003) also applies to fungal food fermentations. This implies ecological niches consisting of suitable substrate, microbial competition, and favourable environmental conditions. It therefore seems quite logical to encounter moulds with optimum growth temperature (T_{opt}) of 20-30 °C in the moderate climates, and those with T_{opt} of 30-40 °C in the humid tropics.

Table 1 presents an overview of selected filamentous fungi, the fermented foods in which they feature, their role in quality improvement, and some recent literature references. In the following sections these cases will be discussed in some detail, with a focus on the scientific questions of recent interest.

Zygomycetes

Among the Zygomycetes, mainly the order of Mucorales is of relevance for this chapter. Four genera, i.e., *Actinomucor, Amylomyces, Mucor,* and *Rhizopus* are of functional importance in a diversity of Oriental fungal fermented food products. The genus *Amylomyces* is considered by some (R. A. Samson, personal communication) as a domesticated form of *Rhizopus*. *Amylomyces rouxii* could not be discriminated from *Rhizopus oryzae* on the basis of 18S-28S rRNA because the amplified sequences were identical (Abe *et al.*, 2004); *A. rouxii* could only be distinguished from *R. oryzae* because of its much higher number of chlamydospores in the aerial and substrate mycelium. On the other hand, the genotypes of *Actinomucor, Rhizopus* and *Mucor* were shown to be distinguishable as separate clusters (Han *et al.*, 2004b).

Actinomucor elegans (Han *et al.*, 2001) and *A. taiwanensis* (Chou *et al.*, 1988) are used as pure culture starters in the manufacture of Chinese fu-ru or sufu (Figure 1a).

Table 1. Overview of filamentous fungi used in food fermentation

Zygomycetes	Species	Foods	Functionality	References
Actinomucor	*A. elegans, A. taiwanensis*	sufu, fu-ru (China)	texture, flavour	Chou *et al.*, 1988; Han *et al.*, 2001
Amylomyces	*A. rouxii*	ragi (Indonesia), marcha (India)	glucose release from starch	Tsuyoshi *et al.*, 2005
Mucor	*M. circinelloides, M. rouxii, M. indicus*	ragi, marcha, tempe (Indonesia), pehtze (China)	enzymic transformation, flavour production	Pedraza-Reyes and Lopez-Romero, 1991; Agranoff and Markham, 1997
Rhizopus	*R. microsporus, R. oligosporus, R. oryzae*	koji (Japan), nuruk (Korea), chu (China), marcha, tempe	texture, enzymic transformations, vitamins	Ginting and Arcot, 2004; Nout and Kiers, 2005
Ascomycetes				
Monascus	*M. pilosus, M. purpureus, M. ruber*	angkak, red yeast rice (China)	colour, flavour, secondary metabolites	Juzlova *et al.*, 1996; Akihisa *et al.*, 2005
Neurospora	*N. sitophila, N. intermedia*	oncom (Indonesia)	texture, flavour, enzymic modifications	Beuchat, 1986
Deuteromycetes				
Aspergillus	*A. oryzae, A. sojae, A. niger*	soy sauce (East Asia)	carbohydrases, proteases, other lytic enzymes	Nout and Aidoo, 2002; Hanya and Nakadai, 2003
	A. glaucus, A. melleus, A. repens, A. candidus	katsuobushi (Japan)	enzymic transformation, flavour production	Campbell-Platt, 1987
Penicillium	*P. glaucum*	katsuobushi		
	P. camemberti	Camembert, Brie (France)	texture, flavour	Leclercq Perlat *et al.*, 2004a
	P. nalgiovense	salami (Europe)	colour, flavour	Fink-Gremmels *et al.*, 1988
	P. roqueforti	Roquefort, blue Stilton, Danablue (Europe)	colour, flavour	Gripon, 2003

The genus *Actinomucor* was described earlier (Benjamin and Hesseltine, 1957); recently *A. taiwanensis* was described as a separate species (Jong and Yuan, 1985; Chou *et al.*, 1988).

The process of preparing sufu or fu-ru starts with the production of soymilk by soaking dehulled soybeans, grinding, sieving and cooking the watery extract. Next, a coagulation step is carried out, by adding salts or acid, in order to obtain a precipitate of mainly soy protein and enclosed lipids. This is collected and pressed to obtain sheets of tofu (soybean curd) of the required moisture content and firmness. After cutting the tofu into cubes (dices) these are inoculated with a suspension of mould spores. Incubation during a few days usually results in a luxuriant mycelial development giving the dices a fluffy appearance. These are now called pehtze, and after flattening the mycelium as a protective skin on the cubes, pehtze is submerged in a maturation mix and left during several months to develop into a flavoursome, soft, cheese-like product. The main functions of the maturation mix are preservation, flavouring and colouring.

Figure 1. Fungal fermented foods (a: sufu; b: men; c: tempe; d: oncom; e: soy sauce; f: Camembert; g: blue-veined cheese).

The preservation is achieved by a combination of salt and alcohol (rice wine may be used), whereas ang-kak (see below) and other ingredients impart specific flavour and colour to the product (Su, 1986; Han *et al.*, 2001). The major function of the moulds in this process is the formation of the protective layer of mycelial biomass surrounding the pehtze cubes, but most importantly, to release several enzymes (Han *et al.*, 2003a) that are responsible for the partial degradation of the protein (Lu *et al.*, 1996), fibre and lipid fractions in pehtze during the maturation. This degradation results in a softening of the texture, solubilization of the dry matter and accumulation of flavour enhancing compounds, such as glycine (Ma *et al.*, 2004) and glutamic acid (Liu and Chou, 1994; Han *et al.*, 2004c). In view of the optimization of industrial sufu-making processes, the response of *A. elegans* to temperature, salt (Han *et al.*, 2003b) and alcohol has been studied. The higher the salt and alcohol levels during the maturation, the slower the enzymatic reactions take place and thus the more maturation time and costs are involved. With the objective of accelerating the maturation, the salt and alcohol levels could be lowered. This is feasible to a level of about 10% alcohol (Chou and Hwan, 1994) in combination with 6% salt; at lower levels the product is susceptible to spoilage by lactic acid bacteria (Han *et al.*, 2004a), as well as survival by pathogens (Shi and Fung, 2000) and enterotoxin formation by *Staphylococcus aureus* (Han *et al.*, 2005).

Amylomyces rouxii is a rather peculiar mould, described by Ellis *et al.* (1976). It finds its importance as a functional component of Oriental traditional starters for alcoholic fermentations. Its main properties of technological importance are the production of amyloglucosidase (Wang *et al.*, 1984), its ability to colonize uncooked rice dough, and its restricted sporulation. Oriental traditional starters for alcoholic fermentations (Hesseltine *et al.*, 1988; Leistner, 1990) are often prepared by mixing powdered milled rice with water to a dough, with the addition of herbs and spices. The dough is portioned into small flattened balls or tablets (Figure 1b) which are dusted with powdered starter from an earlier batch. These inoculated tablets are kept in a warm room during some days where they also gradually dehydrate. The final product is a hard, dry tablet which can be conveniently packed and transported for marketing. The microflora of such starters — Indonesian ragi, Vietnamese men, Indian marcha and numerous others — has been investigated by several researchers (Hesseltine *et al.*, 1988; Tamang and Sarkar, 1995; Tsuyoshi *et al.*, 2005); in principle, three categories of microorganisms can be encountered, viz. starch-degrading fungi (mainly *A. rouxii*, but some starters contain amylolytic yeasts such as *Endomycopsis fibuliger*), alcohol-tolerant yeasts (*Saccharomyces cerevisiae* in particular) and non-functional contaminants such as lactic acid bacteria, *Bacillus* spp., etc. When used in rice wine preparation, rice (either glutinous or non-sticky rice) is soaked, cooked, and the cooled mass is dusted with powdered starter tablets (the level of inoculation needs to be experienced first and depends on the composition and viability of the microflora within the tablet). During incubation at a warm place, a significant liquefaction takes place which results from the degradation of the gelatinized starch. When, after a few days, it is considered that sufficient glucose has been produced to start the alcoholic fermentation, more water is added to submerge the moulded rice, turning the aerobic incubation into anaerobic conditions. This will inhibit the formation of mould biomass and favour yeast fermentation; the yeast being present from the start will ferment as soon as glucose is released, but the most effective alcohol accumulation occurs during this submerged fermentation phase. When the fermentation has stopped, the residual rice and yeast is left to sediment and the supernatant wine is decanted. According to local preference, the wine can be clarified further by filtration, and its shelf-life can be prolonged by fortification, i.e., adding some distilled (rice) alcohol (Kozaki and Uchimura, 1990; Rhee *et al.*, 2003). At a small scale of production, hardly any control of the process is possible and therefore it is not suprising that the yields of glucose and ethanol from rice, as well as important traits such as colour, volatile flavour and taste are rather unpredictable. In traditional rice wine fermentation starters, bacteria — including low numbers (2.6-4.2 log cfu g^{-1}) of lactic acid bacteria — are also present. The fact that the pH of good quality rice wine usually is in the pH range 3.9-4.2 does not necessarily imply the functional contribution of lactic acid bacteria (LAB) to its quality. The pH range indicated above is also found in pure culture experiments with moulds and yeasts and results from the formation of lactic acid — for example by *A. rouxii* (Saito *et al.*, 2004) — and other acidic co-metabolites (by the yeast). If the number of LAB would become higher, the quality of the wine is likely to suffer because of acidity (in poor quality wines we measure pH as low as 3.2). Based on the above, LAB should not be considered as functional flora in rice wine starters, but rather as potential spoilage microorganisms. At an industrial scale, most rice wines (Chinese Shaohing, Japanese Saké, Korean Yakju) are produced with *Aspergillus oryzae* (see below); from 1000 kg polished rice, 3000 litres Saké of 20% v/v ethanol are obtained, representing an almost 100% yield (Nout and Aidoo, 2002). The manufacture of rice wine starters and rice wine constitute two different businesses. There is an increasing demand for the development of defined starters that combine maximum saccharification and alcohol productivity, instead of the traditional tablets of unknown composition and activity.

Mucor spp. such as *M. circinelloides*, *M. indicus*, and *M. rouxii* are encountered in a diversity of fungal fermented food products of the Orient (Tamang and Sarkar, 1995; Agranoff and Markham, 1997; Han *et al.*, 2004b), including starter tablets as well as tempe, a fermented soybean food (see below). *Mucor* spp. grow rapidly and release a range of enzymes including amyloglucosidase, lipases (Chou *et al.*, 1988), proteases (Han *et al.*, 2003a), and carbohydrases (Pedraza-Reyes and Lopez-Romero, 1991). Whereas these are valuable properties that may contribute to the evolution of fermented foods, it appears that in fermentations where *Rhizopus* or *Amylomyces* are present as well, *Mucor* spp. are not the prime movers of the fermentation. They may, however, contribute in other ways, such as formation of certain flavour compounds or fatty acids (Oxlade, 1990; Agranoff and Markham, 1997). In a comparison of *Amylomyces rouxii* and *Mucor circinelloides*, it was found that the latter accumulated glucose less efficiently from rice starch; this was not related to glucoamylase activity but rather to its profuse biomass formation (Dung, 2004).

Rhizopus spp. (Schipper and Stalpers, 1984) of importance in food fermentation are *R. microsporus* and *R. oryzae*. The latter is mesophilic, forms a variety of enzymes particularly starch degrading enzymes and is encountered in a diversity of amylolytic starters for alcoholic fermentations such as koji, nuruk, chu and murcha (Tamang *et al.*, 1996; Nout and Aidoo, 2002; Shrestha *et al.*, 2002); its glucoamylase gene has been brought to expression in *Saccharomyces cerevisiae* to facilitate the direct production of ethanol from raw maize starch (Shigechi *et al.*, 2004). Whereas fungal starch degradation for winemaking is mainly practised in Asia, the use of *Rhizopus* spp. was described as well in a complicated process for making Parakari, an indigenous alcoholic beverage made from cassava in Guyana (Henkel, 2005). *Rhizopus* spp. can produce health-promoting unsaturated fatty acids such as gamma-linolenic acid (GLA) (Liu *et al.*, 2004). *R. oryzae* is also used for soybean fermentations, e.g., in tempe manufacture. On the other hand, *R. microsporus* is thermophilic and prefers temperatures ranges from 30-40 °C. Within this species, varieties are distinguished of which *R. microsporus* var. *oligosporus* (in short: *R. oligosporus*) is best known in relation with the tempe fermentation. Tempe (Figure 1c) originates from Indonesia and is made from cooked seeds (soybeans, cereals or others) or food-processing by-products, by solid substrate fungal fermentation (Nout and Kiers, 2005). In the traditional tempe process, simple methods are employed for the inoculation of the cooked beans. In principle it is possible to use some previously made tempe as inoculum (Ko and Hesseltine, 1979); as tempe contains a considerable load of bacteria, the re-use of tempe as an inoculum incurs the risk of fermentation failure due to bacterial overgrowth. Therefore, professional tempe manufacturers use traditional mould spore concentrates (Samson, 1993). These are, for example, harvested from cooked rice that has been grown with a selected *R. oligosporus* culture, or grown on cooked soybeans between leaves of *Hibiscus tiliaceus* (the waru tree) (Nout *et al.*, 1992). The latter type of starter is widely used, is made by specialized households, and can be purchased in the public markets in Indonesia. For a better control of the fermentation, pure culture spore preparations can be used. These are grown on, for example, cooked rice and stored as dehydrated powders. It was observed that a majority of the spores thus produced are in a state of exogenous dormancy. Using defined media, it was reported earlier (Medwid and Grant, 1984) that a carbon source (e.g., glucose) and nitrogen (amino acids) are necessary to initiate the formation of germ tubes. Recently, it was shown that in addition to glucose, alanine and phosphates contribute to the germ tube formation and further outgrowth of mycelial biomass (Thanh *et al.*, 2005). Some of the interesting properties of *R. oligosporus* in relation to the tempe fermentation are directly linked to this biomass. The characteristic binding of the bean particles by the mycelium results in a considerable stiffness of the tempe cake. The strength of the mycelium can be measured by physical methods (Ariffin *et al.*, 1994) and can be used as an index for fungal growth and quality of tempe. It has been estimated that 5.9% (dry weight basis) of tempe consists of fungal biomass (Sparringa and Owens, 1999). The production and metabolism of such considerable quantities of biomass may easily result in technological problems such as overheating and insufficient supply of oxygen. The traditional, empirical and labour-intensive, tray or bed solid-state fermentation functions well, provided that the depth of the bed and the temperature and ventilation of the environment are in balance. In larger-scale mechanized fermentations, heat and mass transfer can be controlled, especially in mixed fermentors. This has been shown convincingly in rotating drum fermentors (Oostra *et al.*, 2000), cooled either by air or by spraying mist (Nagel *et al.*, 2001), as well as in the agitated bed koji fermentors (Figure 2) used in Japan (Nout and Aidoo, 2002). In the case of tempe, this kind of fermentation implies a departure from the traditional brick-shaped final product, because agitated fermentation results in particulate fermentation products. Nevertheless, mechanized systems could be of interest in tempe fermentations, e.g., when producing novel nutrition ingredients such as tempe flour (Han *et al.*, 1999). Another point of relevance for the acceptability of tempe concerns discolourations caused by enzymatic browning. Phenoloxidase activity, in particular from laccase, has been observed in several fungi such as *Agaricus bisporus* (Wiegant *et al.*, 1992) and *Aspergillus oryzae* (Lertsiri *et al.*, 2003), the activity of the latter fungus being associated with browning of fermented Thai soybean paste. Laccase has also been reported in *R. oligosporus* (McCue *et al.*, 2004), but its role in tempe in relation to browning and quality acceptance has not yet been investigated. An aspect that has not yet attracted much attention is the volatile favour of tempe, in particular the "mushroomy" flavour of freshly fermented tempe. From mushroom research it has been shown that 1-octen-3-ol is one of the major volatiles responsible for the characteristic mushroom smell (Kubickova and Grosch, 1997). It would be of interest to investigate the behaviour of *Rhizopus* spp. in this respect and study the biochemical pathways, precursors and genes involved in key flavour compounds. *Rhizopus* biomass produces a diversity of carbohydrases such as polygalacturonase, endocellulase, xylanase, arabinase, beta-D-glucosidase, alpha-D-galactosidase, beta-D-xylosidase, alpha-L-arabinosidase, and alpha-D-glucosidase (Sarrette *et al.*, 1992) that contribute to the degradation of dietary fibre (non-starch polysaccharides), which mainly consist of arabinogalactans, galactomannans, xylans and pectic substances (Fransen, 1999). This degradation is the cause of a gradual softening (De Reu *et al.*, 1997) of the texture of the fermented product during storage. In addition, the action of proteases, particularly aspartic-(35 kD) and serine (33 kD) protease, each existing in different isoforms (Heskamp and Barz, 1998) causing enzymic protein degradation, results in a strongly improved digestibility of tempe (Kiers *et al.*, 2003). Similar phenomena were observed in tempe made from chickpea (Reyes Moreno *et al.*, 2004) and maize (Cuevas Rodriguez *et al.*, 2004) with concomitant decreases of phytic acid and tannin levels. In addition, the tempe fermentation adds health benefits to the soybean by converting isoflavonoids such as genestein and daidzein into compounds with increased antioxidative capacity such as 3-hydroxyanthranilic acid (Jha *et al.*, 1997; Matsuo *et al.*, 1997; Berghofer *et al.*, 1998). These have been associated with reductions of various types of degenerative diseases. The release of phenolic antioxidants by *R. oligosporus* from isoflavones in soybean (McCue and Shetty, 2003), pineapple waste (Correia *et al.*, 2004b) and soy-guava waste (Correia *et al.*, 2004a) was associated with the considerable β-glucosidase activity of the mould. It has been postulated that the presence of laccase could increase the formation of polymeric phenolics; the latter were shown to inhibit the growth of the peptic ulcer-associated *Helicobacter pylori* (McCue *et al.*, 2004).

Several vitamins (Nout and Kiers, 2005), including folates, mainly 5-formyl-tetrahydrofolate (Ginting and Arcot, 2004), are synthesized during the tempe fermentation. Although a variety of desirable modifications are ascribed to *Rhizopus* spp., there is always a need to ascertain safety of the fermented foods obtained. It was reported that whereas *R. microsporus* can form rhizoxins and rhizonins, the strains of *R. oligosporus* and *R. chinensis* investigated did not produce any of these pharmaceutically active (rhizoxins) or highly toxic (rhizonins A and B) metabolites (Jennessen *et al.*, 2005).

Ascomycetes

In relation with food fermentations, the most significant ascomycetes are the yeasts, especially *Saccharomyces cerevisiae*. There are, however, also a few filamentous fungi that are classified are Ascomycetes, and that have some very interesting properties which make their fermentation products quite appetizing!

The first example is the genus *Neurospora*, with *N. sitophila* and *N. intermedia*. These prefer temperatures of 25-35 °C and produce a very rapidly growing mycelium with large numbers of spores that are very easily detached and spread into the environment. Because of this property, they can spread in laboratory collections and cause havoc! They also can lead to spoilage in bakeries when they contaminate slicing machines. The fermented food of relevance here is oncom (ontjom) (Figure 1d), originating in West-Java and made from peanut press cake, a by-product of peanut oil pressing, by soaking one day, mixing with starchy ingredients such as cassava residues, steaming for about 1 hour, cooling and inoculating with some pre-grown fungal mycelium on the same material. The inoculated dough is shaped in flat rectangular boxes (moulds) obtaining brick-shaped pieces that are covered in banana leaves and incubated during a few days at ambient temperatures (25-30 °C) (Beuchat, 1986). Two types of oncom are distinguished, namely oncom hitam (black oncom) and oncom merah (red oncom) which contain different mycoflora. The yellow-red type contains mainly *Neurospora*, whereas the black oncom contains significant amounts of *Rhizopus* spp. Whereas the black colour of *Rhizopus* sporangiospores is caused by melanoids, several pigments including carotenoids, mainly β-carotene, form the basis of the orange-yellow colour of *Neurospora* (de Fabo *et al.*, 1976).

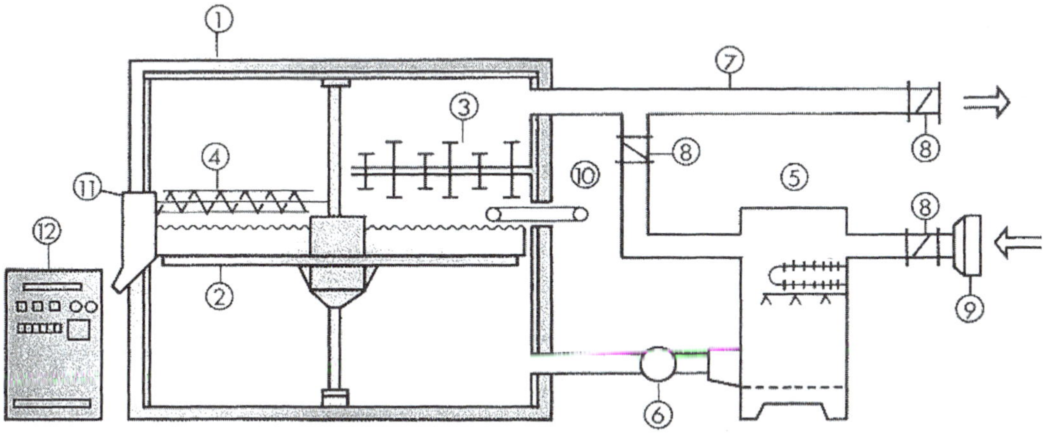

Figure 2. Schematic view of an agitated koji solid-state fermentor.

The flavour of oncom has been described as fruity and somewhat alcoholic; after frying, mince-meat or almond flavours were observed. The enzymic activities (proteases, lipases) contribute to a considerable increase in free fatty acids, and degradation of proteins. Although this does not result in improved protein efficiency ratios, the protein digestibility is improved (Beuchat, 1986) which is of importance for consumers with digestive disorders. Recently, experiments on "oncom-miso" made from soybeans and okara (soymilk extraction residue) demonstrated increased antioxidative and antimutagenic activity, associated with the enzymic release of isoflavone-aglycones (Matsuo, 2004). In contrast with the use of spore-based starters of *Rhizopus,* for example, tempe inoculation, starters for oncom are propagated and maintained by vegetative growth, in a kind of fed-batch solid-state fermentation. Through a moist mixture of peanut-presscake and cassava offal (fibrous residue of cassava starch extraction process), previously overgrown mixture is mixed and incubated. This product will constitute the starter for the next fermentation batch. Although very little controlled experimentation has been done on this fermentation it is presumed that the method of vegetative propagation is needed because the *Neurospora* spores either have limited viability when stored in a dehydrated form, or have a restricted germination ability.

Monascus (*M. ruber, M. pilosus* and *M. purpureus*) is of special interest because of its production of secondary metabolites (Figure 3). Traditionally this organism is used in the production of Chinese "red kojic rice," also referred to as "red-mould rice" and "red-yeast rice." Interestingly, this product has been known in the scientific literature as ang-kak or angka, but in mainland China this name is hardly known. Traditionally, polished rice is soaked overnight, cooked or steamed, cooled and inoculated with spores of *Monascus* spp. Solid-substrate fermentation during approximately one week allows the mould to grow and produce its pigments. The finished fermented product has an attractive red-purple colour and is used as a biocolouring for red sufu (fu-ru), distilled alcoholic beverages, and ceremonial products. Major azaphilone pigments include the orange pigments rubropunctatin and monascorubrin, purple pigments rubropunctamin and monascorubramin, and the yellow pigments ankaflavin and monascin (Pastrana *et al.*, 1995; Akihisa *et al.*, 2005). They are heat-stable over a wide pH-range, and thus of interest as "bio-colorants" in foods. In addition, several other secondary metabolites have been identified such as the furanoisophtalides xanthomonasin A and B, the amino acids (+) and (-) monascumic acid (Akihisa *et al.*, 2005), monascusone A and B (Jongrungruangchok *et al.*, 2004), monacolins (Juzlova *et al.*, 1996) and γ-aminobutyric acid (GABA) (Wang *et al.*, 2003). The flavour of red kojic rice is pleasant: the volatile metabolites (Juzlova *et al.*, 1998) included alcohols, aldehydes, ketones, esters and terpenoid compounds. It was reported earlier (Peters *et al.*, 1993) that in media containing saccharides (glucose) and fatty acids (octanoic acid), the relative toxicity of the fatty acid forced the mould into a detoxification process, oxidising octanoic acid to methyl ketones and secondary alcohols. Recently, major flavour compounds were identified as 3-methyl-1-butanol, ethanol, ethyl acetate, 2-methyl-1-propanol, ethyl butanoate and 3-methylbutyl acetate (Chung *et al.*, 2004). Only after complete detoxification, saccharides were assimilated for fungal metabolism. These properties are of importance for controlled production of singular flavour components. Industrial production of the pigments might be more efficient in liquid submerged fermentations rather than in solid-state fermentations; it was shown that ratios of carbon and nitrogen in liquid media determine the production of specific pigments.

Of recent interest are the health promoting effects of angkak. It was reported (Wang *et al.*, 1997) that during an 8-week trial in a group of 324 hyperlipidemia patients, a daily dose of 1.2 g angkak resulted in significant reductions of serum total cholesterol and low-density cholesterol. Cholesterol lowering ability (Liu *et al.*, 2005) of red rice was ascribed to monacolin K (Wang *et al.*, 2003).

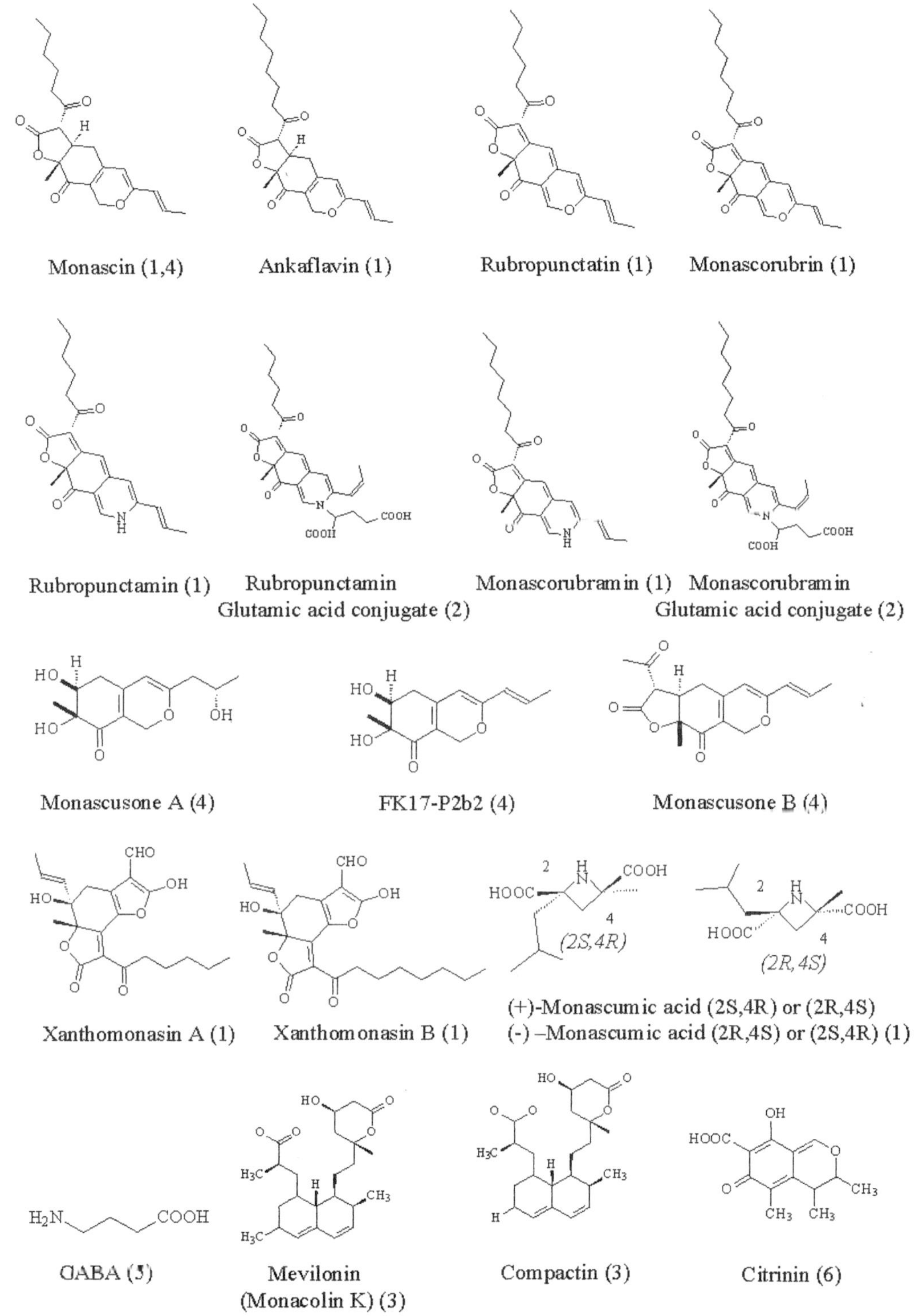

Figure 3. Secondary metabolites of *Monascus* spp. Compiled from literature references. 1: Akihisa *et al.* (2005); 2: Blanc *et al.* (1994); 3: Juzlova *et al.* (1996); 4: Jongrungruangchok *et al.* (2004); 5: Wang *et al.* (2003); 6: Liu *et al.* (2005).

Mevinolin (= monacolin K), compactin and derivates such as pravastatin, and simvastatin are inhibitors of 5-hydroxy-3-methylglutaryl-coenzyme A (HMG-CoA) reductase — a liver enzyme involved in cholesterol production — and can prevent hyperlipidemia (Juzlova *et al.*, 1996; Yang *et al.*, 2005); for human application a daily dose of 5 mg monacolin K has been recommended (Yang *et al.*, 2005). Whereas animal tests with monacolin K did not reveal toxicity as such, a significant transient reduction of cardiac and liver ubiquinone (Coenzyme Q10) levels was reported (Yang *et al.*, 2005) which may have negative effects in the long run if such products are taken as a regular part of the diet. GABA has a hypotensive effect (Wang *et al.*, 2003), and red rice was reported to have anti-inflammatory effects, while it induced antigens (Akihisa *et al.*, 2005) and decreased adipogenic transcription factors (Jeon *et al.*, 2004). Monascusone A had no antimicrobial, or cytotoxic effects (Jongrungruangchok *et al.*, 2004); the toxicology of red rice still needs further clarification. For example, it was observed that in certain fermentation conditions the mycotoxin citrinin may be produced. The risk of mycotoxin formation might seriously jeopardize the use of liquid fermentation-derived *Monascus* pigments as GRAS ingredients. Obviously it is of interest to analyze the traditional red kojic rice from China for the presence of citrinin; we analyzed a limited number of samples from Guang dong, Jiangsu, Hunan, Fujian and Beijing and did not find detectable quantities (detection limit 1 ppb) of citrinin (Han, 2003). Other investigations revealed low levels of citrinin in lipid extracts of red rice which had very low cytotoxic effects (Liu *et al.*, 2005); levels of citrinin in experimental fermentations could be reduced under optimized fermentation conditions (Wang *et al.*, 2003).

Deuteromycetes

Among the Deuteromycetes, the genera *Aspergillus* and *Penicillium* play a dualistic role in food technology. Some of the species used in age-old fermentation processes appear to be closely related to proficient producers of highly toxic and carcinogenic mycotoxins. No wonder why there is so much interest in the aspect of safety of *Aspergillus* and *Penicillium*–derived fermented food products.

Aspergillus oryzae and *A. sojae* are typical industrial moulds that have been used for centuries in the production of koji for the manufacture of soy sauce and miso (Wood, 1982). Whereas DNA fingerprints of *A. oryzae* isolates did not match those of *A. flavus*, *A. parasiticus* or *A. sojae* (Wicklow *et al.*, 2002), all *A. sojae* strains had identical DNA fingerprints and were considered having originated from a common ancestral clonal population, a domesticated form of *A. parasiticus* (Wicklow *et al.*, 2002). *A. sojae* is incapable of forming aflatoxins. AFLR (aflatoxin pathway-specific regulatory gene) was found to be impaired in its ability to activate transcription of aflatoxin biosynthetic genes, as well as being unable to interact with AFLJ (co-activator gene) (Chang, 2004). In traditional products such as Chinese and Japanese koji and soy sauce (Blesa *et al.*, 2004), and Korean Meju, Doenjang and fermented barley (Yang *et al.*, 2004) the mycotoxins ochratoxin and aflatoxins, respectively, could not be detected. Nevertheless, in some Meju and barley samples the presence of aflatoxigenic moulds were detected using multiplex PCR targeted towards 3 genes involved in aflatoxin biosynthesis (Yang *et al.*, 2004). This indicates that even though non-starter aflatoxigenic strains may be present as chance contaminants, there is little chance that these will produce aflatoxins in the fermented product. This may be caused by the food environment, or by microbial competition (Ehrlich *et al.*, 1985; Nout, 1989). Koji is made by soaking soy beans, *Glycine max*, in water, boiling and draining and mixing with ground or crushed roasted wheat. The mixture is placed on trays and mixed with *A. oryzae* or *A. sojae* (tane-koji) and allowed to ferment at about 30 °C for 5 days to form koji. The principal function of the mould is the elaboration and release of a range of hydrolytic enzymes, including amylases, proteases, cellulases, invertases, as well as lipolytic enzymes (Nout and Aidoo, 2002). Its major function is comparable to that of barley malt in brewing technology, i.e., it is a rich source of lytic enzymes. In the production of soy sauce (Figure

1e), the koji is mixed with salt brine (23% w/v) in a ratio of 1:1.5 to make the salt mash or moromi, which undergoes lactic acid bacterial and yeast fermentations for at least one year at ambient temperatures during which colour and flavour develop resulting in quality soy sauce (Nout and Aidoo, 2002). During this process, the carbohydrases degrade wheat starch into fermentable sugars, and proteolytic enzymes degrade soy protein into peptides and other non-protein nitrogenous compounds, such as glutamic acid. Although the moulds are the prime movers in the conversion of soy sauce, other microorganisms, particularly osmotolerant yeasts (*Zygosaccharomyces rouxii*) and halotolerant lactic acid bacteria (*Tetragenococcus halophila)*, are involved. The combination of mixed alcoholic and lactic acid fermentations results in a highly complex mixture of taste and flavour compounds. During the final step of manufacture, the filtered sauce is pasteurized. A number of flavour compounds including alcohols, glycerol, esters, 4-hydroxy-5-methyl-3(3H)-furanone (HMMF), 4-hydroxy-2 (5)-ethyl-5(2)-methyl-3(2H)-furanone (HEMF) and 4-hydroxy-2,5-dimethyl-3(2H)-furanone (HDMF), are formed. Of the furanones, HEMF produced by *Z. rouxii* and *Candida* spp. gives Japanese-type soy sauce its characteristic flavour (Hanya and Nakadai, 2003). This compound is also reported to have antitumor and antioxidative properties (Nagahara *et al.*, 1992; Koga *et al.*, 1998). The industrial scale production of koji is carried out in solid-state fermentations; as mentioned earlier in this chapter, solid-state fermentation presents technological challenges because of the specific behaviour of mycelial fungi. In solid-state, limitations to heat and mass transport rapidly result in gradients of temperature, gas-phase composition, nutrients, water and metabolite levels. Under conditions of decreased water activity, *A. oryzae* forms polyols such as erythritol, arabitol and mannitol as a survival strategy (Blomberg and Adler, 1992; Witteveen and Visser, 1995). Mycelial biomass growing on a surface was shown to consist of a compacted mat and a more open-structured aerial mycelium, the latter making an important contribution to the oxygen uptake and respiratory capacity of *A. oryzae* (Rahardjo *et al.*, 2002). A proteomics approach was used to demonstrate that, compared with submerged (liquid) fermentation, solid-state fermentation conditions result in stronger (enolase, amylase) or exclusive (zinc-finger transcription factor, glucoamylase) expression of genes for key enzymes (Te Biesebeke *et al.*, 2002). These new findings support some of the earlier observations that fungi behave in a specifically different manner when grown on solid substrates.

Aspergillus glaucus, *A. melleus*, *A. repens*, and *A. candidus* have all been reported as functional mycoflora in the fermentation of fish in Japan. The product Katsuobushi is made from Bonito or skipjack tuna flesh (*Katsuwouno pelamis*) which is cut to strips, steamed, and left to dry in a barrel for about 3 weeks. During this period, mould fermentation takes place; mould is scraped off and the fish oven-dried until hard. The presence of *A. ochraceus* (natural contaminant) may lead to mycotoxin contamination. Furthermore, the product contains high levels of histidine, which could be decarboxylated by contaminant flora and increase the risk of histamine poisoning (Campbell-Platt, 1987). This indicates that there is scope for improvement of the microbiological and chemical control of quality and safety of this type of product.

Compared with *Aspergillus*, the genus *Penicillium* is equally important for food and biotechnology. Several *Penicillium* spp. such as *P. italicum* and *P. expansum* cause extensive economic losses as toxigenic spoilage agents in the citrus fruit business. Others, such as *P. chrysogenum*, are widely exploited for their antibiotics production. Only a few species are used as food, particularly *P. camemberti*, *P. nalgiovense*, and the *P. roqueforti* group. *P. camemberti* is used in the manufacture of surface-ripened cheeses such as French Camembert and Brie (Figure 1f). This type of cheese is made by pasteurizing cows' milk, followed by addition of lactic acid bacteria starter (*Lactococcus lactis* and *Streptococcus cremoris*), rennet, and calcium chloride. After coagulation, the curd is cut, transferred to moulds, turned, rubbed with salt, and sprayed with mould spores at the surface of the young cheese. The mould fermentation takes place during 1-4 weeks at 10-

14 °C (Campbell-Platt, 1987). The microbiology of this type of product is complex: in the basic cheese, lactic acid bacteria are essential for flavour, lactose depletion and lactate production. Various yeasts and bacteria are involved in the maturation, along with *P. camemberti*. On the basis of pure culture experiments under aseptic cheesemaking conditions, it was observed that lactate serves as an important carbon source for the energy metabolism of *P. camemberti* (Adour *et al.*, 2004). Whereas bacteria (*Brevibacterium linens*), yeasts (*Kluyveromyces lactis*) and other fungi (*Geotrichum candidum*) contribute to proteolysis, formation of esters (ethyl, butyl, and isoamylacetates) and other volatiles (3-methyl butanol, methyl-3-butanal, 2-octanone), *P. camemberti* has an exclusive contribution to the character of Camembert cheese (Leclercq Perlat *et al.*, 2004a; Leclercq Perlat *et al.*, 2004b). First, it is responsible for the mycelial surface growth. Second, it is the major proteolytic organism releasing ammonia that dominates the flavour and high pH (7.5) in late stages of maturation. Third, it produces volatiles such as styrene, 2-pentanone and 1-octen-3-ol (Husson *et al.*, 2005).

P. nalgiovense is a white sporulating mould that is widely used as a surface growth on fermented meat products (traditional Salami sausages, country cured hams) (Fierro *et al.*, 2004). The safety of this species could be improved if its potential to produce toxic secondary metabolites could be eliminated. There is an interest to develop genetic manipulation tools for "self-cloning," in which genes of a microorganism are cloned within the microorganism itself (Akada, 2002). Revised national guidelines for GM (genetically modified) food (Japan, April 2001) exempt self-cloned bakers' yeast from labeling or treatment as GM yeast (Akada, 2002). This is expected to facilitate the introduction of modified microorganisms on the consumer market (Fierro *et al.*, 2004).

The *P. roqueforti* group can be differentiated into the three species *P. roqueforti*, *P. carneum* and *P. paneum*, using profiles of volatile metabolites (Karlshoj and Larsen, 2005). Whereas *P. roqueforti* spp. cause important economic losses by spoilage of bakery products and ensiled animal feeds, their prime feature is their colour and flavour production in "blue-veined cheeses" such as the French Roquefort, English Stilton, and Danish Danablue (Figure 1g). Roquefort is a blue-veined cheese, with strong flavour strong aroma and creamy consistency. It is prepared from ewe's milk, which is first coagulated with rennet followed by addition of *P. roqueforti* and addition of salt and storage for the maturation. Finally it is pierced just before leaving for the ripening in the cave of Roquefort.

During this maturation, both natural and provoked fermentation will take place. As Roquefort is made with raw milk, interesting natural fermentations of the milk by microflora like *Leuconostoc* or *Geotrichum* will take place, modifying the curd structure and facilitating the growth of the conidia of *P. roqueforti* that were added to the cheese.

Nowadays "Roquefort" is an AOC (Appellation d'Origine Côntrolée) and this name can only be used for a cheese made of milk coming exclusively from the south of France, transformed in a specific process, defined by the law, and ripened in the cave of the city of Roquefort. The maturation takes place in caves where the temperature is naturally regulated; however, this temperature should not be lower than -5 °C (specified by law) and must be lower than 37 °C (maximum temperature for growth of *P. roqueforti*). The holes made in the cheese will ensure a homogenous growth of the mould throughout the product (the strain needs oxygen for a good growth). After 15 days maturation in the cave, the cheese will be packed and transferred to a low temperature room where a slower maturation will continue for at least 3 months. In the blue cheese, *P. roqueforti* plays an important role in the degradation, especially by proteolysis and lipolysis (Gripon, 2003). Indeed in blue cheese, up to 10% of total amino acids are free amino acids and up to 20% of fatty acids are free fatty acids. The latter are particularly important because they will be transformed into methyl ketone, butyric and caproic acids which are responsible for the strong zesty flavour of the blue cheese.

The blue-green colour of the blue-veined cheese is provided by fungal melanins (Figure 4) in the conidia of *P. roqueforti*. These are synthesized by polyketide pathways (Wheeler and Klich, 1995), starting from malonate which is transformed into polyketides, dihydroxynaphtalene and finally into melanins.

Some *P. roqueforti* strains are able to produce mycotoxins (patulin, penicillic acid, PR toxin, roquefortine). During the past years, the risk of mycotoxin contamination of blue-veined cheeses has been the subject of investigations.

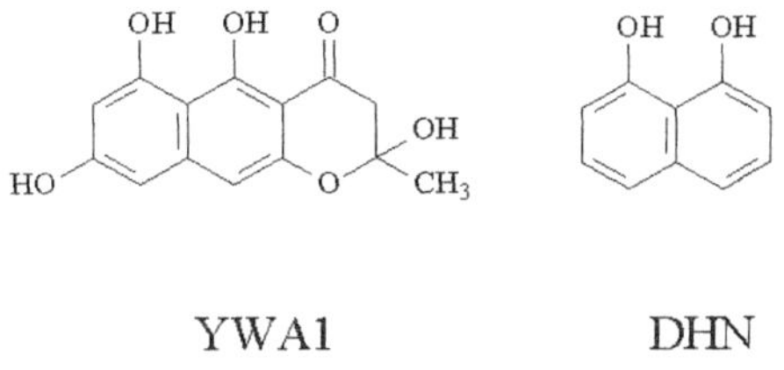

Figure 4. DHN (1,8-dihydroxynaphtalene) and YWA1, intermediates in the melanin biosynthesis of *Penicillium* spp.

Until now, only small traces of roquefortine have been found in cheese inoculated with toxigenic strains (Erdogan and Sert, 2004). It has been assumed that the conditions in cheese (nutrient composition, salt level, pH, etc.) are not favourable to mycotoxin production or stability.

FUTURE CHALLENGES

The ever-improving performance of molecular and analytical techniques offers opportunities to characterize existing food products and to support process innovations. In view of the protection of origin (AOC or certified origin of production) an unequivocal characterization of traditional fermented foods and their microflora will be required; this could be based on combinations of food compositional analysis and metabolite profiles, nucleic acid patterns such as obtained by DGGE and other methods. Innovative processes using non-traditional fermentation conditions — for example by immobilized cells, or in agitated solid-state fermentors — or using pure culture inoculation instead of multi-strain natural fermentations, may invoke changes in secondary metabolite production. In view of maintaining the character of the food, as well as safeguarding the safety of the consumer, the impact of novel processing should be investigated, understood and possibly controlled. Although the aspect of GMO (genetically modified organism) is still sensitive with the general public, safe techniques such as "self-cloning" could be helpful to obtain food-safe fungi that can be guaranteed as "mycotoxin-free." Finally, the fact that fungal fermentation technology is an important source of income in the Asian region raises the question how other regions of the world — particularly the less industrialized African countries — can benefit from this traditional know-how in the development of small- or medium-scale enterprise.

REFERENCES

Abe, A., Sujaya, I. N., Sone, T., Asano, K., and Oda, Y. (2004). Microflora and selected metabolites of potato pulp fermented with an Indonesian starter Ragi tape. Food Technology and Biotechnology 42:169-173.

Adour, L., Couriol, C., and Amrane, A. (2004). The effect of lactate addition on the growth of Penicillium camemberti on glutamate. Journal of Biotechnology 114:307-314.

Agranoff, J., and Markham, P. (1997). Fatty acid components of tempe (and tapeh). In International Tempe Symposium (Sudarmadji, S., Suparmo, S., and Raharjo, S. eds.), Den Pasar, Bali, Indonesia: Indonesian Tempe Foundation, Jakarta, Indonesia, pp. 205- 210.

Akada, R. (2002) Genetically modified industrial yeast ready for application. Journal of Bioscience and Bioengineering 94:536-544.

Akihisa, T., Tokuda, H., Yasukawa, K., Ukiya, M., Kiyota, A., Sakamoto, N., Suzuki, T., Tanabe, N., and Nishino, H. (2005). Azaphilones, furanoisophthalides, and amino acids from the extracts of *Monascus pilosus*-fermented rice (red-mold rice) and their chemopreventive effects. Journal of Agricultural and Food Chemistry 53:562-565.

Ariffin, R., Apostolopoulos, C., Graffham, A., MacDougall, D., and Owens, J. D. (1994). Assessment of hyphal binding in tempe. Letters in Applied Microbiology 18:32-34.

Benjamin, C. R., and Hesseltine, C. W. (1957). The genus *Actinomucor*. Mycologia 49:240-249.

Berghofer, E., Grzeskowiak, B., Mundigler, N., Sentall, W. B., and Walcak, J. (1998). Antioxidative properties of faba bean-, soybean- and oat tempeh. International Journal of Food Sciences and Nutrition 49:45-54.

Beuchat, L. R. (1986). Oncom (fermented peanut press cake). *In* Legume-Based Fermented Foods (Reddy, N. R., Pierson, M. D., and Salunkhe, D. K., eds.), Boca Raton, FL, CRC Press, Inc., U.S.A., pp. 135-144.

Blanc, P. J., Loret, M. O., Santerre, A. L., Pareilleux, A., Prome, D., Prome, J. C., Laussac, J. P., and Goma, G. (1994). Pigments of *Monascus*. Journal of Food Science 59:862-865.

Blesa, J., Soriano, J. M., Molto, J. C., and Manes, J. (2004). Absence of ochratoxin A in soy sauce. International Journal of Food Microbiology 97:221-225.

Blomberg, A., and Adler, L. (1992). Physiology of osmotolerance in fungi. Advances in Microbial Physiology 33:145-212.

Campbell-Platt, G. (1987). Fermented Foods of the World. A dictionary and guide, London: Butterworths.

Chang, P. K. (2004). Lack of interaction between AFLR and AFLJ contributes to nonaflatoxigenicity of *Aspergillus sojae*. Journal of Biotechnology 107:245-253.

Chou, C. C., Ho, F. M., and Tsai, C. S. (1988). Effects of temperature and relative humidity on the growth of and enzyme production by *Actinomucor taiwanensis* during sufu pehtze preparation. Applied and Environmental Microbiology 54:688-692.

Chou, C. C., and Hwan, C. H. (1994). Effect of ethanol on the hydrolysis of protein and lipid during the ageing of a Chinese fermented soya bean curd - sufu. Journal of the Science of Food and Agriculture 66:393-398.

Chung, H. Y., Ma, W. C. J., Kim, J. S., and Chen, F. (2004). Odor-active headspace components in fermented red rice in the presence of a *Monascus* species. Journal of Agricultural and Food Chemistry 52: 6557-6563.

Correia, R. T. P., McCue, P., Magalhaes, M. M. A., Macedo, G. R., and Shetty, K. (2004a). Phenolic antioxidant enrichment of soy flour-supplemented guava waste by *Rhizopus oligosporus*-mediated solid-state bioprocessing. Journal of Food Biochemistry 28:404-418.

Correia, R. T. P., McCue, P., Magalhaes, M. M. A., Macedo, G. R., and Shetty, K. (2004b). Production of phenolic antioxidants by the solid-state bioconversion of pineapple waste mixed with soy flour using *Rhizopus oligosporus*. Process Biochemistry 39:2167-2172.

Cuevas Rodriguez, E. O., Milan Carrillo, J., Mora Escobedo, R., Cardenas Valenzuel, O. G., and Reyes Moreno, C. (2004). Quality protein maize (*Zea mays* L.) tempeh flour through solid state fermentation process. Lebensmittel Wissenschaft und Technologie - Food Science and Technology 37:59-67.

Dalgaard, P., Vancanneyt, M., Vilalta, N. E., Swings, J., Fruekilde, P., and Leisner, J. J. (2003). Identification of lactic acid bacteria from spoilage associations of cooked and brined shrimps stored under modified atmosphere between 0 degrees C and 25 degrees C. Journal of Applied Microbiology 94:80-89.

Dung, N. T. P. (2004). Defined fungal starter granules for purple glutinous rice wine. Ph.D. Thesis, Wageningen University, Wageningen, The Netherlands.

Ehrlich, K., Ciegler, A., Klich, M., and Lee, L. (1985). Fungal competition and mycotoxin production on corn. Experientia 41:691-693.

Ellis, J. J., Rhodes, L. J., and Hesseltine, C. W. (1976). The genus *Amylomyces*. Mycologia 68:131-143.

Erdogan, A., and Sert, S. (2004). Mycotoxin-forming ability of two *Penicillium roqueforti* strains in blue moldy tulum cheese ripened at various temperatures. Journal of Food Protection 67:533-535.

Fabo, E. C. de, Harding, R. W., and Shropshire, W. J. (1976). Action spectrum between 260 nanometers and 800 nanometers for photo induction of carotenoid biosynthesis in *Neurospora crassa*. Plant Physiology 57:440-445.

Fierro, F., Laich, F., Garcia Rico, R. O., and Martin, J. F. (2004). High efficiency transformation of *Penicillium nalgiovense* with integrative and autonomously replicating plasmids. International Journal of Food Microbiology 90:237-248.

Fink-Gremmels, J., Abd-El- Banna, A., and Leistner, L. (1988). Developing mould starter cultures for meat products. Fleischwirtschaft 68:1292-1294.

Fransen, C. T. M. (1999). Structural analysis of soy bean polysaccharides and transgalactosylation products from lactose. PhD. Thesis, Utrecht University, Utrecht, The Netherlands.

Fukushima, D. (1985). Fermented vegetable protein and related foods of Japan and China. Food Reviews International 1:149-209.

Ginting, E., and Arcot, J. (2004). High-performance liquid chromatographic determination of naturally occurring folates during tempe preparation. Journal of Agricultural and Food Chemistry 52:7752-7758.

Gripon, J. C. (2003). Mould-Ripened cheese. *In* Encyclopedia of Dairy Sciences, Vol. 1 (Roginsky, H., Fuquay, J. W., and Fox, P. F., ed.), Academic Press, Amsterdam, The Netherlands, pp. 401-406.

Han, B.-Z. (2003). Characterization and product innovation of sufu, a Chinese fermented soybean food. PhD. Thesis, Wageningen University, Wageningen, The Netherlands.

Han, B.-Z., Kiers, J. L., and Nout, M. J. R. (1999). Solid-substrate fermentation of soybeans with *Rhizopus* spp.: comparison of discontinuous rotation with stationary bed fermentation. Journal of Bioscience and Bioengineering 88:205-209.

Han, B.-Z., Rombouts, F. M., and Nout, M. J. R. (2001). A Chinese fermented soybean food. International Journal of Food Microbiology 65:1-10.

Han, B.-Z., Ma, Y., Rombouts, F. M., and Nout, M. J. R. (2003a). Effects of temperature and relative humidity on growth and enzyme production by *Actinomucor elegans* and *Rhizopus oligosporus* during Sufu Pehtze preparation. Food Chemistry 81:27-34.

Han, B.-Z., Wang, J. H., Rombouts, F. M., and Nout, M. J. R. (2003b). Effect of NaCl on textural changes and protein and lipid degradation during the ripening stage of sufu, a Chinese fermented soybean food. Journal of the Science of Food and Agriculture 83:899-904.

Han, B.-Z., Cao, C.-F., Rombouts, F. M., and Nout, M. J. R. (2004a). Microbial changes during the production of Sufu - a Chinese fermented soybean food. Food Control 15:265-270.

Han, B.-Z., Kuijpers, A. F. A., Thanh, N. V., and Nout, M. J. R. (2004b). Mucoraceous moulds involved in the commercial fermentation of Sufu Pehtze. Antonie van Leeuwenhoek 85:253-257.

Han, B.-Z., Rombouts, F. M., and Nout, M. J. R. (2004c). Amino acid profiles of Sufu, a Chinese fermented soybean food. Journal of Food Composition and Analysis 17:689-698.

Han, B.-Z., Sesenna, B., Beumer, R. R., and Nout, M. J. R. (2005). Behaviour of *Staphylococcus aureus* during Sufu production at laboratory scale. Food Control 16:243-247.

Hanya, Y., and Nakadai, T. (2003). Yeasts and soy products. *In* Yeasts in Food: Beneficial and Detrimental Aspects, (Boekhout, T., and Robert, V., eds.) Hamburg: B. Behr's Verlag GmbH & Co. KG, Germany, pp. 413-428.

Henkel, T. W. (2005). Parakari, an indigenous fermented beverage using amylolytic *Rhizopus* in Guyana. Mycologia 97:1-11.

Heskamp, M. L., and Barz, W. (1998). Expression of proteases by *Rhizopus* species during tempeh fermentation of soybeans. Nahrung - Food 42:23-28.

Hesseltine, C. W., Rogers, R., and Winarno, F. G. (1988). Microbiological studies on amylolytic oriental fermentation starters. Mycopathologia 101 (3):141-155

Husson, F., Krumov, K. N., Cases, E., Cayot, P., Bisakowski, B., Kermasha, S., and Belin, J. M. (2005). Influence of medium composition and structure on the biosynthesis of the natural flavour 1-octen-3-ol by *Penicillium camemberti*. Process Biochemistry 40:1395-1400.

Jennessen, J., Nielsen, K. F., Houbraken, J., Lyhne, E. K., Schnurer, J., Frisvad, J. C., and Samson, R. A. (2005). Secondary metabolite and mycotoxin production by the *Rhizopus microsporus* group. Journal of Agricultural and Food Chemistry 53:1833-1840.

Jeon, T., Hwang, S. G., Hirai, S., Matsui, T., Yano, H., Kawada, T., Lim, B. O., and Park, D. K. (2004). Red yeast rice extracts suppress adipogenesis by down-regulating adipogenic transcription factors and gene expression in 3T3-L1 cells. Life Sciences 75:3195-3203.

Jha, H. C., Kiriakidis, S., Hoppe, M., and Egge, H. (1997). Antioxidative constituents of tempe. *In* International Tempe Symposium (Sudarmadji, S., Suparmo, S., and Raharjo, S., ed.), Den Pasar, Bali, Indonesia: Indonesian Tempe Foundation, Jakarta, Indonesia, pp. 73-84.

Jong, S. C., and Yuan, G. F. (1985). *Actinomucor taiwanensis* sp. nov., for manufacture of fermented soybean food. Mycotaxon 23:261-264.

Jongrungruangchok, S., Kittakoop, P., Yonsmith, B., Bavovada, R., Tanasupawat, S., Lartpornmatulee, N., and ThebtarAnonth, Y. (2004). Azaphilone pigments from a yellow mutant of the fungus *Monascus kaoliang*. Phytochemistry 65:2569-2575.

Juzlova, P., Martinkova, L., and Kren, V. (1996). Secondary metabolites of the fungus *Monascus*: a review. Journal of Industrial Microbiology 16:163-170.

Juzlova, P., Rezanka, T., and Viden, I. (1998). Identification of volatile metabolites from rice fermented by the fungus *Monascus purpureus* (ang-kak). Folia Microbiologica 43:407-410.

Karlshoj, K., and Larsen, T. O. (2005). Differentiation of species from the *Penicillium roqueforti* group by volatile metabolite profiling. Journal of Agricultural and Food Chemistry 53:708-715.

Kiers, J. L., Meijer, J. C., Nout, M. J. R., Rombouts, F. M., Nabuurs, M. J. A., and Meulen, J. van der (2003). Effect of fermented soya beans on diarrhoea and feed efficiency in weaned piglets. Journal of Applied Microbiology 95:545-552.

Ko, S. D., and Hesseltine, C. W. (1979). Tempe and related foods. *In* Microbial Biomass, Vol. 4 (Rose, A. H., ed.), London, Academic Press, U.K., pp. 115-140.

Koga, T., Moro, K., and Matsudo, T. (1998). Antioxidative behaviors of 4-hydroxy-5-methyl-3(3H)-furanone (HMMF), 4-hydroxy-2(or 5)-ethyl-5(or 2)-methyl-3(3H)-furanone and 4-hydroxy-2,5-dimethyl-3(2H)-furanone against lipid peroxidation. Journal of Agricultural and Food Chemistry 46:946-951.

Kozaki, M., and Uchimura, T. (1990). Microorganisms in Chinese starter "Bubod" and rice wine, "tapuy" in the Philippines. Journal of the Brewing Society of Japan 85:818-824.

Kubickova, J., and Grosch, W. (1997). Evaluation of potent odorants of Camembert cheese by dilution and concentration techniques. International Dairy Journal 7:65-70.

Leclercq Perlat, M. N., Buono, F., Lambert, D., Latrille, E., Spinnler, H. E., and Corrieu, G. (2004a). Controlled production of Camembert-type cheeses. Part I: Microbiological and physicochemical evolutions. Journal of Dairy Research 71:346-354.

Leclercq Perlat, M. N., Latrille, E., Corrieu, G., and Spinnler, H. E. (2004b). Controlled production of Camembert-type cheeses. Part II. Changes in the concentration of the more volatile compounds. Journal of Dairy Research 71:355-366.

Leistner, L. (1990). Mould-fermented foods: recent developments. *In* Food Biotechnology 4 (Proceedings of the International Conference on Biotechnology and Food, Hohenheim University, Stuttgart Feb. 20-24, 1989). New York, Marcel Dekker, U.S.A., ISSN 0890-5436, pp. 433-441.

Lertsiri, S., Phontree, K., Thepsingha, W., and Bhumiratana, A. (2003). Evidence of enzymatic browning due to laccase-like enzyme during mash fermentation in Thai soybean paste. Food Chemistry 80:171-176.

Liu, B. H., Wu, T. S., Su, M. C., Chung, C. P., and Yu, F. Y. (2005). Evaluation of citrinin occurrence and cytotoxicity in *Monascus* fermentation products. Journal of Agricultural and Food Chemistry 53:170-175.

Liu, G. Y., Yuan, S., and Dai, C. C. (2004). Factors affecting gamma-linolenic acid content in fermented glutinous rice brewed by *Rhizopus* sp. Food Microbiology 21:299-304.

Liu, Y.-H., and Chou, C. C. (1994). Contents of various types of proteins and water soluble peptides in sufu during ageing and the amino acid composition of taste oligopeptides. Journal of the Chinese Agricultural and Chemical Society 32:276-283.

Lu, J. M., Yu, R. C., and Chou, C. C. (1996). Purification and some properties of glutaminase from *Actinomucor taiwanensis*, starter of sufu. Journal of the Science of Food and Agriculture 70:509-514.

Ma, X., Zhou, X., and Yoshimoto, T. (2004). Purification and properties of a novel glycine amino peptidase from *Actinomucor elegans* and its potential application. Journal of Applied Microbiology 97:985-991.

Matsuo, M. (2004). Low-salt O-miso produced from koji fermentation of oncom improves redox state and cholesterolemia in rats more than low-salt soybean-miso. Journal of Nutritional Science and Vitaminology 50:362-366.

Matsuo, M., Nakamura, N., Shidoji, Y., Muto, Y., Esaki, H., and Osawa, T. (1997). Antioxidative mechanism and apoptosis induction by 3-hydroxyanthranilic acid, an antioxidant in Indonesian food tempeh, in the human hepatoma derived cell line, HUH 7. Journal of Nutritional Science and Vitaminology 43:249-259.

McCue, P., and Shetty, K. (2003). Role of carbohydrate-cleaving enzymes in phenolic antioxidant mobilization from whole soybean fermented with *Rhizopus oligosporus*. Food Biotechnology 17:27-37.

McCue, P., Lin, Y. T., Labbe, R. G., and Shetty, K. (2004). Sprouting and solid-state bioprocessing by *Rhizopus oligosporus* increase the in vitro antibacterial activity of aqueous soybean extracts against *Helicobacter pylori*. Food Biotechnology 18:229-249.

Medwid, R. D., and Grant, D. W. (1984). Germination of *Rhizopus oligosporus* sporangiospores. Applied and Environmental Microbiology 48:1067-1071.

Nagahara, A., Benjamin, H., Storkson, J., Krewson, J., Sheng, K., Liu, W., and Pariza, M. W. (1992). Inhibition of benzo(α)pyrene-induced mouse forestomach neoplasia by a principal flavour component of Japanese-style fermented soy sauce. Cancer Research 52:1754-1756.

Nagel, F. J. J. I., Tramper, J., Bakker, M. S. N., and Rinzema, A. (2001). Model for on-line moisture-content control during solid-state fermentation. Biotechnology and Bioengineering 72:231-243.

Nout, M. J. R. (1989). Effect of *Rhizopus* and *Neurospora* spp. on growth of *Aspergillus flavus* and *A.parasiticus* and accumulation of aflatoxin B1 in groundnut. Mycological Research 93:518-523.

Nout, M. J. R., Martoyuwono, T. D., Bonné, P. C. J., and Odamtten, G. T. (1992). *Hibiscus* leaves for the manufacture of Usar, a traditional inoculum for tempe. Journal of the Science of Food and Agriculture 58:339-346.

Nout, M. J. R., and Aidoo, K. E. (2002). Asian fungal fermented food. *In* Industrial applications Vol. X Osiewacz, H. D., ed.), Berlin-Heidelberg-New York: Springer-Verlag, Germany, pp. 23-47

Nout, M. J. R., and Kiers, J. L. (2005). Tempe fermentation, innovation and functionality: up-date into the 3rd millenium. Journal of Applied Microbiology 98:789-805.

Oostra, J., Tramper, J., and Rinzema, A. (2000). Model-based bioreactor selection for large-scale solid-state cultivation of *Coniothyrium minitans* spores on oats. Enzyme and Microbial Technology 27:652-663.

Oxlade, L. (1990). King's cure-all makes a comeback. Chemistry in Britain 26:813.

Pastrana, L., Blanc, P. J., Santerre, A. L., Loret, M. O., and Goma, G. (1995). Production of red pigments by *Monascus ruber* in synthetic media with a strictly controlled nitrogen source. Process Biochemistry 30:333-341.

Pedraza-Reyes, M., and Lopez-Romero, E. (1991). Detection of nine chitinase species in germinating cells of *Mucor rouxii*. Current Microbiology 22:43-46.

Peters, N., Panitz, C., and Kunz, B. (1993).The influence of carbohydrate dissimilation on the fatty acid metabolism of *Monascus purpureus*. Applied Microbiology and Biotechnology 39:589-592.

Pitt, J. I., and Hocking, A. D. (1985). Fungi and food spoilage. Orlando, Florida: Academic Press Inc.

Rahardjo, Y. S. P., Weber, F. J., Comte, E. P. le, Tramper, J., and Rinzema, A. (2002). Contribution of aerial hyphae of *Aspergillus oryzae* to respiration in a model solid-state fermentation system. Biotechnology and Bioengineering 78:539-544.

Reu, J. C. de, Linssen, V. A. J. M., Rombouts, F. M., and Nout, M. J. R. (1997). Consistency, polysaccharidase activities and non-starch polysaccharides content of soya beans during tempe fermentation. Journal of the Science of Food and Agriculture 73:357-363.

Reyes Moreno, C., Cuevas Rodriguez, E. O., Milan Carrillo, J., Cardenas Valenzuela, O. G., and Barron Hoyos, J. (2004). Solid state fermentation process for producing chickpea (*Cicer arietinum* L) tempeh flour. Physicochemical and nutritional characteristics of the product. Journal of the Science of Food and Agriculture 84:271-278.

Rhee, S. J., Lee, C. Y. J., Kim, K. K., and Lee, C. H. (2003). Comparison of the traditional (Samhaeju) and industrial (Chongju) rice wine brewing in Korea. Food Science and Biotechnology 12:242-247.

Saito, K., Abe, A., Sujaya, I. N., Sone, T., and Oda, Y. (2004). Comparison of *Amylomyces rouxii* and *Rhizopus oryzae* in lactic acid fermentation of potato pulp. Food Science and Technology Research 10:224-226.

Samson, R. A. (1993). The exploitation of moulds in fermented foods. *In* Exploitation of Microorganisms (Jones, D. G., ed.), London, Chapman & Hall, UK, pp. 321-341.

Sarrette, M., Nout, M. J. R., Gervais, P., and Rombouts, F. M. (1992). Effect of water activity on production and activity of *Rhizopus oligosporus* polysaccharidases. Applied Microbiology and Biotechnology 37:420-425.

Schipper, M. A. A., and Stalpers, J. A. (1984). A revision of the genus *Rhizopus*. Studies in Mycology. Centraal Bureau voor Schimmelcultures, Baarn, The Netherlands, pp. 1-34.

Shi, X., and Fung, D. Y. C. (2000). Control of food-borne pathogens during sufu fermentation and aging. Critical Reviews in Food Science and Nutrition 40:399-425.

Shigechi, H., Koh, J., Fujita, Y., Matsumoto, T., Bito, Y., Ueda, M., Satoh, E., Fukuda, H., and Kondo, A. (2004). Direct production of ethanol from raw corn starch via fermentation by use of a novel surface-engineered yeast strain codisplaying glucoamylase and alpha-amylase. Applied and Environmental Microbiology 70:5037-5040.

Shrestha, H., Nand, K., and Rati, E. R. (2002). Microbiological profile of murcha starters and physico-chemical characteristics of poko, a rice based traditional fermented food product of Nepal. Food Biotechnology 16:1-15.

Sparringa, R. A., and Owens, J. D. (1999). Protein utilization during soybean tempe fermentation. Journal of Agricultural and Food Chemistry 47:4375-4378.

Su, Y.-C. (1986). Sufu. *In* Legume-based fermented foods (Reddy, N. R., Pierson, M. D., and Salunkhe, D. K., eds.), CRC Press, Boca Raton, FL, U.S.A., pp. 69-83.

Tamang, J. P., and Sarkar, P. K. (1995). Microflora of murcha: an amylolytic fermentation starter. Microbios 81:115-122.

Tamang, J. P., Thapa, S., Tamang, N., and Rai, B. (1996). Indigenous fermented food beverages of Darjeeling hills and Sikkim: process and product characterization. Journal of Hill Research 9:401-411.

Biesebeke, R. te, Ruijter, G., Rahardjo, Y. S. P., Hoogschagen, M. J., Heerikhuisen, M., Levin, A., Van Driel, K. G. A., Schutyser, M. A. I., Dijksterhuis, J., Zhu, Y., Weber, F. J., Vos, W. M. de, Hondel, C. A. M. J. J. van den, Rinzema, A., and Punt, P. J. (2002). *Aspergillus oryzae* in solid-state and submerged fermentations, Progress report on a multi disciplinary project. FEMS Yeast Research 2:245-248.

Thanh, N. V., Rombouts, F. M., and Nout, M. J. R. (2005). Effect of individual amino acids and glucose on activation and germination of *Rhizopus oligosporus* sporangiospores in tempe starter. Journal of Applied Microbiology 99:1204-1214.

Tsuyoshi, N., Fudou, R., Yamanaka, S., Kozaki, M., Tamang, N., Thapa, S., and Tamang, J. P. (2005). Identification of yeast strains isolated from marcha in Sikkim, a microbial starter for amylolytic fermentation. International Journal of Food Microbiology 99:135-146.

Wang, H. L., Swain, E. W., and Hesseltine, C. W. (1984). Glucoamylase of *Amylomyces rouxii*. Journal of Food Science 49:1210-1211.

Wang, J. J., Lee, C. L., and Pan, T. M. (2003). Improvement of monacolin K, gamma-aminobutyric acid and citrinin production ratio as a function of environmental conditions of *Monascus purpureus* NTU 601. Journal of Industrial Microbiology and Biotechnology 30:669-676.

Wang, J. X., Lu, Z. L., Chi, J. M., Wang, W. H., Su, M. Z., Kou, W. R., Yu, P. L., Yu, L. J., Zhu, J. S., and Chang, J. (1997). Multicenter clinical trial of the serum lipid-lowering effects of a *Monascus purpureus* (red yeast) rice preparation from traditional Chinese medicine. Current Therapeutic Research Clinical and Experimental 58:964-978.

Wheeler, M. H., and Klich, M. A. (1995). The effects of tricyclazole, pyroquilon, phthalide, and related fungicides on the production of conidial wall pigments by *Penicillium* and *Aspergillus* species. Pesticide Biochemistry and Physiology 52:125-136.

Wicklow, D. T., McAlpin, C. E., and Peterson, S. W. (2002). Common genotypes (RFLP) within a diverse collection of yellow-green aspergilli used to produce traditional Oriental fermented foods. Mycoscience 43:289-297.

Wiegant, W. M., Wery, J., Buitenhuis, E. T., and Bont, J. A. M. de (1992). Growth-promoting effect of thermophilic fungi on the mycelium of the edible mushroom *Agaricus bisporus*. Applied and Environmental Microbiology 58:2654-2659.

Witteveen, C. F. B., and Visser, J. (1995). Polyol pools in *Aspergillus niger*. FEMS Microbiology Letters 134:57-62.

Wood, B. J. B. (1982). Soy Sauce and Miso. *In* Economic Microbiology, Volume 7, Fermented Foods (Rose, A. H., ed.), Academic Press, London, U.K., pp. 39-87.

Yang, H. T., Lin, S. H., Huang, S. Y., and Chou, H. J. (2005). Acute administration of red yeast rice (*Monascus purpureus*) depletes tissue coenzyme Q(10) levels in ICR mice. British Journal of Nutrition 93:131-135.

Yang, Z. Y., Shim, W. B., Kim, J. H., Park, S. J., Kang, S. J., Nam, B. S., and Chung, D. H. (2004). Detection of aflatoxin-producing molds in Korean fermented foods and grains by multiplex PCR. Journal of Food Protection 67:2622-2626.

Chapter 18

Fungal protein for food

Ulf Thrane
Center for Microbial Biotechnology, BioCentrum-DTU, Søltofts Plads 221, DK-2800 Kgs. Lyngby, Denmark.

HISTORY OF FUNGAL PROTEIN FOR FOOD

Fungi have been part of the human diet for thousands of years as a food item in itself — especially mushrooms — or as part of fermented foods like yeasts used in bread and beverages. Other well-known fermented products using filamentous fungi are blue and white moulded cheeses, tempe and miso. In the late 1950s searches for new protein sources were initiated as it was estimated that animal protein sources would be insufficient to meet man's requirements for protein. Microbial produced protein, also named single-cell protein (SCP), was in focus and many different microbial sources were investigated, mainly intended for animal feed (Spicer, 1971). Reviews on alternative protein sources and SCP from many different organisms, e.g., bacteria, fungi and algae, have been published elsewhere (Das and Singh, 2004; Özyurt and Deveci, 2004; Sadler, 2004; Villas-Boas *et al.*, 2002). As fungi have a long history of safe use in food and food production, it has been a strategy to use these organisms, especially filamentous fungi, for production of microbial protein for human consumption. Examples on fungal species and some yeasts which have been tested for SCP production intended for food or feed are listed in Table 1.

A general concern on microbial protein from a filamentous fungus, generically named mycoprotein, is the possible contamination of mycotoxins produced by the fungus, as known from food spoilage by fungi (Filtenborg *et al.*, 1996) and from production of enzymes and other food ingredients (Blumenthal, 2004; Mapari *et al.*, 2005). With this in mind, it is alarming that *Aspergillus fumigatus* known as a highly toxigenic and even pathogenic fungus ever has been suggested as a mycoprotein producer (Table 1). The *A. fumigatus* mycoprotein was evaluated as safe in a rat feeding experiment (Khor *et al.*, 1977) and also analysed for aflatoxins, citrinin, ochratoxin, zearalenone, T-2 toxin and sterigmatocystin (Reade and Gregory, 1975). These mycotoxins are not among the toxins known from *A. fumigatus*, and the analyses were negative as well. There have been no further reports on the use of *A. fumigatus* mycoprotein since the abovementioned reports from the late 1970s.

One of the major work horses in biotechnology is *Aspergillus niger* and its use has been evaluated as safe taking into account that ochratoxin A is produced by industrial strains under laboratory conditions (Schuster *et al.*, 2002). During the experimental SCP productions by *A. niger* (Table 1) a possible mycotoxin contamination was never discussed, but despite the wide use of *A. niger* in biotechnology, mycoprotein from this species has never been commercialised. *Geotrichum candidum* is widely used as starter culture in dairy industry for cheese production, but the mycoprotein production (Table 1) has been evaluated to be of less commercial value compared to SCP from *Saccharomyces cerevisiae* and *Candida* species (Table 1) due to their much higher protein per dry weight content (Ziino *et al.*, 1999).

Table 1. Examples of filamentous fungi and yeast tested for production of single cell protein as food or feed

Actinomucor elegans	(Hang, 1976)
Aspergillus fumigatus	(Khor *et al.*, 1977; Reade and Gregory, 1975)
Aspergillus niger	(Christias *et al.*, 1975; Hang, 1976; Oboh *et al.*, 2002; Singh *et al.*, 1991)
Aspergillus oryzae	(Hang, 1976)
Candida lipolytica	(Achremowicz *et al.*, 1977)
Candida tropicalis	(Achremowicz *et al.*, 1977; Christias *et al.*, 1975)
Candida utilis	(Villas-Boas *et al.*, 2003)
Chaetomium globosum	(Hang, 1976)
Fusarium moniliforme[a]	(Christias *et al.*, 1975; Drouliscos *et al.*, 1976)
Fusarium oxysporum	(Sukara and Doelle, 1989)
Fusarium venenatum[b]	(Anderson and Solomons, 1984; Trinci, 1992)
Geotrichum candidum	(Robinson and Smith, 1976; Ziino *et al.*, 1999)
Pestalotiopsis westerdijkii	(Hang, 1976)
Phanerochaete chrysosporium	(Cardoso and Nicoli, 1981)
Rhizopus oligosporus	(Sukara and Doelle, 1989)
Thielavia terrestris[c]	(Bajon *et al.*, 1985; Stevens and Gregory, 1987)
Trichoderma viride	(Hang, 1976; Youssef and Aziz, 1999)

[a] Outdated species epithet (Seifert *et al.*, 2003); [b] Initially identified as *Fusarium graminearum*. Commercial product is Quorn®; [c] Anamorphic state is *Acremonium alabamense* (syn. *Cephalosporium eichhorniae*) (Stevens and Gregory, 1987).

Fusarium moniliforme was found to produce biomass of a high nutritional value when growing on a cheap substrate (Drouliscos *et al.*, 1976; Macris and Kokke, 1978). At that time the authors were concerned about any mycotoxin production, however, based on rat feeding experiments they concluded that no toxins were produced. Since then "*F. moniliforme*" has been split into several closely related species (Nirenberg and O'Donnell, 1998) and the specific epithet "*moniliforme*" is now considered outdated (Seifert *et al.*, 2003). As the SCP producing strains have not been re-identified according to the updated taxonomic schemes it is impossible to tell exactly which species were used. Another *Fusarium* species has gained much more interest as producer of the only commercially available mycoprotein product for human consumption, Quorn®, namely *F. venenatum* (originally identified as *F. graminearum*). A screening for suitable mycoprotein producers in the late 1960s by a British food company, Ranks Hovis McDougall (RHM), resulted in a promising *Fusarium* strain (Anderson and Solomons, 1984; Wilson, 2001). The production strain, *F. venenatum* A3/5, is discussed in details in a following paragraph.

PRODUCTION OF MYCOPROTEIN

Information on the production of mycoprotein used for Quorn® has been obtained from (Rodger, 2001) and the Quorn® homepage (http://www.quorn.com), unless otherwise stated. Mycoprotein is produced in a 50-m tall air-lift fermenter, where rising air bubbles are used to mix the fermentation culture instead of mechanical stirring as in a conventional fermenter. The air-lift or pressure-cycle fermenter design is preferable for high viscous fermentations by filamentous fungi as it implies an improved transfer of oxygen and nutrients, efficient removal of carbon dioxide and reduced generation of heat (Trinci, 1992). The fermentation is started by adding a stock culture of the *F. venenatum* production strain to the sterilized culture broth. After initial batch cultivation, the fermentation is turned into a continuous fermentation where glucose, biotin and mineral salts are pumped in at a constant rate, simultaneously with removal of culture at the same rate. Compressed air and ammonia are added at the bottom of the reactor and the rising air bubbles mix the fermentation culture. All nutrients including oxygen are kept in excess and the carbon dioxide removed from top of the fermenter contains 10% oxygen (Trinci, 1992). The automatic control system ensures a constant environment controlling pH

6.0, temperature 28-30°C, oxygen in excess as well as the rate of inlet and outlet streams, respectively (Rodger, 2001; Wiebe, 2002). The continuous fermentation runs in the 150 m^3 reactor for up to six weeks and produces 300-350 kg biomass/h (Wiebe, 2002).

The outlet stream from the fermenter is collected into a separate tank where the broth and biomass are shock-heated to around 64°C for 20 minutes to reduce the RNA content in the biomass from 10% to below the recommended 2% maximum level. During this process, proteins and cell components also are lost and the net yield of biomass is reduced by around one third. Following this step the biomass is harvested by filtration and concentrated from 1.5% (w/v) to 25-30% (w/v) total solids (Rodger, 2001; Trinci, 1992).

The production strain has been carefully developed and selected by its ability to grow with a sufficient branching pattern, about one branch per 300 μm (Rodger, 2001; Trinci, 1994), so at this stage the biomass consists of mycelium with a fibril structure like meat; however, the cross-linking structures in meat are absent in mycoprotein. Egg albumin is added as binder, together with colorants and flavouring agents (product dependent) and the mixture is formed into the desired shape and heated to gel the product with a meat-like texture. Before freezing for storage the product is shaped as burgers, chunks, sausages, etc., depending on the final usage.

MYCOPROTEIN AS A FOOD PRODUCT

Mycoprotein is a nutritional recommendable product, which when freshly harvested has a protein content of 12% (w/w, wet weight) and a good composition of amino acids; all essential amino acids are present in concentrations comparable to egg (Miller and Dwyer, 2001; Rodger, 2001). The protein digestibility is comparable to beef and soybean, which is even increased in the final formulations (Quorn® products) due to the egg and milk proteins added (Miller and Dwyer, 2001). Mycoprotein has an unsaturated/saturated fatty acid ratio of 4/1 and contains no cholesterol, and also in contrast to animal protein sources, mycoprotein contents 6% (w/w, wet weight) dietary fibre from the mycelium cell wall constituents (Miller and Dwyer, 2001; Rodger, 2001).

In addition to the nutritional functionality, several investigations also has demonstrated clinical functionality of mycoprotein. There is a general agreement between several studies that mycoprotein has a significant effect on appetite, especially satiety (Burley *et al.*, 1993; Turnbull *et al.*, 1993; Williamson *et al.*, 2006). All concluded that subsequent to a mycoprotein meal the amount of food consumed during the following meal was less than to a control meal containing non-mycoprotein (typically chicken). The same effect was seen when tofu was given (Williamson *et al.*, 2006). The appetite variables are closely related to the blood levels of glucose and insulin, and Turnbull and Ward (1995) found that they were reduced after a mycoprotein meal causing reduced appetite. These observations could lead to development of filling, low-energy foods based on mycoprotein for control of body weight and appetite, as well as for diabetes dietary (Turnbull and Ward, 1995; Williamson *et al.*, 2006). Consumption of mycoprotein has also been demonstrated to reduce the total and the low density lipoprotein cholesterol in blood lipids (Rodger, 2001; Turnbull *et al.*, 1990; Turnbull *et al.*, 1992). Finally, it is generally recommended to get a sufficient amount of dietary fibre and as such mycoprotein is a useful source.

In contrast to the positive nutritional value of fungi and fungal protein, it is generally known that some people are sensitive or allergic to an increased level of fungal material in their environment. It is best known from humid and water damaged homes as well as schools and other working environments where a high level of organic dust containing fungal biomass may have adverse effect on workers. In many countries there is general surveillance of the level of *Cladosporium* and *Alternaria* species in outdoor air and the observations are broadcasted to inform the public. Having this in mind, allergic and other adverse reactions to mycoprotein are of concern. There are reported complaints mainly vomiting, stomach cramping and diarrhoea occurring in the hours after eating mycoprotein (Hoff *et al.*, 2003; Jacobson, 2003b). It is difficult to give exact frequencies of complaints as the sources have used different calculation procedures so frequencies vary from 1 per 80,000 (in Switzerland) to 1 per 370,000 (United Kingdom) (Hoff *et al.*, 2003), or as low as 1 per 667,000 products sold (Tee *et al.*, 1993). From 1994 until 2000 the numbers of reported complaints raised from 27 to 89 with a peak of 115 in 1998; however, as the estimated numbers of consumers also raised from 2.25 to 13 millions the frequencies (complaints per consumer) changed gradually from 1 per 83,000 (in 1994) to 1 per 146,000 (in 2000) (Miller and Dwyer, 2001). Based on a telephone survey (N=1004) 5% of the contacted consumers who had eaten mycoprotein (n=396) reported an adverse reaction (Jacobson, 2003b). The 5% corresponds to 20 persons, which is a similar frequency as adverse reactions to milk, peanuts and wheat each tabled as 2% of 1004 persons; however, in the same survey 30 persons (tabled as 3% of 1004) reported reactions to shellfish. Although the data set statistically is insignificant, this telephone survey indicates that sensitivity to mycoprotein is just as frequent as sensitivity to other food items, maybe even less frequent (Peregrin, 2002). On the other hand, the anecdotal case stories are quite scary (Hoff *et al.*, 2003; Jacobson, 2003a; Jacobson, 2003b; Katona and Kaminski, 2002) and should initiate further studies on intolerance, sensitivity and allergic reactions towards fungal proteins to ensure that novel foods are just as safe as well-known foods. Such future activities could also be valuable within the area of fungal contamination in dwellings and indoor air and may add further to the question on whether these reactions in humans are multi-factorial responses as hypothesised in the case of mycoprotein (Tee *et al.*, 1993).

FUSARIUM VENENATUM – THE PRODUCER OF MYCOPROTEIN

As of today the only commercial mycoprotein products for human food are the palette of Quorn® products from Marlow Foods Ltd. based on *Fusarium venenatum* biomass. The biotechnological development of the specific *F. venenatum* strain used for mycoprotein production is well described; however, originally the strain was identified as *F. graminearum* (Trinci, 1994). Since the original Quorn® strain labelled A3/5 (Table 2) has been shown to be a useful host for heterologous protein production (Royer *et al.*, 1995) the importance of the exact identity of this strain has increased. The ATCC 20334 strain (Table 2) was identified as *F. sulphureum* (now *F. sambucinum* (Nirenberg, 1995), *F. crookwellense* or *F. venenatum* by several mycological experts (Yoder and Christianson, 1998). Within this study species-specific DNA primers were developed and the results supported the identity of ATCC 20334 to be *F. venenatum* which was verified by morphological, chemical and phylogenetic data (O'Donnell *et al.*, 1998). *Fusarium venenatum* (Latin for "the poisonous *Fusarium*") was discovered as a new species in 1995 (Nirenberg, 1995) supported by results from an international collaborative polyphasic examination of 41 strains of *F. sambucinum* and related species (Desjardins and Nelson, 1995; Hering and Nirenberg, 1995; Logrieco *et al.*, 1995; Szecsi *et al.*, 1995; Thrane and Hansen, 1995).

There is a general acceptation of the re-identification of A3/5 as *F. venenatum*; however, it is a logical fact that pre-1998 publications and patents on the Quorn® strain will use the epithet *F. graminearum*. Unfortunately many accessible databases (e.g., culture collections and sequence databases) also will use *F. graminearum* for data on the Quorn® strain, as not all are updated as careful as they should be. This means that erroneous information will be retrieved for many years ahead. A major concern in this context is the aspect of mycotoxins as the two species have different mycotoxin pattern. *Fusarium graminearum* is known to produce deoxynivalenol and nivalenol including derivatives hereof, in addition to zearalenone, fusarin C, culmorin and butenolide (Thrane, 2001); whereas *F. venenatum* is known to produce diacetoxyscirpenol (DAS) and several derivatives hereof, nivalenol and fusarenon X (both in trace amounts), butenolide and culmorin (Miller and MacKenzie, 2000; Nielsen and Thrane, 2001; Thrane and Hansen, 1995). Different copies of *F. venenatum* A3/5 have been analysed for mycotoxin production with some deviating results as DAS was detected in ATCC 20334 and DAOM 212262 (a deposit of A3/5) (Miller and MacKenzie, 2000), whereas no DAS could be detected in a culture of NRRL 26139 (a deposit of ATCC 20334) (O'Donnell *et al.*, 1998). The variation between the three copies of what is supposed to be the same strain (Table 2) can partly be explained by different experimental conditions in the two studies, partly by the attenuation in secondary metabolism of *Fusarium* strains after repeated cultivation (Duncan and Bu'Lock, 1985). DAS was also detected in a culture of an aconidial mutant of A3/5 (NRRL 25416) (O'Donnell *et al.*, 1998) and in a culture of a transformant of ATCC 20334 intended for biotechnological use (Miller and MacKenzie, 2000); however, Quorn® products are only to be produced from fermentations of the specific strain ATCC 20334 (Johnstone, 1998).

In three different Quorn® products no mycotoxins were detected at a 0.5 ppm level (O'Donnell *et al.*, 1998) which relates to the fact that conditions for fermentation of *F. venenatum* ATCC 20334 are well outside the conditions for mycotoxin production (Johnstone, 1998). Furthermore, as part of the manufacturer's Quality Control system, samples for mycotoxin analysis are taken at six-hour intervals, and should be free of mycotoxins to release the produced mycoprotein (Johnstone, 1998). In addition to mycoprotein production, *F. venenatum* ATCC 20334 has been used as mother strain for development of mutants blocked in the trichothecene synthesis by deletion of a key gene (*tri5*) (Royer *et al.*, 1999). Some of the mutants were found to be unable to produce trichothecenes in detectable amounts verifying a successful gene deletion (Miller and MacKenzie, 2000). The *tri5* mutants have further been modified for production of enzymes for the food industry (Ahmad *et al.*, 2004; Pedersen and Broadmeadow, 2000).

In these cases extensive toxicological studies were conducted and the authors considered the enzymes as safe for use in the food industry. It is expected that in the future there will be an increasing use of *F. venenatum* strains originating from the original Quorn® strain as biotechnological work horses for production of heterologous proteins for use in many industries.

Table 2. Known strains of *Fusarium venenatum* A3/5 used for mycoprotein production[a]

Strain no.	Origin	Comment
ATCC 20334	A3/5 – the Quorn® strain (Yoder and Christianson, 1998)	= A3/5 = IMI 145425 = NRRL 26139[b]
ATCC PTA-2684	The Quorn® production strain (Rodger, 2001)	=ATCC 20334 = NRRL 26139 (Wiebe, 2002)
DAOM 193459	A3/5 - the Quorn® strain (Yoder and Christianson, 1998)	
DAOM 212262	A3/5 - the Quorn® strain (Yoder and Christianson, 1998)	
IMI 145425	A3/5 - the Quorn® strain (Yoder and Christianson, 1998)	= A3/5 = NRRL 26139[b]
NRRL 25417	A3/5 - the Quorn® strain (Yoder and Christianson, 1998)	= IMI 145425[c]
NRRL 26139	ATCC 20334 (O'Donnell *et al.*, 1998)	= A3/5 = IMI 145425[b]

[a] ATCC, American Type Culture Collection, Fairfax, VA, USA; DAOM, Agriculture Canada and Agri-Food Canada Culture Collection, Ottawa, ON, Canada; IMI, CABI Bioscience Genetic Resource Collection, Egham, Surrey, UK; NRRL, The Agriculture Research Service Culture Collection, National Center for Agricultural Utilization Research, USDA/ARS, Peoria, IL, USA. [b]According to the ATCC database (http://www.atcc.org). [c]Erroneously listed as a mutant of the Quorn® strain by O'Donnell *et al.* (1998).

CONCLUSIONS

Globally there is an increasing consumer request for semi-prepared ready-to-eat food products and examples of such products are the Quorn® products which are marketed in the United Kingdom, Belgium, the Netherlands, Sweden, Switzerland and the United States. Since its introduction it has become a well-established business with an annual retail sale of Quorn® products above U.S.$ 200 million (http://www.quorn.com, April 2006). Over the years there have been numerous reports on adverse effects of mycoprotein products. However, surveys have shown that the frequency of adverse effects is comparable to what is observed by eating other food items, such as shellfish, peanuts and milk. From the available data, Quorn® can be classified as a successful development of novel food, or functional food taking the nutritional characteristics into consideration. From a mycotoxicological point of view, the use of *Fusarium venenatum* is of concern; however, independent studies have shown that the Quorn® strain only produces diacetoxyscirpenol and other mycotoxins in low amounts under optimal laboratory conditions (Miller and MacKenzie, 2000; O'Donnell *et al.*, 1998) (Nielsen and Thrane, unpublished). The surveillance for mycotoxin contamination (Johnstone, 1998) as well as a careful control of the quality of the production strain is crucial control elements to ensure that only mycotoxin free mycoprotein products are marketed.

REFERENCES

Achremowicz, B., Kosikowski, F. V., and Masuyama, K. (1977). Mixed cultures of different yeasts species, yeasts with filamentous fungi in SCP production. I. Production of single cell protein by mixed cultures *Candida lipolytica*, *Candida tropicalis*. Acta Microbiologica Polonica 26:265-271.

Ahmad, S. K., Brinch, D. S., Friis, E. P., and Pedersen, P. B. (2004). Toxicological studies on Lactose Oxidase from *Microdochium nivale* expressed in *Fusarium venenatum*. Regulatory Toxicology and Pharmacology 39:256-270.

Anderson, C., and Solomons, G. L. (1984). Primary metabolism and biomass production from *Fusarium*. Pages 231-250. *In* The applied mycology of *Fusarium* (Moss, M. O., and Smith, J. E., eds.), Cambridge University Press, Cambridge, U.K.

Bajon, A. M., Tsé Hing Yuen, T. L. S., Li Sui Fong, J-C., and Olah, G. M. (1985). Preliminary screening of some fungal strains for protein upgrading of sugar-beet pulp. Biotechnology Letters 7:203-206.

Blumenthal, C. Z. (2004). Production of toxic metabolites in *Aspergillus niger*, *Aspergillus oryzae*, and *Trichoderma reesei*: justification of mycotoxin testing in food grade enzyme preparations derived from the three fungi. Regulatory Toxicology and Pharmacology 39:214-228.

Burley, V. J., Paul, A. W., and Blundell, J. E. (1993). Influence of a high-fiber food (myco-protein*) on appetite: effects on satiation (within meals) and satiety (following meals). European Journal of Clinical Nutrition 47:409-418.

Cardoso, M. B., and Nicoli, J. R. (1981). Single cell protein from the thermotolerant fungus *Phanerochaete chrysosporium* grown in vinasse. I. Production, composition. Nutrition Reports International 24:237-247.

Christias, C., Couvaraki, C., Georgopoulos, S. G., Macris, B., and Vomvoyanni, V. (1975). Protein content, amino acid composition of certain fungi evaluated for microbial protein production. Applied Microbiology 29:250-254.

Das, H., and Singh, S. K. (2004). Useful byproducts from cellulosic wastes of agriculture and food industry - A critical appraisal. Critical Reviews in Food Science and Nutrition 44:77-89.

Desjardins, A. E., and Nelson, P. E. (1995). Sexual fertility of forty *Fusarium* strains from the European *Fusarium sambucinum* project. Mycopathologia 129:149-151.

Drouliscos, N. J., Macris, B. J., and Kokke, R. (1976). Growth of *Fusarium moniliforme* on carob aqueous extract and nutritional evaluation of its biomass. Applied and Environmental Microbiology 31:691-694.

Duncan, J. S., and Bu'Lock, J. D. (1985). Degeneration of zearalenone production in *Fusarium graminearum*. Experimental Mycology 9:133-140.

Filtenborg, O., Frisvad, J. C., and Thrane, U. (1996). Moulds in food spoilage. International Journal of Food Microbiology 33:85-102.

Hang, Y. D. (1976). Fungal treatment of beet waste. Progress in Water Technology 8:325-327.

Hering, O., and Nirenberg, H. I. (1995). Differentiation of *Fusarium sambucinum* Fuckel sensu lato and related species by RAPD PCR. Mycopathologia 129:159-164.

Hoff, M., Trueb, R. M., Ballmer-Weber. B. K., Vieths, S., and Wuethrich, B. (2003). Immediate-type hypersensitivity reaction to ingestion of mycoprotein (Quorn) in a patient allergic to molds caused by acidic ribosomal protein P2. Journal of Allergy and Clinical Immunology 111:1106-1110.

Jacobson, M. F. (2003a). Adverse reactions linked to Quorn-brand foods. Allergy 58:455-456.

Jacobson, M. F. (2003b). Adverse reactions to a new food ingredient. American Journal of Medicine 115:334.

Johnstone, L. (1998). Letter to the Editor. Fungal Genetics and Biology 25:75.

Katona, S. J., and Kaminski, E. R. (2002). Sensitivity to Quorn mycoprotein (*Fusarium venenatum*) in a mould allergic patient. Journal of Clinical Pathology 55:876-877.

Khor G. L., Alexander, J. C., Lumsden, J. H., and Losos, G. J. (1977). Safety evaluation of *Aspergillus fumigatus* grown on cassava for use as an animal feed. Canadian Journal of Comparative Medicine 41:428-434.

Logrieco, A., Peterson, S. W., and Bottalico, A. (1995). Phylogenetic relationship within *Fusarium sambucinum* Fuckel sensu lato, determined from ribosomal RNA sequences. Mycopathologia 129:153-158.

Macris, B. J., and Kokke, R. (1978). Continuous fermentation to produce fungal protein. Effect of growth rate on the biomass yield and chemical composition of *Fusarium moniliforme*. Biotechnology and Bioengineering 20:1027-1035.

Mapari, S. A. S., Nielsen, K. F., Larsen, T. O., Frisvad, J. C., Meyer, A. S., and Thrane, U. (2005). Exploring fungal biodiversity for water-soluble pigments as potential natural food colorants. Current Opinion in Biotechnology 16:231-238.

Miller, J. D., and MacKenzie, S. (2000). Secondary metabolites of *Fusarium venenatum* strains with deletions in the *Tri5* gene encoding trichodiene synthetase. Mycologia 92:764-771.

Miller, S. A., and Dwyer, J. T. (2001). Evaluating the safety and nutritional value of mycoprotein. Food Technology 55:42-47.

Nielsen, K. F., and Thrane, U. (2001). Fast methods for screening of trichothecenes in fungal cultures using gas chromatography-tandem mass spectrometry. Journal of Chromatography A 929:75-87.

Nirenberg, H. I. (1995). Morphological differentiation of *Fusarium sambucinum* Fuckel sensu stricto, *F. torulosum* (Berk. & Curt.) Nirenberg comb. nov. and *F. venenatum* Nirenberg sp. nov. Mycopathologia 129.131-141.

Nirenberg, H. I., and O'Donnell, K. (1998). New *Fusarium* species and combinations within the *Gibberella fujikuroi* species complex. Mycologia 90:434-458.

O'Donnell, K., Cigelnik, E., and Casper, H. H. (1998). Molecular phylogenetic, morphological, and mycotoxin data support reidentification of the Quorn mycoprotein fungus as *Fusarium venenatum*. Fungal Genetics and Biology 23:57-67.

Oboh, G., Akindahunsi, A. A., and Oshodi, A. A. (2002). Nutrient and anti-nutrient contents of *Aspergillus niger*-fermented cassava products (flour and gari). Journal of Food Composition and Analysis 15:617-622.

Özyurt, M., and Deveci, E. Ü. (2004). Conversion of agricultural and industrial wastes for single cell protein production and pollution potential reduction: A review. Fresenius Environmental Bulletin 13:693-699.

Pedersen, P. B., and Broadmeadow, A. (2000). Toxicological studies on *Termomyces lanuginosus* xylanase expressed by *Fusarium venenatum*, intended for use in food. Food Additives and Contaminants 17:739-747.

Peregrin, T. (2002). Mycoprotein: Is America ready for a meat substitute derived from a fungus? Journal of the American Dietetic Association 102:628.

Reade, A. E., and Gregory, K. F. (1975). High-temperature production of protein-enriched feed from cassava by fungi. Applied Microbiology 30:897-904.

Robinson, P. M., and Smith, J. M. (1976). Morphogenesis and growth kinetics of *Geotrichum candidum* in continuous culture. Transactions of the British Mycological Society 66:413-420.

Rodger, G. (2001). Production and properties of mycoprotein as a meat alternative. Food Technology 55:36-41.

Royer, J. C., Christianson, L. M., Yoder, W. T., Gambettta, G. A., Klotz, A. V., Morris, C. L., Brody, H., and Otani, S. (1999). Deletion of the trichodiene synthase gene of *Fusarium venenatum*: two systems for repeated gene deletions. Fungal Genetics and Biology 28:68-78.

Royer, J. C., Moyer, D. L., Reiwitch, S. G., Madden, M. S., Jensen, E. B., Brown, S. H., Yonker, C. C., Johnstone, J. A., Golightly, E. J., Yoder, W. T., and Shuster, J. R. (1995). *Fusarium graminearum* A 3/5 as a novel host for heterologous protein production. Bio/Technology 13:1479-1483.

Sadler, M. J. (2004). Meat alternatives - market developments and health benefits. Trends in Food Science and Technology 15:250-260.

Schuster, E., Dunn-Coleman, N., Frisvad, J. C., and Van Dijck, P. W. (2002). On the safety of *Aspergillus niger* — a review. Applied Microbiology and Biotechnology 59:426-435.

Seifert, K. A., Aoki, T., Baayen, R. P., Brayford, D., Burgess, L. W., Chulze, S., Gams, W., Geiser, D., de Gruyter, J., Leslie, J. F., Logrieco, A., Marasas, W. F. O., Nirenberg, H. I., O'Donnell, K., Rheeder, J. P., Samuels, G. J., Summerell, B. A., Thrane, U., and Waalwijk, C. (2003). The name *Fusarium moniliforme* should no longer be used. Mycological Research 107:643-644.

Singh, A., Abidi, A. B., Agarwal, A. K., and Darmwal, N. S. (1991). Single cell protein production by *Aspergillus niger* and its evaluation. Zentralblatt für Mikrobiologie 146:181-184

Spicer, A. (1971). Synthetic proteins for human and animal consumption. Veterinary Record 89:482-486.

Stevens, C. A., and Gregory, K. F. (1987). Production of microbial biomass protein from potato processing wastes by *Cephalosporium eichhorniae*. Applied and Environmental Microbiology 53:284-291.

Sukara, E., and Doelle, H. W. (1989). Optimization of single cell protein production from cassava starch (*Rhizopus oligosporus*). Acta Biotechnologica 9:99-110.

Szecsi, A., Turoczi, G., and Bordas, B. (1995). Analysis of esterase zymograms of *Fusarium sambucinum* and related species. Mycopathologia 129:165-171.

Tee, R. D., Gordon, D. J., Welch, J. A., and Taylor, A. J. N. (1993). Investigation of possible adverse allergic reactions to mycoprotein ('Quorn'). Clinical and Experimental Allergy 23:257-260.

Thrane, U. (2001). Developments in the taxonomy of *Fusarium* species based on secondary metabolites. Pages 29-49. *In* Fusarium. Paul E. Nelson Memorial Symposium (Summerell, B. A., Leslie, J. F., Backhouse, D., Bryden, W. L., and Burgess, L. W., eds.), APS Press, St. Paul, MN, U.S.A.

Thrane, U., and Hansen, U. (1995). Chemical and physiological characterization of taxa in the *Fusarium sambucinum* complex. Mycopathologia 129:183-190.

Trinci, A. P. J. (1992). Myco-protein: A twenty-year overnight success story. Mycological Research 96:1-13.

Trinci, A. P. J. (1994). Evolution of the Quorn® myco-protein fungus, *Fusarium graminearum* A3/5. Microbiology U.K. 140:2181-2188.

Turnbull, W. H., Leeds, A. R., and Edwards, G. D. (1990). Effect of mycoprotein on blood lipids. American Journal of Clinical Nutrition 52:646-650.

Turnbull, W. H., Leeds, A. R., and Edwards, D. G. (1992). Mycoprotein reduces blood lipids in free-living subjects. American Journal of Clinical Nutrition 55:415-419.

Turnbull, W. H., Walton, J., and Leeds, A. R. (1993). Acute effects of mycoprotein on subsequent energy intake and appetite variables. American Journal of Clinical Nutrition 58:507-512.

Turnbull, W. H., and Ward, T. (1995). Mycoprotein reduces glycemia and insulinemia when taken with an oral-glucose-tolerance test. American Journal of Clinical Nutrition 61:135-140.

Villas-Boas, S. G., Esposito, E., and Mitchell, D. A. (2002). Microbial conversion of lignocellulosic residues for production of animal feeds. Animal Feed Science and Technology 98:1-12.

Villas-Boas, S. G., Esposito, E., and de Mendonca, M. M. (2003). Bioconversion of apple pomace into a nutritionally enriched substrate by *Candida utilis, Pleurotus ostreatus*. World Journal of Microbiology and Biotechnology 19:461-467.

Wiebe, M. G. (2002). Myco-protein from *Fusarium venenatum*: a well-established product for human consumption. Applied Microbiology and Biotechnology 58:421-427.

Williamson, D. A., Geiselman, P. J., Lovejoy, J., Greenway, F., Volaufova, J., Martin, C. K., Arnett, C., and Ortego, L. (2006). Effects of consuming mycoprotein, tofu or chicken upon subsequent eating behaviour, hunger and safety. Appetite 46:41-48.

Wilson, D. (2001). Marketing mycoprotein: The Quorn foods story. Food Technology 55:48-50.

Yoder, W. T., and Christianson, L. M. (1998). Species-specific primers resolve members of *Fusarium* section *Fusarium* - Taxonomic status of the edible "Quorn" fungus reevaluated. Fungal Genetics and Biology 23:68-80.

Youssef, B. M., and Aziz, N. H. (1999). Influence of γ-irradiation on the bioconversion of rice straw by *Trichoderma viride* into single cell protein. Cytobios 97:171-183.

Ziino, M., Lo Curto, R. B., Salvo, F., Signorino, D., Chiofalo, B., and Giuffrida, D. (1999). Lipid composition of *Geotrichum candidum* single cell protein grown in continuous submerged culture. Bioresource Technology 67:7-11.

Chapter 19

Edible mushrooms: from industrial cultivation to collection from the wild

Jacqueline Baar, Gerben Straatsma, Istvan Paradi, and Jos G. M. Amsing
Applied Plant Research, Mushroom Unit, Wageningen University and Research Center, P.O. Box 6042, 5966 AA Horst, The Netherlands.

INTRODUCTION

The domestication and agronomical production of edible mushrooms has a relatively short history compared to that of staple crops and of many vegetables and fruits (Janick, 2005). A method for the cultivation of a Chinese mushroom, shiitake, *Lentinula edodes*, is known from 1313 AD (Wang, 1987). The cultivation of the white button mushroom, *Agaricus bisporus*, was described for the first time by Tournefort (1707). All edible mushroom species that can be cultivated in a system comparable to the industrial cultivation of the white button mushroom are saprobic, being able to use dead organic matter for their growth. Boa (2004) estimates the number of saprobic species that can be cultivated as almost 100. Several valuable edible fungi, in particular ectomycorrhizal species like truffles (*Tuber magnatum* and *Tricholoma melanosporum*), matsutake (*Tricholoma matsutake*), kings boletes (*Boletus edulis*), and chanterelles (*Cantharellus cibarius*) cannot yet be produced agronomically. At best, semi-natural cultivation systems (e.g., truffle orchards) have been developed (Hall *et al.*, 2003a).

INDUSTRIAL CULTIVATION OF WHITE BUTTON MUSHROOMS, *AGARICUS BISPORUS*

White button mushrooms require two substrates for cultivation. A layer of compost nourishes the mycelium and a layer of casing soil covers the compost allowing the mycelium to form fruit bodies. Compost production and the cultivation process have been described in various handbooks (Flegg *et al.*, 1985; van Griensven, 1988; Oei and Maas, 2003; Chang and Miles, 2004). China, the U.S.A. and The Netherlands are the most important producers of white button mushrooms. Productions in these countries are, respectively, 600, 380 and 260 x 10^3 tonnes annually; the statistics from China need to be taken with caution. The following key data characterize the mushroom industry in the Netherlands. Almost 400 kg of mushrooms can be produced per tonne of full grown compost; this amount of compost being filled on a surface area of about 10 m^2. The amounts of "full grown compost" and of "casing soil" used annually are about 700 and 380 x 10^3 tonnes. Due to the short growing cycle, about eight crops of mushrooms per year can be grown in a cultivation room. The net surface area for mushroom growing is over 100 ha. The rest-product of mushroom cultivation is called "spent mushroom substrate," SMS. The amount of SMS produced is about 850 x 10^3 tonnes; this is marketed as a fertilizer in horticultural crops. The number of full-time jobs in primary production is currently about 5000. Key data for the Netherlands may be extrapolated to other countries, although compost-to-mushroom conversion and labour efficiency are considerably lower.

Figure 1. A dense crop of mushrooms (top) is produced by the following scheme of cultivation of *A. bisporus* according to the Dutch system (bottom).

Table 1. Composting and cultivation processes according to the Dutch system. Compost production, Phases I to III.

Process	Location	Temperature (°C)	Inoculation	Duration (days)
Mixing ingredients	hall			1
Phase-I	tunnel	>70		6
Phase-II (pasteurization and conditioning)	tunnel	45	phase-II compost	6
Phase-III (mycelium colonization)	tunnel	24	mushroom spawn	14
Casing soil colonization	cultivation room	24		10
Primordia formation, fruit body growth and cropping	cultivation room	20		30
Cook out	cultivation room	70		1

Process
Innovative processes and systems implemented at a large scale in The Netherlands became a standard in Europe, Australia and in the past years in North America too; the so-called "Dutch system." The Dutch system of composting and cultivation follows a well-defined scheme (Table 1 and Figure 1).

Compost
The first processes concern compost preparation. The ingredients for compost are carbon rich straw, nitrogen rich manure, gypsum and water. Straw and manure are by-products from other agricultural sectors. Straw-bedded horse manure is a popular ingredient because, if available at a relatively short distance around a compost yard, it has a lower price than a mixture of "pure" straw and manure. The three composting processes, Phases I to III, are bulk processes in fermentation rooms called "tunnels" (van Griensven, 1988; Straatsma, 2004). Tunnel floors are perforated with ventilation holes to be able to force air through the compost pile on the floor. Tunnels are well suited to control the composting process. Tunnels for Phase I are constructed and equipped for high temperatures up to 80 °C and low microbial activity. Tunnels for Phase II are differently constructed for moderate temperatures up to 60 °C and high microbial activity. These tunnels are also suited for Phase III. The metabolic processes of composting are heterotrophic; carbon dioxide and heat are produced from organic matter. Heat production drives processing (Straatsma *et al.*, 2000). Excess heat is mainly removed by the evaporation of water and subsequent ventilation. The degradation of organic matter, heat production and water evaporation result in a decrease of the moisture contents of the compost. The moisture and nitrogen contents of the compost affect degradation during processing. The interplay of these factors makes Phases I to III a complex process. A high temperature in Phase I is thought to increase the suitability of plant cell wall material for microbial degradation. After Phase I the substrate is able to retain a high amount of water. The substrate should have the right "structure"; a high bulk density is desired for filling the limited space in a cultivation room and sufficient porosity is required for exchange of oxygen, carbon dioxide and water vapour. After Phase II the substrate needs to be "selective" for mushroom mycelium. Mushroom inoculum must be able to grow well. Therefore, competitors and pathogens of mushroom mycelium must be absent or stay under full control. Selectivity mainly depends on the presence of the thermophilic fungus *Scytalidium thermophilum* that develops in Phase II (Straatsma *et al.*, 1989). *Scytalidium thermophilum* is a widespread fungus and is present in compost ingredients. However, its presence after the high temperature of Phase I is not guaranteed. For this reason finished Phase II compost is added at the start of Phase II as an inoculum (Houdeau *et al.*, 1991, Straatsma *et al.*, 1994). The molecular basis of selectivity provided by *S. thermophilum* has not yet been solved (Straatsma *et al.*, 1993).

Phase III starts with inoculating the compost with mushroom spawn (Flegg *et al.*, 1985; van Griensven, 1988). This inoculum is industrially produced and consists of sterilized grains fully colonized by mushroom mycelium. Colonization of the compost in tunnels is a relatively cheap process. After Phase III full grown compost is emptied from tunnels and filled into shelves or bags in cultivation rooms. During this handling the mycelial coherence within the compost is lost. Handling poses no problem, apparently because hyphal fusions by anastomosis occur easily afterwards. The low risk of handling provides an opportunity for adding "supplements" to the compost just before filling for cultivation. Most common supplements are based on protein-rich soy bean meal (Carroll and Schisler, 1976; Randle, 1983; Gerrits, 1989), which increase mushroom yields, especially in compost having a low nitrogen content.

Fruit body formation and yield
Fruit bodies are not formed directly on the compost. A "casing soil," covering the compost, is required (Flegg *et al.*, 1985; van

Griensven, 1988). The number of fruit bodies being formed depends on the amount of compost and its nutrient status, the water relations in the casing soil, the presence of microorganisms in the casing soil, the timing of the sub-process of "airing" (Table 1, Figure 1) and on its course. At the molecular level, the role of casing soil microorganisms in fruit body formation remains unclear.

Common casing soils contain a high amount of peat. Baar and Konings (2005) investigated whether there are indications if bacteria in peat play a crucial role in fruit-body formation of *A. bisporus*. Recently developed molecular detection techniques were used to describe the diversity of bacteria in three different casing soils and peat at various depth in a large peat land area that is located in the northeastern part of the Netherlands, near the German border. Thirty-two bacterial species was observed in the casing soils with *Pseudomonas* sp., *Bacteriodes* sp. and *Flavobacterium* sp. (both belonging to the Bacteriodetes group) as the dominant ones. In the peat samples, 24 different types of bacteria were observed of which the majority never had been cultured in the laboratory before. Little similarity between the composition of the bacterial population was seen between the peat samples and the casing soils, while partial overlap of the populations was observed within the three sampled casing soils. These differences can be explained by variation in conditions in which the different casing soils were composed and stored. The results of this project suggest that other bacteria than those in peat are important for pinning of mushroom mycelium. These are possibly bacteria that belong to the genus *Pseudomonas* (Baar and Konings, 2005). Kalberer (1990) noted that peat is not a specific prerequisite for fruit body initiation, but is mainly used for its high water holding capacity.

Fruit body initiation is under temperature control too. After sufficient mycelial colonization of the casing layer, the air temperature in the cultivation room is lowered from 24 to below 20 °C, called airing. Small hyphal knots, primordia, are formed on mycelial strands on and inside the casing soil. Some evidence is available that all primordia for the whole cropping period are produced in a relatively short period between the start of airing and the appearance of the first fruit bodies (Flegg, 1978). After harvesting fruit bodies, another "flush" or "break" of fruit bodies, probably developing from already existing primordia, appears within about eight days (Flegg *et al.*, 1985; van Griensven, 1988).

Two major determinants of mushroom yield are nitrogen contents of the compost and water contents of both compost and casing soil. The meaning of micro-structural and biochemical characteristics of plant cell walls and their changes during Phase I and II for growth and exploitation by mushroom mycelium are not clear (Gerrits, 1988; Iiyama *et al.*, 1994; Sharma and Kilpatrick, 2000; Straatsma, 2004).

Pest and pathogens

Within the *A. bisporus* mushroom industry, pests and pathogens occur that can cause significant crop losses in mushroom growing houses (Sahin, 2005). The most important diseases are caused by pathogenic fungi including *Verticillium fungicola* (dry bubble disease), *Trichoderma aggressivum* (green mold) and *Cladobotryum dendroides* (cobweb disease; Gandy, 1985). In addition, *Pseudomonas tolaasii* (bacterial brown blotch) is a bacterial disease in mushroom growth.

V. fungicola causes most of the infections in the Netherlands resulting in considerable damage to the yield of mushrooms. Yield reductions by *V. fungicola* only are estimated at 10 million Euros for the Netherlands only (Baar and Rutjens, personal communication). It is a mycoparasite that attacks *A. bisporus* during its generative period. The symptoms of infection depend on the developmental stage at which *A. bisporus* becomes infected (Ware, 1933; Fekete, 1967; North and Wuest, 1993). Irregularly shaped, light brown necrotic lesions are found on mushrooms that are infected relatively late in their development. On the other hand, the characteristic malformed mushrooms, the so-called "dry bubbles," are formed when the infection occurs at early stages of mushroom development (Fletcher *et al.*, 1986).

The *V. fungicola* mycoparasitism on *A. bisporus* seems to be a complex process (Calonje, 1992; Dragt *et al.*, 1996; Calonje *et al.*, 1997 and 2000; Amey *et al.*, 2003). The infection may be initiated by a non-specific contact between the fungal parasite and the host surface involving hydrophobic interactions. The following stage seems to be comprised to specific interactions between molecules, for instance lectin and carbohydrate which are present on both *A. bisporus* and *V. fungicola* (Bernardo *et al.*, 2004). The result of this is an evident pathological process with the formation of infection structures, such as appressoria, the secretion of extracellular hydrolytic enzymes and penetration of the host by the parasite, and finally necrosis of the mushrooms (Dragt *et al.*, 1996; Calonje *et al.*, 1997; Bernardo *et al.*,2004).

The largest losses in yields of *A. bisporus* were reported when *V. fungicola* spores were applied to *A. bisporus* mycelium that develops on the surface of the casing soils. Insect belonging to the phorids, *Megaselia halterata*, and to the sciarids, *Lycoriella auripila*, are considered to be a major vector of *V. fungicola* (White, 1981; Scheepmaker *et al.*, 1998). Moreover, spores are dispersed by water droplets, people, equipment and dust particles in the air. White (1981) showed that addition of relatively high number of spores could initiate infection of *V. fungicola* in the first flush. A recent study showed that a relatively low number of *M. halterata* flies could introduce sufficient *V. fungicola* spores to initiate an infection by the third flush (Clift *et al.*, 2004).

Another fungal disease, albeit less common is cobweb disease, caused by *C. dendroides*. This fungus grows over the casing soil and *A. bisporus* mycelia as white, webbed, fluffy and sometimes granular mycelia. The mycelia occasionally may have some reddish/orange color. The result of overgrowth of the *A. bisporus* mycelia and fruit bodies is direct decay. This fungal disease is reported from the countries all over the world, including the United States of America, the United Kingdom and The Netherlands.

A major bacterial disease in the cultivation industry of *A. bisporus* is bacterial brown blotch. The damage caused by this disease is estimated for The Netherlands only at 3-5 million Euro per year (Baar, personal communication).

The symptoms of bacterial blotch are discoloration of the caps of the mushrooms and pitting. In 1996 a viral disease, Mushroom Virus X (MVX) was recorded.

Figure 2. Mushroom infected with *Verticillium fungicola* (dry bubble disease).

Nowadays MVX is thought to be responsible for a range of symptoms including bare cropping areas on commercial beds (primordia disruption), crop delay, premature veil opening, off- or brown-colored mushrooms, sporophore malformations and loss of crop yield. All symptoms were associated with loss of yield and/or product quality (Grogan *et al.*, 2003). The mushroom industry in the United Kingdom was severely affected in 2000 and 2001, while MVX has now been reported in other countries, including Ireland and The Netherlands.

The cultivation of shiitake and oyster mushrooms

Shiitake, *Lentinula edodes*, and oyster mushrooms, *Pleurotus ostreatus*, growing on dying wood or fresh logs, are cultivated extensively in fields and woods on their natural substrates. Rather than being litter decomposers like *A. bisporus*, these fungi are also cultivated industrially (Oei and Maas, 2003; Chang and Miles, 2004), but the technology to cultivate these species deviates from that of the cultivation of white button mushrooms. For white button mushrooms, a nutritious substrate is produced that is "selective" for growth of its mycelium. Selectivity by *S. thermophilum* or other microorganisms as in white button mushroom cultivation has no parallel for *L. edodes* and *P. ostreatus*. Therefore, substrates for these mushrooms need artificial selectivity by the eradication of potential competing microorganisms in the substrate by high temperature pasteurization or autoclaving. Straw, wood or sawdust are used as carbon sources. Nitrogen-rich additions are used as agricultural by-products like bran. The requirements for nitrogen for the production of fruit bodies of these fungi seem lower than in the case of *A. bisporus*. For the initiation of primordia formation no special substrate is required and primordia are directly formed on the nutritious substrate. This can be regarded as an advantage because of simplicity of production, but a substrate with a special impact for water-holding capacity is lacking. Especially *L. edodes* and to a lesser extent *P. ostreatus* are vulnerable for microbial contamination and require strictly controlled procedures. This can best be realized in small scale containers like bags or even bottles. Initially, this may seem a disadvantage for bulk production of these species, but small containers seem ideal for handling in automated cultivation/incubation systems. We feel that many opportunities exist to optimize the nutrition within the substrate and to maintain an optimal level of moisture of the substrate.

SEMI-NATURAL CULTIVATION OF EDIBLE MYCORRHIZAL MUSHROOMS

Edible mycorrhizal mushrooms include some of the most expensive and sought after foods of the world, with about 200 species eaten in the Northern Hemisphere and many more potentially edible or not yet discovered (Arnolds, 1995; Yun and Hall, 2004). "European" truffles and "Eastern Asian" matsutake are products of extreme luxury and are as precious as their dry weight in gold. Market prices of mushrooms economically available to a broader public like chanterelles vary from $15 per kg in Germany to $40 per kg in the United States (Smith and Read, 1997; Pilz *et al.*, 1999). The species mentioned are mycorrhizal and most of them are harvested in forests (Hall *et al.*, 2003b). Epigeous species, growing on the soil surface like matsutake and chantarelles, as well as hypogeous species, forming subterranean fruit bodies like truffles, are involved. Collection of edible mushrooms from the wild has suffered a harsh decline over the past 100 years due to various reasons, like destruction and bad management of their forest habitats or environmental pollution (Arnolds, 1995; Hall *et al.*, 2003a and 2003b). Research on the methods for the sustainable cultivation and forest management is therefore highly required to increase production and to protect natural habitats.

In spite of the high and persistent demand, the large-scale or intensive cultivation of mycorrhizal mushrooms is still not resolved (Iwase, 1997; Hall *et al.*, 2003b). Scaling up of cultivation is essentially prevented by the need of mycorrhizal fungi to associate to a host plant for growth and fruit body production. Semi-natural cultivation of important truffles like *Tuber melanosporum* (Périgord black truffle), and *T. uncinatum* (Burgundy truffle) is practiced in so-called "truffle orchards" (Rebiere, 1967; Olivier, 2000; Chevalier *et al.*, 2002, Yun and Hall, 2004). Most truffles on the market presently come from planted fields; however, annual productions reach only 1/5 of the productions a century ago (Olivier, 2000). Semi-natural cultivation systems (e.g., truffle orchards) have the inherent problem of low controllability; therefore, contamination of other, non-productive ectomycorrhizal fungi and environmental, climatic variations cause major problems. Factors posing ecological constraints for the truffles are believed to be (1) the percentage coverage of the canopy (e.g., the amount of sunlight to the forest floor), (2) the thickness of shrub layer, (3) soil moisture (irrigation), (4) temporal distribution of annual precipitation, (5) soil pH, (6) calcium content,

(7) soil temperature, (8) thickness of litter layer, (9) density of trees, and (10) diversity of trees (homogenous or mixed plantations). According to Chevalier (2001), the cultivation technology for *T. uncinatum* is recently improved to a reasonable successful level and Belloli *et al.* (2001) describe this technology adapted for Italian circumstances. Commercial production of the most precious of the truffle species, the white Italian truffle (*T. magnatum*) has not been achieved yet (Olivier, 2000). In some cases, improved management of natural *T. magnatum* areas and establishment of some artificial truffieres at ideal locations succeeded in increased truffle production, but in most cases only fruit bodies of competitive species could be harvested (Gregori, 2002). Some limited success has been achieved also with *T. borchii* and *T. aestivum* (Tanfulli *et al.*, 2001; Vinay and Pirazzi, 2001; Zambonelli *et al.*, 2000 and 2002), and *T. formosanum* (Hu, 2003). Desert truffles like *Terfezia terfezioides* and *T. laveryi* are under study (Yun and Hall, 2004).

Valuable epigeous mycorrhizal species that cannot be cultured yet are *Boletus edulis* and *Cantharellus cibarius* for Europe, and *Tricholoma matsutake* for Eastern Asia. Production of fruit bodies of some other edible species occurred as a side effect of the introduction of mycorrhizal fungi in plantations. Especially pine plantations as exotic trees in South America and Africa produced fruit bodies of *Suillus luteus* and *S. granulatus* (Singer, 1965; Mikola, 1969; Piearce and Ross, 1980; Peredo and Oliva, 1983; Steineck, 1984; Hedger, 1986). Also the edible species *Lactarius deliciosus* fruits quite easily in plantations after mycorrhization of tree seedlings (Poitou *et al.*, 1984; Guinberteau *et al.*, 1990). The first successful fruit body formation of *C. cibarius* occurred in an "artificial" environment, in a greenhouse on *Pinus sylvestris* (Danell and Camacho, 1997). However, transfer of colonized seedlings to the field has not led to controlled fruit body formation up to now (Danell, 2002). Oei and Baar (2002) studied the possibilities to cultivate the edible mycorrhizal fungus *Leccinum duriusculum* on the root systems of white poplar (*Populus alba*). This study resulted in useful information about conditions for cultivation and fruiting of this tasty bolete. From the Southern Hemisphere, Wang *et al.* (2002b) presented the successful cultivation of *Lactarius deliciosus* and *Rhizopogon rubescens* in New Zealand, where inoculated trees produced fruit bodies in experimental plantations. Successful *in vitro* mycorrhizal formation of some Japanese *Lyophyllum, Tricholoma, Suillus, Lactarius* and *Rhizopogon* species were detected in open pot cultures with *Pinus densiflora* seedlings. Basidiocarp formation of *Rhizopogon rubescens* and some *Lactarius* and *Tricholoma* species was also successful (Yamada *et al.*, 2001). The non-edible species *Hebeloma cylindrosporum* seems an interesting species for fundamental research on tree root colonization and fruit body formation since its full life cycle can be controlled under laboratory conditions (Marmeisse *et al.*, 2004).

EDIBLE FUNGAL FRUIT BODIES COLLECTED IN THE WILD

Species numbers and ecological range of species

More than 1,100 different species of wild edible fungi are collected for food in more than 80 countries of the world (Boa, 2004). There are very probably many species that are edible, but not recorded as food or even not yet described scientifically (Hawksworth, 2001). Of this huge variation, only a few mushroom species (e.g., white button mushroom, shiitake, oyster-mushroom or some ectomycorrhizal species, like truffles, cantharelle, matsutake or kings bolete) are used worldwide and the others are only locally important. However, locally known species may have excellent nutritional, culinary or medicinal properties. The great majority of mushroom species are, however, not cultivable in an artificial environment and are used in the local environment where they grow.

Wild edible and non-edible fungi have important roles in ecosystems. Ectomycorrhizal species live in symbiosis with plants (mostly trees) and play an important role in nutrient cycling, development and survival of natural communities. More than 400 species of edible ectomycorrhizal fungal species have been recorded (Wang *et al.*, 2002a). Mycorrhizal fungi can be found all over the world from the taiga to the tropical forests. The symbiotic state is a natural phenomenon for most of the plants worldwide and often certain mushroom and tree species simply cannot live without each other. Also saprobic fungi are important in sustaining the natural nutrient cycling in all ecosystems, being able to decompose died plant material.

In natural systems, management measurements can be taken to enhance production of fruit bodies in natural systems. In *P. sylvestris* stands of different age in The Netherlands, removal of top soil layers significantly increased formation of fruit bodies of edible ectomycorrhizal fungi, including *C. cibarius* and *B. edulis* (Baar and Kuyper, 1998). Furthermore, management measurements in wooded banks of *Quercus robur* can have stimulating effects on the fruit body production. The addition of rain water enhanced fruit body formation of *Boletus edulis* and various *Russula* species (Baar, 2005).

Sustainable harvest of wild edible fungi must be managed in the context of forest management (Boa, 2004). Edible fungi are, as non-wood forest products, only one component of the resources of a forest to be exploited in a sustainable way. Therefore, a balance must be found between fungal biodiversity and the harvesting edible fungi in accordance with the broader (financial) aims of forest management. Regulation of wild collection of mushrooms seems to be inevitable in the long term to avoid overexploitation, whilst research efforts must be dedicated to develop new cultivation methods, e.g., for the commercial production of ectomycorrhizal mushrooms. Protection and management of mycorrhizal associations are especially important, as more and more results show they are one of the main principles of forest development and ecological (nutritional) equilibrium.

Figure 3. *Boletus edulis* in a sixty-year old wooded bank of *Quercus robur* in The Netherlands

Environmental problems of collecting mushrooms in the wild

Intuitively, the picking of mushrooms from the wild is considered to have a negative effect on the survival of the fungal species. However, careful picking for periods of about ten years did not prove to have a negative local effect (Egli *et al.*, 1990; Pilz and Molina, 2002). It is only one functional role of fruit bodies to disseminate spores for the establishment of new mycelia, or perhaps the genetic adaptation of existing mycelia. Additional functional roles in the ecological network are that they provide food for micro- and macrofauna and perhaps their development has an influence on soil organic matter mineralization by the excretion of extracellular enzymes only during this life stage, which might reflect differential resource utilization as observed in some cultivated mushrooms (Flegg *et al.*, 1985).

In certain countries, like China, Korea, Poland, Turkey, the USA and Zimbabwe, harvesting from the wild is a very important commercial business, due to the strong and constant demand from Japan and Europe (Hosford *et al.*, 1997; Fortin and Piche, 2000; Pilz and Molina, 2002; Pilz *et al.*, 2003; Boa, 2004).

Commercial harvesting has side effects like poaching, disturbance by the raking of soil, soil compaction by vehicles, perhaps uncontrolled "management" in people's secret mushroom patches, and even violence among commercial mushroom hunters occurs (Hosford *et al.*, 1997; Pilz and Molina, 2002; Pilz *et al.*, 2003). These forms of disturbance seem to be a greater threat than picking *per se* and demand for an integral forest resource management.

FUTURE CHALLENGES

Sustainable cultivation of the white button mushroom

According to the report "Our common future" (Brundtland, 1987), sustainable development is the "development that meets the needs of the present without compromising the ability of future generations to meet their own needs." Recently Van Calker *et al.* (2005) worked out this definition for Dutch dairy farming, describing aspects and their attributes. They recognize three aspects: economic, social and ecological sustainability, corresponding to the "popular" triple-P approach: People, Planet and Profit. The sustainable production of agricultural produce for an increasing number of people on our planet is an enormous challenge (Smil, 2000; Tilman *et al.*, 2002). Mushroom growing fits into a sustainable agricultural system by recycling by-products of other agricultural sectors and producing its own by-product, spent mushroom substrate, a valuable fertilizer for horticultural crops. However, the position of the mushroom industry within a broader agricultural system has not been optimized and, considered in isolation, the industry is characterized by some specific bottlenecks.

Economic sustainability

The competitive power of the mushroom industry in economically high developed countries is rather weak compared with the emerging mushroom industries in Central and Eastern Europe and China. One option is the development of automatic harvesting for the market of fresh produce (Van Loon, 2003), reducing costs of labour. Other options are optimizing chain management and a more market-oriented production system for just-in-time delivery. Just-in-time delivery reduces costs within the chain and prevents disposal of produce that does not reach a final consumer. Inherent to mushroom production and marketing are the rapid development of fruit bodies within a week with a window for harvesting of about one day and a shelf-life of freshly harvested produce less than a week. Just in time production delivery depend on the development of "precision" cultivation. At present the possibilities to control cropping are insufficient to meet the rapid cultivation process. The shelf-life of fresh produce depends on postharvest development of fruit bodies. A modern approach under study is the down regulation of target genes in postharvest development (Eastwood *et al.*, 2004). Finally, the production of heterologous proteins for pharmacological use in the host *A. bisporus* is considered to give mushroom production an unexpected technological edge and initial research is now done in this area (Zhang *et al.*, 2004).

Ecological sustainability

Compost for cultivation is produced from renewable resources, straw and manure, being by-products of staple crop production and animal husbandry. Despite the positive contribution to sustainability of agriculture as a whole, the use of resources within the mushroom industry needs improvement. Almost 40% of the organic matter in the ingredients is lost during the composting process by degradation. We seriously doubt that this amount of degradation is essential for a good crop of mushrooms. Potentially valuable organic matter is wasted. Moreover, degradation results in the production of large amounts of carbon dioxide, a greenhouse gas. During degradation heat is produced and external energy is used to remove the surplus of biological heat. Process innovations are needed to improve the sustainability of composting for the mushroom industry.

Compost, minerals, by products

Essential substrate characteristics are not fully understood (see above). Individual compost batches can show yield reductions of up to 25% compared to maximal yields, for no clear reason. Accordingly the turnover of substrate into produce is not optimal, resulting in more rest product "spent mushroom substrate," SMS, than in an optimal system. An option would seem to acquire more knowledge on compost composition and degradation by *A. bisporus,* which have been studied for decades, but novel cultivation methods are developing slowly as a consequence of the complexity of the process. Another option is to develop equipment for monitoring crop development in order to adjust cultivation technique to the optimal outcome. There is a need to evaluate obtained data in accurate models of mushroom cultivation which are insufficient at present (Chanter and Thornley, 1978; Chanter, 1979; Straatsma *et al.,* 2000). SMS is useful as a fertilizer in agricultural crops (Wuest *et al.,* 1995), but it contains high levels of salts. This prevents the use of SMS as potting substrate in horticulture. Adaptation of composts, by replacing manure for other nitrogen sources and by lowering the amount of gypsum, to an optimal use of SMS would seem to contribute to sustainability. However, a practical solution is not available yet (Wever *et al.,* 2005).

Casing soil and fruit body initiation

The application of peat in casing soil conflicts with sustainability. Bogs and moors develop so slowly that peat can hardly be seen as a renewable resource. Although widely used, peat is not the ultimate material for full control over fruit body initiation and growth. The time of appearance of fruit bodies as well as the number of fruit bodies that grows out of the soil is insufficiently controlled. If the number is too high, premature ripening and quality loss occurs; low numbers result in low yields. Both time of appearance and number of fruit bodies interact with compost parameters. For labor and marketing planning, full control over fruit body initiation and growth is required. To acquire this control, developmental knowledge of primordia initiation (Flegg, 1978) the outgrowth into fruit bodies (Umar and van Griensven, 1997) and flushes is needed. A beginning understanding at the molecular level is obtained (De Groot *et al.,* 1998). Taken together, alternatives for peat in casing soil (Visscher, 1980; Noble and Dobrovin-Pennington, 2005), together with an improved system for control of fruit body initiation are needed.

Control of pests and pathogens

The pests and pathogens in the cultivation of *A. bisporus* are usually controlled with the use of chemical pesticides. Some of these chemicals are effective while for others less effectiveness is reported. The focus of the governments of a number of European countries is on the reduction of the use of chemical pesticides in order to enable sustainable mushroom cultivation. As a consequence, interest in the development of biological pesticides has grown. Scheepmaker *et al.* (1998) showed that the entomopathogenic nematode species *Steinernema feltiae* was able to reduce the development of the sciarid *Lycoriella auripila.* This has resulted in sustainable control of this pest by commercial availability of the entomopathogenic nematode. In Europe, several more initiatives have been taken to develop more sustainable ways to control pests and pathogens. Also, development of disease resistance of strains of *A. bisporus* is obtained and is an alternative sustainable way of disease control (Kerrigan, 2004; Largeteau *et al.,* 2004). Recent research at Applied Plant Research in The Netherlands showed that wild strains of *A. bisporus* show a wide range of responses to infections of pathogens including *V. fungicola.* Therefore, sources of resistance to *V. fungicola* are available for the development of (partial resistant) strains of *A. bisporus* (Kerrigan, 2004). Very recently, the response of mushrooms to the fungus was studied and 100 uniquely differentially expressed sequences were identified and currently used for functional analyses by RNAi (Costa *et al.,* 2005).

Before everything, prevention of the development of pests and pathogens is the most sustainable way. A system with hygienic managements can prevent development of pests and pathogens. Such a system was recently developed for the Dutch cultivation industry in a project funded by the Dutch ministry of Agriculture and was indicated with "Best practices" (Baar and van Roestel, 2004).

REFERENCES

Amey, R. C., Mills, P. R., Bailey, A., and Foster, G. D. (2003). Investigating the role of a *Verticillium fungicola* beta -1,6-glucanase during infection of *Agaricus bisporus* using targeted gene disruption. Fungal Genetics and Biology 39:264-275.

Arnolds, E. (1995). Conservation and management of natural populations of edible fungi. Canadian Journal of Botany 73:S987-S998.

Baar, J. (2005). Ectomycorrhizal fungi in wooded banks: what is the value? *In* Conference on the biology of Basidiomycete fungi: Basidio2005. HRI / University of Warwick, United Kingdom.

Baar, J., and Konings M. (2005). Bacteria in casing soil: a black box that can be opened using molecular tools? (In Dutch). Report Applied Plant Research, Report numner: 2005-4, 16 pp.

Baar, J., and Van Roestel, A. (2004). Gewasbescherming: "Best practices" (In Dutch). Paddestoelen 10:9.

Baar, J., and Kuyper, T. W. (1998). Restoration of aboveground ectomycorrhizal flora in stands of *Pinus sylvestris* (Scots pine) in The Netherlands by removal of litter and humus. Restoration Ecology 6:227-237.

Belloli, S., Bologna, F., Gregori, G., and Zambonelli, A. (2001). Il tartufo di Fragno (*Tuber uncinatum* Chatin. ecologia a coltivazione. Actes du Ve Congrès International, Science et Culture de la Truffe et des autres Champignons Hypoges Comestibles. 4 au 6 Mars 1999, Aix-en-Provence, France, Federation Française des Trufficulteurs pp. 367-371.

Bernardo, D., Cabo, A. P., Novaes-Ledieu, M., and Mendoza, C. G. (2004). Verticillium disease or "dry bubble" of cultivated mushrooms: the *Agaricus bisporus* lectin recognizes and binds the *Verticillium fungicola* cell wall glucogalactomannan. Canadian Journal of Microbiology 50:729-735.

Boa, E. (2004). Wild edible fungi a global overview of their use and importance to people. Non-wood forest products, No.17, FAO Publishing, Rome.

Brundtland, G. H. (1987). Our Common Future. UN World Commission on Environment and Development (WCED). Oxford University Press.

Calker, K. J. van, Berentsen, P. B. M., Giesen, G. W. J., and Huirne, R. B. M. (2005). Identifying and ranking attributes that determine sustainability in Dutch dairy farming. Agriculture and Human Values 22:53-63.

Calonje, M. (1992). Properties of a hydrophobin isolated from the mycoparasitic fungus *Verticillium fungicola*. Canadian Journal of Microbiology 48:1030-1034.

Calonje, M., Garcia-Mendoza, C., Galan, B., and Novaes-Ledieu, M. (1997). Enzymic activity of the mycoparasite *Verticillium fungicola* on Agaricus bisporus fruit body cell walls. Microbiology Reading 143:2999-3006.

Calonje, M., Garcia-Mendoza, C., Perez-Cabo, A., Bernardo, D., and Novaes-Ledieu, M. (2000). Interaction between the mycoparasite *Verticillium fungicola* and the vegetative mycelial phase of *Agaricus bisporus*. Mycological Research 104:988-992.

Carroll, A. D., and Schisler, L. C. (1976). Delayed release nutrient supplement for mushroom culture. Applied and Environmental Microbiology 31:499-503.

Chang, S. T., and Miles, P. G. (2004). Mushrooms: cultivation, nutritional value, medicinal effect, and environmental impact - 2nd ed, Boca Raton, FL, CRC Press, U.S.A.

Chanter, D. O., and Thornley, J. H. M. (1978). Mycelial growth and the initiation and growth of sporophores in the mushroom crop: a mathematical model. Journal of General Microbiology 106:55-65.

Chanter, D. O. (1979). Harvesting the mushroom crop: A mathematical model. Journal of General Microbiology 115 :79-87.

Chevalier, G. (2001). Du congres de Spoleto a celui d'Aix-en Provence: les avancees en matiere de recherches sur la truffe et la trufficulture en France. Actes du Ve Congrès International, Science et Culture de la Truffe et des autres Champignons Hypoges Comestibles. 4 au 6 Mars 1999, Aix-en-Provence, France, Federation Française des Trufficulteurs, pp. 11-15.

Chevalier, G., Gregori, G., Frochot, H., and Zambonelli A. (2002). The cultivation of the Burgundy truffle. Edible mycorrhizal mushrooms and their cultivation. Proceedings of the Second International Conference on Edible Mycorrhizal Mushrooms, Christchurch, New Zealand, 3-6 July, 2001.

Clift, A., Shamshad, A., and Terras, M. A. (2004). Flies and dry bubble on cultivated mushrooms. Mushroom Science 16:459-464.

Costa, A., Thomas, J., Bailey, A., Foster, G., Challen, M., and Mills, P. (2005). Identificantion and characterisation of *Agaricus bisporus* genes differentially expressed during *Verticillium fungicola* infection. Meeting British Mycological Society. "Exploitation of Fungi," September 5-8, Manchester, 2005.

Danell, E., and Camacho, F. J. (1997). Successful cultivation of the golden cantherelle. Nature 385:303.

Danell, E. (2002). Current research on chanterelle cultivation in Sweden. Edible mycorrhizal mushrooms and their cultivation. Proceedings of the Second International Conference on Edible Mycorrhizal Mushrooms, Christchurch, New Zealand, 3-6 July, 2001.

Dragt, J. W., Geels, F. P., De Bruijn, W. C., and Van Griensven L. J. L. D. van (1996). Intracellular infection of the cultivated mushroom *Agaricus bisporus* by the mycoparasite *Verticillium fungicola* var. *fungicola*. Mycological Research100:1082-1086.

Eastwood, D. C., Sergeant, M. J., Mead, A., and Burton, K. S. (2004). Expression profiles of key genes in *Agaricus bisporus* postharvest development of fruit bodies. Mushroom Science 16:53-58.

Egli, S., Ayer, F., and Chatelain, F. (1990). Die Einfluss des Pilzsammelns auf die Pilzflora. Mycologica Helvetica 3:417-428.

Fekete, K. (1967). Ueber die Morphologie, Biologie und Bekaempfung von *Verticillium malthousei*, einem Parasiten des Kulturchampignons. PhD Thesis 3877, Eidegenossige Technische Hochschule, Zuerich.

Flegg, P. B. (1978). Effect of temperature on sporophore initiation and development in *Agaricus bisporus*. Mushroom Science 10:595-602.

Flegg, P. B., Spencer, D. M., and Wood, D. A. (1985). The biology and technology of the cultivated mushroom. ISBN 0-471-90435-X. Chichester. Wiley.

Fletcher, J. T., White, P. F., and Gaze, R. H. (1986). Mushrooms: pest and disease control. 2nd edition. Intercept, 174 p.

Fortin, J. A., and Piche, Y. (2000). Les champignons forestiers: recolte, commercilisation et conservation de la ressource. CRBF, Universite Laval, Quebec, 199 pp.

Gandy, D. G. (1985). Bacterial and fungal diseases. *In* The biology and technology of the cultivated mushroom. (Flegg, P. B., Spencer, D. M., and Wood, D. A., eds.), ISBN 0-471-90435-X. Chichester [etc.], Wiley, U.K., pp. 261-270.

Gerrits, J. P. G. (1988). Nutrition and compost. *In* The Cultivation of Mushrooms (Griensven, L. J. L. D. van, ed.), Darlington Mushroom Laboratories, Rustington, U.K., pp. 29-72.

Gerrits, J. P. G. (1989). Supplementation of *Agaricus* compost with organic materials with special attention to the uptake of minerals and amino acids. Mushroom Science 12:361-370.

Gregori, G. L. (2002). Problems and expectations with the cultivation of *Tuber magnatum*. Edible mycorrhizal mushrooms and their cultivation. Proceedings of the Second International Conference on Edible Mycorrhizal Mushrooms, Christchurch, New Zealand, 3-6 July, 2001.

Griensven, L. J. L. D. van (1988). The Cultivation of Mushrooms. Darlington Mushroom Laboratories, Rustington, U.K.

Grogan, H. M., Adie, B. A. T, Gaze, R. H., Challen, M. P., and Mills, P. R. (2003). Double-stranded RNA elements associated with the MVX disease of *Agaricus bisporus*. Mycological Research 107:147-154.

Groot, P. W. J. de, Visser, J., Griensven, L. J. L. D. van, and Schaap, P. J. (1998). Biochemical and molecular aspects of growth and fruiting of the edible mushroom *Agaricus bisporus*. Mycological Research 102:1297-1308.

Guinberteau, J., Ducamp, M., Poitou, N., Mamoun, M., and Olivier, J. M. (1990). Ecology of various competitors from an experimental plot of *Pinus pinaster* inoculated with *Suillus granulatus* and *Lactarius deliciosus*. Agriculture Ecosystems and Environment 28:161-165.

Hall, I. R., Stephenson, S. L., Buchanan, P. K., Wang. Y., and Cole, A. L. J. (2003a). Edible and Poisonous Mushrooms of the World. Timber Press Inc., Portland, U.S.A.

Hall, I. R., Wang, Y., and Amicucci, A. (2003b). Cultivation of edible ectomycorrhizal mushrooms. Trends in Biotechnology 21:433-438.

Hawksworth, D. L. (2001). The magnitude of fungal diversity: the 1.5 million species estimate revisited. Mycological Research 105:1422-1432.

Hedger, J. (1986). *Suillus luteus* on the equator. Bulletin of the Britisch Mycological Society 20:53-54.

Houdeau, G., Olivier, J. M., amd Chabbert, B. (1991). Improvement of indoor short composting. Mushroom Science 13:215-220.

Hosford, D., Pilz, D., Molina, D., and Amaranthus, M. (1997). Ecology and management of the commercially harvested american Matsutake mushroom. Gen. Tech. Rep. PNW-GTR-412. USDA Forest Service, Portland. 68 pp.

Hu, H. T. (2003). Cultivation of Ascocarps of *Tuber Formosanum* Hu on *Cyclobalanopsis Glauca* (Thunb.) Oerst in Siliceous Soils by Liming. Abstract. The Fourth International Conference on Mycorrhizae ICOM4, August 10-15.

Iiyama, K., Stone, B. A., and Macauley, B. J. (1994). Compositional changes in compost during composting and growth of *Agaricus bisporus*. Applied and Environmental Microbiology 60:1538-1546.

Iwase, K. (1997). Cultivation of mycorrhizal mushrooms. Food Reviews International 13:431-442.

Janick, J. (2005). History of Horticulture. Purdue University, West Lafayette, IN, U.S.A.

Kalberer, P. P. (1990). Water relations of the mushroom culture *Agaricus bisporus* study of a single break. Scientia Horticulturae 41:277-284.

Kerrigan, R. W. (2004). Trait diversity in wild *Agaricus bisporus*. *In* Science and cultivation of edible and medicinal fungi (Romaine, Keil, Rinker, Royse, eds.), pp. 31-38.

Largeteau, M. L., Rodier, A., Rousseau, T., Juarez del Carmen, S., Vedie, R., and Savoie, J. M. (2004). *Agaricus* susceptibility to *Verticillium fungicola*. Mushroom Science 16:515-523.

Loon, P. C. C. van (2003). Inventarisatie van knelpunten bij de ontwikkeling van een oogstrobot voor champignons. Applied Plant Research, Mushroom Research Unit report 2003-2004.

Marmeisse, R., Guidot, A., Gay, G., Lambilliotte, R., Sentenac, H., Combier, J. P., Melayah, D., Fraissinet-Tachet, L., and Debaud, J. C. (2004) *Hebeloma cylindrosporum* - a model species to study ectomycorrhizal symbiosis from gene to ecosystem. New Phytologist 163:481-498.

Mikola, P. (1969). Mycorrhizal fungi of exotic forest plantations. Karstenia 10:169-176.

Noble, R., and Dobrovin-Pennington, A. (2005). Partial substitution of peat in mushroom casing with fine particle coal tailings. Scientia Horticulturae 104:351-367.

North, L. H., and Wuest, P. J. (1993). The infection process and symptom expression of *Verticillium* disease of *Agaricus bisporus*. Canadian Journal of Plant Pathology 15:74-80.

Oei, P., and Baar, J. (2002). Cultivation of poplar boletes: a step closer? Teelt van harde populierboleet: een stap dichterbij? Coolia 45:167-174.

Oei, P., and Maas, M. (2003). Mushroom cultivation: appropriate technology for mushroom growers - 3rd ed. ISBN 9057821370. Leiden: Backhuys Publishers. 429 pp.

Olivier, J. M. (2000). Progress in the cultivation of truffles. Mushroom Science 15:937-942.

Peredo, H., and Oliva M. (1983). Environmental factors determining the distribution of *Suillus luteus* fructifications in *Pinus radiata* grazing-forest plantations. Plant and Soil 71:367-370.

Piearce, G. D., and Ross, W. K. (1980). The appeal of the pine bolete *Suillus granulatus* as a new Zambian mushroom. International Tree Crops Journal 1:183-197.

Pilz, D., and Molina, R. (2002). Commercial harvests of edible mushrooms from the forests of the Pacific Northwest United States: issues, management, and monitoring for sustainability. Forest Ecology and Management 155:3-16.

Pilz, D., Norvell, L., Danell, E., and Molina, R. (2003). Ecology and management of commercially harvested chantarelle mushrooms. Gen. Tech. Rep. PNW-GTR-576. USDA Forest Service, Portland. 83 pp.

Pilz, D., Smith, J., Amaranthus, M. P., Alexander, S., Molina, R., and Luoma, D. (1999). Mushrooms and timber - Managing commercial harvesting in the Oregon cascades. Journal of Forestry 97:4-11.

Poitou, N., Mamoun, M., Ducamp, M., and Delmas, J. (1984). Apres le bolet granuleux, le lactaire delicieux obtenu en fructification au champ a partir de plants mycorhizes. P.H.M. Revue Horticole 244 :65-68

Randle, P. E. (1983). Supplementation of mushroom composts - a review. Crop Research 23 :51-69.

Rebiere, J. (1967). La truffe du Perigord, sa culture. Fanlac. Periguex. 142 pp.

Sahin, N. (2005). Antimicrobial activity of *Streptomyces* species against mushroom blotch disease pathogen. Journal of Basic Microbiology 45:64-71.

Scheepmaker, J. W. A., Geels, F. P., Rutjens, A. J., Smits, P. H., and Griensven, L. J. L. D. van (1998). Comparison of the efficacy of entomopathogenic nematodes for the biological control of the mushroom pests *Lycoriella auripila* (Sciaridae) and *Megaselia halterata* (Phoridae). Biocontrol Science and Technology 8:277-288.

Sharma, H. S. S., and Kilpatrick, M. (2000). Use of near-infrared spectroscopy to predict potential mushroom (*Agaricus bisporus*) yield of phase II compost. Applied Spectroscopy 54:44-47.

Singer, R. (1965). Die Roehrlinge, Teil 1 Die Boletaceae. Klinkhardt, Bad Heilbrunn.

Smil, V. (2000). Feeding the world : a challenge for the twenty-first century. ISBN 0262194325. Cambridge, Mass., MIT Press. 360 pp.

Smith, S. E., and Read, D. J. (1997). Mycorrhizal symbiosis - 2nd ed. San Diego. Academic Press. 605 pp.

Steineck, H. (1984). Moeglichkeiten fuer den Anbau von Mykorrhizapilzen. Champignon. 273:7-19.

Straatsma, G. (2004). Processing and composition of mushroom compost. Mushroom Science 16:241-246.

Straatsma, G., Gerrits, J. P. G., Augustijn, M. P. A. M., Op den Camp, H. J. M., Vogels, G. D., and Griensven, L. J. L. D. van (1989). Population dynamics of *Scytalidium thermophilum* in mushroom compost and stimulatory effects on growth rate and yield of *Agaricus bisporus*. Journal of General Microbiology 135:751-759.

Straatsma, G., Di Lena, G., Olijnsma, T. W., Op den Camp, H. J. M., and Griensven, L. J. L. D. van (1993). Laboratory media for measuring growth parameters of *Agaricus bisporus* mycelium as influenced by *Scytalidium thermophilum*. Cultivated Mushroom Research Newsletter 1:1-6.

Straatsma, G., Olijnsma, T. W., Gerrits, J. P. G., Amsing, J. G. M., Camp, H. J. M. op den, and Griensven, L. J. L. D. van (1994). Inoculation of *Scytalidium thermophilum* in button mushroom compost and its effect on yield. Applied and Environmental Microbiology 60:3049-3054.

Straatsma, G., Gerrits, J. P.G., Thissen, J. T. N. M., Amsing, J. G. M., Loeffen, H., and Griensven, L. J. L. D. van (2000). Adjustment of the composting process for mushroom cultivation based on initial substrate composition. Bioresource Technology 72:67-74.

Tanfulli, M., Giovanotti, E., Donnini, D., and Baciarelli, F. L. (2001). Analisi della micorrizazione in tartufaie coltivate di *Tuber aestivum* Vittad. e *Tuber borchii* Vittad. Impiantate da oltre 12 anni ambienti pedoclimatici diversi. Actes du Ve Congrès International, Science et Culture de la Truffe et des autres Champignons Hypoges Comestibles. 4 au 6 Mars 1999, Aix-en-Provence, France, Federation Française des Trufficulteurs, pp. 8.480-484.

Tilman, D., Cassman, K. G., Matson, P. A., Naylor, R., and Polasky, S. (2002). Agricultural sustainability and intensive production practices. Nature 418:671-677.

Tournefort, M. (1707). Observations sur la naissance et sur la culture des champignons. Memoires de l'Academie Royale Des Sciences Paris :58-66.

Umar, M. H., and Griensven, L. J. L. D. van (1997). Morphological studies on the life span, developmental stages, senescence and death of fruit bodies of *Agaricus bisporus*. Mycological Research 101:1409-1422.

Vinay, M., and Pirazzi, R. (2001). Realta'ed esigenze per la coltivazione di *Tuber borchii* Vittad. e *Tuber aestivum* Vittad. nell'Italia Centrale. Actes du Ve Congrès International, Science et Culture de la Truffe et des autres Champignons Hypoges Comestibles. 4 au 6 Mars 1999, Aix-en-Provence, France, Federation Française des Trufficulteurs, pp. 7.425-430.

Visscher, H. R. (1980). Substitutes for peat in mushroom casing soil. ISOSC, International society for soilless culture Proceedings 5:395-409.

Wang, Y., Buchanan, P., and Hall, I. R. (2002a). A list of edible ectomycorrhizal mushrooms. Edible mycorrhizal mushrooms and their cultivation. Proceedings of the Second International Conference on Edible Mycorrhizal Mushrooms, Christchurch, New Zealand, 3-6 July, 2001.

Wang, Y., Hall, I. R., Dixon, C., Hance-Halloy, M., Strong, G., and Brass, P. (2002b). The cultivation of *Lactarius deliciosus* (saffron milk cap) and *Rhizopogon rubescens* (shoro) in New Zealand. Edible mycorrhizal mushrooms and their cultivation. Proceedings of the Second International Conference on Edible Mycorrhizal Mushrooms, Christchurch, New Zealand, 3-6 July, 2001.

Wang, Y. C. (1987). Mycology in ancient China. Mycologist 21:59-61.

Ware, W. M. (1933). A disease of cultivated mushrooms caused by *Verticillium malthousei* sp nov. Annals of Botany 47:763-785.

Wever, G., Straatsma, G., and Burg, A. M. M. van (2005). Potential of adapted mushroom compost as a growing medium in horticulture. Acta Horticulturae. In press.

White, P.F. (1981). Spread of the mushroom disease *Verticillium fungicola* by *Megaselia halterata* diptera phoridae. Protection Ecology 3:17-24.

Wuest, P. J., Levanon, D., and Hadar, Y. (1995). Environmental, agricultural and industrial uses for spent mushroom substrate from mushroom farms. American Mushroom Institute, Washington, 156 pp.

Yamada, A., Ogura, T., and Ohmasa, M. (2001). Cultivation of some Japanese edible ectomycorrhizal mushrooms. Edible mycorrhizal mushrooms and their cultivation. Proceedings of the Second International Conference on Edible Mycorrhizal Mushrooms, Christchurch, New-Zealand, 3-6 July, 2001.

Yun, W., snd Hall, I. R. (2004). Edible ectomycorrhizal mushrooms: challenges and achievements. Canadian Journal of Botany 82:1063-1073.

Zambonelli, A., Lotti, M., Giomaro, G., Hall, I. R., and Stocchi, V. (2002). *T. borchii* cultivation: an interesting perspective. Edible mycorrhizal mushrooms and their cultivation. Proceedings of the Second International Conference on Edible Mycorrhizal Mushrooms, Christchurch, New-Zealand, 3-6 July, 2001.

Zambonelli, A., Iotti, M., Rossi, I., and Hall, I. (2000). Interactions between *Tuber borchii* and other ectomycorrhizal fungi in a field plantation. Mycological Research 104:698-702.

Zhang, C., Odon, V., Kim, H. K., Challen, M., Burton, K., Hartley, D., and Elliott T. (2004). Mushrooms for Molecular Pharming. Mushroom Science 16:611-617.

SUBJECT INDEX

G

H

I

M

O

P

Q

R

S